The Cyanobacteria

Molecular Biology, Genomics and Evolution

Edited by Antonia Herrero and Enrique Flores

Caister Academic Press

Caister Scientific Press
Norfolk, UK

www.caister.com

British Library Cataloguing-in-Publication Data
A catalogue record for this book is available from the British Library

ISBN: 978-1-904455-15-8

Printed and bound in Great Britain

Cover: (see also Cover plate)

Background: a surface bloom of the cyanobacterium Nodularia in the southern Baltic Sea in 2004 (courtesy of S. Jansson and B. Bergman). Back, top: electron micrograph of a transversal section of Synechococcus sp. PCC 7942 (courtesy of R. Nevo, Z. Reich, I. Ohad and A. Kaplan). Back, middle: light micrograph of Synechococcus sp. PCC 9321 (courtesy of A. M. Muro-Pastor). Back, bottom: Mesoproterozoic fossil akinete from the 1500 Ma Bil'yakh Group, Siberia (courtesy of A. H. Knoll). Front, top: light micrograph of a natural isolate of a heterocyst-forming cyanobacterium from Mkindo, Tanzania (courtesy of J. E. Frías and A. Herrero). Front, middle: Trichodesmium thiebautii colony (courtesy of B. Bergman). Front, bottom: electron micrograph of a longitudinal section of a filament of Anabaena sp. PCC 7120 with a heterocyst (far right) and vegetative cells (courtesy of I. Maldener). Images are not to scale.

Contents

Contributors

David G. Adams
Faculty of Biological Sciences
University of Leeds
Leeds
UK

d.g.adams@leeds.ac.uk

Hermann Bauwe
Department of Plant Physiology
University of Rostock
Rostock
Germany

Birgitta Bergman
Department of Botany
Stockholm University
Stockholm
Sweden

bergmanb@botan.su.se

Robert E. Blankenship
Departments of Biology and Chemistry
Washington University
St. Louis
MO 63130
USA

blankenship@biology2.wustl.edu

Thomas Börner
Institute of Biology (Genetics)
Humboldt-University
Berlin
Germany

thomas.boerner@rz.hu-berlin.de

Yanina Sarah DeRuyter
Department of Chemistry and Biochemistry
Arizona State University
Tempe, AZ
USA

ysb@asu.edu

Jeff Elhai
Center for the Study of Biological Complexity
Virginia Commonwealth University
Richmond, VA
USA

elhaij@vcu.edu

Enrique Flores
Instituto de Bioquímica Vegetal y Fotosíntesis
Consejo Superior de Investigaciones
Científicas and Universidad de Sevilla
Américo Vespucio 49
E-41092 Seville
Spain

eflores@ibvf.csic.es

Karl Forchhammer
Karl Forchhammer
Lehrstuhl für Mikrobiologie II
Tübingen
Germany

karl.forchhammer@uni-tuebingen.de

Petra Fromme
Department of Chemistry and Biochemistry
Arizona State University
Tempe, AZ
USA

PFromme@asu.edu

Antonia Herrero
Instituto de Boquímica Vegetal y Fotosíntesis
Consejo Superior de Investigaciones
Científicas and Universidad de Sevilla
Seville
Spain

herrero@ibvf.csic.es

Ferran Garcia-Pichel
School of Life Sciences
Arizona State Univesity
Tempe, AZ
USA

ferran@asu.edu

Martin Hagemann
Department of Plant Physiology
University of Rostock
Rostock
Germany

martin.hagemann@uni-rostock.de

Wolfgang R. Hess
University of Freiburg
Institute of Biology II
Experimental Bioinformatics
Freiburg
Germany

wolfgang.hess@biologie.uni-freiburg.de

Shira Kahlon
Department of Plant and Environmental
Sciences
The Hebrew University of Jerusalem
Jerusalem
Israel

Aaron Kaplan
Department of Plant and Environmental
Sciences
The Hebrew University of Jerusalem
Jerusalem
Israel

aaronka@vms.huji.ac.il

Andrew H. Knoll
Department of Organismic and Evolutionary
Biology
Harvard University
Cambridge
MA 02138
USA

aknoll@oeb.harvard.edu

Takao Kondo
Division of Biological Science
Graduate School of Science
Nagoya University
Nagoya
Japan

kondo@biol1.bio.nagoya-u.ac.jp

Michelle Liberton
Department of Biology
Washington University
St. Louis
MO 63130
USA

mliberton@biology2.wustl.edu

Dmitry A. Los
Institute of Plant Physiology
Russian Academy of Sciences
Moscow
Russia

Ignacio Luque
Instituto de Bioquímica Vegetal y Fotosíntesis
Consejo Superior de Investigaciones
Científicas and Universidad de Sevilla
Américo Vespucio 49
E-41092 Seville
Spain

ignacio.luque@ibvf.csic.es

John C. Meeks
Section of Microbiology
University of California
Davis
CA 95616
USA

jcmeeks@ucdavis.edu

Conrad W. Mullineaux
School of Biological and Chemical Sciences
Queen Mary, University of London
Mile End Road
London
UK

c.mullineaux@qmul.ac.uk

Norio Murata
National Institute for Basic Biology
Okazaki 444-8585
Japan

murata@nibb.ac.jp

Alicia M. Muro-Pastor
Instituto de Bioquímica Vegetal y Fotosíntesis
Consejo Superior de Investigaciones
Científicas and Universidad de Sevilla
Américo Vespucio 49
E-41092 Seville
Spain

alicia@ibvf.csic.es

Teruo Ogawa
Institute of Plant Physiology and Ecology
Shanghai Institutes for Biological Sciences
Chinese Academy of Sciences
Shanghai
China

ogawater@xd6.so-net.ne.jp

Himadri B. Pakrasi
Department of Biology
Washington University
St. Louis, MO
USA

pakrasi@wustl.edu

Liang Ran
Department of Botany
Stockholm University
Stockholm
Sweden

ran.liang@botan.su.se

Jason Raymond
Microbial Systems Division
Lawrence Livermore National Laboratory
Livermore
CA 94550
USA

jason.raymond@llnl.gov

Kaarina Sivonen
Department of Applied Chemistry and
Microbiology
Viikki Biocenter
Helsinki University
Finland

kaarina.sivonen@helsinki.fi

Lucas J. Stal
Department of Marine Microbiology
Netherlands Institute of Ecology
Yerseke
The Netherlands

L.Stal@nioo.knaw.nl

Iwane Suzuki
Graduate School of Life and Environmental
Sciences
University of Tsukuba
Tsukuba
Japan

Wesley D. Swingley
School of Life Sciences
Arizona State University
Tempe
AZ 85287
USA

Kazuki Terauchi
Division of Biological Science
Graduate School of Science
Nagoya University
Nagoya
Japan

terauchi@biol1.bio.nagoya-u.ac.jp

C. Peter Wolk
MSU-DOE Plant Research Laboratory and
Department of Plant Biology
Michigan State University
East Lansing, MI
USA

wolk@msu.edu

Xudong Xu
Institute of Hydrobiology
Chinese Academy of Sciences
Wuhan, Hubei
PR China

xux@ihb.ac.cn

Jonathan P. Zehr
Department of Ocean Sciences
University of California
Santa Cruz, CA
USA

zehrj@ucsc.edu

Vladislav V. Zinchenko
Department of Genetics
Moscow State University
Moscow
Russia

Preface

Research in several fields of biology has led to the realization of the immense importance of oxygenic phototrophic prokaryotes, the cyanobacteria, from at least three different perspectives. First, the cyanobacteria, as inventors of oxygenic photosynthesis, have had a unique and key role in the history of the Earth and the evolution of life. Second, these organisms currently contribute substantially to maintaining the biosphere since they are important primary producers at a global scale, with a relevant role in the C and N cycles. Third, the cyanobacteria perform some unique biological processes and can be used as models to study other important biological activities or capabilities.

An unprecedented wealth of data on sequenced cyanobacterial genomes has been obtained recently. Genomics together with novel molecular biology work including structural analysis have resulted in an improvement of our understanding of key topics in the biology of cyanobacteria. This suggests that a new book on the biology of these organisms is timely. The chapters of this book are intended to address some representative topics related to the different areas in which, as noted above, the cyanobacteria are important. In Chapters 1 through 3 we begin with the evolution of cyanobacteria including their origin, phylogeny, and the importance of gene transfer. Comparative genomics is presented to reconstruct the history of cyanobacteria, and the genetic events that have shaped cyanobacterial genome evolution are discussed. Also, molecular phylogenies are combined with analysis of the geological record to infer the course of cyanobacterial evolution, oxygenic photosynthesis development, atmosphere oxygenation, and biosphere reliance on cyanobacterial primary production throughout the Earth's history.

The ecology of cyanobacteria, including a special look at the marine environment, stress responses, the production of active compounds, and the cyanobacterial circadian rhythms are then covered in Chapters 4 through 8. Aspects treated include how the application of molecular biology and genomics to cyanobacterial ecology, together with recent bioinformatics resources and increasing surveying efforts have uncovered an undescribed wealth of genetic diversity among cyanobacteria in nature. They have also enabled a first glimpse at population genetics and community physiology, including analysis of gene expression in nature, and has provided a growing body of information on those cyanobacteria that are fundamental for oceanic primary production. There is also the remarkable finding of a complex reservoir of marine cyanobacterial genes that, embedded in the very abundant cyanophages, can contribute to the high diversity and flexibility observed in some marine cyanobacteria. Genome-wide investigations of gene expression aimed at understanding the mechanisms of perception and reaction to environmental stress, which are involved in the acclimation of cyanobacteria to new environmental situations, are covered. Also covered is the study of the environmentally relevant production by cyanobacteria of bioactive compounds including potent hepatotoxins and neurotoxins, anticancer and antiviral agents, and UV-protectants. Finally, the features of the cyanobacterial circadian oscillator that adapt cellular performance to daily changes in the environment, the first one discovered in a prokaryotic organism and to be reconstructed *in vitro* from its molecular components, are presented.

Then we turn to physiological, bioenergetics and metabolic processes starting from a description of the photosynthetic apparatus while adding two complementary views on the membrane systems and their dynamics in cyanobacteria in Chapters 9 through 11. Oxygenic photosynthesis is the process of conversion of light energy from the sun into chemical energy using water as the source of electrons, which provides energy for most living beings and generates the oxygen that has accumulated in our atmosphere. The structure and function of the protein complexes that catalyze oxygenic photosynthesis are presented here. Also presented is current knowledge on the structure, organization, and biogenesis of the different membrane systems that harbor the photosynthetic and other bioenergetic complexes of cyanobacteria together with recent studies on targeting of the protein complexes to these membrane systems. In Chapters 12 and 13 two crucial assimilatory processes, those of C and N, are addressed including distinct regulatory circuits that tune metabolism to ambient conditions. Two more chapters related to the assimilation of N deal with particularly remarkable processes. In Chapter 14 the development of N_2-fixing heterocysts as a response to the ambiental cue of combined nitrogen limitation is covered. This provides an excellent model for cell differentiation of prokaryotes and the study of one-dimensional pattern formation. Heterocyst-forming cyanobacteria are truly multicellular organisms with specialized-interdependent cells that contribute to the overall performance of the whole. Chapter 15 addresses the importance of cyanobacterial N_2-fixation as the primary way that atmospheric nitrogen is introduced into the oceans. Finally, Chapter 16 presents the symbioses that some cyanobacteria, notably the heterocyst-forming cyanobacteria, can establish with plants, which in a sense closes the circle of our story since, through its involvement in the formation of plastids, symbiosis is at the basis of one of the main contributions of cyanobacteria to the development of life in our planet.

In addressing these topics, the contributors of this book have brought together information gathered from diverse research fields including Earth science, ecology, physiology, and molecular and structural biology. We want to express our recognition and gratitude to the authors for their excellent work, which is the essence of this book. We hope that, globally, the reader will get a sense of the beauty and importance of cyanobacteria.

We, the editors, belong to a school of Spanish biochemists started by Manuel Losada with a long tradition in research on bioenergetics and the assimilation of nitrogen by photosynthetic organisms. We have long considered the utilization of the photosynthetic assimilatory power (ATP and reducing equivalents) as a key aspect of biology. What else is the growth of a cyanobacterium than the most ample utilization of photosynthetic assimilatory power? We wish to thank our teachers and colleagues for many years of inspiring work in this field of biology. Our special recognition goes to the already large number of graduate and post-doctoral students in our laboratory who have been pivotal for the achievements of the Cyanobacterial research group of Seville (CS). Crucial to the performance of our group has also been our post-doctoral work with C. Peter Wolk at Michigan State University, which was devoted to learning genetics of cyanobacteria. This is a good opportunity, with the occasion of his recent seventieth birthday, to show our gratitude to Peter not only for his instruction and continuous support of our research in this field, but also for his decisive contributions to the modern study of cyanobacteria.

We wish to dedicate our work to our children, Marina and Antonio E. Flores Herrero, who have been hearing about cyanobacteria since they were born. We hope that they have acquired a sense of appreciation for these organisms and, most importantly, of the immense power of biology for personal growth.

Antonia Herrero and Enrique Flores
Seville, August 2007

Cyanobacteria and Earth History

1

Andrew H. Knoll

Abstract

The biochemical capacity to use water as the source for electrons in photosynthesis evolved once, in a common ancestor of extant cyanobacteria. The geological record indicates that this transforming event took place early in our planet's history, at least 2450–2320 million years ago (Ma), and possibly much earlier. Geobiological interpretation of Archean (> 2500 Ma) sedimentary rocks remains a challenge; available evidence indicates that life existed 3500 Ma, but the question of when oxygenic photosynthesis evolved continues to engender debate and research. A clear paleontological window on cyanobacterial evolution opened about 2000 Ma, revealing an already diverse biota of blue-greens. Cyanobacteria remained principal primary producers throughout the Proterozoic Eon (2500–543 Ma), in part because the redox structure of the oceans favored photautotrophs capable of nitrogen fixation. Green algae joined blue-greens as major primary producers on continental shelves near the end of the Proterozoic, but only with the Mesozoic (251–65 Ma) radiations of dinoflagellates, coccolithophorids, and diatoms did primary production in marine shelf waters take modern form. Cyanobacteria remain critical to marine ecosystems as primary producers in oceanic gyres, as agents of biological nitrogen fixation, and, in modified form, as the plastids of marine algae.

Introduction

In terms of Earth history, cyanobacteria occupy a privileged position among organisms. As primary producers they play a significant role in Earth's carbon cycle; as N-fixers, they also figure prominently in the nitrogen cycle. Moreover, as the only organisms ever to evolve coupled photosystems that harvest electrons from water and produce dioxygen as a byproduct, they also loom large in our planet's redox history. Molecular phylogenies make predictions about the course of cyanobacterial evolution that can be tested using the geological record. In this chapter, I briefly summarize our current understanding of cyanobacterial phylogeny and then review the fossil record of cyanobacteria, pinning that record as closely as possible to molecular phylogenies. Finally, I consider the geological history of oxygen and the preserved record of primary production through time. Collectively, these three lines of evidence constrain the evolutionary history of cyanobacteria (including plastids) and its relationship to Earth's long term environmental evolution.

Evolutionary pattern: phylogeny

Among bacteria, cyanobacteria exhibit an unusually wide range of morphologies. Not surprisingly, then, traditional taxonomic classifications of blue-greens focused on morphology and development, recognizing five principal groups (e.g. Rippka *et al.*, 1979). Group I (Chroococcales) comprises solitary and colonial unicellular cyanobacteria, for example, *Synechococcus* and *Gloeocapsa*. Group II (Pleurocapsales) consists of unicellular to pseudo-filamentous, thallus-forming cyanobacteria, with cells capable of multiple, as well as binary, fission. Group III (Oscillatoriales) comprises filamentous cyanobacteria without cell differentiation, whereas Group IV (Nostocales) consists of filaments marked by cell differentiation to produce akinetes and heterocysts. Group V (Stigonematales) is comprised of cell-differentiating cyanobacteria with more complex multicellular organization.

Molecular phylogenies of the cyanobacteria (e.g. Giovanonni *et al.*, 1988; Turner *et al.*, 1999; Sánchez-Baracaldo *et al.*, 2005; Tomitani *et al.*, 2006) support some but not all of these groupings. Not surprisingly, the phylogenetic affinities of taxa that share features like cell differentiation or multiple fission find support in molecular analyses, whereas unicells and simple filaments do not constitute monophyletic groupings.

Phylogenies generally agree in placing *Gloeobacter* as the sister group to all other extant cyanobacteria, or as a paraphyletic basal grouping (Turner, 1997). Other clades that show up consistently in molecular phylogenies include a group of unicellular taxa dominated by strains of *Synechecoccus* and *Prochlorococcus*; Group II, baeocyte-forming cyanobacteria; and the *Lyngbya/Phormidium/Plectonema* (LPP) cluster of sheath-forming Group III filaments, along with a few unicellular lineages (Turner *et al.*, 1999; Sánchez-Baracaldo *et al.*, 2005). The phylogeny of Tomitani *et al.* (2006), based on sequence analysis of 16S rRNA, *hetR* and *rbcL* genes and with extensive sampling of multicellular taxa, further supports the monophyly of cyanobacteria that differentiate akinetes and heterocysts: Group V lineages form a clade nested within a paraphyletic Group IV.

Figure 1.1 illustrates phylogenetic relationships among cyanobacteria, as inferred from the molecular analyses cited in the preceding paragraph. As emphasized by Sánchez-Baracaldo *et al.* (2005), the last common ancestor of extant cyanobacteria was a unicellular photoautotroph with coupled photosystems capable of deriving electrons from water. Sánchez-Baracaldo *et al.* (2005) also posit that the earliest cyanobacteria lived in terrestrial or freshwater environments and did not fix nitrogen. Both conclusions are consistent with available phylogenies, but merit further testing based on more extensive sampling of genes and taxa. All phylogenies indicate that simple filaments evolved later in cyanobacterial history, and cell differentiation after that. The mapping of morphology and physiology onto phylogeny, therefore, permits predictions about the evolutionary history of cyanobacteria that can be tested against the direct record of cyanobacterial evolution recorded by sedimentary rocks.

Evolutionary pattern: the fossil record

Preservation and recognition

Relative to other prokaryotes, cyanobacteria commonly have large cells, complex morphology and a high probability of becoming incorporated into the fossil record. Thus, we enjoy the happy circumstance that the bacterial group most centrally relevant to Earth's

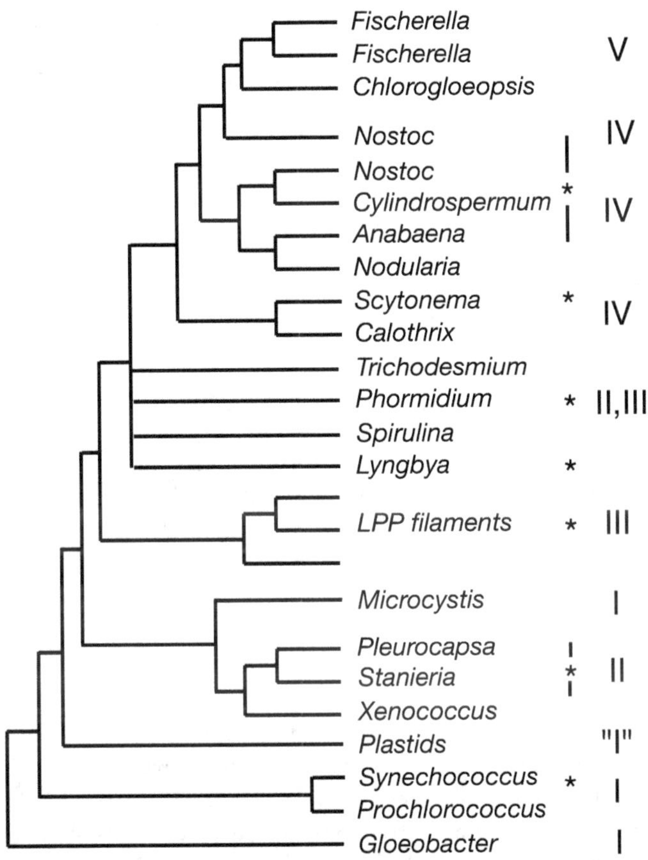

Figure 1.1 Phylogenetic relationships within the Cyanobacteria, based on molecular sequence comparisons, particularly those reported by Sánchez-Baracaldo *et al.* (2005) and Tomitani *et al.* (2006). Roman numerals denote members of Groups I-V, as recognized in morphologically based classification. Asterisks mark morphotypes recorded from 2000–1500 Ma fossil assemblages.

environmental history is the clade most likely to have left a tractable record in sediments (Knoll, 1989, 2003). Not all parts of a cyanobacterium, however, are equally likely to evade post-mortem decay. Experiment and observation show that extracellular sheaths are more likely to preserve than cell walls and cytoplasm (Figure 1.2e; Golubic and Barghoorn, 1977; Bartley, 1996). Thus, the cyanobacterial record is biased toward lineages that form extracellular sheaths and envelopes and toward environments in which such cyanobacteria are or were abundant.

Cyanobacterial trichomes and sheaths can be preserved as organic compressions, pressed along bedding planes in shales (e.g. Butterfield *et al.*, 1994). Such fossils are relatively common in Proterozoic (Table 1.1) successions, but fade to insignificance with the latest Proterozoic evolution of bioturbating animals. More commonly, cyanobacteria have been preserved by early diagenetic mineral precipitation. In Paleozoic (543–251 Ma) and Neoproterozoic (1000–543 Ma) limestones and dolostones, cyanobacterial sheaths are

Table 1.1 The geological time scale as used in this chapter. The fossil record of animal life begins 632 Ma, just before the Proterozoic/Phanerozoic (= Precambrian/Cambrian) boundary.

Phanerozoic Eon (0–542 Ma)
Cenozoic Era (0–65 Ma)
Mesozoic Era (65–251 Ma)
Paleozoic Era (251–542 Ma)
Proterozoic Eon (542–2500 Ma)
Neoproterozoic Era (542–1000 Ma)
Mesoproterozoic Era (1000–1600 Ma)
Paleoproterozoic Era (1600–2500 Ma)
Archean Eon (2500-ca. 4000 Ma)‡

‡ The beginning of the Archean Eon is commonly taken to be the age of oldest known terrestrial rocks, with the time interval between then and the origin of our planet (ca. 4560 Ma) differentiated as the Hadean Eon.

commonly preserved by a thin veneer of carbonate crystals (Riding, 2000; Knoll *et al.*, 1993). In fresh waters, relatively poorly buffered by the total CO_2 system, photosynthetic uptake of ambient carbon dioxide can facilitate carbonate precipitation (Mertz, 1992), but the relevance of this for the interpretation of ancient marine fossils may be limited. In marine environments, carbonate encrustations generally precipitate soon after cell death, in association with bacterial decay (Chaftez and Buczynski, 1992; Bosak and Newman, 2003). Carbonate encrusted sheaths are uncommon in rocks older than about 1000 Ma, not because cyanobacteria were absent but because seafloor precipitation of carbonate crystals obliterated potential fossils (Knoll and Semikhatov, 1998). Although carbonate encrustation fossilizes mat-building filaments, it rarely preserves morphological details that might permit fine-scale taxonomic resolution.

Cyanobacteria can also be preserved by early diagenetic phosphorite (Zhang *et al.*, 1998) or pyrite (Golovenok *et al.*, 1990) precipitation, but these mechanisms are only locally significant. In Precambrian rocks, entombment in early diagenetic silica (SiO_2) constitutes the most abundant and illuminating mode of preservation. Silica deposited in sediments gets redistributed locally during diagenesis to form nodules or discontinuous beds of chert. In the absence of radiolarians, sponges, and diatoms, silica commonly left Precambrian oceans along with carbonate minerals as an early stage evaporite, deposited along the margins of ocean basins. For this reason, the best preserved cyanobacteria occur in early diagenetic chert nodules that formed within peritidal carbonate successions (Maliva *et al.*, 1989). Examples include exceptionally preserved assemblages of the ca. 800 Ma Bitter Springs Formation, Australia (Schopf, 1968; Schopf and Blacic, 1971); the 700–800 Ma Draken Formation, Spitsbergen (Knoll *et al.*, 1991); the ca. 1000 Ma Seryi Klyuch Formation, Siberia (Nagovitsin, 2000); the ca. 1500 Ma Bil'yakh Group, Siberia

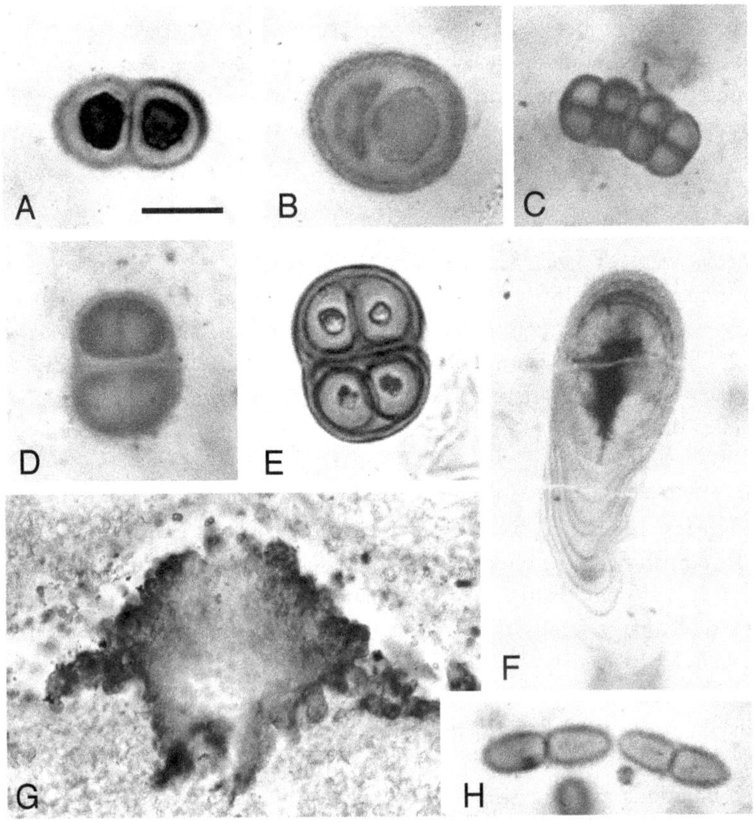

Figure 1.2 Proterozoic fossils interpreted as Group I cyanobacteria, with a partially decayed modern specimen for comparison. A and C, *Sphaerophycus medium*, 700–800 Ma Draken Formation, Spitsbergen; B, *Gloeodiniopsis lamellosa*, ca. 800 Ma Bitter Springs Formation, Australia; D, *Gloeodiniopsis* sp., ca. 1500 Ma Bil'yakh Group, Siberia; E, modern tetrad of *Chroococcus turgidus*, showing that cell contents shrivel during post-mortem decay, whereas extracellular shetahs retain their original geometry; F, *Polybessurus bipartitus*, Draken Formation; G, *Eoentophysalis belcherensis*, Bil'yakh Group; and H, *Eosynechococcus medius*, Draken Formation. Bar in A = 5 µm for A and H; = 10 µm for B–E; = 20 µm for F, and = 50 µm for G.

(Sergeev *et al.*, 1995), and Gaoyuzhuang Formation, China (Zhang, 1981; Seong-Joo and Golubic, 1998, 1999); and the ca. 1650 Ma Amelia Dolomite, Australia (Muir, 1976).

Once again, this mechanism of preservation is not time-invariant. With the Cambrian evolution of siliceous sponges and, especially, radiolarians, silica ceased to leave the oceans as an early stage evaporite, and the microfossil record of tidal mat and coastal cyanobacteria all but disappeared. Similarly, for the first half of Earth history, when much silica left the oceans by adsorption onto ferric hydroxides precipitated from iron-rich waters, silicification of coastal carbonates appears to have been less common and occurred later in diagenesis (Fischer and Knoll, in press). Thus, the structure of the sedimentary record imparts an important pattern to the record of cyanobacterial fossils. Proterozoic microfossils suggest a golden age of cyanobacteria not only because cyanobacteria were abundant and widespread (they were), but because both the earlier processes of iron deposition and the later evolution of bioturbating and biomineralizing eukaryotes closed the principal windows of cyanobacterial preservation. This shifting window of preservation must be kept in mind in any attempt to read evolutionary history from fossil cyanobacteria.

The microfossil record

Since the discovery of silicified fossils in the ca. 800 Ma Bitter Springs Formation (Barghoorn and Schopf, 1965; Schopf, 1968), it has been clear that cyanobacteria-like microfossils are important components of Proterozoic sedimentary rocks (Knoll and Golubic, 1992; Golubic and Seong-Joo, 1999; Sergeev *et al.*, 2002). Over the years, careful comparisons of morphology, development, behavior (as recorded by specific orientation), and ecological distribution (as evaluated by facies relationships) have strengthened the case that diverse and abundant cyanobacteria inhabited Proterozoic oceans.

Fossil cyanobacteria are interpreted on the basis of population-level comparisons with living blue-greens. The earliest example of such an ancient-modern comparison, and still one of the most compelling, is the interpretation of fossil *Eoentophysalis* in terms of its modern counterpart, *Entophysalis* (Figure 1.2g). Hofmann (1976) described large and exceptionally well preserved populations of *Eoentophysalis* from silicified peritidal carbonates of the ca. 2000 Ma Belcher Supergroup, Canada. In a companion paper, Golubic and Hofmann (1976) made a point-by-point comparison of these fossils with species of *Entophysalis* found today in Australia and the Persian Gulf. Size, division pattern, pigmentation, and colony form were all closely similar, as were the morphologies of microbial mats accreted by these populations and their environmental setting. Subsequent research has shown that entophysalid cyanobacteria were principal components of tidal flat microbial mats for at least a billion years, beginning no later than 2000 Ma. Entophysalid fossils are not as common in younger (< 1000 Ma) Proterozoic successions, but this, again, may reflect the changing window through which we view Precambrian fossils rather than evolutionary replacement. Entophysalids remain important today in tidal flat mats, but after ca. 1000 Ma, such environments grew increasingly less important as loci for seafloor carbonate precipitation, decreasing the probability that entophysalid mat populations would be buried and preserved (Kah and Knoll, 1996; Sharma and Sergeev, 2004).

Polybessurus has the opposite distribution, being uncommon in older Proterozoic cherts but abundant in tidal flats formed 1200 Ma and later (Kah and Knoll, 1996). These distinctive fossils consist of coccoidal cells enveloped within multiple envelopes that were elongated in the downward (substrate) direction (Figure 1.2f). Closely similar counterparts related to *Cyanostylon* live today on tidal flats on the Bahamas Banks, where they are associated with the accumulation of fine-grained particulate carbonate (Green *et al.*, 1987). Changing patterns of carbonate deposition through the Proterozoic included an increase in mud generation coincident with the previously noted decline in peritidal seafloor precipitation.

Such fossils join a diverse assemblage of rod-shaped and coccoid microfossils assigned to Group I cyanobacteria with varying levels of confidence (e.g. Schopf, 1968; Knoll *et al.*, 1991; Figure 1.2a–d,h). Many of these fossils are so simple in morphology that attribution to the cyanobacteria must rest principally on assumptions about the likelihood of preservation. A few, however, including large populations of *Gloeocapsa*- and *Gloeothece*-like fossils for which both life cycle and taphonomic alteration are well characterized, suggest strongly that Group I cyanobacteria were abundant and diverse constituents of Proterozoic ecosystems (Knoll and Golubic, 1979; Campbell *et al.*, 1979).

Group II cyanobacteria are represented by ca. 800 Ma *Palaeopleurocapsa* (Figure 1.3a; Knoll *et al.*, 1975), its name underscoring a close similarity to extant *Pleurocapsa*; 1400–1500 Ma epibionts ascribed to the Chaemosiphonales by Seong-Joo *et al.* (1999); and a diversity of endolithic cyanobacteria found within silicified Proterozoic ooids (Figure 1.3b,c). In morphology, development and inferred behavior, endolithic *Eohyella* microfossils bear a close relationship to *Hyella* species that bore into ooids in the present day Bahamas and the Persian Gulf (Green *et al.*, 1988; Al-Thukair and Golubic, 1991). The oldest known hyelloid endoliths occur in ca. 1700 Ma cherts from China (Zhang and Golubic, 1987).

Simple filaments, or trichomes, classically included in Group III, occur throughout the Proterozoic record (Figure 1.4a–e; Mendelson and Schopf, 1992). *Oscillatoria, Lyngbya, Spirulina,* and *Microcoleus*-like fossils are common constituents of silicified assemblages, and the empty sheaths of filaments reasonably related to LPP cyanobacteria occur in both flat microbial laminites and morphologically complex stromatolites. Other bacteria, including both photosynthetic bacteria and sulfur-oxidizers, can produce morphologically similar fossils; nonetheless, the specific and widespread environmental distribution

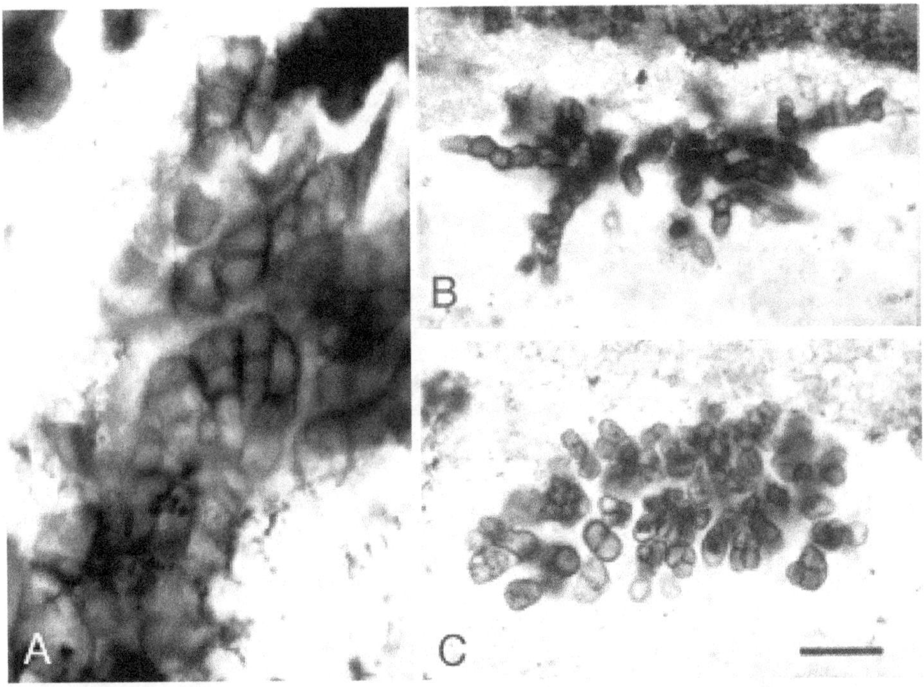

Figure 1.3 Proterozoic fossils interpreted as Group II cyanobacteria. A, *Palaeopleurocapsa wopfnerii*, ca. 800 Ma Skillogalee Formation, Australia; B and C, *Eohyella* spp. from the 700–800 Ma Upper Eleonore Bay Group, Greenland. Bar in C = 15 μm for A and = 50 μm for B and C.

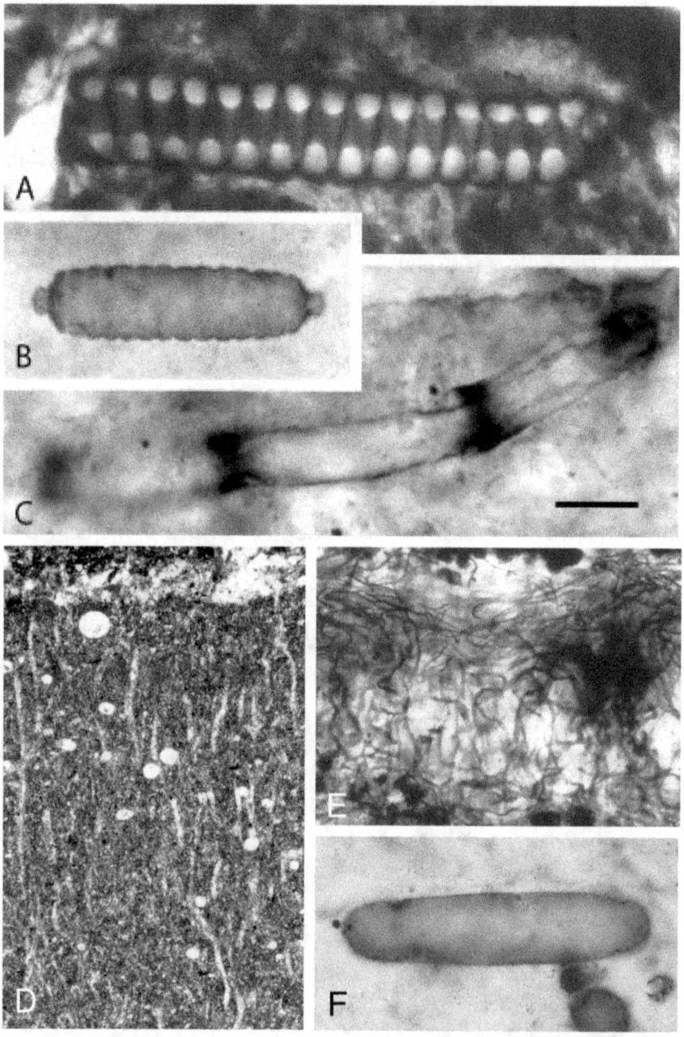

Figure 1.4 Proterozoic fossils interpreted as Group III and IV cyanobacteria. A, *Obruchevella parva*, ca. 600–545 Ma Tinya Formation, Siberia; B, *Filiconstricosus cephalon*, ca. 1500 Ma Bil'yakh Group, Siberia; C, *Circumvaginalis elongatus*, Bil'yakh Group; D, vertically oriented cyanobacterial filaments preserved as molds in a carbonate rock, 700–800 Ma Draken Formation, Spitsbergen—ovoid structures are sections through testate amoebans; E, mat-forming population of *Siphonophycus inornatum*, showing alternating vertical and horizontal orientation of interwoven sheaths, 800–700 Ma Upper Eleonore Bay Group, Greenland; and F, *Archaeoellipsoids grandis*, Bil-yakh Group. Bar in C = 20 μm for A, B and F, = 50 μm for C, = 150 μm for D, and = 250 μm for E.

of these fossil populations supports their general interpretation in terms of Group III cyanobacteria (see Knoll, 1989, for discussion).

What about heterocyst- and akinete-differentiating cyanobacteria? Heterocysts do not preserve well. Indeed, the only widely accepted examples of cyanobacterial thalli with recognizable heterocysts are *Stigonema*-like fossils preserved in the ca. 400 Ma Rhynie Chert, Scotland (Croft and George, 1959). In contrast, akinetes have a relatively high preservation potential, and relatively large, elongate microfossils found widely in Mesoproterozoic (1600–1000 Ma) cherts have been interpreted as nostocalean akinetes (Figure fd; Golubic *et al.*, 1995; Srivastava, 2005). The oldest fossils interpreted with confidence as akinetes occur in ca. 1650 Ma rocks from Australia (Tomitani *et al.*, 2006); the oldest candidate akinetes come from ca. 2100 Ma cherts in Gabon (Amard and Bertrand-Sarfati, 1997). The distinctive, funnel-in-funnel construction of *Scytonema* sheaths is also apparent in mid-Proterozoic cherts from Siberia (Figure 1.4c; Sergeev *et al.*, 1995).

Finally, the endosymbiotic events that spread photoautotrophy through the Eukarya can be viewed cyanocentrically as a major ecological expansion of the cyanobacteria. Molecular clock estimates for the origin of eukaryotic photosynthesis vary, but commonly fall in the mid-Proterozoic (e.g. Yoon *et al.*, 2004). Unambiguous red algae in ca. 1200 Ma cherts (Butterfield, 2000) and possible phycomata of prasinophyte green algae in rocks as old as 1700 Ma (Knoll *et al.*, 2006) are in broad accord with molecular estimates.

Proterozoic rocks clearly preserve an extensive record of early cyanobacteria. 2000–1500 Ma assemblages suggest that the principal features of cyanobacterial diversity were already in place by this time. Unfortunately, for taphonomic reasons outlined above, as well as the simple geological circumstance that older sedimentary rocks are less abundant than younger successions, the trail of microfossils grows cold at about 2000 Ma. Poorly preserved fossils that may be cyanobacteria occur in ca. 2500 Ma cherts from South Africa (Altermann and Schopf, 1995; Klein *et al.*, 1987; Lanier, 1986), but all reports of microfossils in older rocks are subject to ongoing debate (see below). In summary, then, the preservational window through which we view ancient cyanobacteria opened widely about 2000 Ma onto a world in which cyanobacterial radiation had already taken place.

Archean cyanobacteria?

What about the Archean (> 2500 Ma) record? For the past four decades, many in the scientific community have accepted the hypothesis that cyanobacteria evolved early in Earth history, perhaps as early as 3800 Ma. This creates an interesting evolutionary issue, as cyanobacteria fall on a distal branch of the bacterial tree, requiring that major bacterial groups diverged very early in our planet's history. It also introduces a geobiological problem because there is little evidence of persistent oxygen in the atmosphere and surface ocean until 2450–2320 Ma (Bekker *et al.*, 2004; Holland 2006). These two points can, in principle, be reconciled in any one of several ways (see below), but in recent years, an increasingly large number of geobiologists has asked whether the Archean record has been interpreted correctly. As alternative sources of abundant oxygen are unknown, the basal Proterozoic appearance of a stably oxic atmosphere and surface ocean sets a minimum date for cyanobacterial origins (Catling and Clair, 2005). Kopp *et al.* (2005) have proposed that the "Great Oxygenation Event" does indeed record the origin of oxygenic photosynthesis.

Tests of the Kopp *et al.* hypothesis necessarily lie in the geobiological analysis of Archean sedimentary rocks. As noted above, this record is notably poor in interpretable microfossils. Microfossil populations have been reported from cherts as old as ca. 3500 Ma (summarized in Schopf, 2006), but their biogenicity has engendered vigorous debate, focused principally on purported filaments from the Apex Chert of the Warrawoona Group, Australia (Schopf, 1993). Brasier *et al.* (2002, 2005, 2006) argue that microstructures in Warrawoona cherts formed abiologically during diagenesis; Schopf (2006; Schopf *et al.*, 2002) maintains that actual microfossils are preserved. Both parties have trained sophisticated technology on the problem, but without resolution. Raman spectroscopy indicates that the structures contain organic matter (as opposed to graphite), but whether the organics originated as microbial cell walls, biogenic material that coated crystals during diagenesis, or abiotic organic matter cannot be ascertained using this technique (Pasteris and Wopneka, 2003). How the debate will end—and how it will inform reconsideration of other microstructures interpreted as Early Archean microfossils (tabulated in Schopf, 2006) remains to be seen. Perhaps the one existing point of consensus is that even if the Warrawoona and other coeval microstructures are accepted as fossils, they carry little in the way of phylogenetic or physiological information (e.g. Knoll, 2003) and therefore cannot shed light on the question of cyanobacterial antiquity.

Failing resolution through micropaleontology, we must turn to other potential records of ancient cyanobacteria. Stromatolites are laminated structures in (mostly) carbonate rocks that accreted upward from a point or surface on an ancient seafloor or lake bottom (Figure 1.5a). Stromatolites formed on tidal flats can be planar, but many such structures are domal, columnar, or conoidal, reflecting accretion in a broad range of subtidal physical environments. Stromatolites are commonly interpreted in terms of the interaction between

Figure 1.5 Microbial influences on Proterozoic sedimentation. A, columnar stromatolites likely accreted by cyanobacterial mats, ca. 1500 Ma Bil'yakh Group, Siberia; and B, microbially induced sedimentary structures on a rippled sand sandstone bed surface, ca. 548 Ma Nama Group, Namibia—note that the mat has cast the holdfast of an Ediacaran animal in the upper center of the image. Largest column in A is 15 cm cross; coin in B is 2 cm in diameter.

microbial communities and physical sedimentation. Indeed, many geobiologists have historically accepted ancient stromatolites as proxies for cyanobacterial mat builders (e.g. Semikhatov *et al.*, 1979). This interpretation derives from the observation that cyanobacteria are the principal architects of stromatolites in modern lakes and shallow seas (Andres and Reid, 2005). (Eukaryotic algae can play an important role in microbialite accretion, but tend to impart distinctive textures recognizable in ancient microbialites; Feldmann and McKenzie, 1998). Few ancient stromatolites preserve a microfossil record of their builders; thus, the attribution of ancient stromatolites to cyanobacterial mat communities rests on analogy to modern systems. Nonetheless, few would dispute the hypothesis that Proterozoic and Phanerozoic stromatolites are dominated by structures accreted under the influence of cyanobacteria.

As with microfossils, the problem lies with the Archean record, where the assumption that mat communities included cyanobacteria is less easily justified. Moreover, the mechanism by which many Archean stromatolites were built differs from that of most younger Proterozoic examples. Stromatolites can accrete either by the trapping and building of fine-grained carbonate particles (followed by cementation) or by the precipitation of crystalline carbonate minerals. Trapped and bound stromatolites display a range of morphologies and microscopic textures that unambiguously record the activities of microbial mats in their accretion, but many precipitated stromatolites lack these features and can closely resemble laminated tufas or sinters that accrete without the templating influence of microbial mats (Grotzinger and Knoll, 1999). Younger Proterozoic stromatolites are mostly trapped and bound and, therefore, are uncontroversially biogenic. Moving backward in time, however, precipitated stromatolites become more common, constituting a majority of stromatolitic forms in Early Archean successions (Grotzinger and Knoll, 1999). Unambiguously trapped and bound stromatolites are well described from carbonate deposits as old as ca. 3000 Ma (Beukes and Lowe, 1989), and even the oldest known stromatolites, those of the Warrawoona Group, include conoidal structures not likely to accrete in the absence of microbial influence (Allwood *et al.*, 2006). Thus, I read the overall stromatolite record as an indication that benthic microbial communities existed 3500 Ma. Whether cyanobacteria participated in those communities cannot be ascertained from evidence currently in hand. (Note: this does not mean that all Archean stromatolites are best interpreted as biogenic structures; tufa-like precipitates occur, as well, requiring that each new occurrence of stromatolites be carefully and individually evaluated.)

Microbial mats influence the accumulation of sands as well as carbonate sediments. The product of these interactions is not a conspicuous three-dimensional stromatolite but rather a subtle but distinctive pattern on bedding surfaces that is not easily mimicked by purely physical processes (Figure 1.5b; Noffke *et al.*, 2001). Microbially induced sedimentary features have been reported from sandstones as old as 3200 Ma (Noffke *et al.*, 2006), providing independent perspective on the antiquity of benthic microbial communities, if not necessarily cyanobacteria *per se*. Rug-like sedimentary structures called "roll-ups," generated as currents ripped up and then redeposited coherent but unlithified surface layers, also occur in successions as old as ca. 3416 Ma (Tice and Lowe, 2004). In younger rocks, such structures are widely accepted as redeposited microbial mats, and this appears to provide the best explanation for the Early Archean features, as well.

Cyanobacterial antiquity has also been inferred from geochemical data. Carbon isotopic abundances for co-occurring carbonate and organic carbon are consistent with carbon fixation based on ribulose *bis*phosphate carboxylase/oxygenase (Rubisco) throughout recorded Earth history (data summarized in Schopf, 2006). In the absence of corroborating data from fossils or molecular biomarkers, one cannot tell whether the primary producers in question were cyanobacteria, anoxygenic photosynthetic or chemosynthetic bacteria that fix carbon via Rubisco, or microbial communities with other metabolisms that collectively imparted an isotopic signature to organic matter similar to that of Rubsico-bearing autotrophs (Hayes *et al.*, 2006; see below). Recently, Ueno *et al.* (2006) reported isotopic evidence of methanogenic archaeans in Early Archean ecosystems, confirming that by 3500 Ma the carbon cycle was already complex.

Cyanobacteria are principal sources of 2-methylhopanoids with an extended polyhydroxylated side chain (Summons *et al.*, 2006). These lipids, preserved in rocks as old as 2700 Ma (Brocks *et al.*, 1999, 2003), may provide a good record of cyanobacterial antiquity. What remains to be demonstrated, however, is that these compounds are synthesized *only* by cyanobacteria.

Clearly, geobiological discussions of Earth's oldest geologic record must distinguish between evidence for life and evidence of cyanobacteria. There is widespread consensus that life existed 3500 Ma (Knoll, 2003; Schopf, 2006; Brasier *et al.*, 2006), but the phylogenetic and physiological characterization of Early Archean communities is poor. Carbon isotopic evidence from still older (3800 Ma) metamorphosed sedimentary rocks of the Isua succession in southwestern Greenland suggest a still older minimum date for the emergence of life (Rosing, 1999; Fedo *et al.*, 2006).

The Early Archean geobiological record is permissive, and save for the conspicuous absence of geochemical evidence for oxygen, I know of no evidence that is inconsistent with the presence of cyanobacteria 3500–3800 Ma. That stated, few if any data compel us to believe that cyanobacteria originated so early (Figure 1.6). Available observations are equally consistent with a world fueled by anaerobic metabolisms, including photoautotrophy that used ferrous iron or hydrogen as an electron source (e.g. Kappler *et al.*, 2005; Tice and Lowe, 2006). By 2700–2800 Ma, the record is better, and geochemical evidence records microbial carbon and surface cycles somewhat similar to those operating today. Stromatolitic (Buick, 1992) and biomarker (Brocks *et al.*, 1999, 2003) evidence more strongly supports the hypothesis that cyanobacteria existed, although unequivocal rejection of alternate explanations remains difficult.

We are left with an incompletely resolved issue in historical geobiology. Molecular clocks can, in principle, provide independent estimates of cyanobacterial antiquity, but published clock dates for cyanobacterial divergence vary widely and error estimates are large (e.g. Graur and Martin, 2004). Better clock estimates might be obtained using fossils to constrain internal nodes on the cyanobacterial tree (Tomitani *et al.*, 2006), but the court of last resort will continue to be the geologic record.

Toward a geobiological history of cyanobacteria

Life on Earth originated at least 3500 Ma and possibly much earlier. By 2450–2320 Ma, cyanobacteria must have been important primary producers in the world's oceans, and they could have emerged earlier. I confess to a preference for hypotheses in which cyanobacteria

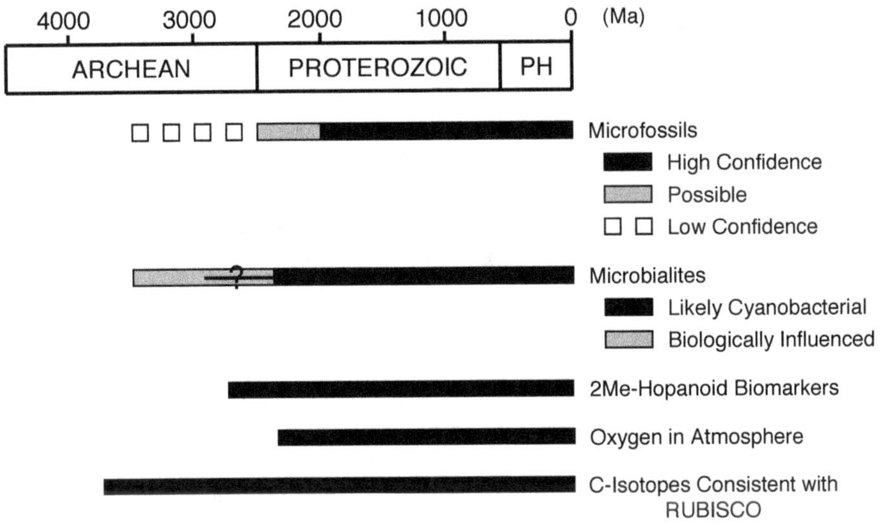

Figure 1.6 Summary of the stratigraphic distribution of fossils and other geobiological features interpreted as evidence for cyanobacteria in Archean and Proterozoic oceans. Cyanobacteria were a major presence in Proterozoic ecosystems; the extent to which they populated Archean oceans remains uncertain. PH = Phanerozoic.

appear on the order of 3000–2700 Ma, significantly before the earliest Proterozoic rise of atmospheric oxygen but well after deposition of the oldest known records of life. This requires that biospheric oxygenation postdate the origin of oxygenic photosynthesis by at least several hundred million years.

Oxygen accumulates in the atmosphere and oceans when production of O_2 by oxygenic photosynthesis exceeds the consumption of oxygen by physical and biological processes. Aerobic respiration is the principal metabolic process that consumes oxygen. Conceivably, aerobic metabolism could have originated in local oxygen oases prior to the global build-up of atmospheric O_2, but aerobic respiration likely assumed global importance only after 2450–2320 Ma, when oxygen gained permanent foothold in the atmosphere. Physical processes that consume oxygen include the oxidation of reduced chemical species in the oceans, predominantly Fe^{++}, and the oxidation of reduced volcanic gases (Hayes and Waldbauer, 2006). Fluxes of reduced materials into the oceans and atmosphere were likely larger early in Earth history. Moreover, hydrogen escape from the upper atmosphere would have been enhanced in the presence of large methane fluxes to the atmosphere, an ecological circumstance suggested by the unusual ^{13}C-depletion of carbon isotopes in Late Archean and early Paleoproterozoic organic matter (Catling and Claire, 2005).

On the other side of the ledger, environmental oxygen levels could have remained low long after the origin of oxygenic photosynthesis if rates of cyanobacterial photosynthesis were limited. Primary production by cyanobacteria might, in principle, have been limited by competition from anoxygenic phototrophs in anoxic oceans, although the strong limitation of electron donors on anoxygenic photosynthesis would seem to favor organisms capable of splitting water. Alternatively, primary production could have been limited in early cyanobacteria that lacked the molecular capacity to fix nitrogen (Grula, 2005); the pattern of phylogenetic spread of nitrogen fixation through the cyanobacteria remains uncertain (Turner *et al.*, 2001; Tomitani *et al.*, 2006), but currently recognized basal lineages do not exhibit this trait (Sánchez-Baracaldo *et al.*, 2005).

Regardless of Archean uncertainties, cyanobacteria made possible the initial oxygenation of the atmosphere and surface oceans, and blue-greens came to dominate primary production and nitrogen fixation in the oceans (see Chapter 15). Of course, cyanobacteria still account for the bulk of biological nitrogen-fixation in the marine realm (Montoya *et al.*, 2004; Davis and McGillicuddy, 2006). Fossils indicate that the major phylogenetic, physiological and morphological groupings of cyanobacteria diverged no later than the Paleoproterozoic Era, although shallow branches on some parts of the cyanobacterial tree indicate continuing diversification within groups.

It is important to note that the initial rise of biospheric oxygen did not result in an ocean and atmosphere with present day redox structure. An increasing volume of geochemical data supports the hypothesis that atmospheric oxygen levels remained relatively low (perhaps 10% of modern levels) until late in the Proterozoic Eon, with oxygen minimum zones in the oceans commonly becoming sulfidic (Brocks *et al.*, 2005, and references therein). Eukaryotic algae emerged during this interval (Knoll *et al.*, 2006), but paleontology and biomarker geochemistry agree that photosynthetic bacteria remained the principal primary producers in many parts of the ocean until about 700–600 Ma (Knoll *et al.*, in press). Indeed, low oxygen levels may have facilitated the continuing ecological dominance of cyanobacteria. In oceans where oxic surface waters were underlain by a sulfidic oxygen-minimum zone, bacterial denitrification and anammox (anaerobic ammonia oxidation) reactions would strongly limit the amount of fixed nitrogen in upwelling waters, favoring cyanobacteria capable of fixing and/or effectively scavenging nitrogen (Anbar and Knoll, 2002; Fennel *et al.*, 2005; Kuypers *et al.*, 2005). Mo, required as a cofactor in the nitrate reductase that makes nitrogen bioavailable to eukaryotes, might further have limited primary production by eukaryotes (Anbar and Knoll, 2002). In sulfidic water masses, Mo is efficiently removed via the formation of insoluble thiomolybdate compounds. It may be no coincidence, therefore, that algae rose to taxonomic and ecological prominence at about the time when the oxygen levels increased to something approaching modern abundances.

Stromatolites and biomarker molecules indicate that cyanobacteria remained important primary producers on shelves of Paleozoic continents. Indeed, it was only with the Mesozoic radiation of dinoflagellates, coccolithophorids, and diatoms that blue-greens ceased to be important players in continental shelf ecology (Falkowski *et al.*, 2004; Knoll *et al.*, in press). In open ocean waters, cyanobacteria remain important as primary producers today.

Conclusions

The cyanobacteria arguably constitute the most important group of organisms ever to appear on our planet. Their capacity to use water as an electron source in photosynthesis freed primary production from earlier limits imposed by the distribution and abundances of alternative electron sources such as H_2S, H_2, and Fe^{2+} (Kharecha *et al.*, 2005). Their ability to couple primary production to nitrogen fixation further increased the biological carrying capacity of the oceans (Grula, 2005). And cyanobacterial production of oxygen gas transformed environments in a way that made novel physiologies and morphologies (including animals) possible. As the photosynthetic partner in the primary endosymbiotic event that introduced photoautotrophy to eukaryotes, cyanobacteria contributed to the environmental transformation of land as well as the sea. The key events in this history took place during the Archean and Proterozoic eons, but their consequences continue to shape nearly every ecosystem on Earth.

Acknowledgments

Research leading to this paper was supported in part by NSF Grant EAR 0420592. I thank S. Golubic, V.N. Sergeev, and A. Tomitani for many helpful discussions.

References

Allwood, A.C., Walter, M.R., Kamber, B.S., Marshall, C.P., and Burch, I.W. (2006). Stromatolite reef from the Early Archaean era of Australia. Nature *441*, 714–718.

Altermann, W., and Schopf, J.W. (1995). Microfossils from the Neoarchean Campbell Group, Griqualand West sequence of the Transvaal Supergroup, and their paleoenvironmental and evolutionary implications. Precambrian Res. *75*, 65–90.

Al-Thukair, A.A., and Golubic, S. (1991). Five new *Hyella* species from the Arabian Gulf. Algol. Stud. *64*, 167–197.

Amard, B., and Bertrand-Sarfati, J. (1997). Microfossils in 2000 Ma old cherty stromatolites of the Franceville Group, Gabon. Precambrian Res. *81*, 197–221.

Anbar, A.D., and Knoll, A.H. (2002). Proterozoic ocean chemistry and evolution: A bioinorganic bridge? Science *297*, 1137–1142.

Andres, M.S., and Reid, R.P. (2006). Growth morphologies of modem marine stromatolites: A case study from Highborne Cay, Bahamas. Sed. Geol. *185*, 319–328.

Barghoorn, E.S., and Schopf, J.W. (1965). Microorganisms from Late Precambrian of central Australia. Science *150*, 337–340.

Bartley, J.K. (1996), Actualistic taphonomy of cyanobacteria: Implications for the Precambrian fossil record. Palaios *11*, 571–586.

Bekker, A., Holland, H.D., Wang, P.L., Rumble, D., Stein, H.J., Hannah, J.L., Coetzee, L.L., and Beukes, N.J. (2004). Dating the rise of atmospheric oxygen. Nature *427*, 117–120.

Beukes, N.J., and Lowe, D.R. (1989). Environmental control on diverse stromatolite morphologies in the 3000 Myr Pongola Supergroup, South-Africa. Sedimentology *36*, 383–397.

Bjerrum, C.J., and Canfield, D.E. (2002). Ocean productivity before about 1.9 Gyr ago limited by phosphorus adsorption onto iron oxides. Nature *417*, 159–162.

Bosak, T., and Newman, D.K. (2003). Microbial nucleation of calcium carbonate in the Precambrian. Geology *31*, 577–580.

Brasier, M.D., Green, O.R., Jephcoat, A.P., Kleppe, A.K., van Kranendonk, M.J., Lindsay, J.F., Steele, A., and Grassineau, N.V. (2002). Questioning the evidence for Earth's oldest fossils. Nature *416*, 76–81.

Brasier, M.D., Green, O.R., Lindsay, J.F., McLoughlin, N., Steele, A., and Stoakes, C. (2005). Critical testing of Earth's oldest putative fossil assemblage from the 3.5 Ga Apex Chert, Chinaman Creek Western Australia. Precambrian Res. 140, 55–102.

Brasier. M., McLoughlin, N., Green, O., and Wacey, D. (2006). A fresh look at the fossil evidence for early Archaean cellular life. Phil. Trans. R. Soc. B. *361*, 887–902.

Brocks, J.J., Logan, G.A., Buick, R., and Summons, R.G. (1999). Archean molecular fossils and the early rise of eukaryotes. Science *285*, 1033–1036.

Brocks, J.J., Buick, R., Summons, R.E., and Logan, G.A. (2003). A reconstruction of Archean biological diversity based on molecular fossils from the 2.78 to 2.45 billion-year-old Mount Bruce Supergroup, Hamersley Basin, Western Australia. Geochim. Cosmochim. Acta *67*, 4321–4335.

Brocks, J.J., Love, G.D., Summons, R.E., Knoll, A.H., Logan, G.A., and Bowden, S.A. (2005). Biomarker evidence for green and purple sulphur bacteria in a stratified Palaeoproterozoic sea. Nature *437*, 866–870.

Buick, R. (1992). The antiquity of oxygenic photosynthesis: evidence from stromatolites in sulfate-deficient Archean lakes. Science *255*, 74–77.

Buick, R., Dunlop, J.S.R., and Groves, D.I. (1981). Stromatolite recognition in ancient rocks: an appraisal of irregularly laminated structures in an early Archaean chert-barite unit from North Pole, Western Australia. Alcheringa 5, 161–181.

Butterfield, N.J. 2000 *Bangiomorpha pubescens* n. gen., n. sp.: implications for the evolution of sex, multicelluarity and the Mesoproterozoic/Neoproterozoic radiation of eukaryotes. Paleobiology *26*, 386–404.

Butterfield, N.J., Knoll, A.H., and Swett, K. (1994). Paleobiology of the Upper Proterozoic Svanbergfjellet Formation, Spitsbergen. Fossils Strata *34*, 1–84.

Canfield, D.E. (1998). A new model for Proterozoic ocean chemistry. Nature *396*, 450–453.

Canfield, D.E., and Raiswell, R. (1999). The evolution of the sulfur cycle: Am. J. Sci. *299*, 697–723.

Canfield, D.E., and Teske, A. (1996). Late Proterozoic rise in atmospheric oxygen concentration inferred from phylogenetic and sulphur-isotope studies. Nature 382, 127–132.

Canfield, D.E., Habicht, K.S., and Thamdrup, B. (2000). The Archean sulfur cycle and the early history of atmospheric oxygen. Science *288*, 658–661.

Catling, D.C., Zahnle, K.J., and McKay, C.P. (2001). Biogenic methane, hydrogen escape, and the irreversible oxidation of early Earth. Science *293*, 839–843.

Catling, D.C., and Claire, M.W. (2005). How Earth's atmosphere evolved to an oxic state: A status report. Earth Planet. Sci. Lett. *237*, 1–20.

Chafetz, H.S., and Buczynski, C. (1992). Bacterially induced lithification of microbial mats. Palaios *7*, 277–293.

Croft, W.N., and George, E.A. (1959). Blue-green algae from the Middle Devonian of Rhynie, Aberdeenshire. Bull. Brit. Mus. (Nat. Hist.) Geology *3*, 341–353.

Davis, C.S., and McGillicuddy, D.J. (2006). Transatlantic abundance of the N_2-fixing colonial cyanobacterium *Trichodesmium*. Science *312*, 1517–1520.

Des Marais, D.J. (1997). Isotopic evolution of the biogeochemical carbon cycle during the Proterozoic Eon. Org. Geochem. *27*, 185–193.

Falkowski, P.G., Katz, M.E., Knoll, A.H, Quigg, A., Raven, J.A., Schofield, O., and Taylor, F.J.R. (2004). The evolution of modern eukaryotic phytoplankton. Science *305*, 354–360.

Farquhar, J., Bao, H.M., and Thiemens, M. (2000). Atmospheric influence of Earth's earliest sulfur cycle. Science *289*, 756–758.

Fedo, C.M., Whitehouse, M.J., and Kamber, B.S. (2006). Geological constraints on detecting the earliest life on Earth: a perspective from the Early Archaean (older than 3.7 Gyr) of southwest Greenland. Phil. Trans. R. Soc. B *361*, 851–867.

Feldmann, M., and McKenzie, J.A. (1998). Stromatolite-thrombolite associations in a modern environment, Lee Stocking Island, Bahamas. Palaios *13*, 201–212.

Fennel, K., Follows, M., and Falkowski, P.G. (2005). The co-evolution of the nitrogen, carbon, and oxygen cycles in the Proterozoic ocean. Am. J. Sci. *305*, 526–545.

Fischer, W.W., and Knoll, A.H. (in press) An Fe shuttle for deep-water SiO_2 in late Archean iron formation. Earth Planet. Sci. Lett.

Giovannoni, S.J., Turner, S., Olsen, G.J., Barnes, S., Lane, D.J., and Pace, N.R. (1988). Evolutionary relationships among cyanobacteria and green chloroplast. J. Bacteriol. *170*, 3584–3592.

Golovenok, V.K., Belova, M.Yu., and Avdeeva, V.I. (1990). Giant *Obruchevella* from Vendina strata of the Siberian Platform. Dokl. Akad. Nauk SSSR *315*, 193–196. (In Russian)

Golubic, S., and Barghoorn, E.S. (1977). Interpretation of microbial fossils with special reference to the Precambrian. In Fossil Algae, E. Flügel, ed. (Berlin: Springer-Verlag), pp. 1–14.

Golubic, S., and Campbell, S.E. (1979). Analogous microbial forms in recent subaerial habitats and in Precambrian cherts: *Gloeothece coerulea* Geitler and *Eosynechococcus moorei* Hofmann. Precambrian Res. *8*, 201–217.

Golubic, S., and Hofmann, H.J. (1976). Comparison of Holocene and mid-Precambrian Entophysalidaceae (Cyanophyta) in stromatolitic algal mats: cell division and degradation. J. Paleontol. *50*, 1074–1082.

Golubic, S., and Seong-Joo, L. (1999). Early cyanobacterial fossil record: preservation, palaeoenvironments and identification. Eur. Phycol. *34*, 339–348.

Golubic, S., Sergeev, N.V., and Knoll, A.H. (1995). Mesoproterozoic *Archaeoellipsoides*: akinetes of heterocystous cyanobacteria. Lethaia *28*, 285–298.

Graur, D., and Martin, W. (2004). Reading the entrails of chickens: Molecular timescales of evolution and the illusion of precision. Trends Genet. *20*, 80–86.

Green, J.W., Knoll, A.H., Golubic, S., and Swett, K. (1987). Paleobiology of distinctive benthic microfossils from the Upper Proterozoic Eleonore Bay Group, central East Greenland. Am. J. Bot. *74*, 928–940.

Green, J.W., Knoll, A.H., and Swett, K. (1988). Microfossils from oolites and pisolites of the Upper Proterozoic Eleonore Bay Group, central East Greenland. J. Paleontol. *62*, 835–852.

Grotzinger, J.P., and Knoll, A.H. (1999). Precambrian stromatolites: Evolutionary milestones or environmental dipsticks? Ann. Rev. Earth Planet. Sci. *27*, 313–358.

Grula, J.W. (2005). Evolution of photosynthesis and the biospheric oxygenation contingent upon nitrogen fixation? Int. J. Astrobiol. *4*, 251–257.

Hayes, J.R., and Waldbauer, J.R. (2006). The carbon cycle and associated redox processes through time. Phil. Trans. R. Soc. B. *361*, 931—950.

Hofmann, H.J. (1976). Precambrian microflora, Belcher Islands, Canada: significance and systematics. J. Paleontol. *50*, 1040–1073.

Hofmann, H.J., Grey, K., Hickman, A.H., and Thorpe, R.I. (1999). Origin of 3.45 Ga coniform stromatolites in Warrawoona Group, Western Australia. Geol. Soc. Am. Bull. *111*, 1256–1262.

Holland, H.D. (2006). The oxygenation of the atmosphere and oceans. Phil. Trans. R. Soc. B *361*, 903—915.

Kah, L., and Knoll, A.H. (1996). Microbenthic distribution of Proterozoic tidal flats: environmental and taphonomic considerations. Geology *24*, 79–82.

Kappler, A., Pasquero, C., Konhauser, K.O., and Newman, D.K. (2005). Deposition of banded iron formations by anoxygenic phototrophic Fe(II)-oxidizing bacteria. Geology *33*, 865–868.

Kharecha, P., Kasting, J., and Siefert, J. (2005). A coupled atmosphere–ecosystem model of the early Archean Earth. Geobiology *3*, 53–76.

Klein, C., Beukes, N.J, and Schopf, J.W. (1987). Filamentous microfossils in the early Proterozoic Transvaal Supergroup: Their morphology, significance, and paleoenvironmental setting. Precambrian Res. *36*, 81–94.

Knoll, A.H. (1989). The paleomicrobiological information in Proterozoic rocks. In Microbial Mats: Physiological Ecology of Benthic Microbial Communities, Y.Cohen and E. Rosenberg, eds. (Washington, DC: ASM Press), pp. 469–485.

Knoll, A.H. (2003). Life on a Young Planet. (Princeton: Princeton University Press).

Knoll, A.H., and Barghoorn, E.S. (1997). Archean microfossils showing cell division from the Swaziland System of South Africa. Science *198*, 396–398.

Knoll, A.H., and Golubic, S. (1979). Anatomy and taphonomy of a Precambrian algal stromatolite. Precambrian Res. *10*, 115–151.

Knoll, A.H., and S. Golubic (1992). Living and Proterozoic cyanobacteria. In Early Organic Evolution: Implications for Mineral and Energy Resources, M. Schidlowski *et al.*, eds. (Berlin: Spring-Verlag), pp. 450–462.

Knoll, A.H., and Semikhatov, M.A. (1998). The genesis and time distribution of two distinctive Proterozoic stromatolitic microstructures. Palaios *13*, 407–421.

Knoll, A.H., Barghoorn, E.S., and Golubic, S. (1975). *Paleopleurocapsa wopfnerii* gen. et sp. nov.: A Late Precambrian alga and its modern counterpart. Proc. Natl. Acad. Sci., USA *72*, 2488–2492.

Knoll, A.H., Fairchild, I.J., and Swett, K. (1993). Calcified microbes in Neoproterozoic carbonates: implications for our understanding of the Proterozoic/Cambrian transition. Palaios *8*, 512–525

Knoll, A.H., Hayes, J.M., Kaufman, A.J., Swett, K., and Lambert, I.B. (1986). Secular variation in carbon isotope ratios from Upper Proterozoic successions of Svalbard and East Greenland. Nature *321*, 832–838.

Knoll, A.H., Javaux, E.J., Hewitt, D., and Cohen, P. (2006). Eukaryotic organisms in Proterozoic oceans. Phil. Trans. R. Soc. B *361*, 1023–1038.

Knoll, A.H., Summons, R.E., Waldbauer, J., and Zumberge, J. (in press). The geological succession of primary producers in the oceans. In The Evolution of Primary Producers in the Sea, P. Falkowski and A.H. Knoll, eds. (New York: Academic Press).

Knoll, A.H., Swett, K., and Mark, J. (1991). The Draken Conglomerate Formation: Paleobiology of a Proterozoic tidal flat complex. J. Paleontol. *65*, 531–569.

Kopp, R.E., Kirschvink, J.L., Hilburn, I.A., and Nash, C.Z. (2005). The Paleoproterozoic snowball Earth: A climate disaster triggered by the evolution of oxygenic photosynthesis. Proc. Natl. Acad. Sci., USA *102*, 11131–11136.

Kuypers, M.M.M., Lavik, G., Woebken, D., Schmid, M., Fuchs, B.M., Amann, R., Jørgensen, B.B., and Jetten, M.S.M. (2005). Massive nitrogen loss from the Bengeula upwelling system through anaerobic ammonium oxidation. Proc. Natl. Acad. Sci. USA *102*, 6478–6483.

Maliva, R.G., Knoll, A.H., and Siever, R. (1989). Secular change in chert distribution: a reflection of evolving biological participation in the silica cycle. Palaios *4*, 519–532.

Mendelson, C.V., and Schopf, J.W. (1992). Proterozoic and selected Early Cambrian microfossils and microfossil-like objects. In The Proterozoic Biosphere: A Multidisciplinary Study, J.W. Schopf, and K. Klein, eds. (Cambridge: Cambridge University Press), pp. 865–952.

Mertz, M.U.M. (1992). The biology of carbonate precipitation by cyanobacteria. Facies *26*, 81–102.

Montoya, J.P., Holl, C.M., Zehr, J.P., Hansen, A., Villareal, T.A., and Capone, D.G. (2004). High rates of N_2 fixation by unicellular diazotrophs in the oligotrophic Pacific Ocean. Nature *430*, 1027–1031.

Muir, M.D. (1976). Proterozoic microfossils from the Amelia Dolomite, McArthur Basin, Northern Territory. Alcheringa *1*, 143–158.

Nagovitsin, K.E. (2000). Silicified microbiotas from the Upper Riphean of the Yenisey Ridge (East Siberia). Siber. Otd., Ross. Akad. Nauk, Geologia Geofizika 2000 (2–3), 7–31.

Noffke, N., Eriksson, K.A., Hazen, R.M., and Simpson, E.L. (2006). A new window into Early Archean life: Microbial mats in Earth's oldest siliciclastic tidal deposits (3.2 Ga Moodies Group, South Africa). Geology *34*, 253–256.

Noffke, N., Gerdes, G., Klenke, T., and Krumbein, W.E. (2001). Microbially induced sedimentary structures: A new category within the classification of primary sedimentary structures. J. Sed. Res. Part A *75*, 649–656.

Riding, R. (2000). Microbial carbonates: the geological record of calcified bacterial-algal mats and biofilms. Sedimentology *47* (Suppl. 1), 179–214.

Rippka, R., Deruelles, J., Waterbury, J.B., Herdman, M., and Stanier, R.Y. (1979). Generic assignments, strain histories and properties of pure cultures of cyanobacteria. J. Gen. Microbiol. *111*, 1–61.

Rosing, M.T. (1999). C^{13}-depleted carbon microparticles in > 3700-Ma sea-floor sedimentary rocks from west Greenland. Science *283*, 674–676.

Rye, R., and Holland, H.D. (1998). Paleosols and the evolution of atmospheric oxygen: A critical review. Am. J. Sci. *298*, 621–672.

Sánchez-Baracaldo, P., Hayes, P.K., and Blank, C.E. (2005). Morphological and habitat evolution in the Cyanobacteria using a compartmentalization approach. Geobiology *3*, 145–166.

Sergeev, V.N., Gerasimenko, L.M., and Zavarzin, G.A. (2002). The Proterozoic history and present state of cyanobacteria. Microbiology *71*, 623–637.

Schopf, J.W. (1968). Microflora of the Bitter Springs Formation, Late Precambrian, central Australia. J. Paleontol. *42*, 651–688.

Schopf, J.W. (1993). Microfossils of the Early Archean Apex Chert: New evidence of the antiquity of life. Science *260*, 640–646.

Schopf, J.W. (2006). Fossil evidence of Archean life. Phil. Trans. Royal Soc., London, *361B*, 861–885.

Schopf, J.W., and Blacic, J.N. (1971). New microorganisms from the Bitter Springs Formation (Late Precambrian) of the north-central Amadeus Basin, Australia. J. Paleontol. *45*, 925–960.

Schopf, J.W., Kudryavtsev, A.B., Agresti, D.G., Wdowiak, T., and Czaja, A.D. (2002). Laser-Raman imagery of Earth's earliest fossils. Nature *416*, 73–76.

Semikhatov, M.A., Gebelein, C.D., Cloud, P., Awramik, S.M, and Benmore, W.C. (1979). Stromatolite morphogenesis: Progress and problems. Can. J. Earth Sci. *16*, 992–1015.

Sergeev, V.N., Gerasimenko, L.M., and Zavarzin, G.A. (2002). The Proterozoic history and present state of cyanobacteria. Microbiology *71*, 623–637.

Sergeev, V.N., Knoll, A.H., and Grotzinger, J.P. (1995). Paleobiology of the Mesoproterozoic Billyakh Group, northern Siberia. Paleontol. Soc. Mem. *39*, 1–37.

Sharma, M., and Sergeev, V.N. (2004). Genesis of carbonate precipitate patterns and associated microfossils in Mesoproterozoic formations of India and Russia—a comparative study. Precambrian Res. *134*, 317–347.

Shen, Y., Buick, R., and Canfield, D.E. (2001). Isotopic evidence for microbial sulphate reduction in the early Archaean era. Nature *410*, 77–81.

Shen Y, Canfield D.E., and Knoll A.H. (2002). Middle Proterozoic ocean chemistry: evidence from the McArthur Basin, northern Australia. Am. J. Sci. *302*, 81–109.

Seong-Joo, L., and Golubic, S. (1998). Multi-trichomous cyanobacterial microfossils from the Mesoproterozoic Gaoyuzhuang Formation, China: Paleoecological and taxonomic implications. Lethaia *31*, 169–184.

Seong-Joo, L., and Golubic, S. (1999). Microfossil populations in the context of synsedimentary micrite deposition and acicular carbonate precipitation: Mesoproterozoic Gaoyuzhuang Formation, China. Precambrian Res. *96*, 183–208.

Seong-Joo, L., Golubic, S., and Verrecchia, E. (1999). Epibiotic relationships in Mesoproterozoic fossil record: Gaoyuzhuang Formation, China. Geology *27*, 1059–1062.

Shen, Y., Knoll, A.H., and Walter, M.R. (2003). Evidence for low sulphate and anoxia in a mid-Proterozoic marine basin. Nature *423*, 632–635.

Srivastava, P. (2005). Vindhyan akinetes: an indicator of Mesoproterozoic biospheric evolution. Origins Life Evol. Biosphere *35*, 175–185.

Summons, R.E., Jahnke, L.L., Hope, J.M., and Logan, G.A. (1999). 2-Methylhopanoids as biomarkers for cyanobacterial oxygenic photosynthesis. Nature *400*, 554–557.

Summons, R.E., Bradley, A.S., Jahnke. L.L., and Waldbauer, J.R. (2006). Steroids, triterpenoids and molecular oxygen. Phil. Trans. R. Soc. B. *361*, 951–968.

Tice, M.M., and Lowe, D.R. (2004). Photosynthetic microbial mats in the 3,416 -Myr-old ocean. Nature *431*, 549–552.

Tice, M.M., and Lowe, D.R. (2006). Hydrogen-based carbon fixation in the earliest known photosynthetic organisms. Geology *34*, 37–40.

Tomitani, A., Knoll, A.H., Cavanaugh, C.M., and Ohno, T. (2006). The evolutionary diversification of cyanobacteria: molecular phylogenetic and paleontolological perspectives. Proc. Natl. Acad. Sci. USA *103*, 5442–5447.

Turner, S. (1997). Molecular systematics of oxygenic photosynthetic bacteria. Plant Systemat. Evol. *11* (suppl.), 13–52.

Turner, S., Haung, T.-C., Chaw, S.-M. (2001). Molecular phylogeny of nitrogen-fixing cyanobacteria. Bot. Bull. Acad. Sin. *42*, 181–186.

Turner, S., Pryer, K.M., Miao, V.P.W., and Palmer, J.D. (1999). Investigating deep phylogenetic relationships among cyanobacteria and plastids by small submit rRNA sequence analysis. J. Euk. Microbiol. *46*, 327–338.

Ueno, Y., Yamada, K., Yoshida, N., Maruyama, S., and Isozaki, Y. (2006). Evidence from fluid inclusions for microbial methanogenesis in the early Archaean era. Nature *440*, 516–519.

Walsh, M.M. (1992). Microfossils and possible microfossils from the Early Archean Onverwacht Group, Barberton Mountain Land, South Africa. Precambrian Res. *54*, 271–293.

Walsh, M.M., and Lowe, D.R. (1999). Modes of accumulation of carbonaceous matter in the early Archean: a petrographic and geochemical study of the carbonaceous cherts of the Swaziland Supergroup. Geol. Soc. Am. Spec. Paper *329*, 115–132.

Yoon, H.S., Hackett, J.D., Ciniglia, C., Pinto, G., and Bhattacharya, D. (2004). A molecular timeline for the origin of photosynthetic eukaryotes. Mol. Biol. Evol. *21*, 809–818.

Zhang, Y. (1981). Proterozoic Stromatolite Microfloras of the Gaoyuzhuang Formation (Early Sinian: Riphean), Hebei, China. J. Paleontol. *55*, 485–506.

Zhang, Y., and Golubic, S. (1987). Endolithic microfossils (cyanophyta) from early Proterozoic stromatolites, Hebei, China. Acta Micropaleontol. Sin. *4*, 1–12.

Zhang, Y., Yin, L., Xiao, S., and Knoll, A.H. (1998). Permineralized fossils from the terminal Proterozoic Doushantuo Formation, China. Paleontol. Soc. Mem. *50*, 1–56.

Insights into Cyanobacterial Evolution from Comparative Genomics

2

Wesley D. Swingley, Robert E. Blankenship, and Jason Raymond

Abstract

Recent high-throughput sequencing has provided DNA sequences at an unprecedented rate, posing considerable analytical challenges, but also offering insight into the genetic mechanisms of adaptation. Here we present a comparative genomics-based approach towards understanding the evolution of these mechanisms in cyanobacteria. Historically, systematic methods of defining morphological traits in cyanobacteria have posed a major barrier in reconstructing their true evolutionary history. The advent of protein, then DNA, sequencing—most notably the use of 16S rRNA as a molecular marker—helped circumvent this barrier and now forms the basis of our understanding of the history of life on Earth. However, these tools have proved insufficient for resolving relationships between closely related cyanobacterial species. The 24 cyanobacteria whose genomes are compared here occupy a wide variety of environmental niches and play major roles in global carbon and nitrogen cycles. By integrating phylogenetic data inferred for hundreds to nearly 1000 protein coding genes common to all or most cyanobacteria, we are able to reconstruct an evolutionary history of the entire phylum, establishing a framework for resolving how their metabolic and phenotypic diversity came about.

Introduction

Cyanobacteria are a highly morphologically diverse group of organisms ranging from single-cells to filaments, some of which are large enough to be visible to the naked eye. Because of this, they were historically termed blue-green algae and classified botanically. Though this method is still used, there has been a push toward genetic characterization of cyanobacteria that is more consistent with prokaryotic systematics (Rippka *et al.*, 1979; Stanier *et al.*, 1978). Because many cyanobacterial families are so closely related, 16S ribosomal RNA (rRNA), which is commonly used to distinguish broad phylogenetic relationships, cannot fully resolve closely related species, even when species are morphologically distinct (Casamatta *et al.*, 2005; Fox *et al.*, 1992).

To compensate for the lack of ribosomal diversity, many studies have turned to the 16S-23S rRNA internal transcribed spacer (ITS) region (Boyer *et al.*, 2001; Brown *et al.*, 2005; Rocap *et al.*, 2002). This region is highly variable in both sequence and length. It contains information responsible for antitermination and rRNA folding and also codes for up to two transfer RNA (tRNA) genes. Although ITS sequences have been useful for differentiating closely related species, there are several issues to consider. Some cyanobacteria have multiple non-identical copies of their ribosomal operons (Boyer *et al.*, 2001). Variations in the length or tRNA coding number of ITS regions within a class of organisms or even a single organism must be dealt with manually, leading to human-error in judgment calls (Boyer *et al.*, 2001; Rocap *et al.*, 2002).

Single proteins conserved across all cyanobacteria have also been used to build phylogenies (Garczarek *et al.*, 2001; Henson *et al.*, 2004). However, no studies thus far have assembled cyanobacterial phylogenies based on the much larger set of all proteins conserved across cyanobacteria. A cyanobacterial phylogeny based on 31 proteins conserved across the entire tree of life was constructed as a part of a large scale tree construction (Ciccarelli *et al.*, 2006). However, this study used only 8 cyanobacterial taxa and the ribosomal proteins used for tree construction did not resolve terminal branches due to high conservation and rate and compositional biases (Hansmann and Martin, 2000).

Despite these limitations, high throughput sequencing has provided a remarkable wealth of genetic data that can be used to decipher organismal evolution. As of this writing, 24 cyanobacterial genomes are either complete or nearly so, and several metagenomics projects are focusing on environments with cyanobacteria as primary producers. Using comparative genomics to extract families of evolutionarily related genes from whole genomes, it becomes possible—though still computationally challenging—to reconstruct the history not only of single genes, but of entire organisms. Though it is clear that complex genetic events like horizontal gene transfer, gene duplication, conversion, and loss are important evolutionary mechanisms (Gogarten *et al.*, 2002), the major route by which prokaryotes share genes is through clonal reproduction; Darwinian descent with modification. Therefore, especially for closely related organisms, at least some subset of genes should share congruent phylogenies (though the extent of this sharing is still stridently debated). This suggests two major goals in reconstructing organismal evolution: to understand how and why the phylogenies of individual genes differ from one another, and to determine whether some core of genes common to a group of organisms are predominantly inherited vertically, thereby tracking the evolution of the organisms in which they are found. Here we develop and discuss methods towards both ends.

Methods

Large scale tree construction is a relatively new technique used to resolve evolutionary relationships between entire genomes. Two basic approaches have been used to tackle this problem. Consensus trees are built by averaging the topology of numerous gene (protein) trees. Concatenated trees are built by stacking alignments end-to-end to build a tree from a single long alignment. Both methods tend to approach a true topology when greater than 20 sequences are used, but concatenation methods tend to reach this topology using fewer sequences and are typically more accurate (Gadagkar *et al.*, 2005).

As with canonical single gene or protein phylogeny, the accuracy of these methods is limited by the ability to select and accurately align homologous sequences. A number of computational tools are available for finding homologous sequences, with the BLAST algorithm/software package being among the most widely used. Simply grouping, aligning, and inferring phylogenies for groups of homologs results in evolutionary trees that are often incorrect, if not indecipherable. This results from the preponderance of duplicated (paralogous) genes present in all genomes, whose inclusion can lead to spurious alignments and evolutionary trees that are difficult to interpret. Therefore, a first step in phylogenetic analysis is to group homologous genes or proteins into orthologs, related by vertical descent and typically performing the same function in different genomes. While simple in principle, selecting orthologs for further analysis is often quite difficult in practice and frequently limits the number of genes included in phylogenetic analysis. The use of thresholds for selecting orthologs is particularly problematic as the cutoffs that work well for one set of homologs—minimum percent identity, length of ungapped/overlapping sequence, or reciprocal closest matches of homologs in 2 or more genomes—may not work well for another (Koski and Golding, 2001; Tatusov *et al.*, 1997; Tatusov *et al.*, 2001). This results in analyses that focus on a narrow set of universally conserved "housekeeping" genes or proteins. Though still potentially informative, phylogenetic analysis of many of these homologs have noted strong biases in composition or evolutionary rates (e.g. ribosomal proteins (Hansmann and Martin, 2000)), as well as clear examples of horizontal gene transfer (e.g. tRNA synthetases (Lamour *et al.*, 1994)). Furthermore, because these proteins are universally conserved, they reveal very little of what's interesting about a particular organism—the phenotypic idiosyncracies that adapt it to its particular niche.

Here we employ a relatively recently developed clustering technique which segregates homologs into topologically distinct families. This method, known as Markov clustering (Enright *et al.*, 2002), begins with either a proximity or distance matrix that can be automatically generated from BLAST results and, in effect, performs a large number of directed walks along the edges of the network (with the probability of traversing a particular edge proportional to the similarity of the two proteins the edge connects). At the end of each run, some subset of "unutilized" edges is pruned, and the procedure is then repeated. This results in discrete clusters of nodes that are highly interconnected, as well as topologically similar (i.e. the nodes within a cluster are similarly connected to one another). Markov clustering is argued to have several advantages over previously developed protein clustering methods (Harlow *et al.*, 2004), but more importantly is quite fast and can be repeated using different parameters whose resulting protein clusters can then be compared based on extent of orthology versus paralogy.

Initial comparisons were done based on all vs. all BLASTP (Altschul *et al.*, 1990), comparing all protein sequences from 26 genomes (24 cyanobacteria, 2 non-cyanobacterial outgroups; see Table 2.1) to one another. A liberal BLAST expectation cutoff of 10^{-4} was used with no restrictions on the number of hits that could be retrieved (note that the default of 500 hits must be changed). Markov clustering (Mcl) was iteratively run on the BLAST results, using inflation parameters ranging from 1.2 to 20.0. The protein family results from each Mcl run were compared to one another based on the number of organisms and the number of paralogs within each protein family. An inflation parameter of 2.8 resulted in 341 protein families present in all 24 cyanobacterial genomes with no obvious paralogs, which were used in subsequent analysis.

Multiple alignments and corresponding Neighbor Joining (NJ) phylogenies for each of the 341 protein families were generated using ClustalW (Thompson *et al.*, 1994) (Gonnet matrix, default gap opening/extension parameters). Note that while the NJ trees are not particularly rigorous, they are useful for ensuing curation of protein families, for example in finding anomalously long branch lengths that can result from inclusion of paralogs.

Table 2.1 Genomes analyzed in this study

Organism	Description	NCBI Accession/in progress
Gloeobacter violaceus	Terrestrial, lacks photosystem components	NC 005125
Thermosynechococcus elongatus	55C optimal growth (hot springs)	NC 004113
Anabaena variabilis	Heterocyst-forming diazotroph	NC 007413
Nostoc punctiforme	Heterocyst-forming diazotroph	IP; Joint Genome Inst.*
Nostoc sp. PCC 7120	Heterocyst-forming diazotroph	NC 003272
Synechococcus elongatus 7942	Freshwater	NC 007604
Synechococcus elongatus 6301	Freshwater	NC 006576
Synechocystis sp. PCC 6803	Freshwater	NC 000911
Synechococcus sp. OS A	Hot spring, diazotroph	NC 007775
Synechococcus sp. OS B′	Hot spring, diazotroph	NC 007776
Acaryochloris marina	Chlorophyll *d*-containing symbiont	IP; Translational Genomics**
Prochlorococcus marinus CCMP1375	Marine, unicellular	NC 005042
Prochlorococcus marinus MED4	Marine, unicellular	NC 005072
Prochlorococcus marinus MIT9312	Marine, unicellular	NC 007577
Prochlorococcus marinus MIT9313	Marine, unicellular	NC 005071
Prochlorococcus marinus NATL2A	Marine, unicellular	NC 007335
Synechococcus sp. CC9605	Marine, unicellular	NC 007516
Synechococcus sp. CC9902	Marine, unicellular	NC 007513
Synechococcus sp. RS9917	Marine, unicellular	IP:NZ AANP00000000#
Synechococcus sp. WH5701	Marine, unicellular	IP;NZ AANO00000000#
Synechococcus sp. WH7805	Marine, unicellular	IP;NZ AAOK00000000#
Synechococcus sp. WH8102	Marine, unicellular	NC 005070
Trichodesmium erythraeum	Marine, filamentous diazotroph	IP;Joint Genome Inst.*
Crocosphaera watsonii	Marine, unicellular diazotroph	IP;Joint Genome Inst.*
Clostridium acetobutylicum	Gram positive (outgroup)	NC 003030
Rhodopseudomonas palustris CGA009	Proteobacterium (outgroup)	NC 005296

*JGI microbial sequencing portal: http://genome.jgi-psf.org/mic_home.html
**Translational genomics sequencing portal: http://genomes.tgen.org/

Manual curation was used to remove paralogs and poorly aligned sequences, and one entire family was deleted from the dataset because of extreme sequence divergence.

Once protein families were aligned and curated, two different methods were used for inferring phylogenies. The first was traditional, single protein phylogeny. Maximum likelihood (ML) trees for each protein family were inferred using Phylip's (Felsenstein, 1989) ProML (JTT probability model, one category of sites with constant rate, with randomized input order), as well as with the quartet-based IQPNNI (Vinh le and von Haeseler, 2004), incorporating a gamma parameter to correct for rate heterogeneity (WAG substitution model, gamma parameter estimated using 4 categories). To help visualize these 340 maximum likelihood phylogenies, a consensus tree was generated using the majority rule heuristic of Phylip's CONSENSE. This approach counts the number of occurrences of specific branching patterns on a tree, and gives an idea of the relative agreement between all trees used to infer the consensus. Two very closely related organisms will be expected to be found together in most or all of the 340 trees, whereas the branching pattern between more distantly related organisms may sometimes differ, due to the fact that some trees are poorly resolved, or to mechanisms like horizontal gene transfer that can create incongruence between two phylogenies. Additionally, the program SplitsTree (Huson, 1998) was used to create a consensus network that allows visualization of particular recurring branching patterns that are in disagreement with the consensus, that may suggest more widespread recombination or gene transfer.

The second approach utilized the sequence concatenation method discussed above. Each of the 340 aligned families was concatenated to produce a single alignment, resulting in 131 786 aligned characters (110 000–115 000 amino acids from each genome). The Shannon information entropy (SEI) was tabulated for each aligned character and corresponding SEI cutoffs between 1.0 and 3.0 were used to produce several filtered alignments from the full concatenation, in order to remove phylogenetically uninformative (or less-informative) positions from the alignment. Phylogenies for all concatenated alignments were reconstructed using Phylip's ProML, as well as the Neighbor Joining method (MEGA3.1 (Kumar *et al.*, 1994), 500 bootstrap replicates, pairwise deletion of sites). Finally, ML trees were also generated using MrBAYES (Huelsenbeck and Ronquist, 2001) (variable/sampled substitution model analyzed over 20 000–30 000 generations in 4 separate runs, with a 1000 generation burn-in; note in all 4 cases topological convergence was achieved within 5000–8000 generations).

Resulting phylogenies were compared to one another using Phylip's CONSENSE, employing the extended majority rule method and both the symmetric (Robinson-Foulds) and branch score distance metrics. Comparisons also included 50 trees comprised of the same cyanobacterial taxa arranged in randomized topologies.

Towards a cyanobacterial species tree

The methods used here provided an unprecedented level of genetic detail into the phylogeny of cyanobacteria. Though two distinct methods were used, one incorporating topological signatures from 340 individual trees, and the other relying on a single, large concatenated dataset, the resultant phylogenies agreed almost exactly with one another. Furthermore, many aspects of the trees generated here agree with the results of previous analyses based on single gene or protein markers such as 16S rRNA, ribosomal proteins, or metabolic proteins such as the Rubisco large subunit and components of the photosynthetic apparatus.

Figure 2.1A shows a consensus tree for the 340 maximum likelihood phylogenies that were inferred. Numbers on (internal) nodes indicate the number of times that particular "split" was observed in the 340 trees. For example, the distinct separation between picophytoplankton—*Prochlorococcus* and marine *Synechococcus*—and other cyanophytes (herein referred to as "canonical cyanophytes," taken to include all cyanobacteria except the marine *Synechococcus*/*Prochlorococcus* clade) was observed in every single tree, as was the clustering of closely related thermophilic *Synechococcus* OS strains A and B′. Other branches were much more poorly supported—the *Acaryochloris marina*/*Thermosynechococcus elongatus* cluster was only observed in a plurality of trees (85). Illustrated in Figure 2.1B are the so-called "split decompositions" resulting from SplitsTree analysis of these phylogenies. This portrays variant but frequently recurring phylogenetic signals as a net rather than a tree, for example the base of the poorly supported *A. marina*/*T. elongatus* clade is shown as something of a tangle, depicting the alternative signal present in many single protein trees.

Figure 2.2 shows the result of a Bayesian maximum likelihood analysis of over 130 000 conserved characters, and shows striking overall congruence with the consensus trees. Because this tree is based on analysis of substitution patterns (as opposed to just topologies), branch lengths are retained, for example showing the relatively long branches indicative of anomalous evolutionary rates in many prochloropyte proteins (discussed below). Another advantage from using concatenated alignments is that they can include protein families with orthologs found outside the cyanobacterial phylum (whereas these non-cyanobacteria were excluded from consensus tree phylogenies to maximize the number of protein families that could be included). Based on inclusion of these outgroups, the *Gloeobacter violaceus* lineage appears to be the most deeply branching among extant cyanobacteria, in agreement with previous findings (Honda *et al.*, 1999; Nakamura *et al.*, 2003). Furthermore, this rooting also suggests an early radiation of some thermophilic cyanobacteria, previously argued against based on analysis of a less diverse set of organisms (Sánchez-Baracaldo *et al.*, 2005).

Phylogenies generated from these 340 protein families highlight a number of relationships that are suggestive of important events during cyanobacterial evolution (and, where noted, are largely consistent with findings based on previous analyses of single genes and proteins):

1 Monophyly—and an apparently singular origin—of heterocyst-forming cyanobacteria (*N. punctiforme*, *Nostoc* sp., *A. variabilis*), and their nested position within filamentous cyanobacteria (*Trichodesmium erythraeum*) (Giovannoni *et al.*, 1988; Honda *et al.*, 1999; Turner *et al.*, 1999).
2 Relatively close relationship between *Synechocystis* sp. PCC 6803 and *Crocosphaera watsonii*, despite marked differences in lifestyle (Dvornyk, 2006; Dyhrman and Haley, 2006).
3 The deeply branching position of G. *violaceus* and thermophilic *Synechococcus* in the rooted tree (Ferris *et al.*, 1996; Giovannoni *et al.*, 1988; Honda *et al.*, 1999).

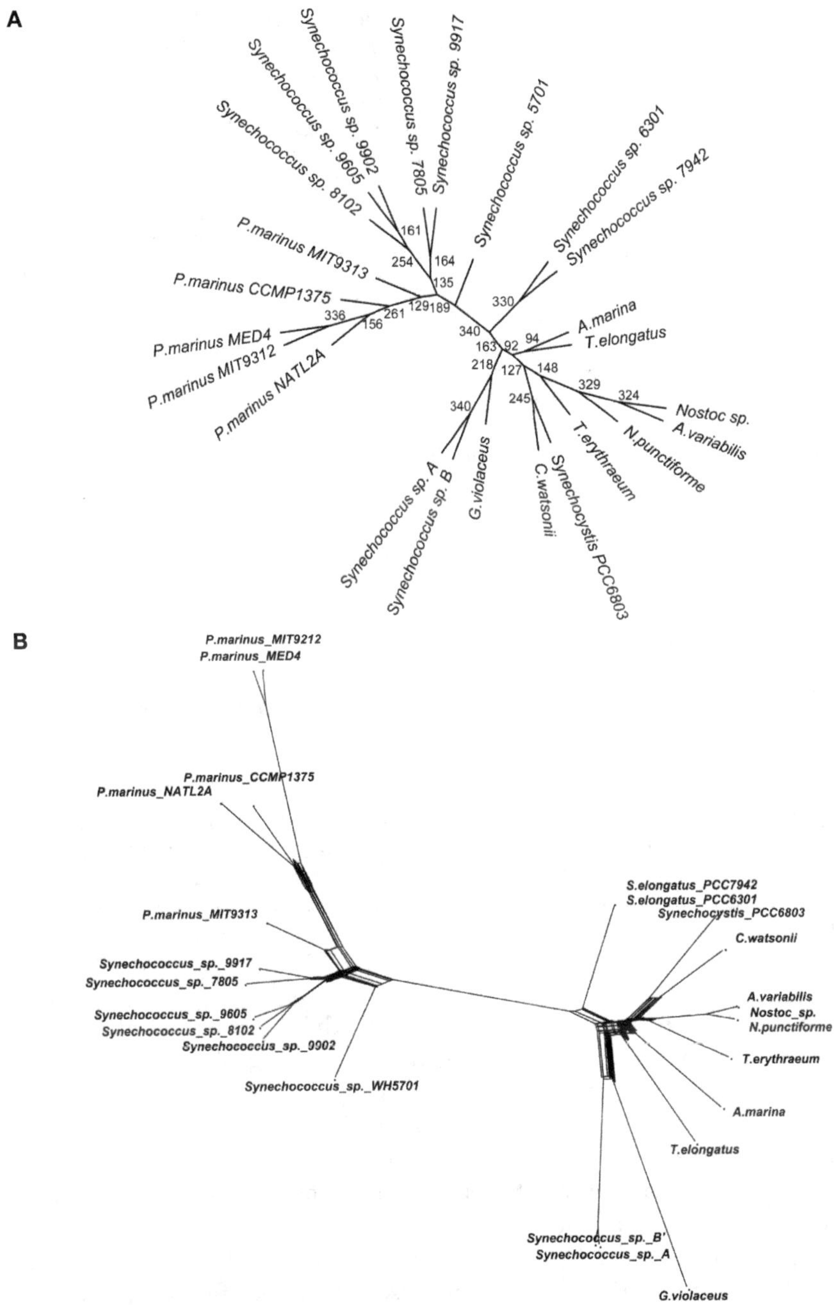

Figure 2.1 A. Consensus phylogeny for 340 protein families common to all cyanobacteria (selected as discussed in the methods). Numbers at each branch indicate the number of times that specific topology was observed in individual protein phylogenies. B. SplitsTree depiction of a phylogenetic "network," based on the same 340 protein families as in 1A. Edges/branches indicate alternative evolutionary signals present in single protein phylogenies, which can result from complex genetic events such as horizontal gene transfer, as well as poorly resolved trees.

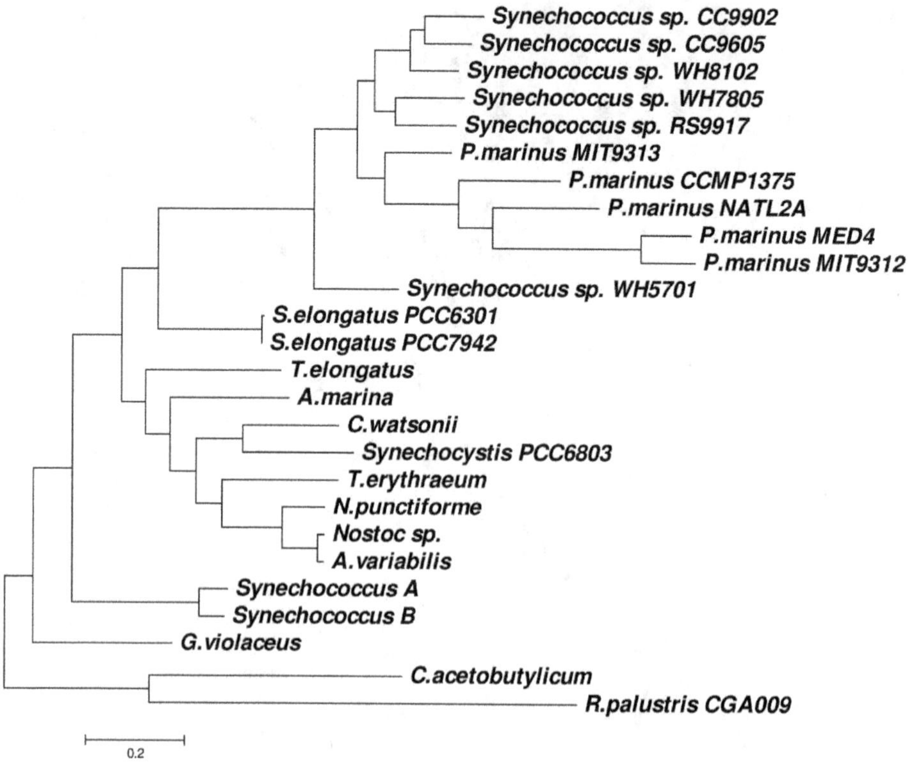

Figure 2.2 Tree resulting from MrBayes analysis of 131 912 characters, based on a concatenation of 340 protein families found in all 24 available cyanobacterial genomes. Scale bar indicates number of amino acid substitutions per site. Further details are given in the methods section. *Synechococcus* A and B refer to *Synechococcus* OS strains A and B′, respectively.

4 The monophyly of the *Prochlorococcaceae* family, which has proven difficult to resolve in some earlier reconstructions (typically due to 16S being too well conserved) (Rocap *et al.*, 2002; Scanlan, 2003; Scanlan and West, 2002).
5 The nested position of the *Prochlorococcaceae* within the marine *Synechococcus*, and the apparently more basal position of *Synechococcus* sp. strain WH 5701, which may be an important modern analog of an ancient transition from freshwater to marine environments (Rocap *et al.*, 2002; Scanlan, 2003; Scanlan and West, 2002).
6 The very deep radiation of freshwater *Synechococcus* (strains PCC 6301 and PCC 7942) relative to the marine clade, and the close relationship of these two strains to one another (Rocap *et al.*, 2002; Scanlan, 2003).

It cannot be understated that any attempt to resolve organismal evolution as a single phylogenetic tree invariably ignores the rich and complex histories of single genes. However, here we emphasize how the inferred history of cyanobacteria at the species level can be compared and contrasted with that of their diverse genetic repertoires, which provides a number of insights and hypotheses into the evolution and adaptation of this diverse phylum.

Nostocales and *Trichodesmium erythraeum* and cyanobacterial N_2 fixation

The phylogeny and diversification of heterocystous cyanobacteria (the three Nostocales included here) has been well studied (Giovannoni *et al.*, 1988; Honda *et al.*, 1999). The sister relationship with non-heterocyst forming filamentous cyanobacteria has previously been noted (e.g. Honda *et al.*, 1999), which generally includes the marine diazotroph *T. erythraeum* among filamentous *Oscillatoria*. Overall, this relationship suggests that heterocyst formation evolved as a specific adaptation within a group of filamentous cyanobacteria, likely driven by nitrogen starvation as the role of heterocysts is primarily suited to creating a cytosolic environment amenable to nitrogen fixation (i.e. expression of O_2-inhibited nitrogenase) (Berman-Frank *et al.*, 2003; Tomitani *et al.*, 2006; see Chapter 14). The paradox of this dual adaptation for oxygenic photosynthesis and anaerobic nitrogen fixation has been well reviewed elsewhere and will not be detailed here, except to note that the formation of anaerobic heterocysts is but one adaptation found among diverse (and paraphyletic) cyanobacterial diazotrophs. In fact, 7 of the 24 cyanobacteria studied here are known to fix nitrogen. *Trichodesmium* species are well-known filamentous marine nitrogen fixers, and use an entirely different mechanism that allows them to perform oxygenic photosynthesis and nitrogen fixation simultaneously (Berman-Frank *et al.*, 2001; see Chapter 15). This involves reduction of molecular oxygen through a number of routes, most notably the Mehler reaction and the use of the respiratory cytochrome oxidase (Kupper *et al.*, 2004). Despite the relatively close relationship suggested by 16S phylogenies and the consensus and concatenated trees inferred here, the mechanistic differences between N_2 fixation in Nostocales and *T. erythraeum* are also reflected in their nitrogenase protein sequences (Berman-Frank *et al.*, 2003). As illustrated in Figure 2.3, the phylogeny of NifD protein sequences shows distinct cyanobacterial clusters that can be (minimally) divided into a) Nif proteins from heterocyst-formers, found distally in the tree and suggesting a relatively recent diversification; b) unicellular, aerobic nitrogen fixers including *Crocosphaera* and *Cyanothece* along with filamentous mat-former *Lyngbya majuscula*; and c) a basal branch comprised of thermophilic *Synechococcus* OS strains A and B′ and *T. erythraeum*. Though variant forms of nitrogenase are known (see Figure 2.3)—specifically the alternative nitrogenases wherein molybdenum at the active site is replaced either by vanadium or iron—so far only *Anabaena variabilis* is known to possess one of these paralogs, expressing a vanadium-dependent nitrogenase under Mo-limited conditions (Thiel, 1993). The discordance between nitrogenase trees and cyanobacteria species trees suggests a complex evolutionary history of this metabolic capability. The energetic cost of nitrogen fixation (minimally 16 mol of ATP per mole of N_2 fixed) no doubt constitutes a strong selection pressure to find alternate sources of fixed nitrogen, manifested as extensive gene loss from various lineages. As has been discussed previously (Raymond *et al.*, 2004) and as is evident, for example by the close relationship of nitrogenase proteins from otherwise distantly related cyanobacteria in Figure 2.3, horizontal gene transfer has also played an important role in the history of N_2 fixation, not only among cyanobacteria but across the tree of life.

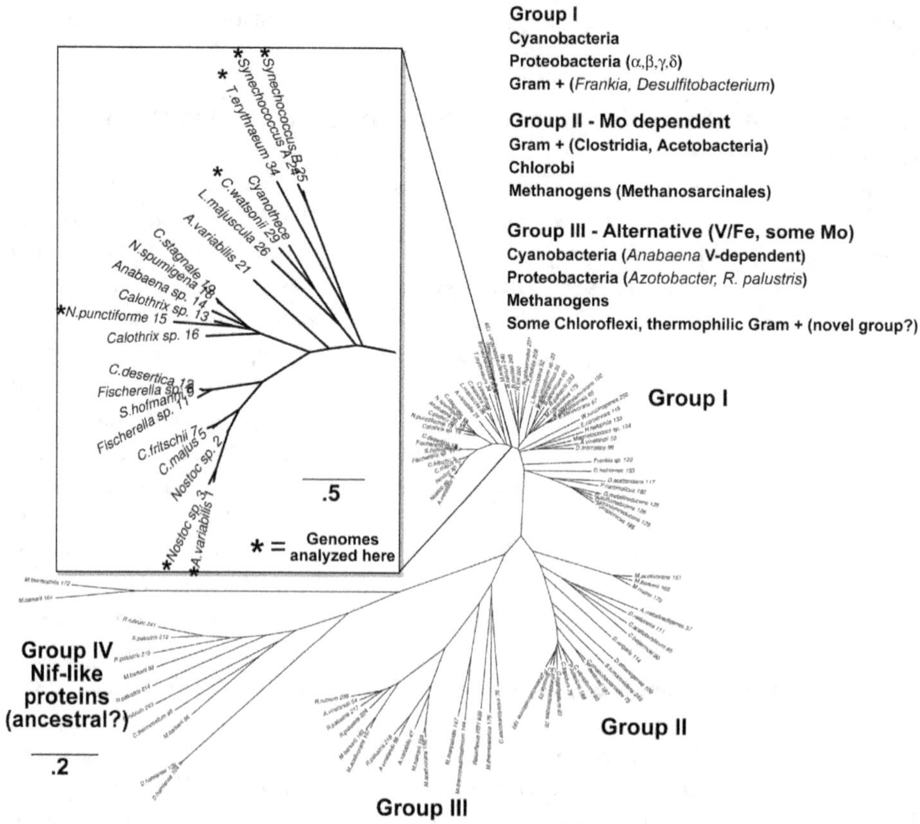

Figure 2.3 Interpreting the distribution and complex history of nitrogen fixation in cyanobacteria, based on cyanobacterial species trees and the phylogeny of specific nitrogenase components. Shown here is the protein phylogeny for NifD, the catalytic, cofactor-containing subunit of the nitrogenase complex. The different phylogenetic groups are indicated (see e.g. Raymond 2004 for more detailed discussion), with cyanobacterial whole genomes from this analysis indicated by *. Scale bars indicate the number of amino acid substitutions per site along each lineage.

***Prochlorococcus*: evolution from a marine *Synechococcus*?**

The genus *Synechococcus* comprises one of the most phylogenetically widespread cyanobacterial genera (along with *Oscillatoria*), mainly due to early botanical classifications that group single-celled cyanobacteria based on a limited number of shared physiological traits. More recently, genetic and genomic analyses have divided this genus into distinct marine and freshwater groups, with marine *Synechococcus* comprising a coherent picophytoplanktonic clade with *Prochlorococcus* species (Hess, 2004; Hess *et al.*, 2001; Honda *et al.*, 1999; Scanlan, 2003; Scanlan and West, 2002). More detailed characterization, integrating genomic signatures, light harvesting pigments, growth modes, and motility have enabled finer scale understanding of niche adaptation and diversification of these organisms (not reviewed here, the reader is referred to Scanlan, 2003; see also Chapter 5).

Notably, marine *Synechococcus* sp. strain WH 5701 is suggested both by previous work (Fuller *et al.*, 2003; Rocap *et al.*, 2003) and by our data to branch basal to the *Prochlorococcus* and *Synechococcus* divergence. As described below, this position of WH 5701 establishes a sequence for the introduction of Chl *b* biosynthesis, either originating or further driving the diversification of the Prochlorococcaceae family. This provides insight into an extremely important event, suggesting a relatively recent diversification of the Prochlorococcaceae that spurred a major shift in global primary production. The requisite synthetase/chlorophyll *a* monooxygenase enzyme has not yet been characterized, and the most likely candidate, identified by Hess *et al.* (2001) and supported by protein family profiles in this analysis, is only distantly related to that found in Chl *b*-containing *Prochlorothrix* and *Prochloron* (Tanaka *et al.*, 1998; Tomitani *et al.*, 1999). Additionally, strain WH 5701 can tolerate both marine and freshwater media and is thus not an obligate marine species (Fuller *et al.*, 2003; Waterbury *et al.*, 1986). The basal position of the freshwater *S. elongatus* strains (PCC 6301 and PCC 7942) further supports the idea of a freshwater ancestor at the base of this diverse radiation, and *Synechococcus* sp. strain WH 5701 may indeed represent a modern analog of a transitional organism in this marine adaptation.

The phylogeny and inferred diversification of the Prochlorococcaceae has been well studied in recent years. *Prochlorococcus* ecotypes are broadly segregated into high light- versus low light-adapted strains with varying chlorophyll *b*/*a* ratios suited to distinct niches in the water column (Hess *et al.*, 2001; Moore *et al.*, 1998; Scanlan and West, 2002). These divisions are phylogenetically well supported, and consensus and concatenated phylogenies generated here show identical topologies as those inferred from 16S rRNA, with the high light-adapted lineages (HL-I:MED4; HL-II:MIT9312) distal and the low-light adapted lineages (LL:NATL2A, 1375/SS120, and MIT9313) more proximal in the *Prochlorococcus* radiation (e.g. Scanlan and West, 2002).

Chlorophyll *b* synthesis

Four distinct cyanobacterial lineages produce alternative chlorophyll pigments. These include the species *Acaryochloris marina*, which produces Chl *d*, and three others, *Prochlorothrix hollandica, Prochloron didemni*, and *Prochlorococcus marinus*, that produce Chl *b*. Each pigment conversion yields a formyl (-CHO) side group-containing chlorophyll, although a vinyl group is converted in Chl *d* rather than the methyl group in Chl *b* synthesis. Similar substrates and products imply similarities in enzymatic mechanisms and possibly evolutionary history. Unfortunately, the apparently complex evolutionary history of Chl *b* biosynthesis only increases the difficulty of finding a similar enzyme for Chl *d* synthesis.

Chl *b* synthesis in plants and green algae is performed by a Rieske-type iron–sulfur containing protein dubbed chlorophyll(ide) *a* oxygenase (CAO), or chlorophyll *b* synthase (Oster *et al.*, 2000; Tanaka *et al.*, 1998). Homologs of CAO are found in the cyanobacterial *Prochloron* and *Prochlorothrix* strains (Tomitani *et al.*, 1999), but not *Prochlorococcus*. *Prochlorococcus* strains appear to synthesize Chl *b* via a novel pathway. An iron–sulfur enzyme was implicated in *Prochlorococcus* Chl *b* synthesis (Hess *et al.*, 2001), though this candidate enzyme has not yet been characterized biochemically.

One possible reason why *Prochlorococcus* does not produce Chl *b* in the same fashion as closely related species is a minor difference in its chlorophyll content. *Prochlorococcus* strains contain divinyl chlorophylls (Chisholm *et al.*, 1992; Chisholm *et al.*, 1988), which have a vinyl rather than ethyl group on ring B, in close proximity to the methyl side group that is formylated during Chl *b* synthesis. It is unclear whether this vinyl side group would interfere with Chl *b* synthesis. *Arabidopsis thaliana* mutants producing divinyl Chl *a* also produced divinyl Chl *b*, albeit at lower than normal concentrations relative to Chl *a* (Nagata *et al.*, 2005; Nakanishi *et al.*, 2005).

With a robust cyanobacterial phylogeny and an orthologous protein library it is possible to identify protein families present in *Prochlorococcus* strains but absent from all other cyanobacteria. This method yields only nine *Prochlorococcus*-unique protein families, including phylogenetically distinct homologs of Photosystem II proteins M and Z (PsbM and PsbZ respectively). One family stands out as containing a putative Rieske-type iron–sulfur binding domain. This protein family shares only 25% identity with a ~200 amino acid segment of the biochemically characterized CAO from *Chlamydomonas reinhardtii* and shares equivalent identities with many oxygenase related proteins from β- and γ-proteobacteria, gram-positive bacteria, cyanobacteria, and various other lineages. Surprisingly, this protein family includes the proteins that were implicated by Hess *et al.* (2001) as putative Chl *b* synthases. The identification of the same protein family via a different bioinformatic approach strengthens support for this assignment, but biochemical studies are still needed to confirm this.

The distinct monophyletic clustering of *Prochlorococcus* strains explains why they would have a different mechanism for Chl *b* biosynthesis. Gene trees have typically placed *Prochlorothrix, Prochloron,* and *Prochlorococcus*—traditionally grouped together as the Prochlorophytes—as evolutionarily distinct, paraphyletic lineages spread around the cyanobacterial tree (Palenik and Haselkorn, 1992; Urbach *et al.*, 1992). Thus, Chl *b* synthase in *Prochlorococcus* is likely the result of convergent evolution or extreme divergence following an ancient horizontal transfer, while the presence of CAO in the other two lineages of Chl *b*-containing cyanobacteria is likely due to horizontal transfer. Interestingly, all Chl *b*-containing cyanobacteria, including *Prochlorothrix* and *Prochloron* share common structural features, such as thylakoid organization and lack of phycobilisomes (Chisholm *et al.*, 1992; Giddings *et al.*, 1980; Hess *et al.*, 1996; Miller *et al.*, 1988), in addition to their pigment composition, suggesting there may be an unknown cause and effect relationship between Chl *b* and photosynthetic apparatus structure.

The origin of *Prochlorococcus* Chl *b* synthesis very likely occurred sometime after the divergence of marine *Synechococcus* and *Prochlorococcus*, or may have promulgated diversification by opening up a new energetic niche. The divergence of marine *Synechococcus* sp. strain WH 5701, which does not contain Chl *b*, provides an important constraint on the advent of *Prochlorococcus* Chl *b* synthesis, defining a relatively narrow window for this gain of function.

Prochlorococcus evolutionary rate

Resolution of tree topology in and around the *Prochlorococcus* clade has historically proven difficult (Fuller *et al.*, 2003; Rocap *et al.*, 2002). This is exemplified in Figure 2.2, where several *Prochlorococcus* lineages are characterized by relatively long branch lengths, indicating higher rates of amino acid substitution. These long branch lengths are also present in single gene and protein phylogenies. With the recent sequencing of several *Prochlorococcus* genomes, it is clear that this difficulty results in large part from genome reduction (Dufresne *et al.*, 2005). Both *Prochlorococcus* and marine *Synechococcus* strains have relatively small genomes, ranging between 1.6 and 3.1 million base pairs. The trend of reduction is most severe in *Prochlorococcus* strains: the more distal lineages in species trees invariably have the most reduced genomes (Dufresne *et al.*, 2005). The terminal species, *Prochlorococcus* sp. strains MED4 and MIT 9312 have the smallest cyanobacterial genomes at 1.65 and 1.71 million basepairs, respectively, and also show the longest branch lengths relative to other cyanobacteria. Such large reduction is atypical of free-living organisms, as all other such cases occur in organisms transitioning from free-living to host-dependent lifestyles.

Previous investigation has postulated that the vast genome reduction in these clades, particularly *Prochlorococcus*, is due to an increase in the rate of DNA replication error (Dufresne *et al.*, 2005). There are several lines of evidence to support this assumption. First, G+C content in these reduced genomes increases as genome size decreases. This is likely caused by an inability to correct A:T to G:C mutations. Dufresne *et al.* (2005) identified the absence of the *ada* gene, which codes for a methylated-DNA–protein-cysteine methyltransferase responsible for repairing alkylated guanine and thymine. The A/G specific DNA glycosylase MutY is also absent from the two terminal *Prochlorococcus* sp. strains MED4 and MIT 9312 (Rocap *et al.*, 2003), although they do include the DNA lyase MutM. Our analysis also verified that the DNA mismatch repair protein MutL is missing from both *Prochlorococcus* and marine *Synechococcus* species, which could account for smaller than average genome sizes in both lineages.

Several enzymes from the *rec* repair family are also missing in various *Prochlorococcus* strains (Dufresne *et al.*, 2005). The DNA helicase RecQ is missing from all *Prochlorococcus* strains except the largest (2.4 mega base pairs) and most basal strain MIT 9313. The two terminal *Prochlorococcus* strains are also the only members missing the single-stranded DNA-specific exonuclease RecJ, which further explains their highly reduced genomes. All marine *Synechococcus* and *Prochlorococcus* strains contain *recO* genes which code for a recombination related protein that did not cluster with other cyanobacterial RecO proteins during our analysis. This evolutionary divergence may represent a loss or gain of function which may affect its interaction with other repair enzymes such as RecF and RecR.

Our cluster analyses also revealed the absence of the UV-damage repair protein RadC in both *Prochlorococcus* and marine *Synechococcus* lineages, both of which contain high-light, open-ocean strains. The successive removal of DNA repair systems in these marine species implies a snowball effect of DNA damage. Each subsequent repair enzyme deactivation increases the DNA error or damage rate, which increases the likelihood of further repair enzyme deactivation. This can clearly be seen in the two terminal *Prochlorococcus marinus* strains MED4 and MIT 9312 that are missing at least two additional repair proteins versus their closest relatives.

Duplicate gene analysis reveals that the gene loss in *Prochlorococcus* and marine *Synechococcus* strains is not merely a random destructive event. Rather, genome reduction selectively removes duplicate protein copies. In all other cyanobacteria and the two bacterial outgroups an average of 25% of all cellular proteins are duplicates (whether functional or not) versus only 10% in the picophytoplankton. Thus, a genome that has been reduced to half-size has lost only 40% of its genomic diversity. This correlation supports the evolutionary selectivity for a duplicate gene loss versus a complete removal of function. It is unclear whether these gene losses decrease the organism's overall metabolic diversity; however, the global success of *Prochlorococcus* would seem to challenge the necessity of these "lost" metabolic pathways.

One positive side effect of the increased mutation rate in the marine cyanobacterial species may be an increase in genetic diversity. The evolutionary divergence of RecO mentioned above is one possible example of this effect. One could speculate that the novel advent of Chl *b* biosynthesis in *Prochlorococcus* is a result of accumulated mutations on a recruited oxygenase.

Thermophilic *Synechococcus* (*elongatus* and Octopus spring A and B′ isolates)

A number of earlier reports suggested that thermophilic cyanobacteria branch early within the phylum (Ward *et al.*, 1998), while a more recent study attempts to argue for a later emergence of thermophilic cyanobacteria (Sánchez-Baracaldo *et al.*, 2005). Our analysis finds strong support for two distinct thermophilic lineages radiating early in the phylum, suggesting that thermophily might indeed have been an ancient trait among cyanobacteria (but not necessarily present within the cyanobacterial common ancestor). Despite their thermophilic lifestyle and overlapping optimal growth ranges, both the *Synechococcus* OS strains A and B′ and *Thermosynechococcus elongatus* lineages appear to be distinct from one another based on the analysis presented here. This is supported by phenotypic and ecological differences, including 16S-based studies as well as the recent discovery of the ability of the *Synechococcus* OS strains A and B′ to fix nitrogen (Steunou *et al.*, 2006).

However, 43 protein families have phylogenies that support a monophyletic group comprised of these three thermophiles, representing a minority disagreement with consensus phylogenies but potentially providing some insight into high temperature adaptation. Importantly, these trees may reflect long-branch attraction between two clades that are both quite different from other organisms in the analysis, and whose genomes no doubt reflect selection in a thermophilic environment. This is exemplified by the PsaC photosystem iron–sulfur protein (protein family 1024), strikingly well conserved in all cyanobacteria—though not yet found in the nearly complete genome of *Acaryochloris marina*—with 92% or greater identity across a diversity ranging from Prochlorales to Nostocales. The two exceptions are *P. marinus* strain MIT 9313 and the clade comprised of *T. elongatus* and *Synechococcus* OS strains A and B′, the latter whose PsaC primary sequences show as little as 85% identity to other cyanobacterial orthologs. Other protein families that exhibit this cluster of thermophiles include enzymes catalyzing steps in lipid, amino acid, heme and carotenoid biosynthesis pathways. While long branch lengths are persistent obstacles to accurately determining phylogenies, the underlying point here is that a number of proteins specifically from these thermophiles are evolving in a way that is quite distinct from their mesophilic cyanobacterial cousins.

Protein family clustering also reveals several features that are unique to these thermophiles. For example, ten protein families—six of which are present in two distinct, tricistronic operons—are well conserved in all three organisms. These protein families are either completely absent from or highly diverged in other cyanobacteria. Moreover, each of these protein families is comprised entirely of unique or conserved hypothetical proteins, having no characterized homologs in any other organism. However, proteins from one of the operons do have nearest (though still divergent) homologs in several other hot spring organisms, including *Thermus thermophilus*, *Aquifex aeolicus*, and *Chloroflexus aurantiacus*, alluding to a possible role in thermal adaptation.

Acaryochloris marina

Similarities have been drawn between *Acaryochloris marina*, the only Chl *d*-producing species, and the Chl *b*-containing cyanobacteria based on their lack of phycobilisomes, appressed thylakoids, Pcb (prochlorophyte Chl *a*/*b* binding-protein) content, and alternative pigment production (Chen *et al.*, 2005; Hu *et al.*, 1999; Marquardt *et al.*, 1999; Miyashita *et al.*, 1996). However, recent work suggests that appressed thylakoids may be dependent on culture conditions (Swingley et al., 2005) and genetic evidence supports only a distant relationship with *A. marina* clustered within the canonical cyanophytes (Miyashita *et al.*, 2003 and this work). This suggests that the physiology of *A. marina* is a novel and perhaps recent adaptation. Using our orthologous family database, we were able to illuminate several issues revolving around this debate.

It was recently suggested that the Isi/Pcb/CP43 family of antenna proteins was laterally transferred across numerous cyanobacterial taxa (Chen *et al.*, 2005). A phylogenetic tree was constructed for large number of these proteins using our protein family library (they all clustered into a single protein family). The resulting tree (not shown) shows a clear distinction between *Prochlorococcus* and canonical cyanophyte proteins.

A. marina and *Nostoc* sp. strain PCC 7120 proteins identified as PcbC proteins are quite distinct from any *Prochlorococcus* Pcb and were perhaps named only by their similarity to sequences in *Prochloron didemni* and *Prochlorothrix hollandica*, neither of which branch within the *Prochlorococcaceae* (Chen *et al.*, 2005; Miyashita *et al.*, 2003). In fact, both *T. erythraeum* and *C. watsonii* (which were not included in the earlier study) contain similar proteins. Further study of this cluster of proteins in Chl *a*-containing cyanobacteria as well as *A. marina* will need to be made to assure confident assignment.

All *Prochlorococcus* Pcb proteins cluster together monophyletically quite distant from cyanobacterial IsiA. Surprisingly, only marine *Synechococcus* sp. strains CC 9605 and CC 9902 (which are nested within the phylum) contain homologs to Pcb/IsiA. The phylogeny of these two proteins fits the expected branching within the gap between *Prochlorococcus* and canonical cyanophytes; however, the *A. marina* proteins branching within the same region do not fit expected phylogenetic patterns. Using several different tree-construction methods this group of proteins occasionally clusters with the IsiA proteins. It is likely that the *A. marina* IsiA-like proteins are highly diverged from typical cyanobacteria to accommodate binding of Chl *d*. This divergence and the associated long branch lengths causes low resolution leading to the potential misinterpretation of a lateral transfer between *Prochlorococcus* and *A. marina*. Our analyses do not support any relationship between *Prochlorococcus* Pcb and *A. marina* IsiA-like proteins.

We also investigated clues in the proteome of *A. marina* and the *Prochlorococcus* species that might explain their similar phycobilisome physiology. Only 11 protein families were missing from *A. marina* and all *Prochlorococcus* species and present in all other canonical cyanophytes (marine *Synechococcus* species were omitted due to their genome similarities with *Prochlorococcus*). Most notably missing from all of these species and present in all other cyanobacteria was the 7.8 kDa phycobilisome linker protein ApcC. The loss of this key linker protein is likely responsible for disrupting the assembly of phycobilisomes in these organisms, even though their genomes contain most other phycobiliprotein and phycobilisome related genes. A weak match (e-value $\sim 10^{-9}$) to the phosphate induced protein YxiE in *Bacillus subtilis* was the only other protein family found in all cyanobacteria except

A. marina and the *Prochlorococcus*. The 9 others (6 of which have no close non-cyanobacterial matches) are also missing in all marine *Synechococcus* species and are not likely to be responsible for any *Prochlorococcus*-specific physiology. No protein families were shared exclusively by *A. marina* and *Prochlorococcus*. However, 2 of the 7 found also in the marine *Synechococcus* are also thought responsible for phosphorous-sequestration related roles—a possible advantage in the phosphate-limited ocean.

There is a puzzling inverse correlation between phycobilisomes and alternate chlorophylls in oxygenic phototrophs. All species of cyanobacteria and algae that contain phycobilisomes contain only Chl *a*. Conversely, cyanobacteria, plant, and algal species that use two types of chlorophyll contain either no phycobilisomes or phycobiliproteins that are not assembled into proper phycobilisome superstructures. It is unclear whether this correlation is a true cause and effect relationship or merely a coincidental occurrence. However, such a clear-cut relationship would suggest an evolutionary pressure driving species to preferentially use one light capture method or the other and not both concurrently.

The disparate origins of cyanobacterial CO_2 fixation

The nature of carbon dioxide fixation amongst the marine *Synechococcus* and *Prochlorococcus* species is a source of great debate. The protein sequences for both large and small ribulose 1,5-bisphosphate carboxylase/oxidase (Rubisco) subunits as well as phosphoribulokinase (PRK) (both key Calvin cycle enzymes) are quite divergent from the canonical cyanophytes. Past analysis has postulated that this is a result of lateral transfer from the proteobacterial lineage (Hess *et al.*, 2001; Shimada *et al.*, 1995). For all three protein phylogenies the proteobacterial sequences lie somewhere in the middle of a long branch connecting canonical cyanophytes to the marine cyanobacteria in a similar fashion to the overall shape of the cyanobacterial tree presented here.

The gene conservation in CO_2-fixation gene clusters reveals a great deal about the history of these genes. A small cluster containing the Calvin cycle enzymes PRK and fructose bisphosphate aldolase is conserved in all marine species, but not in other cyanobacteria. The number of genes separating these two varies between 3 and 5; where 5 genes are present in the most basal *Synechococcus* species, 4 are present in the terminal *Synechococcus* strains CC 9902 and 9605 as well as the basal *Prochlorococcus* species, and only 3 are present in the terminal *Prochlorococcus* strains MIT 9312 and MED4. The loss of these genes is directly related to the genome reduction seen throughout the marine group and clear evidence of genetic flux in this region.

The large gene cluster coding for both Rubisco subunits and several carboxysome (a bacterial CO_2-concentrating structure) related proteins is genomically isolated from the PRK cluster in all marine species (as well as other cyanobacteria). The full cluster, containing six carboxysome genes, is present in all marine species except for the terminal *Prochlorococcus* strains MIT 9312 and MED4—again agreeing with genome reduction—and shares no synteny with any of the sequenced carboxysome-containing proteobacteria. However, this gene cluster is tentatively conserved in both *S. elongatus* species, although the six carboxysome-related proteins coded for are quite divergent from the marine species. The conservation between the marine species and their closest relatives, *S. elongatus*, suggests that the cluster synteny may have been conserved between these species even while the sequences diverged.

Another gene coding for the enzyme ribulose-phosphate 3-epimerase (PPE) further confuses the issue. While it was noted that PPE from MED4 and MIT 9313 were divergent from those in other cyanobacteria (Hess *et al.*, 2001), a full PPE tree reveals a close relationship between the canonical cyanophytes and the marine *Synechococcus* strains as well as the basal *Prochlorococcus* strain MIT 9313. All other *Prochlorococcus* strains are far removed from the canonical cyanophytes in a similar fashion to that seen in the Rubisco and PRK trees. Remarkably, gene synteny is conserved in all of the marine species, with only small insertions in *Synechococcus* strains WH 7805 and CC 9605, further supporting the idea that the divergence between canonical cyanophytes and picophytoplankton is due to vertical inheritance and adaptive evolution in marine *Synechococcus* and *Prochlorococcus*.

Prochlorococcus and marine *Synechococcus* Rubisco is divergent enough from that in other cyanobacteria to warrant its own classification (Badger *et al.*, 2002). However, the perceived similarity of the marine species to proteobacteria could be due to the large amount of genetic drift in the marine species caused by niche adaptation and accelerated substitution resulting from successive loss of repair proteins—as described previously in this chapter. This is likely a problem of not having enough taxa on the tree, which is problematic in Rubisco trees having only three distinct groups ("typical" cyanobacteria, marine cyanobacteria, and proteobacteria) and no obvious root. The lateral transfer of CO_2-fixation genes at three distinct loci is rather improbable and easier explained by a single global event—i.e. the loss of DNA repair enzymes, followed by the transfer of some carbon fixation components into proteobacteria (Badger *et al.*, 2002). The discovery of taxa that fill the gaps between the Calvin cycle-utilizing bacteria will significantly clarify the history of this globally important pathway.

Overall similarities in cyanobacterial and proteobacterial carbon fixation enzymes have led to a number of different explanations for the distribution of the Calvin cycle, including horizontal gene transfer of central components, such as Rubisco, from proteobacteria into cyanobacteria (Delwiche and Palmer, 1996), and vice versa (Badger *et al.*, 2002). These contrasting interpretations have much to do with the extreme divergence seen in many *Prochlorococcus* and marine *Synechococcus* CO_2-fixation enzymes, which are different enough from other cyanobacteria to warrant their own classification (Badger *et al.*, 2002; Tabita, 1999). However, reviewing previously constructed trees, such as those for Rubisco subunits, reveals that the cyanobacteria on these trees show a topology in essence identical to the consensus and concatenated phylogenies inferred here, though with much longer branches leading to the *Prochlorococcus* and marine *Synechococcus* lineages. These long branches make sense given evidence for accelerated evolution in many of these species, as well as selection on CO_2-fixation machinery during adaptation to the marine environment. This revised view suggests that horizontal transfer events have been less frequent than previously assumed, and occurred largely from marine picophytoplankton into a few strains of proteobacteria.

Origin and evolution of the oxygen evolving complex

The most distinctive metabolic innovation by the cyanobacteria is the ability to carry out oxygenic photosynthesis, in which water is utilized as an electron donor and O_2 is produced as a by-product. This metabolic capability has not been achieved by any other class of organisms, except those that inherited it from cyanobacterial ancestors. It is thus a "singular event," which has had a profound effect on the Earth as well as all living things. It is of surpassing interest to understand the evolutionary origins and development of this process, including the origin of the Mn/Ca cluster that forms the heart of the oxygen evolving complex (OEC), as well as the proteins that form the scaffold for the OEC and stabilize it. Another related question is how the two photosytems originated, came to be present in a single organism and became functionally linked. Finally, it is important to identify any intermediate forms of the OEC, alternative electron donors or transitional organisms that may have existed, in order to link up the world of anoxygenic photosynthesis with the oxygenic photosynthesis found in cyanobacteria, algae and plants. These issues are too extensive to review in detail in this chapter, but a few comments will be made and the reader pointed to recent papers that treat the subject in more detail.

The evolutionary origin of the Mn/Ca complex that is the heart of the OEC remains very poorly understood. Various speculations have been made as to its origin as arising from Mn catalase enzymes, Mn inorganic complexes or Mn-containing minerals (Blankenship and Hartman, 1998; Dismukes *et al.*, 2001; Sauer and Yachandra, 2002; Dismukes and Blankenship, 2005), but no strong positive indications have emerged as leading candidates.

The OEC is coordinated primarily by amino acid ligands from the D1 protein from the photosystem II reaction center. In addition to these direct ligands, a series of extrinsic proteins that stabilize the OEC are found in the thylakoid lumen. The most important of the extrinsic OEC proteins is the PsbO protein, also called the 33 kDa protein, which is found in all known oxygenic photosynthetic organisms. It has been extensively analyzed in terms of its structure and evolutionary aspects (De Las Rivas *et al.*, 2004; De Las Rivas and Roman, 2005). In addition to the PsbO protein, most organisms contain two additional extrinsic OEC proteins, although the identity of them is different in cyanobacteria and photosynthetic eukaryotes. Most cyanobacteria contain the PsbU and PsbV proteins, while eukaryotes primarily contain the PsbP and PsbQ proteins, which appear to have been recruited from other sources to take the place of the PsbU and PsbV proteins. In addition to the extrinsic OEC proteins, all photosystem II reaction centers contain as many as twenty or more other protein subunits. The evolutionary origin of most of these subunits is unknown, although some appear to have originated by gene duplication (Raymond and Blankenship, 2004).

The question of how the anoxygenic photosynthetic reaction centers relate to cyanobacterial photosystem I and II is also a poorly understood subject. There is now significant evidence that suggests that all photosynthetic reaction centers derive from a single ancestral complex (Schubert *et al.*, 1998; Sadekar *et al.*, 2006) although the nature of that ancestral complex is not yet well determined and may have been intermediate between the PS I type and PS II type of reaction centers. Another important aspect of the transition from anoxygenic to oxygenic photosynthesis is the change of pigment from the longer wavelength absorbing bacteriochlorophylls to the shorter wavelength absorbing and therefore higher energy chlorophylls (Blankenship and Hartman, 1998).

A recent paper has suggested that the first photosynthetic cells were cyanobacteria, based on analysis of a number of cyanobacterial genomes and a few anoxygenic phototroph genomes (Mulkidjanian *et al.*, 2006). However, this conclusion is contrary to most of what we know about the structural and mechanistic aspects of photosynthesis. In essence, this suggests that the much more sophisticated oxygenic photosynthetic apparatus predates the simpler anoxygenic ones, similar to proposing that an advanced Formula I race car was built before a model T Ford.

Photosynthetic bacteria contain one (anoxygenic phototrophs) or both (oxygenic phototrophs) of two distantly related types of photosynthetic reaction centers. Heliobacteria and green sulfur bacteria have type I reaction centers that transfer electrons through iron–sulfur clusters to cytoplasmic carriers such as ferredoxin. Photosynthetic proteobacteria (in the alpha, beta, and gamma classes) and filamentous anoxygenic phototrophs such as *Chloroflexus* have type II reaction centers with pheophytins and quinones as electron carriers. All oxygenic phototrophs—cyanobacteria, plants, and algae—have both types of reaction centers (photosystems) linked through a complex series of electron transfer intermediates. The distribution of the two types of reaction centers is quite perplexing in light of organismal phylogenies based on e.g. 16S ribosomal RNA sequence analysis, and competing ideas have been proposed in attempt to explain this paraphyletic distribution. The "selective loss hypothesis" of one or the other of the two basic types of photosystems to give rise to the various classes of anoxygenic phototrophs has been proposed many times before. This, along with the "fusion hypothesis," which proposes that lateral gene transfer or genetic fusion of two different organisms, each containing one of the two classes of reaction centers, was the path to the cyanobacteria, has recently been discussed in detail (Blankenship *et al.*, 2006).

Future direction

This chapter is by no means intended as a comprehensive coverage of cyanobacterial evolution, or of the diversity of their pathways and metabolic machinery. Our hope is instead to illustrate potential applications of comparative genomics as an emergent tool for providing unique insight into adaptation and diversification. Often, if not invariably, our gain in understanding from these computational approaches is asymptotically limited by the ability to frame and refine hypotheses by way of additional experiments; computational and experimental biology are simply different, highly complementary sides of the same coin.

Even in the process of writing this chapter, new cyanobacterial sequencing projects have been announced—now totaling 47, according to *genomesonline.org*. While many of these projects focus on model species or legacy organisms, there is increasing emphasis on use of culture-independent techniques to shed light on poorly resolved branches of the tree of life. It is imperative that this emphasis continues. As testament, the 24 remarkably diverse genomes compared here encompass just 4 of the 7 major cyanobacterial lineages enumerated in Honda's well-known study (Honda *et al.*, 1999), barely piercing the depth of Castenholtz's most recent Bergey's survey (Castenholz, 2001). Though the geological record occasionally offers up direct evidence, evolutionary studies are ultimately inferential and are thereby profoundly shaped by our inventory and knowledge of extant organisms and their environments. The ongoing (near exponential) decrease in sequencing cost (Service, 2006) guarantees that the central role of genomics in improving this knowledge will only amplify.

References

Altschul, S.F., Gish, W., Miller, W., Myers, E.W., and Lipman, D.J. (1990). Basic local alignment search tool. J. Mol. Biol. *215*, 403–410.

Badger, M.R., Hanson, D., and Price, G.D. (2002). Evolution and diversity of CO_2 concentrating mechanisms in cyanobacteria. Funct. Plant Biol. *29*, 161–173.

Berman-Frank, I., Lundgren, P., Chen, Y.B., Kupper, H., Kolber, Z., Bergman, B., and Falkowski, P. (2001). Segregation of nitrogen fixation and oxygenic photosynthesis in the marine cyanobacterium Trichodesmium. Science *294*, 1534–1537.

Berman-Frank, I., Lundgren, P., and Falkowski, P. (2003). Nitrogen fixation and photosynthetic oxygen evolution in cyanobacteria. Res. Microbiol. *154*, 157–164.

Blankenship R.E., and Hartman H., (1998). The origin and evolution of oxygenic photosynthesis. Trends Biochem. Sci., *23*, 94–97.

Blankenship, R.E., Sadekar, S., and Raymond, J. (2006). The evolutionary transition from anoxygenic to oxygenic photosynthesis. In Evolution of Aquatic Photoautotrophs, P. Falkowski, and A.N. Knoll, eds. (Amsterdam: Elsevier), pp.39–53.

Boyer, S.L., Flechtner, V.R., and Johansen, J.R. (2001). Is the 16S-23S rRNA internal transcribed spacer region a good tool for use in molecular systematics and population genetics? A case study in cyanobacteria. Mol. Biol. Evol. *18*, 1057–1069.

Brown, M.V., Schwalbach, M.S., Hewson, I., and Fuhrman, J.A. (2005). Coupling 16S-ITS rDNA clone libraries and automated ribosomal intergenic spacer analysis to show marine microbial diversity: development and application to a time series. Environ. Microbiol. *7*, 1466–1479.

Casamatta, D.A., Johansen, J.R., Vis, M.L., and Broadwater, S.T. (2005). Molecular and morphological characterization of ten polar and near-polar strains within the Oscillatoriales (Cyanobacteria). J. Phycol. *41*, 421–438.

Castenholz, R.W. (2001). Phylum BX. Cyanobacteria. Oxygenic photosynthetic bacteria., In Bergey's Manual of Systematic Bacteriology. Volume 1: The Archaea and Deeply Branching and Phototrophic Bacteria, R.W. Castenholz, ed. (New York: Springer-Verlag), pp. 413–439.

Chen, M., Hiller, R.G., Howe, C.J., and Larkum, A.W.D. (2005). Unique origin and lateral transfer of prokaryotic chlorophyll-*b* and chlorophyll-*d* light-harvesting systems. Mol. Biol. Evol. *22*, 21–28.

Chisholm, S.W., Frankel, S.L., Goericke, R., Olson, R.J., Palenik, B., Waterbury, J.B., West-Johnsrud, L., and Zettler, E.R. (1992). *Prochlorococcus marinus* nov. gen. nov. sp.: an oxyphototrophic marine prokaryote containing divinyl chlorophyll *a* and *b*. Arch. Microbiol. *157*, 297–300.

Chisholm, S.W., Olson, R.J., Zettler, E.R., Goericke, R., Waterbury, J.B., and Welschmeyer, N.A. (1988). A novel free-living prochlorophyte abundant in the oceanic euphotic zone. Nature *334*, 340–343.

Ciccarelli, F.D., Doerks, T., von Mering, C., Creevey, C.J., Snel, B., and Bork, P. (2006). Toward automatic reconstruction of a highly resolved tree of life. Science *311*, 1283–1287.

De Las Rivas, J., and Roman A. (2005). Structure and evolution of the extrinsic proteins that stabilize the oxygen-evolving engine. Photochem. Photobiol. Sci., *4*, 1003–1010.

De Las Rivas J., Balsera M., and Barber J. (2004). Evolution of oxygenic photosynthesis: genome-wide analysis of the OEC extrinsic proteins. Trends Plant Science, *9*, 1360–1385.

Delwiche, C.F., and Palmer, J.D. (1996). Rampant horizontal transfer and duplication of rubisco genes in eubacteria and plastids. Mol. Biol. Evol. *13*, 873–882.

Dismukes, G.C., and Blankenship, R.E. (2005). The origin and evolution of photosynthetic oxygen production. In: Photosystem II: The Water/Plastoquinone Oxido-Reductase In Photosynthesis, T. Wydrzynski, and K. Satoh, eds. (Springer, Dordrecht), pp. 683–695.

Dismukes, G.C., Klimov, V., Baranov, S., Kozlov, Y., DasGupta, J., and Tyryshkin A. (2001). The origin of atmospheric oxygen on Earth: The innovation of oxygenic photosynthesis. Proc. Natl Acad. Sci. USA *98*, 2170–2175.

Dufresne, A., Garczarek, L., and Partensky, F. (2005). Accelerated evolution associated with genome reduction in a free-living prokaryote. Genome Biol. *6*, R14.

Dvornyk, V. (2006). Subfamilies of *cpmA*, a gene involved in circadian output, have different evolutionary histories in cyanobacteria. Microbiology *152*, 75–84.

Dyhrman, S.T., and Haley, S.T. (2006). Phosphorus scavenging in the unicellular marine diazotroph *Crocosphaera watsonii*. Appl. Environ. Microbiol. *72*, 1452–1458.

Enright, A.J., Van Dongen, S., and Ouzounis, C.A. (2002). An efficient algorithm for large-scale detection of protein families. Nucleic Acids Res. *30*, 1575–1584.

Felsenstein, J. (1989). PHYLIP—Phylogeny inference package (Version 3.2). Cladistics 5, 164–166.

Ferris, M.J., Ruff-Roberts, A.L., Kopczynski, E.D., Bateson, M.M., and Ward, D.M. (1996). Enrichment culture and microscopy conceal diverse thermophilic *Synechococcus* populations in a single hot spring microbial mat habitat. Appl. Environ. Microbiol. *62*, 1045–1050.

Fox, G.E., Wisotzkey, J.D., and Jurtshuk, P., Jr. (1992). How close is close: 16S rRNA sequence identity may not be sufficient to guarantee species identity. Int. J. Syst. Bacteriol. *42*, 166–170.

Fuller, N.J., Marie, D., Partensky, F., Vaulot, D., Post, A.F., and Scanlan, D.J. (2003). Clade-specific 16S ribosomal DNA oligonucleotides reveal the predominance of a single marine *Synechococcus* clade throughout a stratified water column in the Red Sea. Appl. Environ. Microbiol. *69*, 2430–2443.

Gadagkar, S.R., Rosenberg, M.S., and Kumar, S. (2005). Inferring species phylogenies from multiple genes: concatenated sequence tree versus consensus gene tree. J. Exp. Zool. *304B*, 64–74.

Garczarek, L., van der Staay, G.W.M., Hess, W.R., Le Gall, F., and Partensky, F. (2001). Expression and phylogeny of the multiple antenna genes of the low-light-adapted strain *Prochlorococcus marinus* SS120 (Oxyphotobacteria). Plant Mol. Biol. *46*, 683–693.

Giddings, T.H., Withers, N.W., and Staehelin, L.A. (1980). Supramolecular structure of stacked and unstacked regions of the photosynthetic membranes of *Prochloron* sp., a prokaryote. Proc. Natl. Acad. Sci. USA *77*, 352–356.

Giovannoni, S.J., Turner, S., Olsen, G.J., Barns, S., Lane, D.J., and Pace, N.R. (1988). Evolutionary relationships among cyanobacteria and green chloroplasts. J. Bacteriol. *170*, 3584–3592.

Gogarten, J.P., Doolittle, W.F., and Lawrence, J.G. (2002). Prokaryotic evolution in light of gene transfer. Mol. Biol. Evol. *19*, 2226–2238.

Hansmann, S., and Martin, W. (2000). Phylogeny of 33 ribosomal and six other proteins encoded in an ancient gene cluster that is conserved across prokaryotic genomes: influence of excluding poorly alignable sites from analysis. Int. J. Syst. Evol. Microbiol. *50*, 1655–1663.

Harlow, T., Gogarten, J.P., and Ragan, M. (2004). A hybrid clustering approach to recognition of protein families in 114 microbial genomes. BMC Bioinformatics 5, 45.

Henson, B.J., Hesselbrock, S.M., Watson, L.E., and Barnum, S.R. (2004). Molecular phylogeny of the heterocystous cyanobacteria (subsections IV and V) based on nifD. Int. J. Syst. Evol. Microbiol. *54*, 493–497.

Hess, W.R. (2004). Genome analysis of marine photosynthetic microbes and their global role. Curr. Opin. Biotechnol. *15*, 191–198.

Hess, W.R., Partensky, F., van der Staay, G.W.M., Garcia-Fernandez, J.M., Borner, T., and Vaulot, D. (1996). Coexistence of phycoerythrin and a chlorophyll *a*/*b* antenna in a marine prokaryote. Proc. Natl. Acad. Sci. USA *93*, 11126–11130.

Hess, W.R., Rocap, G., Ting, C.S., Larimer, F., Stilwagen, S., Lamerdin, J., and Chisholm, S.W. (2001). The photosynthetic apparatus of *Prochlorococcus*: Insights through comparative genomics. Photosynthesis Research *70*, 53–71.

Honda, D., Yokota, A., and Sugiyama, J. (1999). Detection of seven major evolutionary lineages in cyanobacteria based on the 16S rRNA gene sequence analysis with new sequences of five marine *Synechococcus* strains. J. Mol. Evol. *48*, 723–739.

Hu, Q., Marquardt, J., Iwasaki, I., Miyashita, H., Kurano, N., Morschel, E., and Miyachi, S. (1999). Molecular structure, localization and function of biliproteins in the chlorophyll *a*/*d* containing oxygenic photosynthetic prokaryote *Acaryochloris marina*. Biochim. Biophys. Acta *1412*, 250–261.

Huelsenbeck, J.P., and Ronquist, F. (2001). MRBAYES: Bayesian inference of phylogenetic trees. Bioinformatics *17*, 754–755.

Huson, D.H. (1998). SplitsTree: analyzing and visualizing evolutionary data. Bioinformatics *14*, 68–73.

Koski, L.B., and Golding, G.B. (2001). The closest BLAST hit is often not the nearest neighbor. J. Mol. Evol. *52*, 540–542.

Kumar, S., Tamura, K., and Nei, M. (1994). MEGA: Molecular Evolutionary Genetics Analysis software for microcomputers. Comput. Appl. Biosci. *10*, 189–191.

Kupper, H., Ferimazova, N., Setlik, I., and Berman-Frank, I. (2004). Traffic lights in *Trichodesmium*. Regulation of photosynthesis for nitrogen fixation studied by chlorophyll fluorescence kinetic microscopy. Plant Physiol. *135*, 2120–2133.

Lamour, V., Quevillon, S., Diriong, S., N'Guyen, V.C., Lipinski, M., and Mirande, M. (1994). Evolution of the Glx-tRNA synthetase family: the glutaminyl enzyme as a case of horizontal gene transfer. Proc. Natl. Acad. Sci. USA *91*, 8670–8674.

Marquardt, J., Schultze, A., Rosenkranz, V., and Wehrmeyer, W. (1999). Ultrastructure and photosynthetic apparatus of *Rhodella violacea* (Porphyridiales, Rhodophyta) grown under iron-deficient conditions. Phycologia *38*, 418–427.

Miller, K.R., Jacob, J.S., Burger-Wiersma, T., and Matthijs, H.C. (1988). Supramolecular structure of the thylakoid membrane of *Prochlorothrix hollandica*: a chlorophyll *b*-containing prokaryote. J. Cell Sci. *91*, 577–586.

Miyashita, H., Ikemoto, H., Kurano, N., Adachi, K., Chihara, M., and Miyachi, S. (1996). Chlorophyll *d* as a major pigment. Nature *383*, 402.

Miyashita, H., Ikemoto, H., Kurano, N., Miyachi, S., and Chihara, M. (2003). *Acaryochloris marina* gen. et sp. nov. (Cyanobacteria), an oxygenic photosynthetic prokaryote containing chl *d* as a major pigment. J. Phycol. *39*, 1247–1253.

Moore, L.R., Rocap, G., and Chisholm, S.W. (1998). Physiology and molecular phylogeny of coexisting Prochlorococcus ecotypes. Nature *393*, 464–467.

Mulkidjanian,, A.Y., Koonin E.V., Makarova, K.S., Mekhedev, S.L., Sorokin, A., Wolf, Y.I., Dufresne, A., Partensky, F., Burd, H., Kaznadzey, D., Haselkorn, R., and Galperin, M.Y. (2006). The cyanobacterial genome core and the origin of photosynthesis. Proc. Natl. Acad. Sci. USA *103*: 13126–13131.

Nagata, N., Tanaka, R., Satoh, S., and Tanaka, A. (2005). Identification of a vinyl reductase gene for chlorophyll synthesis in *Arabidopsis thaliana* and Implications for the evolution of *Prochlorococcus* species. Plant Cell *17*, 233–240.

Nakamura, Y., Kaneko, T., Sato, S., Mimuro, M., Miyashita, H., Tsuchiya, T., Sasamoto, S., Watanabe, A., Kawashima, K., Kishida, Y., Y., Kiyokawa, C., Kohara, M., Matsumoto, M., Matsuno, A., Nakazaki, N., Shimpo, S., Takeuchi, C., Yamada, M., and Tabata, S. (2003). Complete genome structure of *Gloeobacter violaceus* PCC 7421, a cyanobacterium that lacks thylakoids (supplement). DNA Res. *10*, 181–201.

Nakanishi, H., Nozue, H., Suzuki, K., Kaneko, Y., Taguchi, G., and Hayashida, N. (2005). Characterization of the *Arabidopsis thaliana* mutant *pcb2* which accumulates divinyl chlorophylls. Plant Cell Physiol. *46*, 467–473.

Oster, U., Tanaka, R., Tanaka, A., and Rudiger, W. (2000). Cloning and functional expression of the gene encoding the key enzyme for chlorophyll *b* biosynthesis (CAO) from *Arabidopsis thaliana*. Plant J. *21*, 305–310.

Palenik, B., and Haselkorn, R. (1992). Multiple evolutionary origins of prochlorophytes, the chlorophyll *b*-containing prokaryotes. Nature *355*, 265–267.

Raymond, J., and Blankenship, R.E. (2004). The evolutionary development of the protein complement of photosystem 2. Biochim Biophys Acta *1655*: 133– 139.

Raymond, J., Siefert, J.L., Staples, C.R., and Blankenship, R.E. (2004). The natural history of nitrogen fixation. Mol. Biol. Evol. *21*, 541–554.

Rippka, R., Deruelles, J., Waterbury, J.B., Herdman, M., and Stanier, R.Y. (1979). Generic assignments, strain histories and properties of pure cultures of cyanobacteria. J. Gen. Microbiol. *111*, 1–61.

Rocap, G., Distel, D.L., Waterbury, J.B., and Chisholm, S.W. (2002). Resolution of *Prochlorococcus* and *Synechococcus* ecotypes by using 16S-23S ribosomal DNA internal transcribed spacer sequences. Appl. Environ. Microbiol. *68*, 1180–1191.

Rocap, G., Larimer, F.W., Lamerdin, J., Malfatti, S., Chain, P., Ahlgren, N.A., Arellano, A., Coleman, M., Hauser, L., Hess, W.R., Johnson Z.I., Land, M., Lindell, D., Post, A.F., Regala, W., Shah, M., Shaw, S.L., Steglich, C., Sullivan, M.B., Ting, C.S., Tolonen, A., Webb, E.A., Zinser, E.R., and Chisholm, S.W. (2003). Genome divergence in two *Prochlorococcus* ecotypes reflects oceanic niche differentiation. Nature *424*, 1042–1047.

Sadekar, S., Raymond, J., and Blankenship, R.E. (2006). Conservation of distantly related membrane proteins: Photosynthetic reaction centers share a common structural core. Mol. Biol. Evol. *23*, 2001–2007.

Sánchez-Baracaldo, P., Hayes, P.K., and Blank, C.E. (2005). Morphological and habitat evolution in the Cyanobacteria using a compartmentalization approach. Geobiology *3*, 145–165.

Sauer, K., and V. Yachandra. (2002). A possible evolutionary origin for the Mn-4 cluster of the photosynthetic water oxidation complex from natural MnO_2 precipitates in the early ocean. Proc. Natl. Acad. Sci. USA *99*, 8631–8636.

Scanlan, D.J. (2003). Physiological diversity and niche adaptation in marine *Synechococcus*. Adv. Microb. Physiol. *47*, 1–64.

Scanlan, D.J., and West, N.J. (2002). Molecular ecology of the marine cyanobacterial genera *Prochlorococcus* and *Synechococcus*. FEMS Microbiol. Ecol. *40*, 1–12.

Service, R.F. (2006). Gene sequencing. The race for the $1000 genome. Science *311*, 1544–1546.

Shimada, A., Kanai, S., and Maruyama, T. (1995). Partial sequence of ribulose-l,5-bisphosphate carboxylase/oxygenase and the phylogeny of *Prochloron* and *Prochlorococcus* (Prochlorales). J. Mol. Evol. *40*, 671.

Stanier, R.Y., Sistrom, W.R., Hansen, T.A., Whitton, B.A., Castenholz, R.W., Pfennig, N., Gorlenko, V.N., Kondratieva, E.N., Eimhjellen, K.E., Whittenbury, R., *et al.* (1978). Proposal to place nomenclature of cyanobacteria (blue-green-algae) under rules of International Code of Nomenclature of Bacteria. Int. J. Syst. Bacteriol. *28*, 335–336.

Steunou, A.S., Bhaya, D., Bateson, M.M., Melendrez, M.C., Ward, D.M., Brecht, E., Peters, J.W., Kuhl, M., and Grossman, A.R. (2006). *In situ* analysis of nitrogen fixation and metabolic switching in unicellular thermophilic cyanobacteria inhabiting hot spring microbial mats. Proc. Natl. Acad. Sci. USA *103*, 2398–2403.

Swingley, W.D., Hohmann-Marriott, M.F., Le Olson, T., and Blankenship, R.E. (2005). Effect of iron on growth and ultrastructure of *Acaryochloris marina*. Appl. Environ. Microbiol. *71*, 8606–8610.

Tabita, F.R. (1999). Microbial ribulose 1,5-bisphosphate carboxylase/oxygenase: A different perspective. Photosynthesis Res. *60*, 1–28.

Tanaka, A., Ito, H., Tanaka, R., Tanaka, N.K., Yoshida, K., and Okada, K. (1998). Chlorophyll *a* oxygenase (CAO) is involved in chlorophyll *b* formation from chlorophyll *a*. Proc. Natl. Acad. Sci. USA *95*, 12719–12723.

Tatusov, R.L., Koonin, E.V., and Lipman, D.J. (1997). A genomic perspective on protein families. Science *278*, 631–637.

Tatusov, R.L., Natale, D.A., Garkavtsev, I.V., Tatusova, T.A., Shankavaram, U.T., Rao, B.S., Kiryutin, B., Galperin, M.Y., Fedorova, N.D., and Koonin, E.V. (2001). The COG database: new developments in phylogenetic classification of proteins from complete genomes. Nucleic Acids Res. *29*, 22–28.

Thiel, T. (1993). Characterization of genes for an alternative nitrogenase in the cyanobacterium *Anabaena variabilis*. J. Bacteriol. *175*, 6276–6286.

Thompson, J.D., Higgins, D.G., and Gibson, T.J. (1994). CLUSTAL W: improving the sensitivity of progressive multiple sequence alignment through sequence weighting, position-specific gap penalties and weight matrix choice. Nucleic Acids Res. *22*, 4673–4680.

Tomitani, A., Knoll, A.H., Cavanaugh, C.M., and Ohno, T. (2006). The evolutionary diversification of cyanobacteria: molecular-phylogenetic and paleontological perspectives. Proc. Natl. Acad. Sci. USA *103*, 5442–5447.

Tomitani, A., Okada, K., Miyashita, H., Matthijs, H.C.P., Ohno, T., and Tanaka, A. (1999). Chlorophyll *b* and phycobilins in the common ancestor of cyanobacteria and chloroplasts. Nature *400*, 159–162.

Turner, S., Pryer, K.M., Miao, V.P.W., and Palmer, J.D. (1999). Investigating deep phylogenetic relationships among cyanobacteria and plastids by small submit rRNA sequence analysis. J. Eukaryot. Microbiol. *46*, 327–338.

Urbach, E., Robertson, D.L., and Chisholm, S.W. (1992). Multiple evolutionary origins of prochlorophytes within the cyanobacterial radiation. Nature *355*, 267–270.

Vinh le, S., and von Haeseler, A. (2004). IQPNNI: moving fast through tree space and stopping in time. Mol. Biol. Evol. *21*, 1565–1571.

Ward, D.M., Ferris, M.J., Nold, S.C., and Bateson, M.M. (1998). A natural view of microbial biodiversity within hot spring cyanobacterial mat communities. Microbiol. Mol. Biol. Rev. *62*, 1353–1370.

Waterbury, J.B., Watson, S.W., Valois, F.W., and Franks, D.G. (1986). Biological and ecological characterization of the marine unicellular cyanobacterium *Synechococcus*. In Photosynthetic Picoplankton, W.K.W. Li, ed. (Can. Bull. Fish Aquat. Sci.), pp. 71-

Gene Transfer to Cyanobacteria in the Laboratory and in Nature 3

Enrique Flores, Alicia M. Muro-Pastor, and John C. Meeks

Abstract

Horizontal (lateral) gene transfer is a postulated mechanism influencing bacterial evolution. Known mechanisms of DNA transfer into cyanobacteria include genetic transformation and conjugation with *Escherichia coli*, which are widely used in the laboratory with several different cyanobacterial strains. Additionally, direct (likely conjugal) transfer of DNA between cyanobacterial strains has been demonstrated. These transfer mechanisms can represent the basis for genetic recombination in natural populations of cyanobacteria, for which several possible examples have been described, as well as for the horizontal transfer that is deduced for some genes in protein phylogeny studies, which have been made possible by the current availability of numerous complete cyanobacterial genomic sequences.

Introduction

The cyanobacteria represent a group of prokaryotes that are characterized by a well-defined metabolic trait: they perform oxygenic photosynthesis (see Chapter 9). For light absorption in photosynthesis, the cyanobacteria carry pigments that give them characteristic colors. These permit the identification of these organisms in field samples even with the naked eye. Many cyanobacteria can readily be seen in nature, and they are found almost everywhere in our planet (see Chapter 4). Thus, we are confronted with a bacterial group that can be easily identified and that, in phylogenetic trees of core cellular components (i.e. 16S rRNA), form a coherent branch that is well separated from sister bacterial groups (Woese, 1987). However, the cyanobacteria constitute a very diverse group in terms of morphology and metabolic capabilities other than oxygenic photosynthesis (Rippka *et al.*, 1979), which probably reflects billions of years of evolution (see Chapter 1).

Two well-differentiated processes contributing to bacterial diversification are the evolution of vertically inherited genes and gene gain through horizontal (or lateral) gene transfer (Novichkov *et al.*, 2004). Cyanobacterial evolution is thoroughly covered in Chapters 1, 2 and 5 of this book, but here we approach a key aspect for the latter mechanism: DNA transfer. Interestingly, our knowledge of DNA transfer has developed in a context of the need for genetic analysis in modern biological research. In this chapter, we will cover the known processes of DNA transfer in cyanobacteria as demonstrated in the laboratory, and will summarize representative examples illustrating that gene transfer also takes place in nature contributing to the diversity of the genetic pool of these organisms.

Mechanisms of experimental DNA transfer

DNA transfer into cyanobacterial cells has been amply demonstrated in the laboratory, to the point that currently it is a common practice for a number of different laboratory strains. In this context, two historical landmarks were the demonstration of transformation in *Synechococcus elongatus*, formerly known as *Anacystis nidulans* (Shestakov and Khyen, 1970), and of conjugation from *Escherichia coli* to *Anabaena/Nostoc* spp. (Wolk *et al.*, 1984). Additionally, transformation of *S. elongatus* with a heterologous plasmid carrying a transposon, Tn*901*, resulted in endogenous *Synechococcus* plasmids marked with Tn*901* that could be isolated and transferred back to the cyanobacterium (van den Hondel *et al.*, 1980). A very important point is that, in those experiments, DNA transfer was followed by recombination of the incoming material with genomic DNA (Shestakov and Khyen, 1970) or by establishment of replicating plasmids in the recipient (van den Hondel *et al.*, 1980; Wolk *et al.*, 1984). Thus, the basis for genetic analysis was settled, with a wide variety of different experimental approaches having been made possible by these basic DNA transfer procedures.

Transformation in unicellular cyanobacteria

Transformation of the unicellular cyanobacterium *S. elongatus* by incubation of cell suspensions with DNA isolated from antibiotic-resistant strains of the same cyanobacterium is a high-efficiency process, taking place with frequencies of about $2 \cdot 10^{-3}$ per colony-forming unit (CFU) (Shestakov and Khyen, 1970). Such a high frequency of transformation probably relies upon the fact that the incoming DNA bears all the post-replication modifications that are proper to this organism, and indicates that a powerful recombination machinery operates in this cyanobacterium. This methodology has been applied to a number of other unicellular cyanobacteria and, although not every tested strain was successful, two other strains that show robust transformation and have become important organisms for genetic studies of cyanobacteria are *Synechococcus* sp. strain PCC 7002 (formerly, *Agmenellum quadruplicatum*; Stevens and Porter, 1980) and *Synechocystis* sp. strain PCC 6803 (Grigorieva and Shestakov, 1982).

The requirement for gene cloning in molecular biology research encouraged the isolation of cyanobacterial plasmids in order to construct hybrid *E. coli–Synechococcus* plasmids to be used as shuttle vectors between both organisms (see, e.g. Golden and Sherman, 1983). This would allow the cloning of a gene in *E. coli* by standard recombinant DNA techniques and its transfer to the cyanobacterium. Transformation involving replication of the incoming plasmid in the recipient strain takes place, however, with a lower frequency, about 10^{-5} per CFU (Golden and Sherman, 1983), than transformation with chromosomal DNA. Interestingly, higher transformation frequencies (about $5 \cdot 10^{-3}$ per CFU) are obtained if a plasmid that cannot replicate in the cyanobacterium, but that contains cloned DNA from it, is transferred to the cyanobacterium (Tsinoremas *et al.*, 1994). In this case, the incoming DNA is incorporated in the recipient's genome by homologous recombination, and the most efficient type of construct is one in which a gene marker (a selectable gene such as an antibiotic resistance gene) is flanked by cloned contiguous DNA from the recipient. Thus, the frequency of transformation is highly dependent on the fate of the incoming DNA, but it is probably also affected by the efficiency of particular processes (e.g. recombination) and the properties of molecular elements (e.g. DNA construct or plasmid). Whereas homologous recombination appears as a robust mechanism for incorporating DNA in the recipient's genome, the establishment of a replicating plasmid may depend on the particular replicon (origin of replication) that is used.

The mechanism of DNA uptake in better-known natural transformation systems requires proteins related to those involved in the assembly of type IV pili and type II secretion systems (T2SS), functionally coupled to a DNA translocase complex at the cytoplasmic membrane (Chen and Dubnau, 2004). Although the mechanism of transformation in cyanobacteria has not been investigated in detail, it appears to involve breakage of the incoming DNA (Williams and Szalay, 1983). Inspection of cyanobacterial whole genomic sequences allows the identification of putative elements of natural transformation systems (Yura *et al.*, 1999). Additionally, genes encoding Pil proteins for biogenesis of type IV pili, ComA (a membrane-anchored protein that contains DNA binding and endonuclease domains) and ComF (containing a phosphoribosyltransferase domain, a distinguishing feature of other ComF proteins of naturally transformable bacteria) have been demonstrated as needed for transformation of *Synechocystis* sp. strain PCC 6803 (Yoshihara *et al.*, 2001; Okamoto and Ohmori, 2002; Nakasugi *et al.*, 2006).

Endonuclease breakage of the incoming DNA helps to explain some aspects of the frequencies of transformation described above. A linear DNA molecule will require a double recombination event to insure that a fragment of it gets integrated into the genome (Figure 3.1a), because a single recombination would result in a double strand break rendering an open chromosome, which apparently would be lethal for the cyanobacterium. It is clear that such a double recombination event can happen at a substantial frequency in the transformable unicellular cyanobacteria. To be established as an autonomous, replicative genetic element, a linear plasmid will need to be converted back to a circular form. This apparently happens with lower frequency.

Conjugation with heterocyst-forming cyanobacteria

Transformation has not been demonstrated for any filamentous, N_2-fixing cyanobacterium that exhibits heterocyst development, a process of high biological interest (see Chapter 14). However, conjugal transfer of plasmid DNA from *E. coli* to several strains of heterocyst-forming *Anabaena* was attained, promoted by the broad host-range plasmid RP-4 (Wolk *et al.*, 1984). The conjugation system in Gram-negative bacteria requires a type IV secretion system (T4SS) and the presence of a relaxase, which is a site- and strand-specific endonuclease that cuts the strand to be transferred, and of a coupling protein that brings together the relaxosome and the T4SS (Llosa and de la Cruz, 2005). In the intercellular transfer of a plasmid, the relaxase pilots the transfer, pushed by the coupling protein, of a single-stranded DNA through the type IV secretion system into the recipient strain, where the second DNA strand is re-synthesized producing a double-stranded plasmid (Llosa *et al.*, 2002; Gomis-Ruth and Coll, 2006). Once inside the bacterial cell, the plasmid can recombine with the genome of the host (thus integrating itself into a larger replicon) or replicate autonomously if it bears an origin of replication that is functional in the recipient strain.

Mobilizable shuttle vectors used in successful conjugation to cyanobacteria carry replicons for both *E. coli* and the cyanobacterium, contain a site (*bom*, *nic* or *oriT*, for origin of transference) required for mobilization, and are marked by an antibiotic-resistance gene

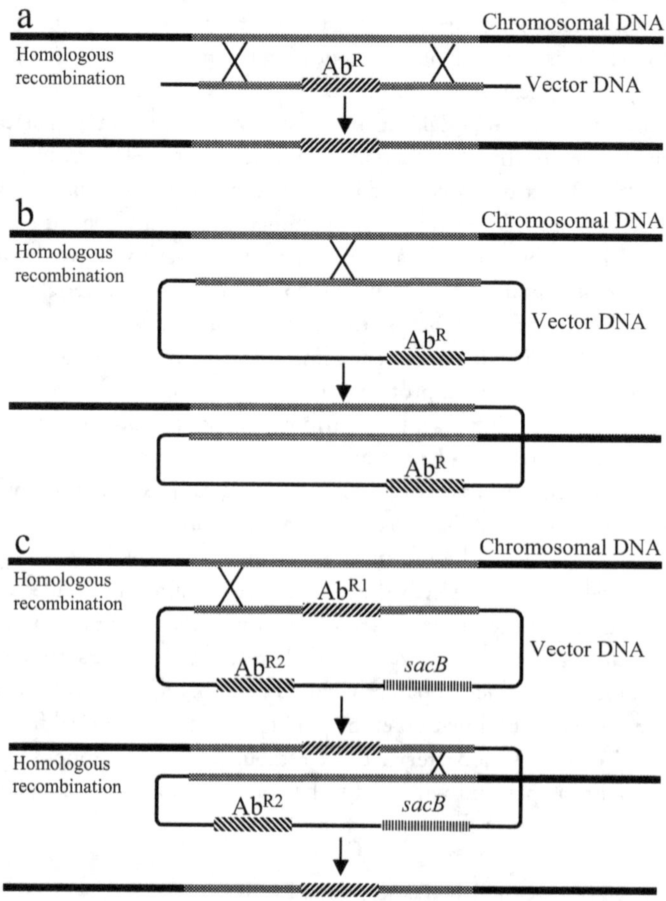

Figure 3.1 Homologous recombination after transformation of unicellular cyanobacteria or conjugation of heterocyst-forming cyanobacteria. After transformation, a linear DNA can recombine with endogenous genomic sequences through a double cross-over event leading to gene replacement (a). (This is necessarily a simplified view of a process that can be complex, since it is unknown whether double or single DNA enters the cytoplasm of the recipient.) After conjugation, a single cross-over leads to integration of the whole plasmid (b, c), and selection for sucrose-resistant clones that experienced a second cross-over event, with concomitant loss of DNA corresponding to the vector portion of the plasmid and some chromosomal DNA, can be done provided the transferred plasmid carries the *sacB* gene (c). This process can result in gene replacement as shown. Ab^R, antibiotic-resistance gene(s). Dotted lines denote chromosomal DNA that has been cloned in cloning vectors. More genetic strategies based on homologous recombination that have been utilized in cyanobacteria are described by Thiel (1994).

that can be selected for in both organisms. The cyanobacterial replicon originally used was the pDU1 plasmid isolated from a *Nostoc* sp. (Wolk *et al.*, 1984). The conjugation system required a helper plasmid carrying the *mob* gene encoding a specific endonuclease that cuts the plasmid to be transferred at the *bom* site. Because cyanobacteria exhibit a diverse and ample restriction enzyme complement, those vectors had sites for *Anabaena* restriction enzymes deleted. Restriction of incoming DNA in the recipient cell has been unraveled as a serious hurdle in conjugation to cyanobacteria (Wolk *et al.*, 1984; Elhai *et al.*, 1997). The conjugation system has been improved throughout the years, and the current protocol includes the use of a helper plasmid that carries methylases corresponding to the restriction/modification systems of a cyanobacterial strain widely used in laboratories worldwide, *Anabaena* sp. strain PCC 7120, in addition to the *mob* gene (Elhai *et al.*, 1997). Frequencies of conjugation up to 10% of the recipient cells have been achieved.

As stated above, in contrast to transformation that renders a piece of linear DNA in the recipient strain, the result of conjugation is a circular plasmid. If the incoming plasmid bears DNA homologous to part of the recipient's genome, a single recombination event will integrate the plasmid into the genome (the chromosome or an endogenous plasmid) of the recipient (Figure 3.1b). However, for some genetic manipulations, such as gene replacement, a double recombination event is necessary. Clones in which a second recombination event (resolution of the plasmid integrate) has happened (Figure 3.1c) can be isolated by positive selection. For example, if the vector portion of the plasmid incorporates the *sacB* gene, which encodes levansucrase and makes sucrose lethal to numerous bacteria, including cyanobacteria, loss of the vector and the *sacB* gene leads to survival in the presence of sucrose (Cai and Wolk, 1990).

The RP-4 based conjugation has been most widely used with *Anabaena* sp. strain PCC 7120, but it has been also shown to work, and utilized, with such important research cyanobacteria as *Nostoc punctiforme* (strain ATCC 29133; Flores and Wolk, 1995), a reference strain in studies of cyanobacterial-plant symbiosis (Meeks and Elhai, 2002), and *Fremyella diplosiphon* (*Tolypothrix* sp. strain PCC 7601; Cobley *et al.*, 1993), a strain widely utilized in studies of complementary chromatic adaptation (Grossman, 2003). Different cyanobacterial strains have their own subtle physiological peculiarities, including varieties of restriction endonucleases, and these lead to differences in optimal conditions for conjugation (Cohen *et al.*, 1998).

Other genetic systems

Although the most widely used cyanobacteria and methods of gene transfer in molecular genetics are those mentioned above, some other cyanobacteria and genetic systems have been utilized. RP-4 based conjugation has been shown to work with high efficiency in *S. elongatus* strain PCC 7942, both with plasmids able to replicate in the recipient (frequency up to 0.5 exconjugants CFU^{-1}) or that integrate in the recipient's genome (10^{-3} exconjugants CFU^{-1}) (Tsinoremas *et al.*, 1994). Promiscuous conjugative plasmid RSF1010 also transfers to unicellular cyanobacteria with different frequencies depending on the strain: to fresh-water strains *S. elongatus* strain PCC 7942 ($5 \cdot 10^{-6}$ exconjugants CFU^{-1}) and *Synechocystis* sp. strain PCC 6803 ($5 \cdot 10^{-4}$ exconjugants CFU^{-1}; Marraccini *et al.*, 1993), and to marine *Synechococcus* spp. ($5 \cdot 10^{-3}$ exconjugants CFU^{-1}; Brahamsha, 1996). The latter is of particular interest given the ecological importance and unique non-flagellar swimming mechanism of the marine *Synechococcus* strains (see Chapter 5).

Electroporation has also been used to transform cyanobacteria, but it has generally been carried out with plasmids developed for conjugation. Electroporation of a pre-methylated pDU1-based plasmid has been shown for *Anabaena* sp. (Thiel and Poo, 1989) and with a similar replicating plasmid for *N. punctiforme* (Summers *et al.*, 1995), as well as of an RSF1010-based vector into *Synechocystis* sp. strain PCC 6803 (Marraccini *et al.*, 1993). Most interestingly, electroporation with a pDU1-based vector has been shown to work with *Croococcidiopsis* sp., the only cyanobacterium of the baeocyte-forming group (taxonomic section II of Rippka *et al.*, 1979) in which gene transfer has been achieved (Billi *et al.*, 2001).

The fate of the incoming DNA in the genetic systems that have been discussed in this section, namely a plasmid that is established as an autonomous replicon or a whole plasmid or a fragment of DNA that is integrated into the recipient's genome, has permitted the development of a wide variety of genetic manipulations in unicellular and heterocyst-forming cyanobacteria (Koksharova and Wolk, 2002). Table 3.1 summarizes some cyanobacteria that are amenable to genetic analysis. However, this is not the only important outcome of the uncovered genetic systems, since the fact that gene transfer can take place by natural transformation and mediated by promiscuous systems of conjugation is of intrinsic interest in the study of the biology of cyanobacteria.

Gene transfer between cyanobacteria

Natural transformation should permit uptake of ambient DNA released by other organisms, including cyanobacteria. Heterospecific transformation in which DNA from a particular unicellular cyanobacterium has been used to transform other different unicellular strains has been reported (Stevens and Porter, 1986). Conjugation should also permit receiving DNA from a broad spectrum of bacteria that can carry the conjugative plasmids. These mechanisms, which can take place for instance in the dense material of a microbial mat, may contribute to cyanobacterial evolution, but is there any evidence of direct transfer of DNA between cyanobacteria?

Conjugal transfer in *Anabaena* strains

Anabaena sp. strain PCC 7120 bears as part of its genome six plasmids ranging in size from 5.5 to 408 kb (Bancroft *et al.*, 1989; Kaneko *et al.*, 2001). The identification of a gene located in one of those plasmids, *nucA*, encoding a sugar-non-specific nuclease, prompted an attempt to test direct gene transfer between two *Anabaena* strains (Muro-Pastor *et al.*, 1994). Transfer of a neomycin resistance marker located in the *nucA* region of the 408-kb α megaplasmid of *Anabaena* sp. strain PCC 7120 to related *Anabaena* sp. strain PCC 7118 was demonstrated. Further analyses determined that the *nucA* region of *Anabaena* sp. strain PCC 7120 (bearing the neomycin resistance marker) had recombined with and replaced the *nucA* region in the corresponding megaplasmid of *Anabaena* sp. strain PCC 7118 (a ca. 510 kb plasmid). The difference in plasmid sizes indicates that the entire strain PCC 7120 plasmid did not replace the strain PCC 7118 plasmid (Muro-Pastor *et al.*, 1994). There is no evidence that genes located outside the α megaplasmid can be transferred. Because the observed transfer of the neomycin resistance marker was not affected by a treatment with DNase, the possibility exists that this transfer is mediated by conjugation, rather than transformation (Muro-Pastor *et al.*, 1994).

The availability of the complete genomic sequence of *Anabaena* sp. strain PCC 7120 (Kaneko *et al.*, 2001) allows inspection of the sequence in order to identify putative cyanobacterial genes whose products could correspond to a bacterial conjugation system. Genes encoding proteins similar to relaxases from mobilizable or conjugative plasmids (see above)

Table 3.1 Some cyanobacteria amenable to genetic analysis, and indication of the gene transfer system available or most frequently utilized with the indicated strain.

Cyanobacterial strain	Gene transfer system
Unicellular strains	
Freshwater	
Synechococcus sp. strain PCC 7942	Transformation
Synechocystis sp. strain PCC 6803	Transformation
Marine	
Synechococcus sp. strain PCC 7002	Transformation
Synechococcus sp. strains WH7803, WH8102, WH8103	Conjugation
Baeocyte-forming	
Chroococcidiopsis spp.	Conjugation/electroporation
Non-heterocyst-forming filamentous strains	
Planktothrix sp., *Plectonema boryanum* M101	Electroporation
Heterocyst-forming strains	
Anabaena sp. strains PCC 7120, ATCC 29413	Conjugation
Fremyella diplosiphon (*Tolypothrix* sp. strain PCC 7601)	Conjugation/electroporation
Nostoc punctiforme strain ATCC 29133	Conjugation/electroporation

have been identified in *Anabaena* plasmids α, δ and γ (Francia *et al.*, 2004; Fernández-López *et al.*, 2006). In the case of plasmid γ, a gene encoding a TrwC-like relaxase appears clustered with a gene encoding a TrwB-like coupling protein (Fernández-López *et al.*, 2006). Further inspection of the plasmid sequences allows the identification of genes encoding a MobM-like relaxase in plasmid ε (*alr9001*), TrwB-like coupling proteins in plasmids α and β (*alr7213* and *alr7539*), as well as proteins similar to other Mob proteins in plasmid β (*all7655, alr7646*). The presence of all these genes in the *Anabaena* genome gives support to the possibility that the observed transfer of the *nucA* region takes place through conjugation.

Genetic recombination in natural populations of cyanobacteria

Population genetic studies of natural samples of several different cyanobacteria have been carried out analyzing in each case a few genetic loci. Allele-specific polymerase chain reaction has helped in these analyses, since it permits investigation of numerous environmental samples. In one study, three loci of the heterocyst-forming cyanobacterium *Nodularia* sp. were investigated in samples from the Baltic Sea. These loci were a non-coding spacer between adjacent copies of the main structural gas vesicle gene *gvpA* (*gvpA*-IGS), the phycocyanin intergenic spacer (PC-IGS) and the rDNA internal transcribed spacer (rDNA-ITS). The analysis unraveled the presence of combinations of alleles suggestive of genetic exchange through horizontal gene transfer, which however was not high enough to produce random association among alleles (linkage equilibrium; Barker *et al.*, 2000).

Another study investigated the genetic diversity of the genus *Arthrospira* using sequences from the phycocyanin operon. Based on this locus, the genus *Arthrospira* appears to consist of three genetically clustered lineages. The distribution of nucleotide substitutions indicates, however, that these three lineages are not the result of a simple cladogenesis characterized by the accumulation of independent substitutions. Instead, the observed clustering appears to result from horizontal transfers of blocks of sequences (Manen and Falquet, 2002). In a study of the microcystin synthetase genes in natural isolates of the unicellular cyanobacterium *Microcystis* sp., a reticulate phylogeny of *mcyA* suggests intragenic recombination in the evolutionary history of this gene, which would have required gene transfer events amongst *Microcystis* spp. (Tanabe *et al.*, 2004). In a final example, multilocus sequence typing of *Microcoleus chthonoplastes* has been used to analyze three genetic loci (*kaiC*, *petB/D*, rDNA-ITS) in individual cyanobacterial filaments from natural microbial mat populations. Non-congruent phylogenetic gene trees for the three loci analyzed suggested the occurrence of horizontal genetic exchange and recombination frequent enough to cause almost random association among alleles (Lodders *et al.*, 2005). Although the mechanisms behind these examples of genetic exchange have not been determined, these studies are consistent with the laboratory studies summarized above demonstrating genetic exchange.

Transduction in cyanobacteria?

Bacteriophages that can infect cyanobacteria, cyanophages, have been described for different cyanobacteria, from heterocyst-formers (see e.g. Hu *et al.*, 1981) to marine unicellular strains (see e.g. Waterbury and Valois, 1993). However, no lysogenic cyanophage has been identified and, thus, no associated genetic transduction has been demonstrated. On the other hand, the recent sequencing of cyanophage genomes has shown the unpredicted presence of cyanobacterial genes in some of these genomes (see Chapter 5). The first described case was the presence of *psbA* and *psbD* genes, which encode core subunits of photosystem II, in a marine *Synechococcus* cyanophage (Mann *et al.*, 2003). The *psbA* gene is transcribed during infection, thus potentially supporting the photosynthetic activity of the host for production of the cyanophage progeny (Clokie *et al.*, 2006). The possible exchange of *psbA* and other photosynthetic genes between *Synechococcus* and *Prochlorococcus* strains via phage intermediates has been suggested, but not yet resolved (Lindell *et al.*, 2004; Zeidner *et al.*, 2005).

Horizontal gene transfer and cyanobacterial evolution

The information about genetic exchange summarized above, which includes well-demonstrated laboratory gene transfer to cyanobacteria, mediated by natural genetic systems of transference, as well as examples of population genetics suggestive of recombination, indicates that horizontal gene transfer can take place in these organisms. The availability of numerous (24 at the time of this writing) sequenced cyanobacterial genomes permits the construction of protein phylogenetic trees, which can be compared to organism phylogenetic trees. Although delineating a definitive cyanobacterial (organismal) phylogenetic tree is still under way (see Chapter 2), major phylogenic associations within the cyanobacteria are clear enough as to detect inconsistencies in some protein phylogenetic trees. The corresponding proteins are obvious candidates for products of horizontally transferred genes.

An extreme case is when a gene that is not commonly found in cyanobacteria is analyzed. Figure 3.2 shows a phylogenetic tree of a Na^+-dependent glutamate transporter, the *gltS* gene product, which apart from *E. coli* has only been characterized in *Synechocystis* sp. strain PCC 6803 (Quintero *et al.*, 2001). The gene is also found in the genomic sequences of *Thermosynechococcus elongatus* and some marine cyanobacteria, namely *Synechococcus* sp. strain WH8102 and *Prochlorococcus* sp. strains MIT9313 and SS120. When a phylogenetic analysis is performed, the proteins from strains PCC 6803 and WH8102 are found clustered with eubacterial GltS proteins, whereas those from *Prochlorococcus, T. elongatus* and a second strain PCC 6803 homologue are clustered with archaeal GltS proteins. Such phylogenetic distribution strongly suggests that *gltS* has been incorporated into cyanobacteria at least in two independent events of horizontal gene transfer.

Another example of horizontal gene transfer from distant organisms to cyanobacteria comes from analysis of threonyl tRNA synthetase (Zhaxybayeva *et al.*, 2006). Threonyl tRNA synthetase from most analyzed cyanobacteria is phylogenetically coherent, in the sense that cyanobacterial proteins cluster together in a phylogenetic tree, but the enzymes from three *Prochlorococcus* strains cluster together with threonyl tRNA synthetase from the γ-proteobacteria. This observation suggests an independent origin of threonyl tRNA synthetase in these *Prochlorococcus* strains and most cyanobacteria (Zhaxybayeva *et al.*, 2006).

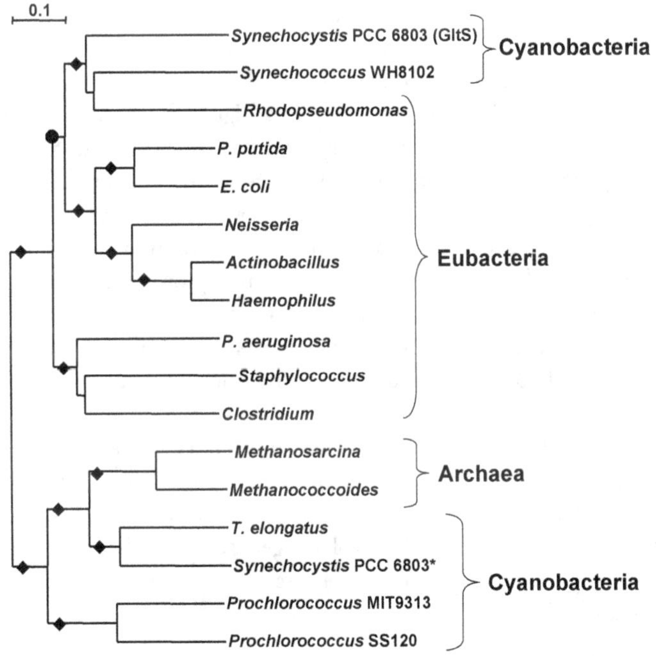

Figure 3.2 Phylogenetic analysis of GltS protein from cyanobacteria, other eubacteria and archaea. A neighbor-joining tree based on ClustalX (1.8) alignments is presented; bootstrap support over 99% (♦) or 86% (•). Size marker, 0.1 substitutions per amino acid residue. The tree was rooted with a distantly related protein, *Escherichia coli* YdfJ (not shown). *A second GltS-like protein in *Synechocystis* sp. PCC 6803.

Horizontal gene transfer has been recently quantified in a phylogenetic analysis of eleven cyanobacterial genomes, with comparison to 168 other bacterial genomes (Zhaxybayeva *et al.*, 2006). In a conservative estimation, the authors concluded that about 50% of the cyanobacterial genes have a history of horizontal gene transfer, either intra-cyanobacterial or between cyanobacteria and other bacterial groups. An important conclusion of this study is that genes from all functional categories (including informational and metabolic genes) can be subjected to horizontal transfer. However, predicted transfer of informational genes is proportionally lower between cyanobacteria and non-cyanobacterial sources than among cyanobacteria, and in favor of metabolic genes. In conclusion, genetic exchange through horizontal gene transfer appears to have played an important role in the evolution of cyanobacteria.

Detection of distant (between cyanobacteria and non-cyanobacterial partners) and related (between different strains of cyanobacteria) gene transfer is of interest. Incorporation of a gene from a phylogenetically distant organism is probably a low-frequency event, since the transferred DNA has to survive a number of barriers (restriction, need for recombination or replication in a heterologous host; Heinemann and Billington, 2004), but may result in the gain of, for instance, a metabolic function. Such transfers are easy to detect in phylogenetic analysis as shown above. An example is the presence in *N. punctiforme* strain ATCC 29133 of a cluster of genes encoding proteins annotated to catabolize hexuronic acids in heterotrophic eubacteria, but in which the proteins appear to have changed their physiological role to the synthesis of a developmental regulatory metabolite under transcriptional control of a symbiotic plant partner (Campbell *et al.*, 2002). These specific genes are absent in the current sequenced genomes of non-symbiotic cyanobacteria.

Transfer between related bacteria is probably a higher frequency phenomenon, since the barriers to transfer are likely less restrictive (similar restriction/modification systems, compatible DNA polymerases for plasmid replication), with different outcomes depending on the transferred gene. First, a new function gained by a bacterium can now spread easily throughout sister strains. Such a scenario has been suggested in the case of proteins for the synthesis of chlorophylls *b* and *d*, and chlorophyll binding proteins in strains that belong to different lineages of the cyanobacterial radiation, but that are physiologically related (Green, 2005). Second, when the transferred gene has an ortholog in the recipient strain, they may simply recombine, as in examples of population genetics given above. This can have the effect of homogenizing the genetic pool of a group of organisms. Consistent with this view, Rudi *et al.* (1998) have presented an evolutionary model for the cyanobacteria that relies on both intergenic and intragenic recombinations resulting in DNA sequence homogenization (concerted evolution) for the modern cyanobacterial species caused by exchange of genetic material for neutrally evolving genes. This exchange of genetic material might explain the apparent stability of cyanobacterial morphological characters over billions of years (Rudi *et al.*, 1998).

Concluding remarks

Laboratory work in molecular genetics, which involves DNA transfer into cyanobacterial cells, environmental studies of genetic recombination and phylogenetic analysis of proteins have all converged into a unitary view that gene transfer takes place in nature and has contributed to shape the cyanobacterial genomes that we can now investigate. However, the quantitative aspects of DNA transfer events in the environment and of establishment of newly incorporated genes into a genome are unknown. It is likely that the extent with which these processes happen is highly variable, depending on the origin of the DNA to be transferred, the microbial community, including cyanobacterial population densities, and the specific ecological habitat. Quantitative documentation of lateral gene transfer could render a novel view of cyanobacterial populations and their continuing dynamic evolution in the biosphere.

Acknowledgments

Work in the authors' laboratories is currently supported by grants number BFU2005–07672 and BFU2004–00872 from Ministerio de Educación y Ciencia, Spain, and MCB 0317104 from the USA National Science Foundation.

References

Bancroft, I., Wolk, C.P., and Oren, E.V. (1989). Physical and genetic maps of the genome of the heterocyst-forming cyanobacterium *Anabaena* sp. strain PCC 7120. J. Bacteriol. *171*, 5940–5948.

Barker, G.L., Handley, B.A., Vacharapiyasophon, P., Stevens, J.R., and Hayes, P.K. (2000). Allele-specific PCR shows that genetic exchange occurs among genetically diverse *Nodularia* (cyanobacteria) filaments in the Baltic Sea. Microbiology *146*, 2865–2875.

Billi, D., Friedmann, E.I., Helm, R.F., and Potts, M. (2001). Gene transfer to the desiccation-tolerant cyanobacterium *Chroococcidiopsis*. J. Bacteriol. *183*, 2298–2305.

Brahamsha, B. (1996). A genetic manipulation system for oceanic cyanobacteria of the genus *Synechococcus*. Appl Environ. Microbiol. *62*, 1747–1751.

Cai, Y., and Wolk, C.P. (1990). Use of a conditionally lethal gene in *Anabaena* sp. strain PCC 7120 to select for double recombinants and to entrap insertion sequences. J. Bacteriol. *172*, 3138–3145.

Campbell, E.L., Wong, F.C.Y., and Meeks, J.C. (2002). DNA binding properties of the HrmR protein of *Nostoc punctiforme* responsible for transcriptional regulation of genes involved in the differentiation of hormogonia. Mol. Microbiol. *47*, 573–582.

Chen, I., and Dubnau, D. (2004). DNA uptake during bacterial transformation. Nat. Rev. Microbiol. *2*, 241–249.

Clokie, M.R., Shan, J., Bailey, S., Jia, Y., Krisch, H.M., West, S., and Mann, N.H. (2006). Transcription of a "photosynthetic" T4-type phage during infection of a marine cyanobacterium. Environ. Microbiol. *8*, 827–835.

Cobley, J.G., Zerweck, E., Reyes, R., Mody, A., Seludo-Unson, J.R., Jaeger, H., Weerasuriya, S., and Navankasattusas, S. (1993). Construction of shuttle plasmids which can be efficiently mobilized from *Escherichia coli* into the chromatically adapting cyanobacterium, *Fremyella diplosiphon*. Plasmid *30*, 90–105.

Cohen, M.F., Meeks, J.C., Cai, Y.A., and Wolk, C.P. (1998). Transposon mutagenesis of heterocyst-forming filamentous cyanobacteria. Methods Enzymol. *297*, 3–17.

Elhai, J., Vepritskiy, A., Muro-Pastor, A.M., Flores, E., and Wolk, C.P. (1997). Reduction of conjugal transfer efficiency by three restriction activities of *Anabaena* sp. strain PCC 7120. J. Bacteriol. *179*, 1998–2005.

Fernández-López, R., Garcillán-Barcia, M.P., Revilla, C., Lázaro, M., Vielva, L., and de la Cruz, F. (2006). Dynamics of the IncW genetic backbone imply general trends in conjugative plasmid evolution. FEMS Microbiol. Rev. *30*, 942–966.

Flores, E., and Wolk, C.P. (1985). Identification of facultatively heterotrophic, N_2-fixing cyanobacteria able to receive plasmid vectors from *Escherichia coli* by conjugation. J. Bacteriol. *162*, 1339–1341.

Francia, M.V., Varsaki, A., Garcillán-Barcia, M.P., Latorre, A., Drainas, C., and de la Cruz, F. (2004). A classification scheme for mobilization regions of bacterial plasmids. FEMS Microbiol. Rev. *28*, 79–100.

Golden, S.S., and Sherman, L.A. (1983). A hybrid plasmid is a stable cloning vector for the cyanobacterium *Anacystis nidulans* R2. J. Bacteriol. *155*, 966–972.

Gomis-Ruth, F.X., and Coll, M. (2006). Cut and move: protein machinery for DNA processing in bacterial conjugation. Curr. Opin. Struct Biol. *16*, 744–752.

Green, B.R. (2005). Lateral gene transfer in the cyanobacteria: chlorophylls, proteins and scraps of ribosomal RNA. J. Phycol. *41*, 449–452.

Grigorieva, G.A., and Shestakov, S.V. (1982). Transformation in the cyanobacterium *Synechocystis* PCC 6803. FEMS Microbiol. Lett. *13*, 367–370.

Grossman, A.R. (2003). A molecular understanding of complementary chromatic adaptation. Photosyn. Res. *76*, 207–215.

Heinemann, J.A., and Billington, C. (2004). How do genomes emerge from genes? ASM News *70*, 464–471.

Hu, N.T., Thiel, T., Giddings, T.H., Jr., and Wolk, C.P. (1981). New *Anabaena* and *Nostoc* cyanophages from sewage settling ponds. Virology *114*, 236–246.

Kaneko, T., Nakamura, Y., Wolk, C.P., Kuritz, T., Sasamoto, S., Watanabe, A., Iriguchi, M., Ishikawa, A., Kawashima, K., Kimura, T., Kishida, Y., Kohara, M., Matsumoto, M., Matsuno, A., Muraki, A., Nakazaki, N. Shimpo, S., Sugimoto, M., Takazawa, M. Yamada, M., Yasuda, M., and Tabata, S. (2001). Complete genomic sequence of the filamentous nitrogen-fixing cyanobacterium *Anabaena* sp. strain PCC 7120. DNA Res. *8*, 205–213; 227–253.

Koksharova, O.A., and Wolk, C.P. (2002). Genetic tools for cyanobacteria. Appl Microbiol. Biotechnol. *58*, 123–137.

Lindell, D., Sullivan, M.B., Johnson, Z.I., Tolonen, A.C., Rohwer, F., and Chisholm, S.W. (2004). Transfer of photosynthesis genes to and from *Prochlorococcus* viruses. Proc. Natl. Acad. Sci. USA. *101*, 11013–11018.

Llosa, M., and de la Cruz, F. (2005). Bacterial conjugation: a potential tool for genomic engineering. Res. Microbiol. *156*, 1–6.

Llosa, M., Gomis-Ruth, F.X., Coll, M., and de la Cruz, F. (2002). Bacterial conjugation: a two-step mechanism for DNA transport. Mol. Microbiol. *45*, 1–8.

Lodders, N., Stackebrandt, E., and Nubel, U. (2005). Frequent genetic recombination in natural populations of the marine cyanobacterium *Microcoleus chthonoplastes*. Environ. Microbiol. *7*, 434–442.

Manen, J.F., and Falquet, J. (2002). The *cpcB-cpcA* locus as a tool for the genetic characterization of the genus *Arthrospira* (Cyanobacteria): evidence for horizontal transfer. Int. J. Syst. Evol. Microbiol. *52*, 861–867.

Mann, N.H., Cook, A., Millard, A., Bailey, S., and Clokie, M. (2003). Marine ecosystems: bacterial photosynthesis genes in a virus. Nature *424*, 741.

Marraccini, P., Bulteau, S., Cassier-Chauvat, C., Mermet-Bouvier, P., and Chauvat, F. (1993). A conjugative plasmid vector for promoter analysis in several cyanobacteria of the genera *Synechococcus* and *Synechocystis*. Plant Mol. Biol. *23*, 905–909.

Meeks, J.C., and Elhai, J. (2002). Regulation of cellular differentiation in filamentous cyanobacteria in free-living and plant-associated symbiotic growth states. Microbiol. Mol. Biol. Rev. *66*, 94–121.

Muro-Pastor, A.M., Kuritz, T., Flores, E., Herrero, A., and Wolk, C.P. (1994). Transfer of a genetic marker from a megaplasmid of *Anabaena* sp. strain PCC 7120 to a megaplasmid of a different *Anabaena* strain. J. Bacteriol. *176*, 1093–1098.

Nakasugi, K., Svenson, C.J., and Neilan, B.A. (2006). The competence gene, *comF*, from *Synechocystis* sp. strain PCC 6803 is involved in natural transformation, phototactic motility and piliation. Microbiology *152*, 3623–3631.

Novichkov, P.S., Omelchenko, M.V., Gelfand, M.S., Mironov, A.A., Wolf, Y.I., and Koonin, E.V. (2004). Genome-wide molecular clock and horizontal gene transfer in bacterial evolution. J. Bacteriol. *186*, 6575–6585.

Okamoto, S., and Ohmori, M. (2002). The cyanobacterial PilT protein responsible for cell motility and transformation hydrolyzes ATP. Plant Cell Physiol. *43*, 1127–1136.

Quintero, M.J., Montesinos, M.L., Herrero, A., and Flores, E. (2001). Identification of genes encoding amino acid permeases by inactivation of selected ORFs from the *Synechocystis* genomic sequence. Genome Res. *11*, 2034–2040.

Rippka, R., Deruelles, J., Waterbury, J.B., Herdman, M., and Stanier, R.Y. (1979). Generic assignments, strain stories and properties of pure cultures of cyanobacteria. J. Gen. Microbiol. *111*, 1–61.

Rudi, K., Skulberg, O.M., and Jakobsen, K.S. (1998). Evolution of cyanobacteria by exchange of genetic material among phyletically related strains. J. Bacteriol. *180*, 3453–3461.

Shestakov, S.V., and Khyen, N.T. (1970). Evidence for genetic transformation in blue-green alga *Anacystis nidulans*. Mol. Gen. Genet. *107*, 372–375.

Stevens, S.E., Jr., and Porter, R.D. (1980). Transformation in *Agmenellum quadruplicatum*. Proc. Natl. Acad. Sci. USA *77*, 6052–6056.

Stevens, S.E., Jr., and Porter, R.D. (1986). Heterospecific transformation among cyanobacteria. J. Bacteriol. *167*, 1074–1076.

Summers, M.L., Wallis, J.G., Campbell, E.L., and Meeks, J.C. (1995). Genetic evidence of a major role for glucose-6-phosphate dehydrogenase in nitrogen fixation and dark growth of the cyanobacterium *Nostoc* sp. strain ATCC 29133. J. Bacteriol. *177*, 6184–6194.

Tanabe, Y., Kaya, K., and Watanabe, M.M. (2004). Evidence for recombination in the microcystin synthetase (*mcy*) genes of toxic cyanobacteria *Microcystis* spp. J. Mol. Evol. *58*, 633–641.

Thiel, T. (1994). Genetic analysis of cyanobacteria. In The Molecular Biology of Cyanobacteria, D.A. Bryant, ed. (Dordrecht, The Netherlands: Kluwer Academic Publishers), pp. 581–611.

Thiel, T., and Poo, H. (1989). Transformation of a filamentous cyanobacterium by electroporation. J. Bacteriol. *171*, 5743–5746.

Tsinoremas, N.F., Kutach, A.K., Strayer, C.A., and Golden, S.S. (1994). Efficient gene transfer in *Synechococcus* sp. strains PCC 7942 and PCC 6301 by interspecies conjugation and chromosomal recombination. J. Bacteriol. *176*, 6764–6768.

van den Hondel, C.A., Verbeek, S., van der Ende, A., Weisbeek, P.J., Borrias, W.E., and van Arkel, G.A. (1980). Introduction of transposon Tn*901* into a plasmid of *Anacystis nidulans*: preparation for cloning in cyanobacteria. Proc. Natl. Acad. Sci. USA. *77*, 1570–1574.

Waterbury, J.B., and Valois, F.W. (1993). Resistance to co-occurring phages enables marine *Synechococcus* communities to coexist with cyanophages abundant in seawater. Appl. Environ. Microbiol. *59*, 3393–3399.

Williams, J.G., and Szalay, A.A. (1983). Stable integration of foreign DNA into the chromosome of the cyanobacterium *Synechococcus* R2. Gene *24*, 37–51.

Woese, C.R. (1987). Bacterial evolution. Microbiol. Rev. *51*, 221–271.

Wolk, C.P., Vonshak, A., Kehoe, P., and Elhai, J. (1984). Construction of shuttle vectors capable of conjugative transfer from *Escherichia coli* to nitrogen-fixing filamentous cyanobacteria. Proc. Natl. Acad. Sci. USA. *81*, 1561–1565.

Yoshihara, S., Geng, X., Okamoto, S., Yura, K., Murata, T., Go, M., Ohmori, M., and Ikeuchi, M. (2001). Mutational analysis of genes involved in pilus structure, motility and transformation competency in the unicellular motile cyanobacterium *Synechocystis* sp. PCC 6803. Plant Cell Physiol. *42*, 63–73.

Yura, K., Toh, H., and Go, M. (1999). Putative mechanism of natural transformation as deduced from genome data. DNA Res. *6*, 75–82.

Zeidner, G., Bielawski, J.P., Shmoish, M., Scanlan, D.J., Sabehi, G., and Beja, O. (2005). Potential photosynthesis gene recombination between *Prochlorococcus* and *Synechococcus* via viral intermediates. Environ. Microbiol. *7*, 1505–1513.

Zhaxybayeva, O., Gogarten, J.P., Charlebois, R.L., Doolittle, W.F., and Papke, R.T. (2006). Phylogenetic analyses of cyanobacterial genomes: quantification of horizontal gene transfer events. Genome Res. *16*, 1099–1108.

Molecular Ecology and Environmental Genomics of Cyanobacteria

4

Ferran Garcia-Pichel

Abstract

The application of molecular biology and genomics to microbial ecology has been a transformative force, making possible the discovery of layer upon layer of complexity in natural communities of microbes. Diversity surveying, community fingerprinting, and functional interrogation of natural populations have become common, enabled by a battery of molecular and bioinformatics techniques. Some were specifically developed for the cyanobacteria, which are reviewed in detail here. The ensuing effects on our views of cyanobacterial ecology have been perhaps less revolutionary, because of the special characteristics of cyanobacteria among microbes, but also significant. We have come to realize that the present taxonomic system is often phylogenetically incorrect; several new cyanobacteria or groups thereof have been discovered, and some established groups have been found to be constructs. Surveying efforts have covered many habitats and have demonstrated that cyanobacterial communities tend to be habitat-specific, and that plenty of undescribed genetic diversity is concealed among morphologically simple types. We can now select among isolates those that are good representatives of natural populations, enabling, among other objectives, ecologically motivated genome sequencing efforts. We have witnessed the first studies addressing population genetics and the blooming of functional studies based on detection of gene expression in Nature. We are entering an era of expansion of the polyphasic approaches that combine molecular, bioinformatics, physiological, and geochemical techniques to study natural communities

Molecular microbial ecology and environmental genomics

Scope and nature of the enterprise

Molecular ecology has been defined as the use of molecular genetic markers to approach ecological and evolutionary questions, as applied to individuals, populations or communities (Carvalho, 1988). Perhaps more simply and broadly, it can be regarded as the application to environmental research of techniques based in molecular biology. Nowhere has this approach been as influential as in microbial ecology, spearheading within the last two decades a true revolution in the study of microorganisms in their natural environment, opening new insights into the complexity and diversity of microbes themselves and into the ecological systems they sustain, as the tools expanded from the use of markers into the study of genomes. Molecular and genomic approaches to microbial ecology can be driven by one of two distinct aims: discovery and inquiry. In discovery-driven applications no particular hypothesis is tested. Rather, a blind search for biological novelty is carried out using surveys of particular genetic markers, or even surveys of unspecified genes. In purpose, this is not unlike the famous naturalist's expeditions of Darwin or von Humboldt. This type of surveys or compilations, which may not appear scientifically as challenging or interesting as hypothesis-driven investigations, have already returned evidence for an overwhelmingly large and apparently unending level of extant microbial diversity. Lucky draws from those molecular expeditions have already resulted in discoveries of importance for ecosystem function, even at the global scale. As these genetic databases grow, so does their potential as tools for hypothesis-driven research. Our ability to recognize patterns in the databases and to use them in the building and testing of hypotheses grows along with database size. All of this will eventually contribute directly to the ultimate goal of establishing generally applicable principles and their boundaries for the entire discipline.

Development

Pioneering work in the early 80s involved laboriously extracting and collecting nucleic acids from field samples to attain quantities that were only large enough for very limited analyses (Stahl *et al.*, 1984; Weller and Ward, 1989). The eventual application of PCR amplification helped circumvent many of the problems associated with the paucity of nucleic acids. The ensuing development of a battery of molecular techniques designed for detection, fingerprinting and functional interrogation of natural populations of microbes allowed molecular microbial ecology to become not only prominent and widespread, but arguably the most significant and influential part of microbial ecology today, an indispensable asset in any comprehensive study of microbes. Early on, techniques to identify microbes *in situ* based on their ribosomal RNA signatures appeared most promising (Olsen *et al.*, 1986; Pace *et al.*, 1986). This was because ribosomal RNAs are universal molecules, present in all organisms, that offered a generic tool; they contained well-conserved regions as well as variable regions, providing recognizable sequence signatures. In addition, rRNA were found in large copy numbers compared to other nucleic acids, which made isolation and sequencing relatively easy. Ribosomal RNA has remained the most widely used molecular genetic marker for microbial studies, even though the list has grown to include other, often functionally meaningful, genes as well. An up-to-date overview of such techniques is beyond the scope of this chapter and can be found in the monograph edited by Osborn and Smith (2005); a large compilation of protocols can be found in the encyclopedic molecular microbial ecology "manual" (Kowalchuk *et al.*, 2004).

Bio-informatic studies of whole genomes, genomic patterns, and collections of genomes, which developed in good measure thanks to improvements in high-throughput DNA sequencing technology, have also found its way into environmental science. The availability of a growing number of fully sequenced microbial genomes, including many cyanobacteria, creates an expanding genomic database with which to interpret sequences obtained directly from the field. While there is a considerable, almost humbling, amount of unknown coding sequences among the cyanobacteria (Meeks *et al.*, 2001), enough is known already to start using genomic information as a source of ecologically relevant hypotheses that can be directly tested (Dyhrman and Haley, 2006). In fact, it is ecologically relevant questions that is driving some of the most recent multistrain cyanobacterial genome sequencing projects with *Prochlorococcus* and thermophilic *Syechococcus*. More directly applied to environmental research is the still nascent field of metagenomics, an attempt to distill functional and structural information from the analyses of genomic information retrieved from heterogenous and often extremely complex collections or ensembles of genomes as they are found in the natural environment. Because these are computationally and technologically demanding enterprises, they are not yet common practice, with only several studies available (Riesenfeld *et al.*, 2004), some of which have yielded data of relevance to cyanobacterial ecology (Venter *et al.*, 2004). In view of what has been achieved so far [see Gillespie *et al.* (2005), for a review of results], it may seem that this approach provides only very modest and unpredictable results, in spite of the large sequencing effort; but the importance of increasing the size of the databases cannot be discounted, and will probably only facilitate future work in the same way that extensive 16S rRNA databases facilitates routine work today.

The need for cultivated anchors

Prior to the introduction of these molecular methods, much of what was known about the diversity and distribution of microbes was derived from studies involving either laboratory-grown cultures or cultivation-based techniques such as plate counts or most-probable-number determinations. But most strains are brought into culture after a strong selection process during enrichment and isolation, and often do not represent the natural population of origin (Garcia-Pichel *et al.*, 1996; Hugenholtz *et al.*, 1998). Molecular methods were perceived by some as a way to reduce, and even substitute the need for the demonstrably biased approach of isolation of pure cultures in the process of determining microbial diversity, distribution, and function in the natural environment (Pace, 1997). But soon enough it became clear that all ecological inference made on the basis of molecular sequence data is based on correlation with identities or functional properties established on pure cultures themselves. The lack of a sufficient number of such cultures that could be used as anchoring points for interpretation of environmental molecular data was underscored by the ever increasing number of branches in the microbial tree of life described solely on the basis of environmental DNA sequences, for which, having no cultivated isolates, we had no idea about their possible role in nature. Perhaps as many as 26 such phyla exist (Rappe and Giovannoni, 2003). This has given way to a true renaissance in the emphasis of cultivation in microbial ecology. Major components of microbial communities in the oceans, that had long been known from molecular surveys, such as the SAR 11 Cluster (Rappe *et al.*, 2002), or some groups of Archaeal marine mesophiles, have only recently been isolated (Koenneke *et al.*, 2005). Cultivation led to unexpected discoveries regarding their metabolism in both cases: those crenarchaeota, now *Nitrosopumilus maritimus*, were found to be chemolithoautotrophic ammonia oxidizers, and the SAR11, now *Pelagibacter ubique*, were found to contain bacteriorhodopsin-like proteins, indicating a potential for photoheterotrophy (Giovannoni *et al.*, 2005). But in turn, molecular techniques can now be used to accompany enrichment and cultivation in order to ensure that the final products of cultivation belong to the field populations they are supposed to represent. Such combination of traditional cultivation and molecular techniques, also known as polyphasic approach, is now the preferred approach in the description of microbial communities.

The ecogenomics toolbox and its application to cyanobacteria

Extraction of nucleic acids from natural cyanobacterial communities

The importance of extraction and purification of nucleic acids from the natural environment varies from the trivial to the crucial, depending on the particular habitat. General protocols are given elsewhere (Osborn and Smith, 2005; Kowalchuk *et al.*, 2004). Planktonic populations in oligotrophic oceans or freshwaters are the simplest cases among the habitats relevant for cyanobacteria. Such samples, after biomass concentration, can be easily extracted with protocols that do not differ significantly from those used with standard laboratory cultures. Natural samples of benthic cyanobacteria (and some bloom-forming plankters) typically contain large amounts of exopolysaccharide which become particularly problematic in microbial mat samples, and need to be precipitated by chemical additives in the extraction buffer (Nübel *et al.*, 1999). The presence of tough extracellular sheaths, particularly in some desiccation-resistant cyanobacteria, may require additional physical treatment for disruption (Garcia-Pichel *et al.*, 2001), or water soaking of the samples prior to extraction (Gaylarde *et al.*, 2005). For soil cyanobacteria, organo-mineral clumping with clays and adsorption of nucleic acids to clay surfaces may represent a problem, and the use of chemical dispersing agents such as pyrophosphate may be called for. In the case of endolithic cyanobacteria, perhaps the toughest samples of all, forceful grinding of the rock may be necessary (Sigler *et al.*, 2003), even though specific mineral-dissolution protocols have been developed for cyanobacterial communities encased in carbonate, phosphate and sulfate minerals (Wade and Garcia-Pichel, 2003). The general idea is always to optimize the protocols and working procedure to allow maximal yield, minimal shearing of the DNA and degradation of the nucleic acid as well as shortest possible time to avoid preferential growth or death of particular community members. Increasing the strength of the physical means of sample disruption from freeze-thaw cycles, to potter grinding, to freeze-grinding or to bead-beating will increase the DNA yield, but also increase the amount of shearing. In extracting nucleic acids from an unfamiliar type of sample, it is always recommended to start with a protocol developed for a sample as similar as possible to the one at hand, and to check for yield and quality of the extracts.

Polymerase chain reaction amplification and allied techniques

In the "polymerase chain reaction," through the use of priming with oligonucleotides of choice, a thermostable DNA polymerase (*Taq* polymerase), and by means of a recurrent switch between temperature regulated priming, extension and denaturation, more than a million copies of particularly targeted DNA sequences can be synthesized *in vitro* in a matter of hours, creating ample amounts of nucleic acid material for further genetic testing (Saiki *et al.*, 1988). Because of the small amounts of template DNA required for amplification, PCR is particularly well suited for application to natural samples of low biomass, making it very valuable to overcome analytical limitation by quantity of nucleic acids. Not only where environmental DNA is limiting, but in all applications, PCR easily allows for a quick transition from extraction to genetic analysis of specific DNA fragments, rapid screening and comparisons of large data sets, replication and statistical treatment, as well as trend analyses over spatial and temporal scales. Not surprisingly then, PCR has become a part of most techniques used in microbial ecology, even though PCR is not without problems. It is sensitive to inhibition by natural substances (metals, acids) that may be present in natural samples, and it may not amplify all possible targets with equal intensity or specificity. The power of PCR is such that routine genetic analyses are rendered possible from microscopic samples consisting of a few cells or filaments of cyanobacteria (Garcia-Pichel *et al.*, 1996; Janson *et al.*, 1999; Lodders *et al.*, 2005).

A variety of PCR modifications have found use in environmental studies, including quantitative-PCR (Q-PCR), reverse-transcriptase PCR (RT-PCR), and nested PCR. In nested PCR, two sequential rounds of PCR are carried out with a different primer set in each; the second pair must prime at a stretch that is within the PCR product of the first reaction. Nested PCR has been used to increase specificity in the overall amplification, and also to improve the minimal concentration of template needed for successful amplification (Zwart *et al.*, 2005). Reverse-transcriptase PCR (RT-PCR) is used to amplify RNA by converting it to cDNA using a viral reverse transcriptase before PCR amplification. Reverse transcriptase PCR is typically used to amplify mRNAs, and thus study gene expression in nature (Steunou *et al.*, 2006; Xu and Tabita, 1996; Zani *et al.*, 2000). If applied to ribosomal RNA analyses, it has been found useful to differentiate between the presence and the activity of a certain microbe. The idea here is that a microbe's presence is gauged by the abundance of its rRNA genes, whereas its activity, or growth rate, will be reflected in the cellular concentration of ribosomes and thus in the concentration of rRNA, which can be amplified from an extract by means of RT-PCR after hydrolyzing the original DNA templates with a deoxyribonuclease treatment. In cyanobacterial research, this approach has served to detect relict populations from actively growing members after environmental changes affecting microbial communities, even before such changes have had time to result in overall shifts in community composition, both in hypersaline (Abed and Garcia-Pichel, 2001) and in hot-spring (Norris *et al.*, 2002) microbial mats. Additionally, because for any given cell, the amount of template obtained as cDNA from rRNA is much larger that that obtained from genomic 16S RNA directly, RT-PCR can be performed to enhance success of PCR amplifications from minute samples, as it has been done to determine the identity of cyanobacteria symbiotic on marine tintinnids and other non-photosynthetic eukaryotes (Foster *et al.*, 2006). Finally, another variation of PCR, Q-PCR has been developed in which the progress of the amplification reaction can be tracked directly (and specifically) by addition of fluorescent probes that report the concentration of double stranded DNA present in the PCR reaction mixture to an online fluorometer. By extrapolation of the exponential phase of amplification, this allows one to quantitatively assess the initial amounts of template copies in a given sample, and adds a quantitative aspect to the technique. Q-PCR is now widely used to determine copy numbers of specific genes. In cyanobacterial ecology it has been used to quantify structural and functional properties of microbial communities: to monitor the waxing and waning of specific cyanobacterial genotypes in populations of freshwater *Synechococcus* in Lake Constance (Becker *et al.*, 2002), and also to monitor gene expression in hot spring cyanobacteria (Steunou *et al.*, 2006).

Genetic markers

A wide variety of genetic markers have been used in studies of cyanobacterial ecology (Table 4.1) that vary from general to narrow in specificity, and from systematic to functional in scope. We will review them in detail below:

Cyanobacteria-specific genetic markers

When choosing genetic markers, specificity for cyanobacteria can be achieved in several ways. The simplest one takes advantage of the large size of some of the supra-cellular colonial structures formed by cyanobacteria, which sometimes are large enough to be physically separated from other microbes in virtually pure stands, either by simple manipulation or by micromanipulation under the microscope. In these cases, and as long as care is taken in surface sterilization of the samples, any appropriate genetic marker can be cyanobacteria-specific. Needless to say that this approach cannot be used for all cyanobacteria, but it is perhaps also true that it has not been used to its fullest possible extent, particularly when implementing microscopic micromanipulation. Appropriately unicyanobacterial, and sometimes axenic samples of cyanobacteria can be easily retrieved in the case of symbionts from lichens, for example; some free-living colonies are axenic in the interior, as is the case in some macroscopic colonies of terrestrial *Nostoc* and aquatic *Aphanothece* (Garcia-Pichel *et al.*, 2002). Micromanipulation has been successfully used for bundle-forming filaments after surface cleaning by dragging on agar (Boyer *et al.*, 2002; Garcia-Pichel *et al.*, 2001; Garcia-Pichel *et al.*, 1998; Garcia-Pichel *et al.*, 1996; Lodders *et al.*, 2005) and for cyanobacterial symbionts inhabiting diatom cells (Janson *et al.*, 1999).

A second approach is to use genes that are exclusively (or virtually so) found in cyanobacteria, and that are also universal (or virtually so) among them. Genes coding for the phycocyanin operon, virtually universal among cyanobacteria, are one example (Neilan *et al.*, 1995; Tillett *et al.*, 2001). Another similar case, if certainly less specific, is the use of the conserved photosynthetic gene *psbA*, coding for the conserved photosystem II D1 protein, which obviously targets all oxygenic phototrophs, namely cyanobacteria and eukaryotic plastids (Zeidner and Béja, 2004; Zeidner *et al.*, 2003).

The third, and most common, consists of targeting a universal gene but achieving specificity through the design of cyanobacteria-specific primers for PCR amplification. This requires a good database on the sequence divergence of the marker among the group and has been used with 16S rRNA and RNA polymerase genes, from which the first is without doubt the most widely used. The sets of primers of Nübel *et al.* (1997) have been designed to target the V3–V4 region of the 16S rRNA gene offering the possibility to use them for a variety of techniques with minor modifications. These and similar primers are not exclusively cyanobacterial, but rather encompass the cyanobacterial line of descent, for they will amplify the alleles from eukaryotic plastids as well. The gene coding for the gamma subunit of RNA polymerase (*rpoC1*), for which it is thought that single copies exist in prokaryotic genomes, has also been used as a cyanobacterial marker using cyanobacteria-specific PCR primers (Palenik, 1994). They have found most use in the study of marine planktonic cyanobacteria (Ma *et al.*, 2004; Palenik, 1994; Toledo and Palenik, 1997), where they have been instrumental in the establishment of genetic niche diversification among marine picoplanktonic populations (Ferris and Palenik, 1998). However, *rpoC1* is probably too conserved a marker to detect geographic patterns in other groups, such as the freshwater *Cylindrospermopsis* (Gugger *et al.*, 2005).

Other relevant genetic markers

Some genetic markers are not specific for cyanobacteria, but become of relevance because of the relevance of cyanobacteria themselves among the target group, or because of their prominence in the target environment. This is the case of nitrogenase genes (*nifH*, particularly), when targeting environments where nitrogen-fixing cyanobacteria are abundant (Omoregie *et al.*, 2004; Yeager *et al.*, 2004), or simply the case of 16S rRNA studies that are not specifically targeted to cyanobacteria, when used in environments where cyanobacteria are numerically important (Nagy *et al.*, 2005; Rothrock and Garcia-Pichel, 2005).

Markers for subgroups and microarrays

Given the large size of the database, subgroup-specific primers have been successfully designed for several cyanobacterial clades or guilds, and even combined into arrays for simultaneous detection (Rudi *et al.*, 2000). Most, but not all (Wilson *et al.*, 2000), are based on ribosomal genes. A recent such effort was designed to simultaneously detect up to 16 groups (Castiglioni *et al.*, 2004). The groups included here were a bit heterogenous, from Antarctic to freshwater temperate forms, as dictated by the availability of consensus sequences in the alignment. Using isolated strains as a test, the arrays reportedly showed very good sensitivity and discriminatory power. In spite of such promising results, routine application in environmental monitoring has not yet been achieved, and the best possibilities for use of microarrays are when used and designed as a habitat-specific tool. On a different level are the guild-specific markers against toxic cyanobacteria that cause public health problems in natural and man-made bodies of water. Major advances in the identification of the non-ribosomal peptide synthases as the biosynthetic pathways of toxins such as microcystin, and the sequencing of the appropriate genomic regions, enabled the development of such markers (Baker *et al.*, 2002; Schatz *et al.*, 2000; Tillett *et al.*, 2001). Phycoerythrin genes can also be used as group-specific markers since not all cyanobacteria contain the red phycobiliprotein, particularly when judiciously used in a certain setting, or with the aid of specific amplification primers, as used for example in marine planktonic communities by Steglich *et al.* (2003). A protocol using a nested-PCR amplication of *rpoC1* specific to *Cylindrospermopsis raciborskii* has been successfully employed to typify a variety of strains of this notoriously problematic toxic freshwater cyanobacterium (Wilson *et al.*, 2000). Specific primers to distinguish *nifH* genes belonging to heterocystous and non-heterocystous forms in the nitrogen-fixing guild of cyanobacteria have also been used (Ben-Porath and Zehr, 1994). Finally, the gene *hetR*, originally described as being involved in heterocyst and akinete differentiation, is present and conserved in all heterocystous cyanobacteria checked. As such it has been successfully used as a marker for the nitrogen-fixing symbionts in studies of host-specificity in the *Richelia*-diatoms symbioses (Janson *et al.*, 1999). The target group of this marker, however is, against initial expectations, not simply the heterocystous cyanobacteria (i.e Nostocales and Stigonematales), since several filamentous non-heterocystous cyanobacteria such as *Trichodesmium* and *Symploca* (i.e. Oscillatoriales) contain genes homologous to the original *hetR* (Lundgren *et al.*, 2005), and yet they are absent from the unicellular cyanobacteria that have been fully sequenced.

Clone libraries: descriptors of community diversity

Cloning environmental DNA into microbial hosts can provide a means to obtain large amounts of single sequences from complex mixtures. It is the expectation that if the clone library is large enough, the distribution of clones in it will be a reflection of the complexity present in the original samples, and can be used as a descriptor of it. Most clone libraries are made using genomic fragments in the thousands of base pairs of length. But libraries of larger clones, containing many genes, known as BAC (for *bacterial artificial chromosome*) libraries are used in metagenomics. Libraries in the size range of hundreds of clones are usually carried out to achieve a moderate level of coverage in most natural communities. The cloned alleles can then be readily sequenced, and their functional relatedness determined based on sequence similarity to known genes, or their phylogenetic relationships reconstructed. In environmental research, clone libraries are usually focused on a single marker, and specificity is achieved with a prior step of PCR amplification. In the case of rRNA libraries, PCR can be avoided if sufficient biomass is present by using a reverse transcriptase step on RNA extracts and cloning cDNAs. The analyses of large clone libraries has led to an increased understanding of the extent of microbial diversity in a variety of different environments (see Pace, 1997, for a review). Many studies using clone library analyses of 16S rRNA genes have contributed to the detection of cyanobacteria, among other microbes, in a wide variety of environments from hot-springs (Fouke *et al.*, 2003; Weller and Ward, 1989) to ocean waters (Ma *et al.*, 2004; Palenik, 1994) and stromatolites (Burns *et al.*, 2004; Papineau *et al.*, 2005), from Antarctic benthic mats (Jungblut *et al.*, 2005; Taton *et al.*, 2003) to sulfur springs (Elshahed *et al.*, 2003) and sponge symbionts (Steindler *et al.*, 2005). The same applies for studies using clone libraries of nitrogen fixation genes from the environment (*nifH*), where in many occasions the contribution of cyanobacteria is significant (Omoregie *et al.*, 2004; Steppe and Paerl, 2005; Yeager *et al.*, 2004), even though they may not have been targeted as a group in particular.

Community fingerprinting

Cloning of 16S rRNA genes, due to its time-intensive nature, is an impractical tool to track microbial population shifts over spatial and temporal scales where a multitude of samples must be processed. In order to rapidly assess succession or spatiotemporal changes in microbial populations, techniques of genetic fingerprinting are used. Here, the various alleles present in a given genetic marker that is universal for the target organismal group are physically separated to reveal explicitly the magnitude of allelic polymorphism, in what constitutes an indirect profile of microbial diversity. Ribosomal genes are also here the most commonly used, but communities can be, and have been, profiled on the basis of other genetic markers such as *nifH*, *rpoB*, *amoA*, and *psbA* (Zeidner and Béja, 2004). Some of these techniques are discussed below.

Lesser techniques

PCR-independent LMW-RNA (low molecular weight-RNA) profiling is the oldest fingerprinting technique, using 5S rRNA and tRNA as genetic markers, analyzed directly after extraction by a separation in a high-resolution polyacrylamide gel electrophoresis; the banding pattern becomes the fingerprint of the community (Bidle and Fletcher, 1995; Hofle and Brettar, 1996). In Orita's single-strand conformation polymorphism (SSCP) fingerprints, one separates a complex allelic mixture of a particular gene by changes in mobility of their single-stranded DNA, which stems from sequence variability, which in turn affects conformation (Orita *et al.*, 1989). Paulsrud and Lindblad have successfully used SCPP of the highly variable $tRNA^{Leu}$ intron region to detect genetic divergence in natural populations of *Nostoc* symbionts of lichens (Paulsrud and Lindblad, 1998). A third technique, RFLP (restriction fragment length polymorphism) profiling, requires PCR amplification of a universal gene, followed by restriction enzyme digestion of the amplified DNA. The digested products are run on an agarose or polyacrylamide gel and the resultant banding patterns constitute a fingerprint, which can be documented and analyzed, with the source of variability being sequence divergence involving potential restriction sites. It is also known as ARDRA (amplified ribosomal DNA restriction analysis) when applied to ribosomal genes in particular. RFLP/ARDRA is widely used in monitoring microbial community changes in a variety of environments. It has been used to probe the genetic diversity of toxic cyanobacterial strains (Neilan *et al.*, 1995) focusing on the phycocyanin locus as marker. ARDRA has been found particularly useful in typifying rather simple mixtures of cyanobacteria (or cyanobacterial sequences), such as groups of cultivated strains (Bolch *et al.*, 1996; Lyra *et al.*, 1997; Margheri *et al.*, 1999; Margheri *et al.*, 2003; Postius *et al.*, 1996). And both ARDRA and RFLP have been used to separate groups of similar sequences in large clone libraries before sequencing, in an attempt to optimize the sequencing effort (Gaylarde *et al.*, 2005; Ma *et al.*, 2004; Postius *et al.*, 1996; Taton *et al.*, 2003).

Terminal restriction fragment length polymorphisms

T-RFLP (Liu *et al.*, 1997) is a variant of RFLP fingerprinting where PCR products are fluorescently labeled during the amplification process prior to enzymatic digestion. These terminally labeled products are subsequently fragmented using restriction enzymes and analyzed using an automated DNA sequencer. The sequencer only detects those DNA fragments that are fluorescently labeled (i.e. terminal) so that a single fragment is detected for each original PCR product. Advantages of T-RFLP profiling include its high resolution through the use of the DNA sequencer, its reproducibility, and its ability to yield quantitative results. T-RFLP analysis does not allow for further characterization of microbial community members via hybridization or direct sequencing, but tentative assignments of identity can be made on the basis of known lengths of the terminal fragments, as determined from isolates or groups of isolates, either newly obtained or from databases. This is a major limitation in comparison with other fingerprinting techniques such as DGGE (see below), even though recent developments may significantly improve identification capacity (Widmer *et al.*, 2006). T-RFLP has been used successfully in some cases to monitor bacterial and cyanobacterial communities in the plankton of reservoirs (Kim *et al.*, 2004), but was neither able to resolve variations with depth in community composition, nor to detect specific phylotypes among the dominant cyanobacteria in intertidal microbial mats using universal primers (Wieland *et al.*, 2003). The combination of group-specific PCR primers with T-RFLP fingerprinting has been used successfully to monitor cyanobacterial infestation of coral reeds directly (Frías-López *et al.*, 2003). AFLP (amplified restriction fragment length polymorphisms) is a related technique, in which a subset of restriction fragments from whole genomic DNA is analyzed. The subset is in this case defined by differential amplification. The advantages are reportedly a higher sensitivity and a much wider genetic basis for analyses than techniques based on single genes. AFLP has found use in comparison of cultivated strains of toxin-producing *Microcystis aeruginosa* (Oberholster *et al.*, 2005).

Denaturing gradient gel electrophoresis (DGGE)

DGGE (Muyzer *et al.*, 1993) is, together with T-RFLP, one of the most widely used genetic fingerprinting techniques. A universal genetic marker for the target group of organisms (typically, but not exclusively, 16S rRNA genes) is PCR amplified to obtain an allelic mixture of fragments of equal length but heterogeneous sequence. This is separated electrophoretically using a polyacrylamide gel containing a linear denaturant gradient, typically composed of urea and formamide. Separation occurs according to melting domains, which is based on the nucleotide composition and secondary structure of the respective gene alleles, causing different ribotypes to stop migration through the gel matrix at different positions along the denaturant gradient. It allows for further analyses of the fingerprints by probe-hybridization (Ovreås *et al.*, 1997; Abed *et al.*, 2002c), or sequencing of specific bands, which is commonly done. This technique can also be performed using a gradient of temperature, rather than of denaturants, to result in similar fingerprinting patterns (named TGGE).

DGGE/TGGE profiling has been applied to many various environments to answer numerous ecologically important questions. In addition to the high-resolution fingerprints (able to distinguish single base pair differences in some cases), individual bands can be excised and subsequently sequenced to determine phylogenetic relationships of the community members, or subjected to taxon-specific probe hybridization for further characterization. Qualitative analyses of DGGE/TGGE fingerprints are possible to calculate ecological parameters such as diversity and richness (Nübel *et al.*, 1999), and community structures can be tracked using statistical analyses such as canonical variate analysis (CVA) (McCaig *et al.*, 2001), non-metric multidimensional scaling (NMDS; van Hannen *et al.*, 1999), and combination of NMDS and similarity analysis (Casamayor *et al.*, 2002; Rothrock and Garcia-Pichel, 2005). DGGE/TGGE is not without its limitations, including a maximally long DNA fragment size of ~700 bp, and an inability to separate certain fragments with similar melting domains but different sequences, in addition to any biases introduced during PCR. Some of the limitations of DGGE are minimized when focusing the analyses on "parts of the community" by using group-specific PCR amplifications; cyanobacteria-specific 16S rRNA-based DGGE being one of the most common and successful (Figure 4.1). Since its first application to the study of microbial mats from hypersaline environments (Nübel *et al.*, 1997), it has been successfully used in a large variety of environments from alpine rocks (Sigler *et al.*, 2003) to paddy fields (Song *et al.*, 2005), pristine estuaries (Geiß *et al.*, 2004) and open sewers (Abed *et al.*, 2002b), and to cyanobacterial soil crusts (Garcia-Pichel *et al.*, 2001), freshwater stromatolites (Elser *et al.*, 2005), intertidal microbial mats (Rothrock and Garcia Pichel, 2005), and even marine mollusk shell endoliths (Chacón *et al.*, 2006), that span a large variety of environments and environmental regions. The fact that the original cyanobacteria-specific primers have retained their specificity and universality in the face of exponentially expanding molecular databases has probably contributed to this widespread use. High-resolution DGGE analyses of cyanobacteria has been achieved by developing fingerprinting aimed at the highly variable ribosomal interspacer locus (Jansen *et al.*, 2003). This has proven useful in discriminating cultivated strains and relatively simple field communities, but is probably too sensitive to be used as a general surveying tool. Other types of DGGE analyses of relevance to cyanobacterial ecology include those developed for oxygenic phototrophs

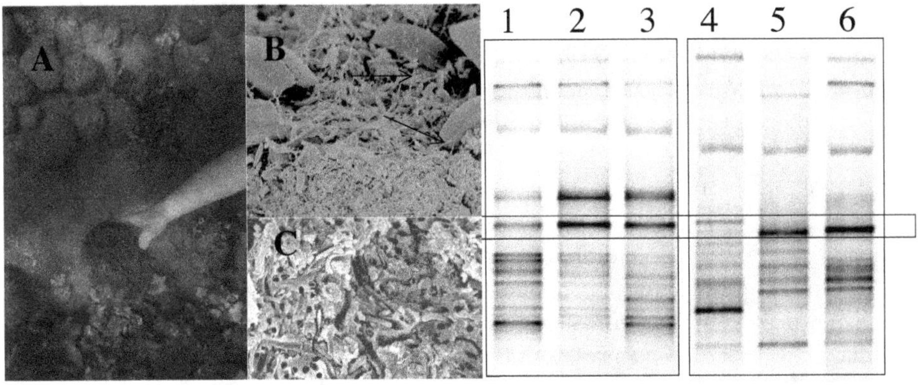

Figure 4.1 (see also Plate 4.1) Fingerprinting of cyanobacterial communities. DGGE fingerprints specific for oxygenic phototrophs done on calcified communities of cyanobacteria found as rolling oncolites (stromatolites) on the bottom of Rio Mezquites, Cuatro Ciénegas, Mexico. Each lane in the gel corresponds to one oncolite, such as that being picked up from the bottom of the river in A. Each band in this community "bar-code" represents one allele of the 16S rRNA found in the complex mixture that has been PCR-amplified from DNA extracts. The bands are separated on the gel on the basis of sequence differences. Strong bands correspond to abundant alleles. The fingerprints allow the immediate realization that oncolites collected at the same site (lanes 1, 2, 3), just meters apart, possess a common pattern of community structure. But shifts in the presence and intensity of particular alleles are also conspicuous as oncolites kilometers downstream are analyzed (lanes 4, 5 and 6). A microscopic investigation of these communities will reveal the presence of surface diatoms (B) and two major morphotypes of cyanobacteria encrusted in the carbonate matrix: one *Calothrix*-like heterocystous filament (B), only the terminal hairs are visible here (arrows) sticking out of the surface, and a thin filamentous, non-heterocystous *Phormidium/Leptolyngbya* (C). Excision, re-amplification and sequencing of the DGGE bands identify those in the upper half of the gel as diatom plastids, the intense band marked by the rectangle as a heterocystous cyanobacterium, and most of the rest as diverse sequences matching best a variety of thin-filamentous cyanobacteria. Data and photographs are from Garcia-Pichel *et al.* (2004) and Elser *et al.* (2005); the DGGE fingerprint is from Wade, B., and Garcia-Pichel, F., unpublished.

through the use of *psbA* amplificates (Zeidner *et al.*, 2003), or those developed for *nifH* (Piceno *et al.*, 1999).

FISH (fluorescent *in situ* hybridization)

FISH is a standard technique in molecular ecology, in which natural, fixed samples are hybridized with fluorescent oligonucletide probes that target ribosomal sequences. These are then examined under fluorescent microscopy (Amann *et al.*, 1995), allowing not only the direct identification and counting of particular groups of bacteria, but also the assessment and identification of their morphology, dynamics, spatial situation within microgradients, or with respect to other types of microorganisms. FISH has not yet found much application in the study of cyanobacteria, in spite of its widespread use with almost every other microbial group. This is probably because of the perceptions that cyanobacteria are sufficiently tagged as a group thanks to the strong autofluorescence of their photosynthetic pigments, and also that any FISH signals may be swamped by this very autofluorescence. The problem with autofluorescence is real, but can be overcome by the use of fluorescence enhancing techniques such as horseradish peroxidase/tyramide labeling of the oligonucleatide probes (Schonhuber *et al.*, 1999), or by using finer separation methodologies such as flow cytometry (Worden *et al.*, 2000). Thus enhanced, FISH has been successfully used to detect an otherwise cryptic cyanobacterium rRNA (Figure 4.2).

Functional assessment

Functional genes: presence and expression

Finding out about the identity, distribution and dynamics of microbes in their natural habitat can only make complete sense if we relate this information to their activity, interactions and biogeochemical consequences. This is what we call the functional aspects of microbial ecology, which can in some cases be addressed with a combination of molecular methods and other specialized techniques such as geochemical measurements, microsensor profil-

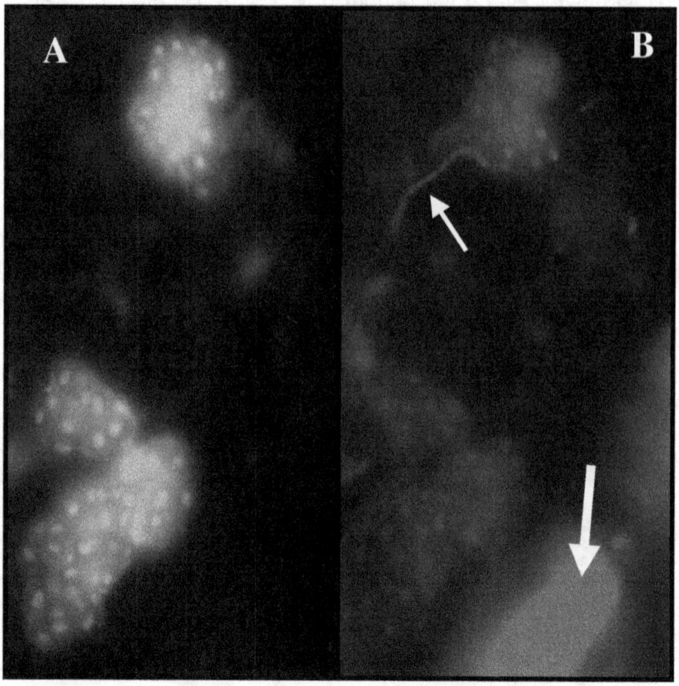

Figure 4.2 (see also Plate 4.2) FISH analysis of a cryptic picobenthic cyanobacterium from microbial mats. A: Fluorescence microscopy of a tiramide-enhanced FISH preparation. B: same frame under autofluorescence by photosynthetic pigments. Note the low levels of fluorescence compared to that of *Microcoleus chthonoplastes* filaments (thick arrow), and the negative hybridization of the probe to small and large filamentous cyanobacteria (thin arrow). Sequences for this previously unidentified cyanobacterium were retrieved from DGGE community fingerprints similar to those in Figure 4.1, and used to design specific oligonucleotide probes. After testing the specificity of the oligonucleotide probes as Southern probes on the DGGE gels, a group of very small unicellular cyanobacteria with low autofluorescence per cell, with the morphology of "*Aphanothece*" could be easily identified as the "missing" cyanobacterium. This picobenthic cyanobacterium has so far eluded cultivation. Data modified from Abed and colleagues (Abed *et al.*, 2002c)

ing, tracer experiments, autoradiography and others. The use of these combined techniques is becoming widespread in environmental research involving cyanobacterial communities. But molecular analyses offer the possibility to interrogate such natural communities with an unprecedented precision, particularly as our genomics databases increase in breadth. The simple detection of functional genes provides some information on potential functional abilities. For example, detection of nitrogen fixation genes in the metagenome by some of the descriptive methods outlined above, proves a potential for nitrogen fixation, even though it does not imply that the gene is being expressed, nor that the enzyme nitrogenase is there in a functional form. Several studies of *nif* genes—see a review by Zehr (Zehr *et al.*, 2003)—offer this sort of information. Functional gene frequency surveys in a Baltic estuary salinity gradient revealed, for example, that glucosyl-glycerol synthesizers (monitored through the marker gene *ggpS*, Table 4.1) were important throughout the gradient (Geiß *et al.*, 2004), contradicting the conventional wisdom gathered through surveys of cultivated cyanobacteria that this osmolyte is prevalent in marine strains (as opposed to trehalose, which is more important in fresh-water isolates). But the first and foremost methodological approach in functional studies is not just the detection of presence, but the detection and quantification of the *expression* of a particular gene. For example, in the same Baltic estuary surveyed for glucosyl-glycerol phosphate synthase genes, Geiß and colleagues also detected the presence of the gene *isiA*, a known marker for iron stress in isolated strains of cyanobacteria (Geiß *et al.*, 2001), in all samples tested. In this case, however, sequences of *isiA* retrieved from the environment were used to construct a consensus probe to carry out Northern hybridization to environmental RNA extracts, thus detecting not only the presence but also the expression of the gene in a qualitative manner (and thereby demonstrating iron stress in the field cyanobacterial populations).

Methodologies and applications

Gene expression methodology can vary in their specificity, sometimes targeting transcripts (mRNA) of particular genetic markers (functional markers, see Table 4.1) but casting a wide net, whereby the diversity of all organisms expressing the gene is encompassed. Alternatively, it can target the transcripts of a single organism or a defined group thereof. But the detection of mRNA, even using powerful techniques such as PCR (RT-PCR in this case) and even nested versions of it, is not a trivial pursuit due to the ephemeral nature of transcripts in the cells, their lack of biochemical stability and, in some cases, their small number of copies (Johnson *et al.*, 2005). In fact, early successful attempts to measure gene expression in natural communities were done on oxygenic phototrophic plankton communities using radio-labeled probes for membrane hybridization against environmental DNA or RNA extracts, and targeted mRNAs of common and abundant proteins such as Rubisco (*rbcL*) (Pichard and Paul, 1991; Pichard and Paul, 1993), for which they could demonstrate a light-dependent transcription in the plankton. Pichard and Paul developed the concept of gene expression per gene dose (mRNA/DNA) as a relative measurement of the level of transcription in the environment. Knowledge of sequence variation in this gene, soon led to the application of group-specific probes to differentiate, for example, between expression by diatoms versus cyanobacteria (Xu and Tabita, 1996).

Also among the physiologically tale-telling genes that have been probed for expression in nature one can count *ftsZ*, which encodes the protein initiating the cell division cycle

Table 4.1 Genetic marker loci used in the study of natural populations of cyanobacteria, either as phylogenetic anchors, populational markers or functional reporters, with and indication of the target groups

Gene or locus		Type of study	Target group	References
rRNA	16S	Diversity fingerprinting	Cyanobacteria/plastids	(Nübel *et al.*, 1999)
rRNA	166–236 ITS	Diversity fingerprinting	Planktonic cyanos	(Laamanen *et al.*, 2002)
STTS	(Tandem repeats)	Diversity fingerprinting	Cycad symbionts (*Nostoc*)	(Zheng *et al.*, 2002)
rpoC1	RNA Polymerase (γ subunit)	Diversity survey	Cyanobacteria (marine plankton)	(Palenik, 1994)
tRNA leu intron		Genetic diversity	Lichen symbionts (*Nostoc*)	(Paulsrud and Lindblad, 1998)
psbA	PSII D1 protein	Diversity survey	Oxygenic phototrophs	(Zeidner *et al.*, 2003)
mcyA	Microcystin synthetase A	Secondary metabolite detection	Toxin (Microcystin) producers	(Tillett *et al.*, 2001)
ggpS	Glucosylglycerol-P synthase	Ecophysiological inference (salinity tolerance)	Plankton	(Geiß *et al.*, 2004)
isiA	Chlorophyll *a*-binding protein	Ecophysiological inference (Fe starvation stress)	Plankton	(Geiß *et al.*, 2004)
hetR	Heterocyst development	Diversity fingerprinting	Heterocystous cyanobacteria	(Janson *et al.*, 1999)
petB/D	Cytochrome b_6f complex components	Diversity/ population genetics	*Microcoleus chthonoplastes*	(Lodders *et al.*, 2005)
kaiC	Circadian oscillator regulator	Diversity/ population genetics	*Microcoleus chthonoplastes*	(Lodders *et al.*, 2005)
cpeB	β-subunit of Phycoerythrin	Microevolution	*Prochlorococcus*	(Steglich *et al.*, 2003)
ftsZ	Cell division initiation protein	Ecophysiological inference (cell division)	*Prochlorococcus*	(Holtzendorff *et al.*, 2002)
ntcA	NH_3-repressed transcriptional activator	Ecophysiological inference (nutrient status)	Marine *Synechococcus*	(Lindell and Post, 2001)

nifH	Nitrogenase iron protein	Ecophysiological inference (nitrogen fixation)	Bacteria	(Omoregie *et al*., 2004)
nifHDK	Nitrogenase (various proteins)	Ecophysiological inference (nitrogen fixation)	Thermophilic *Synechococcus*	(Steunou *et al*., 2006)
cpcB-cpcA IGS	Intergenetic spacer of the phycocyanin operon	Diversity survey	Cyanobacteria (potentially some plastids)	(Neilan *et al*., 1995)
cpcE, cpcF	Phycocyanobilin lyase β subunit	Ecophysiological inference (photosynthesis)	Thermophilic *Synechococcus*	(Steunou *et al*., 2006)
psbB	PSII chlorophyll-binding protein	Ecophysiological inference (photosynthesis)	Thermophilic *Synechococcus*	(Steunou *et al*., 2006)
psaB	PSI reaction center protein	Ecophysiological inference (photosynthesis)	Thermophilic *Synechococcus*	(Steunou *et al*., 2006)
coxA	Cytochrome *c* oxidase subunit I	Ecophysiological inference (respiration)	Thermophilic *Synechococcus*	(Steunou *et al*., 2006)
ldh	L-lactate dehydrogenase	Ecophysiological inference (fermentation)	Thermophilic *Synechococcus*	(Steunou *et al*., 2006)
cydA	Cytochrome *d* oxidase subunit I	Ecophysiological inference (respiration)	Thermophilic *Synechococcus*	(Steunou *et al*., 2006)
acs	Acetyl-coA synthase	Ecophysiological inference (fermentation)	Thermophilic *Synechococcus*	(Steunou *et al*., 2006)
adhE	coA-linked acetaldehyde dehydrogenase/ alcohol dehydrogenase	Ecophysiological inference (fermentation)	Thermophilic *Synechococcus*	(Steunou *et al*., 2006)

Table 4.1 Continued

Gene or locus		Type of study	Target group	References
ackA	Acetate kinase	Ecophysiological inference (fermentation)	Thermophilic *Synechococcus*	(Steunou *et al*., 2006)
pflA	Pyruvate-formate lyase activating enzyme	Ecophysiological inference (fermentation)	Thermophilic *Synechococcus*	(Steunou *et al*., 2006)
pflB	Pyruvate formate lyase	Ecophysiological inference (fermentation)	Thermophilic *Synechococcus*	(Steunou *et al*., 2006)
pdhB	Pyruvate-dehydrogenase b-subunit	Ecophysiological inference (fermentation)	Thermophilic *Synechococcus*	(Steunou *et al*., 2006)

of prokaryotes. Genomic knowledge of the sequence and sequence variation of this gene among marine *Prochlorococcus* have been used by Holtzendorff and colleagues to design probes and monitor its expression in natural populations by means of quantitative RT-PCR (Holtzendorff *et al.*, 2002); they demonstrated, as expected from microscopic observations, the synchronized cell division dynamics of *Prochlorococcus* in nature. Lindell and Post (Lindell and Post, 2001) monitored the expression of a general regulator of nitrogen metabolism in cyanobacteria, *ntcA*, the product of which is a transcriptional activator that mediates negative control by ammonium. Expression of this gene in cultures of marine *Synechococcus* is induced under nutritional stress by lack of nitrogen, and thus is an indicator of nitrogen deprivation. By using a RT-Nested PCR protocol, that was followed during exponential amplification cycles in a sort of Q-PCR manner, they could demonstrate that the levels of *ntcA* mRNA in field populations of *Synechococcus* from a N-enriched site in the gulf of Aqaba were low enough to indicate the preferential use of ammonium as their major source of N (that is, N regenerated *in situ*, as opposed to biologically fixed, upwelled or advected).

The work of Steunou and colleagues (Steunou *et al.*, 2006) represents the most comprehensive study of cyanobacterial gene expression in natural communities. Based on, and spearheaded by, a focused genomic sequencing effort of several thermophilic *Synechococcus* from hot springs, they were able to follow the expression of a large set of genes in natural populations of these microbial mat organisms by Q-PCR during diel cycles. This indicated that *Synechococcus* in their mats undergo a general switch in gene expression at dusk, whereby nitrogenase genes *nifH*, *nifD*, and *nifK* are upregulated as light levels become low and the mat turns anoxic, whereas photosynthetic genes such as *psbB* and *psaB*, as well as *cpcF* and *cpcE* are downregulated, as are some respiration genes such as *cydA*. Interestingly, the switch also involves transcripts for fermentation genes, some of which are downregulated and some upregulated at dusk (see Figure 4.3), indicating the importance of fermentative sugar metabolism in night-time energy generation for oxygen-depleted populations.

Cyanobacteria: a special case

Because of their importance as primary producers, cyanobacteria as a group have been the subject of intense investigation with molecular and genomic techniques, some of which have, as we have seen, been tailored specifically to target them or to deal with the particular idiosyncrasies of this group of oxygenic phototrophs. In some cases the relevance of cyanobacteria in the environment under study itself has indirectly helped expand the baseline of molecular data available for cyanobacteria. But the impact of modern molecular ecological studies on cyanobacterial research has been somewhat different than that on research involving other bacterial groups or on the discipline of microbial ecology as a whole. To begin with, recognition of phylogenetic membership in the group was never a defining problem, since cyanobacteria are in general much easier to recognize as such than other bacteria by mere microscopic, even sometimes macroscopic, observations (Garcia-Pichel, 2000). In general: (i) they are typically large, (ii) they often present clearly defined morphological traits, (iii) their photosynthetic pigment content makes them strongly fluorescent, and (iv) their obligate requirement for sunlight restricts them to surface habitats. At the functional level, because cyanobacteria in nature are photoautotrophs there has been no need to establish their major function in the ecosystem by molecular means, as it certainly was the

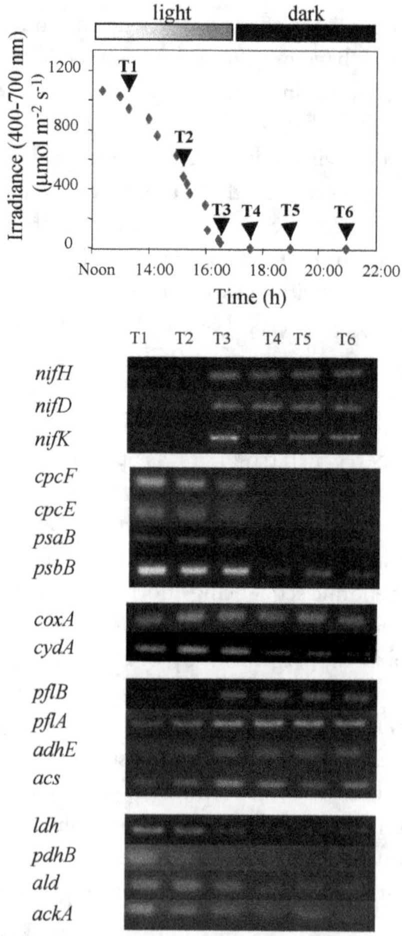

Figure 4.3 Functional analyses of gene expression in natural populations of thermophilic *Synechococcus* from hot spring microbial mats. Transcripts from a variety of genes were probed by RT-PCR along a time course starting in the early afternoon and ending at midnight. The upper graph shows sampling times and light intensity. Genomic analyses of cultured representatives of the major *Synechococcus* populations enabled the choice of genes and the design of specific primers for each. Expression is shown here as agarose gels of the PCR products, revealing a striking switch at dusk. More quantitative assessments—not shown here—were accomplished through Q-PCR. Data is from Steunou *et al.* (2006) and graphs have been recomposed from those graciously provided by Anne-Soisig Steunou. Genetic markers are listed in Table 4.1.

case for other microbial groups. And yet, functional or metabolic aspects secondary to their primary producer role, such as their ability to fix nitrogen, their nutrient status, their toxicity, or the nature of type of nighttime metabolism are still questions that can clearly be best addressed by molecular studies of gene expression.

Before the advent of molecular techniques, cyanobacterial diversity in terms of morphology, ecological distribution, and environmental role, were comparatively well known from over one hundred years of dedicated naturalist's accounts. The contributions of molecular techniques have thus been for the most part incremental rather than transformative. No unexpected habitats have been identified, and no major, deeply branching or radically different groups of cyanobacteria have been detected. The last major discoveries, both members of the picoplankton, pre-dated the molecular revolution or were attained independently of it. The depth of the Phylum Cyanobacteria in the 16S rRNA tree of life has remained virtually unchanged since the very earliest versions (Garcia-Pichel, 2000; Giovannoni *et al.*, 1988; Nelissen *et al.*, 1995; Turner *et al.*, 1999; Wilmotte and Herdman, 2001), with the deepest cyanobacterial node defined (depending on the tree) by terrestrial *Gloeobacter violaceus*, by a unicellular thermophile, by a clade of simple filamentous pseudanabaenas or by the plastid branch (see Chapter 2). No missing links between the phylum of oxygenic phototrophs and its nearest phyla have been revealed. No unexpected metabolic capacities unraveled. And this lack of transformative breakthroughs cannot be attributed to a comparative lack of effort. A simple data-mining exercise, depicted in Figure 4.4, will show that the number of entries for 16S rRNA sequences belonging to the cyanobacteria in Genbank has been increasing exponentially at a rate (base e) of 0.40 per year, not significantly different than the 0.41 at which all bacterial (and Archaeal) sequences have been accreting during the same period. This is of course not to say that our picture of cyanobacterial evolution, diversity, distribution and activity has not improved by the introduction of molecular techniques, for it certainly has. As seen in our review of methodologies and cyanobacterial applications, the molecular studies are many and diverse.

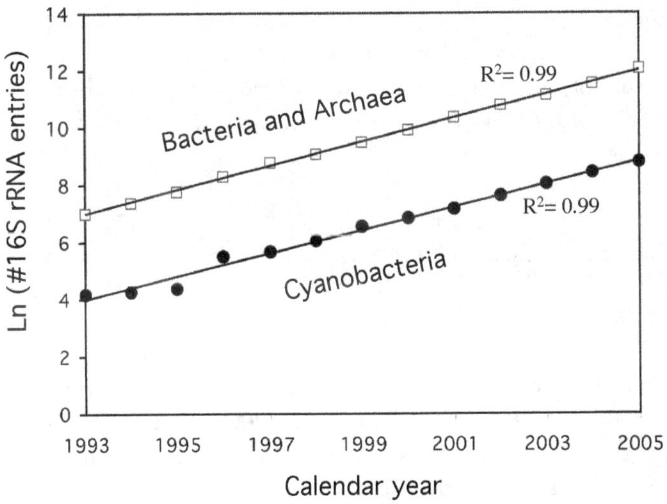

Figure 4.4 Accretion around the turn of the century of unique 16S rRNA gene sequences, including those from cultures or environmental, partial or complete, available publicly at GenBank. The increase has been exponential for all sequences belonging to the Archaea and Bacteria, with an instantaneous growth rate of 0.41. The subset of sequences belonging to the cyanobacteria, while expectedly much smaller, has accreted at essentially the same rate, indicating a proportional research effort

Advances

It is also fair to say that there has been no sustained or systematic effort to integrate the results of independent studies in order to detect emergent distribution patterns or trends, with notable exceptions of perhaps restricted scope (Zehr *et al.*, 2003). For the most part, the information on habitat surveys remains a collection of apparently ever-growing "lists" and functional information is still spotty. At the risk of missing a few of the trees, I will attempt to address the advances here summarizing results from various studies, in such a way that they may start to offer a picture of the forest, even if still coalescing and necessarily blurred.

Cyanobacterial biodiversity and habitat specificity

In general, undescribed cyanobacterial biodiversity has been detected by molecular means in a consistent manner as new environments and habitats have been probed. As mentioned before, however, these have been incremental, rather than transformative finds. Typically novel environmental sequences bore low identity (around 90–93% in terms of 16S rRNA, very seldom below 89%) to those already in the databases (Abed *et al.*, 2003; Chacón *et al.*, 2006; Elser *et al.*, 2005; Gaylarde *et al.*, 2005; Sigler *et al.*, 2003; Steindler *et al.*, 2005), probably representing new genera, according to present microbiological concepts. By contrast, when surveys were carried out on previously well-studied habitats, cyanobacterial sequences retrieved were typically close to, if not statistically undistinguishable from, those obtained before. This is best represented by the case of the extremely halotolerant cyanobacteria, where members within clades originally described from Mexican sites (Garcia-Pichel *et al.*, 1998) were found again on studies of European (Margheri *et al.*, 1999) and Middle Eastern (Sorensen *et al.*, 2005) habitats. The same general trend was found with hot-spring (Papke *et al.*, 2003) and desert crust cyanobacteria (Nagy *et al.*, 2005; Reddy and Garcia-Pichel, 2006). This all indicates that a) cyanobacterial communities are clearly habitat-specific, and b) while there is still much work ahead, the task of attaining a significant level of diversity coverage is certainly within reach in a reasonable time with the typical tools of the trade.

Morphological diversity conceals genetic diversity

In addition, a wealth of studies on the phylogeny of cultured and environmental sequences have revealed that genetic diversity, even at the level of 16S rRNA, is much larger than morphotypical diversity would suggest (Weller *et al.*, 1992; Zheng *et al.*, 2002; Abed *et al.*, 2002b; Ferris *et al.*, 1996; Garcia-Pichel *et al.*, 2001; Garcia-Pichel *et al.*, 1998; Garcia-Pichel *et al.*, 2002; Jungblut *et al.*, 2005; Lau *et al.*, 2005; Miller and Castenholz, 2000; Moore *et al.*, 1998; Song *et al.*, 2005; Taton *et al.*, 2003). This is particularly true when observing complex natural assemblages, where quantitative diversity indices based on 16S rRNA tend to be almost double those obtained by morphological assessment (Nübel *et al.*, 1999). But it must be stressed that this is just an overall trend driven by only some morphotypes. In several cases, a complete coherence has been found between morphological diversification and phylogenetic reconstructions.

The set of morphotypes that conceal comparatively small levels of genetic diversity includes primarily members of the heterocystous cyanobacteria as a group, and in particular genera such as *Nostoc*, *Scytonema*, and the *Calothrix*/*Tolypothrix* group, as well as some Oscillatorian cyanobacteria: *Microcoleus chthonoplastes* (Garcia-Pichel *et al.*, 1996; Lodders *et al.*, 2005), *Microcoleus vaginatus* (Boyer *et al.*, 2002; Garcia-Pichel *et al.*, 2001), *Microcoleus steenstrupii* (Boyer *et al.*, 2002; Nagy *et al.*, 2005), *Trichodesmium* (Capone *et al.*, 1997; Lundgren *et al.*, 2005), *Hydrocoleum* (Abed *et al.*, 2003). A few other morphologically conspicuous types among the unicellular (Chroococcales) or the baeocyte-makers (Pleurocapsales) may indeed prove to be monophyletic and genetically coherent as well, as seems to be the case with *Chroococcidiopsis* (Fewer *et al.*, 2003).

The corollary of these diverse observations seems to be that the level of genetic diversity that a given cyanobacterial morphotype will encompass is variable, such that in simple rods, cocci or filaments, morphology is a bad predictor of relatedness, and in fact these morphotypes are likely to conceal cyanobacteria of divergent origin and wide-ranging, disparate habitat distribution (see also Figure 4.2). Exemplary are the morphogenera *Synechococcus* and *Leptolyngbya/Phormidium* with members from soil, hot spring, marine and freshwater environments clustering around sometimes ill-defined and clearly polyphyletic clades. By contrast cyanobacteria with coherent morphology and phylogeny tend to be those with conspicuous morphological traits, such as those listed above, and notable the truly multicellular cyanobacteria with cell differentiation ability (including apical cell differentiation and filament fragmentation by necridia).

Novel views on macroevolution, phylogeny, and taxonomy

Molecular phylogenetic studies have clearly proven that the traditional five-order model of cyanobacterial taxonomy, in which all taxonomic systems are based in one way or another, is poorly reflective of phylogeny, with 3 out of the 5 orders most likely being polyphyletic (Turner *et al.*, 1999; Garcia-Pichel, 2000; Wilmotte and Herdman, 2001). But also, at a much finer level of phylogenetic resolution, novel, and possibly environmentally relevant cyanobacteria or phylogenetic groups of cyanobacteria have been discovered as diverse environments have been probed, such as the nitrogen fixing marine "*Crocosphaera*" (Zehr *et al.*, 2001), the "*Xeronema*" cluster of desert soil cyanobacteria (Garcia-Pichel *et al.*, 2001) or the *Halomicronema* of microbial mats (Abed *et al.*, 2002a). Some traditional genera, based on simple morphological traits, have been shown to be polyphyletic as well, containing unrelated groups, notably the genus *Microcoleus* (Garcia-Pichel *et al.*, 2001). New phylogenetically coherent clades have been defined on the basis of physiology, rather than morphology: the *Halothece* clade (not yet formally defined as a taxon) and the new genus *Halospirulina*. In both cases it was demonstrated that mere morphological differentiation of genera in classical (microbiological and botanical) taxonomy failed to reflect evolutionary relationships, being here much better described by physiological and biochemical adaptations to salt stress (just like most bacteria are). *Halothece* includes all strains that are unicellular and extremely halotolerant (growing above 120% salt) regardless of morphology (strains in these groups had previously been referred to as *Aphanothece halophytica*, *Cyanothece* spp., and *Dactylococcopsis salina*; Garcia-Pichel *et al.*, 1998). *Halospirulina* encompasses all strains of filamentous tightly coiled cyanobacteria that are extremely halotolerant (Nübel *et al.*, 2000b), well separated evolutionarily from other strains in the morphogenus *Spirulina* (sensu Geitler, 1932) that cannot be differentiated on the basis of morphology. Different investigators have confirmed these results and expanded their geographical distribution (Margheri *et al.*, 1999; Margheri *et al.*, 2003).

Microdiversity, population genetics, and microevolutionary processes

Thanks to the development of molecular studies, we have witnessed the first direct insights into microevolutionary processes of diversification in cyanobacteria, as they may be reconstructed from genetic data obtained from field samples. Molecular studies have confirmed and strengthened the idea of ecotypic differentiation, whereby functional diversity is based on fine-level genetic diversification that is usually overseen in rough surveys of genetic diversity, but is all-important in structuring communities through competition and niche differentiation. Environmentally driven ecotypic differentiation in cyanobacteria is of course not a new concept and had been demonstrated before (see Garcia-Pichel and Castenholz, 1990, for example); the presence of fine-level genetic diversity in several groups of cyanobacteria was also known from molecular diversity studies (see previous sections). The novelty lies in bringing both demonstrations together. The diversification into high-light-adapted and low-light-adapted strains or ecotypes within the marine *Prochlorococcus* constitutes perhaps the best known example, detected genetically using16S rRNA analyses (Moore *et al.*, 1998). Interestingly, now we can start reconstructing molecular events that may have been significant in the process of incipient speciation: studies of the locus *cpeB* (encoding the β-subunit of the red phycobiliprotein) in the same group of picoplankton showed not only that the dual ecotypic split is also reflected in the phylogeny of this protein, but also that in the high-light-adapted clade, *cpeB* was conserved and functional, as opposed to the alleles of *cpeB* in the low-light-adapted clade, where the gene was non-functional (Steglich *et al.*, 2003). Among non-marine strains, excellent microevolutionary reconstructions of thermophily in hot-spring *Synechococcus* have been presented as well (Miller and Castenholz, 2000). Direct, cultivation independent, analyses of population genetics in Baltic and North Sea populations of *Microcoleus chthonoplastes* (a microbial mat builder of worldwide distribution) have shown that these populations sustain a high level of genetic exchange that results in an apparent lack of linkage between the allelic variants in the different populations (Lodders *et al.*, 2005). Similar results could be deduced from filaments of planktonic *Nodularia* in a large bloom (Barker *et al.*, 2000).

Again, the value of cultivation

Finally, if one point has emerged as important, it has been the value of cultivated isolates that can serve as anchors for the interpretation of molecular data. Almost invariably, any molecular studies of cyanobacteria that have gone beyond the descriptive aspect of research, have a history of careful cultivation and characterization behind them. Perhaps not the blind enrichment cultivation of times past, that may have indeed misdirected research efforts, but cultivation that seeks to obtain *bona fide* representatives of field populations and of their diversity. Representatives that can be interrogated physiologically, biochemically and genetically in the laboratory. Explanations of the distribution and abundance of cyanobacteria in hypersaline systems (Nübel *et al.*, 2000a) would have been impossible without much previous work characterizing the physiology of appropriate representatives in the laboratory. The considerable insights into the ecology of thermophilic *Synechococcus* achieved in recent studies, would have been impossible without much previous work identifying the right organisms to cultivate, or without intensive sequencing effort of those isolates as well. Similar arguments could be made for the marine picoplankton and many of the symbiotic cyanobacteria.

Future trends

Integrated studies using community structure investigation, genomic and traditional interrogation of cultivated representatives, as well as genomic and biogeochemical interrogation of field populations should become standard in the study of all cyanobacterial habitats. Of course, our present level of knowledge makes this only possible for some of the well studied habitats. Therefore, the use of molecular techniques is likely to remain in its present dual mode of steady accretion of new descriptive data for many environments and sophisticated analyses focused on a couple of environments or communities (as has been done with marine picoplankton or hot spring systems). Because of their geographical extent, these efforts should be extended to terrestrial communities in arid lands, and possibly also polar and subpolar environments, where community composition and major players are becoming increasingly well-known. But there are no sequencing efforts under way for major terrestrial cyanobacteria such as *Microcoleus vaginatus*, perhaps the third most abundant cyanobacterium on Earth (Garcia-Pichel *et al.*, 2003), nor are there any for psychrophiles. These should probably be the next in line. In the field of cyanobacterial biodiversity, it is in the morphologically simple types where most of the undefined novelty lies, and where most of the future work will need to be directed. Working out a phylogeny and ecological distribution for morphotypes such as the non-heterocystous simple filamentous cyanobacteria (*Leptolyngbya, Phromidium* and allies) will be needed in order to clarify the situation in many environments were they seem to be common.

One final area where one can see a major stumbling stone as genomics and metagenomics attain an increasing share of the field of study, is posed by the large proportions of their genomes, even when easily recognizable as open reading frames, that remain unknown to us. Yet another *renaissance*, this time in traditional genetic and biochemical analysis of mutants for those genes and operons of unknown function, is probably awaiting us, only this time we may be able to probe with the guidance and priorities set by metagenomic studies.

Acknowledgments

This review was written during a sabbatical leave at the CMIMA-CSIC, which was made possible by a grant from the Faculty Mobility Program of the Ministry of Education, Spain, to FGP. I would like to thank Anne-Soisig Steunou and Arthur R. Grossman at Stanford University for the graphs in Figure 4.3, and Carlos Pedrós-Alió for fruitful discussions.

References

Abed, R.M.M., and Garcia-Pichel, F. (2001). Long-term compositional changes after transplant in a microbial mat cyanobacterial community revealed with a polyphasic approach. Environ. Microbiol. 3, 53–62.

Abed, R.M.M., Garcia-Pichel, F., and Hernández-Mariné, M. (2002a). Polyphasic characterization of moderately halophilic, moderately thermophilic, filamentous cyanobacteria of very narrow trichomes: *Halomicronema excentricum gen. novum, sp. nova*. Arch. Microbiol. *177*, 361–370.

Abed, R.M.M., Golubic, S., Garcia-Pichel, F., and Camoin, G. (2003.). Characterization of microbialite forming cyanobacteria in a tropical lagoon; Tikehau Atoll, French Polynesia. J. Phycol. 39, 862–873.

Abed, R.M.M., Safi, N., Köster, J., Rullköter, J., and Garcia-Pichel, F. (2002b). Microbial diversity of a polluted microbial mat and its community changes following degradation of petroleum compounds. Appl. Environ. Microbiol. 68:1674–1683

Abed, R.M.M., Schoenhuber, W., Amann, R.I., and Garcia-Pichel, F. (2002c). Picobenthic cyanobacterial populations revealed by RNA targeted in situ hybridization. Environ. Microbiol. *4*, 375–382.

Amann, R.I., Ludwig, W., and Schleifer, K.-H. (1995). Phylogenetic identification and in situ detection of individual microbial cells without cultivation. Microbiol. Rev. *59*, 143–169.

Baker, J.A., Entsch, B., Neilan, B.A., and McKay, D.B. (2002). Monitoring changing toxigenicity of a cyanobacterial bloom by molecular methods. Appl. Environ. Microbiol. *68*, 6070–6076.

Barker, G.L.A., Handley, B.A., Vacharapiyasophon, P., Stevens, J.R., and Hayes, P.K. (2000). Allele-specific PCR shows that genetic exchange occurs among genetically diverse Nodularia (Cyanobacteria) filaments in the Baltic Sea. Microbiol. *146*, 2865–2875.

Becker, S., Fahrbach, M., Boger, P., and Ernst, A. (2002). Quantitative tracing, by Taq nuclease assays, of a *Synechococcus* ecotype in a highly diversified natural population. Appl. Environ. Microbiol. *68*, 4486–4494.

Ben-Porath, J., and Zehr, J. (1994). Detection and characterization of cyanobacterial *nifH* genes. Appl. Environ. Microbiol. *60*, 880–887.

Bidle, K., and Fletcher, M. (1995). Comparison of free-living and particle-associated bacterial communities in the Chesapeake Bay by stable low-molecular-weight RNA analysis. Appl. Environ. Microbiol. *61*, 944–952.

Bolch, C.J.S., Blackburn, S.I., Neilan, B.A., and Grewe, P.M. (1996). Genetic characterization of strains of cyanobacteria using PCR-RFLP of the *cpcBA* intergenic spacer and flanking regions. J. Phycol. *32*, 445–451.

Boyer, S.I., Johansen, J.R., and Flechter, V.R. (2002). Phylogeny and genetic variance in terrestrial *Microcoleus* (Cyanophyceae) species based on sequence analyses of the 16S rRNA gene and associated 16s-23S ITS region. J. Phycol. *38*, 1222–1235.

Burns, B.P., Goh, F., Allen, M., and Neilan, B.A. (2004). Microbial diversity of extant stromatolites in the hypersaline marine environment of Shark Bay, Australia. Environ. Microbiol. *6*, 1096–1101.

Capone, D.G., Zehr, J.P., Paerl, H.W., Bergman, B., and Carpenter, E.J. (1997). *Trichodesmium*, a globally significant marine cyanobacterium. Science *276*, 1221–1229.

Carvalho, G.R. (1988). Molecular ecology; origins and approach. In Advances in Molecular Ecology, G.R. Carvalho, ed. (Amsterdam, IOS Press), pp. 1–16.

Casamayor, E.O., Massana, R., Benlloch, S., Ovreas, L., Díez, B., Goddard, V.J., Gasol, J.M., Joint, I., Rodríguez-Valera, F., and Pedros-Alio, C. (2002). Changes in archaeal, bacterial and eukaryal assemblages along a salinity gradient by comparison of genetic fingerprinting methods in a multipond solar saltern. Environ. Microbiol. *4*, 338–348.

Castiglioni, B., Rizzi, E., Frosini, A., Sivonen, K., Rajaniemi, P., Rantala, A., Mugnai, M.A., Ventura, S., Wilmotte, A., Boutte, C., Grubisic, S., Balthasart, P., Consolandi, C., Bordoni, R. Mezzelani, A., Battaglia, C., and De Bellis, G. (2004). Development of a universal microarray based on the ligation detection reaction and 16S rRNA gene polymorphism to target diversity of cyanobacteria. Appl. Environ. Microbiol. *70*, 7161–7172.

Chacón, E., Berrendero, E., and Garcia-Pichel, F. (2006). Biogeological signatures of microboring cyanobacterial communities in marine carbonates from Cabo Rojo, Puerto Rico. Sed. Geol. *185*, 215–228.

Dyhrman, S.T., and Haley, S.T. (2006). Phosphorus scavenging in the unicellular marine diazotroph *Crocosphaera watsonii*. Appl. Environ. Microbiol. *72*, 1452–1458.

Elser, J.J., Schampel, J.H., Garcia-Pichel, F., Wade, B.D., Souza, V., Eguiarte, L., Escalante, A., and Farmer, J.D. (2005). Effects of phosphorus enrichment and grazing snails on modern stromatolitic microbial communities. Freshwater Biol. *50*, 1808–1825.

Elshahed, M.S., Senko, J.M., Najar, F.Z., Kenton, S.M., Roe, B.A., Dewers, T.A., Spear, J.R., and Krumholz, L.R. (2003). Bacterial diversity and sulfur cycling in a mesophilic sulfide-rich spring. Appl. Environ. Microbiol. *69*, 5609–5621.

Ferris, M., and Palenik, B. (1998). Niche adaptation in ocean cyanobacteria. Nature *396*, 226–228.

Ferris, M., Ruff-Roberts, A., Kopczynski, E., Bateson, M., and Ward, D. (1996). Enrichment culture and microscopy conceal diverse thermophilic *Synechococcus* populations in a single hot spring microbial mat habitat. Appl. Environ. Microbiol. *62*, 1045–1050.

Fewer, D., Friedl, T., and Büdel, B. (2003). *Chroococcidiopsis* and heterocyst-differentiating cyanobacteria are each other's closest living relatives. Mol. Phylogen. Evol. *23*, 82–90.

Foster, R.A., Collier, J.A., and Carpenter, E.J. (2006). Reverse transcription PCR amplification of cyanobacterial symbiont 16S rRNA sequences from single non-photosynthetic eukaryotic marine planktonic host cells. J. Phycol. *42*, 243–250.

Fouke, B.W., Bonheyo, G.T., Sanzenbacher, B., and Frías-Lopez, J. (2003). Partitioning of bacterial communities between travertine depositional facies at Mammoth Hot Springs, Yellowstone National Park. Can. J. Earth Sci. *40*, 1531–1548.

Frías-Lopez, J., Bonheyo, G.T., Qusheng, J., and Fouke, B.W. (2003). Cyanobacteria associated with coral black band disease in caribbean and Indo-Pacific reefs. Appl. Environ. Microbiol. *69*, 2409–2413.

Garcia-Pichel, F., and Castenholz, R.W. (1990). Comparative anoxygenic photosynthetic capacity in 7 strains of a thermophilic cyanobacterium. Arch. Microbiol. *153*, 344–351.

Garcia-Pichel, F. (2000). Cyanobacteria. In Encyclopedia of Microbiology, J. Lederberg, ed. (San Diego, Academic Press).

Garcia-Pichel, F., Al-Horani, F., Ludwig, R., Farmer, J., and Wade, B. (2004). Balance between calcification and bioerosion in modern stromatolites. Geobiology *2*, 49–57.

Garcia-Pichel, F., Belnap, J., Neuer, S., and Schanz, F. (2003). Estimates of global cyanobacterial biomass and its distribution. Arch. Hydrobiol./Algol. Stud. *109*, 213–228

Garcia-Pichel, F., López-Cortés, A., and Nübel, U. (2001). Phylogenetic and morphological diversity of cyanobacteria in biological soil crusts from the Colorado Plateau. Appl. Environm. Microbiol. *67*, 1902–1910.

Garcia-Pichel, F., Nübel, U., and Muyzer, G. (1998). The phylogeny of unicellular, extremely halotolerant cyanobacteria. Arch. Microbiol. *169*, 469–482.

Garcia-Pichel, F., Prufert-Bebout, L., and Muyzer, G. (1996). Phenotypic and phylogenetic analyses show Microcoleus chthonoplastes to be a cosmopolitan cyanobacterium. Appl. Environ. Microbiol. *62*, 3284–3291.

Garcia-Pichel, F., Wade, B.D., and Farmer, J.D. (2002). Jet-suspended, calcite ballasted cyanobacterial waterwarts in a desert spring. J. Phycol. *38*, 420–428.

Gaylarde, P.M., Crispim, C.A., Neilan, B.A., and Gaylarde, C.C. (2005). Cyanobacteria from Brazilian building walls are distant relatives of aquatic genera. Omics *2005*, 30–42.

Geiß, U., Selig, U., Schumann, R., Steinbruch, R., Bastrop, R., Hagemann, M., and Schoor, A. (2004). Investigations on cyanobacteria diversity in a shallow estuary (Southern Baltic Sea) including genes relevant to salinity resistance and iron starvation acclimation. Environ. Microbiol. *6*, 377–387.

Geiß, U., Vinnemeier, J., Kunert, A., Lindner, I., Gemmer, B., Lorenz, M., Hagemann, M., and Schoor, A. (2001). Detection of the *isiA* gene across cyanobacterial strains: potential for probing iron deficiency. Appl. Environ. Microbiol. *67*, 5247–5253.

Gillespie, D.E., Rondon, M.R., Williamson, L.L., and Handelsman, J. (2005). Metagenomic libraries from uncultured microorganisms. In Molecular microbial ecology, M. Osborn, and C. Smith, eds. (New York, Taylor & Francis), pp. 11–33.

Giovannoni, S.J., Bibbs, L., Cho, J.C., Stapels, M.D., Desiderio, R., Vergin, K.L., Rappe, M.S., Laney, S., Wilhelm, L.J., Tripp, H.J., Mathur, E.J., and Barofsky, D.F. (2005). Proteorhodopsin in the ubiquitous marine bacterium SAR11. Nature *438*, 82–85.

Giovannoni, S.J., Turner, S., Olsen, G.J., Barns, S., Lane, D.J., and Pace, N.R. (1988). Evolutionary relationships among cyanobacteria and green chloroplasts. J. Bacteriol. *170*, 3584–3592.

Gugger, M., Molica, R., Le Berre, B., Dufour, P., Bernard, C., and Humbert, J.-F. (2005). Genetic diversity of *Cylindrospermopsis* strains (Cyanobacteria) isolated from four continents. Appl. Environ. Microbiol. *71*, 1097–1100.

Hofle, M., and Brettar, I. (1996). Genotyping of heterotrophic bacteria from the Central Baltic Sea by use of low-molecular-weight RNA profiles. Appl. Environ. Microbiol. *62*, 1383–1390.

Holtzendorff, J., Marie, D., Post, A.F., Partensky, F., Rivlin, A., and Hess, W.R. (2002). Synchronized expression of *ftsZ* in natural *Prochlorococcus* populations of the Red Sea. Environ. Microbiol. *4*, 644–653.

Hugenholtz, P., Goebel, B.M., and Pace, N.R. (1998). Impact of culture-independent studies on the emerging phylogenetic view of bacterial diversity. J. Bacteriol. *180*, 4765–4774.

Jansen, I., Meima, M., Kardinaal, W.E.A., and Zwart, G. (2003). High-resolution differentiation of cyanobacteria by using rRNA-internal transcribed spacer denaturing gradient gel electrophoresis. Appl. Environ. Microbiol. *69*, 6634–6643.

Janson, S., Wouters, J., Bergman, B., and Carpenter, E.J. (1999). Host specificity in the Richelia-diatom symbiosis revealed by *hetR* gene sequence analyses. Environ. Microbiol. 1: 431–438

Johnson, D.R., Lee, P.K.H., Holmes, V.F., and Alvarez-Cohen, L. (2005). An internal reference technique for accurately quantifying specific mRNAs by real-time PCR with application to the *tceA* reductive dehalogenase gene. Appl. Environ. Microbiol. *71*, 3866–3871.

Jungblut, A.-D., Hawes, I., Mounfort, D., Hitzfeld, B., Dietrich, D.R., Burns, and Neilan, B.A. (2005). Diversity within cyanobacterial mat communities in variable salinty meltwater ponds of McMurdo Ice Shelf, Antarctica. Environ. Microbiol. *7*, 519–529.

Kim, J.B., Moon, M.S., Lee D.H., Lee S.T., Bazzicalupo, M., and Kim, C.K. (2004). Comparative analysis of cyanobacterial communities from polluted reservoirs in Korea. J. Microbiol. *42*, 181–187.

Könneke, M., Bernhard, A.E., de la Torre, J., Walker, C.B., Waterbury, J.B., and Stahl, D.A. (2005). Isolation of a mesophilic autotrophic ammonia-oxidizing marine crenarchaeote. Nature *437*, 543–546.

Kowalchuk, G.A., de Bruijn, F.J., Head, I.M., Akkermans, A.D., and van Elsas, J.D., eds. (2004). Molecular Microbial Ecology Manual, 2nd edition (Heidelberg, Springer Verlag).

Laamanen, M.J., Forström, L., and Sivonen, K. (2002). Diversity of *Aphanizomenon flos-aquae* (Cyanobacterium) populations along a Baltic Sea salinity gradient. Appl. Environm. Microbiol. *68*, 5296–5303.

Lau, E., Nash, C.Z., Vogler, D.R., and Cullings, K.W. (2005). Molecular diversity of cyanobacteria inhabiting coniform structures and surrounding mat in a Yellowstone Hot Spring. Astrobiol. *5*, 83–92.

Lindell, D., and Post, A.F. (2001). Ecological aspects of *ntcA* gene expression and its use as an indicator of the nitrogen status of marine *Synechococcus* spp. Appl. Environ. Microbiol. *67*, 3340–3349.

Liu, W., Marsh, T., Cheng, H., and Forney, L. (1997). Characterization of microbial diversity by determining terminal restriction fragment length polymorphisms of genes encoding 16S rRNA. Appl. Environ. Microbiol. *63*, 4516–4522.

Lodders, N., Stackebrandt, E., and Nübel, U. (2005). Frequent genetic recombination in natural populations of the marine cyanobacterium *Microcoleus chthonoplastes*. Environ. Microbiol. *7*, 434–442.

Lundgren, P., Janson, S., Jonasson, S., Singer, A., and Bergman, B. (2005). Unveiling of novel radiations within *Trichodesmium* cluster by *hetR* gene sequence analysis. Appl. Environ Microbiol. *71*, 190–196.

Lyra, C., Hantula, J., Vainio, E., Rapala, J., Rouhiainen, L., and Sivonen, K. (1997). Characterization of cyanobacteria by SDS-PAGE of whole-cell proteins and PCR/RFLP of the 16S rRNA gene. Arch. Microbiol. *168*, 176–184.

Ma, Y., Jiao, N.Z., and Zeng, Y.H. (2004). Natural community structure of cyanobacteria in the South China Sea as revealed by *rpoC1* gene sequence analysis. Lett. Appl. Microbiol. *39*, 353–358.

Margheri, M.C., Bosco, M., Giovannetti, L., and Ventura, S. (1999). Assessment of the genetic diversity of halotolerant coccoid cyanobacteria using amplified 16S rDNA restriction analysis. FEMS Microbiol. Lett. *179*, 9–16.

Margheri, M.C., Piccardi, R., and Ventura, S. (2003). Genotypic diversity of oscillatoriacean strains belonging to the general *Geitlerinema* and *Spirulina* determined by 16S rDNA. Curr. Microbiol. *46*, 359–364

McCaig, A.E., Glover, L.A., and Prosser, J.I. (2001). Numerical analysis of grassland bacterial community structure under different land management regimens by using 16S ribosomal DNA sequence data and denaturing gradient gel electrophoresis banding patterns. Appl. Environ. Microbiol. *67*, 4554–4559.

Meeks, J.C., Elhai, J., Thiel, T., Potts, M., Larimer, F., Lamerdin, J., Predki, P., and Atlas, R. (2001). An overview of the genome of *Nostoc punctiforme*, a multicellular, symbiotic cyanobacterium. Photosynth. Res. *70*, 85–106.

Miller, S.R., and Castenholz, R.W. (2000). Evolution of thermotolerance in hot spring cyanobacteria of the genus *Synechococcus*. Appl. Environ. Microbiol. *66*, 4222–4229.

Moore, L.R., Rocap, G., and Chisholm, C.S. (1998). Physiology and molecular phylogeny of coexisting Prochlorococcus ecotypes. Nature *393*, 464–467.

Muyzer, G., de Waal, E.C., and Uitterlinden, A.G. (1993). Profiling of complex microbial populations by denaturing gradient gel electrophoresis analysis of polymerase chain reaction-amplified genes coding for 16S rRNA. Appl. Environ. Microbiol. *59*, 695–700.

Nagy, M.L., Pérez, A., and Garcia-Pichel, F. (2005). The prokaryotic diversity of biological soil crusts in the Sonoran Desert (Organ Pipe Cactus National Monument, AZ). FEMS Microbiol. Ecol. *54*, 233–245.

Neilan, B., Jacobs, D., and Goodman, A. (1995). Genetic diversity and phylogeny of toxic cyanobacteria determined by DNA polymorphisms within the phycocyanin locus. Appl. Environ. Microbiol. *61*, 3875–3883.

Nelissen, B., van de Peer, Y., Wilmotte, A., and de Wachter, R. (1995). An early origin of plastids within the cyanobacterial divergence is suggested by evolutionary trees based on complete 16S rRNA sequences. Mol. Biol. Evol. *12*, 1166–1173.

Norris, T.B., McDermott, T.R., and Castenholz, R.W. (2002). The long-term effects of UV exclusion on the microbial composition and photosynthetic competence of bacteria in hot-spring microbial mats. FEMS Microbiol. Ecol. *39*, 193–209.

Nübel, U., Garcia-Pichel, F., Clavero, E., and Muyzer, G. (2000a). Matching molecular diversity and ecophysiology of benthic cyanobacteria and diatoms in communities along a salinity gradient. Environ. Microbiol. *2*, 217–226.

Nübel, U., Garcia-Pichel, F., Kühl, M., and Muyzer, G. (1999). Quantifying microbial diversity: morphotypes, 16S rRNA genes, and carotenoids of oxygenic phototrophs in microbial mats. Appl. Environ. Microbiol. *65*, 422–430.

Nübel, U., Garcia-Pichel, F., and Muyzer, G. (1997). PCR primers to amplify 16S rRNA genes from cyanobacteria. Appl. Environ. Microbiol. *63*, 3327–3332

Nübel, U., Garcia-Pichel, F., and Muyzer, G. (2000b). The halotolerance and phylogeny of cyanobacteria with tightly coiled trichomes (*Spirulina* Turpin) and the description of *Halospirulina tapeticola gen. nov. et sp. nov.* Int. J. System. Evol. Microbiol. *50*, 1265–1277.

Oberholster, P.J., Botha, A.-M., Muller, K., and Cloete, T.E. (2005). Assessment of the genetic diversity of geographically unrelated Microcystis aeruginosa strains using amplified fragment legth polymorphisms (AFLPs). Afric. J. Biotech. *4*, 389–399.

Olsen, G.J., Lane, D.J., Giovannoni, S.J., Pace, N.R., and Stahl, D.A. (1986). Microbial Ecology and Evolution: a ribosomal RNA approach. Ann. Rev. Microbiol. *40*, 337–365.

Omoregie, E.O., Crumbliss, L.L., Bebout, B.M., and Zehr, J.P. (2004). Comparison of diazotroph community structure in *Lyngbya* sp. and *Microcoleus chthonoplastes* dominated microbial mats from Guerrero Negro, Baja, Mexico. FEMS Microbiol. Ecol. *47*, 305–318.

Orita, M., Iwahana, H., Kanazawa, H., Hayashi, K., and Sekiya, T. (1989). Detection of polymorphisms of human DNA by gel electrophoresis as single-strand conformation polymorphisms. Proc. Natl. Acad. Sci. USA *86*, 2766–2770.

Osborn, M., and Smith, C., eds. (2005). Molecular Microbial Ecology (New York, Taylor and Francis).

Øvreås, L., Forney, L., Daae, F., and Torsvik, V. (1997). Distribution of bacterioplankton in meromictic Lake Saelenvannet, as determined by denaturing gradient gel electrophoresis of PCR-amplified gene fragments coding for 16S rRNA. Appl. Environ. Microbiol. *63*, 3367–3373.

Pace, N.R. (1997). A molecular view of microbial diversity and the biosphere. Science *276*, 734–740.

Pace, N.R., Stahl, D.A., Lane, D.J., and Olsen, G.J. (1986). The analyses of microbial communities by ribosomal RNA sequences. Adv. Microbial Ecol. *9*, 1–55.

Palenik, B. (1994). Cyanobacterial community structure as seen from RNA polymerase gene sequence analysis. Appl. Environ. Microbiol. *60*, 3212–3219.

Papineau, D., Walker, J.J., Mojzsis, S.J., and Pace, N.R. (2005). Composition and structure of microbial communities from stromatolites of Hamelin Pool in Shark Bay, Western Australia. Appl. Environ. Microbiol. *71*, 4822–4832.

Papke, R.T., Ramsing, N.B., Bateson, M.M., and Ward, D.M. (2003). Geographical isolation in hot spring cyanobacteria. Environ. Microbiol. 5, 650–659.

Paulsrud, P., and Lindblad, P. (1998). Sequence variation of the tRNALeu intron as a marker for genetic diversity and specificity of symbiotic cyanobacteria in some lichens. Appl. Environ. Microbiol. *64*, 310–315.

Piceno, Y.M., Noble, P.A., and Lovel, R.C. (1999). Spatial and temporal assesment of diazotroph assemblage composition in vegetated salt marsh sediments using denaturing gradient gel electrophoresis analysis. Microb. Ecol. *38*, 157–167.

Pichard, S.L., and Paul, J.H. (1991). Detection of gene expression in genetically engineered microorganisms and natural phytoplankton populations in the marine environment by mRNA analysis. Appl. Environ. Microbiol. *57*, 1721–1727.

Pichard, S.L., and Paul, J.H. (1993). Gene expression per gene dose, a specific measure of gene expression in aquatic microorganisms. Appl. Environ. Microbiol. *59*, 451–457.

Postius, C., Ernst, A., Kenter, U., and Böger, P. (1996). Persistence and genetic diversity among strains of phycoerythrin rich cyanobacteria from the picoplankton of Lake Constance. J. Plankton Res. *18*, 1159–1166.

Rappe, M.S., Connon, S.A., Vergin, K.L., and Giovanonni, S.J. (2002). Cultivation of the ubiquitous SAR11 marine bacterioplankton clade. Nature *418*, 630–631.

Rappe, M.S., and Giovannoni, S.J. (2003). The uncultured microbial majority. Ann. Rev. Microbiol. *57*, 369–394.

Reddy, S., and Garcia-Pichel, F. (2006). The community and phylogenetic diversity of biological soil crusts in the Colorado Plateau studied by molecular fingerprinting and intensive cultivation. Microb. Ecol. *in press*.

Riesenfeld, C.S., Schloss, P.D., and Handelsman, J. (2004). Metagenomics: genomic analysis of microbial communities. Ann. Rev. Genet. *38*.

Rothrock, M.J., and Garcia Pichel, F. (2005). Microbial diversity of benthic mat along a tidal desiccation gradient. Environ. Microbiol. *4*, 593–601.

Rudi, K., Skulberg, O.M., Skulberg, R., and Jakobsen, K.S. (2000). Application of sequence-specific labeled 16S rRNA gene oligonucleotide probes for genetic profiling of cyanobacterial abundance and diversity by array hybridization. Appl. Environ. Microbiol. *66*, 4004–4011.

Saiki, R.K., Gelfand, D.H., Stoffel, S., Scharf, S.J., Higuchi, R., Horn, G.T., Mullis, K.B., and Erlich, H.A. (1988). Primer-directed enzymatic amplification of DNA with a thermostable DNA polymerase. Science *239*, 487–491.

Schatz, D., Eshkol, R., Kaplan, A., Hadas, O., and Sukenik, A. (2000). Molecular monitoring of toxic cyanobacteria. Adv. Limnol. *55*, 45–54.

Schonhuber, W., Zarda, B., Eix, S., Rippka, R., Herdman, M., Ludwig, W., and Amann, R. (1999). *In situ* identification of cyanobacteria with horseradish peroxidase-Labeled, rRNA-targeted oligonucleotide probes. Appl. Environ. Microbiol. *65*, 1259–1267.

Sigler, W.V., Bachoven, R., and Zeyer, J. (2003). Molecular characterization of endolithic cyanobacteria inhabiting exposed dolomite in central Switzerland. Environ. Microbiol. *5*, 618–627.

Song, T., Mårtesson, L., Eriksson, T., Zheng, W., and Rasmussen, U. (2005). Biodiversity and seasonal variation of the cyanobacterial assemblage in a rice paddy field in Fujian, China. FEMS Microbiol. Ecol. *54*, 131–140.

Sorensen, K.B., Canfield, D.E., Teske, A.P., and Oren, A. (2005). Community composition of a hypersaline endoevaporitic microbial mat. Appl. Environ. Microbiol. *71*, 7352–7365.

Stahl, D.A., Lane, J.D., Olsen, J.G., and Pace, N.R. (1984). Analysis of hydrothermal vent-associated symbionts by ribosomalRNA sequences. Science *224*, 409–11.

Steglich, C., Post, A.F., and Hess, W.R. (2003). Analyses of natural populations of Prochlorococcus spp. in the Northern Read Sea using phycoerythrin gene sequences. Environ. Microbiol. *5*, 681–690.☒

Steindler, L., Huchon, D., Avni, A., and Ilan, M. (2005). 16S rRNA phylogeny of associated cyanobacteria. Appl. Environ. Microbiol. *71*, 4127–4131.

Steppe, T.F., and Paerl, H.W. (2005). Nitrogenase activity and nifH expression in a marine intertidal microbial mat. Microb. Ecol. *49*, 315–324.

Steunou, A.-S., Bhaya, D., Bateson, M.M., Melendrez, M.C., Ward, D.M., Brecht, E., Peters, J.W., Kühl, M., and Grossman, A.R. (2006). In situ analysis of nitrogen fixation and metabolic switching in unicellular thermophilic cyanobacteria inhabiting hot spring microbial mats. Proc. Natl. Acad. Sci. USA *103*, 2398–2403.

Taton, A., Grubisic, S., Brambilla, E., De Wit, R., and Wilmotte, A. (2003). Cyanobacterial diversity in natural and artificial microbial mats of Lake Fryxell (McMurdo Dry Valleys, Antarctica): a morphological and molecular approach. Appl. Environ. Microbiol. *69*, 5157–5169.

Tillett, D., Parker, D., and Neilan, B. (2001). Detection of toxigenicity by a probe for the microcystin synthetase A gene (*mcyA*) of the cyanobacterial genus *Microcystis*: comparison of toxicities with 16SrRNA and phycocyanin operon (phycocyanin intergeneric spacer) phylogenies. Appl. Environ. Microbiol. *67*, 2810–2818

Toledo, G., and Palenik, B. (1997). *Synechococcus* diversity in the California current as seen by RNA polymerase gene sequences of isolated strains. Appl. Environ. Microbiol. *63*, 4298–4303.

Turner, S., Pryer, K.M., Miao, V.P., and Palmer, J.D. (1999). Investigating deep phylogenetic relationships among cyanobacteria and plastids by small subunit rRNA sequence analysis. J. Eukaryot. Microbiol. *46*, 327–338.

van Hannen, E.J., Zwart, G., van Agterveld, M.P., Gons, H.J., Ebert, J., and Laanbroek, H.J. (1999). Changes in bacterial and eukaryotic community structure after mass lysis of filamentous cyanobacteria associated with viruses. Appl. Environ. Microbiol. *65*, 795–801.

Venter, J.C., Remington, K., Heidelberg, J.F., Halpern, A.L., Rusch, D., Eisen, J.A., Wu, D., Paulsen, I., Nelson, K.E., Nelson, W., Fouts, D.E., Levy, S., Knap, A.H., Lomas, M.W., Nealson, K., White, O., Peterson, J., Hoffman, J., Parsons, R., Baden-Tillson, H., Pfannkoch, C., Rogers, Y.-H., and Smith H.O. (2004). Environmental genome shotgun sequencing of the Sargasso Sea. Science *304*, 66–74.

Wade, B., and Garcia-Pichel, F. (2003). Evaluation of DNA extraction methods for molecular analysis of microbial communities in modern microbialites. Geomicrobiol. J. *20*, 549–561.

Weller, R., Bateson, M.M., Heimbuch, B.K., Kopczynski, E.D., and Ward, D.M. (1992). Uncultivated cyanobacteria, *Chloroflexus*-like inhabitants, and spirochete-like inhabitants of a hot spring microbial mat. Appl. Environ. Microbiol. *58*, 3964–3969.

Weller, R., and Ward, D.M. (1989). Selective recovery of 16S rRNA sequences from natural microbial communities in the form of cDNA. Appl. Environ. Microbiol. *55*, 1818–1822.

Widmer, F., Hartmann, M., Frey, B., and Kolliker, R. (2006). A novel strategy to extract specific phylogenetic sequence information from community T-RFLP. J. Microbiol. Meth., *in press*.

Wieland, A., Kühl, M., McGowan, L., Fourçans, A., Duran, R, Caumette, P., Garcia de Oteyza, T., Grimalt, J.O., Solé, A., Diestra, E., Esteve, I., and Herbert, R.A. (2003). Microbial mats on the Orkney Islands revisited: microenvironment and microbial community composition. Microb. Ecol. *46*, 371–390.

Wilmotte, A., and Herdman, M. (2001). Phylogenetic relationships among the cyanobacteria based on 16S rRNA sequences. In Bergey's Manual of Systematic Bacteriology. Volume 1: The Archaea and the Deeply Branching and Phototropic Bacteria, D.R.B. Second Edition. G. Garrity, and R.W. Castenholz, eds. (New York, Springer-Verlag), pp. 487–493.

Wilson, K.M., Schembri, M.A., Baker, P.D., and Saint, C.P. (2000). Molecular characterization of the toxic cyanobacterium *Cylindrospermopsis raciborskii* and design of a species-specific PCR. Appl. Environ. Microbiol. 66, 332–338.

Worden, A.Z., Chisholm, S.W., and Binder, B.J. (2000). *In situ* hybridization of *Prochlorococcus* and *Synechococcus* (marine cyanobacteria) spp. with rRNA-targeted peptide nucleic acid probes. Appl. Environ. Microbiol. *66*, 284–289.

Xu, H., and Tabita, F. (1996). Ribulose-1, 5-bisphosphate carboxylase/oxygenase gene expression and diversity of Lake Erie planktonic microorganisms. Appl. Environ. Microbiol. *62*, 1913–1921.

Yeager, C.M., Kornosky, J.L., Housman, D.C., Grote, E.E., Belnap, J., and Kuske, C.R. (2004). Diazotrophic community structure and function in two successional stages of biological soil crusts from the Colorado Plateau and Chihuahuan Desert. Appl. Environ. Microbiol. *70*, 973–983.

Zani, S., Mellon, M.T., Collier, J.L., and Zehr, J.P. (2000). Expression of *nifH* genes in natural microbial assemblages in Lake George, New York, detected by reverse transcriptase PCR. Appl. Environ. Microbiol. *66*, 3119–3124.

Zehr, J.P., Jenkins, B.D., Short, S.M., and Steward, G.F. (2003). Nitrogenase gene diversity and microbial community structure: a cross-system comparison. Environ. Microbiol. *5*, 538–554.

Zehr, J.P., Waterbury, J.B., Turner, P.J., Montoya, J.P., Omoregie, E., Steward, G.F., Hansen, A., and Karl, D.M. (2001). Unicellular cyanobacteria fix N_2 in the subtropical North Pacific Ocean. Nature *412*, 635–638.

Zeidner, G., and Béja, O. (2004). The use of DGGE analyses to explore Eastern Mediterranean and Red Sea marine picophytoplankton assembalges. Environ. Microbiol. *6*, 528–534.

Zeidner, G., Preston, C.M., DeLong, E.F., Massana, R., Post, A.F., Scanlan, D.J., and Béjà, O. (2003). Molecular diversity among marine picophytoplankton as revealed by *psbA* analyses. Environ. Microbiol. *5*, 212–216.

Zheng, W., Song, T., Bao, X., Bergman, B., and Rasmussen, U. (2002). High cyanobacterial diversity in coralloid roots of cycads revealed by PCR fingerprinting. FEMS Microbiol. Ecol. *40*, 215–222.

Zwart, G., Kamst-van Agterveld, M.P., van der Werff-Staverman, I., Hagen, F., Hoogveld, H.L., and Gons, H.J. (2005). Molecular characterization of cyanobacterial diversity in a shallow eutrophic lake. Environ. Microbiol. *7*, 365–377.

Comparative Genomics of Marine Cyanobacteria and their Phages

5

Wolfgang R. Hess

Abstract

At present there are 35 cyanobacteria for which a total genome sequence is available. Twenty-five of these cyanobacteria come from the marine environment. These are 11 *Prochlorococcus* strains, nine marine *Synechococcus* strains, *Trichodesmium erythraeum* IMS101, *Cyanothece* sp. CCY0110, *Nodularia spumigena* CCY9414, *Lyngbya aestuarii* CCY9616 and *Crocosphaera watsonii* WH8501. Several studies have demonstrated how these sequences could be used very successfully to infer important ecological and physiological characteristics of marine cyanobacteria. However, there are many more genome projects currently in progress, amongst those there are further *Prochlorococcus* and marine *Synechococcus* isolates, *Acaryochloris* and *Prochloron*, the N_2-fixing filamentous cyanobacteria *Nodularia spumigena*, *Lyngbya aestuarii* and *Lyngbya majuscula*, as well as bacteriophages infecting marine cyanobacteria. Thus, the growing body of genome information can also be tapped in a more general way to address global problems by applying a comparative approach. Some new and exciting examples of progress in this field are the identification of genes for regulatory RNAs, insights into the evolutionary origin of photosynthesis, or estimation of the contribution of horizontal gene transfer to the genomes that have been analyzed.

Introduction

Life evolved in the ocean and oxygenic photosynthesis appeared first in an ancient cyanobacterium or proto-cyanobacterium (Mulkidjanian *et al.*, 2006), an invention that has transformed the character of our atmosphere and eventually created the oxygen we breathe (Falkowski *et al.*, 2005). Because of the tight molecular interactions between pigments, proteins and lipids the machinery for oxygenic photosynthesis evolved only once on this planet and left its imprint in a set of highly conserved genes (Shi *et al.*, 2005) that can be traced back through evolution by analyzing cyanobacterial genome data (Mulkidjanian *et al.*, 2006). However, marine cyanobacteria are of the highest ecological relevance for our world also today—they are fundamental for oceanic primary production and global carbon cycling, introduce atmospheric nitrogen into the sea and hence they are indispensable for proper functioning of the biosphere. Therefore, genome sequences of cyanobacteria are mainly sequenced for general interest in the environment and ecology, to track the evolution, function and regulation of oxygenic photosynthesis, and sometimes to evaluate their potential for biotechnology. The aim of this chapter is to summarize information and to provide an overview resulting from the genome analysis of almost 30 marine cyanobacteria and phages associated with some of them. Insights into cyanobacterial evolution resulting from comparative genome analysis are dealt with in another chapter of this book (see Chapter 2). Marine cyanobacteria that have been studied so far include tiny unicellular organisms of the *Prochlorococcus/Synechococcus* lineage, unicellular nitrogen fixers such as *Crocosphaera*, as well as filamentous marine cyanobacteria containing considerably more complex genomes. The genomes of some of these cyanobacteria were found to be more dynamic, changing by mutation and gene transfer much more rapidly than originally had been expected. A resource for marine cyanobacterial genes that is even more complex might exist outside their own genomes, embedded in bacteriophage genomes of various types. Recent studies have shown that cyanophages are very abundant in the natural environment and collectively contain a plethora of cyanobacterial genes. Each isolated cyanophage that has been characterized so far harbors a number of cyanobacterial genes in its genome, among them genes that are essential for photosynthesis. Therefore, cyanophages are prime suspects as contributing to the high genome diversity and flexibility observed in some marine cyanobacteria.

Current progress in the sequence analysis of marine cyanobacteria

The first genome sequences that became available for cyanobacteria were from four non-marine strains. Among them was, most prominently, *Synechocystis* sp. PCC 6803 (Kaneko *et al.*, 1996; Kaneko *et al.*, 1996; Kaneko and Tabata, 1997; Kaneko *et al.*, 1995), followed by the analysis of the nitrogen-fixing filamentous *Anabaena (Nostoc)* sp. PCC 7120 (Kaneko *et al.*, 2001) and *Nostoc punctiforme* (Meeks *et al.*, 2001), the thermophilic cyanobacterium *Thermosynechococcus elongatus* BP-1 (Nakamura *et al.*, 2002; Nakamura *et al.*, 2002) and *Gloeobacter violaceus* PCC 7421, the only known cyanobacterium that lacks thylakoids (Nakamura *et al.*, 2003; Nakamura *et al.*, 2003). The pioneering work in this field, genome analysis of *Synechocystis* sp. PCC 6803, laid the foundation for a new generation of experimental work in cyanobacteria in which biochemical, genetic and genome-wide functional studies rely on such genomic information (see Murata and Suzuki, 2006, for a recent review). *Synechocystis* sp. PCC 6803 is a small coccoid and unicellular cyanobacterium, however, it is not closely related to the marine unicellular cyanobacteria such as *Prochlorococcus*. This fact becomes easily visualized if their total genome sequences are compared based on DNA similarity (Figure 5.1). Except strongly conserved ribosomal RNA genes, some tRNA genes and short stretches of DNA in highly conserved photosynthesis genes such as *psbA* or *psaA-psaB*, there is almost no sequence similarity at the DNA level, nor a high level of synteny. As a consequence, well-established model organisms such as *Synechocystis* sp. PCC 6803, *Synechococcus* sp. PCC 7942 or *Anabaena* sp. PCC 7120 are frequently only of limited value for understanding the biology of marine cyanobacteria.

Based on Pulsed Field Gel Electrophoresis, a report in 1999 estimated a genome size of only ~1.8 Mb for the marine cyanobacterium *Prochloroccoccus marinus* SS120 (Strehl *et al.*, 1999). This finding helped to launch a series of sequence analyses in this genus. These analyses were initially designed to gain insight into several problems: how would such an extraordinary small genome be compatible with the fact that *Prochlorococcus* is

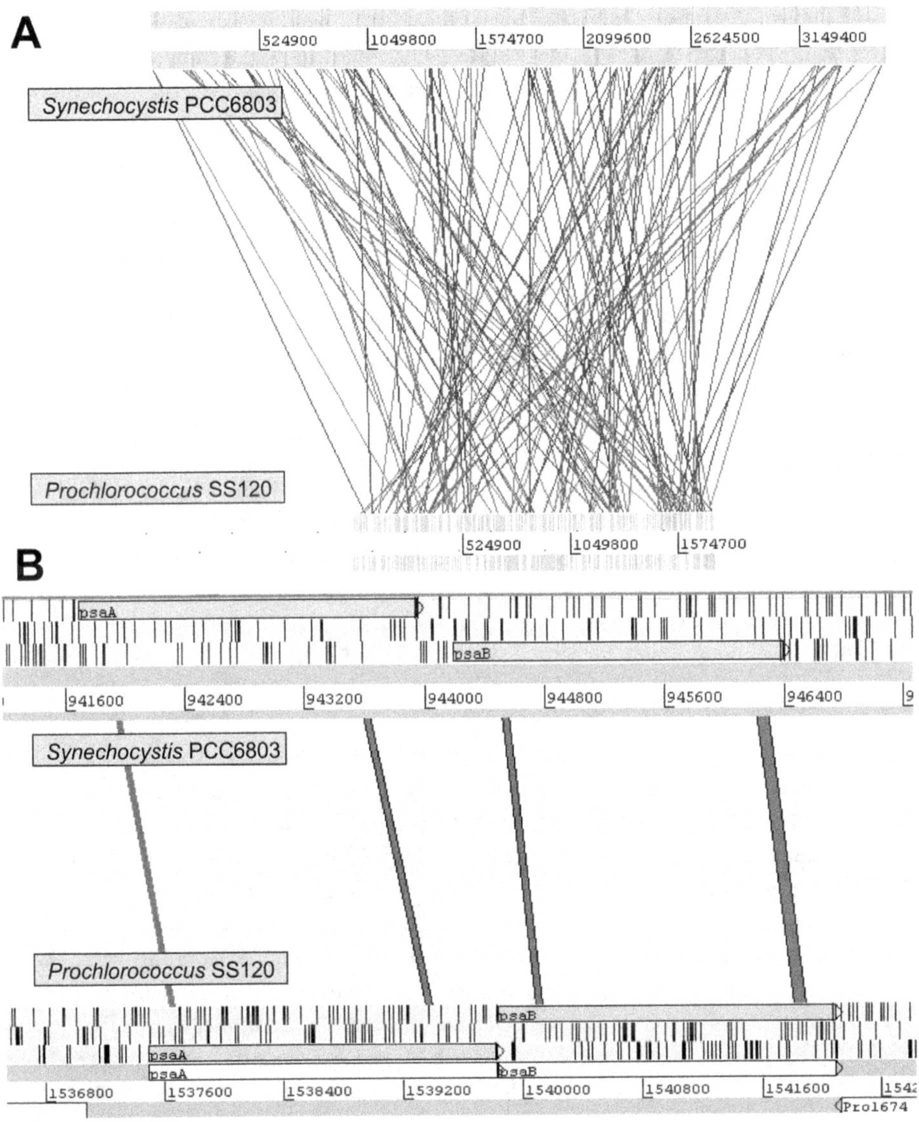

Figure 5.1 (see also Plate 5.1) A. Alignment of the *Synechocystis* sp. PCC 6803 genomic sequence (top) and *Prochlorococcus marinus* SS120 sequence (bottom). The lines connect sequences with a BLASTN E value < 10^{-2}, however, except very few stretches in ultra-conserved genes and some tRNA genes, there is no sequence similarity at the DNA level. There is clearly no synteny in the organization of the two genomes. B. Detail from A, photosynthesis genes such as *psaA-psaB* contain short stretches of conserved nucleotide sequence similarity.

the numerically dominant phytoplankter in tropical and subtropical oceans (Olson *et al.*, 1990; Partensky *et al.*, 1999; Vaulot *et al.*, 1995)? Which genomic information is underpinning a lifestyle of a genus that can account for 21 to 43% of the photosynthetic biomass in oligotrophic oceans, 13 to 48% of the net primary production (Campbell *et al.*, 1997; Campbell *et al.*, 1997; DuRand *et al.*, 2001; Vaulot *et al.*, 1995), and photosynthesizes successfully from the water surface to as deep as 200 m, sometimes reaching abundances greater than 10^5 cells per ml (Partensky *et al.*, 1999)? In other words, how would a limited number of genes sustain the physiology and ecology of an ecologically very successful free-living cyanobacterium?

Confirmation, and another surprise, came with the first analyzed genome sequence from this group, that of *Prochloroccoccus* MED4, finding an even smaller genome, 1.66 Mb, and bearing not more than 1700 genes (Hess *et al.*, 2001). The subsequent detailed comparative analysis of this and three other completed genome sequences from the closely related genera *Prochloroccoccus* and *Synechococcus* revealed that these genomes were not only much smaller but also much more diverse than initially expected, and further indicated that the gene pool of a particular marine cyanobacterium is intimately linked to the respective ecological niche this organism is living in (Dufresne *et al.*, 2003; Palenik *et al.*, 2003; Rocap *et al.*, 2003). The marine ecosystem is not as homogenous as it may seem for the inexperienced observer; thus the genome of a *Prochlorococcus* strain such as MED4, originally isolated from close to the water surface is adapted to thrive under high photon fluxes, whereas that of a strain such as *Prochlorococcus* SS120 which originated from 120 m depth in the Sargasso Sea, is extremely streamlined to live at low light intensities. There are many individual genes which are distinct for the respective environment and life style (Garczarek *et al.*, 2000; Garczarek *et al.*, 2001; Steglich *et al.*, 2005; Steglich *et al.*, 2003) and hence serve as examples for this statement. As a result, these genomes were found to differ by the striking number of hundreds of genes (Hess, 2004). Since these initial analyses, a wealth of new data has provided an unprecedented degree of insight into the genomic underpinnings of diverse marine cyanobacteria's lifestyles. An overview on the current status of genome projects and their distribution alongside the phylogenetic tree is given in Figure 5.2. Right now (April 2007) there are 35 cyanobacteria for which a total genome sequence is available from GenBank (http://www.ncbi.nlm.nih.gov/genomes/MICROBES/microbial_taxtree.html), the Joint Genome Institute (http://www.jgi.doe.gov/) or any of the other public databases. From these, 25 come from marine cyanobacteria and the majority (20) is from the *Prochlorococcus/Synechococcus* group with 11 for *Prochlorococcus* and nine for marine *Synechococcus*. Genome sequences have been analyzed for other, non-marine, *Synechococcus*, such as the well established model strains PCC 6301 and PCC 7942, or the *Synechococcus* strains OS-A and OS-B′, which were isolated from Octopus spring in Yellowstone Park (Allewalt *et al.*, 2006; Steunou *et al.*, 2006), but these are only very distantly related to the marine strains.

Publications which present and discuss the genomic information do not exist yet for all 35 cyanobacteria for which a total genome sequence is currently available from GenBank (see Table 5.1 for details). Clearly, a bottleneck is developing between the technical sequencing capacity and automated annotation on one side, and the interpretation of data in their biological context on the other.

The cyanobacteria for which genome sequences are currently available as shown in Figure 5.2 are only the tip of the iceberg. With the further decrease in sequencing cost, much more genome data will become available in the public databases soon, and only a subfraction of these will be described in individual, specialized publications in the near future. In Table 5.1 an overview is given on all 40 cyanobacterial genome projects, which have been completed or are confirmed to be in a progressed status. The most comprehensive

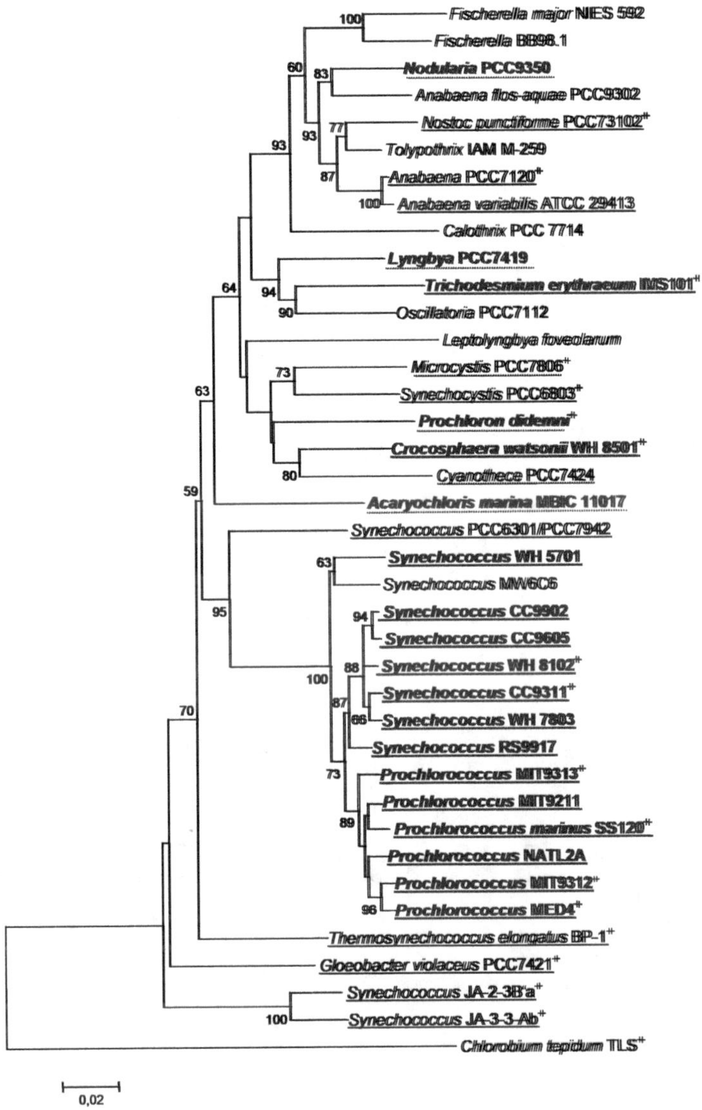

Figure 5.2 Phylogenetic tree of all cyanobacterial strains for which genome sequences are currently available from public databases (underlined), for which at least one publication is available using the genomic information in a substantial way (labeled in superscript^{+}) and/or whose sequences are in progress (underlined by dotted lines). A few additional strains were included for representation of certain phylogenetic groups. The tree is based on a multiple sequence alignment of 16S rRNA sequences analyzed in MEGA3.1 (Kumar *et al.*, 2004). Statistical support values were generated by bootstrapping and are only shown if > 55. Marine strains are in boldface letters.

overview, the Genomes Online database (Liolios *et al.*, 2006), however, lists 65 different cyanobacterial genome projects that are currently ongoing (http://www.genomesonline.org/gold.cgi). For instance, there are 10 additional projects on *Prochlorococcus*, two on marine and five on non-marine *Synechococcus* strains alone. This might give the impression that

Table 5.1 Overview on completed or confirmed cyanobacterial genome projects. A status of inc. means incompleted, cmpl., project has been finished. Remarks: M, marine strain; N, nitrogen fixer; S, symbiontic life style; T, thermophilic; Tx, these cyanobacteria can be toxic to humans; clade number refers to marine *Synechococcus* clades according to Fuller *et al*. (2003). Size gives the number of nucleotides in kb in case of completed projects or is an estimate for the unfinished projects. Under References, publications are given in which the genomic information is presented, interpreted or was used in another substantial way or the sequencing institute is indicated (JCVI, Craig Venter Institute; SBR, Station Biologique de Roscoff; PSU, Pennsylvania State University; TIGR, The Institute for Genome Research).

Organism	Status	Remarks	GC (%)	Size (kb)	GenBank #	References or sequencing institute
Acaryochloris marina MBIC11017	*inc.*	*M, S*		~8000		http://genomes.tgen.org/Acaryochloris.html
Anabaena variabilis ATCC 29413	*cmpl.*	*N*	41.1	6365	CP000117	
Crocosphaera watsonii WH8501	*inc.*	*M, N*	37.1	~6240	AADV00000000	(Dyhrman and Haley, 2006)
Cyanothece CCY0110	*inc.*	*M, N*	37	~5820	AAXW00000000	JCVI
Gloeobacter violaceus PCC 7421	*cmpl.*		62	4659	BA000045	(Nakamura *et al*., 2003; Nakamura *et al*., 2003)
Lyngbya aestuarii CCY9616	*inc.*	*M, N*	41	~7040	AAVU00000000	JCVI
Microcystis aeruginosa PCC 7806	*inc.*	*Tx*		~4800		(Carre-Mlouka *et al*., 2006)
Nodularia spumigena CCY9414	*inc.*	*N,Tx*	42	~5310	AAVW00000000	JCVI
Nostoc (Anabaena) PCC 7120	*cmpl.*	*N*	41.3	6413	NC_003272	(Kaneko *et al*., 2001)
Nostoc punctiforme ATCC 29133(=PCC 73102)	*cmpl.*	*N,S*	41	9020	AAAY02000000	(Meeks *et al*., 2001)
Prochlorococcus MIT9312	*cmpl.*	*M*	31	1709	NC_007577	
Prochlorococcus NATL2A	*cmpl.*	*M*	35	1840	NC_007335	
Prochlorococcus MIT9313	*cmpl.*	*M*	50.7	2410	NC_005071	(Rocap *et al*., 2003)
Prochlorococcus marinus SS120 (=CCMP1375)	*cmpl.*	*M*	36.4	1751	NC_005042	(Dufresne *et al*., 2003)

Prochlorococcus MIT9301	*cmpl.*	*H*	31	1642	NC_009091	JCVI
Prochlorococcus MIT9211	*inc.*	*M*	39	1840	AALP00000000	
Prochlorococcus MIT9303	*cmpl.*	*M*	50	2683	NC_008820	JCVI
Prochlorococcus AS9601	*cmpl.*	*M*	31	1670	NC_008816	JCVI
Prochlorococcus NATL1A	*cmpl.*	*M*	34	1864	NC_008819	JCVI
Prochlorococcus MIT9515	*cmpl.*	*M*	30	1704	NC_008817	JCVI
Prochlorococcus MED4 (=CCMP1986)	*cmpl.*	*M*	30.1	1657	NC_005072	(Rocap *et al.*, 2003)
Prochloron didemni	*inc.*	*M, S*		6000		TIGR
Synechococcus PCC 7942	*cmpl.*		55.4	2696	NC_007604	
Synechococcus PCC 6301	*cmpl.*		55.4	2696	NC_006576	(Sugita *et al.*, 2007)
Synechococcus OS-B′ (JA-2–3B′a(2–13))	*cmpl.*	*T*	58	3046	NC_007776	(Steunou *et al.*, 2006)
Synechococcus OS-A (JA-3–3Ab)	*cmpl.*	*T*	58	2932	NC_007775	(Steunou *et al.*, 2006)
Synechococcus CC9605	*cmpl.*	*M*, clade II	59	2510	NC_007516	
Synechococcus CC9902	*cmpl.*	*M*, clade IV	54	2235	NC_007513	
Synechococcus WH8102	*cmpl.*	*M*, clade III	59.4	2434	NC_005070	(Palenik *et al.*, 2003)
Synechococcus WH7805	*inc.*	*M*, clade VI	57	2620	AAOK00000000	JCVI
Synechococcus WH5701	*inc.*	*M*	65	3043	AANO00000000	JCVI
Synechococcus RS9917 (= RCC556)	*inc.*	*M*, clade VIII	64	2579	AANP00000000	JCVI
Synechococcus RS9916 (= RCC555)	*inc.*	*M*, clade IX	59	2661	AAVA00000000	JCVI

Table 5.1 continued

Organism	Status	Remarks	GC (%)	Size (kb)	GenBank #	References or sequencing institute
Synechococcus BL107	*inc.*	*M*, clade IV	54	2283	AATZ00000000	JCVI
Synechococcus CC9311	*cmpl.*	*M*, clade I	52	2606	NC_008319	(Palenik *et al*., 2006)
Synechococcus RCC307	*inc.*	*M*, calde X	60.2	2370		SBR Genoscope
Synechococcus WH7803	*cmpl.*	*M*, clade V	60.2	2370		SBR Genoscope
Synechococcus PCC 7002	*inc.*			3200		PSU
Synechocystis. PCC 6803	*cmpl.*		47	3573	NC_000911	(Kaneko *et al*., 1996; Kaneko *et al*., 1996; Kaneko and Tabata, 1997; Kaneko *et al*., 1995)
Thermosynechococcus elongatus BP-1	*cmpl.*	*T*	53.9	2593	NC_004113	(Nakamura *et al*., 2002; Nakamura *et al*., 2002)
Trichodesmium erythraeum IMS101	*cmpl.*	*N, M*	34.3	7750	NC_008312	(Dyhrman *et al*., 2006)

simple unicellular cyanobacteria receive a lot of attention. However, another trend is also obvious namely, that genomic sequencing is increasingly spreading towards marine cyanobacteria with a more complex genome organization. The data in Table 5.1 include projects such as for *Trichodesmium erythraeum* IMS101 and *Crocosphaera watsonii* WH8501 with a genome size of 7.75 and 6.24 Mb, respectively. Finally, almost finished projects should be mentioned such as those for the chlorophyll *b*-containing obligate symbiont *Prochloron* sp. (~6 Mb), the chlorophyll *d*-containing *Acaryochloris marina* MBIC 11017 (~8 Mb), and the N_2-fixing filamentous cyanobacteria *Nodularia spumigena* CCY9414 (PCC 9350) and *Lyngbya aestuarii* CCY9616 (PCC 7419) with an estimated genome size of 5.3 and 7.0 Mb, respectively.

In the following, I will briefly summarize some of the scientific highlights and progress that have been found in these diverse marine cyanobacteria, based on their respective genome data.

Prochlorococcus

The initial three genome projects in *Prochlorococcus* MED4, SS120 and MIT9313 demonstrated their small and streamlined genomes, the fact that their gene complement is tuned in accordance with the respective light conditions, the single most important environmental parameter (Dufresne *et al.*, 2003; Palenik *et al.*, 2003; Rocap *et al.*, 2003), and, more generally, that these genomes each differ by hundreds of genes from each other (Hess, 2004). However, if plotted along the 16S rRNA tree, gradients become obvious in three important genome characteristics: in GC content, size, and total gene number. All three decline from MIT9313, the most basally branching strain within the *Prochlorococcus* radiation toward MED4, the most distally evolved strain in this group (Figure 5.2), and this decline can also be seen in the new genomes of strains MIT9211, MIT9312, and NATL2A (Table 5.1 and Figure 5.2). Dufresne *et al.* (2005) were the first who analyzed the underlying evolutionary process and who tried to link the ecologic success of this genus to its genome properties. Their major finding was that the observed genome shrinkage was accompanied by an increased rate of protein evolution in SS120 and even more so in MED4, whereas 16S rRNA sequences seem to evolve clock-like (Dufresne *et al.*, 2005). The separation of the two genera *Prochlorococcus* and *Synechococcus* was estimated to have occurred ~150 million years ago, and the split between high-light- and low-light-adapted *Prochlorococcus* ecotypes (represented by eMED4 and eSS120) around 80 million years ago, assuming an 16S rRNA divergence of 1% per 50 million years. During the intense genome streamlining, several genes were lost which are required for repair of DNA point mutations (Rocap *et al.*, 2003; Dufresne *et al.*, 2005). Although it cannot be excluded that some of these genes were substituted by non-ortholog genes, it appeared likely that these deletions damaged the ability to repair G:C to A:T transitions and thus contributed to the sharp reduction in G+C content. *Prochlorococcus* outcompetes other phytoplankton under highly oligotrophic conditions. Thus, the major driving force for both genome reduction as well as the bias toward an A+T-rich genome would be selection for minimized nutrient requirements within the *Prochlorococcus* radiation: a streamlined genome obviously requires less nitrogen and phosphorus, and an AT base pair contains seven atoms of nitrogen compared to eight in case of a GC; both elements are frequently limiting phytoplankton growth.

Mosaic structure of cyanobacterial genomes

Another important process contributing to *Prochlorococcus* genome evolution was discovered by Coleman *et al.* (2006), who were the first to compare the genomes of two very closely related strains. Both these strains (MED4 and MIT9312) belong to the high-light-adapted ecotype, differ by only 0.8% in their 16S rRNA sequences and their genomes are very similar in GC content and genome size. Yet, the number of strain-specific genes is 236 in MIT9312 and 139 in MED4. Interestingly, the majority of these genes (80 and 74%, respectively) were found to be located in only five major islands, thus, these cyanobacterial genomes resemble the mosaic structure of enterobacterial genomes where they are better known also as pathogenicity islands. In pathogenic bacteria, such islands greatly contribute to the capability for infection and, as the name suggests, pathogenicity. What genes do these islands contain in *Prochlorococcus*? Mainly genes that are differentially expressed and thought to be important for nutrient assimilation, including a cyanate transporter and lyase in MED4 and two transporters, for manganese/iron and amino acids, in MIT9312, whereas other genes might protect from photoinhibition or help to defend against phage infection (Coleman *et al.*, 2006). Why are such potentially useful genes clustered in genomic islands? Again, by analogy to the pathogenicity islands in *E. coli*, Coleman *et al.* suggested that these are hotspots for recombination, arose via lateral gene transfer and continually undergo rearrangement (Coleman *et al.*, 2006). In agreement with this idea, the islands are located in the same position in both genomes, are associated with repeated sequence elements, tRNA genes or pseudogenes, all known as common integration sites for mobile elements in other prokaryotes.

Ecological representation

By means of the *Prochlorococcus* MIT9312 genome sequence still another problem was studied. The huge population of cyanobacteria such as *Prochlorococcus*, estimated to be in the order of 10^{27} cells (Whitman *et al.*, 1998), together with the known genome diversity poses the problem to what extent natural biodiversity can be represented by cultures in the laboratory. A common concern is that cyanobacteria purified in the laboratory may not be typical for the site of sampling but rather represent a minor strain selected for by the respective conditions in the growth chamber (see Chapter 4). The most straightforward way to answer this concern would be to sequence the genome of a single cell directly isolated from the environment. In a process they called "ploning" (for polymerase cloning), Zhang *et al.* (2006) tested the possibility of whole-genome shotgun sequencing of a single cell from *Prochlorococcus* MIT9312. Albeit some technical drawbacks caused by the formation of chimeric DNA molecules, a number of 7484 sequencing reads resulted in an assembly of 477 contigs, containing 62.2% of the MIT9312 genome at least once, which was comparable to the sequence information gathered from the same amount of sequencing in the previous standard shot-gun sequencing of this genome. Subsequent successful ploning of two single wild *Prochlorococcus* cells from a Pacific Ocean sample collected at a depth of 85 m at the Station ALOHA showed their close genetic relatedness with MIT9312, thus confirming that *Prochlorococcus* MIT9312 is typical for an abundant part of the *Prochlorococcus* population in the environment (Zhang *et al.*, 2006). Independently, Coleman *et al.* (2006) came to the same statement by examining 36 large genome fragments with a total length of 1.1 Mb from the Hawaii Ocean Time-Series Station. MIT9312 was found to be the dominant ecotype also in the Atlantic. Based on samples obtained during a meridional transect in the Atlantic Ocean, its abundance was found to exceed the abundance of other ecotypes by two orders of magnitude between 15-N and 15-S (Johnson *et al.*, 2006). Among low light adapted *Prochlorococcus*, the NATL2A strain was found to represent the most abundant ecotype in the environment (Johnson *et al.*, 2006). Since a total genome sequence is available also for the NATL2A strain, it should be possible to infer ecologically meaningful genes also for this ecotype. In the ocean, temperature was suggested as the second most significant factor next to light that correlated with shifts in ecotype abundance (Johnson *et al.*, 2006), thus genes which play a role with regard to light, temperature, nutrients and the protection from viral infection should be expected.

Functional studies

Another aspect that comes with a genome sequence is that it may stimulate systematic biochemical or genetic studies which had not been possible before. Such work is very important also from the point of understanding genome content in its very own sense as, on average, the function of ~40% of all genes in *Prochlorococcus* cannot be inferred from sequence information alone, and similar numbers apply to the other marine cyanobacteria. A substantial step towards understanding the biology of *Prochlorococcus* in all its facets, therefore, has been the introduction of whole genome microarray-based analyses. The analysis of genes induced by P starvation revealed two large gene clusters on the chromosome of *Prochlorococcus* MED4. One of these clusters contains 21 genes, among them the P master regulator *phoB* and most known P-acquisition genes, whereas all except one of the 14 genes in the second cluster are of unknown function and are absent from the genome of *Prochlorococcus* MIT9313 (Martiny *et al.*, 2006). PMM1404, the single known gene in this second cluster, is *hli5* coding for an only 52 amino acids long high light inducible protein (Bhaya *et al.*, 2002). Intriguingly, *hli5* was one of the most strongly induced genes also under high-light stress (Steglich *et al.*, 2006). The genes in this second cluster are located between PMM1403 and PMM1416, which is part of a variable genomic island, discovered by Coleman *et al.* (2006). MED4 originated from surface waters in the northwest Mediterranean Sea, where the P concentration is typically very low. Thus, the presence of previously unknown P-starvation-induced genes in MED4 but not in MIT9313 or several other strains (Martiny *et al.*, 2006) supports the importance of these genes in responding to P limitation. The function of these genes remains to be shown, either if they are involved in a specific response to P starvation or provide an enhanced fitness under other stresses as well.

Three *hli* genes, but not *hli5*, were found to be among the most strongly and rapidly upregulated genes during N stress, possibly to protect the photosystems (Tolonen *et al.*, 2006a). In that study, microarrays were used for an interstrain comparison of how *Prochlorococcus* MED4 responded to changes in ambient nitrogen compared with *Prochlorococcus* MIT9313. Based on the responses of known N-metabolism genes, the results not only suggested that the two strains integrate N and C metabolism in fundamentally different ways, but also identified a number of novel N-regulated genes. Among the most highly upregulated genes during N starvation were several genes of unknown function, some with well-conserved NtcA consensus binding sequences (for instance PMM0958 and PMM0374), suggesting they play a specific role in the N stress response

(Tolonen *et al.*, 2006a). Furthermore, several of these genes are located close to genes with known roles in N metabolism, such as the cyanate transporter in MED4, the nitrite transporter genes or *glnB* in MIT9313, strengthening further their putative function in N-stress response.

Finally, a third paper using microarrays focused on the potential for light-sensing in *Prochlorococcus* MED4 and addressed the genome-wide gene regulation under seven qualitatively and quantitatively different light conditions (Steglich *et al.*, 2006). Two group II sigma factors (PMM1629 and PMM1697) were found to respond in antagonistic ways, thus these two regulators should mediate a major part of the genomic response to light. Since one of these factors, PMM1697, was differentially regulated also in the N-stress experiment, it may be involved in the integration of signals between light- and N-dependent signal transduction chains. Interestingly, about 10% of all genes displayed significant differential expression when shifted from darkness to high white light, indicating the high relevance of light conditions for *Prochlorococcus*. Among the most strongly responding genes were photosynthetic genes and 12 of the 22 *hli* genes, whereas only modest to minor changes were detected for additional five *hli* genes and another five were not affected at all (Steglich *et al.*, 2006). Also in this experiment, several strongly responding genes of unknown function were identified. The most highly upregulated gene (61.2-fold upregulated upon dark to high light transition) was PMM1359. This protein possesses a *doxD*-like domain, thus it has some similarity to a subunit of the terminal quinol oxidase. A possible function of this protein as well as of PMM0336, another light responsive gene that has weak similarities to terminal oxidases, could be in alternative electron transfer routes under high light conditions and thus the function of these genes could be linked intimately to the specific adaptation of *Prochlorococcus* MED4 to grow under high light. Interestingly, PMM1148, the most strongly downregulated gene under high light (11.3 fold reduction) is very highly conserved within the *Synechococcus/Prochlorococcus* lineage (> 60% identity; > 80% similarity), but not found in other cyanobacteria nor any other organism (Steglich *et al.*, 2006). Although the function of PMM1148 is unknown, these findings reinforce the view that results of studies in well-established model strains such as *Synechocystis* sp. PCC 6803 or *Synechococcus* sp. PCC 7942 cannot always be applied to understand the biology of the marine cyanobacteria.

These three functional genomics-type papers demonstrate the high value that the use of microarrays has in *Prochlorococcus*. Future biochemical and computational studies to elucidate the precise function of novel genes in *Prochlorococcus* will clearly benefit from these array experiments. Numerous functional genomics-type studies in this and other marine cyanobacteria will follow. Despite all the genomic information now available for *Prochlorococcus* there is still the disadvantage in working with these cyanobacteria that only a single strain has been amenable to genetic manipulation so far and that at relatively inefficient rates (Tolonen *et al.*, 2006b). However, sometimes experimental work in other systems may help. A shining example for such work was the identification of a novel type of carbonic anhydrase in *Prochlorococcus*. This enzyme is essential for photosynthetic carbon fixation, yet it was not possible to identify a gene during the initial genome annotation work based on homology to known carbonic anhydrases (Hess *et al.*, 2001; Rocap *et al.*, 2003). Following a biochemical approach, So *et al.* (2004) identified in *Halothiobacillus neapolitanus* a novel class of carbonic anhydrase, the ε-class. Using a recombinant protein, they subsequently showed that the carbonic anhydrase of *Prochlorococcus* also belongs to the ε-class. The corresponding gene *csoS3* is part of an operon for carboxysomal shell proteins that was most likely transferred to these marine cyanobacteria via lateral gene transfer from chemolithoautotrophic bacteria. The encoded protein, CsoS3 was unequivocally demonstrated to be the missing carbonic anhydrase in *Prochlorococcus* (So *et al.*, 2004).

Synechococcus

Together with *Prochlorococcus*, cyanobacteria of the genus *Synechococcus* form an essential part of marine picophytoplankton. Marine *Synechococcus* are important contributors to marine primary production and are ubiquitous in the world's oceans as well, although in general they are less abundant than *Prochlorococcus*. They frequently outcompete the latter in total cell counts closer to coastal regions and in less oligotrophic waters. Based on 16S rRNA phylogeny, ten discrete lineages or clades have been differentiated among marine *Synechococcus* of "marine cluster A" ("MC-A"), which is the dominant *Synechococcus* group within the euphotic zone of both open ocean and coastal waters (Fuller *et al.*, 2003). Accordingly, the first marine *Synechococcus* genome sequenced and analyzed, that of the motile strain WH8102 (Palenik *et al.*, 2003), represents one phylogenetic clade (clade III), which may correspond to a certain ecotype, similar to that outlined above for *Prochlorococcus*. To obtain a more complete overview on marine *Synechococcus* clades and ecotypes, several more genome sequences are currently being analyzed (Table 5.1). Compared to *Prochlorococcus*, genomes of marine *Synechococcus* on average are several hundred kb larger in size and contain several hundred more genes. Therefore, the functional information inferred from studying *Synechococcus* genomes has not yet progressed as far as it is for *Prochlorococcus*. Existing studies so far have dealt with two of these genomes, those of WH8102 and CC9311 (McCarren and Brahamsha, 2005; Palenik *et al.*, 2003; Palenik *et al.*, 2006; Six *et al.*, 2005; Su *et al.*, 2003; Su *et al.*, 2006).

Already the availability of one genome sequence has enabled systematic studies, which were not possible previously. Six *et al.* (2005) utilized the information from WH8102 for a genomic-biochemical study of its phycobilisomes. The fact that the majority of phycobiliprotein genes are clustered in an extended genomic region of more than 28 kb led to the description of two novel genes for phycoerythrin-associated linker polypeptides. Finally, it was possible to construct a model of the localization of linker proteins in a phycobilisome rod as deduced from the sequence and biochemical data (Six *et al.*, 2005). Using a comparative genomics-based approach, Su *et al.* (2006) developed a computational model predicting nitrogen assimilation regulatory network biological pathways in *Synechococcus* WH 8102. Combining this model with microarray gene expression data collected under ammonium-rich versus nitrate-rich growth conditions, a tight coordination between the expression of genes involved in nitrogen assimilation and in photosynthesis was found (Su *et al.*, 2006).

As indicated in Table 5.1, at least one genome sequence is in progress or finished for almost all of these clades, hence it will be only a matter of time when we will get a glimpse of ecotypic differences mirrored in the genomes of marine *Synechococcus*. The first genome comparison of this type, between a clade I and a clade III strain, WH8102 and CC9311, revealed a common core of 1730 ORFs (Palenik *et al.*, 2006) representing about 65% of the respective proteome. Of the 2526 protein-coding genes annotated in WH8102, almost 800 have no ortholog in CC9311, illustrating the high genetic diversity among marine *Synechococcus*. Interestingly, a number of these differences can be directly linked to the respective environment these strains were originally isolated from, open ocean in case of WH8102 and a coastal region for CC9311 (from the edge of the California Current after nitrate enrichment and low light incubation; Palenik, 2001). With 11 histidine kinase genes and 17 response regulator genes, CC9311 has nearly twice as many two component systems as WH8102. Since the function of these systems is the regulation of adaptation to fluctuations in the environment (see Chapter 6), CC9311 has a greater capacity to sense and respond to such changes, whereas the small number in WH8102 was interpreted as a genetic adaptation to a relatively constant ecosystem (Palenik *et al.*, 2003). The much larger capacity for uptake and assimilation of metals, especially iron and copper in CC9311, points in the same direction, whereas the lack of several phosphate acquisition systems in CC9311 which are present in WH8102 is consistent with the higher concentration of phosphate in coastal environments.

A major advantage of some of these *Synechococcus* strains compared to *Prochlorococcus* is the availability of an efficient genetic system, a very important technical precondition for performing functional genomics-type studies. The first who have made extensive use of it were McCarren and Brahamsha (2005), who screened WH8102 for a non-motile phenotype upon transposon mutagenesis. All 17 independent mutations causing a loss-of-motility phenotype were located in nine open reading frames, which clustered in three separate regions of the genome. Among them were the previously identified genes *swmA* and *swmB* as well as several genes for transport proteins and a number of glycosyltransferases which are likely surface proteins (McCarren and Brahamsha, 2005). Swimming motility in marine *Synechococcus* is distinct from types of motility in other bacteria. Although not yet understood, studies like the one by McCarren and Brahamsha (2005) will finally solve the enigma of swimming motility. Without a total genome sequence and a functioning genetic system this would hardly be possible.

Crocosphaera and *Trichodesmium*

Crocosphaera as well as *Trichodesmium* belong to the cyanobacterial fraction of oceanic phytoplankton just as *Prochlorococcus* and *Synechococcus*. However, different from these, *Crocosphaera* cells are significantly larger (3 to 10 μm), and its genome is about three times as big (Table 5.1). Although less important for overall marine primary production in total numbers, it can become relevant by introducing a substantial fraction of new nitrogen to the euphotic zone in tropical regions where it occurs (Montoya *et al.*, 2004). As a filamentous cyanobacterium, *Trichodesmium* has a totally different morphology than the unicellular cyanobacteria mentioned above. *Trichodesmium* is perhaps the best-known and best-studied marine diazotroph. It introduces the largest fraction of new nitrogen to the euphotic zone and therefore plays a profound role for the marine biogeochemical cycle with its high abundance and dual roles in carbon and nitrogen fixation (Capone *et al.*, 1997). There are no publications yet that deal specifically with the genomes of *Crocosphaera watsonii* WH8501 or *Trichodesmium erythraeum* IMS101, the strains that have been sequenced. However, the draft versions of both genome sequences are available in Genbank and part of this information has been used in recent publications both for *Trichodesmium* (Castruita *et al.*, 2006; Dyhrman *et al.*, 2006; Sudek *et al.*, 2006) and *Crocosphaera* (Dyhrman and Haley,

2006). With a genome size of 7.75 and 6.24 Mb, respectively, *Trichodesmium erythraeum* IMS101 and *Crocosphaera watsonii* WH8501 have fairly complex genomes (Table 5.1) compared to *Prochlorococcus* and *Synechococcus*.

Both genomes were investigated with regard to their capacity for phosphorus assimilation. Phosphorus has been recognized as a key nutrient that constrains nitrogen fixation in the open ocean (Sanudo-Wilhelmy *et al.*, 2001), at least for *Trichodesmium*. Phosphonates are a class of phosphorus-containing molecules which occur in marine systems and may constitute up to 25% of the dissolved organic phosphorus. There are well-known pathways for phosphonate utilization in bacteria such as *Pseudomonas stutzeri* and *Escherichia coli* via a C–P lyase enzymatic activity that mediates the hydrolysis of a broad range of substrates. However, due to the high stability of the C–P bond typical for phosphonates, they had been considered unavailable for marine cyanobacteria. Therefore it came as a surprise when Dyhrman *et al.* (2006) found a cluster of *phn* genes that is predicted to encode proteins for the transport and hydrolysis of phosphonate compounds by a similar C–P lyase pathway in the genome of *T. erythraeum* IMS101. Reinspection of all available genomes of unicellular cyanobacteria using these data showed that the presence of a C–P lyase is indeed unique to *Trichodesmium* among marine cyanobacteria, whereas a *phnCDE* cluster of putative phosphonate transporter genes is also found in the available *Synechococcus* and *Prochlorococcus* genomes (Moore *et al.*, 2005). Screening of all available *Trichodesmium* species in culture as well as field populations for the presence of *phnD* and *phnJ* genes demonstrated that a C–P lyase is a regular component in this cyanobacterium (Dyhrman *et al.*, 2006). Thus, if no other marine cyanobacteria have *phn* genes, where did the *Trichodesmium* genes come from?

Interestingly, both the sequence identity and genomic arrangement of *phn* genes in *T. erythraeum* IMS101 suggest that it was acquired through a horizontal gene transfer event. The closest relatives are the *phn* gene orthologs in *Thiobacillus denitrificans*, a sulfur-oxidizing obligate chemolithoautotrophic β-proteobacterium (Dyhrman *et al.*, 2006). The *phn* gene cluster constitutes an example for a laterally transferred set of genes that provides a competitive advantage with respect to the utilization of a key nutrient, phosphorus, and may be crucial to explain the global abundance of *Trichodesmium* in many low phosphate environments.

In contrast, analyzing the genome information of *Crocosphaera watsonii* WH8501, the only other marine diazotroph with available genome data, did not find homologs of genes for phosphonate transport and hydrolysis (Dyhrman and Haley, 2006). Instead, *Crocosphaera* does have homologs of *pstSCAB*, coding for a high-affinity phosphate transport system very similar to other marine cyanobacteria thriving in low-phosphate, oligotrophic systems (Moore *et al.*, 2005). However, different from these, *Crocosphaera* seems to have also a low-affinity phosphate permease. With regard to competition for phosphorus these factors may give *Crocosphaera* an ecological niche different from *Prochlorococcus*, *Synechococcus* as well as *Trichodesmium* (Dyhrman and Haley, 2006).

Comparative studies using genome data of marine cyanobacteria

In the preceding paragraphs, I have summarized some of the specific findings that have been made based on genome sequences in different marine cyanobacteria. Instead of using genome sequences from one or two organisms to mine for a specific question, another way to utilize these data is to analyze problems that are more general. Specific insights into cyanobacterial evolution obtained by comparative genome analysis are dealt with in another chapter of this book (see Chapter 2).

Many studies are concerned with the absence or presence of genes that are involved in a particular function, that in turn, may determine, sometimes dramatically, the niche this cyanobacterium may occupy. All the single examples as well as gene numbers mentioned before are protein-coding genes. However, there is a whole class of genes that are not protein-coding. Apart from tRNA and rRNA genes, these genes code for regulatory RNA, a class of molecules regularly overlooked by traditional genome analysis due to the problems involved in finding and annotating them. It has been speculated that even in *E. coli* where ~70 such RNAs are known at present, all major stress responses will include at least one small regulatory RNA as part of the regulon (Gottesman, 2004, 2005). All the more reason we should expect such RNAs contributing significantly to the regulatory system of free-living cyanobacteria. Indeed, one such regulatory RNA was recently found in the cyanobacterium *Synechocystis* sp. PCC 6803 where it regulates the gene expression of *isiA* (Dühring *et al.*, 2006), a gene expressed under iron limitation and oxidative stress (see Chapter 9). Another example in cyanobacteria is the antisense transcript covering the regulatory gene *furA* in *Anabaena* sp. PCC 7120 over its full length (Hernandez *et al.*, 2005). Comparative genome analysis of closely related species is crucial to find this type of regulators and to annotate the respective genes. Frequently, the prediction of genes for regulatory RNAs is limited by the availability of genome information appropriate for this purpose. However, at least in the case of cyanobacteria of the *Prochlorococcus/Synechococcus* lineage this is clearly not the situation (Figure 5.2, Table 5.1). In a screening that made use of this information, several such RNAs were predicted and experimentally verified in *Prochlorococcus* MED4, MIT9313, SS120 and *Synechococcus* WH8102 (Axmann *et al.*, 2005). In addition to regulatory RNAs, also other important regulatory elements were found, such as operon 5′ leader sequences and riboswitches. The responsiveness of some of these regulatory RNAs to environmental perturbations indicated their possible involvement in specific signal transduction processes. Thus, the existence of regulatory RNA in cyanobacteria has been unequivocally shown and in the first few examples we are beginning to understand what their functions are (Axmann *et al.*, 2005; Dühring *et al.*, 2006; Hernández *et al.*, 2005). The availability of many more genome sequences of marine cyanobacteria has now opened the way to unravel the full set of this important type of regulatory molecule.

Another important field of work is the inference of phylogenies based on completed genome sequences. The comparative analysis of 15 complete cyanobacterial genome sequences, among them genomes of eight marine cyanobacteria (five *Prochlorococcus* and three *Synechococcus*), revealed a core set of 1054 protein families called core CyOGS (cyanobacterial clusters of orthologous groups of proteins), which are encoded in at least 14 of them (Mulkidjanian *et al.*, 2006). Interestingly, only 50 of these CyOGS were found to be specific for cyanobacteria and 84 shared by cyanobacteria and photosynthetic eukaryotes. Anoxygenic photosynthetic bacteria possess very few homologs of these 84 photosynthetic CyOGS; furthermore, even these shared genes differ in different bacteria. Therefore, based on analyzing CyOGS in great detail it has been suggested that the first phototrophs on earth might have been anaerobic ancestors of cyanobacteria ("procyanobacteria") that already conducted anoxygenic photosynthesis in a photosystem I-like reaction center, and that photosynthesis spread from these procyanobacteria, to other phyla by lateral gene transfer (Mulkidjanian *et al.*, 2006; see also Chapters 2 and 3).

An obvious question arising from these findings is if it is possible to quantify horizontal gene transfer events within cyanobacteria, as well as between cyanobacteria and other phyla and how much has it contributed to contemporary cyanobacterial genomes? Zhaxybayeva *et al.* (2006) attempted to answer this question by analyzing 1128 protein-coding gene families which are encoded in at least nine from 11 completely sequenced cyanobacterial genomes, including five of marine cyanobacteria (four from the *Prochlorococcus*/marine *Synechococcus* group and that of *C. watsonii*). 700 of these protein-coding gene families were sufficiently resolved and either had cyanobacteria as a coherent group with 80% bootstrap support or had other taxa grouping within cyanobacteria at 80% bootstrap support and thus were found to be phylogenetically informative. From these, only 540 (= 77%) supported cyanobacteria as a coherent group, while 160 (= 23%) either included sequences from other taxa interspersed among cyanobacteria or cyanobacterial sequences were grouping somewhere else, suggesting possible transfer events to or from cyanobacteria (Zhaxybayeva *et al.*, 2006). Thus, cyanobacterial genomes reveal a complex evolutionary history, which cannot be represented by a single strictly bifurcating tree for all genes or even most genes. Interestingly, also in this study, more conflicts were detected within cyanobacteria than between cyanobacteria and other phyla. The latter fact illustrates an intuitively understandable principle: genes become more easily transferred between genomes of more closely related organisms than between those of more distantly related bacteria. However, what are the means by which marine cyanobacteria exchange and transfer genetic material? As with any other bacteria, we know three basic mechanisms: direct uptake of DNA (transformation), conjugation, and last but not least transduction mediated through bacteriophages. It has not been studied yet if any marine cyanobacterium possesses some degree of natural competence for direct DNA uptake. Some but not all have the genes to make pili for conjugation, an important precondition to serve as donors in conjugation. Experiments in a small number of marine *Synechocococcus* (Brahamsha, 1996; McCarren and Brahamsha, 2005), and in a single strain of *Prochlorococcus* (Tolonen *et al.*, 2006b) have been successful at introducing foreign DNA through conjugation. By far the most comprehensive dataset currently exists on marine bacteriophages, some of which may potentially mediate gene transfer though transduction.

Phages of marine cyanobacteria

Viral abundance in the marine ecosystem is stunning: 10^7 virus particles per mL of seawater have repeatedly been measured as a typical concentration (Bergh *et al.*, 1989; Wilhelm and Suttle, 1999). Among them, bacteriophages (bacterial viruses) with a very distinctive morphology, the so-called tailed phages dominate the marine virus population. Since unicellular cyanobacteria are a major component of the total marine bacterial population and constitute a major fraction of total phytoplankton it is likely that many phages infect cyanobacteria in the upper illuminated layers of the oceans. Indeed abundances of up to 10^4 to 10^5 cyanophages per mL have been found (Waterbury and Valois, 1993; Suttle and Chan, 1993).

Cyanophage genomes

Study of marine phage genomes has begun comparatively recently. There are about 500 completed genomes of double-stranded DNA viruses and phages in GenBank; however, of these only 14 infect marine bacteria, among them six cyanophages (P60, S-PM2, P-SSP7, P-SSM2, P-SSM4 and Syn9), as well as eight non-cyanophages, the vibriophages VpV262, KVP40, VP16T, VP16C, K139, VHML; roseophage SIO1; and *Pseudoalteromonas* phage PM2 (for reviews see Mann, 2005; Paul *et al.*, 2002). All six cyanophages for which genome sequences are available had been isolated using unicellular marine cyanobacteria as a host, either *Synechococcus* (P60, Syn9 and S-PM2; Lu *et al.*, 2001; Waterbury and Valois, 1993; Wilson *et al.*, 1993), or *Prochlorococcus* strains (P-SSP7, P-SSM2 and P-SSM4; Sullivan *et al.*, 2003). There are no phage genomes yet for any other marine cyanophage-host system.

All known cyanophages are double-stranded DNA viruses, which belong to one of the three morphologically defined bacteriophage families: Podoviridae, Myoviridae, and Siphoviridae (Sullivan *et al.*, 2003; Suttle and Chan, 1993; Suttle and Chan, 1994; Waterbury and Valois, 1993; Wilson *et al.*, 1993). Morphologically, S-PM2, P-SSM4 and P-SSM2 are most similar to the Myoviridae, whereas P-SSP7 and P60 belong to the Podoviridae (Chen and Lu, 2002; Mann *et al.*, 2005; Sullivan *et al.*, 2005). Excellent model systems exist for both of these phage families: T4 and other T-even phages for the Myoviridae, and T7 for the Podoviridae. Consistent with these observations, the podoviral cyanophage P-SSP7 was found to contain 15 of 26 core T7-like genes and the myoviral cyanophages P-SSM2, P-SSM4 and S-PM2 contain, respectively, 43, 42 and 40 of 75 core T4-like genes (Mann *et al.*, 2005; Sullivan *et al.*, 2005). Genomes of marine cyanophages belonging to the Podoviridae are comparatively small and compact; 44 970 and 47 872 bp were found in case of P-SSP7 and P60, respectively (Chen and Lu, 2002; Sullivan *et al.*, 2005). In contrast, those of myoviral cyanophages are relatively large: 196 280 bp were found for S-PM2, P-SSM4 has 178 249 bp, and P-SSM2 has even a larger genome of 252 401 bp (Mann *et al.*, 2005; Sullivan *et al.*, 2005). These genomes are even larger than the smallest known eubacterial genomes such as that of the psyllid symbiont *Carsonella ruddii* which consists of a chromosome of only 159 662 bp (Nakabachi *et al.*, 2006). An overview on some characteristic features of these phage genomes is given in Table 5.2.

An interesting feature for consideration is the host specificity of cyanophages particularly as a broader host range may indicate a better suitability as potential vectors for horizontal gene transfer via transduction. Among the cyanophages, podoviruses tend to be very host-specific, whereas myoviruses generally have a broader host range, even across genera. Among the sequenced marine cyanophages whose host-ranges have been determined, one podovirus, P-SSP7, is very host-specific infecting only the single high-light-adapted strain *Prochlorococcus* MED4 among 21 different *Prochlorococcus* and *Synechococcus* strains used for testing (Sullivan *et al.*, 2003). In contrast, the myoviruses P-SSM2 and P-SSM4 cross-infected between three and four different *Prochlorococcus* hosts, respectively (Sullivan *et al.*, 2003). Therefore, a connection between host range and phage morphology seems to exist: the podoviruses have a narrow host range, infecting a single to a few closely related, while myoviruses can infect multiple strains of *Synechococcus* or *Prochlorococcus* and some even infect both genera (Waterbury and Valois, 1993; Sullivan *et al.*, 2003; Millard and Mann, 2006).

Table 5.2 Genome-wide characteristics of marine cyanophages

	P-SSP7	P60	S-PM2	P-SSM4	P-SSM2	Syn9
Morphological class	Podoviridae	Podoviridae	Myoviridae	Myoviridae	Myoviridae	Myoviridae
Genome size (bp)	44 970	47 872	196 280	178 249	252 401	176 847
ORFs	54	80	239	198	327	235
G+C%	38.7	53.1	37.8	36.7	35.5	40.65
Host specificity	narrow: *Prochlorococcus* MED4	narrow: *Synechococcus* WH7803	broad: *Synechococcus* WH7803, WH8109 and WH8102, possibly *Prochlorococcus*	broad: *Prochlorococcus* MED4, MIT9215, NATL1A, NATL2A	broad: *Prochlorococcus* MIT9211, NATL1A, NATL2A	broad: *Synechococcus* WH8012 +12 other strains
Cyanobacterial genes	5 (incl. *psbA, hli, nrd, talC*)	4 (incl. *nrdA, nrdB, thyX*)	20 (incl. *psbA, psbD*, 2 *hli*, *cpeT, phoH, cobS*)	14 (incl. *psbA, psbD, pcyA, phoH, pstS, talC, cobS* and *speD*)	47 (incl. 24 genes for lipopoly-saccharide biosynthesis, *psbA*, 6 *hli, petE, petF, pebA, ho1, phoH, pstS, talC, cobS*)	n.d. (incl. *psbA, psbD, petE, phoH, tal, thyX*)
Accession #	AY939843	AF338467	AJ630128	AY940168	AY939844	NC_008296
Reference	(Sullivan *et al.*, 2005)	(Chen and Lu, 2002)	(Mann *et al.*, 2005)	(Sullivan *et al.*, 2005)	(Sullivan *et al.*, 2005)	P. Weigele, W. Pope, G. Hatfull, and R. Hendrix, (unpublished)

Photosynthesis genes in cyanophages

A key feature of all six currently available marine cyanophage genomes is that, in addition to containing typical phage genes, they bear genes specifying proteins whose closest homologues are found in cyanobacteria. Most spectacular, among these host-derived, "cyanobacterial" genes are several photosynthetic genes. For instance, all three sequenced *Prochlorococcus* phage genomes (P-SSP7, P-SSM2 and P-SSM4) as well as *Synechococcus* phage S-PM2 contain the photosynthetic genes *psbA* and *hliP* (Lindell *et al.*, 2004; Mann *et al.*, 2003), coding for the photosystem II reaction center protein D1 and a high-light inducible protein. In addition, the myoviral phages P-SSM4 and S-PM2 also contain a *psbD* gene encoding the photosystem II reaction center protein D2 (Mann *et al.*, 2005; Sullivan *et al.*, 2005). Both reaction center protein genes are absolutely essential for photosystem II and hence are found in all oxygenic photosynthetic organisms.

To get a more comprehensive overview on the prevalence, distribution and evolutionary dynamics of these genes, Sullivan *et al.* (2006) screened 33 cultured cyanophages of known family and host range for the presence of *psbA* and *psbD*. 88% of all screened phage genomes were found to contain *psbA*, and 50% contained both *psbA* and *psbD*. However, such a global number does not allow conclusions on the strain and species-specific distribution of these photosynthetic genes, which in fact were found to be strongly biased: all myoviruses and *Prochlorococcus* podoviruses had a *psbA* gene whereas it was not present in *Prochlorococcus* siphovirus or *Synechococcus* podoviruses (Sullivan *et al.*, 2006). Interestingly, phylogenetic clustering of the myovirus and *Prochlorococcus* podovirus genes suggested that *psbA* and *psbD* have been transferred as whole genes from host to phage in a discrete number of events over the course of evolution (at least four times for *psbA*, and twice for *psbD*), followed by horizontal and vertical distribution among different cyanophages (Sullivan *et al.*, 2006).

Whereas *psbA*, *psbD* and *hliP* are certainly the most spectacular examples, there are several more photosynthetic genes present in these phage genomes (Table 5.2). Among them are genes for components of the photosynthetic electron transfer chain such as *petE* and *petF* (Lindell *et al.*, 2004; Sullivan *et al.*, 2005) which code for the photosynthetic electron carriers plastocyanin and ferredoxin. Another set of genes, involved in the biosynthesis of phycobiliproteins, were also found (Sullivan *et al.*, 2005): *pebA* and *pcyA* code for two enzymes, phycoerythrobilin:ferredoxin oxidoreductase and phycocyanobilin:ferredoxin oxidoreductase, which catalyze the biosynthesis of the chromophores phycoerythrobilin and phycocyanobilin (Frankenberg *et al.*, 2001). There is only one additional enzymatic activity required for phycobilin chromophore biosynthesis, an heme oxygenase that provides the precursor for all phycobiliprotein chromophores by cleaving the heme porphyrin ring system. Indeed, with the *ho1* gene also this gene has been found in phage genomes (Sullivan *et al.*, 2005), indicating that some phages have the full set of genes to make these chromophores. Phycobilin biosynthesis genes are present in all *Prochlorococcus*. Their presence appears obscure because *Prochlorococcus* does not have the intact phycobilisomes characteristic of most cyanobacteria and because the expression level of these genes is low (Hess *et al.*, 1996; Hess *et al.*, 1999). Their conservation in all *Prochlorococcus* (Penno *et al.*, 2000; Steglich *et al.*, 2003, 2005) and additional presence in *Prochlorococcus* phages strongly suggests some important but still unknown function rather than these genes just to be remnants of the evolutionary reduction of the phycobilisome-based antenna. A regulatory gene, *cpeT* (Cobley *et al.*, 2002), completes this list of phycobiliprotein-related cyanobacterial genes found this time in a different cyanophage, in S-PM2 (Mann *et al.*, 2005). Collectively, these results have clearly demonstrated that photosynthesis genes are common in cyanophages (Table 5.2). Only a fraction of phage genes are photosynthesis-related, yet their presence indicates that a significant exchange of genetic material occurs from host to phage, phage to host, and within the phage gene pool (Sullivan *et al.*, 2006).

If photosynthesis genes are such a common constituent of marine cyanophage genomes, are they likely to be functional? The answer to this question is clearly yes. First, the functionality of *psbA* genes is supported by the conservation of amino acid sequences through purifying selection (Lindell *et al.*, 2004; Zeidner *et al.*, 2005). Secondly, there is clear evidence that at least the *psbA* gene becomes expressed from the phage genome during infection (Clokie *et al.*, 2006; Lindell *et al.*, 2005).

Using the podovirus P-SSP7 and *Prochlorococcus* MED4 as a host, Lindell *et al.* (2005) measured that 7–8 hours after infection phage *psbA* mRNA made up 50% of the total *psbA* steady-state transcript levels, compensating for a corresponding drop in host *psbA*-derived mRNA. Moreover, the phage *psbA* gene became not only strongly transcribed but the authors showed unequivocally that the phage-derived mRNA was indeed translated into D1 protein (Lindell *et al.*, 2005). By the end of the latent period, the rapidly turning-over D1 protein of host origin declined to roughly 75% of its maximal level and phage-derived D1 made up about 10% of total D1 in the infected cells. However, the photosystem II efficiency did not decline significantly at all. Therefore, the measured amount of phage D1 appears too low to directly compensate for the observed loss of host D1. Thus, one must take additional factors into account: either the phage D1 protein has higher photochemical efficiency or additional factors help to maintain a highly functional photosystem II. Good candidate proteins to be involved under the latter hypothesis might be phage genome-derived high light-inducible proteins (HLIPs). These could reduce photosystem damage or could facilitate a more rapid reassembly of PSII. Indeed, expression of the P-SSP7 *hli* gene was detected during infection (Lindell *et al.*, 2005). A major conclusion of this study was that replication of the phage genome is dependent on photosynthesis and that phage photosynthesis genes are likely to increase phage fitness by maintaining an appropriate level of important photosynthesis proteins. However, the host-derived, "cyanobacterial" genes are not restricted to photosynthesis. For instance, the myoviruses S-PM2, P-SSM2 and P-SSM4 contain the *phoH* gene involved in phosphate metabolism, and P-SSM2 and P-SSM4 contain with *pstS* a second such gene (Sullivan *et al.*, 2005; Mann *et al.*, 2005). P-SSM2, P-SSM4 and P-SSP7, but not S-PM2, possess an aldolase family gene (*tal*C) that might be involved in alternative routes of carbon metabolism during infection (Sullivan *et al.*, 2005). In contrast, podovirus P60 contains cyanobacterial genes *nrd*A and *nrdB* for ribonucleoside triphosphate reductases A and B, thymidylate synthase, and an unidentified protein which might be involved in the synthesis of phage DNA (Chen and Lu, 2002).

RNA polymerase

Another promiscuous gene, not normally counted as a "host" or "cyanobacterial" gene, however, is the gene for the RNA polymerase. Transcription of phage genes is crucial for phage replication. Therefore, some phages, among them T4-like Myoviridae express proteins to regulate the host RNA polymerase, such as the *gp*55 gene product in S-PM2 which codes for a sigma factor for late transcription (Mann *et al.*, 2005). Podoviridae follow a totally different strategy with regard to this key enzymatic activity, expressing a single-subunit RNA polymerase that is highly active and fundamentally different from the host enzyme. RNA polymerases of this type are widespread in nature. Single-subunit-type RNA polymerases, for instance, transcribe the mitochondrial genome of eukaryotes including humans, and the mitochondria and part of the chloroplast genome of higher plants (Hess and Börner, 1999). Since their discovery, the phylogenetic origin of these genes in eukaryotes has been an enigma. However, there is little doubt about the endosymbiotic origin of mitochondria and chloroplasts from proteobacteria and cyanobacteria, respectively. Thus, an RNA polymerase gene of an endosymbiont-associated phage or a phage gene integrated into a bacterial genome may be the most plausible theory explaining the origin of these RNA polymerases in eukaryotic organelles.

A search for phage-derived RNA polymerase genes in all available genome sequences of marine cyanobacteria revealed an insertion of such a gene in two of them, *Synechococcus* CC9605 and BL107. The sequences are not only closely related to those of podovirus P60 and P-SSP7, but both have a phage-derived integrase (*int*) gene upstream, are present at the same genomic location, and exhibit extensive DNA sequence similarity to each other (Figure 5.3). Whereas the RNA polymerase gene in CC9605 may have become a split pseudogene due to a point mutation that has created an in frame stop-codon, the gene in BL107 has remained a continuous open reading frame (Figure 5.3A). It is intriguing that also in the genome of phage P-SSP7 an integrase gene is located directly upstream of the RNA polymerase gene (Sullivan *et al.*, 2005).

The phylogenetic analysis of these proteins shows the immense sequence diversity present among RNA polymerases of T7-type phages. This diversity is about as big between non-cyanobacterial and cyanobacterial phage proteins as between the latter and plant and algal RNA polymerases (Figure 5.3B). Based on the large extent of sequence diversity one may speculate that it is still possible that one day we will find a polymerase gene that comes close to the phylogenetic origin of organellar RNA polymerases of eukaryotic organisms. With regard to the world of marine cyanobacteria, we may use these genes as another phage signature gene to screen for in new genomes to find possible prophages integrated into host DNA. Still there is no single example for a lysogenic phage infecting marine cyanobacteria, but these and other examples of genome-inserted phage genes suggest that such phages do exist and that they may not be so divergent from some of the already described bacteriophages. Furthermore, the study of bacteriophage RNA polymerase activities will contribute to our understanding of the coordination of phage genome expression with the switching off of most of host metabolism and gene expression while maintaining certain host activities such as photosynthesis.

Concluding remarks

To summarize, the genomes of marine cyanobacteria have been shaped much more by intra- and interphylum gene transfer than originally expected (Zhaxybayeva *et al.*, 2006). In some instances evidence is found for an enhanced mutation rate (Dufresne *et al.*, 2005) and as a result, these genomes are more different from each other, and are much more flexible and dynamic than one might have expected just very few years ago. Cyanophages are found in the natural environment in high numbers and contain a resource of cyano-

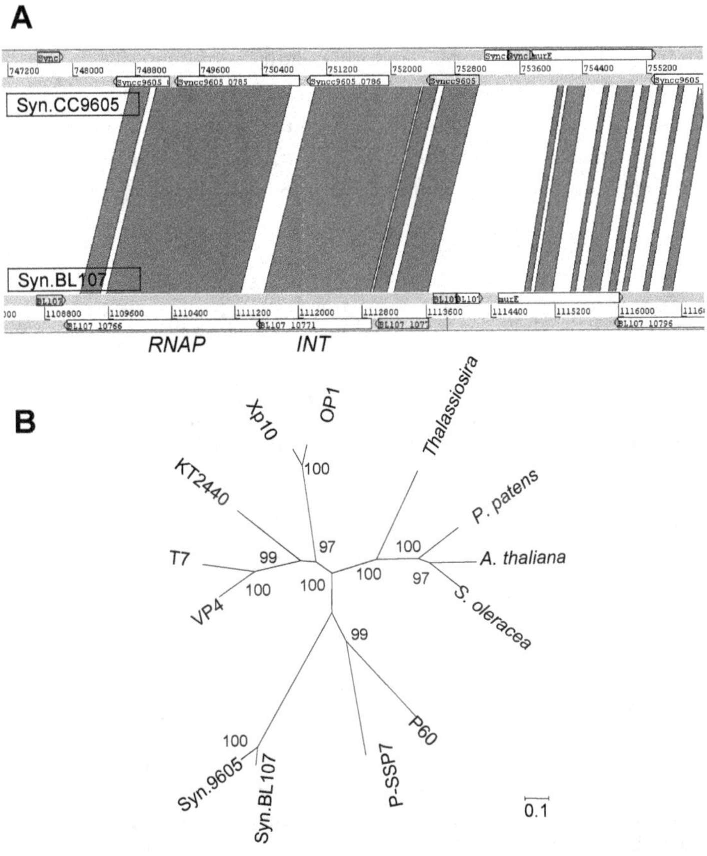

Figure 5.3 (see also Plate 5.3) A. Alignment of part of the *Synechococcus* CC9605 genomic sequence (top) and *Synechococcus* BL107 (bottom) containing genes for a T7-phage-like RNA polymerase (RNAP) and an integrase (INT). The lines connect DNA sequences with a BLASTN E value $< 10^{-2}$. The split of the RNAP gene in two parts in the CC9605 genome is visible, however, the DNA sequence similarity is similar for both parts with the respective gene segment in BL107. B. Phylogenetic analysis of RNAP sequences from cyanobacteria *Synechococcus* CC9605 (Syn.9605) and BL107 (Syn.BL107) and podoviral phages P60 and P-SSP7, non-cyanobacterial phages VP4 (Vibriophage VP4, AAY46276.1), T7 (coliphage, gi|3891671), KT2440 (*Pseudomonas putida* bacteriophage KT2440, NP_744415.1), XP10 (*Xanthomonas oryzae* bacteriophage Xp10, NP_858979.1), OP10 (*Xanthomonas oryzae* phage OP1, YP_453591.1) and eukaryotic RNAPs of the diatom *Thalassiosira pseudonana*, the moss *Physcomitrella patens* (*P. patens*) and the two higher plants *Arabidopsis thaliana* (*A. thaliana*) and *Spinacia oleracea* (*S. oleracea*).

bacterial genes since each cyanophage isolated so far contains a number of cyanobacterial genes in its genome (Lindell *et al.*, 2004; Mann *et al.*, 2005; Mann *et al.*, 2003; Sullivan *et al.*, 2005; Sullivan *et al.*, 2006), among them genes that are essential for photosynthesis. Therefore, future studies need not only to address single genome sequences but will highlight the interplay between the genes found in a certain cyanobacterial chromosome, associated phage(s) and natural populations where both host and phage can be expected to co-occur.

Acknowledgments.

I would like to thank my colleagues, coworkers and friends for sharing unpublished information, for years of fruitful work and collaboration, Debbie Lindell, Dave Scanlan and Claudia Steglich for critical reading. For continuous support I would like to thank the European Union (programs PROMOLEC, MARGENES and Marine Genomics Europe), the Deutsche Forschungsgemeinschaft and the Landesstiftung Baden-Wuerttemberg.

References

Allewalt, J.P., Bateson, M.M., Revsbech, N.P., Slack, K., and Ward, D.M. (2006). Effect of temperature and light on growth of and photosynthesis by *Synechococcus* isolates typical of those predominating in the octopus spring microbial mat community of Yellowstone National Park. Appl. Environ. Microbiol. *72*, 544–550.

Axmann, I.M., Kensche, P., Vogel, J., Kohl, S., Herzel, H., and Hess, W.R. (2005). Identification of cyanobacterial non-coding RNAs by comparative genome analysis. Genome Biol. *6*, R73.

Bergh, O., Borsheim, K.Y., Bratbak, G., and Heldal, M. (1989). High abundance of viruses found in aquatic environments. Nature *340*, 467–468.

Bhaya, D., Dufresne, A., Vaulot, D., and Grossman, A. (2002). Analysis of the *hli* gene family in marine and freshwater cyanobacteria. FEMS Microbiol. Lett *215*, 209–219.

Brahamsha, B. (1996). A genetic manipulation system for oceanic cyanobacteria of the genus *Synechococcus*. Appl. Environ. Microbiol. *62*, 1747–1751.

Campbell, L., Liu, H.B., Nolla, H.A., and Vaulot, D. (1997). Annual variability of phytoplankton and bacteria in the subtropical North Pacific Ocean at Station ALOHA during the 1991–1994 ENSO event. Deep Sea Res. *44*, 167–192.

Campbell, L., Liu, H.B., Nolla, H.A., and Vaulot, D. (1997). Annual variability of phytoplankton and bacteria in the subtropical North Pacific Ocean at Station ALOHA during the 1991–1994 ENSO event. Deep Sea Res. *44*, 167–192.

Capone, D.G., Zehr, J.P., Paerl, H.W., Bergman, B., and Carpenter, E.J. (1997). *Trichodesmium*, a globally significant marine cyanobacterium. Science *276*, 1221–1229.

Carre-Mlouka, A., Mejean, A., Quillardet, P., Ashida, H., Saito, Y., Yokota, A., Callebaut, I., Sekowska, A., Dittmann, E., Bouchier, C., and de Marsac, N.T. (2006). A new rubisco-like protein coexists with a photosynthetic rubisco in the planktonic cyanobacteria *Microcystis*. J. Biol. Chem. *281*, 24462–24471.

Castruita, M., Saito, M., Schottel, P.C., Elmegreen, L.A., Myneni, S., Stiefel, E.I., and Morel, F.M. (2006). Overexpression and characterization of an iron storage and DNA-binding Dps protein from *Trichodesmium erythraeum*. Appl. Environ. Microbiol. *72*, 2918–2924.

Chen, F., and Lu, J. (2002). Genomic sequence and evolution of marine cyanophage P60: a new insight on lytic and lysogenic phages. Appl. Environ. Microbiol. *68*, 2589–2594.

Clokie, M.R., Shan, J., Bailey, S., Jia, Y., Krisch, H.M., West, S., and Mann, N.H. (2006). Transcription of a "photosynthetic" T4-type phage during infection of a marine cyanobacterium. Environ. Microbiol. *8*, 827–835.

Cobley, J.G., Clark, A.C., Weerasurya, S., Queseda, F.A., Xiao, J.Y., Bandrapali, N., D'Silva, I., Thounaojam, M., Oda, J.F., Sumiyoshi, T., and Chu, M.H. (2002). CpeR is an activator required for expression of the phycoerythrin operon (*cpeBA*) in the cyanobacterium *Fremyella diplosiphon* and is encoded in the phycoerythrin linker-polypeptide operon (*cpeCDESTR*). Mol. Microbiol. *44*, 1517–1531.

Coleman, M.L., Sullivan, M.B., Martiny, A.C., Steglich, C., Barry, K., Delong, E.F., and Chisholm, S.W. (2006). Genomic islands and the ecology and evolution of *Prochlorococcus*. Science *311*, 1768–1770.

Dufresne, A., Garczarek, L., and Partensky, F. (2005). Accelerated evolution associated with genome reduction in a free-living prokaryote. Genome Biol. *6*, R14.

Dufresne, A., Salanoubat, M., Partensky, F., Artiguenave, F., Axmann, I.M., Barbe, V., Duprat, S., Galperin, M.Y., Koonin, E.V., Le Gall, F., *et al.* (2003). Genome sequence of the cyanobacterium *Prochlorococcus marinus* SS120, a nearly minimal oxyphototrophic genome. Proc. Natl. Acad. Sci. USA *100*, 10020–10025.

Dühring, U., Axmann, I.M., Hess, W.R., and Wilde, A. (2006). An internal antisense RNA regulates expression of the photosynthesis gene *isiA*. Proc. Natl. Acad. Sci. USA *103*, 7054–7058.

DuRand, M.D., Olson, R.J., and Chisholm, S.W. (2001). Phytoplankton population dynamics at the Bermuda Atlantic Time-series station in the Sargasso Sea. Deep Sea Res. *48*, 1983–2003.

Dyhrman, S.T., Chappell, P.D., Haley, S.T., Moffett, J.W., Orchard, E.D., Waterbury, J.B., and Webb, E.A. (2006). Phosphonate utilization by the globally important marine diazotroph *Trichodesmium*. Nature *439*, 68–71.

Dyhrman, S.T., and Haley, S.T. (2006). Phosphorus scavenging in the unicellular marine diazotroph *Crocosphaera watsonii*. Appl. Environ. Microbiol. *72*, 1452–1458.

Falkowski, P.G., Katz, M.E., Milligan, A.J., Fennel, K., Cramer, B.S., Aubry, M.P., Berner, R.A., Novacek, M.J., and Zapol, W.M. (2005). The rise of oxygen over the past 205 million years and the evolution of large placental mammals. Science *309*, 2202–2204.

Frankenberg, N., Mukougawa, K., Kohchi, T., and Lagarias, J.C. (2001). Functional genomic analysis of the HY2 family of ferredoxin-dependent bilin reductases from oxygenic photosynthetic organisms. Plant Cell *13*, 965–978.

Fuller, N.J., Marie, D., Partensky, F., Vaulot, D., Post, A.F., and Scanlan, D.J. (2003). Clade-specific 16S ribosomal DNA oligonucleotides reveal the predominance of a single marine *Synechococcus* clade throughout a stratified water column in the Red Sea. Appl. Environ. Microbiol. *69*, 2430–2443.

Garczarek, L., Hess, W.R., Holtzendorff, J., van der Staay, G.W., and Partensky, F. (2000). Multiplication of antenna genes as a major adaptation to low light in a marine prokaryote. Proc. Natl. Acad. Sci. USA *97*, 4098–4101.

Garczarek, L., van der Staay, G.W., Hess, W.R., Le Gall, F., and Partensky, F. (2001). Expression and phylogeny of the multiple antenna genes of the low-light-adapted strain *Prochlorococcus marinus* SS120 (Oxyphotobacteria). Plant Mol. Biol. *46*, 683–693.

Gottesman, S. (2004). The small RNA regulators of *Escherichia coli*: roles and mechanisms. Annu. Rev. Microbiol. *58*, 303–328.

Gottesman, S. (2005). Micros for microbes: non-coding regulatory RNAs in bacteria. Trends Genet. *21*, 399–404.

Hernández, J.A., Muro-Pastor, A.M., Flores, E., Bes, M.T., Peleato, M.L., and Fillat, M.F. (2005). Identification of a *furA cis* antisense RNA in the cyanobacterium *Anabaena* sp. PCC 7120. J. Mol. Biol. *355*, 325–334.

Hess, W.R., Partensky, F., van der Staay, G.W., García-Fernández, J.M., Börner, T., and Vaulot, D. (1996). Coexistence of phycoerythrin and a chlorophyll a/b antenna in a marine prokaryote. Proc. Natl. Acad. Sci. USA 93, 11126–11130.

Hess, W.R., Steglich, C., Lichtlé, C., and Partensky, F. (1999). Phycoerythrins of the oxyphotobacterium *Prochlorococcus marinus* are associated to the thylakoid membrane and are encoded by a single large gene cluster. Plant Mol. Biol. 40, 507–521.

Hess, W., and Börner, T. (1999). Organellar RNA polymerases of higher plants. Int. Rev. Cytol. *190*, 1–59.

Hess, W.R. (2004). Genome analysis of marine photosynthetic microbes and their global role. Curr. Opin. Biotechnol. *15*, 191–198.

Hess, W.R., Rocap, G., Ting, C.S., Larimer, F., Stilwagen, S., Lamerdin, J., and Chisholm, S.W. (2001). The photosynthetic apparatus of *Prochlorococcus*: Insights through comparative genomics. Photosynth. Res. *70*, 53–71.

Johnson, Z.I., Zinser, E.R., Coe, A., McNulty, N.P., Woodward, E.M., and Chisholm, S.W. (2006). Niche partitioning among *Prochlorococcus* ecotypes along ocean-scale environmental gradients. Science *311*, 1737–1740.

Kaneko, T., Nakamura, Y., Wolk, C.P., Kuritz, T., Sasamoto, S., Watanabe, A., Iriguchi, M., Ishikawa, A., Kawashima, K., Kimura, T., Kishida, Y., Kohara, M., Matsumoto, M., Matsuno, A., Muraki, A., Nakazaki, N. Shimpo, S., Sugimoto, M., Takazawa, M. Yamada, M., Yasuda, M., and Tabata, S. (2001). Complete genomic sequence of the filamentous nitrogen-fixing cyanobacterium *Anabaena* sp. strain PCC 7120. DNA Res. *8*, 205–213; 227–253.

Kaneko, T., Sato, S., Kotani, H., Tanaka, A., Asamizu, E., Nakamura, Y., Miyajima, N., Hirosawa, M., Sugiura, M., Sasamoto, S., Kimura T., Hosouchi, T., Matsuno, A., Muraki, A., Nakazaki, N., Naruo, K., Okumura, S., Shimpo, S., Takeuchi, C., Wada, T., Watanabe, A., Yamada, M., Yasuda, M., and Tabata, S. (1996). Sequence analysis of the genome of the unicellular cyanobacterium *Synechocystis* sp. strain PCC6803. II. Sequence determination of the entire genome and assignment of potential protein-coding regions. DNA Res. *3*, 109–136.

Kaneko, T., Sato, S., Kotani, H., Tanaka, A., Asamizu, E., Nakamura, Y., Miyajima, N., Hirosawa, M., Sugiura, M., Sasamoto, S., Kimura, T., Hosouchi, T., Matsuno, A., Muraki, A., Nakazaki, N., Naruo, K., Okumura, S., Shimpo, S., Takeuchi, C., Wada, T., Watanabe, A., Yamada, M., Yasuda, M., and Tabata, S. (1996). Sequence analysis of the genome of the unicellular cyanobacterium *Synechocystis* sp. strain PCC6803. II. Sequence determination of the entire genome and assignment of potential protein-coding regions (supplement). DNA Res. *3*, 185–209.

Kaneko, T., and Tabata, S. (1997). Complete genome structure of the unicellular cyanobacterium *Synechocystis* sp. PCC6803. Plant Cell Physiol. *38*, 1171–1176.

Kaneko, T., Tanaka, A., Sato, S., Kotani, H., Sazuka, T., Miyajima, N., Sugiura, M., and Tabata, S. (1995). Sequence analysis of the genome of the unicellular cyanobacterium *Synechocystis* sp. strain PCC6803. I. Sequence features in the 1 Mb region from map positions 64% to 92% of the genome. DNA Res. *2*, 153–166, 191–158.

Kumar, S., Tamura, K., and Nei, M. (2004). MEGA3: Integrated software for Molecular Evolutionary Genetics Analysis and sequence alignment. Brief. Bioinform. *5*, 150–163.

Lindell, D., Jaffe, J.D., Johnson, Z.I., Church, G.M., and Chisholm, S.W. (2005). Photosynthesis genes in marine viruses yield proteins during host infection. Nature *438*, 86–89.

Lindell, D., Sullivan, M.B., Johnson, Z.I., Tolonen, A.C., Rohwer, F., and Chisholm, S.W. (2004). Transfer of photosynthesis genes to and from *Prochlorococcus* viruses. Proc. Natl. Acad. Sci. USA *101*, 11013–11018.

Liolios, K., Tavernarakis, N., Hugenholtz, P., and Kyrpides, N.C. (2006). The Genomes On Line Database (GOLD) v.2: a monitor of genome projects worldwide. Nucleic Acids Res. *34*, D332–334.

Lu, J., Chen, F., and Hodson, R.E. (2001). Distribution, isolation, host specificity, and diversity of cyanophages infecting marine *Synechococcus* spp. in river estuaries. Appl. Environ. Microbiol. *67*, 3285–3290.

Mann, N.H. (2005). The third age of phage. PLoS Biol. *3*, e182.

Mann, N.H., Clokie, M.R., Millard, A., Cook, A., Wilson, W.H., Wheatley, P.J., Letarov, A., and Krisch, H.M. (2005). The genome of S-PM2, a "photosynthetic" T4-type bacteriophage that infects marine *Synechococcus* strains. J. Bacteriol. *187*, 3188–3200.

Mann, N.H., Cook, A., Millard, A., Bailey, S., and Clokie, M. (2003). Marine ecosystems: bacterial photosynthesis genes in a virus. Nature *424*, 741.

Martiny, A.C., Coleman, M.L., and Chisholm, S.W. (2006). Phosphate acquisition genes in *Prochlorococcus* ecotypes: evidence for genome-wide adaptation. Proc. Natl. Acad. Sci. USA *103*, 12552–12557.

McCarren, J., and Brahamsha, B. (2005). Transposon mutagenesis in a marine *Synechococcus* strain: isolation of swimming motility mutants. J. Bacteriol. *187*, 4457–4462.

Meeks, J.C., Elhai, J., Thiel, T., Potts, M., Larimer, F., Lamerdin, J., Predki, P., and Atlas, R. (2001). An overview of the genome of *Nostoc punctiforme*, a multicellular, symbiotic cyanobacterium. Photosynth. Res. *70*, 85–106.

Millard, A.D., and Mann, N.H. (2006). A temporal and spatial investigation of cyanophage abundance in the Gulf of Aqaba, Red Sea. J. Mar. Biol. Ass. UK *86*, 507–515.

Montoya, J.P., Holl, C.M., Zehr, J.P., Hansen, A., Villareal, T.A., and Capone, D.G. (2004). High rates of N_2 fixation by unicellular diazotrophs in the oligotrophic Pacific Ocean. Nature *430*, 1027–1031.

Moore, L.R., Ostrowski, M., Scanlan, D.J., Feren, K., and Sweetsir, T. (2005). Ecotypic variation in phosphorus-acquisition mechanisms within marine picocyanobacteria. Aquat. Microb. Ecol. *39*, 257–269.

Mulkidjanian, A.Y., Koonin, E.V., Makarova, K.S., Mekhedov, S.L., Sorokin, A., Wolf, Y.I., Dufresne, A., Partensky, F., Burd, H., Kaznadzey, D., *et al.* (2006). The cyanobacterial genome core and the origin of photosynthesis. Proc. Natl. Acad. Sci. USA *103*, 13126–13131.

Murata, N., and Suzuki, I. (2006). Exploitation of genomic sequences in a systematic analysis to access how cyanobacteria sense environmental stress. J. Exp. Bot. *57*, 235–247.

Nakabachi, A., Yamashita, A., Toh, H., Ishikawa, H., Dunbar, H.E., Moran, N.A., and Hattori, M. (2006). The 160-kilobase genome of the bacterial endosymbiont *Carsonella*. Science *314*, 267.

Nakamura, Y., Kaneko, T., Sato, S., Ikeuchi, M., Katoh, H., Sasamoto, S., Watanabe, A., Iriguchi, M., Kawashima, K., Kimura, T., Kishida, Y., Kiyokawa, C., Kohara, M., Matsumoto, M., Matsuno, A., Nakazaki, N., Shimpo, S., Sugimoto, M., Takeuchi, C., Yamada, M., and Tabata, S. (2002). Complete genome structure of the thermophilic cyanobacterium *Thermosynechococcus elongatus* BP-1. DNA Res. *9*, 123–130.

Nakamura, Y., Kaneko, T., Sato, S., Ikeuchi, M., Katoh, H., Sasamoto, S., Watanabe, A., Iriguchi, M., Kawashima, K., Kimura, T., Kishida, Y., Kiyokawa, C., Kohara, M., Matsumoto, M., Matsuno, A., Nakazaki, N., Shimpo, S., Sugimoto, M., Takeuchi, C., Yamada, M., and Tabata, S. (2002). Complete genome structure of the thermophilic cyanobacterium *Thermosynechococcus elongatus* BP-1 (supplement). DNA Res. *9*, 135–148.

Nakamura, Y., Kaneko, T., Sato, S., Mimuro, M., Miyashita, H., Tsuchiya, T., Sasamoto, S., Watanabe, A., Kawashima, K., Kishida, Y., Kiyokawa, C., Kohara, M., Matsumoto, M., Matsuno, A., Nakazaki, N., Shimpo, S., Takeuchi, C., Yamada, M., and Tabata, S. (2003). Complete genome structure of *Gloeobacter violaceus* PCC 7421, a cyanobacterium that lacks thylakoids. DNA Res. *10*, 137–145.

Nakamura, Y., Kaneko, T., Sato, S., Mimuro, M., Miyashita, H., Tsuchiya, T., Sasamoto, S., Watanabe, A., Kawashima, K., Kishida, Y., Kiyokawa, C., Kohara, M., Matsumoto, M., Matsuno, A., Nakazaki, N., Shimpo, S., Takeuchi, C., Yamada, M., and Tabata, S. (2003). Complete genome structure of *Gloeobacter violaceus* PCC 7421, a cyanobacterium that lacks thylakoids (supplement). DNA Res. *10*, 181–201.

Olson, R.J., Zettler, E.R., Altabet, M.A., Dusenberry, J.A., and Chisholm, S.W. (1990). Spatial and temporal distributions of prochlorophyte picoplankton in the North Atlantic Ocean. Deep-Sea Res. *37*, 1033–1051.

Palenik, B. (2001). Chromatic adaptation in marine *Synechococcus* strains. Appl. Environ. Microbiol. *67*, 991–994.

Palenik, B., Brahamsha, B., Larimer, F.W., Land, M., Hauser, L., Chain, P., Lamerdin, J., Regala, W., Allen, E.E., McCarren, J., Paulsen I, Dufresne, A. Partensky, F., Webb, E.A., and Waterbury, J. (2003). The genome of a motile marine *Synechococcus*. Nature *424*, 1037–1042.

Palenik, B., Ren, Q., Dupont, C.L., Myers, G.S., Heidelberg, J.F., Badger, J.H., Madupu, R., Nelson, W.C., Brinkac, L.M., Dodson, R.J., Durkin, A.S., Daugherty, S.C., Sullivan, S.A., Khouri, H., Mohamoud, Y., Halpin, R., and Paulsen, I.T. (2006). Genome sequence of *Synechococcus* CC9311: Insights into adaptation to a coastal environment. Proc. Natl. Acad. Sci. USA *103*, 13555–13559.

Partensky, F., Hess, W.R., and Vaulot, D. (1999). *Prochlorococcus*, a marine photosynthetic prokaryote of global significance. Microbiol. Mol. Biol. Rev. *63*, 106–127.

Paul, J.H., Sullivan, M.B., Segall, A.M., and Rohwer, F. (2002). Marine phage genomics. Comp. Biochem. Physiol. B Biochem. Mol. Biol. *133*, 463–476.

Penno, S., Campbell, L., and Hess, W.R. (2000). Presence of phycoerythrin in two strains of *Prochlorococcus* (Cyanobacteria) isolated from the subtropical north Pacific Ocean. J. Phycol. 36, 723–729.

Rocap, G., Larimer, F.W., Lamerdin, J., Malfatti, S., Chain, P., Ahlgren, N.A., Arellano, A., Coleman, M., Hauser, L., Hess, W.R., Johnson Z.I., Land, M., Lindell, D., Post, A.F., Regala, W., Shah, M., Shaw, S.L., Steglich, C., Sullivan, M.B., Ting, C.S., Tolonen, A., Webb, E.A., Zinser, E.R., and Chisholm, S.W. (2003). Genome divergence in two *Prochlorococcus* ecotypes reflects oceanic niche differentiation. Nature *424*, 1042–1047.

Sanudo-Wilhelmy, S.A., Kustka, A.B., Gobler, C.J., Hutchins, D.A., Yang, M., Lwiza, K., Burns, J., Capone, D.G., Raven, JA., and Carpenter, E.J. (2001). Phosphorus limitation of nitrogen fixation by *Trichodesmium* in the central Atlantic Ocean. Nature *411*, 66–69.

Shi, T., Bibby, T.S., Jiang, L., Irwin, A.J., and Falkowski, P.G. (2005). Protein interactions limit the rate of evolution of photosynthetic genes in cyanobacteria. Mol. Biol. Evol. 22, 2179–2189.

Six, C., Thomas, J.C., Thion, L., Lemoine, Y., Zal, F., and Partensky, F. (2005). Two novel phycoerythrin-associated linker proteins in the marine cyanobacterium *Synechococcus* sp. strain WH8102. J. Bacteriol. *187*, 1685–1694.

So, A.K.C., Espie, G.S., Williams, E.B., Shively, J.M., Heinhorst, S., and Cannon, G.C. (2004). A novel evolutionary lineage of carbonic anhydrase (ε class) Is a component of the carboxysome shell. J. Bacteriol. *186*, 623–630.

Steglich, C., Frankenberg-Dinkel, N., Penno, S., and Hess, W.R. (2005). A green light-absorbing phycoerythrin is present in the high-light-adapted marine cyanobacterium *Prochlorococcus* sp. MED4. Environ. Microbiol. 7, 1611–1618.

Steglich, C., Futschik, M., Rector, T., Steen, R., and Chisholm, S.W. (2006). Genome-wide analysis of light sensing in *Prochlorococcus*. J. Bacteriol. *188*, 7796–7806.

Steglich, C., Post, A.F., and Hess, W.R. (2003). Analysis of natural populations of *Prochlorococcus* spp. in the northern Red Sea using phycoerythrin gene sequences. Environ. Microbiol. 5, 681–690.

Steunou, A.S., Bhaya, D., Bateson, M.M., Melendrez, M.C., Ward, D.M., Brecht, E., Peters, J.W., Kuhl, M., and Grossman, A.R. (2006). In situ analysis of nitrogen fixation and metabolic switching in unicellular thermophilic cyanobacteria inhabiting hot spring microbial mats. Proc. Natl. Acad. Sci. USA *103*, 2398–2403.

Strehl, B., Holtzendorff, J., Partensky, F., and Hess, W.R. (1999). A small and compact genome in the marine cyanobacterium *Prochlorococcus marinus* CCMP 1375: lack of an intron in the gene for tRNA(Leu)(UAA) and a single copy of the rRNA operon. FEMS Microbiol. Lett. *181*, 261–266.

Su, Z., Dam, P., Chen, X., Olman, V., Jiang, T., Palenik, B., and Xu, Y. (2003). Computational inference of regulatory pathways in microbes: an application to phosphorus assimilation pathways in *Synechococcus* sp. WH8102. Genome Inform. *14*, 3–13.

Su, Z., Mao, F., Dam, P., Wu, H., Olman, V., Paulsen, I.T., Palenik, B., and Xu, Y. (2006). Computational inference and experimental validation of the nitrogen assimilation regulatory network in cyanobacterium *Synechococcus* sp. WH 8102. Nucleic Acids Res. *34*, 1050–1065.

Sudek, S., Haygood, M.G., Youssef, D.T., and Schmidt, E.W. (2006). Structure of trichamide, a cyclic peptide from the bloom-forming cyanobacterium *Trichodesmium erythraeum*, predicted from the genome sequence. Appl. Environ. Microbiol. *72*, 4382–4387.

Sugita, C., Ogata, K., Shikata, M., Jikuya, H., Takano, J., Furumichi, M., Kanehisa, M., Omata, T., Sugiura, M., Sugita, M. (2007). Complete nucleotide sequence of the freshwater unicellular cyanobacterium *Synechococcus elongatus* PCC 6301. Photosynth. Res. (in press).

Sullivan, M.B., Coleman, M.L., Weigele, P., Rohwer, F., and Chisholm, S.W. (2005). Three *Prochlorococcus* cyanophage genomes: signature features and ecological interpretations. PLoS Biol. *3*, e144.

Sullivan, M.B., Lindell, D., Lee, J.A., Thompson, L.R., Bielawski, J.P., and Chisholm, S.W. (2006). Prevalence and evolution of core photosystem II genes in marine cyanobacterial viruses and their hosts. PLoS Biol. *4*, e234.

Sullivan, M.B., Waterbury, J.B., and Chisholm, S.W. (2003). Cyanophages infecting the oceanic cyanobacterium *Prochlorococcus*. Nature *424*, 1047–1051.

Suttle, C.A., and Chan, A.M. (1993). Marine cyanophages infecting oceanic and coastal strains of *Synechococcus*: Abundance, morphology, cross-infectivity and growth characteristics. Mar. Ecol. Prog. Ser. *92*, 99–109.

Suttle, C.A., and Chan, A.M. (1994). Dynamics and distribution of cyanophages and their effects on marine *Synechococcus* spp. Appl. Environ. Microbiol. *60*, 3167–3174.

Tolonen, A.C., Aach, J., Lindell, D., Johnson, Z.I., Rector, T., Steen, R., Church, G.M., and Chisholm, S.W. (2006a). Global gene expression of *Prochlorococcus* ecotypes in response to changes in nitrogen availability. Mol. Syst. Biol. 1–15.

Tolonen, A.C., Liszt, G.B., and Hess, W.R. (2006b). Genetic manipulation of *Prochlorococcus* MIT9313: GFP expression on an RSF1010 plasmid and Tn5 transposition. Appl. Environ. Microbiol. *72*, 7607–7613.

Vaulot, D., Marie, D., Olson, R.J., and Chisholm, S.W. (1995). Growth of *Prochlorococcus*, a photosynthetic prokaryote, in the equatorial Pacific Ocean. Science *268*, 1480–1482.

Waterbury, J.B., and Valois, F.W. (1993). Resistance to co-occuring phages enables marine *Synechococcus* communities to coexist with cyanophages abundant in seawater. Appl. Environ. Microbiol. *59*, 3393–3399.

Whitman, W.B., Coleman, D.C., and Wiebe, W.J. (1998). Prokaryotes: The unseen majority. Proc. Natl. Acad. Sci. USA *95*, 6578–6583.

Wilhelm, S.W., and Suttle, C.A. (1999). Viruses and nutrient cycles in the sea—Viruses play critical roles in the structure and function of aquatic food webs. Bioscience *49*, 781–788.

Wilson, W.H., Joint, I.R., Carr, N.G., and Mann, N.H. (1993). Isolation and molecular characterization of five marine cyanophages propogated on *Synechococcus* sp. strain WH 7803. Appl. Environ. Microbiol. *59*, 3736–3743.

Zeidner, G., Bielawski, J.P., Shmoish, M., Scanlan, D.J., and Sabehi, G. (2005). Potential photosynthesis gene recombination between *Prochlorococcus* and *Synechococcus* via viral intermediates. Environ. Microbiol. *7*, 1505–1513.

Zhang, K., Martiny, A.C., Reppas, N.B., Barry, K.W., Malek, J., Chisholm, S.W., and Church, G.M. (2006). Sequencing genomes from single cells by polymerase cloning. Nat. Biotechnol. *24*, 680–686.

Zhaxybayeva, O., Gogarten, J.P., Charlebois, R.L., Doolittle, W.F., and Papke, R.T. (2006). Phylogenetic analyses of cyanobacterial genomes: quantification of horizontal gene transfer events. Genome Res. *16*, 1099–1108.

Stress Responses in *Synechocystis*: Regulated Genes and Regulatory Systems

6

Dmitry A. Los, Iwane Suzuki, Vladislav V. Zinchenko, and Norio Murata

Abstract
Genome-wide investigations of gene expression at the transcriptional level in cyanobacteria, using DNA microarrays, have allowed identification of genes whose expression is induced or repressed by various types of environmental stress and also of previously uncharacterized genes that appear to be involved in stress responses. Acclimation to stress begins with perception of stress and transduction of the stress signal. A combination of the systematic mutation of potential sensors and transducers and DNA microarray analysis has led to significant progress in understanding the mechanisms of perception of and reaction to environmental stress in cyanobacteria. In this chapter, we summarize recent progress in the identification of stress-inducible genes and of systems that regulate responses to stress that has been made using DNA microarrays in cyanobacteria and, in particular, in *Synechocystis* sp. PCC 6803.

Introduction

Environmental stresses influence a variety of physiological activities in living organisms. When a change in the environment exceeds a certain threshold level, the activities of some enzymes are inhibited or lost and those of others are enhanced or induced. In response to moderate stress, many organisms activate sets of genes that are specific to the individual type of stress. Specific proteins are synthesized and some of these proteins, in turn, participate in the synthesis of certain stress-specific metabolites. The proteins and metabolites that are synthesized *de novo* in response to stress are important for the acclimation of the organism or cell to the new environment (Figure 6.1).

The first steps in acclimation to environmental stress are the perception of such stress and transduction of the resultant signal. Organisms and/or individual cells appear to be equipped with sensors and signal transducers that perceive and transduce signals that result from changes in the environment. Moreover, the sensors and signal transducers appear to be specific to individual types of environmental stress.

Unicellular cyanobacteria have several features that make them particularly suitable for studies of stress responses at the molecular level. The general features of the plasma membranes and thylakoid membranes of cyanobacterial cells are similar to those of the chloroplasts of higher plants in terms of lipid composition and the assembly of membranes (see Chapters 9 and 10). Therefore, cyanobacteria might be expected to serve as powerful

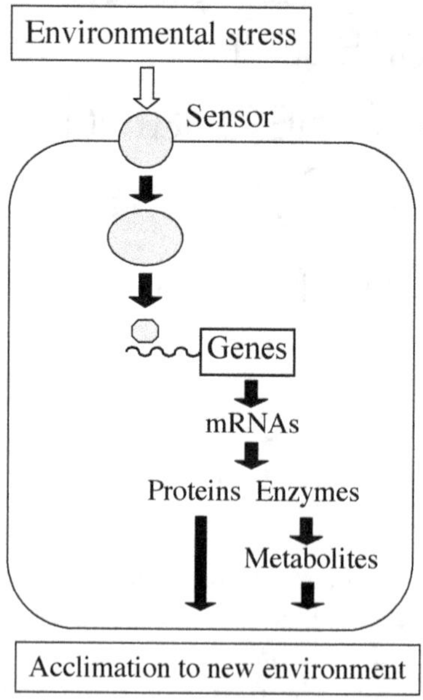

Figure 6.1 A general scheme showing the responses of a cyanobacterial cell to environmental stress. Adapted from Murata and Los (2006) under permission of the publisher.

model systems for studies of the molecular mechanisms of responses and acclimation to environmental stress (Los and Murata, 2004).

The entire nucleotide sequence of the genome of *Synechocystis* sp. PCC 6803 (hereafter, *Synechocystis*) was determined by Kaneko *et al.* (1996), and it was the first sequence of a cyanobacterial genome to be reported. The sequences of a number of cyanobacterial genomes are now available, such as those of *Anabaena* sp. PCC 7120 (hereafter, *Anabaena*; also known as *Nostoc* sp. PCC 7120; Kaneko *et al.*, 2001), *Thermosynechococcus elongatus* BP-1 (Nakamura *et al.*, 2002), *Gloeobacter violaceus* PCC 7421 (Nakamura *et al.*, 2003), *Synechococcus* sp. WH8102 (Palenik *et al.*, 2003), *Synechococcus* sp. PCC 6301 (Sugita M., unpublished; data are available in GenBank under accession no. AP008231), *Prochlorococcus marinus* SS120 (Dufresne *et al.*, 2003), *P. marinus* MED4 and *P. marinus* MIT9313 (Palenik *et al.*, 2003; Rocap *et al.*, 2003), and others (http://www.ncbi.nlm.nih.gov/genomes/lproks.cgi).

Genome sequences provide vast amounts of basic information, which can be exploited for the genome-wide study of gene expression. In 1999, Takara Bio Co. (Ohtu, Japan) initiated the production of a genome-wide cDNA microarray for analysis of gene expression in *Synechocystis*. Their DNA microarray covers 3079 (97%) of the 3165 genes on the chromosome of *Synechocystis* (99 genes for transposases are excluded from this calculation). Moreover, the entire sequences of all four of the plasmids (pSYSM, pSYSX, pSYSA and pSYSG) harbored by *Synechocystis* have been reported (Kaneko *et al.*, 2003). There are 132, 110, 106 and 49 putative genes, respectively, on these plasmids, namely, 397 genes in total. However, these genes are not included in the DNA microarray from Takara Bio Co.

Segment-based DNA microarrays are also available for *Anabaena* (Sato *et al.*, 2004). The complete genome of this cyanobacterium includes 5368 protein-coding genes on the main chromosome of 6.4 Mb (Kaneko *et al.*, 2001). Sato *et al.* (2004) amplified 2407 segments of 3 to 4 kb of *Anabaena* DNA by PCR and spotted them on glass slides for use as microarrays and, more recently, another type of microarray specific for the *Anabaena* genome has been developed that involves spotted oligonucleotides (Ehira and Ohmori, 2006; Higo *et al.*, 2006). This microarray covers a set of regulatory genes in *Anabeana*.

The original results of analysis of patterns of gene expression are accessible at the KEGG expression database (http://www.genome.jp/kegg/expression/).

In this chapter, we summarize recent progress in studies of stress-induced gene expression in *Synechocystis*. These studies, using DNA microarrays and modern methods for the identification of stress-inducible genes and regulatory systems, would not have been possible without the availability of entire genome sequences.

Responses to environmental stress at the transcriptional level

For many years, studies of gene expression at the transcriptional level have exploited Northern blotting, and only very limited numbers of genes were selected for analysis by the research community. The development of methods that included reverse transcriptase-polymerase chain reaction (RT-PCR) significantly broadened the spectrum of investigated genes. Analysis of the expression of all the genes in *Synechocystis* became possible once the complete genome had been sequenced and DNA microarrays became available. DNA microarrays specific for *Synechocystis* cover almost all the genes on one small glass plate, on which cDNA for each gene (or a fragment thereof) is spotted individually. Such a microarray allows the analysis of changes in levels of the transcripts of each gene in response to a particular environmental stress. Moreover, this approach allows us to identify the functions of previously uncharacterized genes that respond to various kinds of stress. Furthermore, in combination with targeted mutagenesis, DNA microarrays allow us to identify genes that are involved in the perception of changes in environmental conditions and in the regulation of stress-dependent expression of downstream genes.

The first successful exploitation of microarrays in studies of cyanobacteria was reported by Suzuki *et al.* (2001), who studied the reaction of *Synechocystis* to cold stress, and by Hihara *et al.* (2001), who investigated changes in gene expression in the same strain in response to light stress.

Cold stress-inducible genes

Cold stress-inducible changes in gene expression were studied in detail at the genome level by our group (Suzuki *et al.*, 2001; Mikami *et al.*, 2002; Inaba *et al.*, 2003). Cold stress was applied by exposing *Synechocystis* cells, which had been grown at 34°C, to 22°C for 20–30 min. We chose only brief exposure to 22°C to allow us to follow the primary events in changes in gene expression in response to cold stress so that we could separate the primary responses from subsequent events. We found more than 100 cold stress-inducible genes by microarray analysis when we set the ratio of transcript levels before and after stress as 1:3 or greater (Suzuki *et al.*, 2001; Inaba *et al.*, 2003)

Table 6.1 shows a list of cold-inducible genes together with the ratios of levels of transcripts before and after cold stress. The most strongly inducible genes are *ndhD2*, which encodes a subunit of NADH dehydrogenase (Dzelzkalns *et al.*, 1994; Ohkawa *et al.*, 2000), and *crhR* (*deadD*), which encodes RNA helicase (Chamot *et al.*, 1999; Yu and Owttrim, 2000). Cold stress also enhances the expression of three *hli* genes, which have been defined as high light-inducible genes (He *et al.*, 2001). The proposed role of CrhR in cold acclimation is the destabilization of the secondary structures of mRNAs, which reverses the cold-induced interruption of the initiation of translation at low temperatures (Yu and Owttrim, 2000).

Table 6.1 includes several genes whose functions have been well characterized, for example, *fus* for elongation factor EF-G; *ycf39* for the quinone-binding protein; *rbpA* for an RNA binding protein (Sato, 1995; Maruyama *et al.*, 1999). However, before we obtained our microarray data, it was difficult to predict that such genes would respond to cold stress. Moreover, there are a number of genes for putative proteins of unknown function that are also strongly induced by cold stress (see the lower part of Table 6.1).

The list of cold-inducible genes (Table 6.1) includes the *desB* and *desA* genes for ω3 and Δ12 fatty acid desaturases, respectively, whose cold inducibility was established in our previous studies by Northern blotting analysis (Los *et al.*, 1997). However, the extent of their induction was low as compared to that of other cold-inducible genes. Thus, genome-wide analysis of gene expression revealed that these important genes, which are responsible for the maintenance of membrane fluidity under cold stress (Vigh *et al.*, 1993; Murata and Los, 1997), do not belong to the group of most strongly cold-induced genes.

Table 6.1 Cold-inducible genes in *Synechocystis* sp. PCC 6803

ORF	Gene	Encoded protein	Ratio of transcript levels
slr1291	*ndhD2*	NADH dehydrogenase I chain M	22.0 ± 7.1
ssl2542	*hliA*	High light-inducible protein	20.7 ± 1.0
slr0083	*crhR*	ATP-dependent RNA helicase	16.3 ± 4.8
ssr2595	*hliB*	High light-inducible protein	10.6 ± 3.2
sll1541		Putative lignostilbene-dioxygenase	6.4 ± 0.8
slr1105	*fus*	Elongation factor EF-G	6.2 ± 0.5
sll0086		Arsenical pump-driving ATPase	5.9 ± 1.0
sll1483		Periplasmic protein	5.5 ± 2.9
ssl1633	*hliC*	High light-inducible protein	5.5 ± 1.0
sll0517	*rbpA*	RNA-binding protein	5.4 ± 1.3
sll0384	*cbiQ*	ABC-type cobalt transport system permease protein	4.9 ± 1.1
slr0400		Putative inorganic polyphosphate/ATP-NAD kinase 1	4.7 ± 0.0
sll2012	*sigD*	RNA polymerase sigma factor	4.4 ± 0.9

Table 6.1 Continued

ORF	Gene	Encoded protein	Ratio of transcript levels
slr0955		Putative tRNA/rRNA methyltransferase	4.4 ± 0.8
slr0401		Spermidine/putrescine transport system protein	4.4 ± 0.1
slr1974		Putative GTP-binding protein	4.4 ± 0.5
slr1254	*crtP*	Phytoene desaturase	4.3 ± 0.6
slr1392	*feoB*	Ferrous iron transport protein B	4.3 ± 0.8
slr0399	*ycf39*	Quinone-binding protein in PS II	4.3 ± 1.2
slr0423	*rlpA*	Rare lipoprotein A	4.1 ± 1.1
sll1441	*desB*	Fatty acid desaturase	4.0 ± 0.0
sll0385	*cbiO*	ABC-type cobalt transport system protein	4.0 ± 0.7
slr1992	*gpx2*	Glutathione peroxidase	3.7 ± 0.3
slr0426	*folE*	GTP cyclohydrolase I	3.7 ± 0.1
sll1770		ABC1-like	3.6 ± 0.5
sll1147		Glutathione S-transferase	3.6 ± 1.3
sll1772	*mutS*	DNA mismatch repair protein	3.6 ± 0.7
slr1881	*natE*	ABC-type neutral amino acid transport system	3.6 ± 0.4
slr1544		Hypothetical protein	20.6 ± 5.7
slr0082		Hypothetical protein	15.4 ± 4.0
slr1927		Hypothetical protein	6.6 ± 0.9
ssr2016		Hypothetical protein	6.2 ± 0.6
sll1611		Hypothetical protein	5.2 ± 0.8
sll1911		Hypothetical protein	5.1 ± 0.5
slr1747		Hypothetical protein	5.0 ± 0.0
sll0494		Hypothetical protein	4.5 ± 0.1
sll0815		Hypothetical protein	4.4 ± 0.5
slr0236		Hypothetical protein	4.4 ± 0.5
slr1677		Hypothetical protein	4.1 ± 1.3
slr1436		Hypothetical protein	4.0 ± 0.6
sll0355		Hypothetical protein	3.9 ± 0.7
slr0616		Hypothetical protein	3.8 ± 1.7
sll0185		Hypothetical protein	3.7 ± 0.6

Cells that had been grown at 34°C were incubated at 22°C for 30 min. Genes with ratios of transcript levels greater than 3.5 are listed. Data adapted from Inaba *et al.* (2003) with inclusion of more recent results.

Hyperosmotic stress-inducible genes

Genome-wide responses of *Synechocystis* cells to hyperosmotic stress were also investigated by our group (Kanesaki *et al.*, 2002; Mikami *et al.*, 2002; Paithoonrangsarid *et al.*, 2004; Shoumskaya *et al.*, 2005; Shapiguzov *et al.*, 2005). Hyperosmotic stress was achieved by exposing *Synechocystis* cells to 0.5 M sorbitol for 15–20 min. Analyses of genome-wide levels of transcripts indicated that hyperosmotic stress induces the expression of 97 genes with a ratio of transcript levels before and after stress of 1:3 or higher (Paithoonrangsarid *et al.*, 2004).

Table 6.2 lists genes whose expression is strongly induced by hyperosmotic stress. These genes includes the so-called heat-shock genes, such as *hspA* for a small heat-shock protein (sHSP), *htp*G for HSP90, *dnaK2* for HSP70, and *dnaJ* for HSP40, as well as *groES*, *groEL1* and *groEL2* that encode chaperonins. Genes induced by hyperosmoric stress also include two genes for proteases, *clpB1* and *htrA*. Three *hli* genes for high light-inducible proteins, which are also cold-inducible (see above), are induced by hyperosmotic stress. In addition, hyperosmotic stress induces the expression of the *sodB* gene for superoxide dismutase and a number of genes involved in the regulation of gene expression, such as *sigB* and *sigD* for sigma factors of RNA polymerase; *hik34*, whose product, histidine kinase 34 (Hik34), controls heat-inducible gene expression (see below), and *spkH* for a putative serine-threonine protein kinase. Alternative sigma factors might facilitate the appropriate adjustment of the transcriptional machinery under stress.

The sets of genes induced by hyperosmotic stress and heat stress overlap significantly, perhaps as a result of the regulatory role of Hik34, which is involved in the perception of hyperosmotic stress, salt stress, and heat stress (see below for details). The effects of both heat stress and hyperosmotic stress might be related to the destabilization of proteins.

Table 6.2 Hyperosmotic stress-inducible genes in *Synechocystis* sp. PCC 6803

ORF	Gene	Encoded protein	Ratio of transcript levels
sll1514	*hspA*	Small heat-shock protein	87.7 ± 1.5
sll1483		Periplasmic protein	39.1 ± 2.2
sll0330	*fabG*	3-Ketoacyl-ACP reductase	32.7 ± 0.6
slr1963		Water-soluble carotenoid protein	17.3 ± 1.0
slr1641	*clpB1*	ClpB protein	17.2 ± 0.1
slr1204	*htrA*	Serine protease HtrA	16.0 ± 0.2
slr0423	*rlpA*	Rare lipoprotein A	14.6 ± 0.3
ssl2542	*hliA*	High light-inducible protein	12.4 ± 0.1
ssr2595	*hliB*	High light-inducible protein	11.4 ± 0.1
sll0306	*sigB*	RNA Polymerase sigma factor	9.7 ± 0.1
slr1516	*sodB*	Superoxide dismutase	9.3 ± 0.1
sll0430	*htpG*	Heat-shock protein 90	9.3 ± 0.3
slr1285	*hik34*	Sensor histidine kinase Hik34	7.5 ± 0.3

ORF	Gene	Encoded protein	Ratio of transcript levels
ssl1633	*hliC*	High light-inducible protein	7.4 ± 0.1
sll0170	*dnaK2*	Heat-shock protein 70	6.8 ± 0.2
sll0005	*spkH*	Serine/threonine kinase	6.1 ± 0.2
slr2076	*groEL1*	60-kDa chaperonin 1	6.0 ± 0.1
slr0093	*dnaJ*	Heat-shock protein 40	5.2 ± 0.4
slr2075	*groES*	10-kDa chaperonin	5.2 ± 0.2
sll0416	*groEL2*	60-kDa chaperonin 2	5.0 ± 0.1
slr0381	*gloA*	Lactoylglutathione lyase	4.6 ± 0.2
sll2021	*sigD*	RNA polymerase sigma factor	4.3 ± 0.2
sll1863		Hypothetical protein	50.4 ± 0.8
sll1862		Hypothetical protein	45.6 ± 1.4
sll0528		Hypothetical protein	40.2 ± 1.5
sll0846		Hypothetical protein	25.1 ± 0.3
slr1544		Hypothetical protein	22.3 ± 0.4
sll1722		Hypothetical protein	12.6 ± 1.2
slr0959		Hypothetical protein	10.1 ± 0.6
sll0939		Hypothetical protein	9.1 ± 0.1
sll1884		Hypothetical protein	8.9 ± 0.1
slr1603		Hypothetical protein	8.9 ± 0.1
slr1915		Hypothetical protein	8.4 ± 0.3
ssr2016		Hypothetical protein	8.1 ± 0.4
ssl2971		Hypothetical protein	7.6 ± 0.2
slr1413		Hypothetical protein	7.1 ± 0.5
slr0967		Hypothetical protein	7.0 ± 0.3
slr0581		Hypothetical protein	6.4 ± 0.2
ssr1853		Hypothetical protein	6.3 ± 0.2
slr1119		Hypothetical protein	6.0 ± 0.3
ssl3446		Hypothetical protein	5.4 ± 0.2
slr0852		Hypothetical protein	5.0 ± 0.2
ssr3188		Hypothetical protein	4.6 ± 0.1
slr0112		Hypothetical protein	4.6 ± 0.1
sll0294		Hypothetical protein	4.5 ± 0.2

Cells that had been grown under normal osmotic conditions were incubated under hyperosmotic conditions (0.5 M sorbitol) for 20 min. Genes with ratios of transcript levels greater than 4 are listed. Data adapted from Paithoonrangsarid *et al*. (2004).

When *Synechocystis* cells are exposed to hyperosmotic stress, their cytoplasmic volume decreases rapidly (Kanesaki *et al.*, 2002; Shapiguzov *et al.*, 2005), causing, perhaps, the tight packing and aggregation of cellular proteins. Thus, the synthesis of chaperones and proteases might be necessary when cells acclimate to hyperosmotic stress. The role of serine-threonine protein kinases and, in particular of the product of the *spkH* gene, in stress responses has not yet been studied, to our knowledge.

There are many genes, unlisted in Table 6.2, whose expression is also enhanced to a lesser extent by hyperosmotic stress. These genes include *rpoA* and *rpoB* for the α and β subunits of RNA polymerase, *ggpS* for glucosylglycerol-phosphate synthase and *crhR*, which encodes an RNA helicase (Paithoonrangsarid *et al.*, 2004).

It should be noted that, in *Synechocystis*, the expression of a relatively small number of genes is induced specifically by hyperosmotic stress. Genes whose expression is strongly induced specifically by hyperosmotic stress are the *fabG* gene for 3-oxoacyl-(acyl-carrier protein) reductase, the *spkH* gene for a serine/threonine protein kinase, and the *gloA* gene for lactoylglutathione lyase (Table 6.2).

Among the hyperosmotic stress-inducible genes, a substantial number of genes encode proteins of as-yet unknown function (Table 6.2, lower part). These genes might play important roles in the acclimation of *Synechocystis* to hyperosmotic stress. Further investigations with genetically engineered mutants are necessary to elucidate the roles of these genes in the response to hyperosmotic stress.

Salt stress-inducible genes

The results of salt stress due to 0.5 M NaCl have been studied in great detail in *Synechocystis* (Kanesaki *et al.*, 2002; Marin *et al.*, 2003; 2004; Shoumskaya *et al.*, 2005). Exposure to salt stress for 15 min induces the expression of approximately 360 genes and represses the expression of approximately 200 genes when the ratio of transcript levels is set at 1:3 or higher and 3:1 or lower, respectively (Kanesaki *et al.*, 2002). According to the timing of maximum induction or repression, salt stress-regulated genes can be divided into four groups (Marin *et al.*, 2004). The first group, which has been the main focus of our attention, includes genes whose expression is only transiently induced or repressed within 10–15 min after the initiation of stress treatment.

Table 6.3 lists genes whose expression is rapidly and strongly induced by salt stress. Resembling hyperosmotic stress, salt stress induces the expression of heat-shock genes, such as *hspA*, *htpG*, *dnaK2*, *dnaJ*, and *groEL2*, and of genes for proteases, such as *clpB1* and *htrA*, as well as the *hli*, *sodB*, *sigB* and *sigD* genes (see previous section for details of these genes).

The set of salt stress-inducible genes is very similar to the set of hyperosmotic stress-inducible genes. The reason for activation of the expression of similar sets of genes might be the rapid decrease in cytoplasmic volume that follows exposure to both types of stress (Kanesaki *et al.*, 2002; Shapiguzov *et al.*, 2005).

Heat stress-inducible genes

Heat-induced gene expression was studied in *Synechocystis* at the genome-wide level by Inaba *et al.* (2003) and Suzuki *et al.* (2005). Upon heat stress, achieved by exposing *Synechocystis* cells, which had been grown at 34°C, to 44°C for 20 min, the levels of expression

Table 6.3 Salt-inducible genes in *Synechocystis* sp. PCC 6803

ORF	Gene	Encoded protein	Ratio of transcript levels
sll1514	*hspA*	Small heat-shock protein	49.7 ± 6.8
slr1641	*clpB1*	ClpB protein	22.3 ± 1.3
sll0306	*sigB*	RNA polymerase sigma factor	20.3 ± 2.2
slr1516	*sodB*	Superoxide dismutase	16.1 ± 2.0
ssr2595	*hliB*	High light-inducible protein	15.1 ± 0.1
ssl3044		Hydrogenase component	14.5 ± 0.0
slr1963		Water-soluble carotenoid protein	14.2 ± 2.2
sll1566	*ggpS*	Glucosylglycerol-phosphate synthase	13.2 ± 2.3
sll0170	*dnaK2*	Heat-shock protein 70	12.0 ± 0.2
sll1167	*pbp*	Penicillin-binding protein 4	11.1 ± 0.5
ssl2542	*hliA*	High light-inducible protein	9.8 ± 0.0
sll1594	*ndhR*	Transcriptional regulator of NdhF3	9.7 ± 0.4
sll1085	*glpD*	Glycerol-3-phosphate dehydrogenase	9.7 ± 0.8
slr1204	*htrA*	Serine protease HtrA	9.5 ± 0.8
slr0093	*dnaJ*	Heat-shock protein 40	9.4 ± 2.0
sll0938		N-Succinyldiaminopimelate aminotransferase	8.5 ± 4.1
sll1621		Membrane protein	8.4 ± 0.8
sll1483		Periplasmic protein	8.0 ± 0.0
sll1797	*ycf21*	Ycf21 gene product	6.2 ± 0.2
slr1916		Esterase	6.1 ± 0.1
ssl1633	*hliC*	High light-inducible protein	4.9 ± 0.1
sll0416	*groEL2*	60-kDa chaperonin 2	5.5 ± 0.8
slr1192		Zinc-containing alcohol dehydrogenase	5.1 ± 0.2
sll2012	*sigD*	RNA polymerase sigma factor	4.9 ± 0.4
slr0095		O-methyltransferase	4.6 ± 0.2
sll1862		Hypothetical protein	152.4 ± 6.4
sll1863		Hypothetical protein	106.5 ± 8.3
sll0528		Hypothetical protein	74.4 ± 4.0
sll0939		Hypothetical protein	35.8 ± 3.1
slr0967		Hypothetical protein	32.3 ± 8.9
slr1544		Hypothetical protein	23.2 ± 0.1
slr1603		Hypothetical protein	22.5 ± 2.1

Table 6.3 Continued

ORF	Gene	Encoded protein	Ratio of transcript levels
slr0959		Hypothetical protein	19.3 ± 1.1
slr1687		Hypothetical protein	16.0 ± 0.4
sll0846		Hypothetical protein	14.6 ± 1.1
slr1704		Hypothetical protein	11.9 ± 3.4
sll1884		Hypothetical protein	11.3 ± 0.8
slr1915		Hypothetical protein	10.3 ± 0.5
sll1722		Hypothetical protein	10.3 ± 0.2
ssr3188		Hypothetical protein	9.4 ± 1.5
ssr2194		Hypothetical protein	8.9 ± 2.1
slr1686		Hypothetical protein	7.3 ± 0.7
ssl2971		Hypothetical protein	7.2 ± 0.6
slr0895		Hypothetical protein	7.2 ± 1.1
sll1652		Hypothetical protein	6.1 ± 0.7
slr0852		Hypothetical protein	5.9 ± 0.1

Cells that had been grown in standard growth medium were incubated in the presence of 0.5 M NaCl for 20 min. Genes with ratios of transcript levels greater than 4.5 are listed. Data adapted from Shoumskaya *et al.* (2005).

of 59 genes rose more than three-fold and those of 58 genes fell more than three-fold (Suzuki *et al.*, 2005). Most of the 59 genes were strongly induced within 10–20 min after an upward shift in temperature from 34°C to 44°C. However, 1 hour after the onset of heat stress the levels of their transcripts declined significantly.

Table 6.4 lists genes whose expression is rapidly and strongly induced by heat stress. The transiently induced genes include *hspA*, *groES*, *groEL1*, *groEL2*, *dnaJ*, *htpG*, and *dnaK2*, which encode proteins that are mostly chaperones. Heat stress also induces the expression of the *clpB1* and *htrA* genes for proteases, the *sigB* gene for a group 2 sigma factor of RNA polymerase, and the *hik34* gene, as well as the *sodB* gene for superoxide dismutase and the *hliC* gene for a high light-inducible protein.

Expression of none of these genes is specific to heat stress; the expression of all these genes is also induced by hyperosmotic stress and salt stress (see above), as well as by oxidative stress, strong visible light and UV-B light (see below). The genes whose expression is specifically induced by heat mainly encode proteins whose functions are, as yet, unknown (*slr0095*, *slr1920*, *slr1232*, and some others).

The levels of expression of most genes that are involved in energy and lipid metabolism, pigment biosynthesis and photosynthesis fall under heat-stress conditions (Suzuki *et al.*, 2005).

Table 6.4 Heat-inducible genes in *Synechocystis* sp. PCC 6803

ORF	Gene	Encoded protein	Ratio of transcript levels
slr1641	*clpB1*	ClpB protease	92.5 ± 9.1
sll1514	*hspA*	Small heat-shock protein	68.1 ± 2.9
slr2075	*groES*	10-kDa chaperonin	44.7 ± 2.3
sll0416	*groEL2*	60-kDa chaperonin 2	42.1 ± 7.7
slr0093	*dnaJ*	Heat-shock protein 40	36.9 ± 3.6
sll0430	*htpG*	Heat-shock protein 90	34.6 ± 5.6
slr0095		*o*-Methyltransferase	24.4 ± 4.6
sll0306	*sigB*	RNA polymerase group 2 s-factor	22.9 ± 0.7
slr1963		Water-soluble carotenoid protein	22.8 ± 0.0
sll0170	*dnaK2*	Heat-shock protein 70	22.3 ± 1.5
slr1285	*hik34*	Histidine kinase 34	18.9 ± 2.6
sll1621		AhpC/TSA family protein	11.0 ± 0.6
ssl3044		Ferredoxin-like protein	10.3 ± 0.0
slr2076	*groEL1*	60-kDa chaperonin 1	8.7 ± 1.5
ssl1633	*hliC*	CAB/ELIP/HLIP superfamily	6.9 ± 0.5
slr1204	*htrA*	Serine protease	6.0 ± 0.7
slr1516	*sodB*	Superoxide dismutase	5.6 ± 0.3
sll1483		Salt-inducible periplasmic protein	5.1 ± 1.1
sll1009	*frpC*	Iron-regulated protein	4.9 ± 0.5
slr1675	*hypA1*	Hydrogenase component protein	4.7 ± 0.1
sll1507		Salt-induced periplasmic protein	4.0 ± 0.6
slr1674		Hypothetical protein	28.3 ± 0.8
sll0528		Hypothetical protein	19.5 ± 0.9
ssl2971		Hypothetical protein	15.1 ± 1.7
sll0939		Hypothetical protein	11.9 ± 0.2
sll0846		Hypothetical protein	11.7 ± 3.0
slr0967		Hypothetical protein	9.8 ± 0.6
slr1915		Hypothetical protein	8.5 ± 0.8
slr0589		Hypothetical protein	7.8 ± 0.5
slr1686		Hypothetical protein	7.6 ± 0.5
slr1597		Hypothetical protein	6.7 ± 1.0
sll1433		Hypothetical protein	6.1 ± 0.3

Table 6.4 Continued

ORF	Gene	Encoded protein	Ratio of transcript levels
sll1680		Hypothetical protein	5.4 ± 0.4
slr1603		Hypothetical protein	5.2 ± 0.0
sll0688		Hypothetical protein	4.4 ± 0.6
slr0364		Hypothetical protein	4.4 ± 0.2
sll1652		Hypothetical protein	4.3 ± 0.1
slr1413		Hypothetical protein	4.2 ± 0.6
slr1712		Hypothetical protein	4.1 ± 0.0

Cells that had been grown 34°C were incubated at 44°C for 20 min. Genes with ratios of transcript levels greater than 4 are listed. Data adapted from Suzuki *et al*. (2005).

Oxidative stress-inducible genes

Oxidative stress-inducible expression of genes in *Synechocystis* has been investigated with DNA microarrays by three research groups (Kobayashi *et al.*, 2004; Singh *et al.*, 2005; Kanesaki *et al.*, 2007). Oxidative stress due to 0.25 mM H_2O_2 strongly induces the expression of 118 genes for proteins of known and as yet unknown function, with ratios of transcript levels of 1:3 and above (Kanesaki *et al.*, 2007).

Table 6.5 lists genes whose expression is strongly induced by oxidative stress due to H_2O_2. These genes include *hspA, dnaJ, clpB1, dnaK2, ctpA, hypA1, sigB, htpG*, and *hik34*, whose expression is also strongly induced by heat stress (Inaba *et al.*, 2003; Suzuki *et al.*, 2005). H_2O_2 also enhances the expression of the three *hli* genes. Furthermore, as mentioned above, the expression of these genes is also strongly induced by salt and hyperosmotic stress.

Oxidative stress induces the expression of the *pgr5* gene for ferredoxin:plastoquinone reductase (Yeremenko *et al.*, 2005), *gifA* and *gifB* genes for glutamine synthase inactivation factors (Garcia-Dominguez *et al.*, 1999), and two genes, *nblA1* and *nblA2*, that are involved in the degradation of phycobilisomes (Baier *et al.*, 2001). However, oxidative stress strongly represses the expression of the majority of genes involved in pigment biosynthesis and in photosynthesis (Singh *et al.*, 2005).

Among the genes whose expression is specifically regulated by H_2O_2, we find the so-called "detoxification" gene, namely, the *aphC* gene for type II peroxiredoxin (Hosoya-Matsuda *et al.*, 2005) and the *perR* gene that encodes the transcription regulator PerR. Expression of two genes, whose products are associated with the scavenging of peroxide, namely, *katG* (*sll1987*) and *tpx* (*sll0755*), is not significantly enhanced under oxidative stress.

Light stress-inducible genes

The influence of strong visible light on gene expression in *Synechocystis* was investigated, using DNA microarrays, by three research groups (Hihara *et al.*, 2001; Huang *et al.*, 2002; Hsiao *et al.*, 2004). Expression of more than 160 genes is enhanced within 15 min (ratio

Table 6.5 Hydrogen peroxide-inducible genes in *Synechocystis* sp. PCC 6803

ORF	Gene	Encoded protein	Ratio of transcript levels
ssr2016	*pgr5*	Ferredoxin:plastoquinone reductase	120.2 ± 25.8
sll1515	*gifB*	Glutamine synthetase inactivating factor	81.3 ± 4.50
sll1514	*hspA*	Small heat-shock protein	76.8 ± 17.4
ssr2595	*hliB*	High light-inducible protein	49.2 ± 5.4
sll1549		Salt-enhanced periplasmic protein	39.4 ± 9.3
ssl1911	*gifA*	Glutamine synthetase inactivating factor	33.2 ± 7.8
ssl2542	*hliA*	High light inducible protein	29.7 ± 7.5
ssl0453	*nblA2*	Phycobilisome degradation protein NblA	29.3 ± 5.5
ssl3044		Hydrogenase component	23.4 ± 2.6
slr0093	*dnaJ*	DnaJ protein, heat-shock protein 40	22.1 ± 5.7
ssl0452	*nblA1*	Phycobilisome degradation protein NblA	18.1 ± 4.3
slr1291	*ndhD2*	NADH dehydrogenase I chain M	18.0 ± 2.3
sll1483		Periplasmic protein	15.2 ± 6.3
slr1641	*clpB1*	ClpB protein	13.7 ± 8.2
sll1621	*ahpC*	Probable alkyl hydroperoxide reductase	13.4 ± 6.0
ssl1633	*hliC*	CAB/ELIP/HLIP superfamily	12.4 ± 2.2
slr1738	*perR*	Transcription factor PerR	12.0 ± 2.2
sll0330		Sepiapterine reductase	9.7 ± 2.6
slr0611	*sds*	Solanesyl diphosphate synthase	9.4 ± 1.5
slr1318	*fecE*	Iron(III) dicitrate transport protein	9.1 ± 2.1
sll0248	*isiB*	Flavodoxin	8.9 ± 2.3
sll0247	*isiA*	Chlorophyll-binding protein	8.2 ± 5.4
sll0990		Formaldehyde dehydrogenase	8.1 ± 1.5
slr1295	*futA1*	ABC-type iron-transport system	7.9 ± 1.2
slr0228	*ftsH*	Cell division protein FtsH homolog	6.9 ± 1.5
slr1963		Water-soluble carotenoid protein	6.7 ± 2.4
sll0170	*dnaK2*	DnaK protein 2, heat-shock protein 70	6.5 ± 1.6
slr0008	*ctpA*	Carboxyl-terminal processing protease	6.4 ± 0.6
slr1675	*hypA1*	Hydrogenase expression/formation protein	6.3 ± 2.4
sll0306	*sigB*	RNA polymerase group 2 sigma factor	6.2 ± 0.8
sll1541		Lignostilbene-alpha, beta-dioxygenase	5.3 ± 0.9
slr0095		O-methyltransferase	5.1 ± 1.2

Table 6.5 Continued

ORF	Gene	Encoded protein	Ratio of transcript levels
sll0992		Esterase	5.0 ± 1.0
slr1544		Hypothetical protein	131.3 ± 11.2
sll0528		Hypothetical protein	74.4 ± 27.3
sll0846		Hypothetical protein	29.6 ± 8.3
slr1687		Hypothetical protein	27.9 ± 4.5
slr0967		Hypothetical protein	26.5 ± 4.7
slr1674		Hypothetical protein	18.9 ± 0.5
sll0939		Hypothetical protein	13.3 ± 3.4
slr1603		Hypothetical protein	13.2 ± 1.9
slr0589		Hypothetical protein	12.1 ± 3.7
sll0185		Hypothetical protein	11.5 ± 1.5
ssl2162		Hypothetical protein	8.6 ± 1.9
slr1686		Hypothetical protein	8.2 ± 2.4
ssl3446		Hypothetical protein	8.1 ± 1.9
ssr2333		Hypothetical protein	7.9 ± 1.2
ssr0692		Hypothetical protein	7.1 ± 1.3
slr1484		Hypothetical protein	7.0 ± 1.3
slr0270		Hypothetical protein	6.5 ± 1.9
sll0157		Hypothetical protein	6.4 ± 1.0
slr1119		Hypothetical protein	5.8 ± 1.2
slr1413		Hypothetical protein	5.7 ± 2.3
sll0249		Hypothetical protein	5.7 ± 2.4
sll0688		Hypothetical protein	5.7 ± 1.4
ssl0832		Hypothetical protein	5.6 ± 0.3
sll1158		Hypothetical protein	5.5 ± 1.8
sll1620		Hypothetical protein	5.4 ± 0.6
slr0818		Hypothetical protein	5.2 ± 0.9

Cells that had been grown under normal conditions were incubated in the presence of 0.25 mM H_2O_2 for 20 min. Genes with ratios of transcript levels greater than 5 are listed. Data adapted from Kanesaki *et al*. (2007) under permission of the publisher.

of transcript levels, 1:2 or greater) after cells are transferred from light at 20 to light at 300 μmol photons m^{-2} s^{-1}(Hihara *et al*., 2001).

Table 6.6 shows that light stress induces the expression of genes for NAD(P)H dehydrogenase, namely, *ndhD2*, *ndhD3*, *ndhF3*, and *ndhK*, and of four genes from two

Table 6.6 Light stress-inducible genes in *Synechocystis* sp. PCC 6803.

ORF	Gene	Encoded protein	Ratio of transcript levels
sll1514	*hspA*	Small heat-shock protein	12.2 ± 0.9
sll1733	*ndhD3*	NAD(P)H dehydrogenase component	11.4 ± 1.6
sll1732	*ndhF3*	NAD(P)H dehydrogenase component	8.2 ± 2.3
slr1641	*clpB1*	ClpB protease	8.2 ± 0.3
slr1291	*ndhD2*	NAD(P)H dehydrogenase component	7.6 ± 0.2
sll1483		Periplasmic protein	7.5 ± 0.4
sll0170	*dnaK2*	Heat-shock protein 70	6.4 ± 0.3
sll0430	*htpG*	Heat-shock protein 90	6.2 ± 0.2
sll1028	*ccmK1*	CO_2-concentrating mechanism protein	6.2 ± 0.1
sll0217		Potential FMN protein	6.0 ± 0.2
ssl3044		Hydrogenase component protein	5.7 ± 0.2
sll1032	*ccmN*	CO_2-concentrating mechanism protein	5.6 ± 0.3
slr0228	*ftsH*	Cell division protein FtsH	5.5 ± 1.0
slr2076	*groEL1*	60-kDa chaperonin 1	5.4 ± 0.2
slr2075	*groES*	10-kDa chaperonin	5.3 ± 0.2
slr1675	*hypA1*	Hydrogenase component protein	5.2 ± 0.3
sll1031	*ccmM*	CO_2-concentrating mechanism protein	4.7 ± 0.1
sll0416	*groEL2*	60-kDa chaperonin 2	4.7 ± 0.1
sll1029	*ccmK2*	CO_2-concentrating mechanism protein	4.6 ± 0.2
slr1604	*ftsH*	Cell division protein FtsH	4.6 ± 0.1
sll1541		Lignostilbene-a,b-dioxygenase	4.4 ± 0.1
slr1516	*sodB*	Superoxide dismutase	4.2 ± 0.2
sll0218		Hypothetical protein	11.3 ± 0.5
sll0528		Hypothetical protein	15.6 ± 0.5
slr1544		Hypothetical protein	9.2 ± 0.7
slr1963		Hypothetical protein	6.1 ± 0.2
sll0846		Hypothetical protein	5.6 ± 0.2
slr0959		Hypothetical protein	5.0 ± 0.2
sll1734		Hypothetical protein	4.5 ± 0.1

Cells that had been grown in light at 20 mmol photons m^{-2} s^{-1} were incubated in light at 300 mmol photons m^{-2} s^{-1} for 15 min. Genes with ratios of transcript levels greater than 4 are listed. Data adapted from Hihara *et al.* (2001) under permission of the publisher.

operons, namely, *ccmK1-ccmK2* and *ccmM-ccmN*. The expression of the *ndh* and *ccm* genes is induced by signals common to exposure of cells to a low concentration of CO_2 and to strong light.

Light stress markedly induces the expression of a set of heat-inducible genes, namely, *hspA, clpB1, dnaK2, htpG, groES, groEL1,* and *groEL2*. Expression of most of the transcripts of heat-shock genes is induced only transiently within 15 min. However, levels of *groES, groEL,* and *groEL2* transcripts remain high even 15 hours after the start of exposure to strong light (Hihara *et al.*, 2001; Huang *et al.*, 2002).

One might expect that light stress, which accelerates the production of reactive oxygen species (ROS), would induce the expression of genes that encode enzymes that scavenge ROS. However, only the expression of the *sodB* gene for superoxide dismutase is enhanced. Levels of transcripts of *katG* for catalase, *gpx1* for glutathione peroxidase, and *tpx* for thioredoxin peroxidase are not significantly affected.

Strong light decreases the expression of the majority of genes whose products are involved in the absorption of light and photochemical reactions. Levels of transcripts of genes for proteins in the oxygen-evolving complex, namely, *psbO, psbU,* and *psbV*, all fall by 50% of control levels within 15 min. However, light stress does not affect levels of transcripts of other genes that encode small subunits of PSII, such as *psbE, psbF,* and *psbK*. Light stress decreased by 5- to 10-fold the levels of transcripts of almost all genes for subunits of PSI (Huang *et al.*, 2002). In particular, the levels of *psaAB* transcripts for the proteins in the reaction center of PSI were 30-fold lower than original levels prior to light stress. By contrast, expression of one of the two *psaK* genes was markedly enhanced within one hour and the level of the transcript remained high throughout the exposure to strong light.

UV light-inducible genes

The effect of UV-B light (280 to 320 nm at 20 µmol photons m^{-2} s^{-1}) on gene expression was studied in *Synechocystis* with custom-made DNA microarrays by Huang *et al.* (2002). The levels of expression of 55 genes rose more than two-fold and that of 44 genes fell more than two-fold during exposure of *Synechocystis* cells to UV-B light for 2 hours.

Table 6.7 lists genes whose expression was strongly induced by UV-B light. UV-B light induced the expression of two genes, *psbA2* and *psbA3*, that encode identical polypeptide precursors to the D1 protein. UV-B light also induced expression of the *ctpA* and *ftsH* genes for proteases that are involved in the maturation and degradation of the D1 protein and of the *nblB* gene, whose product is involved in the degradation of phycobilisomes. Exposure to UV-B light led to induction of the expression of *hli* genes and also of heat-shock genes *hspA, clpB, htpG,* and *dnaK2*.

Iron limitation-inducible genes

Cyanobacteria require various metal ions for their growth, development and function. Singh *et al.* (2003) performed a detailed study, using microarrays, of gene expression under iron-limited conditions. When *Synechocystis* cells are grown in iron-depleted medium for 3 hours, expression of the *isiA* and *isiB* genes for a chlorophyll-binding protein and for flavodoxin, respectively (Kunert *et al.*, 2003), is induced. It is of interest to note that expression of these genes is induced by oxidative stress (see Table 6.5). Expression of several

Table 6.7 UV-B-inducible genes in *Synechocystis* sp. PCC 6803

ORF	Gene	Encoded protein	Ratio of transcript levels
sll1514	*hspA*	Small heat-shock protein	7.8 ± 4.2
ssl2595	*hliB*	High-light-inducible protein	7.3 ± 0.8
slr1641	*clpB*	ClpB protease	6.6 ± 2.1
sll1483		Periplasmic protein	4.7 ± 1.5
slr1311	*psbA2*	Photosystem II D1 protein	4.4 ± 1.5
sll1867	*psbA3*	Photosystem II D1 protein	4.1 ± 0.7
ssl2542	*hliA*	High light-inducible protein	4.1 ± 0.6
slr1687	*nblB*	Phycobilisome degradation protein NblB	3.9 ± 0.6
slr1675	*hypA*	Hydrogenase	3.7 ± 0.9
sll0430	*htpG*	Heat-shock protein 90	3.7 ± 1.4
slr1604	*ftsH*	Cell division protein FtsH	3.6 ± 1.1
sll0170	*dnaK2*	Heat shock protein 70	3.4 ± 0.6
slr0228	*ftsH*	Cell division protein FtsH	3.3 ± 0.9
slr1285	*hik34*	Histidine kinase Hik34	3.1 ± 0.6
sll0528		Hypothetical protein	7.4 ± 2.8
slr1544		Hypothetical protein	6.5 ± 2.5
sll0846		Hypothetical protein	5.6 ± 1.0
slr1603		Hypothetical protein	3.6 ± 0.6
slr1686		Hypothetical protein	3.1 ± 1.3

Cells that had been grown in white light at 25 mmol photons m^{-2} s^{-1} were incubated in UV-B light (280–320 nm) at 20 mmol photons m^{-2} s^{-1} for 2 hours. Genes with ratios of transcript levels greater than 3 are listed. Data adapted from Huang *et al*. (2002) under permission of the publisher.

genes related to iron uptake, such as *slr0513* for periplasmic Fe-binding protein and *fnuA* for ferrichrome-Fe receptor protein, is also induced.

Manganese limitation-inducible genes

Manganese is an essential transition metal in almost all organisms. In addition, it plays a critical role in oxygenic photosynthesis, since a cluster of four Mn atoms in the PSII complex catalyzes the dehydrogenation of water to produce molecular oxygen. The effect of manganese limitation on gene expression was studied with DNA microarrays after *Synechocystis* cells had been incubated in the absence of Mn^{2+} ions for 24 hours (Yamaguchi *et al.*, 2002). Incubation of *Synechocystis* cells in Mn^{2+}-depleted medium significantly enhanced the expression of the *mntA*, *mntB*, and *mntC* genes, which are clustered as a single operon. The product of this operon, *mntCAB*, is an ABC-type permease that mediates high-affinity transport of Mn^{2+} ions in *Synechocystis* (Bartsevich and Pakrasi, 1995; 1996). No other genes appear to be affected by the limitation of manganese.

Phosphate limitation-inducible genes

Phosphate is an essential nutrient, and it is required for the biosynthesis of nucleotides, DNA and RNA, and for the functional regulation of proteins by phosphorylation. However, phosphate is one of the least available nutrients in the environment. Therefore, regulatory mechanisms have developed for the acquisition, storage, and metabolism of phosphate. One such regulatory mechanism involves the induction of the expression of genes for proteins that participate in the uptake of phosphate and/or increase the availability of phosphate under phosphate-limiting conditions. The effect of phosphate limitation on gene expression in *Synechocystis* was studied with microarrays after cells had been incubated in phosphate-depleted medium for 8 hours.

Phosphate limitation strongly induced the expression of the Pho regulon, which includes genes for two types of phosphate transporter (Pst), alkaline phosphatase, and an extracellular nuclease. Both Pst systems, designated Pst1 and Pst2, are encoded by similar clusters of genes. The Pst1 system includes six open-reading frames (*sll0679–sll0684*) in the following order: *sphX-pstS1-pstC1-pstA1-pstB1-pstB1*. The Pst2 system includes four open-reading frames (*slr1247–slr1250*), namely, *pstS2-pstC2-pstA2-pstB2*. The *phoA* gene for alkaline phosphatase and the *nucH* gene for the extracellular nuclease, both of which are involved in increasing the availability of phosphate in the extracellular environment, are located in tandem on the *Synechocystis* genome.

Phosphate limitation also repressed the expression of a number of genes. The expression of the *urtA* gene for the binding protein of urea ABC-type transporter was strongly repressed in response to phosphate limitation. This gene is located at a different site from the phosphate limitation-inducible genes on the genome of *Synechocystis* (Suzuki *et al.*, 2004).

General features of stress-inducible genes

A specific stress induces expression of a specific set of genes. Table 6.8 shows the characteristic distribution of stress-inducible genes according to the nature of the stress. Some genes are induced only by one kind of stress, while other genes are induced in response to several kinds of stress. For example, genes for heat-shock proteins and for proteases respond to various kinds of stress; their expression is induced by heat stress, salt stress, hyperosmotic stress, oxidative stress, UV-B, and light stress, but not by cold stress. Two genes for sigma factors, namely, *sigB* and *sigD*, three *hli* genes, and the *sodB* gene are induced also by various kinds of stress. Nevertheless, the genes induced by cold stress and heat stress are very different with the exception of *hliC*, which turns out to be the only gene, identified to date, that is induced by all types of environmental stress tested. All the other strongly stress-inducible genes listed in Table 6.8 respond to one or even two types of stress. These genes belong to those whose products are involved in gene expression, signal transduction, metabolism, photosynthesis, ROS scavenging and transport of nickel and iron ions (Table 6.8).

Large numbers of genes for hypothetical proteins and for proteins of totally unknown function are affected by environmental stress. These genes represent about half of all the genes on the *Synechocystis* genome. Expression of some of these genes is induced by a specific type of stress, whereas that of some others is induced by several types of stress, as is

Table 6.8 Genes whose expression is strongly induced by five types of environmental stress in *Synechocystis* sp. PCC 6803

			Stress				
ORF	Gene	Encoded protein	Cold	Osmo	Salt	H_2O_2	Heat
Transcription, translation, protein modification							
slr0083	*crhR*	ATP-dependent RNA helicase	+				
slr1105	*fus*	Elongation factor EF-G	+				
sll0517	*rbpA*	RNA-binding protein	+				
sll1594	*ndhR*	Transcriptional regulator of NdhF3			+		
Sigma factors							
sll0306	*sigB*	RNA polymerase sigma factor		+	+		+
sll2012	*sigD*	RNA polymerase sigma factor	+	+	+		
Heat-shock genes							
sll1514	*hspA*	Small heat-shock protein		+	+	+	+
slr0093	*dnaJ*	Heat-shock protein 40		+	+	+	+
sll0170	*dnaK2*	Heat-shock protein 70		+	+		+
sll0416	*groEL2*	60-kDa Chaperonin 2		+	+		+
slr2076	*groEL1*	60-kDa Chaperonin 1		+			+
slr2075	*groES*	10-kDa chaperonin		+			+
sll0430	*htpG*	Heat-shock protein 90		+			+
Proteases							
slr1641	*clpB1*	ClpB protease 1		+	+	+	+
slr1204	*htrA*	Serine protease HtrA		+	+		+
Signal transduction							
slr1285	*hik34*	Sensor histidine kinase Hik34		+			+
sll0005	*spkH*	Serine/threonine kinase		+			
Metabolism							
slr0423	*rlpA*	Rare lipoprotein A	+	+			
sll0330	*fabG*	3-Ketoacyl-ACP reductase		+			
slr1254	*crtP*	Phytoene desaturase	+				
sll1441	*desB*	Fatty acid desaturase	+				

Table 6.8 Continued

ORF	Gene	Encoded protein	Stress				
			Cold	Osmo	Salt	H_2O_2	Heat
slr0426	*folE*	GTP cyclohydrolase I	+				
ssl1911	*gifA*	Glutamine synthetase inactivating factor				+	
sll1515	*gifB*	Glutamine synthetase inactivating factor				+	
sll1566	*ggpS*	Glucosylglycerol phosphate synthase			+		
sll1085	*glpD*	Glycerol-3-phosphate dehydrogenase			+		
Photosynthesis and respiration							
ssl1633	*hliC*	High light-inducible protein	+	+	+	+	+
ssl2542	*hliA*	High light-inducible protein	+	+	+	+	
ssr2995	*hliB*	High light-inducible protein	+	+	+	+	
slr1291	*ndhD2*	NADH dehydrogenase 1 chain M	+			+	
slr0399	*ycf39*	Quinine-binding protein in photosystem II	+				
sll0248	*isiB*	Flavodoxin				+	
sll0247	*isiA*	Chlorophyll-binding protein				+	
ssr2016	*pgr5*	Ferredoxin:plastoquinone reductase				+	
ssl0453	*nblA2*	Phycobilisome degradation protein NblA				+	
ssl0452	*nblA1*	Phycobilisome degradation protein NblA				+	
ROS scavenging							
slr1516	*sodB*	Superoxide dismutase		+	+		+
slr1736	*perR*	Transcription factor PerR				+	
sll1621	*ahpC*	Probable alkyl hydroperoxide reductase				+	
slr1992	*gpx2*	Glutathione peroxidase	+				
Transport							
sll0384	*cbiQ*	ABC-type cobalt transport system	+				

Table 6.8 Continued

ORF	Gene	Encoded protein	Stress: Cold	Osmo	Salt	H_2O_2	Heat
sll0385	*cbiO*	ABC-type cobalt transport system protein	+				
slr1392	*feoB*	Ferrous iron transport protein B	+				
slr1295	*futA1*	ABC-type iron-transport system				+	
Others							
slr1675	*hypA1*	Hydrogenase component protein					+
sll1009	*frpC*	iron-regulated protein					+
slr0611	*sds*	Solanesyl diphosphate synthase				+	
slr0228	*ftsH*	Cell division protein				+	
slr0381	*gloA*	Lactoylglutathione lyase		+			
sll1797	*ycf21*	Ycf21 gene product			+		
sll1167	*pbp*	Penicillin-binding protein 4			+		

For cold stress, cells that had been grown at 34°C were incubated at 22°C for 30 min; for hyperosmotic stress, cells that had been grown under normal osmotic conditions were incubated under hyperosmotic conditions due to 0.5 M sorbitol for 20 min; for salt stress, cells that had been grown in standard growth medium were incubated in medium supplemented with 0.5 M NaCl for 20 min; for oxidative stress, cells that had been grown in normal medium were incubated in medium supplemented with 0.25 mM H_2O_2 for 20 min; and for heat stress, cells that had been grown at 34°C were incubated at 44°C for 20 min.

+ indicates that a respective gene is strongly induced by an individual type of stress; approximately 20 genes whose expression is most strongly induced under each type of stress are taken from Tables 6.1–6.5.

the case for genes for proteins of known function. Thus, much work remains to be done in characterizing all the effects of various stresses on gene expression in *Synechocystis*.

Responses to environmental stress at the protein level

Several studies of responses to environmental stress at the protein level have recently been reported (Fulda *et al.*, 2006; Huang *et al.*, 2006; Slabas *et al.*, 2006). Application of matrix-assisted laser desorption ionization-time of flight (MALDI-TOF) analysis, in combination with large two-dimensional gels, is a powerful technique for the automated identification of proteins and the analysis of their stress inducibility.

Salt stress-inducible proteins

In a study focused on the identification of salt stress-inducible proteins in the periplasmic space of *Synechocystis*, Fulda *et al.* (2000) identified 57 proteins, including six proteases, as well as several enzymes involved in cell wall biosynthesis. However, most of the periplasmic proteins were proteins of unknown function. In particular, salt stress greatly increased levels of Slr1485 and Sll1549, both of which have unknown functions. Genes for all 57 salt stress-inducible proteins in the periplasm encode precursors to periplasmic proteins with signal peptides; 47 of the precursors contain a typical Sec-dependent signal peptide, whereas 10 contain a putative twin-arginine signal peptide (Fulda *et al.*, 2000).

Recent analysis by MALDI-TOF of *Synechocystis* under salt stress (Fulda *et al.*, 2006) revealed that salt stress enhances levels of 55 proteins: 18 salt-inducible proteins appear after salt-stress treatment for 2 hours, whereas synthesis of 38 proteins is induced in cells that have been acclimated to 684 mM NaCl for five days. Salt stress specifically induces the expression of a small number of proteins, such as enzymes involved in the synthesis of glucosylglycerol, whereas synthesis of other salt-inducible proteins is also induced by other kinds of environmental stress. These proteins are heat-shock proteins, such as DnaK2, GroEL1, GroEL2, HspA, and GroES, as well as enzymes that counteract the production of ROS, such as superoxide dismutase and peroxiredoxin.

A detailed analysis of proteins in cytoplasmic membranes that respond to salt stress resulted in the identification of 109 proteins, which corresponded to the products of 66 different genes. Levels of 25 of these proteins were enhanced or reduced under salt stress. Seven of the salt-enhanced proteins that were identified in the plasma membranes of *Synechocystis* are substrate-binding proteins of ABC transporters (Huang *et al.*, 2006), including Slr1740, a putative oligopeptide-binding protein; Sll1450, the nitrate/nitrite-binding protein NrtA; Slr1295, the iron-binding protein FutA1; Slr0529, the glucosylglycerol-binding protein GgtB; Sll0680, the phosphate-binding protein PstS; and Sll1080, a putative nitrate/sulfonate/bicarbonate/taurine-binding protein. These binding proteins include a putative lipoprotein anchor for attachment to the plasma membrane at their amino terminus (Huang *et al.*, 2006).

Heat stress-inducible proteins

We used the MALDI-TOF analysis to examine heat-induced proteins (Slabas *et al.*, 2006). When we exposed *Synechocystis* cells, which had been grown at 34°C, to 44°C for 60 min, we observed, on two-dimensional gels, more than 100 spots that corresponded to 65 different proteins. The most remarkable changes that occurred upon heat shock corresponded to components involved in pathways for the folding, breakdown of aggregates, and degradation of proteins. The small heat-shock protein HspA showed the greatest change, with an increase in level of almost 20-fold. Smaller changes in levels were observed for GroES, GroEL1, GroEL2, HtpG, DnaK, and ClpB, which is an ATP-dependent Clp protease.

Levels of three proteases also rose upon heat shock: oligopeptidase A; the periplasmic protease HhoA; and ClpC, the ATPase subunit of the ClpC/ClpP protease. The levels of peptidyl-prolyl *cis-trans* isomerase (PPIase) also rises in heat-shocked cells.

The activity of the metabolic machinery for protein synthesis increases upon heat shock. Levels of two ribosomal subunits (30S S1 and 50S L6), two tRNA synthetases (the β chain of glycyl-tRNA synthase and aspartyl-tRNA synthase) and four elongation factors (EF-G1, EF-G2, EF-Tu and EF-Ts) are elevated. Levels of five enzymes, which are involved in amino acid biosynthesis as follows, are also enhanced: D-3-phosphoglycerate dehydrogenase (Sera) involved in the metabolism of glycine, serine and threonine; arginosuccinate synthetase (ArgG) involved in the metabolism of aspartate, arginine and proline; aspartate kinase (LysC) involved in the synthesis of glycine, serine, threonine and lysine; and keto-acidreductoisomerase (IlvC) and dihydroxyacid dehydratase (IlvD) both involved in the metabolism of valine, leucine and isoleucine (Slabas *et al.*, 2006).

The PSII complex and, in particular, the oxygen-evolving machinery is the most heat-sensitive component of the photosynthetic apparatus. After heat shock, there is a two-fold increase in the level of the manganese-stabilizing protein, PsbO. Levels of phycobillisome proteins fall upon heat shock and the level of RbcL, the large subunit of ribulose-1,5-bisphosphatase oxygenase (Rubisco) falls to less than half of the control level.

Involvement of two-component regulatory systems in signal transduction during environmental stress

The existence of two-component systems for signal transduction has been well established in eubacteria (Aguilar *et al.*, 2001; Stock *et al.*, 2000). A typical two-component system consists of a histidine kinase (Hik) and a cognate response regulator (Rre). In *Escherichia coli* and *Bacillus subtilis*, genes for the cognate Hik and Rre in each two-component system are, in most cases, located close to each other on the genome. The Hik perceives changes in the environment via its sensor domain and then a histidine residue within the conserved histidine kinase domain is autophosphorylated, with ATP as the donor of the phosphate group (Stock and Da, 2000). The phosphate group is transferred from the Hik to an aspartate residue in the conserved receiver domain of the cognate Rre. Upon phosphorylation, the Rre changes its conformation, and this change allows the Rre to bind to the promoter regions of genes that are located further downstream in the signal-transduction pathway. Two-component systems have been found in prokaryotes including cyanobacteria, fungi, yeasts, plants, and lower animals (Koretke *et al.*, 2000) but not in higher animals. In eukaryotes and, in particular, in higher plants and higher animals, serine/threonine protein kinases are the major sensors and transducers of environmental signals (Widmann *et al.*, 1999).

The availability of the complete sequence of the *Synechocystis* genome (http://www.kazusa.or.jp/cyano/Synechocystis/index.html) has allowed the construction of "gene-knockout" libraries by random mutagenesis with antibiotic-resistance cassettes (Suzuki *et al.*, 2000a) and with transposons (Bhaya *et al.*, 2001) and the determination of the functions of the respective gene products under certain stress conditions. The availability of the sequence of the entire genome allows the identification of the genes and sites that have been mutated.

Another powerful technique, also based on the availability of the complete genome sequence, involves the construction of "knockout" libraries of specific sets of genes by targeted mutagenesis. This technique was used successfully for the identification of many potential sensors and transducers of environmental signals in *Synechocystis* (Suzuki *et al.*, 2000a; Paithoonrangsarid *et al.*, 2004; Shoumskaya *et al.*, 2005). In addition, DNA microarray analysis of the genome-wide expression of genes has allowed examinations of the effects of mutations of specific genes for Hiks and Rres on the global expression of genes.

The *Synechocystis* chromosome includes 44 putative genes for Hiks (Kaneko *et al.*, 1996; Mizuno *et al.*, 1996) and there are three genes for Hiks on its plasmids (Kaneko *et al.*,

2003). There are 42 putative genes for Rres on the chromosome and three on the plasmids. These Hiks and Rres are candidates for sensors and transducers of environmental signals. However, in *Synechocystis*, Hiks and their cognate Rres are not located in close proximity on the genome in most cases, in contrast to their organization in *E. coli* and *B. subtilis*. Therefore, to investigate the specific functions of individual Hiks and Rres, it is necessary to mutate Hiks and Rres individually and then to examine the phenotype of the resultant mutants. We succeeded in completely replacing 44 of the 47 *hik* genes and 42 of the 45 *rre* genes by mutated genes (http://www.kazusa.or.jp/cyano/Synechocystis/mutants/index.html). The *hik* and *rre* genes that we could segregate incompletely were the *hik2, hik11, hik26, rre23, rre25,* and *rre26* genes

The Hiks in *Synechocystis* appear to be involved in the perception and transduction of environmental signals due to low-temperature stress (Suzuki *et al.*, 2000a; 2000b; 2001), high osmolarity (Mikami *et al.*, 2002; Paithoonrangsarid *et al.*, 2004), high concentrations of NaCl (Marin *et al.*, 2003; Shoumskaya *et al.*, 2005), phosphate deprivation (Hirani *et al.*, 2001; Suzuki *et al.*, 2004), excess nickel ions (Lopez-Maury *et al.*, 2002), and manganese limitation (Ogawa *et al.*, 2002; Yamaguchi *et al.*, 2002). In the following sections we will describe the principles of regulation of gene expression by two-component systems and we will provide examples of the way in which we identified Hik-Rre two-component systems as sensors and transducers of environmental signals.

Positive and negative regulation of gene expression

There are, in principle, two types of regulation of stress-inducible gene expression mediated by Hik/Rre systems, namely, positive regulation and negative regulation. In positive regulation (Figure 6.2A), a Hik is inactive under non-stress conditions and, as a result, the corresponding Rre is inactive. Genes that are regulated by this type of two-component system are silent (Figure 6.2A) under non-stress conditions. When cells are exposed to the appropriate environmental stress, the Hik is activated (by phosphorylation in response to the stress) and then the signal is transferred to the cognate Rre. The activated Rre enhances the expression of genes that are silent under non-stress conditions (Figure 6.2A). Most of the stress-inducible regulation of gene expression in *Synechocystis* is associated with this type of regulation (Murata and Suzuki, 2006).

In the negative regulation of the stress-inducible expression of genes (Figure 6.2B), the Hik is active under non-stress conditions and, as a result, the cognate Rre is active. Genes that are regulated by this type of two-component system are repressed under non-stress conditions. When the appropriate environmental stress is applied, the Hik becomes inactive and the Rre also becomes inactive. As a result, genes that are repressed under non-stress conditions are released from repression (Figure 6.2B). Therefore, the levels of expression of genes appear to increase.

Changes in phenotype due to mutations in Hiks and Rres, which reflect the effects of these Hiks and Rres on gene expression, differ between positive regulation and negative regulation. A knockout mutation of either the Hik or the Rre in a two-component system for negative regulation has a marked effect on gene expression under non-stress conditions. The expression of genes under a negatively regulating two-component system is enhanced under non-stress conditions (as described in the following section). Therefore, a specific signal-transduction pathway with a specific Hik and its cognate Rre can be identified with

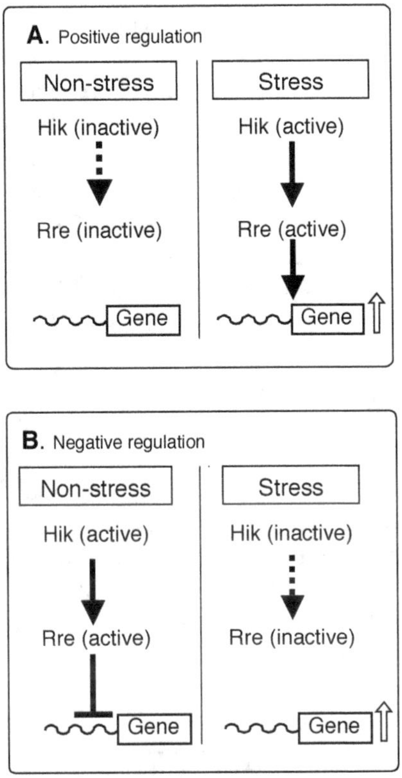

Figure 6.2 Schematic representation of the positive regulation (A) and negative regulation (B) of the stress-inducible expression of genes. Solid arrows indicate signals that activate downstream components and dotted arrows indicate their absence. Inverted T's indicate signals that repress the expression of the downstream genes. Open arrows correspond to the enhancement of gene expression. Adapted from Murata and Suzuki (2006).

relative ease in cases of negative regulation. By contrast, knockout mutation of a Hik and an Rre in a two-component system that operates via positive regulation does not reveal a significant effect on gene expression under non-stress conditions. In this type of two-component system, the identification of the Hik and the Rre in a specific signal-transduction pathway requires the screening of knockout libraries of *hik* and *rre* genes under individual types of stress.

The Hik33-Rre26 system regulates expression of cold-inducible genes

In extensive studies in our laboratory, mutation of the 44 *hik* genes, except *hik20*, *hik27* and *hik34* (see below) did not induce significant changes in gene expression. Therefore, we postulated that the Hiks encoded by these genes might regulate gene expression in a positive manner. To confirm this hypothesis, we screened our library of *hik* mutants under specific kinds of stress.

We had previously identified the histidine kinase Hik33 as a cold sensor in *Synechocystis* (Suzuki *et al.*, 2000a). DNA microarray analysis of Δ*hik33* cells indicated that Hik33 regulates the expression of 21 of 35 cold-inducible genes (Figure 6.3), with ratios of transcript levels higher than 3.0. These 21 genes include *ndhD2, hliA, hliB, hliC, fus, feoB, crtP,* and genes for proteins of unknown function. By contrast, 14 of the 35 cold-inducible genes were not regulated by Hik33. Therefore, we deduced that *Synechocystis* might have another pathway for cold-signal transduction. The genes that are not controlled by Hik33 include *crhR* for RNA helicase and *rbpA1* for a ribosome-binding protein.

To identify the Rre that is located downstream of Hik33, we screened an Rre knockout library by RNA slot–blot hybridization using some of the cold-inducible genes whose expression is controlled by Hik33 as probes. We identified Rre26 as a candidate for the Rre that, with Hik33, constitutes a two-component system for cold-signal transduction (Murata and Los, 2006).

Five two-component systems contribute to the perception and transduction of salt-stress signals

We used DNA microarrays to investigate the impact of a mutation in each Hik on changes in gene expression upon exposure of cells to salt stress. Our comprehensive screening revealed that the inducibility of gene expression by elevated levels of NaCl was significantly affected in Δ*hik16* (*slr1805*), Δ*hik33* (*sll0698*), Δ*hik34* (*slr1285*), and Δ*hik41* (*sll1229*) mutant cells (Marin *et al.*, 2003). In each of these mutants, the expression of several genes was no longer induced by salt or the extent of inducibility by salt was markedly reduced.

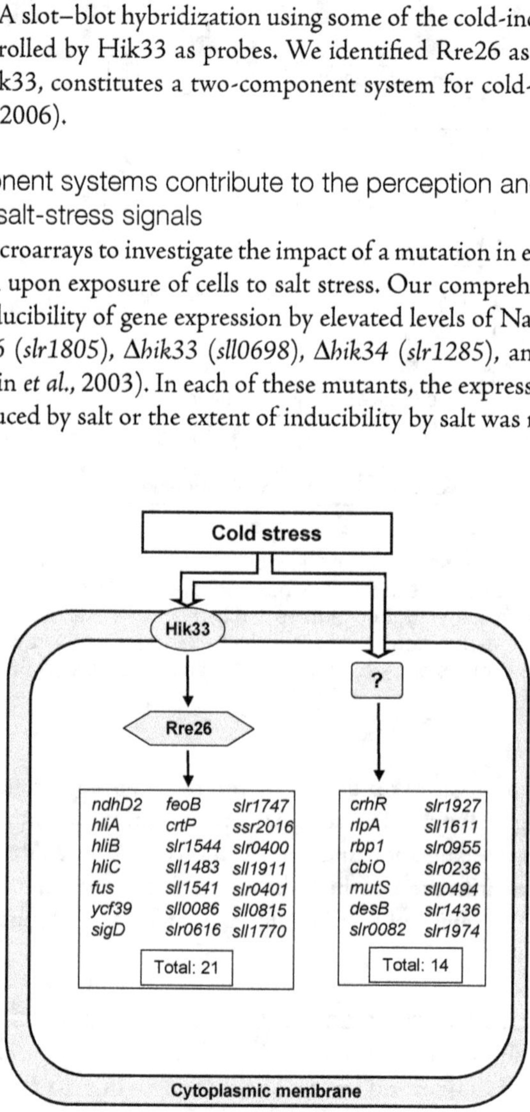

Figure 6.3 A scheme showing the Hik33-Rre26 two-component system that is involved in the transduction of cold stress, as well as the genes that are under the control of this two-component system. Primary signals are represented by open arrows. Selectively regulated genes are shown in boxes. Uncharacterized mechanisms are represented by a question mark. Genes with induction factors higher than 3.0 are included in this scheme. Adapted from Murata and Los (2006) under permission of the publisher.

By screening an Rre knockout library by RNA slot–blot hybridization and with the genome-wide DNA microarray, our group identified four Hik-Rre systems, namely, Hik33-Rre31, Hik10-Rre3, Hik16-Hik41-Rre17 and Hik34-Rre1, that appeared to be involved in the perception of salt stress and transduction of the signal (Shoumskaya *et al.*, 2005). Figure 6.4A shows a hypothetical scheme for the salt signal-transducing systems that involve these Hiks and Rres; the scheme includes the salt-inducible genes that are controlled by individual Hik-Rre two-component systems.

The Hik33-Rre31 system regulates the inducibility by high concentrations of NaCl of the expression of seven genes. Inactivation of either the *hik33* or the *rre31* gene resulted

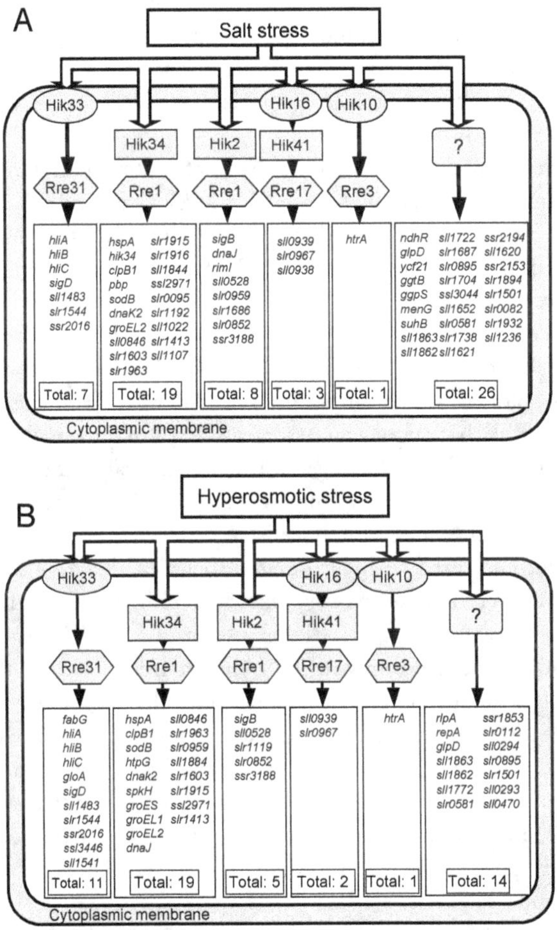

Figure 6.4 Schemes showing the two-component systems that are involved in the transduction of salt stress and hyperosmotic stress, as well as the genes that are under the control of the individual two-component systems. Primary signals are represented by open arrows. Hiks are indicated as ellipses or squares, Rres are indicated as hexagons, and selectively regulated genes are shown in boxes. Uncharacterized mechanisms are represented by question marks. Genes with induction factors higher than 4.0 are included in these schemes. (A) Salt stress; adapted from Shoumskaya *et al.* (2005). (B) Hyperosmotic stress. Adapted from Paithoonrangsarid *et al.* (2004) and Shoumskaya *et al.* (2005).

in the elimination of, or a marked reduction in, the salt stress-inducible expression of these genes. The genes are *hliA, hliB, hliC, sigD* and other as yet uncharacterized genes. Our findings indicate that Hik33 and Rre31 are tightly coupled in salt-signal transduction. The Hik10-Rre3 two-component system regulates the salt stress-inducible expression of only one gene, *htrA*, which encodes a serine protease.

The Hik16-Hik41-Rre17 system regulates the salt-inducible expression of three genes, namely, *sll0939, slr0967*, and *sll0938* (Shoumskaya *et al.*, 2005). Inactivation of *hik16*, of *hik41* or of *rre17* eliminated the expression of these genes, suggesting that Hik16, Hik41, and Rre17 are all essential for the perception of salt stress and for transduction of the signal that controls the expression of these genes. It seems likely that Hik41 acts downstream of Hik16 since Hik41 is a hybrid-type histidine kinase that contains both a signal-receiver domain and a histidine kinase domain. By contrast, Hik16 is a typical Hik with a histidine kinase domain and a potential sensory domain that, hypothetically, spans the membrane seven times. It is also possible that Hik16 and Hik41 might together perceive salt stress, acting as a complex.

The Hik34-Rre1 two-component system regulates the expression of 26 salt stress-inducible genes that include *hspA, dnaK2, groEL2, sodB, clpB1* and other genes of uncharacterized function (Figure 6.4A). The inducible expression of these genes was abolished or significantly reduced by mutation of either Hik34 or Rre1, suggesting the tight coupling of these two components.

The expression of the *sigB, dnaJ*, and *rimI* genes, as well as that of five other salt stress-inducible genes, was controlled by Rre1 but not by Hik34. Screening the library of mutant Hiks by RNA slot–blot hybridization with a probe derived from the *sigB* gene did not allow us to identify a Hik that might act upstream of Rre1 in the signal-transduction pathway that regulates the expression of this gene. These results suggest that such a Hik might be encoded by a gene that cannot be completely replaced by the corresponding mutated gene in Δ*hik2*, Δ*hik11*, and Δ*hik26* mutant cells.

Experiments with a yeast two-hybrid system revealed that Hik2 can associate very tightly with Rre1. Therefore, it seems very likely that Hik2 and Rre1 might constitute a two-component system that regulates the expression of all eight genes in response to salt stress. The studies by our group with DNA microarrays revealed that expression of 26 of the 64 salt stress-inducible genes was not controlled by any of the Hik-Rre two-component systems discussed above. It seems likely that the signals, due to salt stress, that induce the expression of these genes might be perceived by as yet unknown mechanisms that are separate from the Hik-Rre two-component systems. Salt stress might act directly to regulate either the transcription or the stability of the transcripts of these inducible genes.

Five two-component systems contribute to the perception and transduction of hyperosmotic-stress signals

In the literature, hyperosmotic stress and salt stress are frequently regarded as identical forms of stress. However, they are distinctly different. Hyperosmotic stress causes the efflux of water from the cytoplasm, resulting in a decrease in cytoplasmic volume and increased concentrations of cytoplasmic solutes and other components. By contrast, salt stress decreases the cytoplasmic volume only transiently and to a small extent (Kanesaki *et al.*, 2002), while rapidly increasing cytoplasmic concentrations of Na^+ and Cl^- ions (Allakhverdiev *et al.*, 2002) via an influx of NaCl through Na^+/K^+ and Cl^- channels.

Earlier studies in our laboratory demonstrated that perception of hyperosmotic stress involves a histidine kinase, Hik33, and other unknown components (Mikami *et al.*, 2002), and that the set of hyperosmotic stress-inducible genes differs from the set of salt stress-inducible genes (Kanesaki *et al.*, 2002). However, we found subsequently that both types of stress appear to share common sensors and response regulators (Paithoonrangsarid *et al.*, 2004).

Systematic mutagenesis of almost all the *hik* and *rre* genes in *Synechocystis*, in combination with the genome-wide analysis of gene expression at the transcriptional level, allowed the identification of four Hik-Rre systems, namely, Hik33-Rre31, Hi34-Rre1, Hik16-Hik41-Rre17, and Hik10-Rre3, and another potential two-component system, namely Hik2-Rre1, for the perception and transduction of hyperosmotic-stress signals. The expression of most of the hyperosmotic stress-inducible genes is controlled by these Hik-Rre systems (Paithoonrangsarid *et al.*, 2004). Figure 6.4B shows the signal-transduction pathways that operate when cells are exposed to hyperosmotic stress, as well as the hyperosmotic stress-inducible genes whose expression is controlled by the individual Hik-Rre systems.

The Hik33-Rre31 system regulates the inducible expression of 11 genes. These 11 genes include *hliA*, *hliB*, *hliC* and *sigD*, which are induced by salt stress under the control of the Hik33-Rre31 system, and, in addition, *fabG*, *gloA*, *ssl3446* and *sll1541*, which are specifically induced by hyperosmotic stress (Figure 6.4; Paithoonrangsarid *et al.*, 2004; Shoumskaya *et al.*, 2005). The Hik10-Rre3 system regulates the hyperosmotic stress-inducible expression of only the *htrA* gene, as in the case of the salt inducibility of gene expression. The Hik16-Hik41-Rre17 system regulates the hyperosmotic stress-inducible expression of two genes, namely, *sll0939* and *slr0967*. The Hik34-Rre1 two-component system regulates the expression of 19 hyperosmotic stress-inducible genes. However, this set of inducible genes is not the same as the set of salt stress-inducible genes that are under the control of the Hik34-Rre1 system. For example, expression of the *htpG*, *spkH*, *groEL1*, and *groES* genes is induced by hyperosmotic stress but to a much lesser extent by salt stress, whereas *pbp* and some other uncharacterized genes are induced by salt stress but not by hyperosmotic stress (Shoumskaya *et al.*, 2005). The expression of the *sigB* gene and of four other hyperosmotic stress-inducible genes is controlled by Rre1, but not by Hik34. It seems likely that Hik2 and Rre1 might constitute a two-component system that regulates the expression of *sigB* and four other genes in response to hyperosmotic stress, a situation similar to that encountered during salt stress and described in the previous section.

The studies with DNA microarrays revealed that expression of 14 of the 48 hyperosmotic stress-inducible genes was not controlled by any of the five Hik-Rre two-component systems discussed above. It is likely that the signals due to hyperosmotic stress that induce the expression of these genes might be perceived by as yet unknown mechanisms that are distinct from the Hik-Rre two-component systems (Paithoonrangsarid *et al.*, 2004; Shoumskaya *et al.*, 2005).

In *Synechocystis*, salt stress and hyperosmotic stress appear to be perceived by identical sets of Hik-Rre systems, which regulate rather similar sets of genes, albeit to different extents. However, it is also clear that identical Hik-Rre systems may also control expression of different sets of genes under hyperosmotic stress and salt stress (Figure 6.4). Typical examples are the Hik33-Rre31 and Hik34-Rre1 pairs mentioned above. These findings cannot be explained in terms of the currently accepted concept of two-component systems. For example, although Hik33 might perceive both salt stress and hyperosmotic stress and transduce these signals to Rre31, the current concept of two-component systems does not explain how the Hik33-Rre31 system controls the differential expression of the two different sets of genes under different types of stress. It seems reasonable to postulate the presence of some unknown factor(s) that provides each two-component system with the strict specificity that is related to the specific nature of the stress.

Hik34 is involved in the control of the heat-stress response

Screening of our Hik knock-out library under non-stress conditions using DNA microarrays allowed us to identify Hik34 as an important contributor to the regulation of expression of heat-shock genes and to thermotolerance in *Synechocystis* (Suzuki *et al.*, 2005). Mutation of the *hik34* gene enhanced the levels of transcripts of a number of heat shock genes that included *htpG* and *groESL1* (Figure 6.5). Furthermore, overexpression of the *hik34* gene repressed the expression of these heat-shock genes. In addition, Hik34-mutant cells survived incubation at 48°C for 3 hours, while wild-type cells and cells with mutations in other Hiks failed to do so. However, mutation of the *hik34* gene had only an insignificant effect on the global expression of genes during incubation of the mutant cells at 44°C for 20 min. Among 59 heat stress-inducible genes with ratios of transcript levels greater than 1:3, mutation of the *hik34* gene markedly decreased the heat inducibility of only two genes, namely, *htpG* and *slr1963*. Recombinant Hik34 expressed in *E. coli* was autophosphorylated *in vitro* at physiological temperatures but not at elevated temperatures, such as 44°C. Thus, it seems likely that Hik34 is involved in the negative regulation of the expression of certain heat-shock genes that might be related to thermotolerance in *Synechocystis*.

Hik33 is a major contributor to signal transduction during oxidative stress

Screening of the library of Hik mutant cells by monitoring changes in the genome-wide expression of genes upon exposure to 0.25 mM H_2O_2 for 20 min revealed that four histidine kinases, namely, Hik34, Hik16, Hik41, and Hik33, regulate gene expression under oxidative stress. These Hiks regulate the expression of 26 of the 77 H_2O_2-inducible genes with ratios of transcript levels 1:4 or greater (Kanesaki *et al.*, 2006). Hik34, which was identified as a sensor/transducer of signals due to salt (Shoumskaya *et al.*, 2005), hyperosmotic (Paithoonrangsarid *et al.*, 2004), and heat (Suzuki *et al.*, 2005) stress, also regulates the expression of the H_2O_2-inducible *htpG* gene. Hik16 and Hik41 together regulate the expression of *sll0967* and *sll0939*. Hik33 is the main contributor to the regulation of H_2O_2-inducible gene expression and regulates the induction of expression of 22 of the 26 H_2O_2-inducible genes that are under the control of histidine kinases. Hik33 regulates the expression of *ndhD2*, of *hli* genes, of *pgr5* for ferredoxin:plastoquinone reductase, of *nblA1*

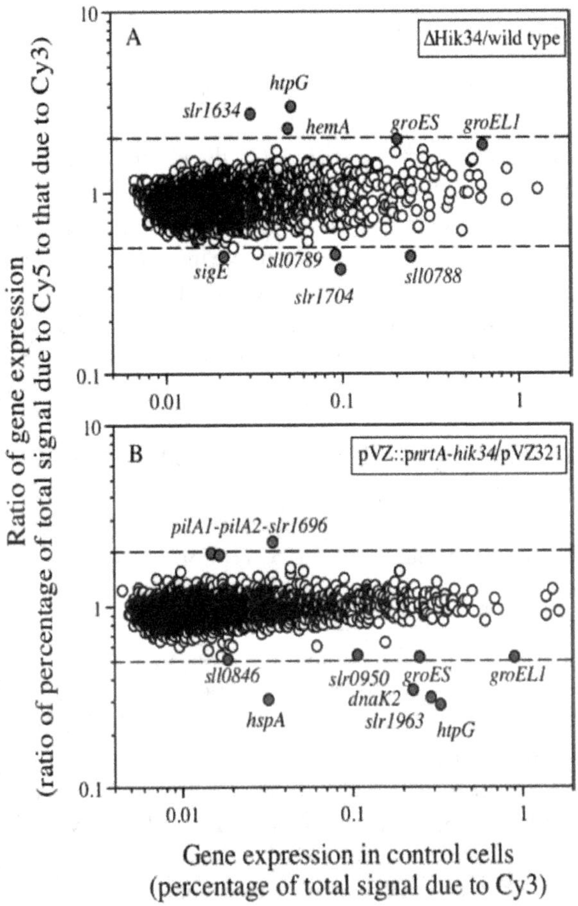

Figure 6.5 DNA microarray analysis of changes in gene expression under non-stress conditions upon mutation (A) and overexpression (B) of the *hik34* gene in *Synechocystis* cells. Adapted from Suzuki *et al.* (2005); see the original paper for full details.

and *nblA2* for proteins involved in phycobilisome degradation, and of several other genes (Kanesaki *et al.*, 2006).

Several Hiks are involved in the perception and transduction of light-stress signals

The potential sensors of light in *Synechocystis* are histidine kinases with phytochrome-like features (Kehoe and Grossman, 1994; 1996). The *Synechocystis* genome contains seven *hik* genes, namely, *hik35* (*slr0473*), *hik3* (*sll1124*), *hik32* (*sll1473*/*sll1475*), *hik1* (*slr1393*), *hik44* (*slr1212*), *hik24* (*slr1969*), and *hik18* (*sll0041*), that exhibit different degrees of similarity to plant genes for phytochromes.

When the *hik35* (*cph1*) gene was expressed in *E. coli* cells that synthesized phycocyanobilin, the *E. coli* cells produced an adduct with a red/far-red photoreversible signature that is typical of phytochrome (Hughes *et al.*, 1997). The *rre27* (*rcp1* or *slr0474*) gene is located adjacent to the *hik35* gene (Yeh *et al.*, 1997). Insertional inactivation of *hik35* in *Synechocystis* impaired the growth of mutant cells under continuous far-red light (Fiedler *et al.*, 2004).

We used microarrays in an attempt to identify the molecular targets of Hik35 (Cph1) activity (Hubschmann *et al.*, 2005). In wild-type and two lines of phytochrome-mutant cells, one-fourth of all 3165 putative genes responded to light. Red light predominantly enhanced the expression of genes whose products are involved in transcription, translation, and photosynthesis, whereas far-red light enhanced the levels of transcripts of genes whose expression is induced by various kinds of stress. The absence of Hik35 (Cph1) altered the light-dependent expression of approximately 20 genes. Hence, light receptor(s) different from this two-component system might trigger global red/far-red alterations in the profile of gene expression.

It does not seem likely that all of the Hiks mentioned above are typical phytochromes in terms of red/far-red reversibility. Indeed, Hik3 seems to be involved in blue light-dependent growth (Wilde *et al.*, 1997), and Hik44 (Etr1) binds ethylene (Sineshchekov *et al.*, 1998). The potential involvement in the perception of light-stress signals of Hik32, which is intact in a strain from the Pasteur Culture Collection but is disrupted by an insertion in a strain from Dupont (Okamoto *et al.*, 1999), Hik1, and Hik24 merits further investigation. The involvement of Hik33 in the regulation by light of gene expression has also been suggested (Hsiao *et al.*, 2004; Tu *et al.*, 2004).

In the filamentous cyanobacterium *Calothrix* sp. PCC 7601, Jorissen *et al.* (2002) identified CphA-RcpA and CphB-RcpB as two-component light-sensing phytochrome-like systems. These two regulatory systems are both homologous to Hik35-Rre27 in *Synechocystis*. In *Anabaena*, Vogeley *et al.* (2004) and Sineshchekov *et al.* (2005) identified cyanobacterial rhodopsin that has biochemical and biophysical characteristics suggestive of involvement in the regulation of chromatic adaptation.

The Hik27-Rre16 system negatively regulates gene expression in response to manganese limitation

Screening of our libraries of Hik- and Rre-mutant cells under non-stress conditions revealed that Hik27 and Rre16 constitute a two-component system for the perception of manganese deficiency and transduction of the resultant signal (Yamaguchi *et al.*, 2002). Using DNA microarrays, we compared gene expression in Δ*hik27* mutant cells with that in wild-type cells during growth under non-stress conditions, namely, in BG-11 medium that contained 9 μM Mn^{2+} ions. Marked changes, with ratios of transcript levels greater than 1:10, due to mutation of the *hik27* gene (*slr0640*) or the *rre16* gene (*slr1837*), were recognized only in the expression of three genes, namely, *mntC*, *mntA*, and *mntB*, which constitute the *mntCAB* operon that encodes subunits of the ABC-type Mn^{2+} transporter. These results suggest that the Hik27-Rre16 two-component system might be active under non-stress conditions and repress the expression of the *mntCAB* operon and, moreover, that disappearance of this signal, due to inactivation of Hik27 or Rre16, might allow the *mntCAB* operon to be expressed. However, mutation of *hik27* or *rre16* abolished the Mn^{2+} deficiency-induced expression of the *mntCAB* operon (Yamaguchi *et al.*, 2002). Ogawa *et al.* (2002), working independently and using traditional methods, identified the same two-component system as the sensor and signal transducer of Mn^{2+} deficiency.

The Hik7-Rre29 system regulates gene expression in response to phosphate limitation

Hirani *et al.* (2001) discovered that the *hik7* (*sll0337* or *sphS*) and *rre29* (*slr0081* or *sphR*) genes in *Synechocystis* encode proteins that are homologous, respectively, to PhoR and PhoB of *E. coli*, to PhoR and PhoP of *B. subtilis*, and to SphS and SphR of *Synechococcus* sp. PCC 7942. They reported that the Hik7-Rre29 two-component system induces the expression of the *phoA* gene for alkaline phosphatase in response to phosphate limitation.

DNA microarray analysis of the genome-wide expression of genes in *Synechocystis* revealed that the expression of 12 genes was strongly induced while that of only one gene was strongly repressed by phosphate limitation (Suzuki *et al.*, 2004). The expression of all the phosphate limitation-inducible genes was completely eliminated upon inactivation of either Hik7 or Rre29. The response regulator Rre29 binds to the upstream flanking regions of three genes at repetitive PyTTAAPyPy(T/A)-like sequences. Our observations suggest that the Hik7-Rre29 two-component system might be the only system for the perception and transduction of the phosphate-limitation signal in *Synechocystis* (Suzuki *et al.*, 2004).

The Hik30-Rre33 system regulates gene expression in response to excess Ni^{2+} ions

Lopez-Maury *et al.* (2002) identified the Hik30-Rre33 (or NrsS-NrsR) two-component system that regulates transcription of the nickel-resistance operon *nrsBACD*. The *hik30-rre33* operon is located near the *nrsBACD* operon for subunits of the ABC-type Ni^{2+} ion transporter. Knockout mutation of either the *hik30* or the *rre33* gene completely abolishes the Ni^{2+}-induced transcription of the *nrsBACD* operon. As a result, cells fail to survive when there are high concentrations of Ni^{2+} ions in the growth medium. It is probable that Hik30 binds a Ni^{2+} ion and is then autophosphorylated. The phosphate group is then transferred to Rre33, which binds to the intergenic regions of *nrsRS* and *nrsBACD*, which are located close to each other on the chromosome, stimulating transcription of the latter operon. This example, therefore, represents a positive type of regulatory system. However, to examine whether the Hik30-Rre33 system regulates only the *nrsBACD* operon or also regulates other operons and/or genes, it is necessary to use microarray analysis to investigate global patterns of gene expression under appropriate conditions.

Future perspectives in studies of two-component regulatory systems

Hik33 was first identified as a potential sensor of cold stress (Iwane *et al.*, 2001), and then of hyperosmotic stress (Mikami *et al.*, 2002). Further studies revealed that it is also involved in responses to salt stress (Marin *et al.*, 2003), oxidative stress (Kanesaki *et al.*, 2006), and light stress (Hsiao *et al.*, 2004). By comparing the stress-inducible genes that are under the control of Hik33 under individual types of stress, our group discovered that Hik33 regulates the expression of certain genes in a stress-specific manner. Hik33 regulates the expression of *fus*, *ycf39*, *feoB*, and *crtP* as cold-specific genes; of *fabG* and *gloA* as hyperosmotic stress-specific genes; and of *pgr5*, *nblA1* and *nblA2* as oxidative stress-specific genes. The Hik34-Rre1 system regulates the expression of *sll1107* as a salt stress-specific gene and of *htpG* as a hyperosmotic stress-specific gene (Figure 6.4).

The findings discussed above suggest that the perception of salt stress and hyperosmotic stress by the Hik-Rre systems is complex and that salt stress and hyperosmotic stress are perceived as distinct signals by the individual Hik-Rre systems. This hypothetical scenario cannot be interpreted by reference to the currently accepted scheme for two-component systems. Therefore, we are confident that as-yet unidentified components are present and that these components endow the sensory systems with the respective activities that allow the systems to induce responses specific to each individual type of stress.

Other potential sensors and transducers of environmental signals

Serine/threonine protein kinases, tyrosine protein kinases and protein phosphatases

While eubacteria, including cyanobacteria, use two-component systems for many types of signal transduction, eukaryotes exploit serine/threonine protein kinases (Ser/Thr kinases), tyrosine protein kinases (Tyr kinases), and protein phosphatases for similar purposes. The presence of genes for Ser/Thr kinases and phosphatases in cyanobacteria was revealed only after the complete sequence of *Synechocystis* genome had become available (Zhang *et al.*, 1998). The putative proteins were identified by comparison of deduced amino acid sequences encoded by open-reading frames with known amino acid sequences of eukaryotic protein kinases.

Among 12 putative genes for Ser/Thr kinases in *Synechocystis*, seven encode proteins that belong to the PKN2 subfamily of Ser/Thr kinases and five encode proteins that belong to the ABC1 subfamily of Ser/Thr kinases (Leonard *et al.*, 1998; Shi *et al.*, 1998). The genes for kinases of the PKN2 type are designated *spkA, spkB, spkC, spkD, spkE, spkF* and *spkG* and those for kinases of the ABC1 type are designated *spkH, spkI, spkJ, spkK* and *spkL* (CyanoBase; http://www.kazusa.or.jp/cyano/). Only three gene products have been functionally characterized to date (Zhang *et al.*, 2005). SpkA and SpkB appear to be involved in the control of cell motility (Kamei *et al.*, 2001; 2003), and SpkE is probably involved in the regulation of nitrogen metabolism (Galkin *et al.*, 2003).

A recent study with DNA microarrays demonstrated connections among the activity of SpkA, the genome-wide expression of certain genes, the formation of thick pili, and cell motility (Panichkin *et al.*, 2006). It is likely that, under non-stress conditions, SpkA activates expression of the putative *pilA9-pilA10-pilA11-slr2018* operon and inactivates expression of the *pilA5-pilA6* and *pilA1-pilA2* operons. Electron microscopy revealed that the SpkA activity is essential for the formation of thick pili. It seems likely that SpkA is a regulator of the expression of these putative operons, whose products are regulators of the formation of thick pili, which are, in turn, essential for cell motility. The molecular mechanism responsible for the contribution of SpkB to cell motility remains to be clarified.

Systematic analysis of the regulation of gene expression using DNA microarrays has not been applied to the functional characterization of the other Ser/Thr kinases, Tyr kinases and protein phosphatases. Much work remains to be done.

Sigma factors and transcription factors that control the activities of RNA polymerases

The initiation of transcription, mediated by the RNA polymerase holoenzyme, is the main determinant of gene expression in eubacteria. Cyanobacterial RNA polymerases are complexes of a core enzyme, composed of α_2, β, β', γ, and ω subunits, as well as a sigma (σ) factor (Tuominen *et al.*, 2003). The sigma factor is responsible for the recognition of promoter sequences (Wosten, 1998).

The *Synechocystis* genome contains nine open-reading frames that encode putative sigma factors (Kaneko *et al.*, 1996). All of these sigma factors belong to the σ^{70} family. Sequence homology suggests that the *sigA* gene encodes the primary sigma factor (PSF); the *sigB*, *sigC*, *sigD*, and *sigE* genes encode group 2 sigma factors; and the *sigF*, *sigG*, *sigH*, and *sigI* genes encode group 3 sigma factors. Attempts to characterize the functional role of each sigma factor have been recently made. However, systematic analysis using DNA microarrays has not been performed in efforts to characterize these sigma factors.

A search for and classification of DNA-binding transcription factors in *Synechocystis* was conducted using the sequence of the genome and a set of bioinformatic tools (Kryazhov and Zinchenko, unpublished). Fifty-seven genes for transcription factors that account for 1.7% of all the genes in the *Synechocystis* genome are found. Transcription factors contain the DNA-binding domains of seven families of factors. In total, forty-four transcription factors have a winged helix or the C-terminal effector domain of bipartite response regulators.

Several transcription factors have been functionally characterized. The role of a LysR-type regulator of transcription, NdhR (Sll1594), was studied systematically with DNA microarrays (Wang *et al.*, 2004a). NdhR negatively regulates the expression of its own gene, as well as that of the *ndhF3* and *ndhD3* genes for NDH-1 (see Chapter 12). NdhR also regulates the expression of the *nhaS1* (*slr1727*) gene for one of the Na^+/H^+ antiporters.

HrcA (Sll1670) encodes an ortholog of a protein that negatively regulates the expression of the heat-shock *grpE-dnaK-dnaJ* and *groESL* operons for chaperonins in *B. subtilis* (Hecker *et al.*, 1996). In *Synechocystis*, HrcA only regulates expression of the *groESL1* operon and the *groEL2* gene in response to heat stress (Nakamoto *et al.*, 2003), suggesting that expression of other heat-shock genes might be regulated by different mechanisms. SufR (Sll0088) functions as a repressor of the *sufBCDS* operon, which is involved in biogenesis of the iron–sulfur cluster in PSI (Wang *et al.*, 2004b). ArsR (Sll1957) represses the expression of the *arsBHC* operon, which consists of three genes: *arsB* for a putative arsenite and antimonite carrier; *arsH* for a protein of unknown function; and *arsC* for a putative arsenate reductase that confers arsenic resistance (Lopez-Maury *et al.*, 2003). However, no systematic analysis using DNA microarrays has been performed to examine the roles of these transcription factors.

Conclusion

Substantial progress has been made in the elucidation of the molecular mechanisms of stress responses in cyanobacteria. This progress has been made possible mainly by the availability of complete genome sequences and of DNA microarrays.

Nonetheless, many questions remain to be answered. Genome-wide transcription networks and their controlling mechanisms await detailed studies and characterization. Several two-component regulatory systems have been identified and assigned to various acclimatory processes, such as responses to cold, heat, salt, and osmotic stress. Nevertheless, signal-transduction pathways in cyanobacteria seem to be much more complicated than simple interactions between a sensory kinase, a cognate response regulator, and the corresponding promoter regions of downstream genes. One of the most important goals should now be the identification and characterization of those factors that provide cyanobacterial two-component systems with specificity to individual types of stress. The functional characterization of Ser/Thr kinases and transcription factors is also important. A combination of physiological, biochemical, and genetic approaches should lead to a deeper understanding of the mechanisms of stress responses in lower and higher organisms.

Acknowledgments

The authors acknowledge Drs. Yu Kanesaki and Kalyanee Paithoonrangsarid for their assistance during preparation of the manuscript. This work was supported, in part, by a grant (no. 13854002) to N.M. and I.S. from the Japan Society for the Promotion of Science. It was also supported by grants from the Russian Foundation for Basic Research (nos. 06–04–48581 and 05–04–50883) to D.A.L and to V.V. Zinchenko, and by a grant from the "Molecular and Cell Biology Program" of the Russian Academy of Sciences to D.A.L. It was also supported, in part, by the Japan-Russia Research Cooperative Program to I.S. and D.A.L. from the Japan Society for the Promotion of Science and the Russian Foundation for Basic Research. The authors gratefully acknowledge the support of the NIBB Cooperative Research Program on the Stress Tolerance of Plants.

References

Aguilar, P.S., Hernandez-Arriaga, A.M., Cybulski, L.E., Erazo, A.C., and de Mendoza, D. (2001). Molecular basis of thermosensing: a two-component signal transduction thermometer in *Bacillus subtilis*. EMBO J. *20*, 1681–1691.

Allakhverdiev, S.I., Nishiyama, Y., Miyairi, S., Yamamoto, H., Inagaki, N., Kanesaki, Y., and Murata N. (2002). Salt stress inhibits the repair of photodamaged photosystem II by suppressing the transcription and translation of *psbA* genes in *Synechocystis*. Plant Physiol. *130*, 1443–1453.

Baier, K., Nicklisch, S., Grundner, C., Reinecke, J., Lockau, W. (2001). Expression of two *nblA*-homologous genes is required for phycobilisome degradation in nitrogen-starved *Synechocystis* sp. PCC 6803. FEMS Microbiol Lett. *195*, 35–39.

Bartsevich, V.V., and Pakrasi, H.B. (1995). Molecular identification of an ABC transporter complex for manganese: analysis of a cyanobacterial mutant strain impaired in the photosynthetic oxygen evolution process. EMBO J. *14*, 1845–1853.

Bartsevich, V.V., and Pakrasi, H.B. (1996). Manganese transport in the cyanobacterium *Synechocystis* sp. PCC 6803. J. Biol. Chem. *271*, 26057–26061.

Bhaya, D., Takahashi, A., Shahi, P., and Grossman, A.R. (2001). Novel motility mutants of *Synechocystis* strain PCC 6803 generated by *in vitro* transposon mutagenesis. J. Bacteriol. *183*, 6140–6143.

Chamot, D., Magee, W.C., Yu, E., and Owttrim, G.W. (1999). A cold shock-induced cyanobacterial RNA helicase. J. Bacteriol. *181*, 1728–1732.

Dufresne, A., Salanoubat, M., Partensky, F., Artiguenave, F., Axmann, I.M., Barbe, V., Duprat, S., Galperin, M.Y., Koonin, E.V., Le Gall, F., Makarova, K.S., Ostrowski, M., Oztas, S., Robert, C., Rogozin, I.B., Scanlan, D.J., Tandeau de Marsac, N., Weissenbach, J., Wincker, P., Wolf, Y.I., and Hess, W.R. (2003). Genome sequence of the cyanobacterium *Prochlorococcus marinus* SS120, a nearly minimal oxyphototrophic genome. Proc. Natl. Acad. Sci. USA *100*, 10020–10025.

Dzelzkalns, V.A., Obinger, C., Regelsberger, G., Niederhauser, H., Kamensek, M., Peschek, G.A., and Bogorad, L. (1994). Deletion of the structural gene for the NADH-dehydrogenase subunit 4 of *Synechocystis* 6803 alters respiratory properties. Plant Physiol. *106*, 1435–1442.

Ehira, S., and Ohmori, M. (2006). NrrA, a nitrogen-responsive response regulator, facilitates heterocyst development in the cyanobacterium *Anabaena* sp. strain PCC 7120. Mol. Microbiol. *59*, 1692–1703.

Fiedler, B., Broc, D., Schubert, H., Rediger, A., Boerner, T., and Wilde, A. (2004). Involvement of cyanobacterial phytochromes in growth under different light qualities and quantities. Photochem. Photobiol. *79*, 551–555.

Fulda, S., Huang, F., Nilsson, F., Hagemann, M., and Norling, B. (2000). Proteomics of *Synechocystis* sp. strain PCC 6803. Identification of periplasmic proteins in cells grown at low and high salt concentrations. Eur. J. Biochem. *267*, 5900–5907.

Fulda, S., Mikkat, S., Huang, F., Huckauf, J., Marin, K., Norling, B., and Hagemann, M. (2006). Proteome analysis of salt stress response in the cyanobacterium *Synechocystis* sp. strain PCC 6803. Proteomics *6*, 2733–2745.

Galkin, A.N., Mikheeva, L.E., and Shestakov, S.V. (2003). Insertional inactivation of genes encoding eukaryotic-type serine/threonine protein kinases in the cyanobacterium *Synechocystis* sp. PCC 6803. Mikrobiologiia *72*, 64–69.

Garcia-Dominguez, M., Reyes, J.C., and Florencio, F.J. (1999). Glutamine synthetase inactivation by protein-protein interaction. Proc. Natl. Acad. Sci. USA *96*, 7161–7166.

He, Q., Dolganov, N., Bjorkman, O., and Grossman, A.R. (2001). The high light-inducible polypeptides in *Synechocystis* PCC 6803. Expression and function in high light. J. Biol. Chem. *276*, 306–314.

Hecker, M., Schumann, W., and Volker, U. (1996). Heat-shock and general stress response in *Bacillus subtilis*. Mol. Microbiol. *19*, 417–428.

Higo, A., Katoh, H., Ohmori, K., Ikeuchi M., and Ohmori, M. (2006). The role of a gene cluster for trehalose metabolism in dehydration tolerance of the filamentous cyanobacterium *Anabaena* sp. PCC 7120. Microbiology. *152*, 979–987.

Hihara, Y., Kamei, A., Kanehisa, M., Kaplan, A., and Ikeuchi, M. (2001). DNA microarray analysis of cyanobacterial gene expression during acclimation to high light. Plant Cell *13*, 793–806.

Hihara, Y., Sonoike, K., Kanehisa, M., and Ikeuchi, M. (2003). DNA microarray analysis of redox-responsive genes in the genome of the cyanobacterium *Synechocystis* sp. strain PCC 6803. J. Bacteriol. *185*, 1719–1725.

Hirani, T.A., Suzuki, I., Murata, N., Hayashi, H., and Eaton-Rye, J.J. (2001). Characterization of a two-component signal transduction system involved in the induction of alkaline phosphatase under phosphate-limiting conditions in *Synechocystis* sp. PCC 6803. Plant Mol. Biol. *45*, 133–144.

Hosoya-Matsuda, N., Motohashi, K., Yoshimura, H., Nozaki, A., Inoue, K., Ohmori, M., and Hisabori, T. (2005). Anti-oxidative stress system in cyanobacteria. Significance of type II peroxiredoxin and the role of 1-Cys peroxiredoxin in *Synechocystis* sp. strain PCC 6803. J. Biol. Chem. *280*, 840–846.

Hsiao, H.Y., He, Q., Van Waasbergen, L.G., and Grossman, A.R. (2004). Control of photosynthetic and high light-responsive genes by the histidine kinase DspA: negative and positive regulation and interactions between signal transduction pathways. J. Bacteriol. *186*, 3882–3888.

Huang, F., Fulda, S., Hagemann, M., and Norling, B. (2006). Proteomic screening of salt stress-induced changes in plasma membranes of *Synechocystis* sp. strain PCC 6803. Proteomics *6*, 910–920.

Huang, L., McCluskey, M.P., Ni, H., and LaRossa, R.A. (2002). Global gene expression profiles of the cyanobacterium *Synechocystis* sp. strain PCC 6803 in response to irradiation with UV-B and white light. J. Bacteriol. *184*, 6845–6858.

Hubschmann, T., Yamamoto, H., Gieler, T., Murata, N., and Boerner, T. (2005). Red and far-red light alter the transcript profile in the cyanobacterium *Synechocystis* sp. PCC 6803: impact of cyanobacterial phytochromes. FEBS Lett. *579*, 1613–1618.

Hughes, J., Lamparter, T., Mittmann, F., Hartmann, E., Gartner, W., Wilde, A., and Boerner, T. (1997). A prokaryotic phytochrome. Nature *386*, 663.

Inaba, M., Suzuki, I., Szalontai, B., Kanesaki, Y., Los, D.A., Hayashi, H., and Murata, N. (2003). Gene-engineered rigidification of membrane lipids enhances the cold inducibility of gene expression in *Synechocystis*. J. Biol. Chem. *278*, 12191–12198.

Jorissen, H.J., Quest, B., Remberg, A., Coursin, T., Braslavsky, S.E., Schaffner, K., Tandeau de Marsac, N., and Gartner, W. (2002). Two independent, light-sensing, two-component systems in a filamentous cyanobacterium. Eur. J. Biochem. *269*, 2662–2671.

Kamei, A., Yuasa, T., Orikawa, K., Geng, X.X., and Ikeuchi, M. (2001). A eukaryotic-type protein kinase, SpkA, is required for normal motility of the unicellular cyanobacterium *Synechocystis* sp. strain PCC 6803. J. Bacteriol. *183*, 1505–1510.

Kamei, A., Yoshihara, S., Yuasa, T., Geng, X., and Ikeuchi, M. (2003). Biochemical and functional characterization of a eukaryotic-type protein kinase, SpkB, in the cyanobacterium *Synechocystis* sp. PCC 6803. Curr. Microbiol. *46*, 296–301.

Kaneko, T., Sato, S., Kotani, H., Tanaka, A., Asamizu, E., Nakamura, Y., Miyajima,N., Hirosawa, M., Sugiura, M., Sasamoto, S., Kimura, T., Hosouchi, T., Matsuno, A., Muraki, A., Nakazaki, N., Naruo, K., Okumura, S., Shimpo, S., Takeuchi, C., Wada, T., Watanabe, A., Yamada, M., Yasuda, M., and Tabata, S. (1996). Sequence analysis of the genome of the unicellular cyanobacterium *Synechocystis* sp. strain PCC6803. II. Sequence determination of the entire genome and assignment of potential protein-coding regions. DNA Res. *3*, 185–209.

Kaneko, T., Nakamura, Y., Wolk, C.P., Kuritz, T., Sasamoto, S., Watanabe, A., Iriguchi, M., Ishikawa, A., Kawashima, K., Kimura, T., Kishida, Y., Kohara, M., Matsumoto, M., Matsuno, A., Muraki, A., Nakazaki, N., Shimpo, S., Sugimoto, M., Takazawa, M., Yamada, M., Yasuda, M., and Tabata, S. (2001). Complete genomic sequence of the filamentous nitrogen-fixing cyanobacterium *Anabaena* sp. strain PCC 7120. DNA Res. *8*, 205–213.

Kaneko, T., Nakamura, Y., Sasamoto, S., Watanabe, A., Kohara, M., Matsumoto, M., Shimpo, S., Yamada, M., and Tabata, S. (2003). Structural analysis of four large plasmids harboring in a unicellular cyanobacterium, *Synechocystis* sp. PCC 6803. DNA Res. *10*, 221–228.

Kanesaki, Y., Suzuki, I., Allakhverdiev, S.I., Mikami, K., and Murata, N. (2002). Salt stress and hyperosmotic stress regulate the expression of different sets of genes in *Synechocystis* sp. PCC 6803. Biochem. Biophys. Res. Commun. *290*, 339–348.

Kanesaki, Y., Yamamoto, H., Paithoonrangsarid, K., Shoumskaya, M., Suzuki, I., Hayashi, H., and Murata, N. (2007). Histidine kinases play important roles in the perception and signal transduction of H_2O_2 in the cyanobacterium *Synechocystis*. Plant J. *49*, 313–324.

Kehoe, D.M., and Grossman, A.R. (1994). Complementary chromatic adaptation: photoperception to gene regulation. Semin. Cell Biol. *5*, 303–313.

Kehoe, D.M., and Grossman, A.R. (1996). Similarity of a chromatic adaptation sensor to phytochrome and ethylene receptors. Science *273*, 1409–1412.

Kobayashi, M., Ishizuka, T., Katayama, M., Kanehisa, M., Bhattacharyya-Pakrasi, M., Pakrasi, H.B., and Ikeuchi, M. (2004). Response to oxidative stress involves a novel peroxiredoxin gene in the unicellular cyanobacterium *Synechocystis* sp. PCC 6803. Plant Cell Physiol. *45*, 290–299.

Koretke, K.K., Lupas, A.N., Warren, P.V., Rosenberg, M., and Brown, J.R. (2000). Evolution of two-component signal transduction. Mol. Biol. Evol. *17*, 1956–1970.

Kunert, A., Vinnemeier, J., Erdmann, N., and Hagemann, M. (2003). Repression by Fur is not the main mechanism controlling the iron-inducible *isiAB* operon in the cyanobacterium *Synechocystis* sp. PCC 6803. FEMS Microbiol. Lett. *227*, 255–262.

Leonard, C.J., Aravind, L., and Koonin, E.V. (1998). Novel families of putative protein kinases in bacteria and archaea: evolution of the "eukaryotic" protein kinase superfamily. Genome Res. *8*, 1038–1047.

López-Maury, L., Garcia-Dominguez, M., Florencio, F.J., and Reyes, J.C. (2002). A two-component signal transduction system involved in nickel sensing in the cyanobacterium *Synechocystis* sp. PCC 6803. Mol. Microbiol. *43*, 247–256.

López-Maury, L., Florencio, F.J., and Reyes, J.C. (2003). Arsenic sensing and resistance system in the cyanobacterium *Synechocystis* sp. strain PCC 6803. J. Bacteriol. *185*, 5363–5371.

Los, D.A., and Murata, N. (2004). Membrane fluidity and its roles in the perception of environmental signals. Biochim. Biophys. Acta *1666*, 142–157.

Los, D.A., Ray, M.K., and Murata, N. (1997). Differences in the control of the temperature-dependent expression of four genes for desaturases in *Synechocystis* sp. PCC 6803. Mol. Microbiol. *25*, 1167–1175.

Marin, K., Suzuki, I., Yamaguchi, K., Ribbeck, K., Yamamoto, H., Kanesaki, Y., Hagemann, M., and Murata, N. (2003). Identification of histidine kinases that act as sensors in the perception of salt stress in *Synechocystis* sp. PCC 6803. Proc. Natl. Acad. Sci. USA *100*, 9061–9066.

Marin, K., Kanesaki, Y., Los, D.A., Murata, N., Suzuki, I., and Hagemann, M. (2004). Gene expression profiling reflects physiological processes in salt acclimation of *Synechocystis* sp. strain PCC 6803. Plant Physiol. *136*, 3290–3300.

Maruyama, K., Sato, N., and Ohta, N. (1999). Conservation of structure and cold-regulation of RNA-binding proteins in cyanobacteria: probable convergent evolution with eukaryotic glycine-rich RNA-binding proteins. Nucleic Acids Res. *27*, 2029–2036.

Mikami, K., Kanesaki, Y., Suzuki, I., and Murata, N. (2002). The histidine kinase Hik33 perceives osmotic stress and cold stress in *Synechocystis* sp. PCC 6803. Mol. Microbiol. *46*, 905–915.

Mizuno, T., Kaneko, T., and Tabata, S. (1996). Compilation of all genes encoding bacterial two-component signal transducers in the genome of the cyanobacterium *Synechocystis* sp. strain PCC 6803. DNA Res. *3*, 407–414.

Murata, N., and Los, D.A. (1997). Membrane fluidity and temperature perception. Plant Physiol. *115*, 875–879.

Murata, N., and Los, D.A. (2006). Histidine kinase Hik33 is an important participant in cold signal transduction in cyanobacteria. Physiol. Plant. *126*, 17–27.

Murata, N., and Suzuki, I. (2006). Exploitation of genomic sequences in a systematic analysis to access how cyanobacteria sense environmental stress. J. Exp. Bot. *57*, 235–247.

Nakamoto, H., Suzuki, M., and Kojima, K. (2003). Targeted inactivation of the *hrcA* repressor gene in cyanobacteria. FEBS Lett. *549*, 57–62.

Nakamura, Y., Kaneko, T., Sato, S., Ikeuchi, M., Katoh, H., Sasamoto, S., Watanabe, A., Iriguchi, M., Kawashima, K., Kimura, T., Kishida, Y., Kiyokawa, C., Kohara, M., Matsumoto, M., Matsuno, A., Nakazaki, N., Shimpo, S., Sugimoto, M., Takeuchi, C., Yamada, M., and Tabata, S. (2002). Complete genome structure of the thermophilic cyanobacterium *Thermosynechococcus elongatus* BP-1. DNA Res. *9*, 123–130.

Nakamura, Y., Kaneko, T., Sato, S., Mimuro, M., Miyashita, H., Tsuchiya, T., Sasamoto, S., Watanabe, A., Kawashima, K., Kishida, Y., Kiyokawa, C., Kohara, M., Matsumoto, M., Matsuno, A., Nakazaki, N., Shimpo, S., Takeuchi, C., Yamada, M., and Tabata, S. (2003). Complete genome structure of *Gloeobacter violaceus* PCC 7421, a cyanobacterium that lacks thylakoids. DNA Res. *10*, 137–145.

Ogawa, T., Bao, D.H., Katoh, H., Shibata, M., Pakrasi, H.B., and Bhattacharyya-Pakrasi, M. (2002). A two-component signal transduction pathway regulates manganese homeostasis in *Synechocystis* 6803, a photosynthetic organism. J. Biol. Chem. *277*, 28981–28986.

Ohkawa, H., Pakrasi, H.B., and Ogawa, T. (2000). Two types of functionally distinct NAD(P)H dehydrogenases in *Synechocystis* sp. strain PCC 6803. J. Biol. Chem. *275*, 31630–31634.

Okamoto S, Ikeuchi M, Ohmori M. (1999). Experimental analysis of recently transposed insertion sequences in the cyanobacterium *Synechocystis* sp. PCC 6803. DNA Res. *6*, 265–273.

Paithoonrangsarid, K., Shoumskaya, M.A., Kanesaki, Y., Satoh, S., Tabata, S., Los, D.A., Zinchenko, V.V., Hayashi, H., Tanticharoen, M., Suzuki, I., and Murata, N. (2004). Five histidine kinases perceive osmotic stress and regulate distinct sets of genes in *Synechocystis*. J. Biol. Chem. *279*, 53078–53086.

Palenik, B., Brahamsha, B., Larimer, F.W., Land, M., Hauser, L., Chain, P., Lamerdin, J., Regala, W., Allen, E.E., McCarren, J., Paulsen, I., Dufresne, A., Partensky, F., Webb, E.A., and Waterbury, J. (2003). The genome of a motile marine *Synechococcus*. Nature *424*, 1037–1042.

Panichkin, V.B., Arakawa-Kobayashi, S., Kanaseki, T., Suzuki, I., Los, D.A., Shestakov, S.V., and Murata, N. (2006). Serine/threonine protein kinase, SpkA, in *Synechocystis* sp. PCC 6803 is a regulator of expression of three putative *pilA* operons, formation of thick pili, and cell motility. J. Bacteriol. *188*, 7696–7699.

Rocap, G., Larimer, F.W., Lamerdin, J., Malfatti, S., Chain, P., Ahlgren, N.A., Arellano, A., Coleman, M., Hauser, L., Hess, W.R., Johnson, Z.I., Land, M., Lindell, D., Post, A.F., Regala, W., Shah, M., Shaw, S.L., Steglich, C., Sullivan, M.B., Ting, C.S., Tolonen, A., Webb, E.A., Zinser, E.R., and Chisholm, S.W. (2003). Genome divergence in two *Prochlorococcus* ecotypes reflects oceanic niche differentiation. Nature *424*, 1042–1047.

Sato, N. (1995). A family of cold-regulated RNA-binding protein genes in the cyanobacterium *Anabaena variabilis* M3. Nucleic Acids Res. *23*, 2161–2167.

Sato, N., Ohmori, M., Ikeuchi, M., Tashiro, K., Wolk, C.P., Kaneko, T., Okada, K., Tsuzuki, M., Ehira, S., Katoh, H., Okamoto, S., Yoshimura, H., Fujisawa, T., Kamei, A., Yoshihara, S., Narikawa, R., Hamano, T., Tabata, S., and Kuhara, S. (2004). Use of segment-based microarray in the analysis of global gene expression in response to various environmental stresses in the cyanobacterium *Anabaena* sp. PCC 7120. J. Gen. Appl. Microbiol. *50*, 1–8.

Shapiguzov, A., Lyukevich, A.A., Allakhverdiev, S.I., Sergeyenko, T.V., Suzuki, I., Murata, N., and Los, D.A. (2005). Osmotic shrinkage of cells of *Synechocystis* sp. PCC 6803 by water efflux via aquaporins regulates osmostress-inducible gene expression. Microbiology. *151*, 447–455.

Shoumskaya, M.A., Paithoonrangsarid, K., Kanesaki, Y., Los, D.A., Zinchenko, V.V., Tanticharoen, M., Suzuki, I., and Murata, N. (2005). Identical Hik-Rre systems are involved in perception and transduction of salt signals and hyperosmotic signals but regulate the expression of individual genes to different extents in *Synechocystis*. J. Biol. Chem. *80*, 21531–21538.

Sineshchekov, V., Hughes, J., Hartmann, E., and Lamparter, T. (1998). Fluorescence and photochemistry of recombinant phytochrome from the cyanobacterium *Synechocystis*. Photochem. Photobiol. *67*, 263–267.

Sineshchekov, O.A., Trivedi, V.D., Sasaki, J., and Spudich, J.L. (2005). Photochromicity of *Anabaena* sensory rhodopsin, an atypical microbial receptor with a *cis*-retinal light-adapted form. J. Biol. Chem. *280*, 14663–14668.

Shi, L., Potts, M., and Kennelly, P.J. (1998). The serine, threonine, and/or tyrosine-specific protein kinases and protein phosphatases of prokaryotic organisms. A family portrait. FEMS Microbiol. Rev. *22*, 229–253.

Singh, A.K., McIntyre, L.M., and Sherman, L.A. (2003). Microarray analysis of the genome-wide response to iron deficiency and iron reconstitution in the cyanobacterium *Synechocystis* sp. PCC 6803. Plant Physiol. *132*, 1825–1839.

Singh, A.K., Li, H., Bono, L., and Sherman, L.A. (2005). Novel adaptive responses revealed by transcription profiling of a *Synechocystis* sp. PCC 6803 Δ*isiA* mutant in the presence and absence of hydrogen peroxide. Photosynth. Res. *84*, 65–70.

Slabas, A.R., Suzuki, I., Murata, N., Simon, W.J., and Hall, J.J. (2006). Proteomic analysis of the heat shock response in *Synechocystis* PCC 6803 and a thermally tolerant knockout strain lacking the histidine kinase 34 gene. Proteomics 6, 845–864.

Stock, A.M., Robinson, V.L., and Goudreau, P.N. (2000). Two-component signal transduction. Annu. Rev. Biochem. 69, 183–215.

Stock, J., and Da, R.S. (2000). Signal transduction: response regulators on and off. Curr. Biol. *10*, R420-R424.

Suzuki, I., Los, D.A., Kanesaki, Y., Mikami, K., and Murata, N. (2000a). The pathway for perception and transduction of low-temperature signals in *Synechocystis*. EMBO J. *19*, 1327–1334.

Suzuki, I., Los, D.A., and Murata, N. (2000b). Perception and transduction of low-temperature signals to induce desaturation of fatty acids. Biochem. Soc. Trans. *28*, 628–630.

Suzuki, I., Kanesaki, Y., Mikami, K., Kanehisa, M., and Murata, N. (2001). Cold-regulated genes under control of the cold sensor Hik33 in *Synechocystis*. Mol. Microbiol. *40*, 235–244.

Suzuki, I., Kanesaki, Y., Hayashi, H., Hall, J.J., Simon, W.J., Slabas, A.R., and Murata, N. (2005). The histidine kinase Hik34 is involved in thermotolerance by regulating the expression of heat shock genes in *Synechocystis*. Plant Physiol. *138*, 1409–1421.

Suzuki, S., Ferjani, A., Suzuki, I., and Murata, N. (2004). The SphS-SphR two component system is the exclusive sensor for the induction of gene expression in response to phosphate limitation in *Synechocystis*. J. Biol. Chem. *279*, 13234–13240.

Tu, C.J., Shrager, J., Burnap, R.L., Postier, B.L., and Grossman, A.R. (2004). Consequences of a deletion in *dspA* on transcript accumulation in *Synechocystis* sp. strain PCC 6803. J. Bacteriol. *186*, 3889–3902.

Tuominen, I., Tyystjarvi, E., and Tyystjarvi, T. (2003). Expression of primary σ factor (PSF) and PSF-like sigma factors in the cyanobacterium *Synechocystis* sp. strain PCC 6803. J. Bacteriol. *185*, 1116–1119.

Valladares, A., Montesinos, M.L., Herrero, A., and Flores, E. (2002). An ABC-type, high-affinity urea permease identified in cyanobacteria. Mol. Microbiol. *43*, 703–715.

Vigh, L., Los, D.A., Horvath, I., and Murata, N. (1993). The primary signal in the biological perception of temperature: Pd-catalyzed hydrogenation of membrane lipids stimulated the expression of the *desA* gene in *Synechocystis* PCC 6803. Proc. Natl. Acad. Sci. USA *90*, 9090–9094.

Vogeley, L., Sineshchekov, O.A., Trivedi, V.D., Sasaki, J., Spudich, J.L., and Luecke, H. (2004). *Anabaena* sensory rhodopsin: a photochromic color sensor at 2.0 A. Science *306*, 1390–1393.

Wang, H.L., Postier, B.L., and Burnap, R.L. (2004a). Alterations in global patterns of gene expression in *Synechocystis* sp. PCC 6803 in response to inorganic carbon limitation and the inactivation of *ndhR*, a LysR family regulator. J. Biol. Chem. *279*, 5739–5751.

Wang, T., Shen, G., Balasubramanian, R., McIntosh, L., Bryant, D.A., and Golbeck, J.H. (2004b) The *sufR* gene (*sll0088* in *Synechocystis* sp. strain PCC 6803) functions as a repressor of the *sufBCDS* operon in iron–sulfur cluster biogenesis in cyanobacteria. J. Bacteriol. *186*, 956–967.

Widmann, C., Gibson, S., Jarpe, M.B., and Johnson, G.L. (1999). Mitogen-activated protein kinase: conservation of a three-kinase module from yeast to human. Physiol. Rev. *79*, 143–180.

Wilde, A., Churin, Y., Schubert, H., and Boerner, T. (1997). Disruption of a *Synechocystis* sp. PCC 6803 gene with partial similarity to phytochrome genes alters growth under changing light qualities. FEBS Lett. *406*, 89–92.

Wosten, M.M. (1998). Eubacterial sigma factors. FEMS Microbiol. Rev. *22*, 127–150.

Yamaguchi, K., Suzuki, I., Yamamoto, H., Lyukevich, A., Bodrova, I., Los, D.A., Piven, I., Zinchenko, V., Kanehisa, M., and Murata, N. (2002). A two-component Mn^{2+}-sensing system negatively regulates expression of the *mntCAB* operon in *Synechocystis*. Plant Cell *14*, 2901–2913.

Yeh, K.C., Wu, S.H., Murphy, J.T., and Lagarias, J.C. (1997). A cyanobacterial phytochrome two-component light sensory system. Science *277*, 1505–1508.

Yeremenko, N., Jeanjean, R., Prommeenate, P., Krasikov, V., Nixon, P.J., Vermaas, W.F., Havaux, M., and Matthijs, H.C. (2005). Open reading frame *ssr2016* is required for antimycin A-sensitive photosystem I driven cyclic electron flow in the cyanobacterium *Synechocystis* sp. PCC 6803. Plant Cell Physiol. *46*, 1433–1436.

Yu, E., and Owttrim, G.W. (2000). Characterization of the cold stress-induced cyanobacterial DEAD-box protein CrhC as an RNA helicase. Nucleic Acids Res. *28*, 3926–3934.

Zhang, C.C., Gonzalez, L., and Phalip, V. (1998). Survey, analysis and genetic organization of genes encoding eukaryotic-like signaling proteins on a cyanobacterial genome. Nucleic Acids Res. *26*, 3619–3625.

Zhang, C.C., Jang, J., Sakr, S., and Wang, L. (2005). Protein phosphorylation on Ser, Thr and Tyr residues in cyanobacteria. J. Mol. Microbiol. Biotechnol. *9*, 154–166.

Bioactive Compounds Produced by Cyanobacteria

7

Kaarina Sivonen and Thomas Börner

Abstract

Cyanobacteria produce a large variety of bioactive compounds, including substances with anticancer and antiviral activity, UV protectants, specific inhibitors of enzymes, and potent hepatotoxins and neurotoxins. Only a few biosynthetic pathways have been elucidated. So far genes have been identified for several bioactive proteins, "ribosomal" and "non-ribosomal" peptides, and peptide–polyketide hybrid molecules. Potential functions of bioactive compounds for the producing cells and evolutionary aspects are discussed, and methods for the detection of cyanobacterial toxins and harmful cyanobacteria are described.

Cyanobacteria as producers of toxins and bioactive compounds

Cyanobacteria produce a wide variety of toxins that cause animal poisonings and human health risk. They also produce a large number of bioactive compounds that have potential as drug leads and cell reagents. The separation of these cyanobacterial products into toxins and bioactive compounds is somewhat artificial and has historical roots. Such toxins as cyanobacterial hepato- and neurotoxins for example, are used as research tools (cell biological reagents) and are sold by reagent suppliers. Many of the screens for cyanobacterial bioactive compounds are targeted towards their effects on mammalian cells (e.g. anticancer compounds), in other words they are often screens for cytotoxins. Cyanobacteria produce bioactive compounds that are not only effective on eukaryotic cells/organisms, but also act on viruses as well as other bacteria/cyanobacteria. Cyanobacteria as autotrophic organisms are very versatile and grow in diverse environments, which may partly explain the variety of bioactive compounds produced by these organisms. Increasingly, large existing culture collections are being used to study these fascinating organisms and their bioactive compounds.

Cyanobacterial toxins

Cyanotoxins include very potent toxins. They have caused poisonings of wild and domestic animals and are health risks for human beings due to the combination of toxicity of the compounds and the large mass occurrences of cyanobacteria producing these toxins. Many of the cyanotoxins seem to be associated with planktonic cyanobacteria that occur in fresh and brackish waters and less frequently in marine environments, yet benthic and terrestrial cyanobacteria may also be toxic. Toxin-producing cyanobacteria are frequently those which form mass occurrences in water. These occurrences are often referred to as water blooms and appear as greenish scums (or red-brownish in the case of red-pigmented strains) floating on surface water. Mass occurrences of such cyanobacteria as *Cylindrospermopsis* and *Planktothrix* may also occur in deeper water and are less obvious.

Microcystins and nodularins—cyanobacterial hepatotoxins

Worldwide, the cyanobacterial toxins most frequently found in freshwater blooms are the cyclic heptapeptide microcystins, whereas in brackish waters, the cyclic pentapeptide nodularin is common. These toxins are very potent hepatotoxins. In mouse bioassays these toxins in acute doses cause death within a few hours through liver injury and internal hemorrhages. On autopsy a large, dark, blood-engorged liver is found; the effect on the liver is thus obvious and the name "hepatotoxin" well deserved. These toxins have been identified from planktonic *Anabaena, Microcystis, Oscillatoria/Planktothrix, Nostoc, Anabaenopsis* (Sivonen and Jones, 1999), from which *Microcystis* is the most common producer globally followed by *Planktothrix* and *Anabaena*. In addition, microcystins have more recently been discovered in *Radiocystis feernandoi* in Brazil (Lompardo *et al.*, 2006) and in marine *Synechococcus* from Salton Sea, California (Carmichael and Li, 2006). Microcystins have further been reported from benthic cyanobacteria. Mats formed by *Oscillatoria limosa* and *Phormidium konstantinosum* in alpine pastures in Switzerland caused cattle poisonings and were shown to contain microcystins (Mez *et al.*, 1997). Mats in Antarctica were found to contain microcystins and nodularins (Hitzfeld *et al.*, 2000), and reservoirs in California showed microcystin-LR containing *Phormidium* (Izaguirre and Neilan, 2004). Most recently, *Anabaena subcylindrica, Anabaena variabilis, Nostoc spongiaforme, Plectonema boryanum,* and *Phormidium corium* strains originating from mats in the Nile River or irrigation canals were shown to produce microcystins (Mohamed *et al.*, 2006). There are also a few cases where terrestrial cyanobacteria such as *Hapalosiphon* (Prinsep *et al.*, 1992a) and *Nostoc* (Oksanen *et al.*, 2004), have been implicated as microcystin producers. Nodularin originates conclusively from *Nodularia spumigena*, but the analogous compound motuporin (in which L-Val replaces L-Arg in nodularin) has been characterized from the marine sponge *Theonella swinhoei* (de Silva *et al.*, 1992).

The general structure of microcystins is cyclo(D-alanine^1-X^2-D-MeAsp^3-Z^4-Adda^5-D-glutamate^6-Mdha^7), in which X and Z are variable L-amino acids, D-MeAsp^3 is D-*erythro*-ß-methyl aspartic acid, and Mdha is N-methyldehydroalanine (Figure 7.1). The amino acid Adda, (2S,3S,8S,9S)-3-amino-9-methoxy-2,6,8-trimethyl-10-phenyldeca-4,6-dienoic acid, is the novel structure in cyanobacterial hepatotoxins (Botes *et al.*, 1984, 1985; Rinehart *et al.*, 1988). Microcystins were first isolated from the cyanobacterium *Microcystis aeruginosa* and named according this producer (Carmichael *et al.*, 1988). The chemical structures of microcystins vary ([M+H]+ 910–1117); currently, over 80 different varieties of microcystins have been reported in the scientific literature. Variations occur in all amino acids but are most frequently found among the L-amino acids at positions 2 and 4, while demethylation of amino acids is found at positions 3 and/or 7.

In brackish waters such as the Baltic Sea and in saline lakes and estuaries, the cyanobacterium *Nodularia spumigena* produces nodularin. The chemical structure of nodu-

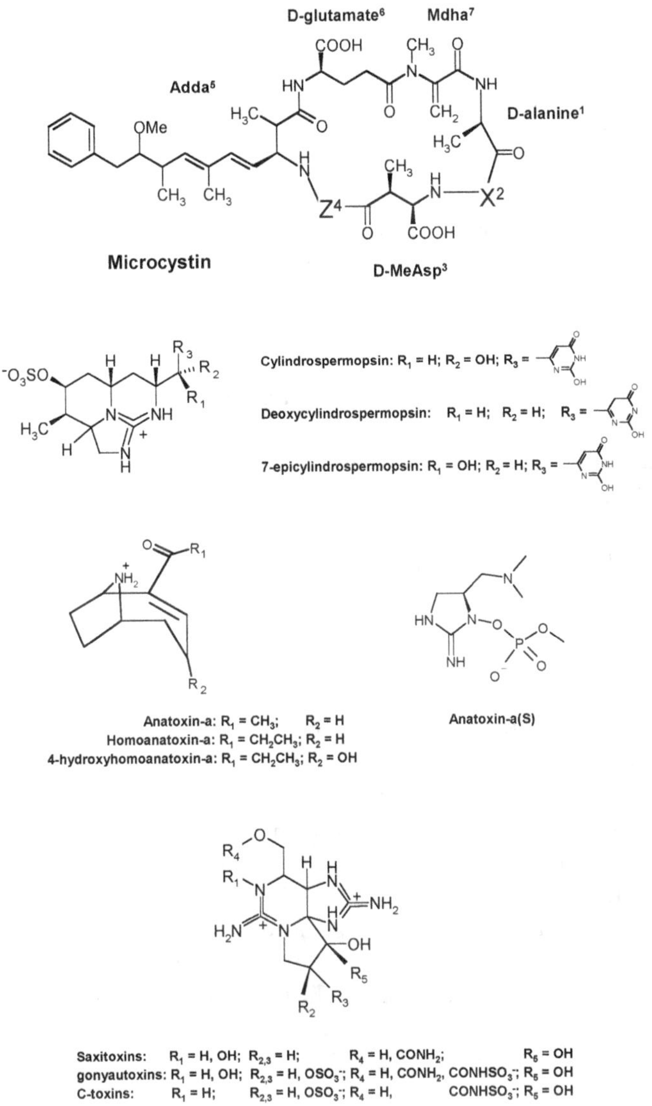

Figure 7.1 Structures of the major cyanobacterial toxins: general structure of microcystin, cylindrospermopsins, anatoxin-a (and variants), anatoxin-a(S), and saxitoxin(s).

larin is cyclo(D-MeAsp[1]-L-arginine[2]-Adda[3]-D-glutamate[4]-Mdhb[5]), in which Mdhb is 2-(methylamino)-2-dehydrobutyric acid. Nodularin was first identified from a bloom sample originating from brackish water in New Zealand (Rinehart *et al.*, 1988) and the following year the same compound was found in bloom samples and *Nodularia* strains originating from the Baltic Sea (Sivonen *et al.*, 1989). Only a few variants of nodularin are found in nature: the most common are the ones with D-Asp[1] in place of D-MeAsp[1] and DMAdda[3] in place of Adda[3] (Namikoshi *et al.*, 1994).

Toxicities of microcystins and nodularins vary (LD_{50} values vary from 50 μg kg^{-1}, i.p. [= intraperitoneal injection] mice to practically non-toxic), microcystin-LR and nodularin being the most potent. The non-toxic variants of both of these toxins include the 6Z-stereoisomer of $Adda^3$ and esterified carboxylic group in glutamate as well as the linear variants of these compounds, which most likely are break-down products (Harada *et al.*, 1990 a,b; Rinehart *et al.*, 1994 Choi *et al.*, 1993; Bourne *et al.*, 1996). Both microcystins and nodularins are serine/threonine-specific protein phosphatase (1 and 2 A) inhibitors (MacKintosh *et al.*, 1990; Yoshizawa *et al.*, 1990). The correct configuration of these molecules is thus needed for effective inhibition (Rudolph-Böhner *et al.*, 1994; Barford and Keller, 1994; Goldberg *et al.*, 1995; Annila *et al.*, 1996).

Cylindrospermopsins—cyanobacterial hepato- and cytotoxins

Cylindrospermopsin is a cyclic guanidine alkaloid with a molecular weight of 415 (Ohtani *et al.*, 1992; Figure 7.1). It is a tropically or subtropically occurring cyanotoxin produced by *Cylindrospermopsis raciborskii* (Hawkins *et al.*, 1985, 1997), *Umezakia natans* (Harada *et al.*, 1994), *Aphanizomenon ovalisporum* (Banker *et al.*, 1997; Shaw *et al.*, 1999), *Aphanizomenon flos-aquae* (Preußel *et al.*, 2006), *Raphidiopsis curvata* (Li *et al.*, 2001), and *Anabaena bergii* (Schembri *et al.*, 2001; Fergusson *et al.*, 2003). In addition to being a hepatotoxin, cylindrospermopsin also affects the kidneys, spleen, thymus, and heart. Pure cylindrospermopsin has an LD_{50} in mice (i.p.) of 2.1 mg/kg at 24 hours and 0.2 mg/kg at 5–6 days (Ohtani *et al.*, 1992). The toxin is a protein synthesis inhibitor (Runnegar *et al.*, 1994, 1995; Terao *et al.*, 1994). Recently, new structural variants of cylindrospermopsin have been isolated from an Australian strain of C. *raciborskii.* One variant was identified as deoxycylindrospermopsin, which seems to be practically non-toxic (Norris *et al.*, 1999), whereas the structural variant 7-epicylindrospermopsin isolated from *Aph. ovalisporum* was toxic (Banker *et al.*, 2001). In tropical and subtropical waters of Australia, cylindrospermopsin has caused health problems via drinking water supplies (Hawkins *et al.*, 1985). This toxin has also been reported in New Zealand (Stirling and Quilliam, 2001), Japan (Harada *et al.*, 1994), Thailand (Li *et al.*, 2001), Israel (Banker *et al.*, 1997) and Europe (Fastner *et al.*, 2003; Preußel *et al.*, 2006). The C. *raciborskii* strains isolated from lakes in France, Germany, and Portugal were toxic but did not contain cylindrospermopsins (Bernard *et al.*, 2003; Fastner *et al.*, 2003; Saker *et al.*, 2003).

Cyanobacterial neurotoxins

Cyanobacterial neurotoxins are not as common as hepatotoxins. These neurotoxins are very rapidly acting poisons and have been reported most frequently in North America, Europe, and Australia. In mouse bioassays death by respiratory arrest occurs within two to 30 minutes. Three types of cyanobacterial neurotoxins, anatoxin-a, anatoxin-a(S), and saxitoxins are known. Recently, the occurrence of a neurotoxic amino acid, β-methylamino-L-alanine (BMAA) was reported in a wide variety of cyanobacterial strains (Cox *et al.*, 2005). In marine *Trichodesmium* blooms from the Virgin Islands, a neurotoxic factor was reported that was neither anatoxin-a nor anatoxin-a(S), yet the structure of the compound has remained unknown (Hawser *et al.*, 1991).

Anatoxin-a is a small (MW=165) alkaloid, a secondary amine, 2-acetyl-9-azabicyclo(4–2–1)non-2-ene (Figure 7.1, Devlin *et al.*, 1977). It was the first cyanobacterial toxin whose structure was fully determined. This toxin mimics the effect of acetyl choline and causes rapid death by respiratory arrest. Anatoxin-a is shown to be produced by *Anabaena flos-aquae, Anabaena* spp. (*flos-aquae/lemmermannii/circinalis-* group), *A. planktonica, Aphanizomenon, Cylindrospermum, Oscillatoria* (Sivonen and Jones, 1999), *Planktothrix rubescens* (Viaggiu *et al.*, 2005), benthic *Oscillatoria* (Edwards *et al.*, 1992), and *Phormidium flavosum* (Gugger *et al.*, 2005). Anatoxin-a is perhaps the most common cyanobacterial neurotoxin and it is particularly abundant in North America and Europe. Recently, it was reported to occur in low amounts in *Arthrospira fusiformis* originating from certain lakes in Kenya (Ballot *et al.*, 2005). Structural variants seem to be rare. Homoanatoxin-a, (MW 179) is an anatoxin-a homologue. It was first isolated from an *Oscillatoria formosa* strain in Norway (Skulberg *et al.*, 1992). It has a propionyl group at C-2 instead of the acetyl group in anatoxin-a (Skulberg *et al.*, 1992). The LD_{50} values of anatoxin-a and homoanatoxin-a are similar, 200–250 μg kg^{-1} (Devlin *et al.*, 1977; Carmichael *et al.*, 1990; Skulberg *et al.*, 1992). Seventy-six axenic strains in the Pasteur Culture Collection were analyzed, and co-occurrence of anatoxin-a and homoanatoxin-a in two *Oscillatoria* strains were reported as well as the presence of anatoxin-a in three *Oscillatoria* strains (Araoz *et al.*, 2005). The simultaneous production of anatoxin-a and homoanatoxin-a as well as 4-hydroxyhomoanatoxin–a was reported in *Raphidiopsis mediterranes* (Namikoshi *et al.*, 2003). The 4-hydroxyhomoanatoxin has an OH group in lieu of hydrogen in C4.

Anatoxin-a(S) is a phosphate ester of a cyclic N-hydroxyguanine (MW 252, Figure 7.1), originally identified from an *Anabaena flos-aquae* strain NRC 525–17 (Matsunaga *et al.*, 1989). Anatoxin-a(S) is an anticholinesterase, and the characteristic signs of its poisonings in mice include salivation (thus the S in its name). The LD_{50} value of anatoxin-a(S) is 20 μg/kg (i.p. mouse) (Carmichael *et al.*, 1990). There are no structural variants of anatoxin-a(S). The toxin seems to occur only rarely. Anatoxin-a(S) was first reported in North America where it has caused animal poisonings (Mahmood *et al.*, 1988). Later it was also found in Denmark, produced by *A. lemmermannii* (Henriksen *et al.*, 1997; Onodera *et al.*, 1997a).

Saxitoxins are a group of carbamate alkaloid neurotoxins, which are either non-sulfated (saxitoxins—STX), singly sulfated (gonyautoxins—GTX), or doubly sulfated (C-toxins) (Figure 7.1). In addition, decarbamoyl variants and several related new toxins such as "*Lyngbya wollei* toxins" have been identified (Sivonen and Jones, 1999). Saxitoxins block nerve-cell sodium channels. The most potent of these toxins are saxitoxins and neosaxitoxins (LD_{50} i.p. in mice 10 μg kg^{-1}). Saxitoxins were originally isolated and identified from marine dinoflagellates. These dinoflagellate saxitoxins (PSPs) have caused human poisonings owing to their ability to concentrate in shellfish (Anderson, 1994). In cyanobacteria, saxitoxins have been found in *Anabaena circinalis, Aphanizomenon flos-aquae, Aph. issatschenkoi, Cylindrospermopsis raciborskii, Lyngbya wollei,* and *Planktothrix sp.* (Lagos *et al.*, 1999; Sivonen and Jones, 1999; Pereira *et al.*, 2000; Pomati *et al.*, 2000; Fereira *et al.*, 2001; Dias *et al.*, 2002; Nogueira *et al.*, 2004). The first report of saxitoxins came from New Hampshire, where *Aphanizomenon flos-aquae* was the toxin producer (Ikawa *et al.*, 1982; Mahmood and Carmichael, 1986). The largest and widest mass occurrences of saxitoxin containing the cyanobacteria *Anabaena circinalis* have been reported in Australia, where they caused animal poisonings (Humpage *et al.*, 1994; Negri *et al.*, 1995, 1997). Later, the benthic cyanobacterium *Lyngbya wollei* was shown to produce decarbamoylsaxitoxins and six new variants of saxitoxins (Onodera *et al.*, 1997b). In Europe, saxitoxins have been reported in freshwater in Portugal (e.g. Pereira *et al.*, 2000) and Finland (Rapala *et al.*, 2005).

Dermatotoxins and other marine biotoxins

Benthic cyanobacteria such as *Lyngbya, Oscillatoria,* and *Schizothrix* have been reported to produce toxins causing severe dermatitis among swimmers who came into contact with cyanobacteria in Hawaii and Okinawa (Moikeha and Chu, 1971; Hashimoto *et al.*, 1976). The toxins of *Lyngbya* were identified as aplysiatoxins and debromoaplysiatoxins which are potent tumor promoters and protein kinase C activators (Mynderse *et al.*, 1977; Fujiki *et al.*, 1990). Lyngbyatoxin-a identified from another *Lyngbya* strain caused dermatitis and severe oral and gastrointestinal inflammation (Cardellina *et al.*, 1979).

Further cyanobacterial bioactive compounds

Cyanobacteria are increasingly valued as a prolific source of new bioactive compounds for the biotechnological and biomedical industries. Marine environments and especially benthic cyanobacteria seem to be among the best sources for new bioactive compounds. The greatest interest has focused on screening for anticancer compounds. Terrestrial cyanobacteria screening has revealed some interesting antiviral compounds and yielded promising anticancer leads. On the other hand, research on planktonic cyanobacteria from fresh and brackish waters has found a vast array of linear and cyclic peptides other than hepatotoxic microcystins and nodularins. These compounds were not toxic in a mouse bioassay. They were first found by screening for serine protease-inhibiting compounds (prevention of blood clotting). Small alkaloids originating from cyanobacteria may also have biomedical potential.

Compounds from marine environments

Many new compounds and families of compounds were identified from natural assemblages of *Lyngbya* sp. (mostly *L. majuscula*) collected from tropical waters. These organisms grow in abundance, attached to coral reefs, to rocks or form mats in shallow water. Most of the compounds isolated and identified were cyclic or linear peptides or lipopeptides, and many of them were cytotoxic and antineoplastic.

A selection of these compounds is presented here (see the examples in Figure 7.2). One structurally novel antimitotic and antiproliferative compound, curacin A (Figure 7.2), was identified from a Caribbean collection of *Lyngbya majuscula* (Gerwick *et al.*, 1994: Nagle *et al.*, 1995). From the same material, barbamide, a chlorinated compound with molluscidal activity (Orjala and Gerwick, 1996; Figure 7.2) and ichthyotoxic compounds such as malyngamide H were identified (Orjala *et al.*, 1995). *Lyngbya majuscula* isolates from Jamaica contained the sodium-channel blocking and fish-toxic compounds, jamaicamides (Edwards *et al.*, 2004; Figure 7.2). Several depsipeptides having antitumor activity and disrupting cellular microfilament networks were isolated from natural assemblages of *L. majuscula/Schizothrix calcicola* or *L. majuscula* near Guam, e.g. dolastatin 12, lyngbyastatin 1 (Harrigan *et al.*, 1998b), and lyngbyabellins (Luesch *et al.*, 2000).

Other benthic, marine cyanobacterium rich of bioactive compounds is *Symploca* sp. collected near Guam and Palau and found to produce several linear peptides (Harrigan

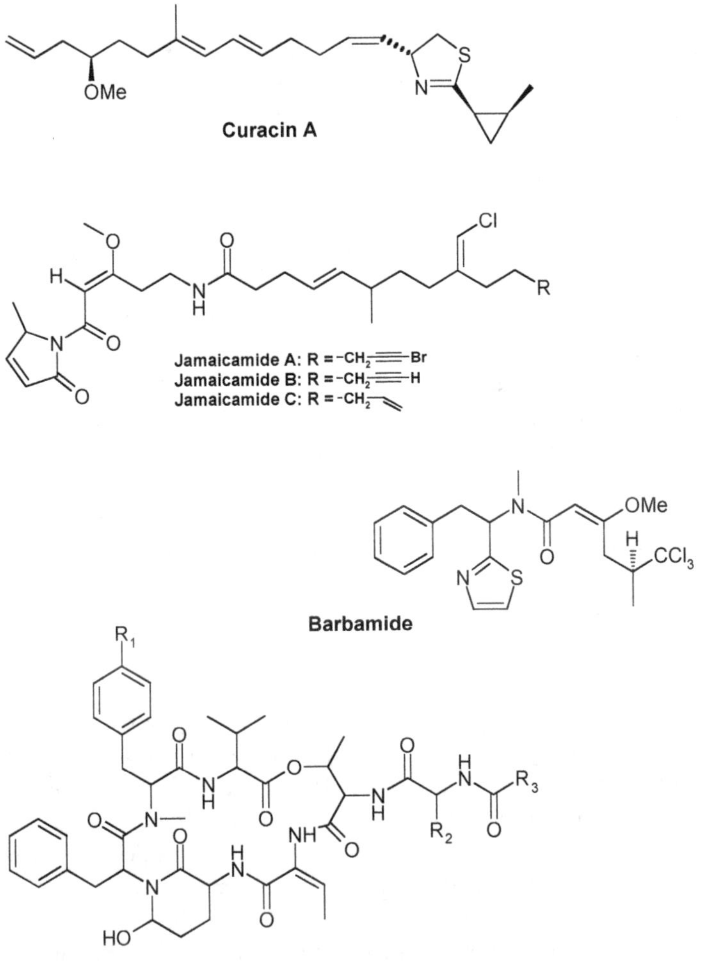

Somamide A: R_1 = OH; R_2 = $CH_2CH_2SOCH_3$; R_3 = $(CH_2)_4CH_3$
Somamide B: R_1 = OH; R_2 = $CH_2CH_2CONH_2$; R_3 = $(CH_2)_2CH_3$
Symplostatin 2: R_1 = OH; R_2 = $CH_2CH_2SOCH_3$; R_3 = $CH(CH(CH_3)CH_2CH_3)NHCO(CH_2)_2CH_3$
Dolastatin 13: R_1 = H; R_2 = $CH(CH_3)_2$; R_3 = $CH(OCH_3)CH_2OH$

Figure 7.2 Structures of cyanobacterial bioactive compounds from marine cyanobacteria *Lyngbya majuscula* and *Symploca* sp: curacin A, jamaicamides, barbamide, and compounds related to dolastatin 13.

et al., 1998a) as well as cyclic peptides (Harrigan *et al.*, 1999) and cytotoxic alkaloids (William *et al.*, 2004). Interestingly, *Symploca* sp. also synthesizes several analogs and an identical compound to dolastatin 10 (Luesch *et al.*, 2001). Dolastatin 10, an anticancer drug in phase II clinical trials, was originally purified from the marine sea hare *Dolabella auricularia* (Mooberry *et al.*, 2003). The concentrations in cyanobacteria were much higher, suggesting that the sea hare had gained this compound by feeding on cyanobacteria. Similarly, many compounds found in *Lyngbya* (e.g. dolastatin 12, 16) were originally reported to occur in lower amounts in marine tunicates (Nogle and Gerwick, 2002). Also somamide A, B isolated from *Lyngbya* (Nogle *et al.*, 2001) and symplostatin 2 isolated from *Symploca* (Harrigan *et al.*, 1999) closely resemble dolastatin 13 (Figure 7.2), which is a cytostatic agent originally obtained from the Indian Ocean mollusk *Dolabella auricularia* (sea hare) (Pettit *et al.*, 1989), but these too are likely products of cyanobacteria.

Symbiotic cyanobacteria may also produce interesting bioactive compounds. Several compounds originally isolated from tunicates *Lissoclinum patella*/*Lissoclinum bistratum* were later verified as products of the symbiotic cyanobacterium *Prochloron* sp. These compounds include several octa- and hexapeptides with thiazole- and oxazole-rings. They show cytotoxic, antineoplastic properties as well as multidrug-resistance reversing activity (Ireland *et al.*, 1982; Schmidt *et al.*, 2004, 2005; Hampley *et al.*, 1992). Similar compounds have also been identified from freshwater cyanobacteria, e.g. *Nostoc* (Jüttner *et al.*, 2001; Banker and Carmeli, 1998), and terrestrial cyanobacteria such as *Westiellopsis* (Princep *et al.*, 1992b), *Oscillatoria* (Admi *et al.*, 1996), and *Stigonema* (Oringo *et al.*, 1996). These compounds were reported as anticyanobacterial/antialgal and toxic to rotifers and sea urchin embryos.

Recently, neuro-apoptogenic and blood platelet-targeting bioactivities were reported in benthic cyanobacteria isolated on the Portuguese coast (Selhem *et al.*, 2005). Similarly, benthic cyanobacteria isolated on the northern coast of the Baltic Sea contained cytotoxic and apoptogenic bioactivities (Herfindal *et al.*, 2005; Surakka *et al.*, 2005).

Bioactive compounds from terrestrial cyanobacteria

Interesting bioactive compounds have also been found from terrestrial cyanobacteria. The best examples are the cryptophycins originating from the isolated Indian *Nostoc* sp. strain GSV224 and the Scottish *Nostoc* sp. strain ATCC 53789 (Golakoti *et al.*, 1994; Figure 7.3). Cryptophycins, originally described as antifungal compounds, were later recognized as efficient antitumor depsipeptides. Cryptophycins are peptolides and include over 25 natural variants and several synthetic/semi-synthetic compounds (Golakoti *et al.*, 1994, 1995; Subbaraju *et al.*, 1997; Chagaty *et al.*, 2004). Cryptophycin 1 is a cytotoxin, binds to tubulin, and disrupts tubulin dynamics, resulting in apoptosis of tumor cells (Panda *et al.*, 1997; Liang *et al.*, 2005). Due to its stability, cryptophycin-52 entered clinical trials but produced only weak antitumor activity. Currently, new generations of stable cryptophycins were synthesized and shown to be effective against many tumors including the multidrug resistant types (Liang *et al.*, 2005). In addition to cryptophycins, *Nostoc* sp. ATCC 53789 produces two weak cytotoxic heptapeptides, nostocyclopeptides A1 and A2, possessing a unique imino linkage (Golakoti *et al.*, 2001; Figure 7.3). *Nostoc* sp. GSC224 produces minor amounts of nostopeptolides (A1, A2 and A3), which have nine amino acid residues including 4-methylproline and L-leucylacetate (Golakoti *et al.*, 2000; Figure 7.3).

Interesting compounds have also been described from terrestrial *Scytonema* (e.g. Moore *et al.*, 1986). The pigment scytonemin from *Scytonema* was found to be a kinase inhibitor (Stevenson *et al.*, 2002). Previously, toxins from *Lyngbya* were shown to be protein kinase C activators (Mynderse *et al.*, 1977).

Two recent examples of antiviral compounds have been isolated from cyanobacteria: cyanovirin-N from *Nostoc ellipsosporum* (Boyd *et al.*, 1997) and scytovirin from *Scytonema varium* (Bokesch *et al.*, 2003). Cyanovirin-N is a novel 11 kDa protein, which inactivates the human immunodeficiency virus (HIV). It binds to N-linked, high-mannose oligosac-

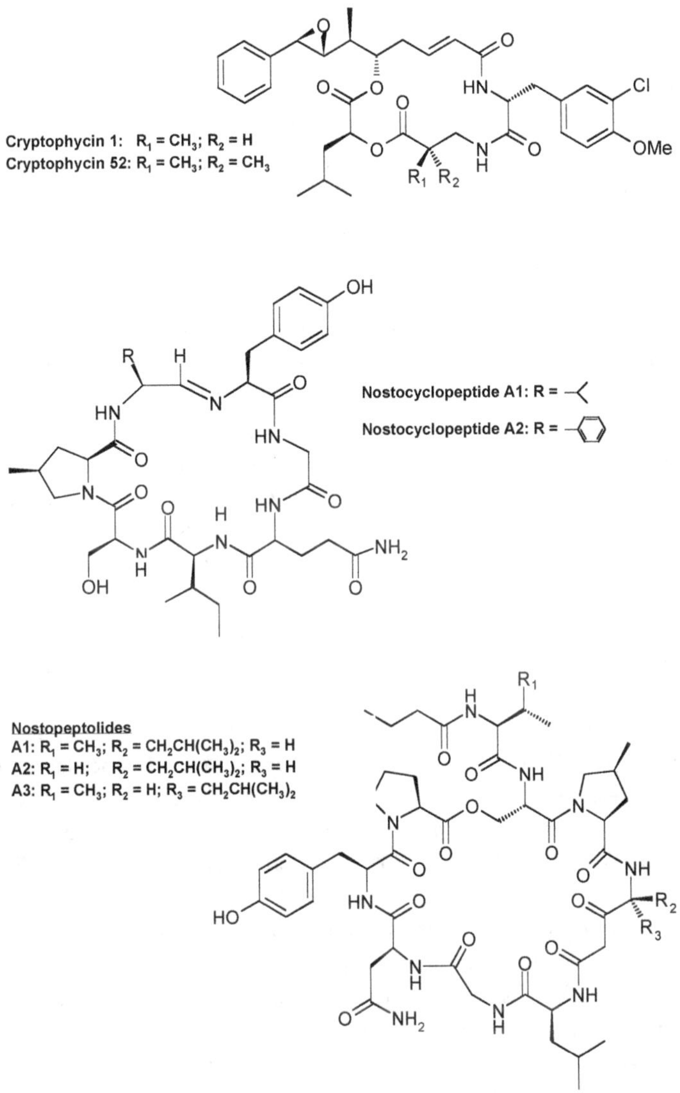

Figure 7.3 Cyanobacterial bioactive compounds from terrestrial cyanobacteria *Nostoc* sp.: cryptophycins, nostocyclopeptides, and nostopeptolides.

charides on the viral envelope (Bolmstedt *et al.*, 2001). Cyanovirin-N is under development as an antiviral agent, thanks to its effectiveness against HIV, its non-toxicity to human cells, and its persistence (Bewley *et al.*, 1998). Scytovirin is a protein that acts similarly to cyanovirin-N, but is less efficient in inactivation of viruses (Bokesch *et al.*, 2003).

Compounds from planktonic fresh and brackish water cyanobacteria

Planktonic cyanobacteria from fresh and brackish water environments, even the same strains, that contain cyanobacterial hepatotoxins, may also produce a wide variety of linear and cyclic peptides. The common compounds include depsipeptides, such as cyanopeptolins, D-lysine, and ureido bond-containing peptides such as anabaenopeptins, linear peptides such as aeruginosins and microginins as well as tricyclic microviridins. There are several reports of other peptides in freshwater cyanobacteria. For example the cyclic peptide nostophycin was isolated from a microcystin-producing *Nostoc* strain (Fujii *et al.*, 1999; Sivonen *et al.*, 1992), which contained a β-amino acid similar to the Adda in microcystins. Kawaguchipeptin B, a cyclic undecapeptide from *Microcystis aeruginosa*, showed antibacterial activity (Ishida *et al.*, 1997).

Depsipeptides containing a unique residue, 3-amino-6-hydroxy-2-piperidone (Ahp), form a large group of cyanobacterial peptides commonly called cyanopeptolides (these peptides also go by such names as aeruginopeptins, anabaenopeptilides, microcystilides, micropeptins, nostocyclin, nostopeptin, oscillapeptilides and oscillapeptins) (Figure 7.4). Altogether, 56 structurally different compounds of this class were reported in fresh water cyanobacteria (see the list in Rouhiainen, 2004). They were first found in *Microcystis aeruginosa* (Harada *et al.*, 1993; Martin *et al.*, 1993). Currently, they have been isolated and identified from several strains of freshwater *Anabaena, Microcystis, Nostoc,* and *Oscillatoria/Planktothrix* (e.g. Fujii *et al.*, 1996, 2000, 2002; Kaya *et al.*, 1996; Okino *et al.*, 1997; Shin *et al.*, 1995). Similar peptides (nine structurally different compounds) have also been identified from terrestrial (*Scytonema, Microchaete,* Matern *et al.*, 2001; Lee *et al.*, 1994) and marine cyanobacteria (*Lyngbya, Symploca,* Nogle *et al.*, 2001; Harrigan *et al.*, 1999). Many of these compounds are serine protease inhibitors (Namikoshi and Rinehart, 1996; Weckesser *et al.*, 1996).

Hexapeptides containing D-lysine and a unique ureido linkage are most frequently called anabaenopeptins (but are also referred to as ferintonic acids, nodulapeptins, or oscillamides) (Figure 7.4). They are also produced by various cyanobacterial species: *Anabaena, Aphanizomenon, Microcystis, Nodularia,* and *Planktothrix/Oscillatoria* (e.g. Fujii *et al.*, 1996, 1997, 2002; Murakami *et al.*, 1997a, 2000; Williams *et al.*, 1996). In addition, similar compounds have been found in a terrestrial *Schizothrix* sp. (Reshef and Carmeli, 2002). Some of these peptides seem to be protease inhibitors (Namikoshi and Rinehart, 1996).

Aeruginosins (microcin, spumigins) (Figure 7.4) are serine protease inhibitors produced by *Microcystis areuginosa/viridis, Oscillatoria/Planktothrix agardhii,* and *Nodularia spumigena* (e.g. Fujii *et al.*, 1997; Matsuda *et al.*, 1996; Murakami *et al.*, 1994; Shin *et al.*, 1997). Typically, these linear peptides contain the unusual amino acid, 2-carboxy-6-hydroxyoctahydroindole (= Choi) or its variant (Fujii *et al.*, 1997; Murakami *et al.*, 1994; Matsuda *et al.*, 1996; Shin *et al.*, 1997).

Microginins (Figure 7.4), which are linear peptides, were isolated from *M. aeruginosa* (Okino *et al.*, 1993), *Planktothrix/Oscillatoria agardhii* (oscillaginin, Sano and Kaya, 1997), and *Nostoc* sp. (nostoginin, Ploutno and Carmeli, 2002). They are characterized by having the amino acid Adha (3-amino-(2-hydroxy) decenoid acid). Recently, two types of unusual linear peptides, aeruginoguanidines (Ishida *et al.*, 2002) and aeruginosamide (Lawton *et al.*, 1999), were reported in *Microcystis aeruginosa*.

Microviridins were first isolated from *Microcystis viridis* (Ishitsuka *et al.*, 1990) and later from *Microcystis aeruginosa* (Okino *et al.*, 1995; Rohrlack *et al.*, 2003), *Planktothrix/Oscillatoria agardhii* (Shin *et al.*, 1996; Fujii *et al.*, 2000), and *Nostoc minutum* (Murakami *et al.*, 1997b). These compounds are depsipeptides having the special structural feature of a tricyclic ring system (Figure 7.4). These compounds are not toxic in mouse bioassay but may have significant ecological effects on grazers (Rohrlack *et al.*, 2003).

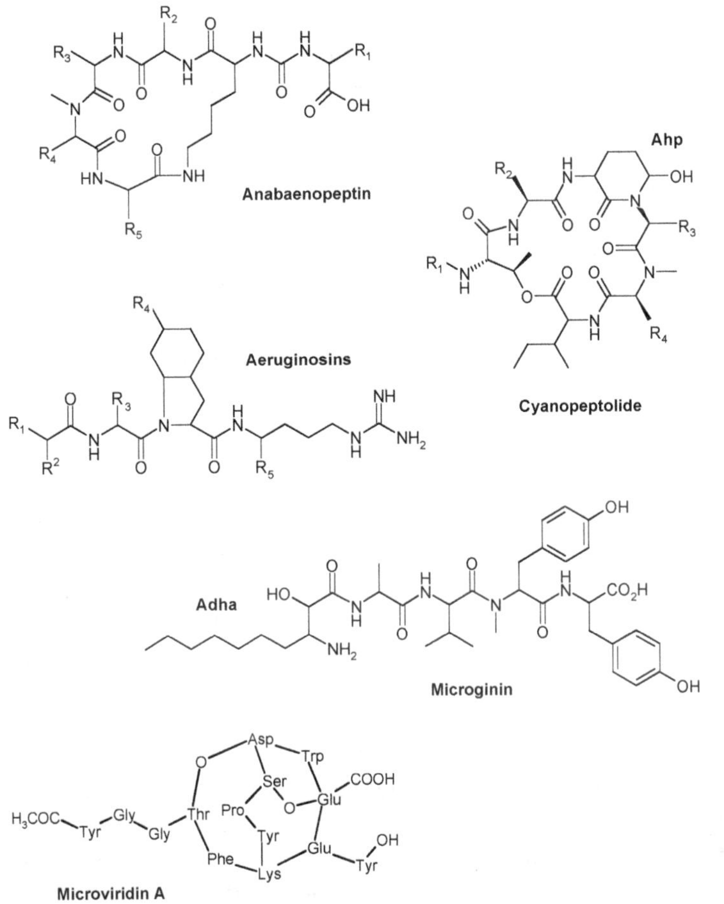

Figure 7.4 Structures of cyanobacterial bioactive compounds from fresh and brackish water cyanobacteria: anabaenopeptin, cyanopeptolide, aeruginosins, microginin and microviridin A.

Cyanobacteria produce a wide variety of alkaloids in addition to the known neurotoxins. One example with pharmacological interest is nostocarboline, a new cholinesterase inhibitor identified in the freshwater isolate *Nostoc* 78–12A (Becher *et al.*, 2005). The effects of this compound were comparable to galanthamine, an approved drug for Alzheimer's disease.

Biosynthesis and biosynthetic gene clusters of cyanobacterial bioactive compounds

The first attempts to solve the biosynthesis of cyanobacterial toxins and bioactive compounds included the feeding experiments with precursors labeled to follow their incorporation into the carbon skeleton of these compounds. This research gave useful information about biosynthesis and was later used in combination with genetic studies to solve the biosynthesis of a few cyanobacterial metabolites (e.g. Chang *et al.*, 2002, 2004). The occurrence of unusual amino acids in cyanobacterial peptides and the common incorporation of acetyl units in many cyanobacterial compounds led researchers to investigate the involvement of multienzyme complexes such as non-ribosomal peptide synthetase (NRPSs) and polyketide synthases (PKSs) in the peptide/lipopeptide production of cyanobacteria. The approach has been successful, and currently, biosynthesis and the genes involved in the biosynthesis of several cyanobacterial compounds are known. See the more detailed information about the cyanobacterial NRPS and PKS systems in the recent article by Welker and von Döhren (2006).

Non-ribosomal peptide synthetases are common in bacteria (e.g. *Brevibacillus brevis*, a producer of gramicidin) and lower eukaryotes (e.g. *Tolypocladium niveus, a* producer of cyclosporins) and are responsible for the production of several biomedically important cyclic and linear peptides (von Döhren *et al.*, 1997; Marahiel *et al.*, 1997). PKS systems are common in bacteria and have been intensively investigated among bacteria such as *Streptomycetes*, a prolific source of antibiotics and other bioactive compounds (Hopwood, 1997). Characteristically, many non-ribosomal peptides of cyanobacteria are products of integrated systems where both NRPS and PKS activities are needed.

Gene clusters of combined NRPS and PKS systems

The biosynthesis of the cyanobacterial hepatotoxins, microcystins (Moore *et al.*, 1991) and nodularin (Choi *et al.*, 1993; Rinehart *et al.*, 1994) was first studied by following the incorporation of labeled precursors into the toxins by NMR. These studies indicated that the Adda amino acid derived from phenylalanine and acetate. Methyldehydrobutyrine in nodularin was possibly formed from threonine, its methyl group coming from methionine. The dehydroamino acid in microcystin was suggested to be formed from serine rather than from threonine (Rinehart *et al.*, 1994).

Dittmann *et al.* (1997) were the first to show by knock-out mutagenesis that peptide synthetase genes were involved in the production of a cyanobacterial bioactive compound. Their study demonstrated that one gene cluster was responsible for the production of all microcystin variants in the strain *Microcystis aeruginosa* PCC 7806. It is now well established that microcystins and nodularins are produced by a combination of the NRPS and PKS systems (Börner and Dittmann, 2005). The biosynthetic gene clusters have been fully sequenced from *Microcystis* (Nishizawa *et al.*, 1999/2000; Tillett *et al.*, 2000), *Planktothrix* (Christiansen *et al.*, 2003), *Anabaena* (Rouhiainen *et al.*, 2004), *Nodularia* (Moffitt and Neilan, 2004), and *Nostoc* (Fewer, D., L. Rouhiainen and K. Sivonen, unpublished results).

In microcystin biosynthesis the peptide synthetase genes (*mcyA, mcyB, mcyC*) code for the synthesis of the pentapeptide part [(Me)Dha-Arg-X-(Me)Asp-Z] of microcystins. The genes *mcyD, mcyG,* and *mcyE* code for the synthesis of Adda (3-amino-9-methoxy-2,6,8-trimethyl-10-phenyldeca-4,6-dienoic acid). The activation of glutamic acid and its combination with Adda is also encoded in *mcyE*. The genes *mcyF, mcyI,* and *mcyJ* have a tailoring function. The *mcyF* gene was shown to encode an aspartate racemase (Sielaff *et al.*, 2003). The targeted mutation of *mcyJ* in *Planktothrix agardhii* revealed that it codes for an O-methyltransferase responsible for the methylation of Adda (Christiansen *et al.*, 2003). McyH is possibly involved in the transport of microcystin (Pearson *et al.*, 2004). There are differences between the gene clusters of the different microcystin producers; the order of operons and genes varies between the genera. In all analyzed heterocystous cyanobacteria (*Anabaena, Nostoc* and *Nodularia*), the gene order follows the co-linearity rule of peptide synthetases and its products (Figure 7.5).

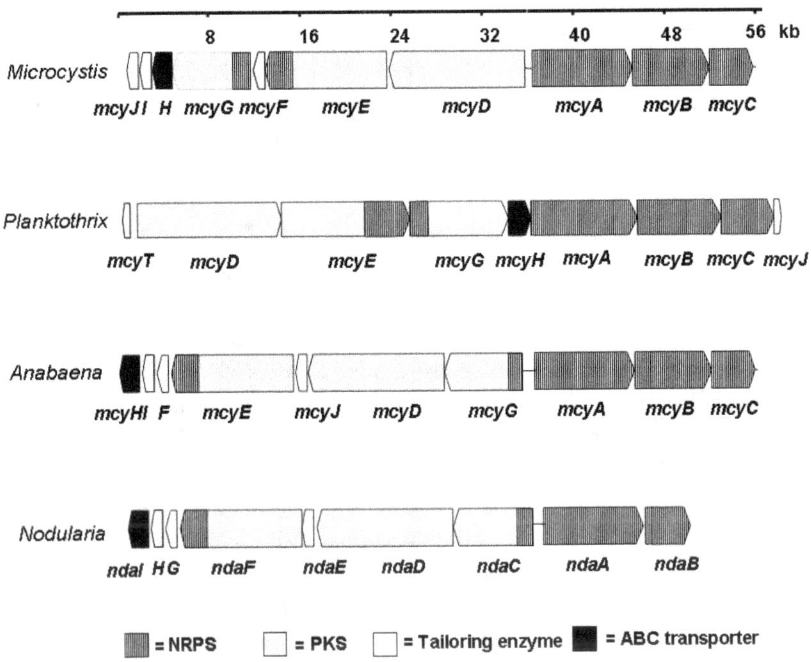

Figure 7.5 Schematic representation of the *mcy* gene clusters coding for biosynthesis of microcystins in *Anabaena, Microcystis* and *Planktothrix* and the *nda* gene cluster coding for the biosynthesis of nodularin in *Nodularia spumigena* (based on data from Tillett *et al.*, 2000; Christiansen *et al.*, 2003; Moffitt and Neilan, 2004; Rouhiainen *et al.*, 2004).

Other examples of combined PKS/NRPS systems involved in biosynthesis of cyanobacterial metabolites come from the marine *Lyngbya majuscula* strain L19. The biosynthesis of barbamide in that strain was first investigated by feeding experiments, which indicated that the compound is derived from acetate, L-phenylalanine, L-leucine, and L-cysteine. The gene cluster of 26 kb containing 12 ORFs, PKS, and NRPS clusters and tailoring genes was co-linear with the barbamide structure. The synthesis involved an unusual trichloroleucine starter unit. The activation of the corresponding amino acids by the adenylation domains of the NRPSs was verified by ATP-[^{32}P]PPi exchange assays (Chang *et al.*, 2002).

More recently, the biosynthesis of curacin A in the same strain was investigated using labeled substrate incorporation followed by NMR analysis and subsequent identification of a 64-kb putative curacin A gene cluster. The feeding experiment revealed incorporation of 10 acetate units and one cysteine as well as two S-adenosyl methionine-derived methyl groups. Consequently, the gene cluster consists of one NRPS unit and multiple PKS units; it is co-linear with the curacin A structure (Chang *et al.*, 2004).

A *Lyngbya majuscula* strain isolated from Jamaica was shown to produce neurotoxic lipopeptides (jamaicamides), the biosynthesis of which was studied by precursor feeding and subsequent gene cluster characterization. The cluster comprises 58 kb and 17 open reading frames. The order of modules is co-linear with the product. It is the largest mixed PKS/NRPS gene cluster thus far identified in cyanobacteria (Edwards *et al.*, 2004).

Nostopeptilide biosynthesis is encoded in a cluster of 40 kb in *Nostoc* sp. GSV224, a mixed NRPS-PKS system that is co-linear with the product (Hoffmann *et al.*, 2003). The gene cluster contains a PKS unit (*nosB*) and several NRPS and tailoring genes. Two of the genes, *nosE* and *nosF*, code for dehydrogenase and reductase activities that are involved in the synthesis of the unusual amino acid, 4-methylproline (Luesch *et al.*, 2003).

NRPS gene clusters

The gene cluster for the synthesis of anabaenopeptilides (cyanopeptilides) in *Anabaena* strain 90 was the first fully characterized, cyanobacterial biosynthetic system comprised solely of NRPSs. Its function was verified by knock-out mutagenesis (Rouhiainen *et al.*, 2000). A 29-kb region in the *Anabaena* 90 genome contained three genes (*apdA, apdB,* and *apdD*) coding for seven modules of non-ribosomal peptide synthetases. Other three genes (*apdC, apdE,* and *apdF*) code for a halogenase, a methyltransferase, and a reductase, respectively. The order of the functional units coded for by the NRPS genes was co-linear with the sequence of the amino acids in the anabaenopeptilides. The gene *apdC* was the first halogenase gene reported in cyanobacteria. The formyltransferase domain found in the N-terminus of ApdA was the first reported in an NRPS and showed low similarity to the bacterial methionyl-tRNA formyltransferases. Also the 33-kb nostocyclopeptilide gene cluster of *Nostoc* sp. ATCC 53789 codes for NRPS and tailoring functions, and the order of modules and domains is co-linear with the structure of nostocyclopeptides (Becker *et al.*, 2004).

Cylindrospermopsins

The structure of the alkaloid cylindrospermopsin suggested a polyketide origin for this toxin (Moore *et al.*, 1993). The feeding experiments indicated that the starter unit for the polyketide chain in cylindrospermopsin is guanidinoacetic acid formed by amidination of glycine (Burgoyne *et al.*, 2000). Candidate PKS and NRPS genes, possibly involved in cylindrospermopsin production, have been partially identified from *Cylindrospermopsis raciborskii* (Schembri *et al.*, 2001) and *Aphanizomenon ovalisporum* (Shalev-Avlon *et al.*, 2002), but not verified either by gene knock-out experiments or biochemically.

Ribosomal cyclic peptides in cyanobacteria

From the structure of patellamides one may suspect that these compounds are also products of non-ribosomal peptide synthesis. Recently, however, evidence was presented for a ribosomal synthesis of these compounds by *Prochloron didemni* (Schmidt *et al.*, 2005). The *pat* gene cluster consists of seven genes for patellamide A and C as well as sequence motifs that most likely direct the cyclization of the peptides. Similar gene clusters have been found in the genomes of the marine bloom forming *Trichodesmium erythraeum* (Sudek *et al.*, 2006) and the freshwater strains *M. aeruginosa* PCC 7806 (E. Dittmann, personal communication) and *Anabaena* sp. 90 (unpublished results by the Sivonen laboratory).

Neurotoxins

Feeding studies with labeled compounds have been carried out with strains producing the most common cyanobacterial neurotoxins. However, the genes involved have thus far remained unknown. An *Aphanizomenon flos-aquae* strain was used to study the biosynthesis of saxitoxins (Shimizu *et al.*, 1984). The proposed new pathway for neosaxitoxin biosynthesis involved the condensation of an acetate unit, or its derivative, to the amino group bearing an α-carbon of arginine or an equivalent, and a subsequent loss of the carboxyl carbon and imidazole ring formation on the adjacent carbonyl carbon. The origins of all the carbons in the toxin alkaloid ring system were established, and the side-chain carbon was found to be derived from methionine (Shimizu, 1986).

Anatoxin-a is related structurally to the tropane class of alkaloids found in higher plants. Based mainly on ^{14}C labeled precursors and enzymatic studies, Gallon *et al.* (1990, 1994) suggested that the biosynthesis of anatoxin-a is similar to that of tropanes. However, the research by Hemscheidt *et al.* (1995b) with ^{13}C labeling experiments and ^{13}C NMR indicated that the carbon skeleton of anatoxin-a was derived from acetate and glutamate and that C-1 of glutamic acid is retained during the transformation of anatoxin-a rather than being lost by decarboxylation, and thus, it does not support the tropane alkaloid theory.

Moore *et al.* (1992, 1993) concluded that all the carbons of anatoxin-a(S) are derived from amino acids. Three methyl carbons arise from L-methionine or other donors to the tetrahydrofolate C1 pool, and C-2, C-4, C-5, and C-6 carbons of the toxin originate from L-arginine (Moore *et al.*, 1992, 1993). (2*S*,4*S*)-4-hydroxyarginine was found to be the intermediate in the biosynthesis of anatoxin-a(S) from L-arginine (Hemscheidt *et al.*, 1995a).

Potential functions

Cyanobacteria need a high input of energy to replicate their large gene clusters encoding polyketide synthases and non-ribosomal peptide synthetases and to synthesize these metabolites (Welker and von Döhren, 2006). Therefore, their biosynthesis should be advantageous to the cell. In spite of the wealth of data that has been published on the occurrence and structures of bioactive compounds in cyanobacteria, our knowledge about their function remains very limited. Typically, these substances are produced by certain cyanobacterial strains and missing in others. Producers and non-producers are often closely related suggesting that possession or lack of the respective compound might have no major impact on either growth or fitness of the cells. Non-ribosomal peptides, polyketides, and other bioactive compounds are secondary metabolites and may not be needed by the cell for survival and growth.

There are also "ribosomal" peptides with structural properties and bioactivities similar to non-ribosomal peptides. The patellamides, like many non-ribosomal peptides and other natural products, contribute to the great variety of bioactivities known from cyanobacteria (see above). Patellamides are produced by some *Prochloron* spp. strains that live in symbiosis with didemnid ascidians. They exhibit cytotoxic activity (Schmidt *et al.*, 2005). Although the importance of patellamides for symbiosis has been discussed in the literature, it should be noted that other symbiotic *Prochloron* strains do not synthesize those peptides (Schmidt *et al.*, 2005). Closely related cytotoxic ribosomal peptides or genes for their biosynthesis were found in some non-symbiotic *Microcystis* strains (Ishida *et al.*, 2000) and in a *Trichodesmium erythraeum* strain (Schmidt *et al.*, 2005), and similar "cyclamides" occur in many different cyanobacteria (see p.166; Welker and von Döhren, 2006). With few exceptions, functions of cyanobacterial bioactive compounds (secondary metabolites) remain obscure and subject to speculation. In the following section, different approaches to the functional roles of bioactive compounds are discussed with a focus on one of the most widely studied family of metabolites, the microcystins.

Role of metabolites as allelochemicals and toxins

Secondary metabolites produced by bacteria, eukaryotic microorganisms, fungi, plants, and animals often serve as allelochemicals in the suppression of growth of competitors, play a role in the defense against predators, or protect from abiotic stresses. The allelopathic effects of cyanobacterial secondary metabolites have been demonstrated on natural competitors (which, above all, are other cyanobacteria and algae insofar freshwater and marine strains are concerned, and bacteria and other microorganisms as well as fungi in terrestrial environments). However, these effects have usually been shown under laboratory conditions and with the use of isolated compounds (often at concentrations that rarely occur in the environment) or undefined cyanobacterial extracellular products (e.g. Keating, 1978; Flores and Wolk, 1986; Klein *et al.*, 1995; von Elert and Jüttner, 1997; Smith and Doan, 1999; Ishida and Murakami, 2000; Kearns and Hunter, 2000, 2001; Singh *et al.*, 2001; Kaya *et al.*, 2002; Schagerl *et al.*, 2002; Hu *et al.*, 2004; Suikkanen *et al.*, 2004; for a detailed discussion see Babica *et al.*, 2006). Interestingly, competing algae and cyanobacteria may have an effect on the production of microcystins as observed by co-cultivation or treatment with a spent culture medium (Kearns and Hunter, 2000; Sukenik *et al.*, 2002; Vardi *et al.*, 2002; Jang *et al.*, 2006), thus supporting the idea of an allelopathic role for these toxins and suggesting a communication via allelochemicals between cyanobacteria and competitors. The allelopathic effects of *Nodularia spumigena* on cryptophytes and diatoms (Suikkanen *et al.*, 2004, 2005) were due to metabolites as yet unknown and not to the toxin nodularin (Suikkanen *et al.*, 2006). Allelochemicals directed against competitors need to act outside the producing cells, i.e. they are usually secreted. An active export of secondary metabolites by cyanobacteria has not been reported thus far. The gene clusters coding for the biosynthetic enzymes of non-ribosomal peptides in cyanobacteria often include a gene for a putative ABC transporter protein as shown for genome regions coding for the microcystin synthetases in Figure 7.5 (Börner and Dittmann, 2005; unpublished data). In the case of the microcystin synthetase cluster, the amino acid sequence of this putative transport protein resembles ABC exporters more than importers, suggesting a role in the secretion of microcystins (Pearson *et al.*, 2005). However, a function as peptide transporter has so far not been experimentally demonstrated for any of these cyanobacterial proteins.

Through immunodetection in combination with electron microscopy, microcystins have been located in the inner part of the cells (nucleoplasm, thylakoids, and major inclusions). Only minor amounts were observed in the cell wall, periplasm, or sheath (Shi *et al.*, 1995; Young *et al.*, 2005; Gerbersdorf, 2006). When the distribution of microcystins between cells and surrounding medium was compared, most, and in some cases all, of the peptides were found to be cell-bound (Rapala *et al.*, 1997; Wiedner *et al.*, 2003; Tonk *et al.*, 2005). These data are not contradictory, but they also do not support the idea of a role for microcystins as allelochemicals.

Cell-bound cyanobacterial toxins, growth inhibitors, etc. may act as deterrents of and weapons against predators, i.e. against planktonic herbivores. Microcystins and other metabolites have been shown to adversely affect and even kill potential predators like

Daphnia and copepods (e.g. Burns, 1987; Benndorf and Henning, 1989; DeMott *et al.*, 1991; Gilbert, 1996; Rohrlack *et al.*, 1999, 2004). Grazers prefer eukaryotic algae such as small chlorophytes and cryptophytes to cyanobacteria as food. A large size due to the formation of filaments or colonies by cyanobacteria is regarded as better protection from herbivorous grazers than are toxins. However, consumption of filamentous and colonial cyanobacteria by (macro)zooplankton seems to be common in lakes with dense populations of cyanobacteria (Work and Havens, 2003). It is not size alone that makes many cyanobacteria a less preferred food. Herbivorous zooplankton may also be deterred from cyanobacteria by as yet unknown metabolites, e.g. components of the sheath. Rohrlack *et al.* (1999) have demonstrated that it is not microcystin that deters *Daphnia* from grazing on single-cellular toxic *Microcystis*, i.e. the substances causing toxicity and deterrence are not identical.

It remains a challenge to demonstrate the influence of cyanobacterial bioactive metabolites as a deterrent in the natural environment (e.g. Benndorf and Henning, 1989; Hanazato, 1991; Ferrao-Filho *et al.*, 2002; Vardi *et al.*, 2002). Although it is tempting to speculate that the stronger or weaker inhibitory effects and the toxicity of certain cyanobacterial secondary metabolites are related to the function of the respective compound for the producing cell, final evidence for this hypothesis is still lacking. This is true not only for the well-investigated microcystins and other non-ribosomal peptides and polyketides but also for the intensively studied saxitoxins, for which a primary function as a toxin is questionable (Llewellyn, 2006).

Response to environmental factors

Clues to the function of cyanobacterial bioactive compounds may be obtained from searching the conditions under which the compounds are produced. Many secondary metabolites are produced by bacteria and eukaryotes to protect the cells from abiotic stresses. Among the few cyanobacterial secondary metabolites with rather well-defined functions are substances that serve the protection of cells from osmotic stresses, including high salt concentrations in the environment and radiation by UV light; their synthesis is induced by the type of stress (Shibata, 1969; Garcia-Pichel and Castenholz, 1993; Gröniger *et al.*, 2000; Marin *et al.*, 2002; Bultel-Poncé *et al.*, 2004; Wright *et al.*, 2005).

Microcystins are synthesized and present in the cells in all phases of growth (cf. Sivonen and Jones, 1999; Kardinaal and Visser, 2005). Microcystins share this feature with many primary metabolites, in contrast to the many secondary metabolites that bacteria produce, accumulate, or secrete only in the stationary phase or under certain environmental conditions. Non-ribosomal synthesis, presence only in some strains, and apparent lack of a role in primary metabolism, however, clearly classify microcystins as secondary metabolites. Nevertheless, factors such as light, pH, temperature, nitrogen, phosphorous and iron have been shown to affect the content of microcystins per biomass, biovolume, or cell in the field and in laboratory studies (for literature and further discussion, see Kardinaal and Visser, 2005). In most, but not all studies the highest microcystin production occurred under conditions that also supported optimal growth (Sivonen and Jones, 1999), leading to the hypothesis that environmental cues affect microcystin production indirectly via their effects on growth (Orr and Jones, 1998; Long *et al.*, 2001). This hypothesis may be an oversimplification, since different *Microcystis* strains may respond in opposite ways to light and phosphorous (Hesse and Kohl, 2001), and light at least was shown to act very fast on *mcy* genes (see below).

The idea of a function of microcystins connected to light is supported by studies of gene activity and subcellular localization. A marked, light-induced accumulation of *mcy* transcripts and Mcy proteins was observed in *M. aeruginosa* PCC 7806 (Kaebernick *et al.*, 2000; Börner and Dittmann, 2005). Low white light (16 μE $m^{-2} s^{-1}$) was sufficient to increase gene expression as was blue and red light of the same intensity. Enhanced transcript levels could already be observed after a few minutes of illumination, arguing for a direct effect of light on the transcription of *mcy* genes (Kaebernick *et al.*, 2000). More recently, a several-fold, light-induced (white, red, blue) increase in microcystin content per cell was measured by immunodetection of the toxin in cryosections of cells from *M. aeruginosa* PCC 7806 under the electron microscope. Medium light intensities of 25 μE $m^{-2} s^{-1}$ (white, red, blue) were found to be most effective in increasing the microcystin content compared to cells cultured under low (5 μE $m^{-2} s^{-1}$) and high light (75 μE $m^{-2} s^{-1}$) (Gerbersdorf, 2006). Light seems not to affect microcystin synthesis and content under all conditions of growth, since cell density was found to interfere with light effects (Kaebernick *et al.*, 2000; Gerbersdorf, 2006).

Microcystins, like non-ribosomal peptides that serve as siderophores (Crosa and Walsh, 2002), have an affinity to metal cations (Humble *et al.*, 1994). Toxic and non-toxic *Microcystis* strains responded differently to variable concentrations of iron in the medium, leading to speculation about a function of these compounds in iron homeostasis as an iron chelator (Utkilen and Gjolme, 1995; Lyck *et al.*, 1996). Most recently, this idea received additional support with a demonstration of the binding of Fur to *Microcystis mcy* promoters *in vitro* (Martin-Luna *et al.*, 2006). Fur (ferric-uptake regulator) is an iron-dependent repressor of transcription in many bacteria (Escolar *et al.*, 1999), but it also has roles in the response to acidic stress and oxidative stress (Hall and Foster, 1996; Nunoshiba *et al.*, 1999). Any effect of iron on *Microcystis* Fur *in vivo* and on *mcy* transcription has yet to be shown. Interestingly, oxidative stress drastically reduced *mcy* transcript levels in *M. aeruginosa* PCC 7806 (Kaebernick *et al.*, 2000). Whether this effect is connected to regulation by Fur is unknown.

An entirely different role of microcystins and other peptides is suggested by the observation that their presence in the medium may be sensed by *Microcystis* cells (Schatz *et al.*, 2007). Lysis of *Microcystis* cells induced the accumulation of McyB, an essential subunit of the microcystin synthetase, and of microcystin within the remaining intact cells. This effect is likely due to the release of microcystins and other non-ribosomal peptides into the medium since it was shown that addition of microcystin (also of microginin or micropeptin) to the medium enhanced drastically the McyB levels in *Microcystis* cells. These results suggest that microcystin may act as infochemical for the cells and as autoinducer of its biosynthesis (Schatz *et al.*, 2007).

Analysis of mutants

Mutants with defective biosynthesis of bioactive compounds ought to be ideal subjects in which to search for the function(s) of these metabolites. Provided a strain is accessible to genetic manipulation, genes can be inactivated ("knocked out" or "knocked down") by directed mutagenesis, resulting in insertional disruption or deletion of the target gene. In the case of microcystins, such mutants could be generated by directed mutagenesis of *mcy* genes in *M. aeruginosa* and *P. agardhii* strains (Dittmann *et al.*, 1997; Nichizawa *et al.*, 2000; Christiansen *et al.*, 2003; Pearson *et al.*, 2004). The lack of all microcystins in an *mcyB* mutant of *M. aeruginosa* PCC 7806, a well-investigated producer of different microcystins (Dittmann *et al.*, 1997), had no effect on growth of the mutant cells under different light conditions as compared to the microcystin-containing wild-type cells (Hesse *et al.*, 2001). One reason for the normal growth of *mcyB* mutants under laboratory conditions might be that microcystins and other cyanobacterial bioactive compounds play a role with relevance for the producer only under certain conditions in the natural environment, a typical feature of many secondary metabolites. However, even under natural conditions, the lack of microcystins due to mutation of *mcy* genes does not seem to reduce markedly the fitness of a mutant compared to wild types. Kurmayer *et al.* (2004) reported that among *Planktothrix* genotypes, up to 21% of so-called "inactive microcystin genotypes" of *P. rubescence* may occur in Austrian lakes. "Inactive microcystin genotypes" are defined as cell clones that contain *mcy* genes, yet do not produce microcystins. More recently it was shown that the "inactive microcystin genotypes" represent mutants with deletions and transposon insertions, respectively, in *mcy* genes. Since identical *mcy* mutations were found in different lakes and in related, though clearly different, strains, these mutants seem to have co-existed with microcystin-producing wild-type genotypes for long periods of time (Christiansen *et al.*, 2006).

It may be argued that the adverse effects potentially caused by the lack of a certain metabolite could be compensated for by synthesis of one or more compounds with identical or similar function. Indeed, cyanobacterial strains often synthesize a bunch of different, though structurally related, bioactive substances that might well have similar functions and thus be able to substitute for each other (Repka *et al.*, 2004). Also the wild-type strain *Microcystis* sp. PCC 7806 that was used for generation of *mcy* mutants not only produces microcystins, but also cyanopeptolins (Martin *et al.*, 1993) and aeruginosin (M. Welker and N. Tandeau de Marsac, personal communication). These peptides are present in the microcystin-lacking cells of an *mcyB* mutant (Dittmann *et al.*, 1997). If microcystins and cyanopeptolins and/or aeruginosins were to have similar functions, the lack of one of them might easily be tolerated. Moreover, microcystin producers and non-producers usually co-occur in their natural environments. The situation is similar for other bioactive compounds (Welker and von Döhren, 2006). Hence, it could be sufficient that some of the cells synthesize a specific substance if the joint effects of several compounds were decisive for the fitness of populations. Nevertheless, toxic strains may have advantages over non-toxic ones. Schatz *et al.* (2005) observed the superiority of two toxic, microcystin-producing strains over two genetically related, non-toxic "subcultures" in lake and growth experiments.

Although an effect of the incapability of synthesizing microcystins in growth and fitness of genetically designed mutants has not been described so far, analyses of mutants with impaired biosynthesis of bioactive compounds may nevertheless pave the way to a better understanding of the physiological or ecological function of these substances. Comparisons of the proteome of a mutant lacking all microcystins and microcystin-producing wild-type cells of PCC 7806 have revealed distinct differences. Among the proteins exhibiting altered accumulation due to knock-out mutation of a single *mcy* gene are MrpA and MrpB (homologues of quorum-sensing controlled *Rhizobium* proteins; Dittmann *et al.*, 2001), and a lectin, microvirin, that specifically binds to mannans on the surface of *Microcystis* cells. Microvirin is similar to the antiviral protein cyanovirin-C from *Nostoc ellipsosporum* (Kehr *et al.*, 2006). Like microcystin biosynthesis, the synthesis of these proteins is also affected by light (Dittmann *et al.*, 2001; Kehr *et al.*, 2006). Lectins are important for cell-cell interactions. One may, therefore, speculate about roles in the formation of colonies or in the interaction with other microorganisms. *M. aeruginosa* cells are always closely associated with other bacteria in field samples. Removal of the accompanying bacteria leads to slower growth of *Microcystis* and lower rates of CO_2-fixation (e.g. Pearl and Milly, 1996).

Other observations suggest a link between microcystins and photosynthesis. Enzymes of the Calvin cycle are among the proteins with altered accumulation in *mcy* mutants (Y. Zilliges and E. Dittmann, personal communication). A role for microcystin in the fixation of CO_2 and in photorespiration as an inhibitor of ribulose-1,5 bisphosphate carboxylase/oxygenase (Rubisco) was previously suggested by Jaehnichen *et al.* (2001). An inhibition of the expression of Rubisco by ocadaic acid was shown for maize and discussed as an indirect effect of the inhibition of protein phosphatases (Sheen, 1993). Although microcystins, like ocadaic acid, are specific inhibitors of protein phosphatases 1 and 2A, the data obtained with maize cannot be interpreted as support for the similar effect of microcystins on cyanobacteria, since bacterial protein phosphatases, including cyanobacterial ones, are not inhibited by microcystins (Shi *et al.*, 1999). Nevertheless, the hypothesis for a role for microcystins in CO_2 fixation and/or photorespiration gained ground with the recent observation that microcystins were not randomly distributed over the cell, but rather showed a preferential clustering with polyphosphate bodies and carboxysomes in cryosections of *Microcystis* sp. PCC 7806 cells (Gerbersdorf, 2006). Carboxysomes contain Rubisco and other enzymes of the Calvin cycle (Badger *et al.*, 2006). The microcystin-free *mcyB* mutant of strain PCC 7806, though not affected in growth, was found to have a higher phycocyanin/chlorophyll ratio under light-limiting conditions compared to the wild type (Hesse *et al.*, 2002). It should be noted in this context that microcystins were also found to be associated with the thylakoid membranes (Shi *et al.*, 1995; Young *et al.*, 2005).

Considering all the evidence, there are several implications for intra- and extracellular functions of microcystins in metabolism or cell-cell interactions. Yet hard evidence is lacking for any of the suggested functions. In the near future, more detailed molecular, biochemical, and (eco)physiological studies can be expected to provide further insights.

Evolutionary aspects

Studies of the evolution of biosynthetic pathways for bioactive compounds in cyanobacteria are in the beginning stages. Many of the metabolites are specific for cyanobacteria, i.e. the evolution of the respective pathway must have taken place within the cyanobacteria rather than in other organisms followed by "import" via lateral (horizontal) gene transfer (LGT). There are, however, several cyanobacterial secondary metabolites that are structurally related to compounds produced by other organisms. Well-known examples are the saxitoxins that are synthesized by cyanobacteria and marine dinoflagellates (Llewellyn, 2006), and several cyanobacterial peptides, including anabaenopeptins, cyanopeptolins, microcystins, nodularins, and cyclamides, that have structural homologues in some marine invertebrates and sponges (see pp 164–166; Welker and von Döhren, 2006). These marine eukaryotes live in symbiosis with bacteria. As an alternative to LGT, the possibility has been considered that the bioactive compounds are synthesized by (cyano)bacterial symbionts rather than by their eukaryotic hosts. At least in the case of the patellamides, the cyanobacterial symbionts were indeed shown to harbor the biosynthetic genes (Schmidt *et al.*, 2005).

A characteristic feature of many cyanobacterial secondary metabolites is their patchy distribution, i.e. identical or structurally related compounds occur in several, but not all genera, within the genus in several but not all species, and within the species in several but not all strains (genotypes, clones) (see Welker and von Döhren, 2006, for a recent discussion of the situation with non-ribosomal peptides and polyketides). Microcystins, for example, have been detected in the cyanobacterial genera *Anabaena*, *Anabaenopsis*, *Hapalosiphon*, *Microcystis*, *Nostoc*, *Planktothrix*, *Phormidium*, *and Synechococcus* (see p. 160). Identical structures of bioactive compounds in different genera together with patchy distribution might be regarded as indications of a distribution of the biosynthetic genes among cyanobacteria by LGT. In the case of microcystins, enough sequence data were available to investigate the evolution of the *mcy* gene cluster. Rantala *et al.* (2004) compared phylogenetic trees of housekeeping genes and conserved regions of three NRPS genes, *mcyA*, *mcyD*, and *mcyE*, which are essential for the biosynthesis of microcystins. Interestingly, identical trees from all genes were obtained, i.e. the *mcy* genes co-evolved with the housekeeping genes (Figure 7.6). This data rules out an LGT of the investigated *mcy* genes *between genera*. Other important findings of that study were that microcystin synthetase genes are very ancient, and that the non-toxic strains have lost these genes during evolution. This finding also implies that individual strains of cyanobacteria may have retained these genes, and new toxin producers among the cyanobacteria currently regarded as non-producing species may still be found. This study further predicted that nodularin is a shorter variant of microcystins (cf. Moffitt and Neilan, 2004). Although these conclusions can only be drawn for the investigated *mcy* sequences, there is good reason to assume that they are valid for all *mcy* genes. Moreover, Jungblut and Neilan (2006), by analyzing the phylogeny of the aminotransferase domain of McyE, observed congruence between the evolution of this domain and 16S rRNA.

The modular structure of NRPS and PKS and the composition of modules from domains with distinct enzymatic activities suggest specific paths of evolution. Homologous

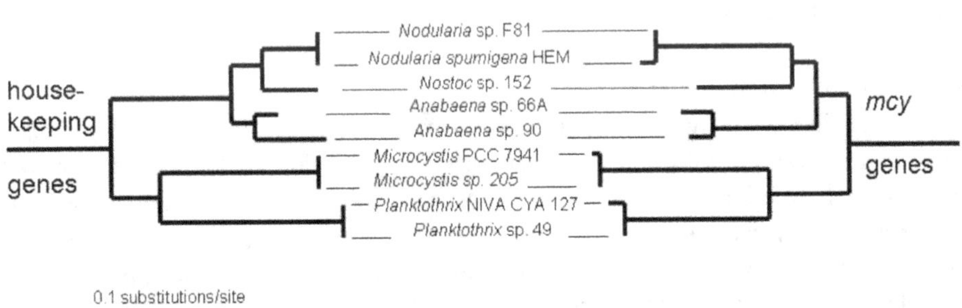

Figure 7.6 Sequence analysis of housekeeping genes (*16Srrn*, *rpoC1*) and of genes coding for the microcystin synthetase (*mcyA, mcyD, mcyE*) result in nearly identical phylogenetic trees indicating their coevolution (Rantala *et al.*, 2004).

recombination should be an efficient tool for generating new modules with new combinations of domains, new genes with new combinations of modules, or even new gene clusters with new combinations of genes. Sequence comparison of different modules and domains indeed led to the conclusion that homologous recombination contributed to the evolution of the *mcy* gene clusters in the different genera (Christiansen *et al.*, 2003; Mikalsen *et al.*, 2003; Rouhiainen *et al.*, 2004), possibly based on an intraspecific, interstrain exchange of sequences by LGT (Tanabe *et al.*, 2004). Nature has obviously used mechanisms for the evolution of secondary metabolism that are now copied by experimenters aiming at the generation of new bioactive compounds (Grünewald and Marahiel, 2006; Hill, 2006; Jenke-Kodoma *et al.*, 2006).

Cyanobacterial toxins: problems and detection methods

Problems caused by toxic cyanobacteria

Mass occurrences of toxic cyanobacteria are common worldwide (Sivonen and Jones, 1999). In temperate regions toxic blooms are seasonal, occurring mostly in late summer and autumn. In warmer climates the blooms may persist almost year around. Toxic mass occurrences of cyanobacteria in drinking water or in important recreational areas cause health risks for human beings (see e.g. Kuiper-Goodman *et al.*, 1999; Codd *et al.*, 2005). The WHO has established a provisional guideline value of 1 μg/l for microcystin-LR concentration in drinking water. Once systematic surveys revealed the common occurrence of toxic cyanobacteria in water reservoirs, studies of the removal of toxins and toxic cells in water treatment became a high priority among researchers and water companies. Currently, there are efficient methods for removing both toxins and toxic cells during water treatment procedures (Hoeger *et al.*, 2005). It is also well known that children especially may be in danger when swimming in the waterblooms. Thus, in most countries swimming in lakes containing apparent blooms is prohibited, and beaches are closed during the bloom period. Toxic blooms may also harm aquaculture. Cyanobacterial toxins are water soluble and do not accumulate in lipid layers as does, for example DDT. The highest hepatotoxin concentrations are found in the liver; thus, it is recommended that livers and intestines of fish and water fowl should not be used for human consumption during the bloom period. Cyanobacterial toxins have caused numerous animal poisonings of both domestic and wild animals (Ressom *et al.*, 1994). Recent research has shown that cyanotoxins may also be found in cyanobacterial dietary supplements (Dietrich and Hoeger, 2005).

Detection methods for toxins

In earlier studies, bioassays (mostly using mice) were the only way to detect cyanobacterial toxins. Later, purification and structural characterization of the most widespread cyanotoxins allowed for the development of efficient methods such as enzyme-linked immunosorbent assay (ELISA), protein phosphatase inhibition assay (PPIA), high-performance liquid chromatography (HPLC), and liquid chromatography—mass spectrometry (LC/MS) for detection of cyanotoxins. ELISA is available for cyanobacterial hepatotoxins and saxitoxins, whereas PPIA is used to detect microcystin and nodularins. ELISA and PPIA are sensitive and fast screening methods (Harada *et al.*, 1999; Rapala *et al.*, 2002;

Hilborn *et al.*, 2005), but they do not discriminate sufficiently between structurally related compounds. Matrix-assisted laser desorption/ionization time-of-flight (MALDI-TOF) mass spectrometry has been proven as a fast chemical screening method for detection of cyanobacterial peptide toxins and bioactive compounds, including different isoforms (e.g. Erhard *et al.*, 1997; Fastner *et al.*, 2001; Welker *et al.*, 2004). For accurate identification and quantification of the individual toxins, methods such as HPLC combined with UV, FT, or mass detection are used (Harada *et al.*, 1999; Sivonen, 2000; Meriluoto *et al.*, 2004; Meriluoto and Codd, 2005). These detection methods have yielded important information about environmentally occurring concentrations of these compounds for risk assessment.

Detection of toxic cyanobacteria

Previously, identification of cyanobacterial strains as toxin producers required isolation of the organisms and subsequent mass cultivation and testing. It soon became evident that toxin producers cannot be identified by microscopy alone since toxic and non-toxic strains of the same species with identical phenotype exist and occur together in blooms (Ohtake *et al.*, 1989; Vezie *et al.*, 1998). Molecular taxonomy based on 16S rRNA and other common cyanobacteria genes also failed to separate toxic strains from non-toxic ones (e.g. *Microcystis* and *Planktothrix*, Lyra *et al.*, 2001). On the other hand, determination of the toxin does not help to identify the producer, since the toxins are produced by different species and genera (except nodularin which is produced only by *Nodularia spumigena*). Development of the first molecular methods for detecting toxic strains became possible only with knowledge of the genes involved in microcystin and nodularin production.

Potential toxin (microcystin)-producing strains in samples can be detected by conventional PCR based on microcystin synthetase genes (*mcy* genes). Most of the primers are designed and used for detection of *Microcystis aeruginosa* strains (Tillett *et al.*, 2001; Pan *et al.*, 2002; Baker *et al.*, 2002; Nonneman and Zimba, 2002; Kurmayer *et al.*, 2002, 2003; Via-Ordorika *et al.*, 2004; Ouahid *et al.*, 2005; Hotto *et al.*, 2005). Studies based on colonies collected directly from lakes and analyzed by PCR showed co-existence of toxic and non-toxic colonies in European lakes (Kurmayer *et al.*, 2002, 2003; Via-Odorika *et al.*, 2004). In these lakes the presence of *mcy* genes was associated most frequently with the *M. aeruginosa* morphotype, not with the *M. wesenbergii*, thereby supporting earlier findings made using strain isolations. Large colonies more frequently contained microcystin than small ones (Kurmayer *et al.*, 2002, 2003; Via-Odorika *et al.*, 2004). Also in *Planktothrix*, the distribution of *mcy* genes correlated to a certain extent with the phenotype: red-pigmented *P. rubescens* contained *mcy* genes much more frequently than green-pigmented *P. agardhii* (Kurmayer *et al.*, 2004; Mbedi *et al.*, 2005). Five percent of the strains in Lake Irrsee and 21% of those in Lake Mondsee (both in Austria) bore mutations in *mcy* genes, leading to an inability to produce microcystins (Kurmayer *et al.*, 2004; Christiansen *et al.*, 2006). General primers to identify all potential microcystin producers have been developed (Neilan *et al.*, 1999; Hisbergues *et al.*, 2003; Rantala *et al.*, 2004; Jungblut and Neilan, 2006). Hisbergues *et al.* (2003) identified microcystin producers and their genus by digesting the PCR-product with restriction enzymes. Rantala *et al.* (2004, 2006) successfully used *mcyE* primers and obtained the PCR product from all known microcystin and nodularin producers and designed genus-specific primers. Simultaneous occurrence of several potential microcystin producers in Finnish lakes seems common (e.g. Vaitomaa *et al.*, 2003; Rantala *et al.*, 2006). Interestingly, toxic cyanobacteria may thrive especially in nutrient-rich lakes (Rantala *et al.*, 2006). The protection of water and the restoration of lakes may thus be effective in reducing toxic cyanobacterial blooms.

Analyses of a high number of *Microcystis* strains and populations have revealed only a few genotypes containing the *mcy* genes but lacking microcystins, possibly due to mutations in the large gene cluster (Kaebernick *et al.*, 2001; Tillett *et al.*, 2001; Mikalsen *et al.*, 2003). This sparsity is in contrast to studies of *Planktothrix* strains and filaments collected from Alpine lakes, where *mcy* mutants that are unable to synthesize any microcystin occur rather frequently (Kurmayer *et al.*, 2004; Christiansen *et al.*, 2006).

Molecular methods for detecting producers of cylindrospermopsins have been developed based on polyketide synthase (PKS) and peptide synthetase (PS) gene clusters present only in cylindrospermopsin-producing strains, but verification of the involvement of these genes in cylindrospermopsin biosynthesis is still lacking (Schembri *et al.*, 2001; Fergusson and Saint, 2003). Fergusson and Saint (2003) developed a multiplex PCR assay for *Cylindrospermopsis raciborskii*, *Aphanizomenon ovalisporum*, and *Anabaena bergii* based on the PKS and PS genes, which correlated well with the production of cylindrospermopsin.

Lack of information about biosynthesis and the biosynthetic genes of other toxins such as the cyanobacterial neurotoxins prevents the use of molecular methods to detect the producers of these toxins.

Real-time PCR yields quantitative information and may answer the question of which organism is a major producer of toxins in a given sample. Very few applications of this method for detection of cyanotoxin-producing cyanobacteria have been reported so far (Foulds *et al.*, 2002; Kurmayer and Kutzenberger, 2003; Vaitomaa *et al.*, 2003; Rinta-Kanto *et al.*, 2005). The prerequisite for this method, namely, that the biosynthetic genes are known, restricts its current use to microcystin and nodularin producers. The combination of real-time PCR with environmental analyses may open the door to the identification of factors promoting toxic blooms especially. Also the design of strategies for the restoration of lakes may profit from quantitative information about toxic cyanobacterial populations.

Use of DNA chips (microarrays) is another attractive method for monitoring, since large amounts of data can be created fast and the method can be automated. Currently, the most often used microscopic identification of cyanobacteria in environmental samples is based on subjective criteria, requires a long training period, and is labor intensive. DNA chip technology can identify all the cyanobacteria, and the toxin producers for which probes are available. At present, applications of DNA chip technology for detection of cyanobacteria are scarce. Rudi *et al.* (2000) developed a microarray based on 16S rRNA genes for a few groups of cyanobacteria and the array was used to detect cyanobacteria in eight Norwegian lakes. Another example of the development of microarray for detection of cyanobacteria comes from an EU-project MIDI-CHIP (http://www.cip.ugl.ac.be/midichip/; Castiglioni *et al.*, 2004). The prototype microarray was based on 16S rRNA genes and detected 19 major cyanobacteria groups. These arrays have been developed further to include a few, more important phylogenetic groups as well as the detection of microcystin/nodularin-producing cyanobacteria (MIDI-CHIP project, unpublished results).

Future prospects

Whole genome sequencing projects include many cyanobacterial strains (e.g. http://www.kazusa.or.jp/cyano/; http://genome.jgi-psf.org/; http://www.moore.org/microgenome/). Most recently, toxin-producing, cyanobacteria genome-sequencing projects have been initiated: two microcystin-producing strains, *Microcystis aeruginosa* (Genoscope, the Pasteur Institute, N. Tandeau de Marsac) and *Anabaena* sp. strain 90 (K. Sivonen with the Beijing Genomics Institute, China), and a nodularin-producing *Nodularia spumigena* strain (Moore Foundation/Venter Institute, initiated by L. Stal, NIOO, NL). Cyanobacteria such as toxic *Planktothrix*, *Cylindrospermopsis* and all of the neurotoxin producers, which form ecologically relevant and harmful mass occurrences, still remain to be sequenced. The same is true for the cyanobacteria producing biomedically important compounds.

Detailed annotation of these whole genomes is likely to reveal genes associated with toxin/bioactive compound production and their regulation. Comparison of these genomes will yield information about the metabolic versatility of these organisms and the differences between the various producers of toxins. Knowledge of genome sequences makes it possible to design arrays to study the gene expression of these organisms under various conditions. Gene knockout experiments with planktonic cyanobacteria have been difficult to accomplish. However, a few mutants of *Microcystis* (Dittmann *et al.*, 1997; Nishizawa *et al.*, 2000; Pearson *et al.*, 2004), *Anabaena* (Rouhiainen *et al.*, 2000), and *Planktothrix* (Christiansen *et al.*, 2003) have been generated. They give hope for the successful genetic manipulation and directed mutagenesis needed in the future to identify the large number of gene clusters already sequenced from different cyanobacterial strains. The whole genome sequences of the toxin-producing and bioactive compound-producing cyanobacteria will pave the way for proteomic research into these organisms. The combination of genome sequencing with studies of transcriptomes, proteomes, and metabolomes is likely to reveal key aspects of the biology of these important organisms in the near future.

Acknowledgments

Grants (nos. 214457 and 211494) from the Academy of Finland to KS and collaboration with the EU-projects MIDI-CHIP and PEPCY are gratefully acknowledged. We are also grateful for Dr. Jouni Jokela (Univ. of Helsinki) for making Figures 7.1–7.4.

References

Admi, V., Afek, U., and Carmeli, S. (1996). Raocyclamides A and B, novel cyclic hexapeptides isolated from the cyanobacterium *Oscillatoria raoi*. J. Nat. Prod. 59, 396–399.

Anderson, D.M. (1994). Red tides. Scientific American *August*, 52–58.

Annila, A., Lehtimäki, J., Mattila, K., Eriksson, J.E., Sivonen, K., Rantala, T.T., and Drakenberg, T. (1996). Solution structure of nodularin—an inhibitor of serine/threonine specific protein phosphatases. J. Biol. Chem. *271*, 16695–16702.

Araoz, R., Nghiem, H.O., Rippka, R., Palibroda, N., Tandeau de Marsac N., and Herdman, M. (2005). Neurotoxins in axenic oscillatorian cyanobacteria: coexistence of anatoxin-a and homoanatoxin-a determined by ligand-binding assay and GC/MS. Microbiology *151*, 1263–1273.

Babica, P., Bláha, L., and Marsálek, B. (2006). Exploring the natural role of microcystin—a review of effects on photoautotrophic organisms. J. Phycol. 42, 9–20.

Badger, M.R., Price, G.D., Long, B.M., and Woodger, F.J. (2006). The environmental plasticity and ecological genomics of the cyanobacterial CO_2 concentrating mechanism. J. Exp. Bot. 57, 249–265.

Baker, J.A., Entsch, B., Neilan, B.A., and McKay, D.B. (2002). Monitoring changing toxigenicity of a cyanobacterial bloom by molecular methods. Appl. Environ. Microbiol. *68*, 6070–6076.

Banker, P.D., and Carmeli, S. (1998). Tenuecyclamides A-D, cyclic hexapeptides from the cyanobacterium *Nostoc spongiaeforme* var *tenue*. J. Nat. Prod. *61*, 1248–1251.
Banker, P.D., Carmeli, S., Hadas, O., Teltsch, B., Porat, R., and Sukenik, A. (1997). Identification of cylindrospermopsin in *Aphanizomenon ovalisporum* (Cyanophyceae) isolated from Lake Kinneret, Israel. J. Phycol. *33*, 613–616.
Banker, P.D., Carmeli, S., Werman, M., Teltsch, B., Porat, R., and Sukenik, A. (2001). Uracil moiety is required for toxicity of the cyanobacterial hepatotoxins cylindrospermopsin. J. Toxicol. Environ. Health, Part A *62*, 281–288.
Ballot, A., Krienitz, L., Kotut, K., Wiegand, C., and Pflugmacher, S. (2005). Cyanobacteria and cyanobacterial toxins in the alkaline crater lakes Sonachi and Simbi, Kenya. Harmful Algae *4*, 139–150.
Barford, D., and Keller, J. (1994). Co-crystallization of the catalytic subunit of the serine/threonine specific protein phosphatase 1 from human in complex with microcystin LR. J. Mol. Biol. *235*, 763–766.
Becher, P.G., Beuchat, J., Gademann, K., and Jüttner, F. (2005). Nostocarboline: isolation and synthesis of a new cholinesterase inhibitor from *Nostoc* 78–12A. J. Nat. Prod. *68*, 1793–1795.
Becker, J.E., Moore, R.E., and Moore, B.S. (2004). Cloning, sequencing, and biochemical characterization of the nostocyclopeptide biosynthetic gene cluster: molecular basis for imine macrocyclization. Gene *325*, 35–42.
Benndorf, J., and Henning, M. (1989). *Daphnia* and toxic blooms of *Microcystis aeruginosa* in Bautzen Reservoir (GDR) Int. Revue ges. Hydrobiol. *74*, 233–248.
Bernard, C., Harvey, M., Briand, J.F., Biré, R., Krys, S., and Fontaine, J.J. (2003). Toxicological comparison of diverse *Cylindrospermosis raciborskii* strains: evidence of liver damage caused by a French C. *raciborskii* strain. Environ. Toxicol. *18*, 176–186.
Bewley, C.A., Gustafson, K.R., Boyd, M.R., Covell, D.G., Bax, A., Clore, G.M., and Gronenborn, A.M. (1998). Solution structure of cyanovirin-N, a potent HIV-inactivating protein. Nature Structural Biology *5*, 571–578.
Bokesch, H.R., O'Keefe, B.R., McKee, T.C., Pannell, L.K., Patterson, G.M.L., Gardellina II, R.S., Sowder, R.C., Turpin, J., Watson, K., Buckheit Jr., R.W., and Boyd, M.R. (2003). A potent novel anti-HIV protein from the cultured cyanobacterium *Scytonema varium*. Biochemistry *42*, 2578–2584.
Bolmstedt, A.J., O'Keefe, B.R., Shenoy, R.S., Mcmahon, J.B., and Boyd, M.R. (2001). Cyanoviridin-N defines a new class of antiviral agent targeting N-linked, high-manose glycans in an oligosaccharide-specific manner. Mol. Pharmacol. *59*, 949–954.
Botes, D.P., Tuinman, A.A., Wessels, P.L., Viljoen, C.C., and Kruger, H. (1984). The structure of cyanoginosin-LA, a cyclic heptapeptide toxin from the cyanobacterium *Microcystis aeruginosa*. J. Chem. Soc., Perkin Transactions I, 2311-2318.
Botes, D.P., Wessels, P.L., Kruger, H., Runnegar, M.T.C., Santikarn, S., Smith, R.J., Barna, J.C.J., and Williams, D.H. (1985). Structural studies on cyanoginosins-LR, -YR, -YA, and -YM, peptide toxins from *Microcystis aeruginosa*. J. Chem. Soc., Perkin Transactions I, 2747-2748.
Bourne, D., Jones, G.J., Blakeley, R.L., Jones, A., Negri, A.P., and Riddles, P. (1996). Enzymatic pathway for the bacterial degradation of the cyanobacterial cyclic peptide toxin microcystin LR. Appl. Environ. Microbiol. *62*, 4086–4094.
Boyd, M.R., Gustafson, K.R., Mcmahon, J.B., Shoemaker, R.H., O'Keefe, B.R., Gulakowski, T.R.J., Wu, L., Rivera, M.I., Laurencot, C.M., Currens, M.J., Gardellina II, R.S., Buckheit Jr., R.W., Nara, P.L., Pannell, L.K., Sowder II, R.C., and Henderson, L.E. (1997). Discovery of cyanovirin-N, a novel human immunodeficiency virus-inactivating protein that binds viral surface envelope glycoprotein gp 120: potential applications to microbicide development. Antimicrobial Agents Chemotheraphy *42*, 1521–1530.
Börner, T., and Dittmann, E. (2005). Molecular biology of cyanobacterial toxins. In Harmful Cyanobacteria, J. Huisman, H.C.P. Matthijs, and P.M. Visser, eds. (Dordrecht, The Netherlands: Springer). pp. 25–40.
Bultel-Poncé, V., Felix-Theodose, F., Sarthou, C., Ponge, J.-F., and Bodo, B. (2004). New pigments from the terrestrial cyanobacterium *Scytonema* sp. collected on the Mitaraka inselberg, French Guyana. J. Nat. Prod. *67*, 678–681.
Burgoyne, D.L., Hemscheidt, T.K., Moore, R.E., and Runnegar, M.T. (2000). Biosynthesis of cylindrospermopsin. J. Org. Chem. *65*, 152–156.
Burns, W.C. (1987). Insights into zooplankton-cyanobacteria interactions derived from enclosure studies. N. Zealand J. Mar. Freshw. Res. *21*, 477–482.

Cardellina II, J.H., Marner, F.-J., and Moore, R.E. (1979). Seaweed dermatitis, structure of lyngbyatoxin A. Science *204*, 193–195.

Carmichael, W.W., Beasley, V.R., Bunner, D.L., Eloff, J.N., Falconer, I., Gorham, P., Harada, K.-I., Krishnamurthy, T., Yu, M.-J., Moore, R.E., Rinehart, K., Runnegar, M., Skulberg, O.M., and Watanabe, M. (1988). Naming of cyclic heptapeptide toxins of cyanobacteria (blue-green algae). Toxicon 26, 971-973.

Carmichael, W.W., and Li, R. (2006). Cyanobacteria toxins in the Salton Sea. Saline Systems. 2, 5. (doi: 190.1186/1745–1488–2–5).

Carmichael, W.W., Mahmood, N.A., and Hyde, E.G. (1990). Natural toxins from cyanobacteria (blue-green algae). In Marine Toxins, Origin, Structure and Molecular Pharmacology, vol. 418, S. Hall, and G. Strichartz, eds. (Washington, DC: American Chemical Society), pp. 87–106.

Castiglioni, B., Rizzi, E., Frosini, A., Sivonen, K., Rajaniemi, R., Rantala, A., Mugnai, M.A., Ventura, S., Wilmotte, A., Boutte, C., Grubisic, S., Balthasart, B., Consolandi, C., Bordoni, R., Mezzelani, A., Battaglia, C., and De Bellis, G. (2004). Development of a universal microarray based on the ligation detection reaction and 16S rRNA gene polymorphism to target diversity of cyanobacteria. Appl. Environ. Microbiol. *70*, 7161–7172.

Chaganty, S., Golakoti, T., Heltzel, C., Moore, R.E., and Yoshida, W.Y. (2004). Isolation and structure determination of cryptophycins 38, 326 and 327 from the terrestrial cyanobacterium *Nostoc* sp. GSV 224. J. Nat. Prod. *67*, 1403–1406.

Chang, Z., Flatt, P.M., Gerwick, W.H., Nguyen, V.-A., Willis, C.L., and Sherman, D.H. (2002). The barbamide biosynthetic gene cluster: a novel marine cyanobacterial system of mixed polyketide synthase (PKS)-non-ribosomal peptide synthetase (NRPS) origin involving an unusual trichloroleucyl starter unit. Gene *296*, 235–247.

Chang, Z., Sitachitta, N., Rossi, J.V., Roberts, M.A., Flatt, P.M., Jia, J., Sherman, D.H., and Gerwick, W.H. (2004). Biosynthetic pathway and gene cluster analysis of curacin A, an antitubulin natural product from the tropical marine cyanobacterium *Lyngbya majuscula*. J. Nat. Prod. *67*, 1356–1367.

Choi, B.W., Namikoshi, M., Sun, F., Rinehart, K.L., Carmichael, W.W., Kaup, A.M., Evans, W.R., and Beasley, V.R. (1993). Isolation of linear peptides related to the hepatotoxins nodularin and microcystins. Tetrahedron Letters *34*, 7881–7884.

Christiansen, G., Fastner, J., Erhard, M., Börner, T., and Dittmann, E. (2003). Microcystin biosynthesis in *Planktothrix*: genes, evolution, and manipulation. J. Bacteriol. *185*, 564–572.

Christiansen, G., Kurmayer, R., Liu, Q., and Börner, T. (2006). Transposons inactivate the biosynthesis of the nonribosomal peptide microcystin in naturally occurring *Planktothrix* spp. Appl. Environ. Microbiol. 72, 117–123.

Codd, G.A., Morrison, L.F., and Metcalf, J.S. (2005). Cyanobacterial toxins: risk management for health protection. Toxicol. Appl. Pharmacol. *203*, 264–272.

Cox, P.A., Banack, S.A., Murch, S.J., Rasmussen, U., Tien, G., Bidigare, R.R., Metcalf, J.S., Morrison, L.F., Codd, G.A., and Bergman, B. (2005). Diverse taxa of cyanobacteria produce β-*N*-methylamino-L-alanine, a neurotoxic amino acid. Proc. Natl. Acad. Sci. USA *102*, 5074–5078.

Crosa, J.H., and Walsh, J.T. (2002). Genetics and assembly line enzymology of siderophore biosynthesis in bacteria. Microbiol. Mol. Biol. Rev. *66*, 223–249.

DeBernhardi, R., and Giussani, G. (1990). Are blue-green algae a suitable food for zooplankton? An overview. Hydrobiologia *200*, 29–41.

DeMott, W.R., Zhang, Q.-X., and Carmichael, W.W. (1991). Effects of toxic cyanobacteria and purified toxins on the survival and feeding of a copepod and three specis of *Daphnia*. Limnol. Oceanogr. *36*, 1346–1357.

de Silva E.D., Williams, D.E., Andersen, R.J., Klix, H., Holmes, C.F.B., and Allen, T.M. (1992). Motuporin, a potent protein phosphatase inhibitor isolated from the Papua New Guinea sponge *Theonella swinhoei* Gray. Tetrahedron Lett. 33, 1561–1564.

Devlin, J.P., Edwards, O.E., Gorham, P.R., Hunter, M.R., Pike, R.K., and Stavric, B. (1977). Anatoxin-a, a toxic alkaloid from *Anabaena flos-aquae* NCR-44h. Can. J. Chem. 55, 1367–1371.

Dias, E., Pereira, P., and Franca, S. (2002). Production of paralytic shellfish toxins by *Aphanizomenon* sp. LMECYA 31 (cyanobacteria). J. Phycol. *38*, 705–712.

Dietrich, D., and Hoeger, S. (2005). Guidance values for microcystins in water and cyanobacterial supplement products (blue-green algal supplements): a reasonable or misguided approach? Toxicol. Appl. Pharmacol. *203*, 273–289.

Dittmann, E., Neilan, B.A., Erhard, M., von Döhren, H., and Börner, T. (1997). Insertional mutagenesis of a peptide synthetase gene that is responsible for hepatotoxin production in the cyanobacterium *Microcystis aeruginosa* PCC 7806. Mol. Microbiol. *26*, 779–787.

Dittmann, E., Erhard, M., Kaebernick, M., Scheler, C., Neilan, B.A., and Börner, T. (2001). Altered expression of two light-dependent genes in a microcystin-lacking mutant of *Microcystis aeruginosa* PCC 7806. Microbiology *147*, 3113–3119.

Edwards, C., Beattie, K.A., Scrimgeour, C.M., and Codd, G.A. (1992). Identification of anatoxin-a in benthic cyanobacteria (blue-green algae) and associated dog poisonings at Loch Insh, Scotland. Toxicon *30*, 1165–1175.

Edwards, D.J., Marquez, B.L., Nogle, L.M., McPhail, K., Goeger, D.E., Roberts, M.A., and Gerwick, W.H. (2004). Structure and biosynthesis of the jamaicamides, new mixed polyketide-peptide neurotoxins from the marine cyanobacterium *Lyngbya majuscula*. Chem. Biol. *11*, 817–833.

Erhard, M., von Döhren, H., and Jungblut, P. (1997). Rapid typing and elucidation of new secundary metabolites of intact cyanobacteria using MALDI-TOF mass spectrometry. Nature Biotechnol. *15*, 906–909.

Escolar, L., Perez-Martin, J., and de Lorenzo, V. (1999). Opening the iron box: transcriptional metalloregulation by the Fur protein. J. Bacteriol. *181*, 6223–6229.

Fastner, J., Erhard, M., and von Döhren, H. (2001). Determination of oligopeptide diversity within a natural population of *Microcystis* spp. (cyanobacteria) by typing single colonies by matrix-assisted laser desorption ionization-time of flight mass spectrometry. Appl. Environ. Microbiol. *67*, 5069–5076.

Fastner, J., Heinze, R,. Humpage, A.R., Mischke, U., Eaglesham, G.K., and Chorus, I. (2003). Cylindrospermosin occurrence in two German lakes and preliminary assessment of toxicity and toxin production of *Cylindrospermopsis raciborskii* (cyanobacteria) isolates. Toxicon *42*, 313–321.

Fergusson, K.M., and Saint, C.P. (2003). Multiplex PCR assay for *Cylindrospermopsis raciborskii* and cylindrospermopsin-producing cyanobacteria. Environ. Toxicol. *18*, 120–125.

Ferrao-Filha, A.S., Domingos, P., and Azevedo, S.M.F.O. (2002). Influence of *Microcystis aeruginosa* KÜTZING bloom on zooplankton populations in Jacarepaguá Lagoon (Rio de Janeiro, Brazil). Limnol. Ecol. Manag. Inland Wat. *32*, 295–308.

Ferreira, F.M., Franco Soler, J.M., Fidalgo, M.L., and Fernandez-Vila, P. (2001). PSP toxins from *Aphanizomenon flos-aquae* (cyanobacteria) collected in the Crestuma-Lever reservoir (Douro river, northern Portugal). Toxicon *39*, 757–761.

Flores, E. and Wolk, C.P. (1986). Production, by filamentous, nitrogen-fixing cyanobacteria, of a bacteriocin and of other antibiotics that kill related strains. Arch. Microbiol. *145*, 215-219.

Foulds, I.V., Granacki, A., Xiao, C., Krull, U.J., Castle, A., and Horgen, P.A. (2002). Quantitation of microcystin-producing cyanobacteria and *E. coli* in water by 5′-nuclease PCR. J. Appl. Microbiol. *93*, 825–834.

Fujii, K., Harada, K., Suzuki, M., Kondo, F., Ikai, Y., Oka, H., Carmichael, W.W., and Sivonen, K. (1996). Occurrence of novel cyclic peptides together with microcystins from toxic cyanobacteria, *Anabaena* species. In Harmful and Toxic Algal Blooms. T. Yasumoto, Y. Oshima, and Y. Fukuyo, eds. (Paris: Intergovernmental Oceanographic Commission of UNESCO), pp. 559–562.

Fujii, K., Sivonen, K., Adachi, K., Noguchi, K., Sano, H., Hirayama, K., Suzuki, M., and Harada, K.-I. (1997). Comparative study of toxic and non-toxic cyanobacterial products: novel peptides from toxic *Nodularia spumigena* AV1. Tetrahedron Lett. *38*, 5525–5528.

Fujii, K., Sivonen, K., Kashiwagi, T., Hirayama, K., and Harada, K.-I. (1999). Nostophycin, a novel cyclic peptide from the toxic cyanobacterium, *Nostoc* sp. 152. J. Org. Chem. *64*, 5777–5782.

Fujii, K., Sivonen, K., Naganawa, E., and Harada, K.-I. (2000). Non-toxic peptides from toxic cyanobacteria, *Oscillatoria agardhii*. Tetrahedron *56*, 725–733.

Fujii, K., Sivonen, K., Nakano, T., and Harada, K.-I. (2002). Structural elucidation of cyanobacterial peptides encoded by peptide synthetase genes in *Anabaena* species. Tetrahedron *58*, 6863–6871.

Fujiki, H., Suganuma, M., Suguri, H., Yoshizawa, S., Takagi, K., Nakayasu, M., Ojika, M., Yamada, K., Yasumoto, T., Moore, R.E., and Sugimura, T. (1990). New tumor promoters from marine natural products. In Marine Toxins, Origin, Structure and Molecular Pharmacology, vol. 418. S. Hall, and G. Strichartz, eds. (Washington D.C.: American Chemical Society), pp. 232–240.

Gallon, J.R., Chit, K.N., and Brown, E.G. (1990). Biosynthesis of the tropane-related cyanobacterial toxin anatoxin-a: role of ornithine decarboxylase. Phytochemistry *29*, 1107–1111.

Gallon, J.R., Kittakoop, O., and Brown, E.G. (1994). Biosynthesis of anatoxin-a by *Anabaena flos-aquae*: examination of primary enzymic steps. Phytochemistry *35*, 1195–1203.

Gerbersdorf, S.U. (2006). An advanced technique for immuno-labelling of microcystins in cryosectioned cells of *Microcystis aeruginosa* PCC 7806 (cyanobacteria): Implementations of an experiment with varying light scenarios and culture densities. Toxicon *47*, 218–228.

Gerwick, W.H., Proteau, P.J., Nagle, D.G., Hamel, E., Blokhin, A.J., and Slate, D.L. (1994). Structure of curacin A, a novel antimitotic, antiproliferative, and brine shrimp toxic natural product from the marine cyanobacterium *Lyngbya majuscula*. J. Org. Chem. *59*, 1243–1245.

Gilbert, J.J. (1996). Effect of temperature on the response of planktonic rotifers to a toxic cyanobacterium. Ecology *77*, 1174–1180.

Golakoti, T., Ohtani, I., Patterson, G.L.M., Moore, R.E., Corbett, T.H., Valeriote, F.A., and Demchik, L. (1994). Total structures of cryptophycins, potent antitumor depsipeptides from the blue-green alga *Nostoc* sp. strain GSV 224. J. Am. Chem. Soc. *116*, 4729–4737.

Golakoti, T., Ogino, J., Heltzel, C.E., Husebo, T.L., Jensen, C.M., Larsen, L.K., Patterson, G.M.L., Moore, R.E., Mooberry, S.L., Corbett, T.H., and Valeriote, F.A. (1995). Structure determination, conformational analysis, chemical stability studies and antitumor evaluation of the cryptophycins. Isolation of 18 new analogs from *Nostoc* sp. strain GSV 224. J. Am. Chem. Soc. *117*, 12030–12049.

Golakoti, T., Yoshida, W.Y., Chaganty, S., and Moore, R.E. (2000). Isolation and structures of nostopeptolides A1, A2 and A3 from the cyanobacterium *Nostoc* sp. GVS 224. Tetrahedron *56*, 9093–9102.

Golakoti, T., Yoshida, W.Y., Chaganty, S., and Moore, R.E. (2001). Isolation and structure determination of nostocyclopeptilides A1 and A2 from the terrestrial cyanobacterium *Nostoc* sp. ATCC53789. J. Nat. Prod. *64*, 54–59.

Goldberg, J., Huang, H.B., Kwon, Y.G., Greengard, P., Nairn, A.C., and Kuriyan, J. (1995). Three-dimensional structure of the catalytic subunit of protein serine/threonine phosphatase-1. Nature *376*, 745–753.

Gröniger, A., Sinha, R.P., Klisch, M., and Häder, D.P. (2000). Photoprotective compounds in cyanobacteria, phytoplankton and macroalgae—a database. J. Photochem. Photobiol. B: Biology *58*, 115–122.

Grunewald, J., and Marahiel, M.A. (2006). Chemoenzymatic and template-directed synthesis of bioactive macrocyclic peptides. Microbiol. Mol. Biol. Rev. *70*, 121–146.

Gugger, M., Lenoir, S., Berger, C., Ledreux, A., Druart, C.-J., Humbert, J.-F., Guette, C., and Bernard, C. (2005). First report in a river in France of the benthic cyanobacterium *Phormidium favosum* producing anatoxin-a associated with dog neurotoxicosis. Toxicon *45*, 919–928.

Hall, H.K., and Foster, J.W. (1996). The role of Fur in the acidic tolerance response of *Salmonella typhimurium* is physiologically and genetically separable from its role in iron acquisition. J. Bacteriol. *178*, 5683–5691.

Hambley, T.W., Hawkins, C.J., Lavin, M.F., van der Brenk, A., and Watters, D.J. (1992). Cycloxazolone: a cytotoxic cyclic hexapeptide from the ascidian *Lissoclinium bistratum*. Tetrahedron *48*, 341–348.

Hanazato, T. (1991). Interrelations between *Microcystis* and *Cladocera* in the highly eutrophic Lake Kasumigaura, Japan. Int. Revue ges. Hydrobiol. *76*, 21–36.

Harada, K., Mayumi, T., Shimada, T., Suzuki, M., Kondo, F., and Watanabe, M.F. (1993). Occurrence of four depsipeptides, together with microcystins from toxic cyanobacteria. Tetrahedron Lett. *34*, 6091–6094.

Harada, K.-I., Kondo, F., and Lawton, L.A. (1999). Laboratory analysis of cyanotoxins. In Toxic Cyanobacteria in Water. A Guide to their Public Health Consequences, Monitoring and Management. I. Chorus, and J. Bartram, eds. (London: The World Health Organization. E and FN Spoon), pp. 369–405.

Harada, K.-I., Ohtani, I., Iwamoto, K., Suzuki, M., Watanabe, M.F., Watanabe, M., and Terao, K. (1994). Isolation of cylindrospermopsin from a cyanobacterium *Umezakia natans* and its screening method. Toxicon *32*, 73–84.

Harada, K.-I., Ogawa, K., Matsuura, K., Murata, H., Suzuki, M., Watanabe, M.F., Itezono, Y., and Nakayama, N. (1990a). Structural determination of geometrical isomers of microcystins LR and RR from cyanobacteria by two-dimensional NMR spectroscopic techniques. Chem. Res. Toxicol. *3*, 473–481.

Harada, K.-I., Matsuura, K., Suzuki, M., Watanabe, M.F., Oishi, S.A., Dahlem, M., Beasley, V.R., and Carmichael, W.W. (1990b). Isolation and characterization of the minor components associated with microcystins-LR and –RR in the cyanobacterium (blue-green algae). Toxicon *28*, 55–64.

Harrigan, G.G., Luesch, H., Yoshida, W.Y., Moore, R.E., Nagle, D.G., Paul, V.J., Mooberry, S., Corbett, T.H., and Valeriote, F.A. (1998a). Symplostatin 1: a dolastatin 10 analogue from the marine cyanobacterium *Symploca hydnoides*. J. Nat. Prod. *61*, 1075–1077.

Harrigan, G.G., Yoshida, W.Y., Moore, R.E., Nagle, D.G., Park, P.U., Biggs, J., Paul, V.J., Mooberry, S.,T. Corbett, H., and Valeriote F.A. (1998b). Isolation, structure determination, and biological activity of dolastatin 12 and lyngbyastatin 1 from *Lyngbya majsucula/Schizothrix calcicola* cyanobacterial assemblages. J. Nat. Prod. *61*, 1221–1225.

Harrigan, G.G., Luesch, H., Yoshida, W.Y., Moore, R.E., Nagle D.G., and Paul, V.J. (1999). Symplostatin 2: a dolastatin 13 analogue from the marine cyanobacterium *Symploca hydnoides*. J. Nat. Prod. *62*, 655–658.

Hashimoto, Y., Kamiya, H., Yamazato, K., and Nozawa, K. (1976). Occurrence of a toxic blue-green alga inducing skin dermatitis in Okinawa. In Animal, Plant, and Microbial Toxins, Vol. 1, A. Ohsaka, K. Hayashi, and Y. Sawai, eds. (New York: Plenum Publishing), pp. 333–338.

Hawkins, P.R., Runnegar, M.T.C. Jackson, A.R.B., and Falconer, I.R. (1985). Severe hepatotoxicity caused by the tropical cyanobacterium (blue-green alga) *Cylindrospermopsis raciborskii* (Woloszynska) Seenaya and Subba Raju isolated from a domestic water supply reservoir. Appl. Environ. Microbiol. *50*, 1292–1295.

Hawkins, P.R., Chandrasena, N.R., Jones, G.J., Humpage, A.R., and Falconer, I.R. (1997). Isolation and toxicity of *Cylindrospermopsis raciborskii* from an ornamental lake. Toxicon *35*, 341–346.

Hawser, S.P., Codd, G.A., Carpenter, E.J., and Capone, D.G. (1991). A neurotoxic factor associated with the bloom-forming cyanobacterium *Trichodesmium*. Toxicon *29*, 277–278.

Hemscheidt, T., Rapala, J., Sivonen, D.L., and Moore, R.E. (1995a). Biosynthesis of anatoxin-a(s). (2*S*,4*S*)-4-hydroxyarginine as an intermediate. J. Chem. Soc. Chem. Commun. 205–206.

Hemscheidt, T., Rapala, J., Sivonen, K., and Skulberg, O.M. (1995b). Biosynthesis of anatoxin-a in *Anabaena flos-aquae* and homoanatoxin-a in *Oscillatoria formosa*. J. Chem. Soc. Chem. Commun. 1361–1362.

Henriksen, P., Carmichael, W.W., An, J., and Moestrup, Ø. (1997). Detection of an anatoxin-a(s)-like anticholinesterase in natural blooms and cultures of cyanobacteria/blue-green algae from Danish lakes and in the stomach contents of poisoned birds. Toxicon *35*, 901–913.

Heresztyn, T., and Nicholson, B.C. (1997). Nodularin concentrations in Lakes Alexandrina and Albert, South Australia, during bloom of the cyanobacterium (blue-green alga) *Nodularia spumigena* and degradation of the toxin. Environ. Toxicol. Water Qual. *12*, 273–281.

Herfindal, L., Oftedal, L., Selheim, F., Wahlsten, M., Sivonen, K., and Døskeland, S.O. (2005). A high proportion of Baltic Sea bentic cyanobacterial isolates contain apoptogens able to induce rapid death of isolated rat hepatocytes. Toxicon *46*, 252–260.

Hesse, K., and Kohl, J.G. (2001). Effects of light and nutrient supply on growth and microcystin content of different strains of *Microcystis aeruginosa*. In Cyanotoxins—Occurrence, Causes, Consequences. I. Chorus, ed. (Berlin: Springer), pp. 104–114.

Hilborn, D.E., Carmichael, W.W., Yuan, M., and Azevedo, S.M.F.O. (2005). A simple colorimetric method to detect biological evidence of human exposure to microcystins. Toxicon *46*, 218–221.

Hill, A.M. (2006). The biosynthesis, molecular genetics and enzymology of the polyketide-derived metabolites. Nat. Prod. Rep. *23*, 256–320.

Hisbergues, M., Christiansen, G., Rouhiainen, L., Sivonen, K., and Börner, T. (2003). PCR-based identification of microcystin-producing genotypes of different cyanobacterial genera. Arch. Microbiol. *180*, 402–410.

Hitzfeld, B.C., Lampert, C.S., Spaeth, N., Mountfort, D., Kaspar, H., and Dietrich, D.R. (2000). Toxin production in cyanobacterial mats from ponds on the McMurdo Ice Shelf, Antarctica. Toxicon *38*, 1731–1748.

Hoeger, S.J., Hitzfeld, B.C., and Dietrich, D.R. (2005). Occurrence and elimination of cyanobacterial toxins in drinking water treatment plants. Toxicol. Appl. Pharmacol. *203*, 231–242.

Hoffmann, D., Hevel, J.M., Moore, R.E., and Moore, B.S. (2003). Sequence analysis and biochemical characterization of the nostopeptolide A biosynthetic gene cluster from *Nostoc* sp. GSV224. Gene *311*, 171–180.

Hopwood, D.A. (1997). Genetic contributions to understanding polyketide synthases. Chem. Rev. *97*, 2465–2497.

Hotto, A., Satchwell, M., and Boyer, G. (2005). Seasonal production and molecular characterization of microcystins in Oneida Lake, New York, USA. Environ. Toxicol. *20*, 243–248.

Hu, Z.-Q., Liu, Y.-D., and Li, D.-H. (2004). Physiological and biochemical analyses of microcystin-RR toxicity to the cyanobacterium *Synechococcus elongatus*. Environ. Toxicol. *19*, 571–577.

Humble, A., Gadd, G.M., and Codd, G.A. (1994). Polarographic analysis of the interactions between cyanobacterial microcystin (hepatotoxins) variants and metal cations. In Abstracts VIII International Symposium on Phototrophic Prokaryotes, (Florence), p. 82.

Humpage, A.R., Rositano, J., Bretag, A.H., Brown, R., Baler, P.D., Nicholson, B.C., and Steffensen, D.A. (1994). Paralytic shellfish poisons from Australian cyanobacterial blooms. Aust. J. Mar. Freshwater Res. *45*, 761–771.

Ikawa, M., Wegener, K., Foxall, T.L., and Sasner, J.J. (1982). Comparison of the toxins of the blue-green alga *Aphanizomenon flos-aquae* with the *Gonyaulax* toxins. Toxicon *20*, 747–752.

Ireland, C.M., Durso Jr., A.R., Newman, R.A., and Hacker, M.P. (1982). Antineoplastic cyclic peptides from marine tunicate *Lissoclinum patella*. J. Org. Chem. *47*, 1807–1811.

Ishida, K., Matsuda, H., Murakami, M., and Yamaguchi, K. (1997). Kawaguchipeptin B, an antibacterial cycic undecapeptide from the cyanobacterium *Microcystis aeruginosa*. J. Nat. Prod. *60*, 724–726.

Ishida, K., Matsuda, H., Okita, Y., and Murakami, M. (2002). Aeruginoguanidines 98-A-98-C: cytotoxic unusual peptides from the cyanobacterium *Microcystis aeruginosa*. Tetrahedron *58*, 7645–7652.

Ishida, K., and Murakami, M. (2000). Kasumigamide, an antialgal peptide from the cyanobacterium *Microcystis aeruginosa*. J. Org. Chem. *65*, 5898–5900.

Ishitsuka, M.O., Kusumi, T., Kakisawa, H., Kaya, K., and Watanabe, M.M. (1990). Microviridin A novel tricyclic depsipeptide from the toxic cyanobacterium *Microcystis viridis*. J. Am. Chem. Soc. *112*, 8180–8182.

Izaguirre, G., and Neilan, B.A. (2004). Benthic *Phormidium* species that produces microcystin-LR, isolated from three reservoirs in Southern California. Sixth International Conference on Toxic Cyanobacteria. Abstract book p. 50.

Jang, M.-H., Ha, K., Jung, L.-M., and Takamura, N. (2006). Increased microcystin production of *Microcystis aeruginosa* by indirect exposure of nontoxic cyanobacteria: Potential role in the development of *Microcystis* bloom. Bull. Environ. Contam. Toxicol. *76*, 957–962.

Jenke-Kordoma, H., Börner, T., and Dittmann, E. (2006). Natural biocombinatorics in the polyketide synthetase genes of the actinobacterium *Streptomyces avermitilis*. PLoS Comput. Sci. *2*, e132.

Jungblut, A.-D., and Neilan, B.A. (2006). Molecular identification and evolution of the cyclic peptide hepatoxins, microcystin and nodularin, synthetase genes in three orders of cyanobacteria. Arch. Microbiol. *185*, 107–114.

Jüttner, F., Todorova, A.K., Walch, N., and von Philipsborn, W. (2001). Nostocyclamide M: a cyanobacterial cyclic peptide with allelopathic activity from *Nostoc* 31. Phytochemistry *57*, 613–619.

Kaebernick, M., Rohrlack, T., Christofferssen, K., and Neilan, B.A. (2001). A spontaneous mutant of microcystin biosynthesis: genetic characterization and effect on *Daphnia*. Environ. Microbiol. *3*, 669–679.

Kardinaal, W.E.A., and Visser, P.M. (2005). Dynamics of cyanobacterial toxins. Sources of variability in microcystin concentrations. In Harmful Cyanobacteria, J. Huisman, H.C.P. Matthijs, and P.M. Visser, eds. (Dordrecht, The Netherlands: Springer), pp. 41–63.

Kaya, K., Sano, T., Beattie, K.A., and Codd, G.A. (1996). Nostocyclin, a novel 3-amino-6-hydroxy-2-piperidone-containing cyclic depsipeptide from the cyanobacterium *Nostoc* sp. Tetrahedron Lett. *37*, 6725–6728.

Kaya, K., Mahakhant, A., Keovara, L., Sano, L., Kubo, T., and Takagi, H. (2002). Spiroidesin, a novel lipopeptide from the cyanobacterium *Anabaena spiroides* that inhibits cell growth of the cyanobacterium *Microcystis aeruginosa*. J. Nat. Prod. *65*, 920–921.

Kearns, K.D., and Hunter, M.D. (2000). Green algal extracellular products regulate anti-algal toxin production in a cyanobacterium. Environ. Microbiol. 2, 291–297.

Kearns, K.D., and Hunter, M.D. (2001). Toxin-producing *Anabaena flos-aquae* induces settling of *Chlamydomonas reinhardii*, a competing motile alga. Microbial Ecol. 42, 80–86.

Kehr, J.C., Zilliges, Y., Springer, A., Disney, M.D., Ratner, D.D., Bouchier, C., Seeberger, H.P., Tandeau de Marsac, N., and Dittmann, E. (2006). A mannan binding lectin is involved in cell-cell attachment in a toxic strain of *Microcystis aeruginosa*. Mol. Microbiol. 59, 893–906.

Klein, D., Daloze, D., Braekman, J.C., Hoffmann, L., and Demoulin, V. (1995). New hapaindoles from the cyanophyte *Hapalosiphon laingii*. J. Nat. Prod. 58, 1781–1785.

Kuiper-Goodman, T., Falconer, I., and Fitzgerald, J. (1999). Human health aspects. In Toxic cyanobacteria in water. A Guide to their Public Health Consequences, Monitoring and Management. I. Chorus and J. Bartram, eds. (London: The World Health Organization. E and FN Spoon), pp. 113–153.

Kurmayer, R., and Kutzenberger, T. (2003). Application of real-time PCR for quantification of microcystin genotypes in a population of the toxic cyanobacterium *Microcystis* sp. Appl. Environ. Microbiol. *69*, 6723–6730.

Kurmayer, R., Dittmann, E., Fastner, J., and Chorus, I. (2002). Diversity of microcystin genes within a population of the toxic cyanobacterium *Microcystis* spp. in Lake Wannsee (Berlin, Germany). Microb. Ecol. *43*, 107–118.

Kurmayer, R., Christiansen, G., and Chorus, I. (2003). The abundance of microcystin-producing genotypes correlates positively with colony size in *Microcystis* sp. and determines its microcystin net production in Lake Wannsee. Appl. Environ. Microbiol. *69*, 787–795.

Kurmayer, R., Christiansen, G., Fastner, J., and Börner, T. (2004). Abundance of active and inactive microcystin genotypes in populations of the toxic cyanobacterium *Planktothrix* spp. Environ. Microbiol. *6*, 831–841.

Kurmayer, R., Christiansen, G., Gumpenberger, M., and Fastner, J. (2005). Genetic identification of microcystin ecotypes in toxic cyanobacteria of the genus *Planktothrix*. Microbiology *151*, 1525–1533.

Lagos, N., Onodera, H., Zagatto, P.A., Andrinolo, D., Azevedo, S.M., and Oshima, Y. (1999). The first evidence of paralytic shellfish toxins in the fresh water cyanobacterium *Cylindrospermopsis raciborskii*, isolated from Brazil. Toxicon *37*, 1359–1373.

Lawton L.A, Morris, L.A., and Jaspars, M. (1999). A bioactive modified peptide, aeruginosamide, isolated from the cyanobacterium *Microcystis aeruginosa*. J. Org. Chem. *64*, 5329–5332.

Lee, A.Y., Smitka, T.A., Bonjouklian, R., and Clardy, J. (1994). Atomic structure of the trypsin-A90720A complex: a unified approach to structure and function. Chem. Biol. *1*, 113–117.

Li, R., Carmichael, W.W., Brittain, S., Eaglesham, G.K., Shaw, G.S., Mahakhant, A., Noparatnaraporn, W., Yongmanitchai, K., Kaya, K., and Watanabe, W.W. (2001). Isolation and identification of the cyanotoxin cylindrospermopsin and deoxy-cylindrospermopsin from a Thailand strain of *Cylindrospermopsis raciborskii* (Cyanobacteria). Toxicon *39*, 973–980.

Liang, J., Moore, R.E., Moher, E.D., Munroe, J.E., Al-awar, R.S., Hay, D.A., Varie, D.L., Zhang, T.Y., Aikins, J.A., Martinelli, M.J., Shih, C., Ray, J.E., Gibson, L.L., Vasudevan, V., Polin, L., White, K., Kushner, J., Simpson, C., Pugh, S., and Corbett, T.H. (2005). Cryptophycins-309, 249 and other cryptophycin analogs: preclinical efficacy studies with mouse and human tumors. Investigational New Drugs *23*, 213–224.

Llewellyn, L.E. (2006). Saxitoxin, a toxic marine natural product that targets a multitude of receptors. Nat. Prod. Rep. 23, 200–220.

Lombardo, M., Pinto, F.C.R., Vieira, J.M.S., Honda, R.Y., Pimenta, A.M.C., Bemquerer, M.P.L., Carvalho, R., and Kiyota, S. (2006). Isolation and structural characterization of microcystin-LR and three minor oligopeptides simultaneously produced by *Radiocystis fernandoi* (Chroococcales, Cyanobacteriae): A Brazilian toxic cyanobacterium. Toxicon *47*, 560–566.

Long, B.M., Jones, G.J., and Orr, P.T. (2001). Cellular microcystin content in N-limited *Microcystis aeruginosa* can be predicted from growth rate. Appl. Environ. Microbiol. *67*, 278–283.

Luesch, H., Yoshida, W.Y., Moore, R.E., and Paul, V.J. (2000). Isolation and structure of the cytotoxin lyngbyabellin B and absolute configuration of lyngbyapeptin A from the marine cyanobacterium *Lyngbya majuscula*. J. Nat. Prod. *63*, 1437–1439.

Luesch, H., Moore, R.E., Paul, V.J., Mooberry, S.L., and Corbett, T.H. (2001). Isolation of dolastatin 10 from the marine cyanobacterium *Symploca* species VP642 and total stereochemistry and biological evaluation of its analogue symplostatin 1. J. Nat. Prod. *64*, 907–910.

Luesch, H., Hoffmann, D., Hevel, J.M., Becker, J.E., Golakoti, T., and Moore, R.E. (2003). Biosynthesis of 4-methylproline in cyanobacteria: cloning of *nosE* and *nosF* genes and biochemical characterization of the encoded dehydrogenase and reductase activities. J. Org. Chem. *68*, 83–91.

Lyck, S., Gjølme, N., and Utkilen, H. (1996). Iron starvation increases toxicity of *Microcystis aeruginosa* CYA 228/1 (Chroococcales, Cyanophyceae). Phycologia 35, 120–124.

Lyra, C., Suomalainen, S., Gugger, M., Vezie, C., Sundman, P., Paulin, L., and Sivonen, K. (2001). Molecular characterization of planktic cyanobacteria of *Anabaena, Aphanizomenon, Microcystis* and *Planktothrix* genera. Int. J. System. Evol. Microbiol. *51*, 513–526.

MacKintosh, C., Beattie, K.A., Klumpp, S., Cohen, P., and Codd, G.A. (1990). Cyanobacterial microcystin-LR is a potent and specific inhibitor of protein phosphatase 1 and 2A from both mammals and higher plants. FEBS Lett. *264*, 187–192.

Mahmood, N.A., and Carmichael, W.W. (1986). Paralytic shellfish poisons produced by the freshwater cyanobacterium *Aphanizomenon flos-aquae* NH-5. Toxicon 24, 175–186.

Mahmood, N.A., and Carmichael, W.W. (1987). Anatoxin-a(S), an anticholinesterase from the cyanobacterium *Anabaena flos-aquae* NCR-525–17. Toxicon 25, 1221–1227.

Mahmood, N.A., Carmichael, W.W., and Pfahler, D. (1988). Anticholinesterase poisonings in dogs from a cyanobacteriunm (blue-green alga) bloom dominated by *Anabaena flos-aquae*. Am. J. Vet. Res. *49*, 500–503.

Marahiel, M.A., Stachelhaus, T., and Mootz, H.D. (1997). Modular peptide synthetases involved in nonribosomal peptide synthesis. Chem. Rev. *97*, 2651–2673.

Marin, K., Huckauf, J., Fulda, S., and Hagemann, M. (2002). Salt-dependent expression of glucosylglycerol-phosphate synthase, involved in osmolyte synthesis in the cyanobacterium *Synechocystis* sp. strain PCC 6803. J. Bacteriol. *184*, 2870–2877.

Martin, C., Oberer, L., Ino, T., Konig, W.A., Busch, M., and Weckesser, J. (1993). Cyanopeptolins, new depsipeptides from the cyanobacterium *Microcystis* sp. PCC 7806. J. Antibiot. *46*, 1550–1556.

Martin-Luna, B., Sevilla, E., Hernandez, J.A., Bes, M.T., Fillat, M.F., and Peleato, M.L. (2006). Fur from *Microcystis aeruginosa* binds *in vitro* promoter regions of the microcystin biosynthesis gene cluster. Phytochemistry *67*, 876–881.

Matern, U., Oberer, L., Falchetto, R.A., Erhard, M., König, M. Herdman, W.A., and Weckesser, J. (2001). Scyptolin A and B, cyclic depsipeptides from axenic cultures of *Scytonema hofmanni* PCC 7110. Phytochemistry *58*, 1087–1095.

Matsuda, H., Okino, T., Murakami, M., and Yamaguchi, K. (1996). Aeruginosins 102-A and B, new thrombin inhibitors from the cyanobacterium *Microcystis viridis* (NIES-102). Tetrahedron *52*, 14501–14506.

Matsunaga, S., Moore, R.E., Niemczura, W.P., and Carmichael, W.W. (1989). Anatoxin-a(s), a potent anticholinesterase from *Anabaena flos-aquae*. J. Am. Chem. Soc. *111*, 8021–8023.

Mbedi, S., Welker, M., Fastner, J., and Wiedner, C. (2005). Variability of the microcystin synthetase gene cluster in the genus *Planktothrix* (Oscillatoriales, Cyanobacteria). FEMS Microbiol. Lett. *245*, 299–306.

Meriluoto J., and Codd, G.A., eds. (2005). TOXIC: Cyanobacterial Monitoring and Cyanotoxin Analysis. (Turku: Åbo Akademi University Press).

Meriluoto, J., Karlsson, K., and Spoof, L. (2004). High-throughput screening of ten microcystins and nodularins, cyanobacterial peptide hepatotoxins, by reversed-phase liquid chromatography-electro spray ionization mass spectrometry. Chromatographia 59, 291–298.

Mez, K., Beattie, K.A., Codd, G.A. Hanselmann, K., Hauser, B., Naegeli, H., and Preisig, H.R. (1997). Identification of a microcystin in benthic cyanobacteria linked to cattle deaths on alpine pastures in Switzerland. Eur. J. Phycol. 32, 111–117.

Mikalsen, B., Boison, G., Skulberg, O.M., Fastner, J., Davies, W., Gabrielsen, T.M., Rudi, K., and Jakobsen, K.S. (2003). Natural variation in the microcystin synthetase operon *mcyABC* and impact on microcystin production in *Microcystis* strains. J. Bact. *185*, 2774–2785.

Moffitt, M.C., and Neilan, B.A. (2004). Characterization of the nodularin synthetase gene cluster and proposed theory of the evolution of cyanobacterial hepatotoxins. Appl. Environ. Microbiol. *70*, 6353–6362.

Mohamed, Z.A., El-Sharouny, H.M., and Ali, W.S.M. (2006). Microcystin production in benthic mats of cyanobacteria in the Nile River and irrigation canals, Egypt. Toxicon *38*, 1759–1771.

Moikeha, S.N., and Chu, G.W. (1971). Dermatitis-producing alga *Lyngbya majuscula* Gomont in Hawaii. II. Biological properties of the toxic factor. J. Phycol. *7*, 8–13.

Mooberry, S.L., Leal, R.M., Tinley, T.L., Luesch, H., Moore, R.E., and Corbett, T.H. (2003). The moleclar pharmacology of symplostatin 1: a new antimitotic dolastatin 10 analog. Int. J. Cancer *104*, 512–521.

Moore, R.E., Patterson, G.M.L., Mynderse, J.S., and Barchi, Jr. J. (1986). Toxins from cyanophytes belonging to the Scytonemataceae. Pure Appl. Chem. *58*, 263–271.

Moore, R.E., Lu Chen, J., Moore, B.S., and Patterson, G.M.L. (1991). Biosynthesis of microcystin-LR. Origin of the carbons in the Adda and Masp units. J. Am. Chem. Soc. *113*, 5083–5084.

Moore, B.S., Ohtani, I., de Koning, C.B., Moore, R.E., and Carmichael, W.W. (1992). Biosynthesis of anatoxin-a(s). Origin of the carbons. Tetrahedron Lett. *33*, 6595–6598.

Moore, R.E., Ohtani, I., Moore, B.S., de Konig, C.B., Yoshida, W.Y., Runnegar, M.T.C., and Carmichael, W.W. (1993). Cyanobacterial toxins. Cazzetta Chimica Italiana *123*, 329–336.

Murakami, M., Okita, Y., Matsuda, H., Okino, T., and Yamaguchi, K. (1994). Aeruginosin 298-A, a thrombin and trypsin inhibitor from the blue-green alga *Microcystis aeruginosa* (NIES-298). Tetrahedron Lett. *35*, 3129–3132.

Murakami, M., Shin, H.J., Matsuda, H., Ishida, K., and Yamaguchi, K. (1997a). A cyclic peptide, anabaenopeptin B, from the cyanobacterium *Oscillatoria agardhii*. Phytochemistry *44*, 449–452.

Murakami, M., Sun, Q., Ishida, K., Matsuda, H., Okino, T., and Yamaguchi, K. (1997b). Microviridins, elastase inhibitors from the cyanobacterium *Nostoc minutum* (NIES-26). Phytochemistry *45*, 1197–1202.

Murakami, M., Suzuki, S., Itou, Y., Kodani, S., and Ishida, K. (2000). New anabaenopeptins, potent carboxypeptidase-A inhibitors from the cyanobacterium *Aphanizomenon flos-aquae*. J. Nat. Prod. *63*, 1280–1282.

Mynderse, J.S., Moore, R.E., Kashiwagi, M., and Norton, T.R. (1977). Antileukemia activity in the Oscillatoriaceae, isolation of debromoaplysiatoxin from *Lyngbya*. Science *196*, 538–540.

Nagle, D.G., Geralds, R.S., Yoo, H.-D., Gerwick, W.H., Kim, T.-S., Nambu, M., and White, J.D. (1995). Absolute configuration of curacin A, a novel antimitotic agent from the tropical marine cyanobacterium *Lyngbya majuscula*. Tetrahedron Lett. *36*, 1189–1192.

Namikoshi, M., Choi, B.W., Sakai, R., Sun, F., Rinehart, K.L., Carmichael, W.W., Evans, W.R., Cruz, P., Munro, M.H.G., and Blunt, J.W. (1994). New nodularins, a general method for structure assignment. J. Org. Chem. *59*, 2349–2357.

Namikoshi, M., Murakami, T., Watanabe, M.F., Oda, T., Yamada, J., Tsujimura, S., Nagai, H., and Oishi, S. (2005). Simultaneous production of homoanatoxin-a, anatoxin-a, and a new non-toxic 4-hydroxyhomoanatoxin-a by the cyanobacterium *Raphidiopsis mediterranea* Skuja. Toxicon *42*, 533–538.

Namikoshi, M., and Rinehart, K.L. (1996). Bioactive compounds produced by cyanobacteria. J. Ind. Microbiol. *17*, 373–384.

Negri, A.P., Jones, G.J., and Hindmarsh, M. (1995). Sheep mortality associated with paralytic shellfish poisoning toxins from the cyanobacterium *Anabaena circinalis*. Toxicon *33*, 1321–1329.

Negri, A.P., Jones, G.J., Blackburn, S.I., Oshima, Y., and Onodera, H. (1997). Effect of culture and bloom development and of sample storage on paralytic shellfish poisons in the cyanobacterium *Anabaena circinalis*. J. Phycol. *33*, 26–35.

Neilan, B.A., Dittmann, E., Rouhiainen, L., Bass, R.A., Schaub, V., Sivonen, K., and Börner, T. (1999). Nonribosomal peptide synthesis and toxigenicity of cyanobacteria. J. Bact. *181*, 4089–4097.

Nishizawa, T., Asayama, M., Fujii, K., Harada, K.-I., and Shirai, M. (1999). Genetic analysis of the peptide synthetase genes for a cyclic heptapeptide microcystin in *Microcystis* spp. J. Biochem. *126*, 520–529.

Nishizawa, T., Ueda, A., Asayama, M., Fujii, K., Harada, K.-I., Ochi, K., and Shirai, M. (2000). Polyketide synthase gene coupled to the peptide synthetase module involved in the biosynthesis of the cyclic heptapeptide microcystin. J. Biochem. *127*, 779–789.

Nogle, L.M., and Gerwick, W.H. (2002). Isolation of four new cyclic depsipeptides, antanapeptins A-D, and dolastatin 16 from a Madagascan collection of *Lyngbya majuscula*. J. Nat. Prod. *65*, 21–24.

Nogle, L.M., Williamson, R.T., and Gerwick, W.H. (2001). Somamides A and B, two new depsipeptide analogues of dolastatin 13 from a Fijian cyanobacterial assemblage of *Lyngbya majuscula* and *Schizothrix* species. J. Nat. Prod. *64*, 716–719.

Nogueira, I.C.G., Pereira, P., Dias, E., Pflugmacher, S., Wiegand, C., Franca, S., and Vasconcelos, V.M. (2004). Accumulation of paralytic shellfish toxins (PTS) from the cyanobacterium *Aphanizomenon issatschenkoi* by the cladoceran *Daphnia magna*. Toxicon *44*, 773–780.

Norris, R.L., Eaglesham, G.K., Pierens, G., Shaw, G.R., Smith, M.J., Chriswell, R.K., Seawright, A.A., and Moore, M.R. (1999). Deoxycylindrospermopsin, an analog of cylindrospermopsin from *Cylindrospermopsis raciborskii*. Environ. Toxicol. *14*, 163–165.

Nonneman, D., and Zimba, P.V. (2002). A PCR-based test to assess the potential for microcystin occurrence in channel catfish production ponds. J. Phycol. *38*, 230–233.

Ogino, J., Moore, R.E., Patterson, G.M.L., and Smith, C.D. (1996). Dendroamines, new cyclic hexapeptides from a blue-green alga. Multidrug-resistance reversing activity of dendroamide A. J. Nat. Prod. *59*, 581–586.

Ohtake, A., Shirai, M., Aida, T., Mori, N., Harada, K.-I., Matsura, K., Suzuki, M., and Nakano, M. (1989). Toxicity of *Microcystis* species isolated from natural blooms and purification of the toxin. Appl. Environ. Microbiol. *55*, 3202–3207.

Ohtani, I., Moore, R.E., and Runnegar, M.T.C. (1992). Cylindrospermopsin, a potent hepatotoxin from the blue-green alga *Cylindrospermopsis raciborskii*. J. Am. Chem. Soc. *114*, 7941–7942.

Okino, T., Matsuda, H., Murakami, M., and Yamaguchi, K. (1993). Microginin, an angiotensin-converting enzyme inhibitor from the blue-green alga *Microcystis aeruginosa*. Tetrahedron Lett. *34*, 501–504.

Okino, T., Matsuda, H., Murakami, M., and Yamaguchi, K. (1995). New microviridins, elastase inhibitors from the blue-green alga *Microcystis aeruginosa*. Tetrahedron *51*, 10679–10686.

Okino, T., Qi, S., Matsuda, H., Murakami, M., and Yamaguchi, K. (1997). Nostopeptins A and B, elastase inhibitors from the cyanobacterium *Nostoc minutum*. J. Nat. Prod. *60*, 158–161.

Oksanen, I., Jokela, J., Fewer, D., Wahlsten, M., Rikkinen, J., and Sivonen, K. (2004). Discovery of rare and highly toxic microcystins from lichen associated cyanobacterium *Nostoc* sp. strain IO-102-I. Appl. Environ. Microbiol. *70*, 5756–5763.

Onodera, H., Oshima, Y., Henriksen, P., and Yasumoto, T. (1997a). Confirmation of anatoxin-a(s) in the cyanobacterium *Anabaena lemmermannii* as the cause of bird kills in Danish lakes. Toxicon *35*, 1645–1648.

Onodera, H., Satake, M., Oshima, Y., Yasumoto, T., and Carmichael, W.W. (1997b). New saxitoxin analogues from the freshwater filamentous cyanobacterium *Lyngbya wollei*. Natural Toxins *5*, 146–151.

Orjala, J., and Gerwick, W.H. (1996). Barbamide, a chlorinated metabolite with molluscidal activity from the Caribbean cyanobacterium *Lyngbya majuscula*. J. Nat. Prod. *59*, 427–230.

Orjala, J., Nagle, D., and Gerwick, W.H. (1995). Malyngamide H, an ichthyotoxic amide possessing a new carbon skeleton from the Caribbean cyanobacterium *Lyngbya majuscula*. J. Nat. Prod. *58*, 764–768.

Orr, P.T., and Jones, G.J. (2001). Relationship between microcystin production and cell division rates in nitrogen-limited *Microcystis aeruginosa* cultures. Limnol. Oceanogr. *43*, 1604–1614.

Ouahid, Y., Pérez-Silva, G., and del Campo, F.F. (2005). Identification of potentially toxic environmental *Microcystis* by individual and multiple PCR amplifications of specific microcystin synthetase gene regions. Environ. Toxicol. *20*, 235–242.

Pan, H., Song, L., Liu, Y., and Börner, T. (2002). Detection of hepatotoxic *Microcystis* strains by PCR with intact cells from both culture and environmental samples. Arch. Microbiol. *178*, 421–427.

Panda, D., Himes, R.H., Moore, R.E., Wilson, L., and Jordan, M.A. (1997). Mechanism of action of the unusually potent microtubule inhibitor cryptophycin 1. Biochemistry *36*, 12948–12953.

Pearl, H.W., and Millie, D.F. (1996). Physiological ecology of toxic aquatic cyanobacteria. Phycologia *35*, 160–167.

Pearson, L.A., Hisbergues, M., Börner, T., Dittmann, E., and Neilan, B.A. (2004). Inactivation of an ABC transporter gene, *mcyH*, results in loss of microcystin production in the cyanobacterium *Microcystis aeruginosa* PCC 7806. Appl. Environ. Microbiol. *70*, 6370–6378.

Pereira P., Onodera, H., Andrinolo, D., Franca, S., Araujo, F., Lagos, N., and Oshima, Y. (2000). Paralytic shellfish toxins in the freshwater cyanobacterium *Aphanizomenon flos-aquae*, isolated from Montargil reservoir, Portugal. Toxicon *38*, 1689–1702.

Pettit, G.R., Kamano, Y., Herald, C.L., Dufresne, C., Cerny, R.L., Herald, D.L., Schmidt, J.M., and Kizu, H. (1989). Isolation and structure of the cytostatic depsipeptide dolastatin 13 from the sea hare *Dolabella auricularia*. J. Am. Chem. Soc. *111*, 5015–5017.

Ploutno, A., and Carmeli, S. (2002). Modified peptides from a water bloom of the cyanobacterium *Nostoc* sp. Tetrahedron *58*, 9949–9957.

Pomati, F., Sacci, S., Rossetti, C., Giovannardi, S., Onodera, H., Oshima, Y., and Neilan, B.A. (2000). The freshwater cyanobacterium *Planktothrix* sp. FP1: molecular identification and detection of paralytic shellfish poisoning toxins. J. Phycol. *36*, 553–562.

Portwich, A., and Garcia-Pichel, F. (1999). Ultraviolet and osmotic stresses induce and regulate the synthesis of mycosporines in the cyanobacterium *Chlorogloeopsis* PCC 6912. Arch. Microbiol. *172*, 187–192.

Preußel, K., Stüken, A., Wiedner, C., Chorus, I., and Fastner, J. (2006). First report on cylindrospermopsin producing *Aphanizomenon flos-aquae* (Cyanobacteria) isolated from two German lakes. Toxicon *47*, 156–162.

Prinsep, M.R., Caplan, F.R., Moore, R.E., Patterson, G.M.L., Honkanen, R.E., and Boynton, A.L. (1992a). Microcystin-LA from a blue-green alga belonging to the Stigonematales. Phytochemistry *31*, 1247–1248.

Prinsep, M.R., Moore, R.E., Levine, I.A., Patterson, G.M.L. (1992b). Westiellamide, a bistratamide-related cyclic peptide from the blue-green alga *Westiellopsis prolifica*. J. Nat. Prod. 55, 140–142.

Rantala, A., Fewer, D.P., Hisbergues, M., Rouhiainen, L., Vaitomaa, J., Börner, T., and Sivonen, K. (2004). Phylogenetic evidence for the early evolution of microcystin synthesis. Proc. Natl. Acad. Sci. USA *101*, 568–573.

Rantala, A., Rajaniemi-Wacklin, P., Lyra, C., Lepistö, L., Rintala, J., Mankiewicz-Boczek, J., and Sivonen, K. (2006). Detection of microcystin-producing cyanobacteria in Finnish lakes with genus-specific microcystin synthetase gene E (*mcyE*) PCR and associations with environmental factors. Appl. Environ. Microbiol. *72*, 6101–6110.

Rapala, J., Sivonen, K., Lyra, C., and Niemelä, S.I. (1997). Variation of microcystins, cyanobacterial hepatotoxins, in *Anabaena* spp. as a function of growth stimuli. Appl. Environ. Microbiol. *63*, 2206–2012.

Rapala, J., Erkomaa, K., Kukkonen, J., Sivonen, K., and Lahti, K. (2002). Detection of microcystins with protein phosphatase inhibition assay, high-performance liquid chromatography-UV detection and enzyme-linked immunosorbent assay. Comparison of methods. Anal. Chim. Acta *466*, 213–231.

Rapala J., Robertson, A., Negri, A.P., Berg, K.A., Tuomi, P., Lyra, C., Erkomaa, K., Lahti, K., Hoppu, K., and Lepistö, L. (2005). First report of saxitoxin in Finnish lakes and possible associated effects on human health. Environ. Toxicol. *3*, 331–340.

Repka, S., Koivula, M., Harjunpää, V., Rouhiainen, L., and Sivonen, K. (2004). Effects of phosphate and light on growth of and bioactive peptide production by the cyanobacterium *Anabaena* strain 90 and its anabaenopeptilide mutant. Appl. Environ. Microbiol. *70*, 4551–4560.

Reshef, V., and Carmeli, S. (2002). Schizopeptin 791, a new anabaenopeptin-like cyclic peptide from the cyanobacterium *Schizothrix* sp. J. Nat. Prod. *65*, 1187–1189.

Ressom, R., Soong, F.S., Fitzgerald, J., Turczynowicz, L., El Saadi, O., Roder, D., Maynard, T., and Falconer, I. (1994). Health Effects of Toxic Cyanobacteria (Blue-Green Algae) (Canberra: National Health and Medical Research Council, Australian Government Publishing Service).

Rinehart, K.L., Harada, K.-I., Namikoshi, M., Chen, C., Harvis, C.A., Munro, M.H.G., Blunt, J.W., Mulligan, P.E., Beasley, V.R., Dahlem, A.M., and Carmichael, W.W. (1988). Nodularin, microcystin, and the configuration of Adda. J. Am. Chem. Soc. *110*, 8557–8558.

Rinehart, K.L., Namikoshi, M., and Choi, B.W. (1994). Structure and biosynthesis of toxins from blue-green alga (cyanobacteria). J. Appl. Phycol. *6*, 159–176.

Rinta-Kanto, J.M., Ouellette, A.J.A., Boyer, G.L., Twiss, M.R., Bridgeman, T.B., and Wilhelm, S.W. (2005). Quantification of toxic *Microcystis* spp. during the 2003 and 2004 blooms in Western Lake Erie using quantitative real-time PCR. Environ. Sci. Technol. *39*, 4198–4205.

Rohrlack, T., Christoffersen, K., Hansen, P.E., Zhang, W., Czarnecki, O., Henning, M., Fastner, J., Erhard, M., Neilan, B.A., and Kaebernick, M. (2003). Isolation, characterization, and quantitative analysis of microviridin J, a new *Microcystis* metabolite toxic to *Daphnia*. J. Chem. Ecol. *29*, 1757–1770.

Rouhiainen, L. (2004). Characterization of *Anabaena* Cyanobacteria: Repeated Sequences and Genes Involved in Biosynthesis of Microcystins and Anabaenopeptilides. Thesis. (Helsinki: University of Helsinki) (pdf available at http://ethesis.helsinki.fi/julkaisut/maa/skemi/vk/rouhiainen/).

Rouhiainen, L., Paulin, L., Suomalainen, S., Hyytiäinen, H., Buikema, W., Haselkorn, R., and Sivonen, K. (2000). Genes encoding synthetases of cyclic depsipeptides, anabaenopeptilides, in *Anabaena* strain 90. Mol. Microbiol. *37*, 156–167.

Rouhiainen, L., Vakkilainen, T., Lumbye Siemer, B., Buikema, W., Haselkorn, R., and Sivonen, K. (2004). Genes coding for hepatotoxic heptapeptides (microcystins) in the cyanobacterium *Anabaena* strain 90. Appl. Environ. Microbiol. *70*, 686–692.

Rudi, K., Skulberg, O.M., Kulberg, K., and Jakobsen, K.S. (2000). Application of sequence-specific labeled 16S rRNA gene oligonucleotide probes for genetic profiling of cyanobacterial abundance and diversity by array hybridisation. Appl. Environ. Microbiol. *66*, 4004–4011.

Rudolph-Böhner, S., Mierke, D.F., and Moroder, L. (1994). Molecular structure of cyanobacterial tumour-promoting microcystins. FEBS Lett. *349*, 319–323.

Runnegar, M.T., Kong, S.-M., Zhong, Y.-Z., Ge, J.-E., and Lu, S.C. (1994). The role of glutathione in the toxicity of a novel cyanobacterial alkaloid cylindrospermopsin in cultured rat hepatocytes. Biochem. Biophys. Res. Commun. *201*, 235–241.

Runnegar, M.T., Kong, S.-M., Zhong, Y.-Z., and Lu, S.C. (1995). Inhibition of reduced glutathione synthesis by cyanobacterial alkaloid cylindrospermopsin in cultured rat hepatocytes. Biochem. Pharmacol. *49*, 219–225.

Saker, M.L., Nogueira, I.C.G., Vasconcelos, V.M., Neilan, B.A., Eaglesham, G.K., and Pereira, P. (2003). First report and toxicological assessment of the cyanobacterium *Cylindrospermopsis raciborskii* from Portuguese freshwaters. Ecotoxicol. Environ. Safety *55*, 243–250.

Sano, T., and Kaya, K. (1997). A 3-amino-10-chloro-2-hydroxydecanoic acid-containing tetrapeptide from *Oscillatoria agardhii*. Phytochemistry *44*, 1503–1505.

Schagerl, M., Unterreider, I., and Angeler, D.G. (2002). Allelopathy among cyanoprokaryota and other algae originating from lake Neusiedlersee (Austria). Internat. Rev. Hydrobiol. *87*, 365–374.

Schatz, D., Keren, Y., Hasa, O., Carmeli, S., Sukenik, A., and Kaplan, A. (2005). Ecological implications of the emergence of non-toxic subcultures from toxic *Microcystis* strains. Environ. Microbiol. *7*, 798–805.

Schatz, D., Keren, Y., Vardi, Y., Sukenik, A., Carmeli, S., Börner, T., Dittmann, E., and Kaplan, A. (2007). Towards clarification of the biological role of microcystins, a family of cyanobacterial toxins. Environ. Microbiol., in press.

Schembri, M.A., Neilan, B.A., and Saint, C.P. (2001). Identification of genes implicated in toxin production in the cyanobacterium *Cylindrospermopsis raciborskii*. Environ. Toxicol. *16*, 413–421.

Singh, D.P., Tyagi, M.B., Kumar, A., Thakur, J.K., and Kumar, A. (2001). Antialgal activity of a hepatotoxin-producing cyanobacterium, *Microcystis aeruginosa*. World J. Microbiol. Biotechnol. *17*, 15–22.

Shibata, K. (1969). Pigments and a UV-absorbing substance in corals and blue-green algae living in the Great Barrier Reef. Plant Cell Physiol. *10*, 325–335.

Shimizu, Y., Norte, M., Hori, A., Genenah, A., and Kobayashi, M. (1984). Biosynthesis of saxitoxin analogues: the unexpected pathway. J. Am. Chem. Soc. *106*, 6433–6434.

Shimizu, Y. (1986). Toxigenesis and biosynthesis of saxitoxin analogues. Pure Appl. Chem. *58*, 257–262.

Schmidt, E.W., Sudek, S., and Haygood, M.G. (2004). Genetic evidence supports secondary metabolic diversity in *Prochloron* spp., the cyanobacterial symbiont of a tropical ascidian. J. Nat. Prod. *67*, 1341–1345.

Schmidt, E.W., Nelson, J.T., Rasko, D.A., Sudek, S., Eisen, J.A., Haygood, M.G., and Ravel, J. (2005). Patellamide A and C biosynthesis by a microcin-like pathway in *Prochloron didemni*, the cyanobacterial symbiont of *Lissoclinum patella*. Proc. Natl. Acad. Sci. USA *102*, 7315–7320.

Selheim, F., Herfindal, L., Martins, R., Vasconcelos, V., and Døskeland, S.O. (2005). Neuro-apoptogenic and blood platelet targeting toxins in benthic marine cyanobacteria from the Portuguese coast. Aquatic Toxicol. *74*, 294–306.

Shalev-Alon, G., Sukenik, A., Livnah, O., Schwarz, R., and Kaplan, A. (2002). A novel gene encoding amidinotransferase in the cylindrospermopsin producing cyanobacterium *Aphanizomenon ovalisporum*. FEMS Microbiol. Lett. *209*, 87–91.

Shaw, G.R., Sukenik, A., Livne, A., Chriswell, R.K., Smith, M.J., Seawright, A.A., Norris, R.L., Eaglesham, G.K., and Moore, M.R. (1999). Blooms of the cylindrospermopsin containing cyanobacterium, *Aphanizomenon ovalisporum* (Forti), in newly contructed lakes, Queensland, Australia. Environ. Toxicol. *14*, 167–177.

Shi, L., Carmichael, W.W, and Kennedy, P.J. (1999). Cyanobacterial PPP family protein phosphatases possess multifunctional capabilities and are resistant to microcystin-LR. J. Biol. Chem. *268*, 7632–7635.

Shi, L., Carmichael, W.W., and Miller, I. (1995). Immuno-gold localization of hepatotoxins in cyanobacterial cells. Arch. Microbiol. *163*, 7–15.

Shin, H.J., Murakami, M., Matsuda, H., Ishida, K., and Yamaguchi, K. (1995). Oscillapeptin, an elastase and chymotrypsin inhibitor from the cyanobacterium *Oscillatoria agardhii* (NIES-204). Tetrahedron Lett. *36*, 5235–5238.

Shin, H.J., Murakami, M., Matsuda, K., and Yamaguchi, K. (1996). Microviridins D-F, serine protease inhibitors from the cyanobacterium *Oscillatoria agardhii* (NIES-204). Tetrahedron *52*, 8159–8168.

Shin, H.J., Matsuda, H., Murakami, M., and Yamaguchi, K. (1997). Aeruginosins 205A and B, serine protease inhibitory glycopeptides from the cyanobacterium *Oscillatoria agardhii* (NIES-205). J. Org. Chem. *62*, 1810–1813.

Sielaff, H., Dittmann, E., Tandeau de Marsac, N., Bouchier, C., von Döhren, H., Börner, T., and Schwecke, T. (2003). The *mcy*F gene of the microcystin biosynthetic gene cluster from *Microcystis aeruginosa* encodes an aspartate racemase. Biochem. J. *373*, 909–916.

Sivonen, K. (2000). Freshwater cyanobacterial neurotoxins: ecobiology, chemistry and detection. In Seafood and Freshwater Toxins. L.M. Botana, ed. (New York: Marcel Dekker, Inc.), pp. 567–582.

Sivonen, K., and Jones, G.J. (1999). Cyanobacterial toxins. In Toxic Cyanobacteria in Water: a Guide to Public Health. Significance, Monitoring and Management, I. Chorus, and J. Bertram, eds. (London: The World Health Organization. E. and F.N. Spon), pp. 41–111.

Sivonen, K., Kononen, K., Carmichael, W.W., Dahlem, A.M.,. Rinehart, K., Kiviranta, J., and Niemelä, S.I. (1989). Occurrence of the hepatotoxic cyanobacterium *Nodularia spumigena* in the Baltic Sea and the structure of the toxin. Appl. Environ. Microbiol. 55, 1990–1995.

Sivonen, K., Namikoshi, M., Evans, W.R., Färdig, M., Carmichael, W.W., and Rinehart, K.L. (1992). Three new microcystins, cyclic heptapeptide hepatotoxins, from *Nostoc* sp. strain 152. Chem. Res. Toxicol. 5, 464–469.

Skulberg, O.M., Carmichael, W.W., Anderson, R.A., Matsunaga, S., Moore, R.E., and Skulberg, R. (1992). Investigations of a neurotoxic Oscillatorialean strain (cyanophyceae) and its toxin. Isolation and characterization of homoanatoxin-a. Environ. Toxicol. Chem. *11*, 321–329.

Smith, G.D., and Doan, N.T. (1999). Cyanobacterial metabolites with bioactivity against photosynthesis in cyanobacteria, algae and higher plants. J. Appl. Phycol. *11*, 337–344.

Stevenson, C.S., Capper, E.A., Roshak, A.K., Marquez, B., Grace, K., Gerwick, W.H., Jacobs, R.S., and Marshall, L.A. (2002). Scytonemin—a marine natural product inhibitor of kinases key in hyperproliferative inflammatory diseases. Inflamm. Res. *51*, 112–114.

Stirling, D.J., and Quilliam, M.A. (2001). First report of the cyanobacterial toxin cylindrospermopsin in New Zealand. Toxicon *39*, 1219–1222.

Subbaraju, G., Golakoti, V.T., Patterson, G.M.L., and Moore, R.E. (1997). Three new cryptophycins from *Nostoc* sp. GSV 224. J. Nat. Prod. *60*, 302–305.

Sudek, S., Haygood, M.G., Youssef, D.T.A., and Schmidt, E.W. (2006). Structure of trichamide, a cyclic peptide from the bloom-forming cyanobacterium *Trichodesmium erythraeum*, predicted from the genome sequence. Appl. Environ. Microbiol. *72*, 4382–4387.

Suikkanen, S., Engström-Öst, J., Jokela, J., Sivonen, K., and Viitasalo, M. (2006). Allelopathy of Baltic Sea cyanobacteria: no evidence for the role of nodularin. J. Plankton Res. *28*, 543–550.

Suikkanen, S., Fistarol, G.O., and Granéli, E. (2004). Allelopathic effects of the Baltic cyanobacteria *Nodularia spumigena, Aphanizomenon flos-aquae* and *Anabaena lemmermannii* on algal monocultures. J. Exp. Mar. Biol. Ecol. *308*, 85–101.

Suikkanen, S., Fistarol, G.O., and Granéli, E. (2005). Effects of cyanobacterial allelochemicals on a natural plankton community. Mar. Ecol. Prog. Ser. *287*, 1–9.

Sukenik, A., Eshkol, R., Livne, A., Hadas, O., Rom, M., Tchernov, D., Vardi, A., and Kaplan, A. (2002). Inhibition of growth and photosynthesis of the dinoflagellate *Peridinium gatunense* by *Microcystis* sp. (cyanobacteria): a novel allelopathic mechanism. Limnol. Oceanogr. *47*, 1656–1663.

Surakka, A., Sihvonen, L.M., Lehtimäki, J.M., Wahlsten, M., Vuorela, P., and Sivonen, K. (2005). Benthic cyanobacteria from the Baltic Sea contain cytotoxic *Anabaena, Nodularia* and *Nostoc* strains and an apoptosis inducing *Phormidium* strain. Environ. Toxicol. *20*, 285–292.

Tanabe, Y., Kaya, K., and Watanabe, M.M. (2004). Evidence for recombination in the microcystin synthetase (*mcy*) genes of toxic cyanobacteria *Microcystis* spp. J. Mol. Evol. *58*, 633–641.

Terao, K., Ohmori, S.,Igarashi, K., Ohtani, I., Watanabe, M.F., Harada, K.-I., Ito, E., and Watanabe, M. (1994). Electron microscopic studies on experimental poisoning in mice induced by cylindrospermopsin isolated from the blue-green alga *Umezakia natans*. Toxicon *32*, 833–843.

Tillett, D., Parker, D.L., and Neilan, B.A. (2001). Detection of toxigenity by a probe for the microcystin synthetase A gene (*mcyA*) of the cyanobacterial genus *Microcystis*: comparison of toxicities with 16S rRNA and phycocyanin operon (phycocyanin intergenic spacer) phylogenies. Appl. Environ. Microbiol. *67*, 2810–2818.

Tillett, D., Dittmann, E., Erhard, M., von Döhren, H., Börner, T., and Neilan, B.A. (2000). Structural organization of microcystin biosynthesis in *Microcystis aeruginosa* PCC 7806: an integrated peptide-polyketide synthase system. Chem. Biol. *7*, 753–764.

Tonk, L., Visser, P.M., Christiansen, G., Dittmann, E., Snelder, E.O.F.M., Wiedner, C., Mur, L.C., and Huisman, J. (2005). The microcystin composition of the cyanobacterium *Planktothrix agardhii* changes toward a more toxic variant with increasing light intensity. Appl. Environ. Microbiol. *71*, 5177–5181.

Utkilen, H., and Gjølme, N. (1995). Iron-stimulated toxin production in *Microcystis aeruginosa*. Appl. Environ. Microbiol. *61*, 797–800.

Vaitomaa, J., Rantala, A., Halinen, K., Rouhiainen, L., Tallberg, P., Mokelke, L., and Sivonen, K. (2003). Quantitative real-time-PCR for determination of microcystin synthetase E copy numbers for *Microcystis* and *Anabaena* in lakes. Appl. Environ. Microbiol. *69*, 7289–7297.

Vardi, A., Schatz, D., Beeri, K., Motro, U., Sukenik, A., Levine, A., and Kaplan, A. (2002). Dinoflagellate-cyanobacterium communication may determine the composition of phytoplankton assemblage in a mesotrophic lake. Curr. Biol. *12*, 1767–1772.

Vezie, C., Brient, L., Sivonen, K., Betru, G., Lefeuvre, C.-J., and Salkinoja-Salonen, M. (1998). Variation of microcystin content of cyanobacterial blooms and isolated strains in Lake Grand-Lieu (France). Microbial Ecology *35*, 126–135.

Viaggiu, E., Melchiorre, S., Volpi, F., Di Corcia, A., Mancini, R., Garibaldi, L., Crichigno, G., and Bruno, M. (2004). Anatoxin-a toxin in the cyanobacterium *Planktothrix rubescens* from a fishing pond in Northern Italy. Environ. Toxicol. *19*, 191–197.

Via-Ordorika, L., Fastner, J., Kurmayer, R., Hisbergues, M., Dittmann, E., Komarek, J., Erhard, M., and Chorus, I. (2004). Distribution of microcystin-producing and non-microcystin-producing *Microcystis* sp. in European freshwater bodies: detection of microcystins and microcystin genes in individual colonies. System. Appl. Microbiol. *27*, 592–602.

von Döhren, H., Keller, U., Vater, J., and Zocher, R. (1997). Multifunctional peptide synthetases. Chem. Rev. *97*, 2675–2705.

von Elert, E., and Jüttner, F. (1997). Phosphorus limitation and not light controls the extracellular release of allelopathic compounds by *Trichormus doliolum* (Cyanobacteria). Limnol. Oceanogr. *42*, 1796–1802.

Weckesser, J., Martin, C., and Jakobi, C. (1996). Cyanopeptolins, depsipeptides from cyanobacteria. System. Appl. Microbiol. *19*, 133–138.

Welker, M., and von Döhren, H. (2006). Cyanobacterial peptides—Nature's own combinatorial biosynthesis. FEMS Microbiol. Rev. *30*, 530–563.

Welker, M., Christiansen, G., and von Döhren, H. (2004). Diversity of coexisting *Planktothrix* (cyanobacteria) chemotypes deduced by mass spectral analysis of microcystins and other oligopeptides. Arch. Microbiol. *182*, 288–298.

Wiedner, C., Visser, P.M., Fastner, J., Metcalf, J.S., Codd, G.A., and Mur, L.R. (2003). Effects of light on the microcystin content of *Microcystis* strain PCC 7806. Appl. Environ. Microbiol. *69*, 1475–1481.

Williams, D.E., Craig, M., Holmes, C.F.B., and Andersen, R.J. (1996). Ferintoic acids A and B, new cyclic hexapeptides from the freshwater cyanobacterium *Microcystis aeruginosa*. J. Nat. Prod. *59*, 570–575.

Williams, P.G., Yoshida, W.Y., Moore, R.E., and Paul, V.J. (2004). Micromide and guamamide: cytotoxic alkaloids from species of the marine cyanobacterium *Symploca*. J. Nat. Prod. *67*, 49–53.

Work, K.A., and Havens, K.E. (2003). Zooplankton grazing on bacteria and cyanobacteria in a eutrophic lake. J. Plankton Res. *25*, 1301–1307.

Wright, D.J., Smith, S.C., Joardar, V., Scherer, S., Jervis, J., Warren, A., Helm, R.F., and Potts, M. (2005). UV irradiation and dessication modulate the three-dimensional extracelluar matrix of *Nostoc commune* (cyanobacteria). J. Biol. Chem. *280*, 40271–40281.

Yoshizawa, S., Matsushima, R., Watanabe, M.F., Harada, K.-I., Ichihara, A., Carmichael, W.W., and Fujiki, H. (1990). Inhibition of protein phosphatases by microcystin and nodularin associated with hepatotoxicity. J. Cancer Res. Clin. Oncol. *116*, 609–614.

Young, F.M., Thomson, T., Metcalf, J.S., Lucocq, J.M., and Codd, G.A. (2005). Immunogold localization of microcystins in cryosectioned cells of *Microcystis*. J. Struct. Biol. *151*, 206–214.

The Cyanobacterial Circadian Clock and the KaiC Phosphorylation Cycle

8

Kazuki Terauchi and Takao Kondo

Abstract

The circadian clock is an endogenous timing mechanism in living organisms that coordinates their lives with environmental changes. Cyanobacteria are the simplest organisms that are known to exhibit circadian rhythms, and they have become one of the most successful model organisms for circadian biology. Although the circadian clock in cyanobacteria has the same fundamental features as that in eukaryotes, its individual components that have been identified to date are unique. The molecular mechanism of the cyanobacterial clock is different from that described for the eukaryotic clock. The clock core in cyanobacteria is the KaiC phosphorylation cycle.

Introduction

Circadian rhythms are endogenous biological oscillations that have been ubiquitously observed in various physiological processes in organisms ranging from cyanobacteria to green plants and humans. As the name suggests, circadian rhythms are biological oscillations with periods of about a day (*circa*, about; *diem*, day). A living cell adapts to the daily changes in its environment. As such, the circadian clock that generates the biological oscillations has been thought of as a system that is adaptive to daily environmental alterations in light, temperature and humidity (Pittendrigh, 1993). Three intrinsic properties have come to define these clocks: (1) persistence of rhythms with a period of about 24 hours in the absence of external signals; (2) entrainment of the phase by changes in environmental cues, such as light/dark and temperature signals; and (3) stability of the free-running period at different temperature conditions (temperature compensation). Temperature compensation is the remarkable property of the circadian clock. Temperature has little effect on the pace of the timekeeping mechanism at different constant temperatures, in contrast to simple biochemical reactions whose rates are affected by temperature changes. Recent molecular and genetic studies on circadian clocks in many organisms have identified and characterized several molecular factors that contribute to generation of circadian rhythms. Cyanobacteria are the simplest organisms to exhibit circadian rhythms, and they have become one of the most successful model organisms for circadian biology.

This chapter focuses on the molecular mechanisms of circadian rhythm generation in cyanobacteria. We first give a concise description of the physiological aspects of circadian rhythms in cyanobacteria. Two decades ago, the circadian clock had been believed to be restricted to eukaryotes. Cyanobacteria are the first prokaryotes found to have a circadian clock within the cell, which divides more than once per day. Second, we discuss the *kaiABC* gene cluster, which is essential for cyanobacterial circadian oscillation. Third, we consider the input and output pathways that support the cellular circadian system. Key proteins that function in both pathways have been identified. Fourth, we describe the behavior of Kai proteins and their structural analysis. Last, we report on the accomplishment of the *in vitro* recapitulation of the clock by reconstituting the KaiC phosphorylation cycle with three Kai proteins and ATP in a test-tube.

Circadian rhythms in cyanobacteria

Discovery of the cyanobacterial circadian clock

The circadian clocks were initially assumed to exist only in eukaryotes. However, in the 1980s several circadian regulations were observed in cyanobacterial physiology (reviewed by Golden *et al.*, 1997). One example of physiological processes in cyanobacteria that occur in a circadian manner is nitrogen fixation, which is temporally separated from photosynthesis. Nitrogenase is a key enzyme for nitrogen fixation converting dinitrogen to ammonia and extremely sensitive to oxygen generated by photosynthesis. Cyanobacteria have developed several strategies to protect the enzyme from oxygen (see Chapters 14 and 15). In some filamentous non-heterocystous and unicellular cyanobacteria, temporal separation has been proposed for the two incompatible activities. They perform photosynthesis during the day and nitrogen fixation at night. Robust daily oscillation of nitrogenase activity has been reported in a marine unicellular cyanobacterium under constant conditions (Mitsui *et al.*, 1986). Huang and coworkers have also reported the circadian rhythm of nitrogenase in a unicellular strain *Synechococcus* sp. RF-1 (Grobbelaar *et al.*, 1986). The initial discovery of *bona fide* circadian rhythms of physiological processes, including nitrogen fixation, cell division and amino acid uptake, was made in cyanobacterial species that had not been manipulated genetically.

Physiological significance of circadian rhythms in cyanobacteria

Circadian clocks are widely believed to improve survival adaptation to environmental changes (Pittendrigh, 1993). As described above, the circadian clock could regulate photosynthesis to be active during the day and nitrogen fixation at night in some unicellular cyanobacteria. However, there was no experimental evidence revealing the merit of circadian control for adaptation to environmental change. To address this, competition experiments were performed using wild-type and mutant strains that exhibit different period lengths in their rhythms. The strains had the same growth rate in single-strain culture, apparently showing that there is no growth advantage or disadvantage in having different circadian periods. However, when each of the strains was mixed with another, and they were grown together under light/dark cycle, strains that had a circadian period similar to that of the light/dark cycle were favored under competition (Ouyang *et al.*, 1998). This result supports the notion of a selective advantage to circadian timing in cyanobacteria.

DNA microarray analysis has provided obvious evidence for the temporal separation of general biochemical pathways by the circadian clock in the non-diazotrophic unicellular cyanobacterium *Synechocystis* sp. PCC 6803 (Kucho *et al.*, 2005). This strain expresses the genes required for respiration at night in a circadian fashion, peaking at subjective night. On the other hand, expression of many other clock-controlled genes, including photosynthetic genes, peaks at subjective daytime (Kucho *et al.*, 2005). These results provided crucial insights into the effect of the circadian clock on cyanobacterial cellular physiology.

Coordination of circadian rhythms and cell division cycles

Many cyanobacteria are able to grow fast with doubling times of less than 24 hours under optimal conditions. The question arises as to how the timing circuits for cell division and circadian rhythms are coordinated. There had been a dogma among clock researchers that the circadian clock cannot function in cells that divide more than once in the circadian period (Edmunds, 1988). However, cyanobacteria were demonstrated to display circadian rhythms even though they divide more than once per a day (Mori *et al.*, 1996; Kondo *et al.*, 1997). Moreover, the cell division cycle in cyanobacteria is gated by the clock, although the clock operates independently of cell division (Mori *et al.*, 1996).

The gene expression rhythms were monitored from a single cyanobacterial cell with self-bioluminescence, indicating that the circadian clock is a property of individual cells (Mihalcescu *et al.*, 2004). When cyanobacterial cells divided, their daughter cells remain synchronized with each other. Moreover, when two microcolonies with different circadian phases were observed together, all the cells from both colonies oscillated with essentially the same period while the period was specific to each colony. These results confirmed that the intracellular rhythms are more precise and more robust than had ever been expected, since intercellular communication seems to be negligible (Mihalcescu *et al.*, 2004).

Discovery of clock genes *kaiABC*

Utility of the bioluminescence reporter system

The four researchers, Ishiura, Golden, Johnson and Kondo, utilized a genetic system to monitor rhythms in the cyanobacterial circadian clock. A unicellular fresh-water strain, *Synechococcus elongatus* PCC 7942, was selected as an experimental organism. This strain is not diazotrophic, but it is naturally transformable by exogenously added DNA. Reporter genes can be exogenously introduced and expressed to monitor promoter activity (Figure 8.1). The *psbA1* gene, encoding a PSII reaction center protein, is known to be strongly expressed in this cyanobacterium (Li and Golden, 1993). A promoterless *luxAB* gene encoding bacterial luciferase subunits from *Vibrio harveyi* was fused to the promoter of *psbA1* and integrated at a specific targeting region (neutral site I) in the chromosome of the cyanobacterium. The resulting reporter strain exhibited a persistent circadian bioluminescence oscillation under constant light conditions (Kondo *et al.*, 1993). Moreover, the phase of the rhythms could be reset by a single dark pulse in a phase-dependent manner, and the period length was not sensitive to change in temperature. Thus, the cyanobacterial bioluminescence rhythm was found to satisfy the three fundamental criteria for circadian oscillations established in eukaryotes (Kondo *et al.*, 1993). The reporter strain of *Synechococcus elongatus* PCC 7942 has, thus, become the simplest model organism for studying the circadian clock.

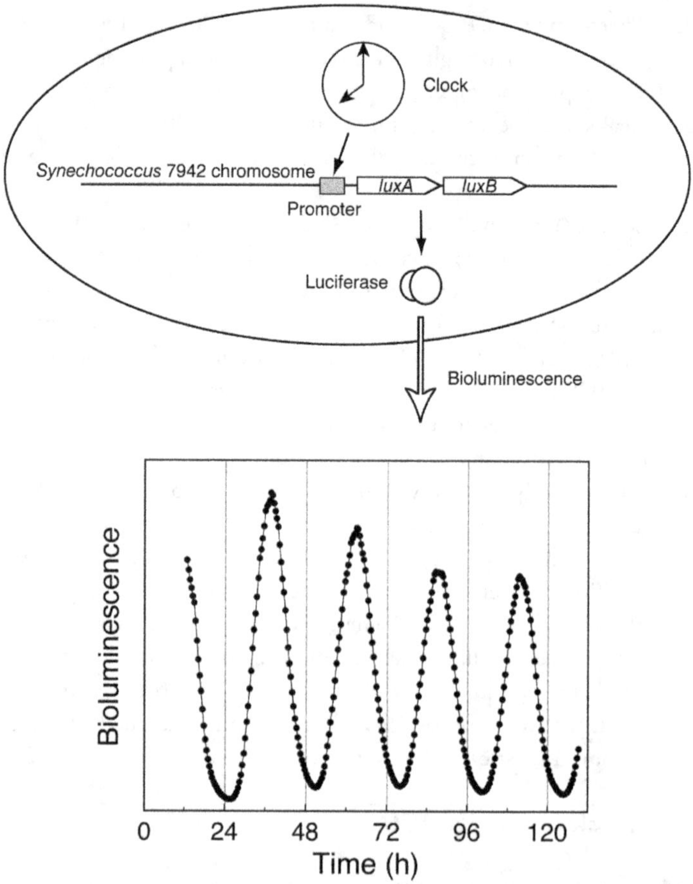

Figure 8.1 Circadian rhythm of *Synechococcus elongatus* PCC 7942 under continuous light conditions. The luciferase operon (*luxAB*) *of Vivrio harveyi* was fused to a promoter and integrated in the chromosome of *Synechococcus elongatus* PCC 7942. The reporter strain that had been entrained previously to 12-h dark period was monitored under continuous light conditions. The bioluminescence from the cyanobacterial culture oscillated with a period of about 24 hours.

The *kaiABC* gene cluster

More than 150 000 EMS-mutagenized clones of *Synechococcus elongatus* PCC 7942 were screened by monitoring bioluminescence from isolated colonies on an agar plate. A wide range of clock mutants were isolated, with altered period lengths ranging from 16 to 60 hours, including arrhythmic mutants (Kondo *et al*., 1994). By genetic complementation of the mutants, a gene cluster composed of three genes, *kaiA*, *kaiB* and *kaiC*, was cloned (Ishiura *et al*., 1998). The genes were named after the meaning of "cycle" in Japanese. All mutations leading to short-period, long-period and arrhythmic phenotypes were mapped to these three genes. The *kaiA*, *kaiB*, and *kaiC* genes encode 284, 102 and 519 amino-acid proteins, respectively. *kaiB* and *kaiC* form an operon, while *kaiA* is independently transcribed (Ishiura *et al*., 1998). Expressions of both *kaiA* and *kaiBC* oscillate in a circadian fashion under continuous light conditions. Disruption of any of these three genes abolished circadian rhythms, indicating that all three *kai* genes are essential for circadian function (Ishiura *et al.*, 1998). Thus, the *kai* gene cluster has a crucial function for the circadian oscillation in *Synechococcus elongatus* PCC 7942.

Evolutionary aspects of *kai* genes

The three *kai* genes are essential components for the circadian clock mechanisms in *Synechococcus elongatus* PCC 7942. Complete genome analysis of a number of cyanobacterial species has shown that the *kai* genes are ubiquitous among cyanobacteria, with the exception of *Gloeobacter violaceus* PCC 7421 (Nakamura *et al.*, 2003). The bioluminescence reporter system has been successfully employed to find circadian rhythms of gene expression in other cyanobacterial species, such as *Synechocystis* sp. PCC 6803 (Aoki *et al.*, 1995) that is widely used in photosynthesis studies, the unicellular thermophilic species *Thermosynechococcus elongatus* BP-1 (Onai *et al.*, 2004) and the non-heterocystous filamentous diazotrophic strain *Plectonema boryanum* (Terauchi *et al.*, 2005). These data suggest that most cyanobacteria have a circadian clock and *kai* genes are essential for the cyanobacterial circadian clocks.

Conserved homologs of KaiB and KaiC have been found in the complete genomes of some archaea and bacteria (Leipe *et al.*, 2000; Dvornyk *et al.*, 2003). A phylogenic analysis of the KaiC sequence proposed that ancestral KaiC was derived from a RecA-like gene by gene duplication, initially moved into Archaea, and then horizontally transferred into ancestral cyanobacteria (Leipe *et al.*, 2000). Circadian rhythmicity has not been established in archaea to date, but rhythmic gene expression with a period of 20.5 hours using a luciferase reporter gene system has been shown in a purple bacterium, *Rhodobacter sphaeroides* (Min *et al.*, 2005).

Cyanobacteria are believed to be an ancestor of chloroplasts in plants according to the endosymbiotic theory. Many genes encoded by chloroplast genomes in the green algae *Chlamydomonas* and in higher plants have been shown to be expressed under the control of circadian clocks (Hwang *et al.*, 1996; Nakahira *et al.*, 1998). The *kai* genes might be expected to have moved into plant cells together with other cyanobacterial genes. However, to date, *kai*-related genes have not been found in any eukaryotes, including plastid genomes of higher plants whose whole entire nucleotide sequences have been determined. Kai oscillators might have been lost from the plastid even if a transfer occurred, due to the dramatic difference between prokaryote and eukaryotic gene regulation (Kondo and Ishiura, 1999). Moreover, key proteins identified in other model organisms do not share any sequence similarity with Kai proteins, despite the fact that the circadian clock in cyanobacteria has the same fundamental features as that in eukaryotes. Thus, it is likely that circadian clocks have developed independently several times in different evolutionary lineages.

Oscillator networks in cellular circadian clocks

Transcriptional control in circadian timing

The *kai* genes are expressed from two promoters, one upstream of *kaiA* and one upstream of *kaiBC*. Overexpression of the *kaiC* gene completely suppressed its own promoter activity, while *kaiA* overexpression enhanced *kaiBC* transcription. Disruption of *kaiA* lowered *kaiBC* promoter activity. In fact, the *kaiBC* operon is upregulated by the KaiA protein and downregulated by the KaiC protein. Thus, a circadian feedback model was proposed similar to that which has been characterized in eukaryotes (Ishiura *et al.*, 1998). Recently, the KaiC phosphorylation cycle has been reconstituted *in vitro* (see below), indicating that the primary pacemaker of the cyanobacterial circadian clock is the biochemical process involving the three Kai proteins rather than feedback regulation of *kaiBC* expression (Tomita *et al.*, 2005; Nakajima *et al.*, 2005). This mechanism would make the feedback regulation of *kaiBC* expression unnecessary. However, the feedback regulation is still recognized to be essential for the post-translational cycle to function physiologically in the cyanobacterial cell. The transcriptional feedback would be a reinforcer rather than the primary mechanism of the circadian cycle.

As mentioned above, overexpression of *kaiC* dramatically reduced *kaiBC* promoter activity (Ishiura *et al.*, 1998). This effect was not only on the *kaiBC* promoter but also on almost all gene promoters examined in *Synechococcus elongatus* PCC 7942 (Nakahira *et al.*, 2004). This suggests that the KaiC protein primarily coordinates genomewide gene expression, including that of its own *kaiBC* promoter. Moreover, a promoter derived from *E. coli* was able to control the expression of *kaiBC* in *kaiBC*-inactivated cyanobacterial strains, resulting in the restoration of normal circadian rhythms (Xu *et al.*, 2003; Nakahira *et al.*, 2004). Specific regulation of the *kaiBC* promoter is not likely to be essential for driving the circadian rhythm of *kaiBC* expression. In fact, mutations or deletions in the upstream region of *kaiBC* do not change the rhythmicity or response to circadian feedback regulation, suggesting that the upstream region of *kaiBC* is not a target for feedback regulation by Kai proteins (Kutsuna *et al.*, 2005).

A random promoter trap analysis in *Synechococcus elongatus* PCC 7942 revealed that all promoter activities examined were rhythmic (Liu *et al.*, 1995). While 80% of clones of *Synechocystis* sp. strain PCC 6803 were identified by this method to have circadian oscillation (Aoki *et al.*, 2002), a DNA microarray analysis of this strain revealed that only ~10% of genes showed circadian rhythm (Kucho *et al.*, 2005). This discrepancy could be due to the differences between the two methods, such as sensitivity and time resolution. It should also be noted that the microarray method monitors the amount of accumulated mRNA, while the promoter trap method monitors transcriptional activity (Kucho *et al.*, 2005). Thus, the question of how many genes are under the control of the circadian clock system in cyanobacteria remains unsettled.

The control mechanism of global gene expression by Kai protein oscillators remains to be elucidated. Since KaiC belongs to the RecA protein superfamily, which contains proteins involved in DNA recombination, it was proposed that KaiC might recognize chromosomal DNA (Mori and Johnson, 2001). KaiC was reported to weakly bind to DNA despite the lack of any known DNA-binding motif (Mori *et al.*, 2002). These data led to the hypothesis that KaiC may affect chromosomal state to regulate genomewide gene expression. Recently, a chromosome compaction was reported to exhibit *kaiC*-dependent circadian rhythms (Smith and Williams, 2006). The degree of chromosome compaction might inhibit the transcriptional machinery from accessing promoter elements. Circadian control of DNA topology or superhelicity may be involved in genomewide control by the clock.

Input pathway

Circadian rhythms in cyanobacterial cells require the input mechanisms that synchronize the clock with the environment and also the output pathways that control temporal behaviors (Figure 8.2). To date, two-component regulatory systems have been shown to be involved in both pathways in *Synechococcus elongatus* PCC 7942.

The *cikA* (circadian input kinase) gene, which encodes a sensor histidine kinase, has been identified to control the input pathway of the circadian oscillator (Schmitz *et al.*, 2000). The *cikA* gene was first identified from a transposon-insertion mutant that had a subtle defect in light-responsive regulation of its photosynthesis genes. The mutant had a clearer defect in circadian rhythms of gene expression (Schmitz *et al.*, 2000). CikA mutants had a shortened period length and reduced amplitude of oscillation, and they also failed to reset the phase of their circadian rhythm. Interestingly, cells of *cikA* mutants are

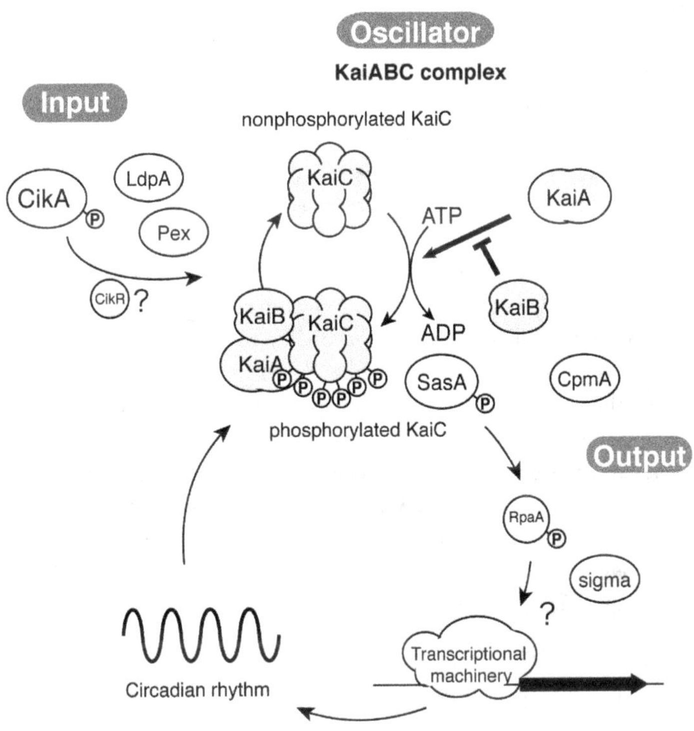

Figure 8.2 Overview of the relationships among the key clock components in cyanobacteria. The KaiC phosphorylation cycle is the pacemaker of the cyanobacterial circadian rhythm. KaiA enhances the autokinase activity of KaiC, whereas KaiB functions as an attenuator of KaiA-enhanced phosphorylation. Environmental signals are transmitted to the central pacemaker via the CikA histidine kinase and an unidentified cognate response regulator, CikR, in input pathways that set the clock to the local time each day. LdpA and Pex, described in the text, are thought to be involved in the input pathway. SasA is a histidine kinase that physically interacts with the Kai proteins; phosphorylated KaiC stimulates SasA autophosphorylation. RpaA is a cognate response regulator, and phosphorylated RpaA activates target genes that remain undefined. The group 2 sigma factors (Tsinoremas *et al.*, 1996) and CpmA (circadian phase modifier) (Katayama *et al.*, 1999) also help to transmit temporal information to downstream genes.

abnormally long and rod-shaped, suggesting a defect in cell division (Miyagishima *et al.*, 2005).

CikA contains three domains, a histidine kinase, a GAF domain and a pseudo-receiver domain (Schmitz *et al.*, 2000). The CikA protein belongs to a bacteriophytochrome family (Montgomery and Lagarias, 2002). Bacteriophytochromes are bacterial proteins that are homologous to plant-photoreceptor phytochromes, which bind a bilin cofactor at a GAF domain and have a light-dependent histidine kinase activity. *In vitro* assays demonstrated that CikA showed autophosphorylation activity that was regulated by other domains (Mutsuda *et al.*, 2003). The GAF domain of CikA does not have a conserved cysteine or histidine residue that binds covalently to a bilin-type chromophore, and CikA protein purified from cyanobacterial cells did not attach a cofactor (Mutsuda *et al.*, 2003). CikA may operate with a cognate response regulator termed CikR, but it has yet to be identified (Zhang *et al.*, 2006). The pseudo-receiver domain in CikA cannot be phosphorylated by the CikA histidine kinase domain (Mutsuda *et al.*, 2003), and it negatively regulated the histidine kinase activity *in vivo*, indicating that the pseudo-receiver domain is not likely to play a role as a cognate response regulator (Zhang *et al.*, 2006).

Light-dependent protein A (LdpA) is another component of the input pathway of the cyanobacterial circadian clock (Katayama *et al.*, 2003; Ivleva *et al.*, 2005). Unlike CikA mutants, the strains that lack LdpA still reset the clock in response to a dark pulse. LdpA was predicted to be a metallo-protein that carries two iron–sulfur clusters (Katayama *et al.*, 2003). UV-visible absorption and electron paramagnetic resonance analyses revealed that LdpA has the expected clusters that are redox sensitive (Ivleva *et al.*, 2005). Moreover, LdpA is sensitive to the redox state of the cell, and it co-purifies with KaiA, CikA, and SasA (see below). LdpA affects CikA abundance according to light intensity and the sensitivity of CikA to respond to cellular redox state. These data suggest that LdpA is a component of the clock complex that is able to sense the redox state of the cell (Ivleva *et al.*, 2005).

Another component of the input pathway is the *pex* gene (period extender) (Kutsuna *et al.*, 1998). Overproduction of Pex extends circadian periods and decreases amplitude of the rhythms. The disruption of *pex* gene increases the expression of *kaiA*, suggesting that *pex* might play a role in repression of *kaiA* expression (Kutsuna *et al.*, 1998). Recently the expression of *pex* has been reported to increase in darkness and to be required to delay the clock (Takai *et al.*, 2006a).

Output pathway

The pathway by which the Kai oscillator transfers temporal information to downstream gene expression is important for the circadian biological activity in cyanobacteria (Figure 8.2). A pair of proteins, SasA and RpaA, plays a key role in the output pathway. Yeast two-hybrid screening for KaiC-associating proteins identified a gene encoding a sensory histidine kinase gene, *sasA* (*Synechococcus* adaptive sensor) (Iwasaki *et al.*, 2000). Although the *sasA* gene was originally isolated from *Synechococcus elongatus* PCC 7942 by screening of two-component-type histidine kinases in an *E. coli* background, its physiological functions were unknown (Nagaya *et al.*, 1993). Disruption of the *sasA* gene dramatically reduced the expression of *kaiBC* and clock-controlled genes, some of which became completely arrhythmic, indicating that SasA functions in the output pathway of the clock.

Interestingly, the presumed sensory input domain of SasA resembles a domain found in KaiB and appears sufficient for association with KaiC (Iwasaki *et al.*, 2000).

Recently, a cognate response regulator of SasA was identified by systematic genome-wide disruptions of response regulator genes in *Synechococcus* sp. PCC 6301, which is a strain closely related to *Synechococcus elongatus* PCC 7942 (Takai *et al.*, 2006b). Disruption of one gene, *rpaA* (regulator of phycobilisome-associated), dramatically affected the expression of *kaiBC* and the clock-controlled genes, producing a phenotype similar to that seen with the loss of SasA. Moreover, KaiC activated *in vitro* SasA autophosphorylation activity (Smith and Williams, 2006; Takai *et al.*, 2006b) and enhanced phosphotransfer from SasA to RpaA (Takai *et al.*, 2006b). Interestingly, SasA autophosphorylation and RpaA phosphorylation were modulated depending on the phase of the KaiC phosphorylation cycle, which was reconstituted *in vitro* by the three Kai proteins and ATP (see below) (Takai *et al.*, 2006b). These results suggested that SasA and RpaA form a two-component phosphorelay signaling pathway that mediates between KaiC phosphorylation rhythm and global transcription. The identification of RpaA targets will be an important next step.

Kai proteins

A single amino acid substitution in each Kai protein shortened or lengthened the period length or abolished a circadian rhythm (Ishiura *et al.*, 1998), suggesting that biochemical analyses of Kai proteins will be essential for elucidating the molecular mechanism of the clocks. There is no sequence similarity between each Kai protein and any known clock proteins in other organisms. Moreover, the Kai proteins do not contain any known functional motifs except for the Walker-motifs in KaiC that are conserved in many ATP/GTP-binding proteins (Ishiura *et al.*, 1998).

KaiC protein

KaiC contains two internally duplicated domains (CI and CII), each of which has a set of ATP-binding motifs (Ishiura *et al.*, 1998), and it belongs to the superfamily of bacterial DNA helicases/recombinases (Leipe *et al.*, 2000). Like other members of this family, KaiC forms a stable hexamer upon binding of ATP, and electron microscopy analysis revealed a ring structure with a central pore of the KaiC hexamer (Mori *et al.*, 2002; Hayashi *et al.*, 2003). The ATP-binding activity of KaiC was demonstrated *in vitro*, and genetic disruption of nucleotide-binding sites of KaiC severely impaired circadian *kaiBC* expression profiles (Nishiwaki *et al.*, 2000). It has also been demonstrated that two ATP molecules bind to one KaiC molecule and that the ATP-binding motif in CI is responsible for KaiC hexamerization, whereas the CII motif is involved in KaiC phosphorylation (Hayashi *et al.*, 2004; Hayashi *et al.*, 2006).

KaiC has an autokinase activity *in vitro* (Nishiwaki *et al.*, 2000) and shows a circadian phosphorylation rhythm *in vivo* (Iwasaki *et al.*, 2002). Mass spectrometry analysis determined the presence of two phosphorylation residues on KaiC, serine 431 and threonine 432 (Nishiwaki *et al.*, 2004). Site-direct mutagenesis of these sites completely abolished the circadian phosphorylation rhythm, formation of the clock complexes and *kaiBC* expression rhythm in cyanobacterial cells, indicating that KaiC phosphorylation is essential for circadian function in cyanobacteria. KaiC protein accumulation oscillates in cyanobacterial cells. Protein synthesis assays using ^{35}S-methionine revealed that the rate of KaiC synthesis changes in a circadian manner (Imai *et al.*, 2004). As the half-life of the KaiC protein *in vivo* is ~10 hours, KaiC is likely to be a relatively stable protein (Imai *et al.*, 2004).

The crystal structure of the full-length KaiC protein from *Synechococcus elongatus* PCC 7942 revealed that KaiC forms a homo-hexamer in the shape of two parallel-stacked rings of CI and CII with a central pore (Pattanayek *et al.*, 2004). It also revealed that the 12 ATPγs molecules were localized in the interfaces between protomers in both the CI and CII domains. The structure of KaiC helped to identify three autophosphorylation sites at threonine 426, serine 431, and threonine 432 near the ATP binding site of the CII domain (Pattanayek *et al.*, 2004; Xu *et al.*, 2004). Threonine 426 and serine 431 are likely to share a phosphate molecule that shuttles between these two residues. However, as described above, mass spectrometry analysis revealed only two phosphorylation residues, serine 431 and threonine 432, on KaiC (Nishiwaki *et al.*, 2004).

A combined structural study of the three Kai proteins will improve the understanding of the molecular mechanism of the cyanobacterial circadian clock. Recently site-directed mutagenesis in combination with an electron microscopy study revealed that KaiA binds to the CII half domain of the KaiC hexamer (Pattanayek *et al.*, 2006). To date, no consistent structural studies on the associations between the three Kai proteins have been reported.

KaiA and KaiB

KaiB and KaiC protein levels oscillate, whereas KaiA accumulates constitutively in *Synechococcus elongatus* PCC 7942 (Xu *et al.*, 2000; Kitayama *et al.*, 2003). The biochemical functions of KaiA and KaiB have not been well characterized due to the fact that any of these proteins has known functional motifs. Biochemical analysis *in vivo* and *in vitro* demonstrated that KaiA and KaiB modulate the KaiC phosphorylation status. KaiA enhances an autokinase activity of KaiC (Iwasaki *et al.*, 2002; Williams *et al.*, 2002), whereas KaiB functions as an attenuator of KaiA-enhanced phosphorylation of KaiC *in vitro* and *in vivo* (Kitayama *et al.*, 2003; Xu *et al.*, 2003). The KaiC phosphorylation level is a function of the antagonistic interplay between KaiA and KaiB.

KaiA is composed of two domains, an amino-terminal domain and a carboxy-terminal domain. The carboxy-terminal domain is highly conserved among cyanobacteria, while there is limited sequence conservation in the amino-terminal domain of KaiA. KaiA from some cyanobacterial strains, such as *Anabaena* sp. PCC 7120, contain a truncated form that lacks the amino-terminal domain. NMR analysis revealed that the structure of the KaiA amino-terminal domain is similar to that of a pseudo-receiver domain of bacterial response regulators although it lacks the conserved aspartate residue necessary for phosphorylation (Williams *et al.*, 2002). The carboxy-terminal domain is responsible for dimer formation and binding to KaiC to enhance KaiC phosphorylation (Williams *et al.*, 2002; Vakonakis *et al.*, 2004; Ye *et al.*, 2004; Uzumaki *et al.*, 2004). The structure of the carboxy-terminal domain of KaiA from *Thermosynechococcus elongatus* BP-1, which was solved by X-ray crystallography (Uzumaki *et al.*, 2004) and NMR (Vakonakis *et al.*, 2004), revealed that the overall fold of KaiA is that of a four-helix bundle. A full-length KaiA from *Anabaena* sp. PCC 7120 was also solved and found to have essentially the same structure (Garces *et al.*, 2004).

The amino acid sequence of KaiB is well conserved among cyanobacteria. To date, the crystal structures of KaiB have been solved from *Anabaena* sp. PCC 7120 (Garces *et al.*, 2004), *Synechocystis* sp. PCC 6803 (Hitomi *et al.*, 2005) and *Thermosynechococcus elongatus* BP-1 (Iwase *et al.*, 2005). Although KaiB proteins from these three strains share a high sequence similarity, they have different oligomeric structures. KaiB was reported to form a dimer in *Anabaena* sp. PCC 7120 but a tetramer in *Synechocystis* sp. PCC 6803 and *Thermosynechococcus elongatus* BP-1. The KaiB structures from *Thermosynechococcus elongatus* BP-1 and *Anabaena* sp. PCC 7120 were also studied in solution using gel-filtration chromatography, chemical cross-linking and analytical ultracentrifugation (Iwase *et al.*, 2005). The results supported the tetrameric structure of KaiB from both strains. On the other hand, gel-filtration chromatography demonstrated that KaiB from *Synechococcus elongatus* PCC 7942 is a dimer (Kageyama *et al.*, 2006).

Kai complex

The cellular content of KaiA is kept constant at about 500 molecules per cell, and the abundance levels of KaiB and KaiC oscillate in the course of a circadian cycle around an average level of 10 000 molecules per cell (Kitayama *et al.*, 2003). KaiA and KaiC are localized in the cytosolic fractions, while KaiB was present in both membrane and cytosolic fractions (Kitayama *et al.*, 2003). KaiA, KaiB and KaiC proteins interact with each other in all possible combinations *in vitro* and in cyanobacterial cells (Iwasaki *et al.*, 1999; Kageyama *et al.*, 2003; Kitayama *et al.*, 2003). Three Kai proteins and a histidine kinase, SasA, form a heteromultimeric protein complex in a circadian manner in *Synechococcus elongatus* PCC 7942 (Kageyama *et al.*, 2003).

The complex formation profile of Kai proteins *in vivo* was elucidated (Kitayama *et al.*, 2003). As KaiC accumulates in cells during the day, KaiA enhances KaiC autokinase activity without a stable binding to KaiC. The phosphorylated KaiC protein apparently forms a tight complex with KaiA. Then, KaiB is proposed to be released from the membrane into the cytosol to interact with the KaiA-KaiC complex to form the larger clock complex (Kitayama *et al.*, 2003). The dynamic circadian behavior of Kai protein complex formation could regulate the KaiC phosphorylation cycle.

KaiC phosphorylation cycle in vitro

TTO model of cyanobacteria

The consensus model in cyanobacteria was similar to eukaryotic models, which share a common feature: a transcriptional and translational oscillatory process (TTO) based on negative feedback regulation of clock genes (Young and Kay, 2002). In the TTO model, clock genes are transcribed and translated into clock proteins, which repress their own expression. In *Synechococcus elongatus* PCC 7942, the negative feedback regulation of the expression of the *kaiBC* operon by Kai proteins was proposed to form this type of transcriptional negative feedback (Ishiura *et al.*, 1998). KaiC has been found to negatively regulate the expression of the *kaiBC* operon, while KaiA positively influences *kaiC* expression (Ishiura *et al.*, 1998).

However, the TTO model in cyanobacteria had to be revised due to a breakthrough study using cyanobacterial cell cultures in continual darkness. *Synechococcus elongatus* PCC

7942 is an obligate photoautotrophic organism, and general gene transcription totally depend on energy derived from photosynthesis. In constant darkness, mRNA synthesis almost stops due to the lack of energy input. Thus, the clock studies in *Synechococcus elongatus* PCC 7942 have been done with the cells grown under continuous light conditions. If the TTO model is true for cyanobacteria, no oscillation should have been generated in cells grown in constant darkness. However, when cyanobacterial cells were incubated in constant darkness after transferring from constant light, surprisingly, the circadian rhythm of KaiC phosphorylation persisted even though *kaiC* mRNA levels were undetectable and KaiC protein content was almost constant (Tomita *et al.*, 2005). The phosphorylation rhythm oscillated even in the presence of inhibitors of transcription and translation. These results indicated that KaiC phosphorylation oscillation is independent of *de novo* transcriptional and translational processes. KaiC phosphorylation oscillation in constant darkness also retains the property of temperature compensation *in vivo*. The period of KaiC phosphorylation oscillation was lengthened in the KaiC mutant strain, in which the period of the *kaiBC* expression rhythm was also lengthened (Tomita *et al.*, 2005). These data also support the notion that the KaiC phosphorylation cycle is the fundamental pacemaker of the cyanobacterial clock.

Even more astonishing was the accomplishment of *in vitro* reconstitution of the KaiC phosphorylation cycle with purified KaiA, KaiB, and KaiC proteins incubated in the presence of ATP (Nakajima *et al.*, 2005). These exciting findings suggest that self-sustained circadian oscillation is generated by rhythmic phosphorylation of KaiC rather than by a transcriptional-translational feedback loop (Nakajima *et al.*, 2005; Tomita *et al.*, 2005).

Reconstitution of the KaiC phosphorylation cycle

It was a dream for researchers in chronobiology to reconstitute a 24-hour oscillation in a test tube. Our group first demonstrated it in 2005. We incubated three Kai proteins with ATP in a test tube and showed that the KaiC phosphorylation state oscillates with a period of about 24 hours without damping *in vitro* (Nakajima *et al.*, 2005). The experimental procedure is quite simple, that is, the purified recombinant KaiC protein was incubated with KaiA and KaiB in the buffer containing 1 mM ATP at 30°C (Figure 8.3). Aliquots of the reaction mixture were collected at regular intervals and subjected to SDS-polyacrylamide gel electrophoresis to evaluate the phosphorylation state of KaiC. KaiC protein is detected as two bands in the gel; the upper and lower bands correspond to the phosphorylated and unphosphorylated form, respectively. The ratio of phosphorylated KaiC to total KaiC is determined by densitometry of each band, and the ratio is plotted as a function of the incubation time. Thus, it was found that the ratio oscillates in a circadian manner.

The period length of the KaiC phosphorylation cycle was temperature-compensated between 25 and 35°C, and the thermal sensitivity of the period was 1.1, which is consistent with what has been reported for gene expression rhythm *in vivo* (Nakajima *et al.*, 2005). Moreover, analysis using KaiC mutant proteins revealed that the periods of the KaiC phosphorylation cycle *in vitro* were in good agreement with those of the respective mutant rhythms *in vivo* (Nakajima *et al.*, 2005), suggesting that the period is an inherent feature of the amino acid sequence of the KaiC protein. Thus, the clock in a test tube satisfied the intrinsic properties of circadian clocks apart from entrainment. These findings clearly prove that KaiC phosphorylation is the true circadian pacemaker in cyanobacteria.

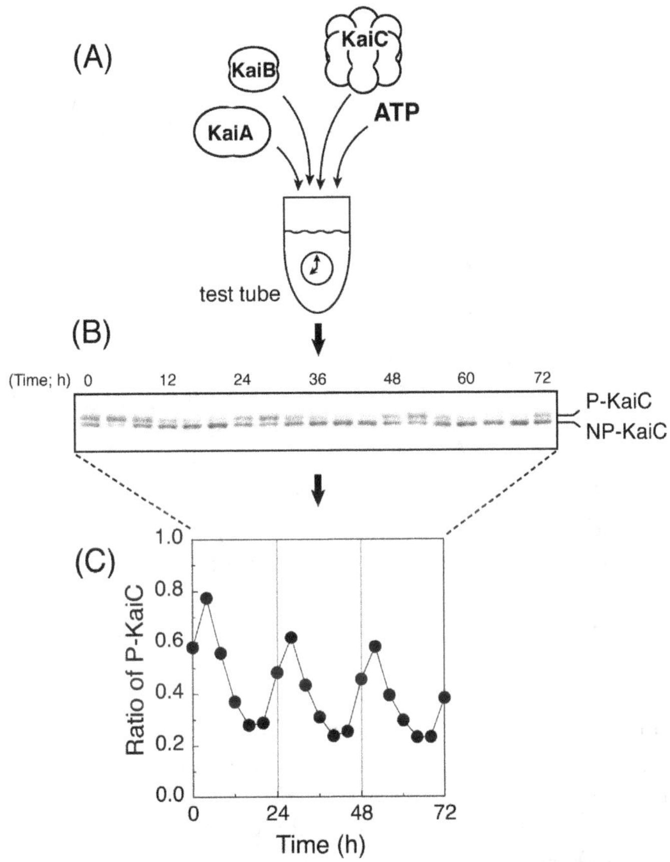

Figure 8.3 KaiC phosphorylation cycle *in vitro*. (A) Recombinant KaiC protein (3.5 µM) was incubated with KaiA (1.2 µM) and KaiB (3.5 µM) in the presence of ATP (1 mM) at 30°C. (B) Aliquots of the reaction mixture were collected every 4 hours and subjected to SDS-polyacrylamide gel electrophoresis. The upper and lower bands correspond to phosphorylated (P-KaiC) and unphosphorylated KaiC (NP-KaiC), respectively. (C) NIH image software was used to perform densitometric analysis of the data in (B). The ratios of P-KaiC to total KaiC are plotted as a function of incubation time.

The KaiA, KaiB and KaiC protein concentrations in the standard reaction mixture for the reconstitution of the KaiC phosphorylation cycle are 1.2, 3.5 and 3.5 µM, respectively, which is inconsistent with the cellular concentrations of KaiB and KaiC that are estimated at 10–30 µM and that of KaiA at 0.5–1 µM (Kageyama *et al.*, 2006). The ratio of the three proteins in the reconstitution mixture is not likely to coincide with the physiological ratio. This fact may be attributed mainly to the intracellular localization of the Kai proteins (Kageyama *et al.*, 2006). The *in vitro* KaiC phosphorylation rhythm was dependent on the relative concentrations of KaiA or KaiB to KaiC (Kageyama *et al.*, 2006). Lowering the KaiA concentration to two-thirds of the standard concentration extended the period and to one-fourth abolished it. Reducing the KaiB concentration to one-third also abolished the rhythm (Kageyama *et al.*, 2006).

The molecular mechanism of the KaiC phosphorylation cycle *in vitro* has now been studied. Recently, the temporal profile of complex formation between the three Kai proteins was analyzed by gel-filtration chromatography (Kageyama *et al.*, 2006). KaiA actively and repeatedly binds KaiC during the phosphorylation process, and the phosphorylated KaiC induced KaiC association with KaiB, which inactivates KaiA to accelerate the dephosphorylation of KaiC. Reduction of phosphorylated KaiC results in KaiB dissociation from KaiC. These protein association/dissociation processes repeat every 24 hours in the KaiC phosphorylation cycle *in vitro*.

Future directions

The KaiC phosphorylation cycle reconstituted *in vitro* has been shown to be the pacemaker in the cyanobacterial circadian clock. However, the molecular mechanism of the three key features of the circadian clock remains to be elucidated. We are currently investigating how the KaiC phosphorylation cycle attains the circadian characteristics. Further detailed studies, such as biochemical, physical and structural studies are needed for the elucidation of those characteristics. One of the features of the circadian clock, entrainment, remains to be investigated *in vitro* in the KaiC phosphorylation cycle. A single amino acid substitution in KaiC resulted in a defective phase shifting to dark pulses *in vivo* (Kiyohara *et al.*, 2005), which may provide a potential for entrainment in the KaiC phosphorylation cycle *in vitro*.

To date, it has been discovered that the KaiC phosphorylation cycle is the only pacemaker to generate circadian rhythms within the cyanobacterial cell. There is no evidence for the existence of more than one oscillation pacemaker in cyanobacteria. However, it should be noted that the KaiC mutant protein K294H is not phosphorylated, yet the cyanobacterial strain with this mutation exhibited the bioluminescence circadian rhythm, although irregularly (Hayashi *et al.*, 2004). The temporal coordination of oscillator components, including the KaiC phosphorylation cycle, is essential for generating a robust and precise circadian clock system in the cell. The details of the molecular mechanism for the regulation of global gene expression in cyanobacterial cells by the KaiC phosphorylation cycle also remains to be studied.

Acknowledgments

We are grateful to Dr. Y. Fujita (Nagoya University) for critical reading of the manuscript and helpful discussions. Work in the authors' laboratory is currently supported by Grants-in-aid from the Ministry of Education, Culture, Sports, Science and Technology of Japan (15GS0308 and 14COEEA01).

Reference

Aoki, S., Kondo, T., and Ishiura, M. (1995). Circadian expression of the *dnaK* gene in the cyanobacterium *Synechocystis* sp. strain PCC 6803. J. Bacteriol. *177*, 5606–5611.

Aoki, S., Kondo, T., and Ishiura, M. (2002). A promoter-trap vector for clock-controlled genes in the cyanobacterium *Synechocystis* sp. PCC 6803. J. Microbiol. Methods *49*, 265–274.

Dvornyk, V., Vinogradova, O., and Nevo, E. (2003). Origin and evolution of circadian clock genes in prokaryotes. Proc. Natl. Acad. Sci. USA *100*, 2495–2500.

Edmunds, L.N. Jr. (1988). Cellular and Molecular Bases of Biological Clocks (Springer-Verlag, New York).

Golden, S.S., Ishiura, M., Johnson, C.H., and Kondo, T. (1997). Cyanobacterial circadian rhythms. Annu. Rev. Plant Physiol. Mol. Biol. *48*, 327–354.

Garces, R.G., Wu, N., Gillon, W., and Pai, E.F. (2004). *Anabaena* circadian clock proteins KaiA and KaiB reveal a potential common binding site to their partner KaiC. EMBO J. *23*, 1688–1698.

Grobbelaar, N., Huang, T.C., Lin, H.Y., and Chow, T.J. (1986). Dinitrogen-fixing endogenous rhythm in *Synechococcus* RF-1. FEMS Microbiol. Lett. *37*, 173–177.

Hayashi, F., Itoh, N., Uzumaki. T., Iwase, R., Tsuchiya, Y., Yamakawa, H., Morishita, M., Onai, K., Itoh, S., and Ishiura, M. (2004). Roles of two ATPase-motif-containing domains in cyanobacterial circadian clock protein KaiC. J. Biol. Chem. *279*, 52331–52337.

Hayashi, F., Iwase, R., Uzumaki, T., and Ishiura, M. (2006). Hexamerization by the N-terminal domain and intersubunit phosphorylation by the C-terminal domain of cyanobacterial circadian clock protein KaiC. Biochem. Biophys. Res. Commun. *348*, 864–872.

Hayashi, F., Suzuki, H., Iwase, R., Uzumaki, T., Miyake, A., Shen, J.R, Imada, K., Furukawa, Y., Yonekura, K., Namba, K., and Ishiura, M. (2003). ATP-induced hexameric ring structure of the cyanobacterial circadian clock protein KaiC. Genes Cells *8*, 287–296.

Hitomi, K., Oyama T., Han, S., Arvai, A.S., and Getzoff, E.D. (2005). Tetrameric architecture of the circadian clock protein KaiB. A novel interface for intermolecular interactions and its impact on the circadian rhythm. J. Biol. Chem. *280*, 19127–19135.

Hwang, S., Kawazoe, R., and Herrin, D.L. (1996). Transcription of *tufA* and other chloroplast-encoded genes is controlled by a circadian clock in *Chlamydomonas*. Proc. Natl. Acad. Sci. USA *93*, 996–1000.

Imai, K., Nishiwaki, T., Kondo, T., and Iwasaki, H. (2004). Circadian rhythms in the synthesis and degradation of a master clock protein KaiC in cyanobacteria. J. Biol. Chem. *279*, 36534–36539.

Ishiura, M., Kutsuna, K., Aoki, S., Iwasaki, H., Andersson, C.R., Tanabe, A., Golden, S.S., Johnson, C.H., and Kondo, T. (1998). Expression of a gene cluster *kaiABC* as a circadian feedback process in cyanobacteria. Science *281*, 1519–1523.

Ivleva, N.B., Bramlett, M.R., Lindahl, P.A., and Golden, S.S. (2005). LdpA: a component of the circadian clock senses redox state of the cell. EMBO J. *24*, 1202–1210.

Iwasaki, H., Nishiwaki, T, Kitayama, Y., Nakajima, M., and Kondo, T. (2002). KaiA-stimulated KaiC phosphorylation in circadian timing loops in cyanobacteria. Proc. Natl. Acad. Sci. USA *99*, 15788–15793.

Iwasaki, H., Taniguchi, Y., Ishiura, M., and Kondo, T. (1999). Physical interactions among circadian clock proteins KaiA, KaiB and KaiC in cyanobacteria. EMBO J. *18*, 1137–1145.

Iwasaki, H., Williams, S.B., Kitayama, Y., Ishiura, M. Golden, S.S., and Kondo, T. (2000). A *kaiC*-interacting sensory histidine kinase, SasA, necessary to sustain robust circadian oscillation in cyanobacteria. Cell *101*, 223–233.

Iwase, R., Imada, K., Hayashi, F., Uzumaki, T., Morishita, M., Onai, K., Furukawa, Y., Namba, K., and Ishiura, M. (2005). Functionally important substructures of circadian clock protein KaiB in a unique tetramer complex. J. Biol. Chem. *280*, 43141–43149.

Kageyama, H., Kondo, T., and Iwasaki, H. (2003). Circadian formation of clock protein complexes by KaiA, KaiB, KaiC, and SasA in cyanobacteria. J. Biol. Chem. *278*, 2388–2395.

Kageyama, H., Nishiwaki, T., Nakajima, M., Iwasaki, H., Oyama, T., and Kondo, T. (2006). Cyanobacterial circadian pacemaker: Kai protein complex dynamics in the KaiC phosphorylation cycle *in vitro*. Mol. Cell *23*, 161–171.

Katayama, M., Kondo, T., Xiong, J., and Golden, S.S. (2003). *ldpA* encodes an iron–sulfur protein involved in light-dependent modulation of the circadian period in the cyanobacterium *Synechococcus elongatus* PCC 7942. J. Bacteriol. *185*, 1415–1422.

Katayama, M., Tsinoremas, N.F., Kondo, T., and Golden, S.S. (1999). *cpmA*, a gene involved in an output pathway of the cyanobacterial circadian system. J. Bacteriol. *181*, 3516–3524.

Kitayama, Y., Iwasaki, H., Nishiwaki, T., and Kondo, T. (2003). KaiB functions as an attenuator of KaiC phosphorylation in the cyanobacterial circadian clock system. EMBO J. *22*, 2127–2134.

Kiyohara, Y.B., Katayama, M., and Kondo, T. (2005). A novel mutation in *kaiC* affects resetting of the cyanobacterial circadian clock. J. Bacteriol. *187*, 2559–2564.

Kondo, T., and Ishiura, M. (1999). The circadian clocks of plants and cyanobacteria. Trends Plant Sci. *4*, 171–176.

Kondo, T., Mori, T., Lebedeva, N.V., Aoki, S., Ishiura, M., and Golden, S.S. (1997). Circadian rhythms in rapidly dividing cyanobacteria. Science 275, 224–227.

Kondo, T., Strayer, C.A., Kulkarni, R.D., Taylor, W., Ishiura, M., Golden, S.S., and Johnson, C.H. (1993). Circadian rhythms in prokaryotes: luciferase as a reporter of circadian gene expression in cyanobacteria. Proc. Natl. Acad. Sci. USA *90*, 5672–5676.

Kondo, T., Tsinoremas, N.F., Golden, S.S., Johnson, C.H., Kutsuna, S., and Ishiura, M. (1994). Circadian clock mutants of cyanobacteria. Science *266*, 1233–1236.

Kucho, K., Okamoto, K., Tsuchiya, Y., Nomura, S., Nango, M., Kanehisa, M., and Ishiura, M. (2005). Global analysis of circadian expression in the cyanobacterium *Synechocystis* sp. strain PCC 6803. J. Bacteriol. *187*, 2190–2199.

Kutsuna, S., Kondo, T., Aoki, S., and Ishiura, M. (1998). A period-extender gene, *pex*, that extends the period of the circadian clock in the cyanobacterium *Synechococcus* sp. strain PCC 7942. J. Bacteriol. *180*, 2167–2174.

Kutsuna, S., Nakahira, Y., Katayama, M., Ishiura, M., and Kondo, T. (2005). Transcriptional regulation of the circadian clock operon *kaiBC* by upstream regions in cyanobacteria. Mol. Microbiol. *57*, 1474–1484.

Leipe, D.D., Aravind, L., Grishin, N.V., and Koonin, E.V. (2000). The bacterial replicative helicase DnaB evolved from a RecA duplication. Genome Res. *10*, 5–16.

Li, R., and Golden, S.S. (1993). Enhancer activity of light-responsive regulatory elements in the untranslated leader regions of cyanobacterial *psbA* genes. Proc. Natl. Acad. Sci. USA *90*, 11678–11682.

Liu, Y., Tsinoremas, N.F., Johnson, C.H., Lebedeva, N.V., Golden, S.S., Ishiura, M., and Kondo, T. (1995). Circadian orchestration of gene expression in cyanobacteria. Genes Dev. *9*, 1469–1478.

Mihalcescu, I., Hsing, W., and Leibler, S. (2004). Resilient circadian oscillator revealed in individual cyanobacteria. Nature *430*, 81–85.

Min, H., Guo, H., and Xiong, J. (2005). Rhythmic gene expression in a purple photosynthetic bacterium, *Rhodobacter sphaeroides*. FEBS Lett. *579*, 808–812.

Mitsui, A., Kumazawa, S., Takahashi, A., Ikemoto, H., Cao, S., and Arai, T. (1986). Strategy by which nitrogen-fixing unicellular cyanobacteria grow photoautotrophically. Nature *323*, 720–722.

Miyagishima, S.Y, Wolk. C.P., and Osteryoung, K.W. (2005). Identification of cyanobacterial cell division genes by comparative and mutational analyses. Mol. Microbiol. *56*, 126–143.

Montgomery, B.L., and Lagarias, J.C. (2002). Phytochrome ancestry: sensors of bilins and light. Trends Plant Sci. *7*, 357–366.

Mori, T., Binder, B., and Johnson, C.H. (1996). Circadian gating of cell division in cyanobacteria growing with average doubling times of less than 24 hours. Proc. Natl. Acad. Sci. USA *93*, 10183–10188.

Mori, T., and Johnson, C.H. (2001). Circadian programming in cyanobacteria. Semin. Cell Dev. Biol. *12*, 271–278.

Mori, T., Saveliev, S.V., Xu, Y., Stafford,W.F., Cox, M.M., Inman, R.B., and Johnson, C.H. (2002). Circadian clock protein KaiC forms ATP-dependent hexameric rings and binds DNA. Proc. Natl. Acad. Sci. USA *99*, 17203–17208.

Mutsuda, M., Michel, K.P., Zhang, X., Montgomery, B.L., and Golden, S.S. (2003). Biochemical properties of CikA, an unusual phytochrome-like histidine protein kinase that resets the circadian clock in *Synechococcus elongatus* PCC 7942. J. Biol. Chem. *278*, 19102–19110.

Nagaya, M., Aiba, H., and Mizuno, T. (1993). Cloning of a sensory-kinase-encoding gene that belongs to the two-component regulatory family from the cyanobacterium *Synechococcus* sp. PCC 7942. Gene *131*, 119–124.

Nakahira, Y., Baba, K., Yoneda, A., Shiina, T., and Toyoshima, Y. (1998). Circadian-regulated transcription of the *psbD* light-responsive promoter in wheat chloroplasts. Plant Physiol. *118*, 1079–1088.

Nakahira, Y., Katayama, M., Miyashita, H., Kutsuna, S., Iwasaki, H., Oyama, T., and Kondo, T. (2004). Global gene repression by KaiC as a master process of prokaryotic circadian system. Proc. Natl. Acad. Sci. USA *101*, 881–885.

Nakajima, M., Imai, K., Ito, H., Nishiwaki, T., Murayama, Y., Iwasaki, H., Oyama, T., and Kondo, T. (2005). Reconstitution of circadian oscillation of cyanobacterial KaiC phosphorylation *in vitro*. Science *308*, 414–415.

Nakamura, Y., Kaneko, T., Sato, S., Mimuro, M., Miyashita, H., Tsuchiya, T., Sasamoto, S., Watanabe, A., Kawashima, K., Kishida, Y., Kiyokawa, C., Kohara, M., Matsumoto, M., Matsuno, A., Nakazaki, N., Shimpo, S., Takeuchi, C., Yamada, M., and Tabata, S. (2003). Complete genome structure of *Gloeobacter violaceus* PCC 7421, a cyanobacterium that lacks thylakoids. DNA Res. *10*, 137–145.

Nishiwaki, T., Iwasaki, H,. Ishiura, M., and Kondo, T. (2000). Nucleotide binding and autophosphorylation of the clock protein KaiC as a circadian timing process of cyanobacteria. Proc. Natl. Acad. Sci. USA *97*, 495–499.

Nishiwaki, T., Satomi, Y., Nakajima, M., Lee, C., Kiyohara, R., Kageyama, H., Kitayama, Y., Temamoto, M., Yamaguchi, A., Hijikata, A., Go, M., Iwasaki, H., Takao, T,., and Kondo, T. (2004). Role of KaiC phosphorylation in the circadian clock system of *Synechococcus elongatus* PCC 7942. Proc. Natl. Acad. Sci. USA *101*, 13927–13932.

Onai, K., Morishita,M., Itoh, S., Okamoto, K., and Ishiura, M. (2004). Circadian rhythms in the thermophilic cyanobacterium *Thermosynechococcus elongatus*: compensation of period length over a wide temperature range. J. Bacteriol. *186*, 4972–4977.

Ouyang, Y., Andersson, C.R., Kondo, T., Golden, S.S., and Johnson, C.H. (1998). Resonating circadian clocks enhance fitness in cyanobacteria. Proc. Natl. Acad. Sci. USA *95*, 8660–8664.

Pattanayek, R., Wang, J., Mori, T., Xu, Y., Johnson, C.H, and Egli, M. (2004). Visualizing a circadian clock protein: crystal structure of KaiC and functional insights. Mol. Cell *15*, 375–388.

Pattanayek, R., Williams, D.R., Pattanayek, S., Xu, Y., Mori, T., Johnson, C.H., Stewart, P.L., and Egli, M. (2006). Analysis of KaiA-KaiC protein interactions in the cyanobacterial circadian clock using hybrid structural methods. EMBO J. *25*, 2017–2028.

Pittendrigh, C.S. (1993). Temporal organization: reflections of a Darwinian clock-watcher. Annu. Rev. Physiol. *55*, 17–54.

Schmitz, O., Katayama, M., Williams, S.B., Kondo, T., and Golden, S.S. (2000). CikA, a bacteriophytochrome that resets the cyanobacterial circadian clock. Science *289*, 765–768.

Smith, R.M., and Williams, S.B. (2006). Circadian rhythms in gene transcription imparted by chromosome compaction in the cyanobacterium *Synechococcus elongatus*. Proc. Natl. Acad. Sci. USA *103*, 8564–8569.

Takai, N., Ikeuchi, S., Manabe, K., and Kutsuna, S. (2006a). Expression of the circadian clock-related gene *pex* in cyanobacteria increase in darkness and is required to delay the clock. J. Biol. Rhythms *21*, 235–244.

Takai, N., Nakajima, M., Oyama, T., Kito, R., Sugita, C., Sugita, M., Kondo, T., and Iwasaki, H. (2006b). A KaiC-associating SasA-RpaA two-component regulatory system as a major circadian timing mediator in cyanobacteria. Proc. Natl. Acad. Sci. USA *103*, 12109–12114.

Terauchi, K., Katayama, M., Fujita, Y., and Kondo, T. (2005). Circadian rhythm of the facultative cyanobacterium *Plectonema boryanum*. In Photosynthesis: Fundamental Aspects to Global Perspectives, A. van der Est and B. Diner, eds. (Lawrence, KS: Allen Press), vol. 2, pp. 729–730.

Tomita, J., Nakajima, M., Kondo, T., and Iwasaki, H. (2005). No transcription-translation feedback in circadian rhythm of KaiC phosphorylation. Science *307*, 251–254.

Tsinoremas, N.F., Ishiura, M., Kondo, T., Tanaka, K., Takahashi, H., Johnson, C.H., and Golden, S.S. (1996). A sigma factor that modifies the circadian expression of a subset of genes in cyanobacteria. EMBO J. *15*, 2488–2495.

Uzumaki, T., Fujita, M., Nakatsu, T., Hayashi, F., Shibata, H., Itoh, N., Kato, H., and Ishiura, M. (2004). Crystal structure of the C-terminal clock-oscillator domain of the cyanobacterial KaiA protein. Nat. Struct. Mol. Biol. *11*, 623–631.

Vakonakis, I., Sun, J., Wu, T., Holzenburg, A., Golden, S.S., and LiWang, A.C. (2004). NMR structure of the KaiC-interacting C-terminal domain of KaiA, a circadian clock protein: implications for KaiA-KaiC interaction. Proc. Natl. Acad. Sci. USA *101*, 1479–1484.

Williams, S.B., Vakonakis, I., Golden, S.S., and LiWang, A.C. (2002). Structure and function from the circadian clock protein KaiA of *Synechococcus elongatus*: a potential clock input mechanism. Proc. Natl. Acad. Sci. USA *99*, 15357–15362.

Xu, Y., Mori, T., and Johnson, C.H. (2000). Circadian clock-protein expression in cyanobacteria: rhythms and phase setting. EMBO J. *19*, 3349–3357.

Xu, Y., Mori, T., and Johnson, C.H. (2003). Cyanobacterial circadian clockwork: roles of KaiA, KaiB and the *kaiBC* promoter in regulating KaiC. EMBO J. 22, 2117–2126.

Xu, Y., Mori, T., Pattanayek, R., Pattanayek, S., Egli, M., and Johnson, C.H. (2004). Identification of key phosphorylation sites in the circadian clock protein KaiC by crystallographic and mutagenetic analyses. Proc. Natl. Acad. Sci. USA *101*, 13933–13938.

Ye, S., Vakonakis, I., Ioerger, T.R., LiWang, A.C., and Sacchettini, J.C. (2004). Crystal structure of circadian clock protein KaiA from *Synechococcus elongatus*. J. Biol. Chem. *279*, 20511–20518.

Young, M.W., and Kay, S.A. (2001). Time zones; A comparative genetics of circadian clocks. Nat. Rev. Genet. *2*, 702–715.

Zhang, X., Dong, G., and Golden, S.S. (2006). The pseudo-receiver domain of CikA regulates the cyanobacterial circadian input pathway. Mol. Microbiol. *60*, 658–668.

Molecular Structure of the Photosynthetic Apparatus

9

Yanina Sarah DeRuyter and Petra Fromme

Abstract

The main process of conversion of light energy from the sun into chemical energy is oxygenic photosynthesis. It is the process that provides all higher life on earth with energy. All oxygen in the atmosphere is evolved by this process, which was invented 2.8 billion years ago by the ancestors of cyanobacteria. Cyanobacteria are even nowadays very important members of the global ecosystem, and contribute up to 30% of the yearly oxygen production on earth. This chapter describes the structure and function of the protein complexes that catalyze the first steps of the energy conversion. Light is captured by antenna complexes and transferred to two large bio-solar systems, photosystem I and II, which catalyze the transmembrane charge separation. This drives the photosynthetic process and provides the energy for production of the high-energy substrate ATP and reduced hydrogen in the form of NADPH. The photosystems are functionally coupled by the cytochrome b_6f complex, the membrane-intrinsic plastoquinone pool and lumenal electron carriers. The reactions of the electron transport chain lead to an electrochemical proton gradient, which drives synthesis of ATP by the molecular motor, the ATP synthase. The structures of the complexes are described and discussed in respect to the function and evolution of the photosynthetic apparatus.

Introduction

Oxygenic photosynthesis is the main process on earth that converts solar energy into chemical energy. All higher life on earth depends on this process, since all the oxygen in the atmosphere is produced by oxygenic photosynthesis.

The reaction starts with capturing of the light by antenna systems, which transfer the excitation energy to reaction centers that, in turn, catalyze light-induced electron transfer across the thylakoid membrane. Oxygenic photosynthetic organisms such as plants, green algae and cyanobacteria contain two photosystems, photosystem I (PSI) and photosystem II (PSII), which function in series. They can be regarded as a nano-bio-solar system, as they covert light energy into electrical energy. The two reaction centers are functionally coupled by a mobile pool of plastoquinone molecules, the cytochrome b_6f complex and the lumenal soluble electron carriers—plastocyanin or cytochrome c_6. Figure 9.1 shows an overview of the electron transfer chain of oxygenic photosynthesis.

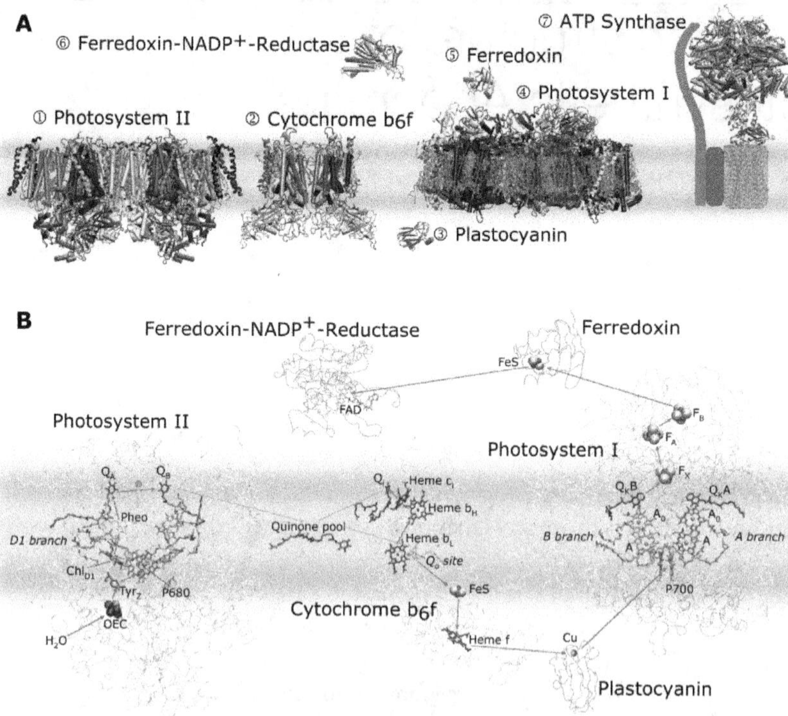

Figure 9.1 (see also Plate 9.1) Structures of the protein complexes involved in the primary reactions of oxygenic photosynthesis, with corresponding Protein Data Bank (PDB) accession codes. A, Structure of the complexes. B, Cofactors of the electron transport chain. Left to right: photosystem II (Ferreira *et al.*, 2004, 1S5L), the cytochrome b_6f complex (Kurisu *et al.*, 2003 1VF5, Stroebel *et al.*, 2003, 1Q90), photosystem I (Fromme *et al.*, 2001; Jordan *et al.*, 2001 1JB0) and the ATP synthase (Abrahams *et al.*, 1994, 1BMF; Stock *et al.*, 1999, 1QO1) as well as the soluble electron carriers plastocyanine, (Redinbo *et al.*, 1993 2PLT), cytochrome c_6 (Kerfeld *et al.*, 1995, 1CYI; Frazao *et al.*, 1995, 1CTJ), ferredoxin (Adman *et al.*, 1976, no PDB code; Morales *et al.*, 1999, 1CZP), flavodoxin (Watenpaugh *et al.*, 1972, 1FX1) and the ferredoxin: $NADP^+$ oxidoreductase (Bruns and Karplus, 1995, 1FNB).

PSII catalyzes light-driven electron transfer from water to plastoquinone (PQ). In each charge separation event, one electron is extracted from the Mn-containing oxygen evolving complex, and oxygen is evolved after 4 positive charges have been accumulated. During this process, 4 H^+ are released into the lumen of the thylakoids. The electron is transferred from P680* across the membrane by a chain of electron carriers to PQ at the Q_B-side. After 2 charge separation events, PQ is doubly reduced, takes up two protons from the stroma and leaves PSII as PQH_2. A PQ from the PQ-pool re-fills the Q_B site. PQH_2 diffuses through the membrane to the cytochrome b_6f complex, which reduces the soluble electron carrier plastocyanin (PC), a soluble protein located in the lumen of the thylakoids. In some cyanobacteria, cytochrome c_6 replaces plastocyanin. The two protons of PQH_2 are released into the lumen and two additional protons are pumped across the membrane by the b_6f complex. The reduced plastocyanin leaves the b_6f complex and docks to PSI.

PSI captures light energy by a large antenna system and transfers it to the center of the complex. When the excitation energy reaches P700, P700* is formed. An electron is ejected and transferred across the membrane to the terminal FeS cluster F_B, located at the stromal side of the PSI complex. From F_B the electron is transferred to ferredoxin. Under iron deficiency, flavodoxin can replace ferredoxin as a soluble electron carrier. The electron transfer process is completed by reduction of $P700^+$ at the lumenal side by plastocyanin/cytochrome c_6. Ferredoxin or flavodoxin finally transfers the electron to the ferredoxin: $NADP^+$ oxidoreductase (FNR), which reduces $NADP^+$ to NADPH.

The electron transfer reactions build up an electrochemical proton gradient, which is then used by ATP synthase for the synthesis of ATP from ADP and P_i. In the dark reactions, the high energy products ATP and reduced hydrogen in the form of NADPH (or ferredoxin) are used to fix CO_2 and build up glucose in the Calvin cycle, among other anabolic reactions.

Antenna complexes

Phycobilisomes

The highly efficient job of capturing light energy, for which chlorophyll molecules contained in the photosynthetic reaction centers are responsible, is in fact not enough to supply all the needed light. This is because chlorophyll (Chl) *a* does not absorb light across the entire spectrum, but has narrow absorbance bands at approximately 430–440 nm and again at 670 nm. This spectrum may be tuned to some small degree by the protein environment but, largely, leaves the organism "blind" to the rest of the visible spectrum—which it is not able to utilize. To overcome this hurdle, photosynthetic organisms develop auxiliary antenna complexes that absorb in parts of the spectrum not covered by chlorophyll absorbance. This is crucial in some environments, such as sea water, where red light is generally extinguished after about 2 m of depth, and the greatest light transmittance is at approximately 500 nm, a wavelength at which chlorophyll simply does not strongly absorb. Antenna proteins can be generally organized into two types: the water-soluble group and the membrane-bound type, which among others includes the antenna portions of PsaA and PsaB from PSI and CP43 and CP47 of PSII (which will be discussed later).

Higher plants use large protein complexes (LHCI and LHCII) that contain an array of Chl*a* and Chl*b* within a specific protein domain. Some marine organisms, such as diatoms and brown algae, utilize xanthophyll, Chl*a* and Chl*c* within a light-harvesting complex. Other algal species and cyanobacteria, however, have partially resolved this problem by utilizing the phycobilisome (PBS), a water-soluble antenna complex of fluorescent, pigmented proteins associated mainly with PSII. Depending on the light intensity and wavelength, cyanobacteria can also undergo state transitions and transfer the PBS to PSI. The PBS differs markedly in structure from other antenna-complexes, but works very efficiently to capture light in some parts of the spectrum missed by the chlorophylls contained in the photosystems. Since the PBS is located on the thylakoid membrane, on the stromal side of PSII (and PSI), it is then no great chore to pass on the light's energy to the photosystem itself.

The phycobilisome, on examination, is a very interesting structure. It consists of a base, attached to the thylakoid membrane, and several rods that protrude from it in a fan-like arrangement, giving it either a semi-elliptical or a semi-circular overall shape (see Figure 9.2). The base consists of three cylinders, which are actually each made up of two smaller cylinders stacked together—each of these being a trimer or double-stacked trimer (hexamer) of allophycocyanin (Figure 9.3). The rods feature a similar arrangement of stacked trimers or hexamers, but show much more heterogeneity. The stacked disks here may be phycoerythrin, phycocyanin or another type of protein, as discussed below, depending on the organism. Although each protein directly absorbs light, it will also either pass on or absorb energy to or from one of the other proteins. Energy transfer always follows the order of:

Phycoerythrin→Phycocyanin→Allophycocyanin→Photosystem chlorophyll

As we have already stated, the phycobilisome consists of pigmented proteins—the phycobiliproteins (PBPs). These are relatively large proteins, responsible for as much as 50% of protein mass within a cell, and containing several chromophores that both give the proteins their fluorescence and result in extremely efficient fluorochromes with an ef-

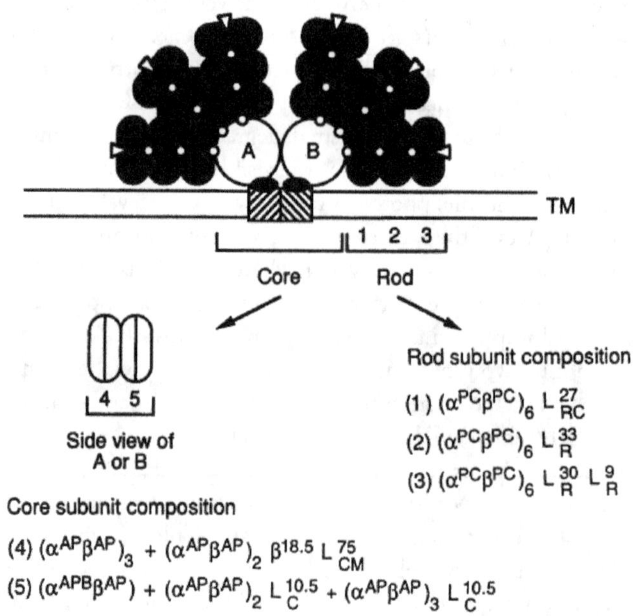

Figure 9.2 A schematic representation of the phycobilisome's fan-like arrangement of the rods, with their pigmented protein stacks, radiating from the core—composed of three short stacks of protein. Components, bottom to top—Reaction center: cross-hatched ovals within the membrane. Proteins anchoring the phycobilisome to the reaction center: small solid ovals. A, B: core subunits, with their composition detailed below the diagram. 1, 2, 3: rod hexamer subunits, with their composition detailed below the diagram. Linker peptides associated with the rod hexamers are designated "L," and shown in the diagram as open circles between the individual rod subunits. Protein composition detail—AP: allophycocyanin (Brejc *et al.*, 1995), PC (corresponds to PhyC): phycocyanin (Schirmer *et al.*, 1985). The small linker peptide found in the core is not shown here.

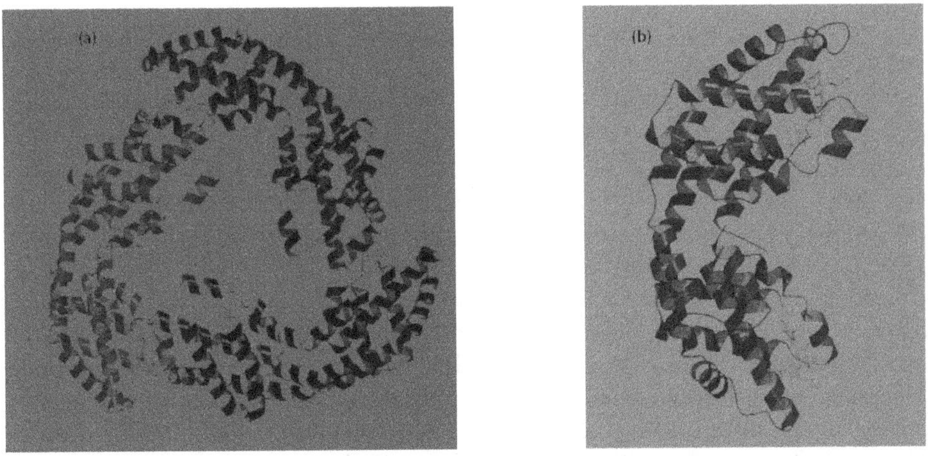

Figure 9.3 (a) A ribbon diagram of the allophycocyanin trimer from *Spirulina platensis* (Brejc *et al.*, 1995). Note the clear delineation of the three monomers. (b) Ribbon diagram of one of the monomeric units from the structure at left. Note the α-helical architecture of the protein, further dimeric nature of the subunit, and the chromophore moieties associated with each of the dimer's monomers (Cogdell and Lindsay, 2000).

ficiency of upward of 90%. In addition to these, there are non-pigmented "linker" proteins that not only help to stabilize the physical structure of the PBS, but also participate in efficient transfer of energy between PBPs and on to the photosystem (Grossman, 2003). PBPs absorb light between 450 nm and 660 nm, with each protein having its own particular absorbance spectrum due to its own chromophore. Phycoerythrin (PE), phycocyanin (PhyC) and allophycocyanin (AP) are the predominant types of PBPs in cyanobacteria, with phycoerythrocyanin (PEC) replacing PE in certain organisms (Grossman, 2003; Grossman *et al.*, 1993).

Each protein of the above four is distinctly colored, and, in fact, may contribute to the overall blue-green or red-green appearance of the organism in which it was first discovered. Frequently, the designation letter in front of the PBP's name gives a clue to its origin—"R" for "red algae," "C" for "cyanobacteria." However, it is now known that the specific PBP may not be specific to a particular organism or even taxon, and that there is a variety of spectral types of PBPs, based on the relative numbers of chromophores present in any given protein. An organism may produce the needed protein at the appropriate time in its lifetime, as, for example, during chromatic adaptation, and therefore the letter designations should really only be used to distinguish the individual PBP within its family.

For example, a particular variety of PE is bright orange-red, with orange fluorescence. It was first found in red algae, and has not been found in other taxa. It is often referred to by its full name, R-PE ("R" denoting "rhodophyta"). PE is also found in the B-PE (pink with orange fluorescence), Y-PE (pink with yellow fluorescence) and PE-566 (violet with red fluorescence) forms. Y-PE is, in fact, of cyanobacterial origin, while PE-566 is found in cryptomonads. PEC is lavender, with red fluorescence. In this spirit, C-PhyC is intensely blue, with a red fluorescence, while R-PhyC is purple-blue with a red fluorescence. There are also several forms of AP, a cyan-blue-green protein, which vary between organisms

(Glazer, 1982). Table 9.1 summarizes the physical and spectral properties of commonly found phycobiliproteins.

Chromatic adaptation itself is the response to a particular light environment, in which the organism modifies the protein content of the PBS to reflect the most efficient profile for the greatest absorbance of light energy in the particular situation. Under green light, there will be a predominance of the red-colored PE, while under red light, of the blue-colored PhyC. Only those cyanobacteria that are capable of synthesizing PE can adapt to varying light conditions (Houmard and de Marsac, 1988).

As we have already seen, PBPs are monomers within disk or flat cylinder-like trimers that can also be double-stacked to form hexamers which, in turn, are stacked like rolls of coins either within the core cylinders of the PBS or within its rods. However, on a more detailed level, structurally, each PBP is not actually a single contiguous entity. Instead, it is composed of subunits labeled α or β, with a respective mass of (generally) 15–20 kDa and 17–22 kDa. Sometimes, a third subunit is present. Each of the subunits may be associated with different chromophores, which results in the unique spectral properties of each protein. The subunits are largely α-helical structures, with folding motifs resembling that of myoglobin, but with individual folds, arrangements of helices and chromophores that are quite unique (Cogdell and Lindsay, 2000).

Table 9.1 Spectral properties and chromophores of common phycobiliproteins. Note that some PBPs have more than one type of chromophore. These may interact with each other as well as the surrounding protein, affecting each other's absorbance spectra and producing colors that will slightly differ from that of each chromophore alone. Chromophore abbreviations: PBB—phycocyanobilin, PEB—phycoerythrobilin, PUB—phycourobilin, PCV—phycobiliviolin. (*)—content varies

Phycobiliprotein	Color	Absorbance max.	Fluorescence	Chromophore
Allophycocyanin	Cyan	652 nm	660 nm	α—PCB β—PCB
Phycoerythrocyanin	Lavender	575 nm	625 nm	α—PCB β—PBV
C-phycocyanin	Blue	620 nm	647 nm	α—PCB β—2 PCB
R-phycocyanin	Purple-blue	555 nm 617 nm	637 nm	α—PCB β—PCB, PEB
R-phycoerythrin	Red-orange	492 nm 565 nm	578 nm	α—2 PEB β—2 or 3PEB, PUB γ—2PEB, 2PUB*
B-phycoerythrin	Pink	545 nm	575 nm	α—PEB × 2 β—PEB × 2 γ—3PEB, 2PUB*
Y-phycoerythrin	Pink	495 nm 545 nm	563 nm	α—PUB β—PUB
Phycoerythrin-566	Violet	566 nm	617 nm	α—PEB β—PEB

The chromophores associated with each subunit of a PBP are the actual source of its color and spectral properties. They are all linear tetrapyrroles, resembling unfolded porphyrin rings, and are in an extended conformation within the PBP, being linked to the protein via thioether linkages through one of their rings (usually A, but sometimes both A and D). This is illustrated in Table 9.2. The chromophores are isomers of each other, with varying absorbance maxima. There is some resemblance between these chromophores and the photoreceptor phytochrome of higher plants.

Table 9.2 Spectral properties and host proteins of PBP chromophores, the phycobilins (Grossman *et al.*, 1993)

Chromophore	Absorbance maximum
Phycocyanobilin 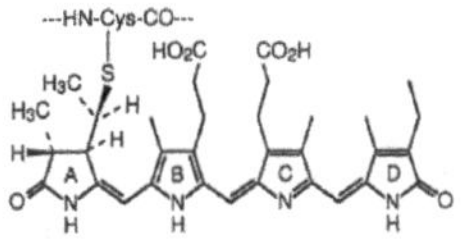 Phycocyanobilin (Amax = 590-670 nm)	590–670 nm
Phycoerythrobilin 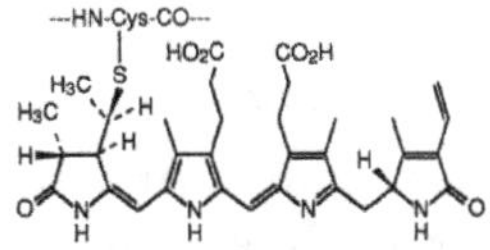 Phycoerythrobilin (Amax = 535-567 nm)	535–567 nm
Phycourobilin 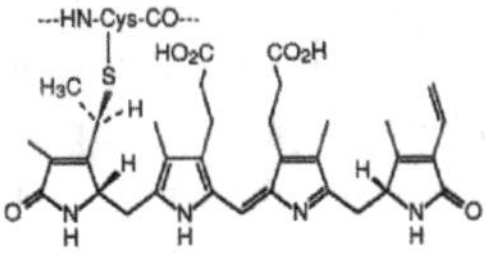 Phycourobilin (Amax = 498 nm)	498 nm
Phycobiliviolin (cryptoviolin) 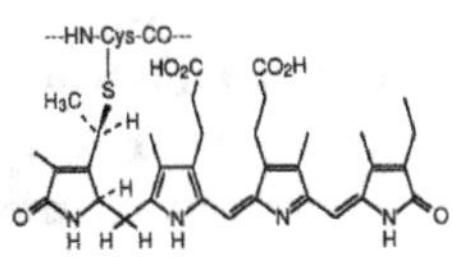 Phycobiliviolin (Amax = 568 nm)	568 nm
Bilin 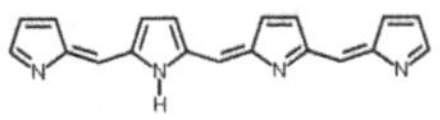	697 nm

As seen in Table 9.2, phycocyanobilin and phycoerythrobilin have a tunable range of absorbance. These are considered the two major chromophores, and one (or both) of them is always present in PBPs. In addition, one of the other three (minor) chromophores may be present.

One interesting point is the maximal absorbance of phycourobilin, at 498 nm. This is quite near the maximal transmission wavelength of seawater, 500 nm. In marine organisms, replacement of phycoerythrobilin by phycourobilin will result in a blue shift of PBS absorbance and, of course aid in optimization of light harvesting efficiency in the marine environment.

Genetically, the phycobilisome is a product of several genes. Those encoding PhyC and AP (*cpcBA* and *apcAB*, respectively) have been cloned from the alga *Cyanophora paradoxa* and several cyanobacteria such as *Synechococcus* sp. PCC 7002. The "A" and "B" in the gene name refer to the α or β subunit of the protein. PE is a product of the *cpeBA* genes, which have been isolated from the cyanobacteria *Synechocystis* sp. PCC 6701, PCC 6803 and others. In all cases, the genes for both subunits are cotranscribed (Grossman *et al.*, 1993).

Having discussed the production of the phycobilisome, we should also mention its destruction or disassembly, as a last note. This is a visually dramatic process, and leads to the ability of an organism to recycle protein components for survival during stress conditions. In fact, mutants unable to break down their PBSs during nutrient deprivation have been found to survive only in low light (Grossman *et al.*, 2001)! The PBS is disassembled in a very strict order under stress conditions such as nitrogen or sulfur starvation. Synthesis of PBPs is repressed during such times, but restored on re-addition of the limiting nutrient. Once disassembly begins, the terminal rod hexamers are degraded first, along with their associated linker proteins, until an entire rod is broken down to its constituents. An entire PBS will be degraded within 24 hours, but, once again, will be rapidly re-synthesized once favorable conditions are restored.

Pcb and *IsiA*

In addition to the membrane extrinsic phycobilisomes, cyanobacteria also have membrane intrinsic antenna complexes, which all belong to a class of proteins with 6 transmembrane helices, called the Pcb family. All proteins of this family share sequence and structural homology with the CP43 protein of PSII and, to a much smaller extent, structural similarity with the N-terminal helices of PsaA and PsaB of PSI.

Those cyanobacteria that contain phycobilisomes express the protein IsiA (which belongs to the Pcb family) under conditions of iron deficiency. *Synechocystis* sp. PCC 6803, *Synechococcus* sp. PCC 7942 and PCC 7002 as well as *Thermosynechococcus elongatus* and *Thermosynechococcus vulcanus* belong to this group of cyanobacteria. Conversely, in a second group of cyanobacteria, which do not contain phycobilisomes, Pcb proteins are the major antenna proteins of the photosynthetic apparatus, and are expressed permanently. This group includes different strains of *Prochlorococcus* and *Arcaryochloris*. Most of these strains contain also different *pcb* genes. The number varies from 1 gene for *Prochlorococcus* sp. MED4 (which lives close to the ocean surface) to as many as 8 genes for *Prochlorococcus* sp. SS120, which can be found up to 200 m deep in the water column.

IsiA

The expression of IsiA is most prominent under iron deficiency, but has also been reported under oxidative stress (Michel and Pistorius, 2004; Yousef *et al.*, 2003) and salt stress (Geiss *et al.*, 2001). The protein is also expressed in a mutant that lacks the PSI subunits PsaF/PsaJ (Jeanjean *et al.*, 2003) (which is interesting, as the docking of phycobilisomes to PSI is mediated by PsaF), and in mutants that lack the electron donor to PSI, cytochrome c_6 (Ardelean *et al.*, 2002). Two functions have been proposed for IsiA: light capturing for PSI (Melkozernov *et al.*, 2003) and prevention of photodamage (Sandstrom *et al.*, 2001). In 2001, two groups discovered that IsiA forms a ring of 18 IsiA proteins surrounding PSI (Bibby *et al.*, 2001; Boekema *et al.*, 2001) (see Figure 9.4, left panel). The IsiA protein contains Chl*a*. No other type of chlorophyll has been found in any IsiA complex. With an estimated number of 16 chlorophylls per IsiA complex, addition of the IsiA to the PSI trimer increases the cross section of Chl*a* molecules per P700 from 96 Chl/P700 in the PSI trimer to about 130 Chl*a*/P700 in the PSI-IsiA complex. Functionally, the IsiA antenna is tightly coupled to the core of PSI (Melkozernow *et al.*, 2003). The first panel of Figure 9.4 shows the structure of the PSI-IsiA complex as revealed by electron microscopy (Boekema *et al.*, 2001). No crystal structure of the PSI-IsiA complex has yet been determined.

The 18-ring surrounding one PSI trimer is not the only form in which IsiA is present in the cell. Depending on the growth conditions of the cells under iron deficiency, individual IsiA molecules can be the dominant form of the protein, with incomplete rings surrounding PSI being formed. Occasionally, even empty rings and double rings of IsiA exist (Yeremenko *et al.*, 2004). No IsiA-PSII complex has been discovered so far.

Pcb

The Pcb proteins show high sequence similarity to the IsiA protein. Each of them consists of 6 transmembrane helices. However, despite their high sequence and structural similarity, the group of Pcb proteins is diverse, as they show much variation in their pigment composition. They bind Chl*a*/*b* in *Prochloron didemni* and *Prochlorotrix* (Hiller and Larkum, 1985; Chen and Bibby, 2005), Chl*d* in *Acaryochloris* and for *Prochlorococcus marinus*, divinyl-Chl*a*/*b* has been reported (Partensky *et al.*, 1999; Helfrich *et al.*, 1999). Pcb proteins can

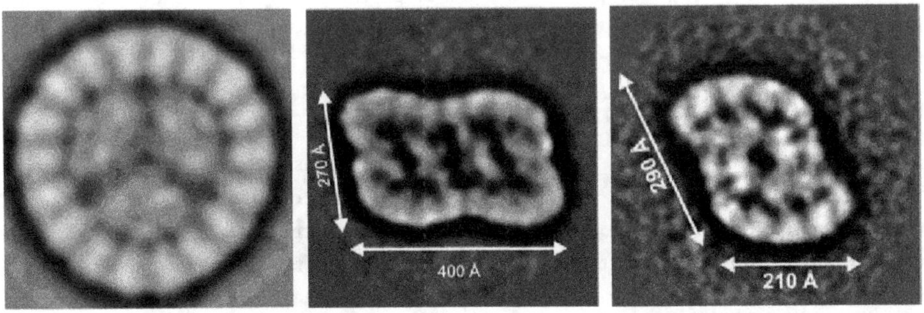

Figure 9.4 Electron micrographs of IsiA and Pcb complexes with photosystems I and II. Left, PSI IsiA complex (from Boekema *et al.*, 2001). Center and right, Pcb-PSII supercomplexes with varying numbers of Pcb proteins associated with dimeric PSII RCs (Chen and Bibby, 2005).

form 18-meric rings surrounding the PSI trimer, producing a structure with the same structural features as the IsiA-PSI complex, but varying in spectral properties according to the variation in chlorophyll molecules.

In contrast to IsiA, which binds exclusively to PSI, Pcb proteins also serve as antenna complexes for PSII. Since these bacteria lack phycobilisomes, specialized Pcb protein serve as the sole peripheral antenna for PSII. When an organism contains several different Pcb proteins, each Pcb protein binds either to PSI or PSII (Bibby *et al.*, 2003). The proteins can bind both to the monomeric and the dimeric form of PSII. 10 Pcb proteins are found in the monomeric Pcb-PSII complex, whereas 16 Pcb proteins are bound to the dimeric PSII complex (see Figure 9.4, center and right-side images) (Chen and Bibby, 2005).

Light capturing by the Pcb protein must be very efficient. One particular photosynthetic organism, *Prochlorococcus* sp. SS120, which is adapted to a very dark environment by living up to 200 m deep in the water column, contains 8 *pcb* genes!

Photosystem II: the site of water splitting

Functional overview

Light-driven electron transfer from water to plastoquinone is catalyzed by PSII. This protein complex is unique, in that it provides the whole photosynthetic process with electrons by extracting them from water:

$$H_2O \rightarrow O_2 + 4H^+ + 4e^-$$

Oxygen is evolved as a by-product and provides the basis for respiration, which is the energy source for all higher non-photosynthetic organisms, including humans. Light for the photoreactions in PSII is captured by the phycobilisomes and the core antenna of PSII, which contains 35 chlorophylls and 11 carotenoids. Excitation energy is transferred from the antenna into the center of the complex where the electron transfer chain is located. Excitation of P680, which may be one or more of the 4 chlorophylls in the reaction center (RC) core of PSII, leads to a charge separation across the photosynthetic membrane. From P680, the electron is transferred to pheophytin (Pheo) and the tightly bound plastoquinone Q_A. The mobile plastoquinone, Q_B, serves as the terminal electron acceptor in PSII. After two subsequent electron-transfer steps, the doubly reduced Q_B^{2-} binds two protons and leaves the binding pocket as PQH_2. The empty pocket is then refilled by a plastoquinone from the PQ pool.

In each round of the photocycle, $P680^+$ is re-reduced by extracting one electron from the oxygen evolving complex (OEC) consisting of 4 Mn ions and 1 Ca ion (hence referred to as the Mn cluster). A redox-active tyrosine, TyrZ, functions as the intermediate electron carrier between $P680^+$ and the Mn cluster. Once four subsequent electron transfer steps have been performed, and four positive charges are accumulated, one molecule of oxygen is released. The proton release pattern and the mechanism of catalytic reaction, including the oxidation states of the Mn cluster are a vividly discussed and still-unsolved problem.

PSII is severely prone to photodamage, mainly due to the high redox potential of $P680^+$ of 1.1 V. Very likely, the damage of the core protein of PSII, D1, is caused by side reactions, such as the direct oxidation of the protein by $P680^+$, or by the formation of the ^{3}P680 triplet and, subsequently, highly reactive singlet oxygen. The life-time of D1 depends on light intensity, and the protein must be replaced every half an hour in bright light. Both cyanobacterial PSII and plant PSII undergo D1 damage. Assuming that PSII is already 2.8 billion years old, and that cyanobacteria and plants separated 1.5 billion years ago, it is astonishing that Nature has not been able to solve the problem of photodamage in PSII, and that the time- and energy-consuming repair cycle of PSII is still the only way to deal with the problem (Baena-Gonzalez and Aro, 2002; see Chapter 11).

Structure of PSII

PSII has the same dimeric oligomeric structure in plants and cyanobacteria. It consists of 17 protein subunits, which bind 35 to 40 chlorophylls and 8 to 12 carotenoids. The thermophilic cyanobacterium *T. elongatus* was the source of the first crystals of PSII that were able to split water (Zouni *et al.*, 1998), and led to the first X-ray structural model of the intact PSII complex at 3.8 Å resolution (Zouni *et al.*, 2001). The crystals were competent in oxygen evolution (Zouni *et al.*, 2000) and their protein content has been analyzed in detail (Zouni *et al.*, 2001). This model provided the first insight into the structure of the water-splitting PSII. Recently, several structures of PSII have been published at 3.7 to 3.0 Å resolution, based on the same crystals (Fromme *et al.*, 2002; Ferreira *et al.*, 2004; Biesiadka *et al.*, 2004; Loll *et al.*, 2005). Furthermore, one new crystal structure has been determined of PSII from *T. vulcanus* at 3.7 Å resolution (Kamiya and Shen, 2003).

The protein subunits in PSII

Thermophilic PSII from *T. elongatus* consists of 17 protein subunits. These subunits are named according to their genes: PsbA to PsbO, PsbU, PsbV and PsbX. For historical reasons, the two major central subunits that form the RC core, PsbA and PsbD, are still named D1 and D2, and the large antenna proteins PsbB and PsbC are named CP43 and CP47. One monomeric unit of PSII contains fourteen membrane intrinsic subunits and three lumenal subunits, PsbO, PsbU and PsbV (cytochrome c_{550}), which do not contain transmembrane α-helices. Figure 9.5 shows the 3.0 Å structure of the dimeric PSII. A top view, with the arrangement of transmembrane helices, is shown in Figure 9.6C. The central core of each PSII monomer harbors the electron transfer chain. It consists of a cluster of 2 × 5 transmembrane helices, assigned to the protein subunits D1 (PsbA) and D2 (PsbD). The arrangement of these subunits resembles the structure of subunits L and M of the reaction center of purple bacteria (PbRC) (Deisenhofer *et al.*, 1984; Deisenhofer *et al.*, 1995) and, to a lesser extent, the structure of the C-terminal domain of the large subunits PsaA and PsaB of PSI (Jordan *et al.*, 2001). A comparison of the structure of the transmembrane helices of PSII and PSI is shown in Figures 9.6B and C. The inner core in PSII and the C-terminal domain of PSI, which form the respective RC cores, consist of 10 transmembrane helices surrounding the electron transfer chain. PSII and the PbRC (not shown here) represent the class II reaction centers (see Figure 9.15). The RC helices of the class II centers form a more open "S" shaped conformation, whereas they surround the electron transfer chain like a fence in PSI. The structural similarity between the central core of PSI and PSII is remarkable, as the sequence identity is less than 15%. The structural comparison shows that all photo-reaction centers have evolved from a common ancestor as previously proposed (Blankenship, 1992; Schubert *et al.*, 1998).

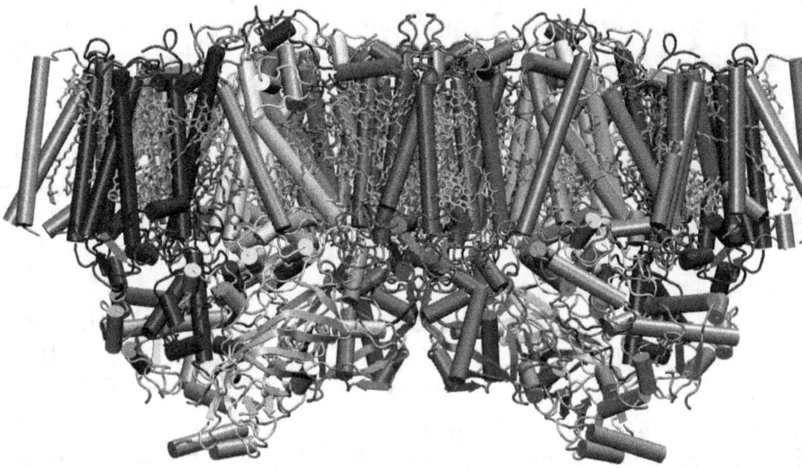

Figure 9.5 (see also Plate 9.5) Structure of photosystem II, picture prepared from PDB file 2AXT (Loll *et al.*, 2005). Yellow—D1; orange—D2; red—CP47; blue—CP43; green—PsbO; all other subunits shown in grey. Cofactors: green—chlorophyll; orange—carotenoids; grey—lipids and detergents.

The central core of D1 and D2 is flanked by the antenna proteins CP47 (PsbB) and CP43 (PsbC). Each of these subunits consists of six transmembrane helices. Comparison with the structure of PSI (Figures 9.6B and C) reveals that the arrangement of the transmembrane helices of these proteins is very similar to the arrangement of the transmembrane helices in the N-terminal part of the PsaA/PsaB in PSI.

All other small membrane intrinsic proteins are located peripheral to the central core, and show no homology between PSI and PSII, so they may have been added at a later stage of evolution. Two helices, which are located in close vicinity to helix A of D2, are assigned to the two proteins PsbE and PsbF, constituting the membrane-bound cytochrome b_{559}. The structural assignment of the small subunits is still under debate. The structural model at 3.5 Å (Ferreira *et al.*, 2004), as shown in Figure 9.6C, contains assignments of all small subunits, whereas the authors of the publication of the structural model at the higher resolution of 3.0 Å (Loll *et al.*, 2005) have raised questions about this assignment and left several helices unassigned.

PSII also contains three extrinsic subunits, located at the lumenal side of the core complex: the 33 kDa protein (PsbO), the 12 kDa protein (PsbU) and cytochrome c_{550} (PsbV). The latter two subunits are unique to cyanobacteria, whereas PsbO is also present in the PSII complex of higher plants. PsbO has a β-barrel structure with a long "lip" pointed towards the Mn cluster. PsbU forms the "tip" of the lumenal hump of PSII. PsbV contains a heme group, but the function of this heme is still unknown, as it is too far away from the electron transport chain (ETC) to be directly involved in electron transport. This subunit shows strong sequence and structural similarity to the soluble cytochrome c_6 that serves as electron donor to PSI and the cytochrome that serves as electron donor to the RC in purple bacteria. This similarity has led to the recent hypothesis that cytochrome c_{550} might represent the original electron donor to the ancient PSII RC, which could have been trapped by a mutation and bound tightly to the RC core without being able to donate

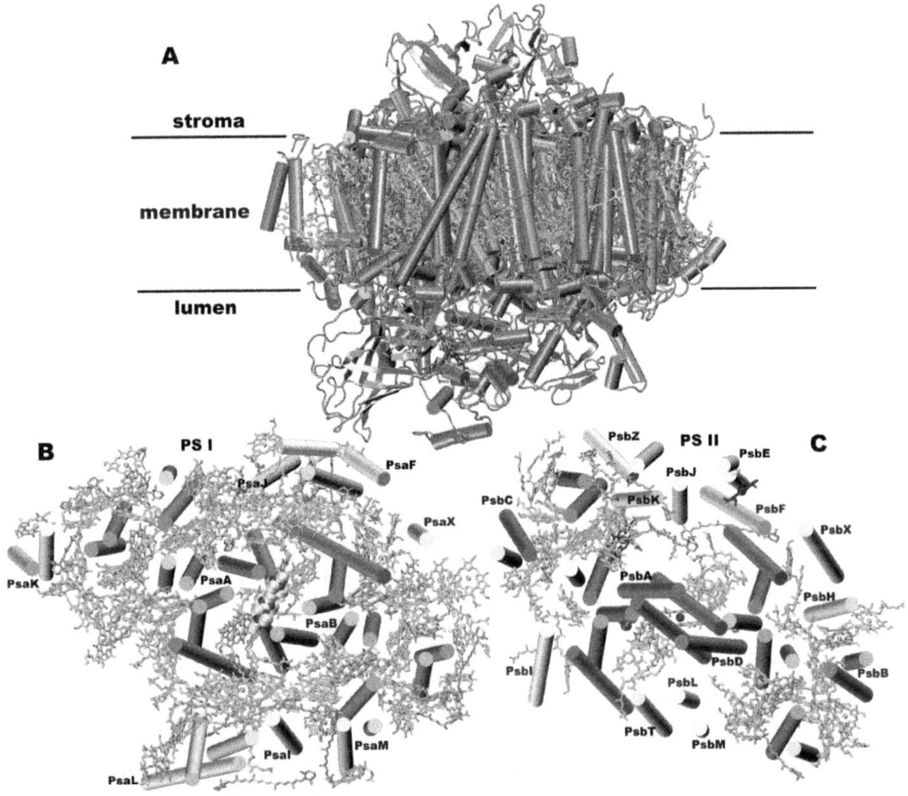

Figure 9.6 (see also Plate 9.6) Comparison of the structure of photosystems I and II. The pictures are derived from the PDB files for PSI, 1JB0 (Jordan *et al.*, 2001) and PSII, 1S6L (Ferreira *et al.*, 2004). A, view parallel to the membrane plane. PSI proteins are shown in gold, PSII in silver, chlorophylls are depicted in green and carotenoids in blue. B, view of PSI from the stromal side onto the membrane plane, the subunits are labeled according to their genes. C, view of PSII from the stromal side of the membrane, the subunits are labeled according to their genes. Note that assignment of small subunits of PSII is still under discussion (see Loll *et al.*, 2005). For clarity only transmembrane helices are shown in B and C.

electrons to the RC. This mutant, impaired by the docking of the cytochrome, might have lead to the development of the OEC. With the cytochrome kept away from P680, P680 might have reduced some metal ions in the near vicinity instead as a starting point for the development of the OEC (Grotjohann and Fromme, 2004).

The electron transport chain of PSII

The core of PSII harbors the cofactors of the ETC, which is comprised of four Chl*a* molecules, two pheophytins, two plastoquinones, two tyrosines (the redox active tyrosine TyrZ and TyrD) and the 4Mn-1Ca cluster. Figure 9.7 shows the position and orientation of the cofactors of the ETC (Ferreira *et al.*, 2004). The acceptor side of PSII, with its Pheo, quinones (Q_A and Q_B) and non-heme iron resembles the structural arrangement of the ETC in PbRC. This result is not astonishing, as the physical chemical properties of these

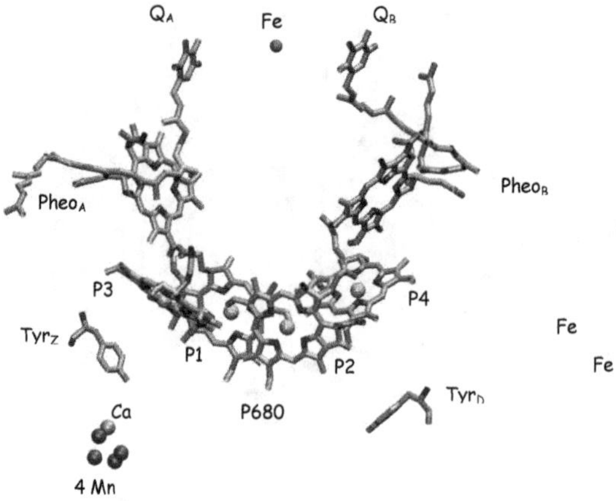

Figure 9.7 (see also Plate 9.7) Electron transfer chain of photosystem II as seen at 3.57 Å resolution, from the PDB file 1S6L (Ferreira *et al.*, 2004).

cofactors are quite similar in PSII and PbRC, and shows that PSII and the PbRC have a common ancestor. The donor site of the ETC, which differs between the PbRC and PSII will be now discussed in more detail.

The primary electron donor: P680

One of the major differences between PSII and the PbRC is the redox potential of the primary electron donors. P680 has the highest redox potential reported for any cofactor in biological system – 1.1 V. This unique redox potential is a prerequisite for water oxidation. One of the most exciting questions is which of the four chlorophylls located in the center of the D1/D2 core represent(s) P680. The chlorophylls are arranged in two symmetrically related pairs, P1/P2 and P3/P4. P1 and P2 are oriented perpendicular to the membrane plane. The center-to-center distance between the chlorophylls varies between the different structures, and is 8.2 Å in Ferreira *et al.* (2004) and 7.6 Å in Loll *et al.* (2005). The chlorophylls are more separated from each other in PSII than the chlorophylls in PSI's P700 or the PbRC's special pair, indicating that the two chlorophylls are not strongly excitonically coupled and therefore may be regarded as chlorophyll monomers. The second pair of chlorophylls (P3/P4) is located at a distance of ≈10 Å from the first pair of chlorophylls. The plane of the chlorin head group of these chlorophylls is tilted at an angle of 30° to the membrane plane.

The answer about the nature of P680 may depend on the redox state of P680, i.e. $P680^{+\bullet}$, the triplet state $^3P680$ and the excited state P680* may be located on different chlorophylls. The cationic radical $P680^{+\bullet}$ is very likely located on P1, because the distance between $P680^{+\bullet}$ and TyrZ has been determined by EPR investigations to be in the range of 7.9 ± 0.2 Å (Lakshmi *et al.*, 1999) (see also Zech *et al.*, 1997; Zech *et al.*, 1999), which fits only for P1. The excited state P680* may be delocalized—at least at room temperature—among all four chlorophylls, as none of these chlorophylls are oxidized by P1 instead of the TyrZ (redox potential 1.0 V). Thereby all chlorophylls may have a high redox potential (between 1.0 and 1.3 V), and may be able to perform the initial charge separation.

The triplet state of P680, ^{3}P680, may be localized on P3, because optical spectroscopy investigations on partially oriented samples and single crystals have provided evidence that the triplet state ^{3}P680 is located on a chlorophyll molecule with an inclination of 30° to the membrane plane (van Mieghem, 1991; Kammel *et al.*, 2003). The triplet state must be primarily formed from the state $P680^{+\bullet}$ located on P1 by intersystem crossing. Therefore, P4 can be excluded as a possible candidate, and P3 remains as the most probable molecule for the location of ^{3}P680.

What causes the very positive redox potential of $P680^{+\bullet}$? The binding pocket for P680, as for P700 in PSI, is very hydrophobic, but several partially positively charged amino acids in the vicinity of P680 may destabilize the cation and increase the redox potential as discussed in (Grotjohann and Fromme, 2004). Another cause for the high redox potential may be the close proximity of Trp191, which is π-stacked with the D2-His197 ligand for P2 (Biesiadka *et al.*, 2004; Keilty *et al.*, 2001).

The water-oxidizing complex

PSII has the unique ability to oxidize water to O_2 and 4 H^+, catalyzing the process by a cluster of four Mn ions and one Ca ion. The first structure of PSII, at 3.8 Å resolution, showed, for the first time, the location and shape of the Mn cluster (Zouni *et al.*, 2001). Four Mn atoms were fitted into a papaya-shaped structure, implying a 3+1 organization of the four Mn atoms of the cluster. The subsequent structural models, at improved resolution, confirmed this general structural arrangement (Kamiya and Shen, 2003; Ferreira *et al.*, 2004; Biesiadka *et al.*, 2004; Loll *et al.*, 2005). However, the electron density of the Mn cluster has not been improved in these structural models and, even in the structure with the highest nominal resolution, 3 Å, is still a featureless blob (Figure 9.8). This does not allow assignment of individual Mn atoms or the visualization of bridging oxygen molecules and the substrate water molecules. Therefore, the structure of the Mn cluster is still unsolved and the mechanism of water oxidation a matter of lively discussion.

Ca^{2+} and Cl^- also play an important role in the process of water splitting. Location of the Ca^{2+} was identified for the first time in the structure of PSII at 3.5 Å resolution (Ferreira *et al.*, 2004). The Ca^{2+} is clustered together with the three Mn atoms, thereby

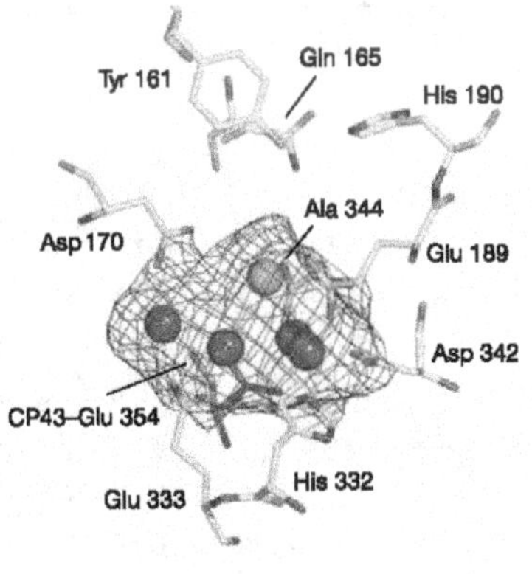

Figure 9.8 (see also Plate 9.8) Electron density of the Mn cluster at 3 Å resolution, from PDB file 2AXT (Loll *et al.*, 2005).

possibly forming a distorted cubane structure consisting of three Mn and one Ca ion—with the fourth Mn being more distal to the distorted cube. This model of the 4Mn-1Ca cluster is in agreement with EPR and XAFS data. The ligands to the Mn cluster are still under discussion, because they differ slightly in the different X-ray structures and also do not agree in all points with mutagenesis studies. Asp170, Glu189, His332, Glu333, Asp342 and Ala344 of D1 and Glu354 of CP43 have been proposed as ligands for the Mn cluster (Ferreira *et al.*, 2004; Loll *et al.*, 2005). However, as all current structures are only at medium resolution and suffer from X-ray damage of the Mn cluster during data collection, the structures differ significantly in the distance and side chain conformation of the potential ligands.

The final answers to the multitude of questions and uncertainties may be provided in the future by an improved X-ray structure in combination with spectroscopic and mutagenesis studies.

The antenna of PSII

A total of 42 cofactors had been identified at 3.8 Å resolution: 26 antenna chlorophylls bound to CP47/43, six chlorophylls coordinated by D1/D2 as well as the Mn cluster, one non-heme iron, located at the D1/D2 interface on the stromal side of the membrane and two heme groups belonging to cytochrome b_{559} and cytochrome c_{550}, respectively. In the most recent structure of PSII (Loll *et al.*, 2005), 35 chlorophylls, 11 β-carotene molecules and 14 lipids were identified. All X-ray structural models show very similar locations for the chlorophylls.

Figures 9.6B and C compares the chlorophylls assigned in the structural model at 3.6 Å with the chlorophyll arrangement in PSI. The striking difference between the antenna system in PSI and PSII is the lack of the central antenna domain in PSII. Whereas more than 50 chlorophylls surround the reaction center domain carrying the electron transport chain in PSI, the central core in PSII contains only two chlorophylls ($Chl_{Z(D1)}$ and $Chl_{Z(D2)}$). This lack of a central antenna domain may be responsible for the lower efficiency of the excitation energy transfer in PSII compared to PSI. The absence of the central antenna domain may be the price PSII has to pay for its ability to use water as an unlimited electron source, which leads to the process of photodamage and repair of the D1 protein. This process may have hindered the fusion of the RC domain and the antenna domain in PSII, which is still at an older evolutionary stage than PSI with regard to the antenna function.

Cytochrome b_6f complex: coupling of electron- and proton-transfer

Between PSII and PSI, also in the thylakoid membrane, lies another large membrane-bound protein complex—the cytochrome b_6f complex. "Between" does not necessarily imply a physical location, although it is located within proximity to both PSI and PSII. What it does imply, however, is the cytochrome b_6fs role as the functional intermediary between the two photosystems. That is, on its way through the photosynthetic pathway, the electron extracted from water by the water-splitting complex of PSII is transferred by a mobile plastoquinone from PSII to cytochrome b_6f. It is then moved internally through the cofactors of the cytochrome b_6f and, finally, handed off by the cytochrome *f* subunit of the cytochrome b_6f complex to one of the lumenal electron carrier proteins, such as PC or cytochrome c_6, which carries the electron to PSI. It is in this way that the complex fulfills its role as a redox protein. Further, this complex pumps protons across the membrane, thereby contributing to the formation of the electrochemical gradient that drives ATP synthesis.

Historically, cytochrome b_6f has been of intense interest; not only as one of the chief participants in photosynthesis, or a model for study of oxidoreductase proteins—but also as a good example of a ubiquitous binding site, in terms of the cytochrome b_6f's ability to interact with both PC and cytochrome c_6 on the lumenal side (Schlarb-Ridley *et al.*, 2002). As far back as 1981, Hurt and Hauska discussed a basic characterization of its polypeptide architecture (Hurt and Hauska, 1981). A breakthrough in the knowledge about the cytochrome b_6f complex was achieved in 2003, when two groups independently determined the crystal structures of the cytochrome b_6f complex from the cyanobacterium *Mastigocladus laminosus* (Kurisu *et al.*, 2003) and the green algae *Chlamydomonas rheinhardtii* (Stroebel *et al.*, 2003). The structure determination of the cytochrome b_6f complex completed the trio of high-resolution membrane-integral photosynthetic protein structures available at the time, which also included PSI and PSII.

Like its photosystem partners, the cytochrome b_6f is a multiprotein complex. The native dimer, as seen in Figure 9.9, has a molecular weight of 217 kDa. The cytochrome b_6f complex is a large structure, measuring 90 × 55 Å on the stromal side and 120 × 75 Å on the lumenal side. The transmembrane domains plus the extrinsic protein together extend 100 Å in "height"—i.e. perpendicular to the membrane plane (Kurisu *et al.*, 2003). The deep brown color of the complex is mostly due to the pigments bound within the complex.

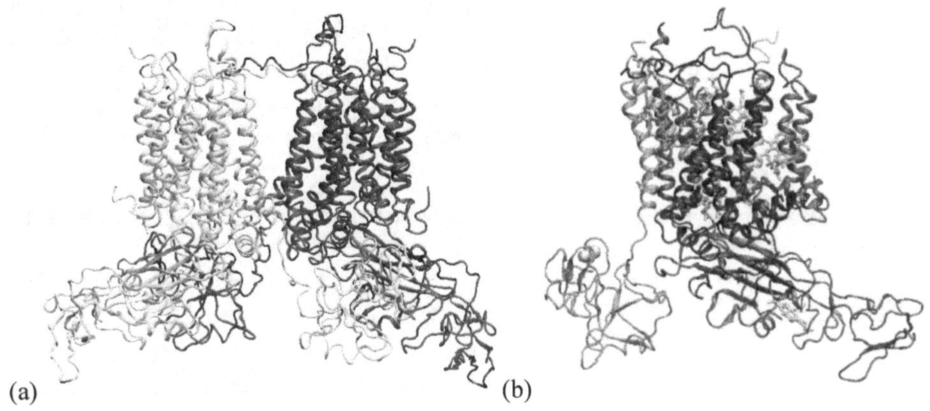

Figure 9.9 (see also Plate 9.8) (a) The dimeric nature of the cytochrome b_6f complex. Note that the Rieske protein is actually located on the opposite side of the axis of symmetry from the rest of its partners. This is a case of "domain swapping," and is thought to contribute to stability of the dimer. (b) The multisubunit nature of a cytochrome b_6f monomer, along with the cofactors belonging to each subunit. The monomer is rotated slightly with relation to its purple analogue in picture (a), to better show the complex assembly of subunits and cofactors. Subunits, coded by color—Blue: cytochome b_6. Red: subunit IV. Gray: cytochrome *f*. Orange: Rieske iron–sulfur protein. Yellow: PetL. Tan: PetM. Silver: PetG. Green: PetN. Cofactors, coded by color—Chla: green. Carotenoid: orange. 2Fe-2S cluster: yellow and green spheres. Lipids: cyan. Hemes: yellow. The structure given here is drawn from the RCSB/PDB coordinate file 1VF5, for the cytochrome b_6f from *Mastigocladus laminosus* (Kurisu *et al.*, 2003).

In general, the architecture of each of the two constituent monomers further consists of eight subunits and several cofactors ligated by the protein. Four of these are considered the "large" subunits, and include cytochrome *f*, the Rieske iron–sulfur protein (ISP), cytochrome b_6 and the somewhat prosaically named subunit IV (SubIV). The remaining component proteins are often grouped as the four small, hydrophobic subunits and include PetG, PetL, PetM and PetN. Subunits and cofactors are briefly summarized in Table 9.3. We will discuss the details of each as they pertain to the cytochrome b_6f's structure.

In order to better understand the complex's structure, it is necessary to discuss briefly the steps in the process this enzyme catalyzes. For clarity, we will defer to convention in terminology—since this protein participates in movement of charge across a membrane, some publications (Kurisu *et al.*, 2003) refer to the stromal side of the membrane as the *n*-side (negative), where proton uptake occurs, and to the lumenal side, where proton extrusion takes place, as the *p*-side (positive). The molecule that transfers electrons and protons from PSII to the cytochrome b_6f complex is a mobile PQ molecule in its doubly reduced form with two protons added—plastoquinol or PQH_2. However, unlike in the photosystem proteins, the two electrons do not take the same path. The first electron moves through the *high-potential* transfer chain. ISP protein and cytochrome *f*, which are both part of the cytochrome b_6f complex and are located on the lumenal side, mediate this transfer, which results in extrusion of two protons to the lumen. The second electron is transferred by two *b*-type hemes to the stromal side of the membrane. Alternatively, it may also be moved on the anionic plastosemiquinol if cyclic electron transfer is taking place. In either case, an extra proton is drawn from the stromal side and either released to the lumen as in the first step, or taken up by the plastoquinol. This "Q-cycle" is structurally facilitated by the presence of two large central cavities, one each on the stromal and lumenal sides of the complex, which will be further discussed later, and may be compared to the electron transfer cycle in cytochrome bc_1. However, although the functional and structural similarity exists, cytochrome b_6f and the cytochrome bc_1 complex are only distantly related in terms of structure—the iron–sulfur proteins are homologous, and cytochrome b_6f's

Table 9.3 A summary of cytochrome b_6f subunits and their cofactors. Data from PDB file 1VF5

Segment	Identity	Mass (kDa)	Cofactor association
A	Cytochrome b_6	24.7	2 *b*-type hemes 1 c_i-type heme (heme "x") Chl*a* tail
B	Subunit IV	17.5	Chl*a* head
C	Cytochrome *f*	32.3	*c*-type heme
D	Rieske Fe-S protein	19.3	2Fe-2S cluster
E	Protein PetL	3.8	Carotenoid
F	Protein PetM	3.5	Carotenoid
G	Protein PetG	4	Carotenoid.
H	Protein PetN	3.3	

cytochrome b_6 and SubIV are reminiscent of the N- and C-terminal parts of cytochrome bc_1's cytochrome *b* subunit. While cytochrome *f* and cytochrome c_1 share a common function, they are not structurally related. There is also no structural homology for the small subunits PetG, PetL, PetM and PetN.

Cytochrome *f*

With an apparent molecular mass of 32 kDa, cytochrome *f* is the largest of the 4 "large" subunits. However, it also has the largest membrane-extrinsic domain. This is seen in Figure 9.9 as the large moiety, projecting into the lumen. Largely consisting of β-sheets, this domain contains the *c*-type heme, which is covalently bound within the protein. The heme itself is ligated by the first 25 residues—the N-terminus—of the protein. The ligating sequence also includes the signature heme-binding peptide, Cys-X-X-Cys-His, of *c*-type cytochromes. In the cyanobacterium *Phormidium laminosum* these are residues 21–25. His25 ligates the heme iron atom, which is also coordinated by the side-chain amino group of Tyr1 (Kurisu *et al.*, 2003). This ligation by the N-terminal residue is an unusual situation for *c*-type cytochromes. The protein contains only one transmembrane helix. The C-terminal domain thereby acts as the membrane anchor of cytochrome *f*. Thus, it can be said that the large extrinsic domain of cytochrome *f* is flanked on the C-terminal end by the membrane anchoring helix and on the N-terminal end by the heme-binding motif.

The cyanobacterial cytochrome *f* bears a high sequence identity (62% in the case of *P. laminosum* and turnip cytochrome *f*) to plant cytochrome *f*, but there are other similarities that bear mention. The ligands to a chain of highly ordered, hydrogen-bonded water molecules (Kurisu *et al.*, 2003) are extremely well conserved between cyanobacterial cytochrome *f* and plants. In fact, 7 of the 10 residues (His25, Ala25, Val60 and 7 other nearby residues) associated with the water chain are completely invariant in the 22 known cytochrome *f* sequences.

Since a buried group of water molecules is, indeed, a rarity in soluble proteins (mostly due to the high entropic cost associated with such an arrangement), it is worthwhile to consider the possible purpose of this structure. Since structural stability in a soluble domain is usually achieved through de-solvation of hydrophobic side-chains, the waters are unlikely to serve a "stabilizing" purpose. Therefore, their presence must serve some functional purpose. It is possible that the water chain might serve as a "guide-wire" for protons exiting the complex. Alternatively, the water may act as a "dielectric well," binding a proton upon reduction of the heme and thereby stabilizing the system electrostatically. Recently, it has been shown that site-directed mutation of residues forming hydrogen bonds with the water molecules resulted in impaired electron transfer function *in vivo*, which underscores the importance of the waters' presence (Kurisu *et al.*, 2003). Thus, the waters' function is not fully known at this time, and remains an interesting question.

In general, the structure of cytochrome *f* is very well conserved. Of the 249 residues of the extrinsic domain, 63 are invariant—and most of these are involved to some degree with the heme. In fact, the very positive midpoint potential of cytochrome *f* (290–360 mV) is thought to be a result of the heme's protein environment. Another invariant cluster is associated with the water chain, providing evidence that the water molecules are somehow critical to the cytochrome *f*'s function. A third invariant cluster is actually near the transmembrane helix, which suggests that there may be some highly conserved form of interaction between cytochrome *f* and other proteins of the complex. A surprising observation may be made regarding the interior of the large membrane-extrinsic domain—although it is highly conserved across species in terms of its hydrophobic profile, the similarity ends there, and sequence similarity is not as pronounced as it is in the more invariant regions. Within the dimer, although the two cytochrome *f* subunits do not come into direct contact with each other, they form a bowl-like structure. This has been suggested to function as a shield for movement of the two Rieske proteins (see below), so that there are no interfering interactions within the limited space (Kurisu *et al.*, 2003).

Cytochrome b_6

An interesting property of this 24.7 kDa subunit is that it is actually the gene product of two genes—*petB* and *petD*. PetB is therefore the N-terminal part of the protein, while PetD is the C-terminal part. The two parts are analogous to the cytochrome *b* component of cytochrome bc_1. Cytochrome b_6 non-covalently binds two *b*-type hemes, which are often referred to by their absorbance maxima—b_{562} and b_{565}. Here, we will refer to them as b_n and b_p, with the subscript referring to whether the particular heme is closer to the *n* (stromal) or *p* (lumenal) side of the membrane. The two are also sometimes called b_H (H = high potential) and b_L (L = low potential), respectively. The hemes are axially ligated by two His residues each, with the four histidines being highly conserved. Cytochrome b_6, unlike other large subunits, is almost entirely membrane-integral, with four trans-membrane helices (labeled A-D). The helices are not strictly perpendicular to the membrane, but are tilted. This produces a cleft or groove in the membrane surface, which is thought to aid in transfer of a plastoquinone between the two PQ binding sites.

Heme "x": heme c_i

Perhaps one of the most interesting recent findings about the cytochrome b_6f complex is the identification of heme c_i (also called heme "*x*"). It is located near the stromal side of the membrane, near the location of proton uptake and plastoquinone reduction. Its binding site, Q_i, opens out into the hydrophobic groove in the dimer interface. Heme c_i itself appears to screen heme b_H from the surface. The heme is covalently bonded to the protein through a single thioether bond between one of its vinyl groups and a cytochrome b_6 Cys35 sulfur, which distinguishes it as a *c*-type heme. However, so far no common consensus sequence such as CXnCH or AXnCH has been found, and Cys35 is not part of any signature sequence.

As originally thought, this heme has proved to be unusual in a number of ways. Its center iron is not directly axially ligated by any residue on one side, but is instead thought to be coordinated by either a water or a hydroxide moiety. The other side of the ring appears to "line" the active site pocket, and no axial ligand is visible. This combination of a weak fifth axial ligand and missing sixth ligand produces a high-spin heme. One face of the ring plane is close to the edge of both heme b_p and helices A and D, while the other is in contact with both the central cavity and helix E of SubIV.

It is not clear what purpose heme c_i may serve. It does not appear to participate in the Q cycle. Cyanobacteria predominantly use the cyclic pathway at physiological CO_2 concentrations or low light levels. It is possible, therefore, that heme c_i is involved with such a process. This is supported by the fact that this heme not only blocks PQ access to heme

b_H, but is itself accessible from the stroma (direct access to Q_i is only prevented by Asp35 of SubIV and Arg95 of PetN). It may further be postulated that the positive potential on the stromal side of cytochrome b_6f would be extremely favorable to docking anionic ferredoxin on this side of the complex near heme c_i (Kurisu *et al.*, 2003).

Rieske iron–sulfur protein

The high-potential Rieske protein is a product of the *petC* gene, and has a mass of 19.3 kDa. Most notably, it is the site of residence for the 2Fe-2S cluster, bound within the signature sequence Cys-X-His-$(X)_{15-17}$-Cys-X-X-His. Within this sequence, one of the iron atoms is coordinated by two Cys residues, while the other is coordinated by the two histidines through the N^{δ} atoms. The domain binding the cluster is reminiscent of an incomplete β-barrel. However, it is also remarkable for both its location and its conformation. Unlike the rest of the monomer to which it belongs, it is located diametrically across the axis of symmetry from the rest of the monomer's subunits, and the FeS cluster is part of the ET chain of the neighboring monomer. This is thought to confer a particular stability on the structural dimer. Further, its prominent, luminally positioned, membrane-extrinsic 2Fe-2S binding domain makes up the C-terminal two-thirds of the protein. The remaining N-terminal one-third consists of a 22 amino acid hydrophobic domain followed by an 18 amino acid glycine-rich domain. In cyanobacteria, the two domains are separated by a Lys residue (Kurisu *et al.*, 2003). An interesting feature is the mobility of the soluble domain, which may be able to swing, as if attached to its transmembrane anchor by a "hinge," bridging the gap between heme b_p (sometimes called heme b_L) and cytochrome f during the process of electron transfer. The idea for this movement came from the bc_1 complex, where the related FeS protein can be found in different positions within the complex. However, this has not been experimentally shown for the cytochrome b_6f complex.

Subunit IV

This subunit, although it is the least massive of the four "large" subunits, is also almost entirely membrane-integral, with 3 transmembrane helices (labeled E-G). For some time, this 19.3 kDa subunit was notable as being the only one of the "large" subunits not to contain any cofactors. However, this proved to be a misconception (Kurisu *et al.*, 2003). Instead, the structures have shown that the Chl*a* molecule associated with cytochrome b_6f is ligated by SubIV via the chlorophyll's tetrapyrrole ring. The ring is located in the center of the membrane, between helices F and G of SubIV, but it is not exactly clear what acts as the axial ligand. Weak electron density points to the possibility of a carbonyl group of Gly136 in SubIV being a ligand, but the distance is too great for direct interaction with the chlorophyll. It is possible, therefore, that a water molecule acts as the axial ligand. The phytol chain projects towards helix C of cytochrome b_6, and has been suggested as the reason for Chl*a* co-purifying with cytochrome b_6 due to hydrophobic interactions of the phytol tail with the protein.

Several studies have been conducted in an attempt to determine or, at least, clarify the role of the chlorophyll within a non-light-harvesting complex. One possibility was that the chlorophyll acted as a sort of sensor, connecting an interacting partner with one of plastoquinone binding sites, but other hypotheses cannot, at the present, be ruled out. One of the biggest questions is the potential ability of an excited triplet-state chlorophyll molecule to produce singlet oxygen and, therefore, generate toxic oxygen species. Normally, the presence of a chlorophyll would be mitigated by a nearby carotenoid located within 4 Å of the chlorophyll and acting as an energy sink. However, the single carotenoid molecule found in cytochrome b_6f is very likely too far to function in this role. In an attempt to explain this phenomenon, a recent study suggested that here, the particular arrangement of local protein structure affords specific protection by shortening the excited state lifetime of the chlorophyll by a factor of 20–25. The study further postulated that the excited state lifetime was shortened due to an electron exchange transfer with a nearby tyrosine (Kurisu *et al.*, 2003).

Internal cavity

An extremely important feature of the cytochrome b_6f dimer is the so-called central "quinone binding cavity." It is in this cavity that lipophilic quinone pool plastoquinones may dock to the cytochrome b_6f complex. There are two of these cavities, located on each side of the trans-membrane interface and measuring 30 × 15 × 25 Å ($h \times d \times w$). The 25-Å width measurement is taken at the base of the n-side interface. Both monomers contribute to the construction of the cavities, with helices C, D and F of one monomer combining with helices A and E and the ISP helix of the other monomer. The N-terminus of cytochrome b_6 forms the floor along with space-filling lipid head groups. Short peripheral helices connecting helices C and D of cytochrome b_6 and the C-terminus of the ISP transmembrane helix form the roof. Plastoquinone reduction occurs on the stromal side of the complex. Helices C and F form a small 11 × 12 Å "portal" in the wall of each cavity, leading to a lumenal side pocket (the Q_p pocket) where plastoquinone oxidation takes place. Plastoquinones may pass through the portal at a rate of about 100 per second.

The four small subunits and their cofactors

These four small subunits, PetG, PetL, PetM and PetN, are unique to the cytochrome b_6f complex, and find no homologue in the complex's closest structural relative, cytochrome bc_1. However, each possesses a single helix, labeled after the specific subunit, and takes part in interaction with cofactors and, possibly, other subunit proteins. For example, while the PetM subunit does not participate in electron transfer directly, mutants lacking this subunit demonstrate negative effects on the levels of other energy-transduction protein complexes (Kurisu *et al.*, 2003). An interesting idea is that certain specific features of cytochrome b_6f, such as the PC-binding site of cytochrome f, the chlorophyll, the carotenoid and the bundle of small subunit helices all appear to be facing the same direction, out of the complex. This opens the possibility that they are all intended to interact with some other participants in the photosynthetic chain. However, at this point this has yet to be ascertained.

It is between the transmembrane helices of PetL and PetM, also coming into contact with the B, E and PetG helices, that the cytochrome b_6f carotenoid—9-cis β-carotene—is found. One of the cyclohexene rings is deeply imbedded near heme c_i and reaches towards helix A of cytochrome b_6, while the other is presumed to be somewhere near the protein-lipid interface. The remainder of the carotenoid is positioned at an oblique angle to both the plane of the membrane and the ring plane of the chlorophyll. While the function of the carotenoid, like that of the chlorophyll, is not known at the present, it has been suggested that the carotenoid might also play the role of a sensor between some interacting partner and one of the quinone binding sites. Another possibility, although a much less exciting one, is that it is simply there to fill structure gaps the way bound lipids often do in membrane protein environments. Three endogenous lipids are, in fact, present in the cytochrome b_6f—one, a sulphoquinovosyldiacylglycerol, is positioned stromally, so that its sulphonate group interacts with Lys272 on the C-terminal end of cytochrome *f*s transmembrane helix. Two others, likely monogalactosydildiacylglycerol, are located on the lumenal side. Two well-defined acyl chains are imbedded in the protein and interact with the carotenoid in one case, and with the chlorophyll in the other. Other acyl chains are located at the protein-lipid interface.

Plastocyanin and cytochrome c_6: lumenal soluble electron carriers

As we have seen, electrons generated by photoexcitation pass from PSII to the cytochrome b_6f complex via the PQ pool in the thylakoid membrane. However, when leaving the cytochrome b_6f complex, the electron temporarily leaves the membrane environment, and travels through the lumen via another type of carrier. That is, it is shuttled from the membrane-bound cytochrome b_6f (cytochrome *f*) complex to PSI by either PC or cytochrome c_6 (see Figure 9.10). These are small, soluble electron carrier proteins present on the lumenal side of the thylakoid membrane, which participate in the movement of electrons in the photosynthetic ETC.

Having received an electron from cytochrome *f*, the reduced redox carrier proteins freely diffuse on the lumenal side of the thylakoid membrane. However, within the limited domain that it traverses between cytochrome *f* and PSI, the carrier does not travel randomly. Its goal, the binding site on PSI, actually guides its approach through electrostatic or hydrophobic interaction. In fact, the nature of the interaction varies between species (see Table 9.4). In some cyanobacteria, such as *Phormidium laminosum*, the protein is mostly guided into place and "docked" using a hydrophobic relationship with PSI. In the same bacterium, once the protein is docked, a re-arrangement or "locking" occurs prior

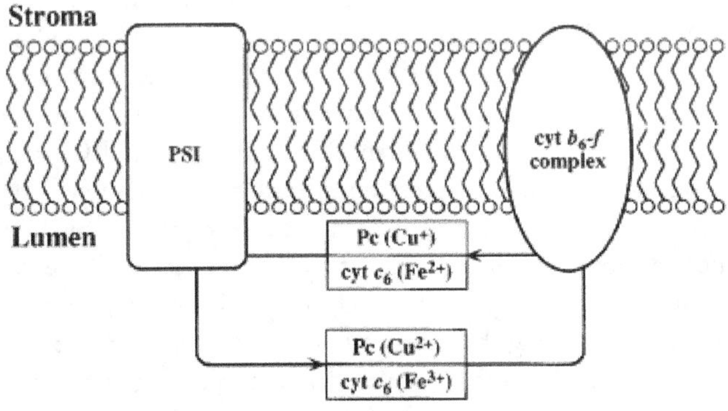

Figure 9.10 A scheme of electron transfer between cytochrome b_6f and photosystem I (Navarro *et al.*, 1997).

Table 9.4 Isoelectric points and character of interaction with PSI, by organism. Character of interaction with PSI (cytochrome c_6/plastocyanin): +, attractive. –, repulsive. 0, neutral (de la Rosa *et al.*, 2002)

	Isoelectric point (pI)		
Organism	Cytochrome c_6	Plastocyanin	Interaction with PSI
Cyanobacteria			
Prochlorothrix hollandica	~8	8.4	+/0
Synechocystis sp. PCC 6803	5.6	5.5	–/–
Phormidium laminosum	5.1	5.0	–/–
Anabaena sp. PCC 7119	9.0	8.8	+/+
Pseudanabaena sp. PCC 6903	8.0	8.3	+/+
Green Alga			
Monoraphidium braunii	3.6	3.7	+/+
Plant			
Spinacea oleracea	n/a	4.2	+

to transfer of the electron. In other cyanobacteria, the docking might be guided more by electrostatics; in which case the interaction would be more *attractive*. In any case, once the carrier finds and docks in a groove on the lumenal side of PSI, it donates its electron to the photo-oxidized chlorophyll dimer, $P700^+$. Having completed the "hand-off" of the electron, the carrier dissociates from PSI and diffuses back into the lumen, and the cycle starts anew (Kurisu *et al.*, 2003).

While higher plants and many cyanobacteria utilize solely PC, in some green algae and cyanobacteria copper-containing PC may be functionally substituted by the heme-bearing cytochrome c_6. In cyanobacteria capable of utilizing either PC or cytochrome c_6 as the lumenal shuttle, it has been observed that PC expression increases with increasing levels of copper in the environment, while cytochrome c_6 levels concurrently decrease (Merchant and Bogorad, 1986; Varley *et al.*, 1995). Some cyanobacteria, such as *T. elongatus*, contain only cytochrome c_6 as the lumenal electron carrier, as they lack the gene for PC. In natural environments, PC appears to be the default carrier for green algae. However within and among cyanobacteria, regulation of these two proteins differs between species. Both *Synechocystis* sp. PCC 6803 and *Anabaena* sp. PCC 7937 control PC production at the transcriptional level—in *Anabaena*, expression of both the PC and cytochrome c_6 genes is controlled via transcription initiation with at least one required *trans*-acting factor, but there are no identified copper-dependent regulatory proteins or target sites for these (Bovy *et al.*, 1992). *P. laminosum* also appears to produce cytochrome c_6 in response to copper deficiency. It is worth mentioning that these two redox carriers may not be alone in their task—a third possible carrier has been postulated. Although the subject is still sporadically investigated, certain studies show that suppression of PC synthesis by copper depletion at the same time as suppressing cytochrome c_6 via gene deletion has had no effect on the progress of photosynthetic processes. This raises the question of how the electron might be transferred without the mediation of either PC or cytochrome c_6! A protein named cytochrome c_{6a} and found also in the green alga *Chlamydomonas reinhardtii*, may serve this function. This protein resembles cytochrome c_6, but it still remains somewhat of a mystery whether it can replace cytochrome c_6 in its function (Wastl *et al.*, 2004).

Although PC and cytochrome c_6 look strikingly different (see Figures 9.11 and 9.12), their function is the same—this suggests that they may possess some common elements. In fact, a property very important to the ability to dock correctly is the electrostatic profile of each protein—and in this respect, the two are very similar. If the two proteins (from the same organism) are aligned according to their dipole moments, then regions of similar surface potential become apparent (Frazao *et al.*, 1995). This leads to the two being acidic in green algae (just like PC in higher plants), but the pair may be basic, neutral or acidic in different cyanobacteria. The pI points of the two are similar within the organism, but also vary—in parallel—between species as described in Table 9.4 (Hervas *et al.*, 1995)!

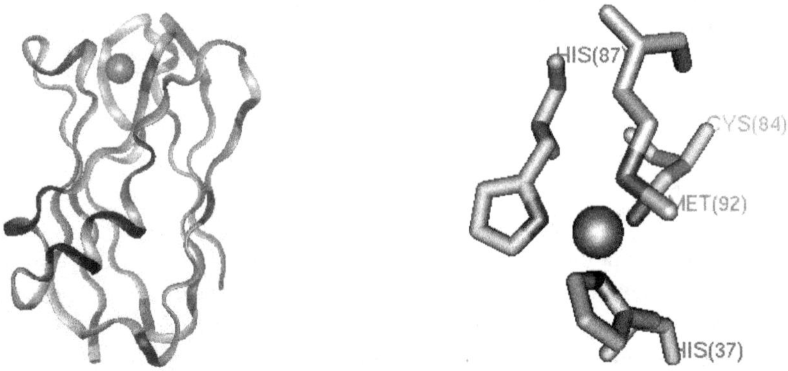

Figure 9.11 Left: ribbon model of plastocyanin from *Phormidium laminosum* (PDB database accession code 1BAW) (Bond *et al.*, 1999). Right: copper coordination site from spinach plastocyanin (PDB code 1AG6) (Xue *et al.*, 1998).

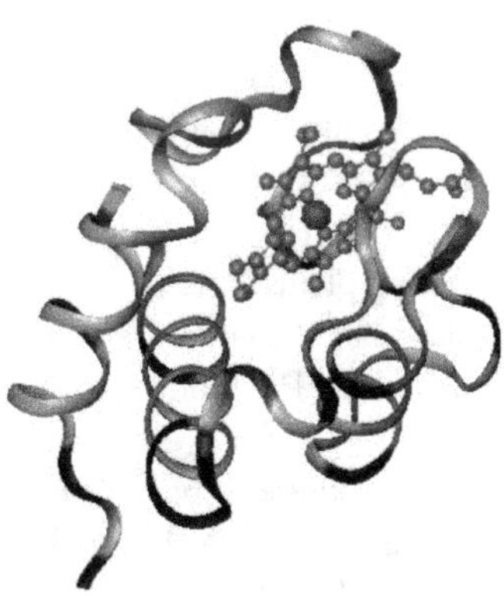

Figure 9.12 Cytochrome c_6 from *Anabaena variabilis* (PDB code 1CTJ) (Frazao *et al.*, 1995). Note the drastic difference in structure from plastocyanin, given in Figure 9.11.

Plastocyanin

PC belongs to a family of blue copper proteins collectively known as the "cupredoxins." This is a reference to the similarity of their role to another electron transfer family of proteins—the iron-containing ferredoxins. The structure of PC is highly conserved among both higher plants and cyanobacteria. The protein has a mass of 11.3 kDa, with a pI that varies between and within both plants and cyanobacteria. In plants, it varies from 4.0 to 4.2. In cyanobacteria, the variability is much greater, from a moderate 5.2 in *P. laminosum* to 8.4 in *Anabaena variabilis* (Schlarb-Ridley *et al.*, 2002).

PC is a Type I cupredoxin, or electron-carrying, redox-active blue copper protein, and possesses a single copper ion (in a distorted trigonal planar environment with a weak fourth axial ligand) at its "northern" end. The metal is located near, but beneath, the surface of the protein and is the moiety responsible for the actual reduced or oxidized state, with a mid-point redox potential of approximately 322 mV.

Type I classification refers to the spectroscopic characteristics of this group of proteins, which show unusual features in their EPR spectra. Their high g and low A values suggest that the unpaired electron in the complex is much more delocalized than in other Cu(II) complexes. The delocalization extends over the sulfur atom of a Cys, to which the Cu is covalently bonded.

The "blue" characterization originates from the fact that the protein possesses an intense absorption band in the visible spectrum, at $\gamma_{max} \approx 600$, due to a (Cys)$S\gamma\pi \rightarrow$Cu(II)(d_{x2-y2}) ligand-to-metal-charge-transfer (LMCT). Presence of the copper ion confers this deep blue color on the protein in its oxidized state.

The copper-coordinating Cys is part of another distinctive feature of these proteins—the strong coordination of the Cu(II) ion by the Nδs of two nearby histidines and a Sγ of a cysteine, forming a trigonal plane (Figure 9.11, right panel). This is referred to as the N_2S plane. The Cu-S_{Cys} bond distance is longer than that of the Cu-Ns, at approximately 2.15 Å, but this is still shorter than that commonly seen in coordination compounds (Harding, 2001).

PC's surface near the Cu site shows a hydrophobic patch with some positive potential. This potential is likely due mostly to the oxidized Cu ion, since, when the Cu(II) is reduced, the potential essentially disappears (Durell *et al.*, 1990). It has been suggested that this patch is the site of the PC's interaction with its redox partners. Charge distribution over the protein's surface varies considerably between species. In higher plants, a negatively charged (acidic) patch is found on the remote face, around Tyr83 (plant sequence numbering). In cyanobacteria, negative and positive charge (acidic and basic residues, respectively) is distributed more evenly (Schlarb *et al.*, 1999).

Although modeling studies using plant proteins have suggested residues that might be involved in docking and interaction of PC with cytochrome b_6f, work is still in progress to determine the nature of cyanobacterial PC binding to cytochrome b_6f and PSI. In particular, two Asp residues at positions 44 and 45 (according to numbering from the plant protein) have been implicated as being quite important to proper complex formation. It is suggested that, in cyanobacteria, PsaA and PsaB are chiefly responsible for providing a shallow groove into which PC may dock (Schubert *et al.*, 1997). Electron transfer to PC from cytochrome b_6f is believed to occur through His92 (87 in corresponding plant structure numbering), one of the ligands coordinating the Cu atom.

Cytochrome c_6

In organisms capable of surviving in copper-restricted environments, the task of shuttling electrons from the cytochrome b_6f complex to PSI falls to another small, water-soluble protein. This is cytochrome c_6—a type I, *c*-type (heme-containing) cytochrome, which is produced by eukaryotic algae like *Monoraphidium braunii* and some cyanobacteria, such as *Anabaena* sp. PCC 7119 and *Synechocystis* sp. PCC 6803. This particular type of cytochrome is characterized by the axial ligands of the iron within the heme ring—a His and a Met. In fact, in some cyanobacterial species, it is the sole luminal electron carrier. As previously mentioned, in some cyanobacteria, such as *T. elongatus*, the gene for PC is lacking. Thus, these organisms are restricted to using only cytochrome c_6. The kinetics of the carriers' interaction with their partners are the same for the functionally interchangeable proteins within the same species (Frazao *et al.*, 1995). It has been suggested that cytochrome c_6 may be the more ancient electron donor, and that PC may have evolved in a convergent evolution after the earth's atmosphere became oxygenic and Fe was depleted from the ocean (de la Rosa *et al.*, 2002).

Cytochrome c_6 contains 83–90 residues, and has a mass of about 9.3–10 kDa. Its α-band exhibits a maximum peak at 553 nm, and its redox potential is between 335 and 390 mV. Complex filamentous cyanobacteria such as *A. variabilis* have a cytochrome c_6 with a pI as high as 9.3. However, unicellular cyanobacterial species have pI values ranging anywhere from 3.8 to 5.2. Of course, as already mentioned, PC and cytochrome c_6 from the same organisms will have very close pI values, since they will have to interact with the same membrane proteins. It has been observed, subsequently, that these changes in pI values between species have been accompanied by complementary changes in charged residues in the region of cytochrome *f* where redox carriers are purported to dock. It is interesting to speculate whether PSI might have undergone similar alterations in its region for redox carrier docking as the cytochrome b_6f complex!

Although cytochrome c_6 is functionally interchangeable with PC, structurally it differs markedly (see Figures 9.11 and 9.12). Wrapped around a partially exposed heme group, the protein is folded into four helices and three loops. The iron atom in the heme is axially ligated by a histidine and a methionine. The heme itself, of which only 6% is exposed to the solvent, resides in a hydrophobic pocket and is bordered by 23 amino acids. Of the 23, 13 are conserved or very conservatively replaced in other cytochromes of the same type. Of these 13, 4 are surface residues, and are invariant—Phe11, Trp64, Tyr79 and Trp89.

Photosystem I: transmembrane electron transfer to ferredoxin

PSI, like PSII and cytochrome b_6f, is a membrane-bound participant in the photosynthetic chain. As we have already seen, it is to the lumenal face of this photosystem that PC and cytochrome c_6 dock. PSI is a light-driven oxidoreductase—after it oxidizes the lumenal electron carrier by accepting an electron, it in turn reduces ferredoxin on the stromal side (which we will discuss in more detail later) by passing the electron to it in the normal sequence of events (Fromme *et al.*, 2001). That particular process supplies electrons for subsequent reduction of $NADP^+$ to NADPH by the FNR (Fromme *et al.*, 2001). It also serves generation of a proton electrochemical gradient across the thylakoid membrane, which ultimately allows synthesis of ATP from ADP and P_i. PSI possesses several features, both structural and functional, that make it an extremely interesting subject of study.

Indeed, as we shall see, there are still ongoing debates regarding some of its functional properties—even though it has been the subject of exhaustive study for many years, since its discovery 45 years ago (Duysens *et al.*, 1961)!

Structural overview

The first crystals of cyanobacterial PSI were obtained from *P. laminosum* as far back as 1987 by the group of Robert Ford (Ford *et al.*, 1987), but no crystal structure was determined from this organism. Instead, all structural information about cyanobacterial Photosystem I is based on crystals from the thermophilic cyanobacterium *T. elongatus*. This organism was originally isolated from hot springs in Japan, and has been renamed several times. The first crystal structure was determined in 1993 (Krauss *et al.*, 1993), when the organism was named *Synechococcus* sp. An improved crystal structure was published at 4 Å resolution in 1996 (Krauss *et al.*, 1996) followed by another structure, at atomic resolution—2.5 Å—in 2001 (Jordan *et al.*, 2001) (see Figures 9.6A and B and Figures 9.13, 9.14, 9.16 and 9.17). At that time, the cyanobacterium was named *Synechococcus elongatus*. In 2002, after the complete genome of this organism was sequenced (Nakamura *et al.*, 2002), the cyanobacterium was again renamed and is now called *Thermosynechococcus elongatus*.

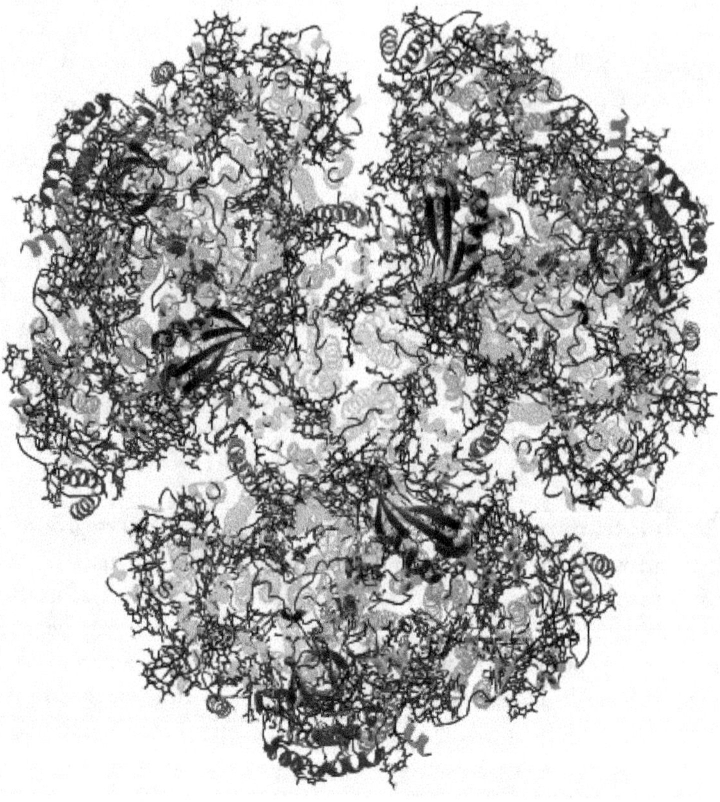

Figure 9.13 (see also Plate 9.13) PSI trimer from the cyanobacterium *Thermosynechococcus elongatus*, seen from the stromal side onto the membrane plane, PDB 1JB0 (Jordan *et al.*, 2001).

One of the properties that will astonish any casual observer, of course, is the immense size of PSI. As the largest membrane protein ever crystallized in its entirety, the PSI trimer weighs in at a heavy 356000 Da per monomer in the naturally existing trimer. This puts the total mass of the complex, including all inorganic and organic co-factors, at more than 1 million Daltons! When viewed at a normal to the membrane plane, from the lumenal side, PSI presents an impressive, attractive sight, and looks much like a three-leaf clover—as seen in Figure 9.13. Each of the ellipsoidal monomers measures approximately 130 Å long and 90 Å wide at its widest point. Measured in thickness through the membrane plane, the trans-membrane plus extramembranous regions total about 90 Å. While the stromal-side extramembranous subunits contribute about 35 Å to this thickness, the lumenal-side projecting regions are not as pronounced as on the stromal side, extending 15 Å into the thylakoid lumen. The remaining 40 Å belongs to the transmembrane region. A model of one of the three monomers of PSI is given in Figure 9.14.

PSI is not alone in the family of photosynthetic oxidoreductase proteins. The protein sequences of subunits analogous to similar structures in other organisms appear to be well conserved. In fact, the homodimeric reaction centers from *Heliobacterium mobilis* and *Chlorobium* have some sequence similarity to PSI and the hydrophobicity plot shows similarity to PsaB and (to a lesser extent) to PsaA from PSI from cyanobacteria, which shows that all class I reactions centers may have been evolved from a common ancestor (Blankenship, 1992; Blankenship and Hartman, 1998; Schubert *et al.*, 1998). Further, all cyanobacterial and plant PSI sequences of PsaA and PsaB show a high degree of identity. For example, the PsaB subunit from *T. elongatus* PSI bears an approximately 86% sequence identity to that of *P. laminosum* (Hart, 2005). This certainly bespeaks a high degree of conservation within the PSI structure. The sequence homology is reflected in the structural

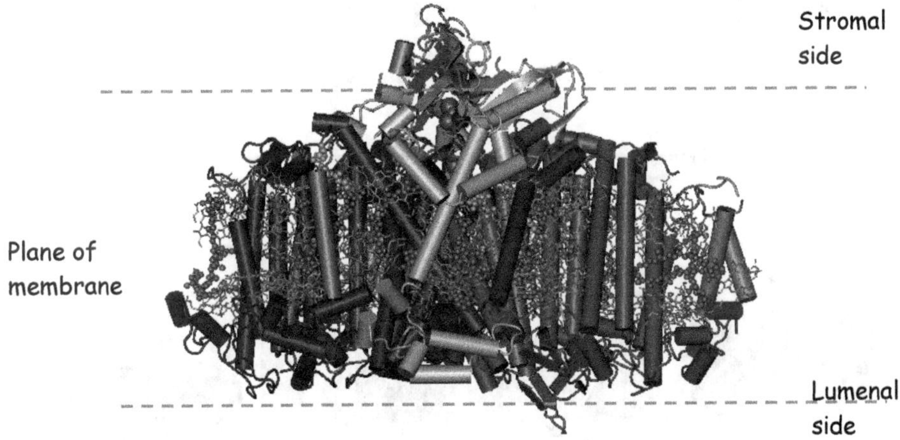

Figure 9.14 (see also Plate 9.14) A model of photosystem I from *Thermosynechococcus elongatus*, drawn from the coordinates of the structure deposited with the Brookhaven Protein Data Bank under accession code 1JB0 (Jordan *et al.*, 2001). Ribbons color-coded by subunits. PsaA—red, PsaB—blue, PsaC—green, PsaD—yellow, PsaE—cyan, PsaF—gray, PsaI—mauve, PsaJ—ochre, PsaK—pink, PsaL—periwinkle, PsaM—tan, PsaX—orange. Co-factors: chlorophyll—green, carotenoids—orange bead chains, phylloquinones—yellow bead chains, lipids—gray bead chains, iron–sulfur clusters—large magenta spheres.

similarity, as the core structure of plant PSI (Ben-Shem *et al.*, 2003) is very similar to the structure of cyanobacterial PSI (Jordan *et al.*, 2001; Jolley *et al.*, 2005).

Reaction center types may be subdivided by type—class I and class II. The particular type is determined by the nature of the terminal electron receptors. This is graphically illustrated in Figure 9.15. As we have already discussed, PSII is a class II RC, since it utilizes plastoquinones in this role. However, class I RCs employ iron–sulfur clusters to receive electrons from the transfer chain cofactors. Having 3 4Fe-4S clusters therefore designates PSI as a class I reaction center.

A product of the *psa* gene family, PSI consists of 11 well-characterized, large and medium subunits plus an additional small, less well-elucidated protein. This extra 12th protein is known as PsaX and has been identified in two thermophilic cyanobacteria by N-terminal sequencing, but its 10-kDa sequence is not yet known (Ikeuchi *et al.*, 1991). Nine of the subunits, including PsaX, are transmembrane, while three are extra-membrane and reside on the stromal side of the complex. It should be recalled that plant PSI contains 14 subunits. However, while plants have the additional subunits PsaG, PsaH, PsaN and PsaO (Knoetzel *et al.*, 2002), they do not possess PsaM.

In addition, there are 127 cofactors per single monomer of PSI that participate in tasks varying from electron transfer to photo-protection. These contribute approximately 30% of the total mass of PSI. We will group the protein subunits and the cofactors by location and function, and discuss the significance of each.

PsaA and PsaB

These are the largest subunits of the RC, with a mass of 83175 Da for PsaA and 82905 Da for PsaB, and a respective sequence length of 755 and 740 residues. The two subunits together make up the core of the RC, and, because they are similar but not identical, render the RC a *heterodimer*. It should be remembered that the entire functional PSI is

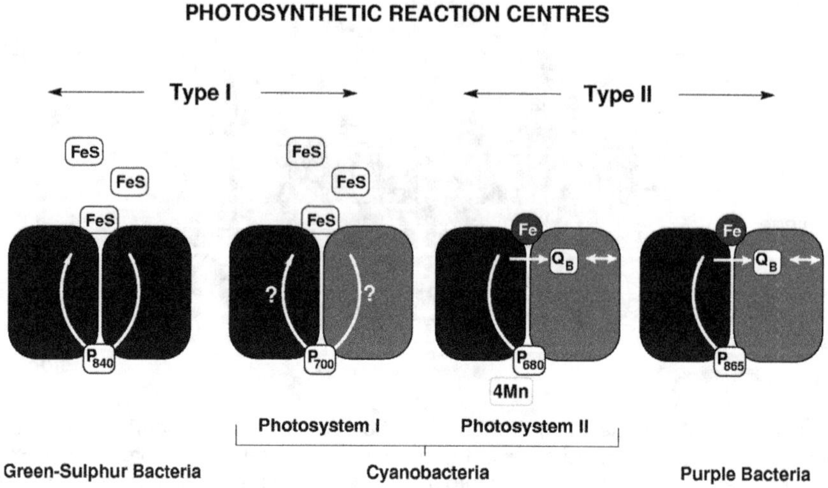

Figure 9.15 A schematic representation of reaction center types, designated according to the nature of the terminal electron acceptor. Figure kind courtesy of Dr. A. Van der Est, Brock University, Canada.

usually found in the cyanobacterial thylakoid membrane as a *trimer* of these individual *heterodimeric* RCs. The PSI core is not only a reaction center—it is a photosystem. A photosystem contains a reaction center core, which harbors the ETC and a core antenna system. In PSI, the core subunits PsaA and PsaB are joint antenna-RC proteins. Both subunits lie immersed within the membrane, with loops and helix-end regions protruding into the stromal or lumenal space. The loop regions are quite long and are important, as they make up some 30% of the PsaA and PsaB mass.

The two subunits each consist of 11 transmembrane helices (6, a–f, in the N-terminal antenna domain and 5, g–k, in the C-terminal RC domain). The helices of the C-terminal domain form a motif resembling two inter-clasped hands or intertwined semi-circles if viewed from above. Helices e and f form a bridge through a cylindrically distorted antenna and RC domain region. At the interface of the two domains lies a large portion of the antenna cofactors, which we will address separately. The 5 helices labeled g to k are closely associated with the ETC. A notable difference between the subunits exists on the lumenal side, where PsaA loops are more pronounced than the loops of PsaB. In fact, it is these two subunits that are implicated in exclusively binding the lumenal electron carrier proteins in a docking site located in a cleft made by PsaA and PsaB on the flat under-surface of PSI. Unlike in plants, in which docking of plastocyanin is aided by PsaF as well as PsaA and PsaB, deletion of PsaF has no influence on the binding character of plastocyanin in cyanobacteria. Therefore, it is only the cleft made up of the large core subunits that provides a binding site. Within this cleft, helices i and j of each of A and B form the base of the "hollow." The loop connecting helices j and k contains a helix that may provide the docking site for lumenal electron carriers. Two conserved Trp residues, A655 and B631, are located here. Their indole rings participate in a π-stacking arrangement whose orientation, pointing into the supposed binding site, may be implicated in either direct interaction with or even electron transfer from the donating lumenal carrier, such as plastocyanin, to the accepting oxidized $P700^{+\bullet}$ dimer—the "keystone" of the ETC.

On the lumenal side of the RC core, several short loops, helices and sheets not similar in A and B combine to produce a slightly corrugated, relatively flat and planar surface that shields the ETC and other cofactors from the aqueous phase of the thylakoid lumen.

Perhaps the greatest significance of the core subunits is the housing of the electron transfer chain itself, which we will discuss below. It should be noted, however, that the first FeS cluster, a nearly cube-shaped 4Fe-4S cluster termed F_x, is "clasped" by the loops connecting helices h and i of PsaA and PsaB. These loops contain the PCDGPGRGGTCD motif, which is the most highly conserved sequence in PSI, and it includes the cysteine residues that directly coordinate the Fe atoms of the cluster.

Curiously, one of the major differences between PSII and PSI lies here, in these two large core proteins, and the way they ligate the antenna cofactors. While the PSII RC subunits D1 and D2 only ligate 2 antenna chlorophylls, the C-terminal domain of the PSI core subunits coordinates approximately half of all antenna chlorophylls, as seen in Figure 9.6B. In addition, the 12 (collective) N-terminal helices, a to f, of PsaA and PsaB coordinate 28 and 26 antenna chlorophylls, respectively, while the C-terminal domains, encompassing helices g through k of PsaA and PsaB, not only enfold the ETC cofactors, but are also involved in respectively coordinating 12 and 13 antenna chlorophylls. However, an interesting fact is that of these 79 antenna chlorophylls (Chl*a*), only 48 are ligated by helical regions. 30 are coordinated by loop region side-chains, and one is even ligated by the phosphodiester group of lipid III (a negatively charged phospholipid).

A final note should be made regarding the lipids present in PSI. All 4 of these are located within the PsaA/PsaB core, with one placed at the interface with PsaX: lipids I, III and IV are negatively charged phospholipids, and lipid II is a neutral galactolipid. The head groups of all four reside closer to the stromal side of the complex, and the two fatty acid chains of each intertwine between transmembrane helices, stretching toward the middle of the membrane.

PsaC, PsaD, and PsaE

As may be seen in Figure 9.6B and Figure 9.14, the entire arrangement of "clasping hands" constructed of PsaA and PsaB, containing the ETC and cradling the first of the 3 Fe-S clusters, Fe_X, is "capped" by 3 stromal subunits. The PsaC protein is an 80-residue, 8652 Da subunit that resides atop the PsaA/PsaB heterodimer and, together with the larger PsaD (138 aa, 15223 Da) and smaller PsaE (75 aa, 8240 Da) polypeptides, shields the exposed loop and helical regions of PsaA and PsaB.

PsaC's function is to house the remaining 2 of the 3 Fe-S clusters—F_A (ligated by Cys21, 38, 51 and 54) and F_B (ligated by Cys11, 14, 17 and 58), since it contains the well-conserved sequence CXXCXXCXXXCP, repeated twice. Due to sequence similarity and the presence of the two iron–sulfur clusters, PsaC resembles bacterial ferredoxin with its two 4Fe-4S clusters—but varies from it in the N- and C-terminal structure as well as a 10-residue extension. The C-terminus of PsaC interacts with both PsaA/PsaB and PsaD, and is crucial for proper complex assembly (Jolley *et al.*, 2006). At the base of the three subunits, atop the loops belonging to PsaA, is the suggested ferredoxin-docking site.

The resemblance of PsaC to 2[4Fe-4S] ferredoxin is strengthened if we consider the fact that PsaC, PsaD and PsaE are implicated in ferredoxin docking. PsaD consists of a large antiparralel 4-stranded β-sheet connected by a short loop to the only helix of the subunit, and followed by a 2-stranded sheet. This motif is followed by a clamp, constructed of residues D95-D123, which drape over PsaC like a "clamshell," and exert a stabilizing influence required for correct orientation of PsaC. Residues D124 to 129 interact with PsaB through extensive H-bonding. The C-terminus of PsaD is stabilized by close interaction with the N-terminus of nearby PsaL.

Although we can define the roles of PsaC and PsaD in a rather direct manner, PsaE presents a challenge. A rather small subunit, it is nevertheless significant in cyclic electron transport as well as interaction with ferredoxin. This compact protein takes the shape of a 5-stranded, antiparallel β-sheet, with the loop between strands 3 and 4 being quite flexible. This loop appears to change its conformation considerably during PSI assembly, as it is involved in interactions with both PsaA/PsaB and PsaC. On the other hand, the loop between PsaE's strands 2 and 3 extends toward the suggested ferredoxin-docking site. PsaE also interacts with PsaF, as well as with the C-terminus of PsaD.

PsaI, PsaL, and PsaM

These three small subunits, with masses of 4.3 kDa, 16.6 kDa and 3.4 kDa respectively, belong to a group of small structural subunits located on the periphery of the reaction center core and surrounding the PsaA/PsaB heterodimer. Located deep inside, around the axis of the PSI trimer, far from the membrane interface, they act to stabilize the trimeric structure of the complex, as well as house some of the antenna complex chlorophylls and several carotenoids. The chlorophyll molecules, Chl*a*, are ligated via their central Mg^{2+} ions either directly by protein sidechains or, within the protein environment of individual subunits, by present water molecules.

PsaI, although very small, is crucial for stabilization of PsaL, and is the only one of the three that does not coordinate any chlorophyll molecule. However, it does interact with the adjacent PSI monomer and provides hydrophobic stabilizing contacts with nearby carotenoid molecules.

PsaL is a structurally very important subunit, since it participates in trimerization of the entire PSI functional trimer and forms most of the contacts between the monomers. The subunit contains a stromal N-terminal loop with 3 β-strands and 1 α-helix. 3 transmembrane helices are connected by a short loop on the lumenal side between helices 1 and 2, and another on the stromal side between 2 and 3. The last small helix becomes the lumenal C-terminus. A Ca^{2+} ion is ligated by this protein, with the participation of PsaA, between adjacent monomers. It has been suggested that the Ca^{2+} is necessary for proper trimer formation. Like PsaI, PsaL provides hydrophobic contacts for carotenoids, but also ligates 3 antenna Chl*a* molecules. The formation of a trimer appears to provide extended light harvesting properties—it has been shown that significant transfer of excitation energy takes place between monomeric units (Sener *et al.*, 2004).

PsaM contains only one transmembrane helix with its N-terminus in the lumen and C-terminus in the stroma. It is the smallest subunit of PSI, and exists only in cyanobacteria. The 2.5 Å structure finally established its location as the interface between monomers, near PsaI and PsaB. For such a tiny subunit, however, it is rather a busy one—it coordinates one of the antenna Chl*a* molecules as well as makes hydrophobic contacts with a nearby carotenoid. In addition, the chlorophyll coordinated by PsaM is important for excitation energy transfer between monomers, and is functionally closely coupled to the neighboring monomer.

PsaF, PsaJ, and PsaK

Unlike PsaI, M and L, which are located at the monomer-monomer interface, PsaF, PsaJ and PsaK (respective masses 15 kDa, 4.4 kDa and 8.5 kDa) lie distally to the trimeric axis—and are exposed to external interactions such as detergents and the membrane environment.

PsaF contains an N-terminal lumenal domain, a single transmembrane helix and two small hydrophobic helices. The second of these, termed F-h, enters the stromal side of the membrane only to end after the first 1/3 of the membrane's thickness, and continue as the last of the three short helices, F-i, which re-emerges on the stromal side and is followed by the C-terminus, which interacts with PsaE. PsaF does not ligate any antenna Chl*a* Mg^{2+}, but does form several contacts with carotenoid molecules and the phytyl tails of chlorophylls. Unlike plant PsaF, cyanobacterial PsaF is not implicated in interactions with lumenal carrier proteins like PC, but may instead interact with external antenna proteins—the above-discussed phycobilisomes.

Located close to PsaF and originating with a stromal-side N-terminus, PsaJ consists of a single transmembrane helix and ends with its C-terminus in the lumen. In spite of its small size, this simple subunit provides ligands for 3 antenna Chl*a* molecules—as well as some contacts for carotenoids.

Somewhat larger than PsaJ, the PsaK subunit is constructed of 2 transmembrane helices and is located at the extreme periphery of the PSI monomer, in close proximity to PsaA, as seen in Figure 9.14 and Figure 9.6B. This contributes to its being the least ordered of the PSI subunits in the 2.5 Å X-ray structure—in fact, its original structure was modeled with alanines. The structural model of PsaK relied on the fact that helix-connecting loops, if they carry a strong positive charge, will not be moved through the hydrophobic membrane environment during insertion and assembly. Therefore, the loop connecting the two helices was assigned to the stromal side. PsaK is not near and does not make any contacts with any other of the non-core subunits, and, in fact, PsaA is its only neighbor. However, PsaK still coordinates 2 antenna Chl*a*s and provides contacts for carotenoids.

PsaX

Its name befits this small (3.8 kDa), somewhat mysterious peripheral subunit, which is located near PsaF and has been identified by its N-terminal sequence in *S. vulcanus* and *A. variabilis*. Unfortunately, no gene for this protein has yet been sequenced. What is known is that it contains 35 residues. Six are located stromally, in a very flexible region, at the N-terminus, and are followed by a single transmembrane helix. It is not certain whether this particular subunit is a peculiarity of thermophilic organisms, such as those in which it has been identified.

Cofactors

Having described the structure and function of the individual protein subunits of PSI, we can now examine the cofactors that are the "powerhouse" and the "engine" of the photosystem. These are the antenna, composed of 90 antenna chlorophylls, 22 carotenoids, 4 lipids and the ETC consisting of 6 Chl*a* molecules, 2 phylloquinones and 3 4Fe-4S clusters. The 4 lipids belonging to each of the PSI monomers have already been discussed along with PsaA and PsaB. In addition, there are several water molecules within the protein's structure—some of these are responsible for coordinating antenna Chl*a* molecules through their central Mg^{2+} ions. We will confine our discussion to the physical and molecular structure of the co-factor arrangement, as photosystem function will be explored in more detail elsewhere.

Electron transfer chain

The ETC is the functional heart of PSI. The ETC is located at the very center of the reaction center core, at the "interlocked" region of subunits PsaA and PsaB. Actually, the ETC is bifurcated—and consists of two diametrically opposed branches, labeled A and B—but each of the two branches includes co-factors from each of PsaA and PsaB, as there is a crossing-over of the branches at ecA1 and ecB1 (shown in Figure 9.16). The electron proceeds up one of the branches, from P700 to the uppermost Fe-S cluster, F_B. Much experimentation has recently been devoted to the identity of the branch preferred for conduction of the electron (Guergova-Kuras *et al.*, 2001; Cohen *et al.*, 2004; Ramesh *et al.*, 2004; Li *et al.*, 2006). However, the current prevailing opinion is that although both branches are active, the kinetics of branch B are 10 times faster—at least in green algae.

The use of the two branches may differ between organisms. Another point worth making is regarding the conservation of residues ligating the ETC's cofactors. All residues involved in coordinating the Mg^{2+} of ETC chlorophylls or in H-bonding or otherwise ligating other co-factors are very strictly conserved between species—from cyanobacteria to higher plants. This is strong evidence for the essential nature of the specific residues, and their role in enhancing the specific reduction potentials within the ETC.

At the base of the ETC, we find P700. This is a pseudo-heterodimer of Chl*a* on the B side, and its C13′ epimer, Chl*a*′ on the A side, whose overlapping rings are arranged in parallel (with apparent π-π stacking), and whose combined plane is perpendicular to the membrane as seen in Figure 9.16. The two chlorophylls are ligated by His B660 and His A680, respectively, and are separated by 6.3 Å between their Mg^{2+} ions. It is here that charge separation takes place.

Above the P700 dimer is a second pair of Chl*a* molecules, commonly referred to as A. These are separated by 12 Å center-to-center, and are the "odd" cofactors of the ETC, whose ligating subunit is PsaA on the B side, and PsaB on the A side. The actual coordination of the Mg^{2+} ion, in each case, is provided by a water molecule H-bonded to an Asn on each side (A branch—Asn B591, B branch—Asn A604). It is presumed that this pair is directly implicated in electron transfer from P700 to the third pair of Chl*a* (A_0).

The third pair of chlorophylls in the ETC, also known as A_0, is considered to be the primary electron acceptor and is located fairly close to the previous pair, about 3.8 Å above. A particularly unique property of this pair, and one that moves against Hard/Soft Acid/Base theory, is that the central Mg^{2+} ions of both chlorophylls are axially ligated by an uncommon donor—the sulfur atoms of nearby Met residues, A688 and B668. It is

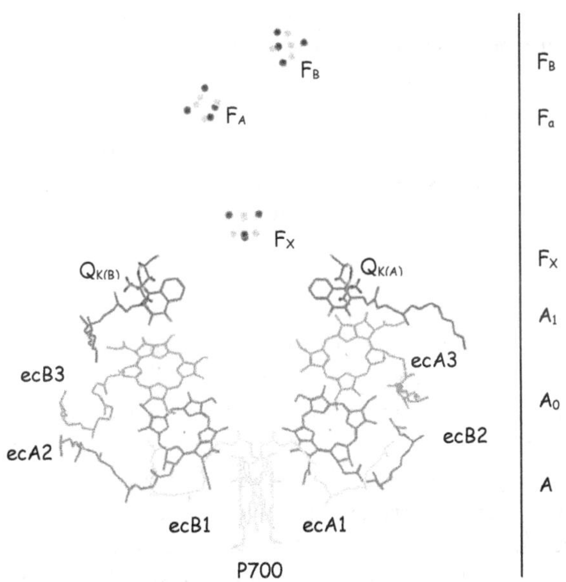

Figure 9.16 Diagram of the bifurcated electron transfer chain of *Thermosynechococcus elongatus*, note that the 3 Fe-S clusters do not show a twofold symmetry. Drawn from coordinates of structure 1JB0 (Jordan *et al.*, 2001). Note that the phylloquinones are H-bonded to the backbone of nearby Leu residues.

questionable but possible that A_0's extremely low redox potential is caused by this very weak interaction with a "mismatched" axial ligand.

Following the three pairs of chlorophylls is another pair of cofactors, positioned 8.6 Å above their A_0 partners—the secondary electron acceptors, phylloquinones, Q_{KA} and Q_{KB} (Vitamin K_1), commonly called A_1. Unlike the plastoquinone in PSII, these phylloquinones are firmly bound in place through hydrophobic interactions of their long phytyl tails with the surrounding protein, and are H-bonded to the backbone nitrogens of nearby Leu residues, A722 and B706. The absence of a second H-bond from the other phylloquinones carbonyl oxygen may account for the lower-than-expected redox potential in comparison to other reaction centers. The quinone rings are situated very close to and participate in π-stacking with two highly conserved Trp residues in their respective binding pockets—Trp A697 and Trp B677.

The iron–sulfur clusters F_X, F_A and F_B

The three 4-iron–4-sulfur clusters constitute the last set of cofactors within the PSI ETC, F_B is the terminal acceptor. The electron passed up the chain to the phylloquinones is transferred to F_X, which is the rate-limiting step of the ETC. The electron transfer from F_X to F_A and then F_B is very fast, after which the electron is passed on to ferredoxin.

As previously mentioned, F_X is ligated by two loops, one each from PsaA and PsaB, and is capped by PsaC, which contains the remaining two clusters. The clusters are cube-like, with alternating Fe and S atoms at the apexes of the cube. The Fe atoms of F_X are ligated by the Cys S atoms of A578, A587, B565 and B574. Nearby are 4 Pro residues—together, they form a flat surface that may help protect F_X from dioxygen (Golbeck, 2003). It appears that F_X may play a structural role in addition to that of electron transport—it has been found that, in the absence of F_X, PsaC will not bind to the reaction center core. Therefore, F_X is necessary for proper assembly of the complex as well as for electron transport.

F_A and F_B are contained strictly inside PsaC, with the Fe atoms of F_A ligated by S atoms from cysteines C20, C47, C50 and C53 for F_A, and of F_B by cysteines C10, C13, C16 and C57 (Golbeck, 2003). Transfer from F_X to these two clusters proceeds very quickly—faster, in fact, than from the phylloquinones to F_X. Their role is understood to be "vectoring" the electron from F_X to ferredoxin or flavodoxin.

Antenna chlorophylls

In total, each of the PSI trimer's three monomers contains 95 Chl*a* molecules and 1 Chl*a*′, a C13 epimer of Chl*a*. Six actually belong to the ETC, and will be discussed separately, but the remaining 90 make up the peripheral antenna system—their job is to harvest light and funnel the energy to fuel the transmembrane charge separation. The arrangement of chlorophylls and carotentoids in PSI is shown in Figure 9.17. The distances of 7–16 Å between antenna chlorophylls support the Förster (resonance) mechanism of fast excitation energy transfer. That is, a transition dipole resulting from decay of excitation of one Chl*a* can induce a transition dipole on a neighboring Chl*a* if there is some overlap between the fluorescence spectrum of the first molecule and the absorption spectrum of the neighboring one.

Of 90 antenna chlorophylls, 79 molecules are associated with PsaA and PsaB, and are generally grouped by their location in the reaction center. In the central domain, 43 Chl*a*

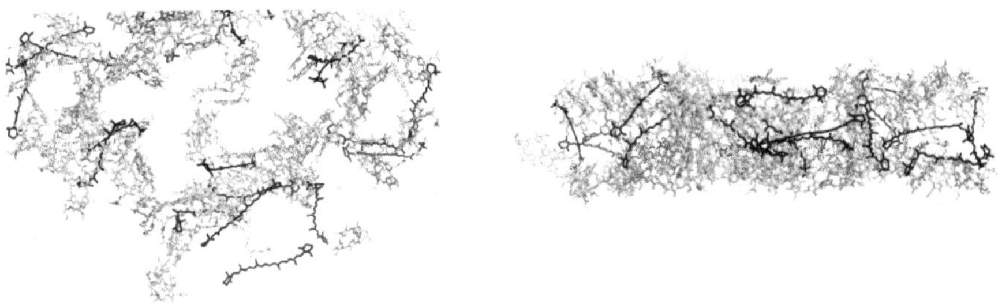

Figure 9.17 The Chl*a* molecules (in gray) and carotenoids (in black), including the ETC, associated with PSI: left, top view, from the stromal side. Right, side view, in the membrane plane. Coordinates drawn from *Thermosynechococcus elongatus* PSI structure 1JB0 (Jordan *et al.*, 2001).

molecules are well distributed among the inner transmembrane helices, and appear scattered throughout the entire thickness of the membrane. However, in the two "peripheral" domains bound by PsaA or PsaB, each containing 18 Chl*a*s, there is a clear organization of the chlorophylls into two layers—close to the lumenal surface and close to the stromal surface. Most of these Chl*a*s are axially coordinated through their Mg^{2+} ions by His imidazole N atoms. However, there are exceptions where Gln, Asp, Glu or Tyr side chain O atoms, backbone carbonyl or even water oxygens participate in the coordination.

Of the previously mentioned 79 Chl*a* molecules, 2 are worthy of extra attention. These are Chl*a*40, ligated by PsaA (A40), and Chl*a*39, ligated by PsaB (B39). Although the 90 antenna chlorophylls are generally well-separated by a distance of at least 18 Å from the 6 chlorophylls in the ETC, A40 and B39 may act as structural or functional linkages between the ETC and the antenna, since they are located approximately 11 Å on the A side and about 13 Å on the B side from the third pair of ETC chlorophylls, ecA3 and ecB3. The chlorophylls of the ETC will thereby serve as energy conduits between P700 and the antenna complex. They therefore function in the dual roles of both participating in excitation energy and electron transfer!

Of the 90 antenna chlorophylls, 11 remain to be mentioned. 10 of these are associated with the small subunits PsaJ, PsaK, PsaL, PsaM and PsaX, as we previously discussed. These 10 chlorophylls are ligated, as in PsaA and PsaB, by His nitrogens or Gln, Asp, Glu, Tyr (hydroxyl), backbone carbonyl or water oxygens. An eleventh Chl*a* is, however, axially coordinated by a phosphatidylglycerol (Lipid III). Likely, the various coordination environments and the coupling between the Chls contribute to "tuning" the absorption spectrum of individual chlorophylls. In fact, within PSI, there are specific chlorophylls (usually termed "red"), which absorb at a longer wavelength than the rest of the antenna, but their location is still under discussion—there are between 9 and 11 of these "red" chlorophylls in *T. elongatus*, for example. The bathochromic shift may be due to several factors: excitonic interactions, "buckling" of the normally planar chlorin ring of the chlorophyll's porphyrin system, charge effects, local protein interactions and even local environmental dielectric constant variations. Excitonic coupling is the most probable of these, and it is on this basis that individual "red" chlorophylls may be identified.

Antenna carotenoids

As well as functioning as the traditionally understood photoprotective "energy sinks," capable of accepting energy from excited chlorophylls and dissipating it as heat, PSI carotenoids also contribute to both structural stabilization of the complex and light harvesting. We can begin to appreciate the amazing construction of the photosystem when we see that carotenoids' absorption spectrum (450–570 nm) is quite different from that of Chl*a*. That is, carotenoids cover that range of wavelengths where chlorophyll is ineffectual and, therefore, lend some flexibility to the photosystem's light requirements. In addition, in their role as photoprotectors, carotenoids can quench Chl*a* excited triplet states via a "charge-transfer" mechanism and prevent formation of singlet oxygen.

The 22 carotenoid molecules identified in *T. elongatus* PSI demonstrate some variations in their structure: 17 are in the all-trans conformation, 2 are 9-cis, and three more contain one of 9,9′-cis, 9,13′-cis and 13-cis bonds (see Figure 9.17). Some of these isomers have been found in other cyanobacteria. They have all been modeled on β-carotene.

Within the *T. elongatus* PSI, there are 6 groupings of carotenoids. Clusters 1–4 are associated with PsaA and PsaB, with cluster 1 also interacting with PsaK, and contain 3, 3, 2 and 2 carotenoids, respectively. Clusters 5 and 6 each have 6 molecules. Cluster 5 interacts with PsaA, PsaB, PsaF and PsaJ and is located peripherally, on the outside of the complex. Cluster 6 is near the trimeric axis and participates in hydrophobic contacts with PsaA, PsaB, PsaI, PsaL and PsaM. Clusters 1–4 are arranged in rough C_2 symmetry, with 1 and 3 located approximately opposite to 2 and 4.

All the PSI carotenoids are ideally situated to fulfill their tasks, being located within approximately 3.6 Å (Van der Waals contact) of 60 chlorophyll porphyrin rings. This, as well as the occasional π-π stacking of carotenoids with chlorophylls, provides for efficient energy transfer from carotenoids to the rest of the antenna as well as for antenna chlorophyll absorption band tuning.

Ferredoxin and flavodoxin: stromal electron carriers

We have followed the electron through PSI, tracing its path through the reaction center up to the three 4Fe-4S clusters of PSI. The terminal cluster, F_B, is the final cofactor in PSI's ETC and, in conjunction with its partner, F_A, works to provide a pathway for the electron from F_X out to the stromal side of the membrane. We can now follow it to the stromal electron acceptor proteins, ferredoxin or flavodoxin, which receive the electron from PSI and carry it to the ferredoxin/$NADP^+$ oxidoreductase (Fromme *et al.*, 2001)—a gateway out of the photosynthetic pathway.

Ferredoxin

As we have already seen in our discussion of PSI, PsaC together with PsaD and PsaE form a stromal "hump" on the stromal side of PSI. We have noted that at the base of these three subunits, on top of the loops formed by PsaA, is the postulated ferredoxin docking site. We have also seen that PsaD is implicated in correct docking of ferredoxin, and that PsaE is significant in interaction of PSI with ferredoxin. We can therefore form the summary concept that ferredoxin is a soluble, globular protein that approaches PSI from the stromal side of the thylakoid membrane, docks to its upper surface and interacts with the three stromal subunits, PsaC, PsaD and PsaE. To clarify this, it is possible that, in plants, FNR may dock to PsaE. There is also some evidence from cyanobacterial mutants for a weak interaction of FNR with PSI via PsaE. On receiving an electron from F_B, it dissociates from PSI and diffuses into the stromal cytosol where it eventually encounters the oxidoreductase to which it passes the electron. With this basic knowledge in hand, we may now examine the structure of ferredoxin.

As a group, the soluble ferredoxin-type proteins are involved in a multitude of electron transfer reactions, due to the variable oxidation states of iron atoms found in their constituent iron–sulfur clusters. It is on the basis of the iron–sulfur clusters that they are also classified, because the number of the iron ions can vary from 2 to 4. We will focus on the "plant-type" ferredoxin, which is found in plants, algae and cyanobacteria and is characterized by a 2Fe-2S cluster (Fish *et al.*, 2005). The cluster itself, its protein environment and the associated three-dimensional architecture are highly conserved features of all ferredoxins within a particular type. Here, each of the two Fe ions is approximately tetrahedrally coordinated by four sulfur ligands—two inorganic sulfurs (which also bridge the irons) and two Cys residues. This type of ferredoxin typically possesses a mid-point potential (E_m) of around –0.42 eV, making reduced ferredoxin one of the strongest soluble natural reducing agents, although still not as strong as PSI! Ferredoxins from thermophilic organisms appear to be very thermostable. It has been suggested that factors contributing to this are a salt bridge network at the protein's C-terminus, and a flexible loop between strands b1 and b2 that might increase accessible hydrophobic surface area (Fish *et al.*, 2005). Since there are over 70 known ferredoxin structures, and at least 20 of them plant-type (the type used by cyanobacteria), we will focus on the ferredoxin from *M. laminosus*, a thermophilic cyanobacterium, the structure of which was published in 2005 (Figure 9.18).

Ferredoxin is the gene product of the light-induced *fedI* gene (Mazouni *et al.*, 2003). It is a relatively small soluble protein, consisting of 100 residues, with a molecular mass of approximately 10.6 kDa. It is constructed in a β-grasp motif around a central core, and composed of 5 antiparallel β-sheet strands (β1-β3, β5, β7) with an α-helix above the β-sheet motif, as seen in Figure 9.18.

The iron–sulfur cluster itself is located quite close to the surface, in a "solvent-exposed" pocket. The pocket is constructed of the [helix α1-strand β3] loop, which contains three of the four iron-ligating cysteines—Cys41, Cys46 and Cys49. The last cysteine, Cys79, is located in the [strand β5-strand β6] loop.

Ferredoxin possesses a very acidic electrostatic profile. We may therefore deduce that both long-range and short-range "steering" of ferredoxin towards its binding site on PSI are likely dominated by electrostatics. In fact, surface charge mutations on ferredoxin from *S. elongatus* have shown that adding negative charges increased ferredoxin affinity for the binding site, while adding positive charges decreased it. Although no structure of a PSI-ferredoxin complex has been published to date, it has been conjectured that, on docking, ferredoxin will assume an orientation that will minimize the distance between its 2Fe-2S cluster and cluster F_B of PSI (Setif *et al.*, 2002).

Flavodoxin

Just as specific interactions are required between PSI and its lumenal electron donors, so are equally great specificity and efficiency required for interactions between PSI and its stromal electron acceptors. However, once again PSI is able to retain interaction specificity

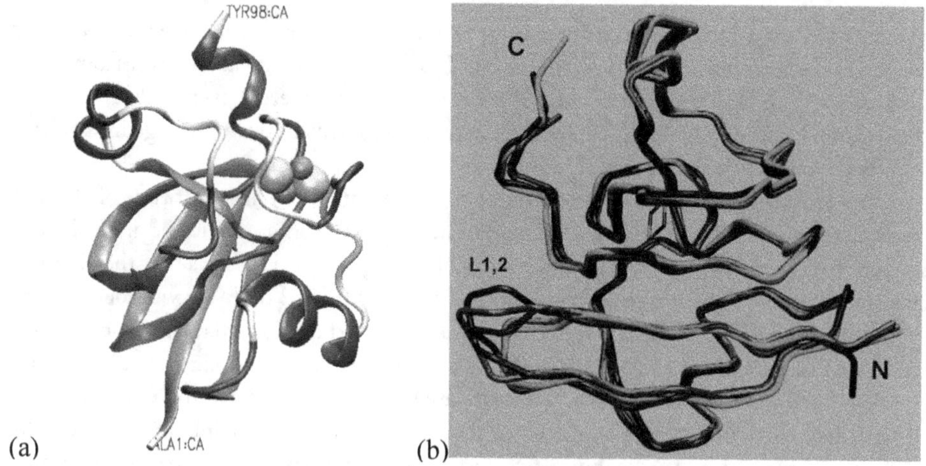

Figure 9.18 (a) A model of the ferredoxin from thermophilic cyanobacterium *Mastigocladus laminosus*, structure 1RFK in the PDB. Note the inner core, framed by the β-sheet structure (Fish *et al.*, 2005). The iron–sulfur cluster is drawn as dark and light spheres, clasped by a loop "clamp." (b) A superposition of the 1RFK structure on structures from *Anabaena* sp. PCC 7119, *Spirulina platensis*, *Synechocystis* sp. PCC 6803, and *Spinacia oleracea*. Note the identical fold and the flexible L1,2 loop region, which is thought to be partially responsible for thermostability of the protein (Fish *et al.*, 2005). The salt bridge, also thought to contribute to protein stability, consists of Lys93 and its two bridges with Glu90 and Glu96—8 and 2 residues away, respectively, from the C-terminus. Structure (b) from (Fish *et al.*, 2005).

while demonstrating adaptability and allowing more than one protein to fulfill a particular role. We have seen that copper-deprivation in some organisms allows them to synthesize cytochrome c_6 to act instead of PC as the lumenal electron carrier. A similar mechanism is in place for the stromal carriers—in conditions of iron deprivation, the iron-containing ferredoxin is replaced by its functional but iron-less analogue, flavodoxin. This allows electron transfer to continue in the event that iron is not available. Flavodoxins are reminiscent of their partners, the ferredoxins, in a number of ways. They are found in a widely varying group of organisms, from strict and facultative anaerobes to cyanobacteria, eukaryotic algae and other higher organisms. In photosynthetic bacteria, this product of the *isiB* gene (Leonhardt and Straus, 1989) functionally replaces ferredoxin in iron-deficient conditions (Burkhart *et al.*, 1995). Flavodoxin, like ferredoxin, is a small, highly acidic protein—with a mass of 19–23 kDa and being anywhere from about 130 to 175 residues in length. Here, we'll refer to flavodoxin from *Anabaena* sp. PCC 7120, which is a "long-chain" flavodoxin with 169 residues, weighing 18.8 kDa (PDB accession code 1RCytf) (see Figure 9.19).

Structurally, flavodoxin consists of a five-stranded parallel β-sheet central core, flanked by two α-helices on one side and 3 on the other, with a 20-residue "lobe" inserted in the middle of the β_5 strand. The lobe does not appear to disrupt hydrogen bonding between the strands. A low-potential cofactor flavin mononucleotide (FMN), responsible for the redox properties on the protein is non-covalently bound at one end of the sheet, in a binding pocket of highly conserved structure, with its dimethyl benzene edge exposed to solvent (Rao *et al.*, 1992).

The FMN is the central feature of flavodoxin, in that it allows flavodoxin to exist in three oxidation states: *ox* (fully oxidized), *sq* (semiquinone; one-electron reduced), and

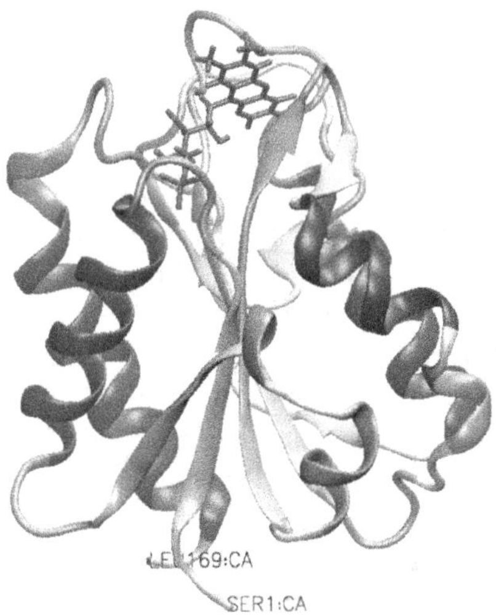

Figure 9.19 A model of the *Anabaena* sp. PCC 7120 flavodoxin, structure 1RCytf in the PDB (Burkhart *et al.*, 1995). N and C termini are labeled. Note the quinone-type flavin cofactor, and the very general similarity of the structure to that of ferredoxin—β-sheet accompanied by α-helices, although here the β-sheet is "flanked" by helices on either side. The two also differ in that the flavodoxin is parallel instead of being antiparallel as in ferredoxin, so that the cofactor binding motif is not a separate domain as it is in ferredoxin.

hq (hydroquinone; two-electron reduced). In flavodoxin, it alternates between the *sq* and *hq* states. The *ox/sq* reduction potential of long-chain flavodoxins (E_2) is similar to that for free FMN in solution at pH 7.0 (−238 mV). However the potential for the *sq/hq* couple commonly seen *in vivo* (E_l) is lower by a factor of 2½. This produces a net effect of well-separated potentials E_2 and E_l, and a sizeable energy gap upon oxidation or reduction, with the stabilized *sq* form favored over the *hq* form (Burkhart *et al.*, 1995). In summary, when the cofactor binds to the apoprotein, FMN's midpoint potential is drastically altered, making the FMN semiquinone much more stable. Flavodoxin is therefore to behave as a one-electron transfer center, that, *in vivo*, cycles between the semiquinone and the fully reduced forms (Medina *et al.*, 1999).

ATP synthase: molecular motor for ATP synthesis

ATP synthase (F_0F_1) is a molecular motor that is present in almost all living organisms. It is one of the most important proteins in energy transduction, catalyzing the synthesis of ATP from ADP and P_i, and driven by the transmembrane electrochemical potential of protons. Despite its outstanding importance, no structure of the intact enzyme has yet been determined. The main architecture of this molecular machine is similar in all organisms (Figure 9.20; for recent reviews, see Dimroth *et al.*, 2006; Muller *et al.*, 2005; Fillingame *et al.*, 2003; Walker and Dickson, 2006). The complex consists of two structurally and functionally distinct domains: a membrane-intrinsic proton translocation system (the F_0 part) and a membrane-extrinsic catalytic domain (the F_1 part). Both entities are structurally and functionally coupled by at least two stalks. The enzyme works as a molecular motor. Ion flux through the F_0 part drives rotation of the rings of subunit *c*, coupled by the central stalk to the rotation of the gamma subunit, which drives ATP synthesis at the catalytic sites.

Structure of F_1

The membrane-extrinsic F_1 part contains the six nucleotide binding sites. F_1 consists of a hexamer of three α- and three β-subunits, with the γ-subunit located asymmetrically in the middle of the hexamer. Three catalytic nucleotide-binding sites are located on the β-subunits, whereas the α-subunits contain three regulatory nucleotide-binding sites. The structure of the main part of the F_1 domain was first solved by X-ray structure analysis in 1994 by the group of J. Walker (Abrahams *et al.*, 1994). In the initial structure of Abrahams *et al.*, the three catalytic binding sites on the β-subunits had different nucleotide occupancy (see Figure 9.21A). The site nearest the γ-subunit was empty, one site was occupied by ADP and P_i and one site (which is closed and not accessible from the medium) was occupied by an ATP analogue. The structure of F_1 strongly supported the binding change mechanism proposed by Boyer in the 1980s (Kohlbrenner and Boyer, 1983). Boyer and Kohlbrenner proposed a binding change mechanism, in which the three sites alternately change their binding affinity for ATP, ADP and P_i within a cooperative three-step mechanism. One site is always occupied by ATP, one by ADP and P_i and the third site is empty. The γ-subunit was supposed to rotate in the middle of the $\alpha_3\beta_3$ hexamer, thereby inducing cooperative changes in the binding affinity. They also proposed that the major energy-consuming step is the release of pre-formed ATP, i.e. the opening of the binding pocket (see Figure 9.21B). Rotation of γ was unequivocally shown and visualized by the elegant experiments of Yoshida and coworkers (Noji *et al.*, 1997). In 1997, Walker and Boyer received the

Figure 9.20 (see also Plate 9.20) A schematic structural diagram of ATP synthase. Structure for F1 was modified after 2F43 (Chen *et al.*, 2006). *c*-ring represents the structure of the c11 ring from *Ilyobacter tartaricus* taken from Meier *et al.* (2005); PDB accession code 1YCE.

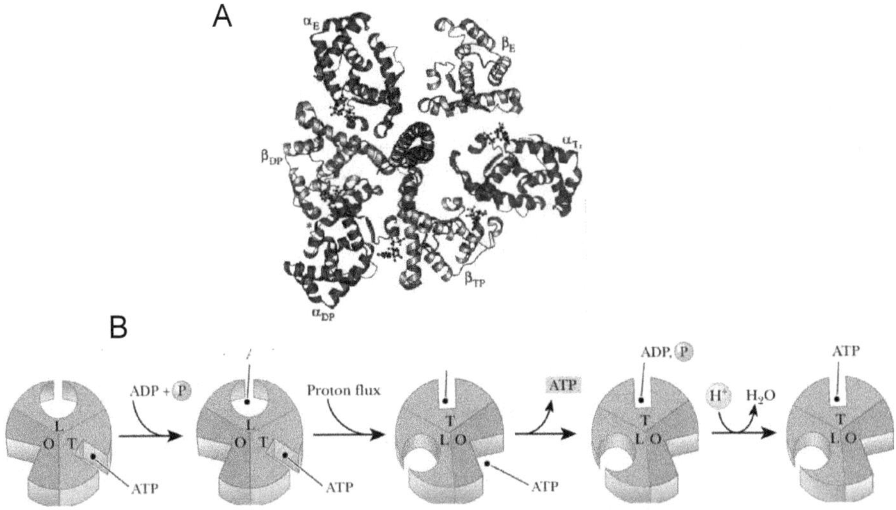

Figure 9.21 (see also Plate 9.21) A, structure of the F_1 part of the ATP synthase from beef heart mitochondria as determined by Abrahams *et al.* in 1994 (Figure from Abrahams *et al.*, 1994). B, binding change mechanism as proposed by Kohlbrenner and Boyer (1983).

Nobel prize for their work. This brilliant work led to a further effort to obtain structural and functional information about the other F_1 subunits and the F_0 domain of the enzyme. Since determination of the initial structure of F_1 from bovine heart, multiple structures of F_1 from different organisms have been solved by X-ray crystallography, among them the structure of F_1 from rat liver (Chen *et al.*, 2006; Bianchet *et al.*, 1998), spinach (Groth and Pohl, 2001), *E. coli* (Hausrath *et al.*, 2001) and the $\alpha_3\beta_3$ complex from a thermophilic bacterium (Shirakihara *et al.*, 1997).

No structure is yet available of the intact F_0F_1 enzyme. The first direct structural information on the proton translocating part and its coupling to F_0 was derived from a 3.9 Å X-ray structural model of a yeast-ATP synthase sub-complex, consisting of three α- and three β-subunits, as well as one γ and one δ subunit (homologous to the ε subunit of *E. coli*) together with a ring of 10 *c* subunits (Stock *et al.*, 1999).

The two stalks of the ATP synthase

The head of the ATP synthase is coupled to the proton translocating F_0 part by two stalks (Wilkens and Capaldi, 1998; Karrasch and Walker, 1999; Bottcher and Graber, 2000; Gaballo *et al.*, 2002). The subunits in these stalk regions differ between organisms, as eukaryotic mitochondrial ATP synthases are much more complex than the prokaryotic (including cyanobacteria) and chloroplast enzymes (Rubinstein *et al.*, 2003). The central stalk plays the role of an energy coupling rotatory element, whereas the second stalk, located asymmetrically on one side of the complex, has been suggested to function as a stator fixing the α/β hexamer. It has also been suggested that subunits γ and ε are part of the rotor, whereas subunit δ belongs to the "stator," together with two copies of subunit *b*. Subunit *b* was shown to reveal an α-helical content of 80%, whereas 14% of the structure consists of β-turns (Greie *et al.*, 2000). Very recently, a large subcomplex of the proteins involved in the formation of the second stalk of bovine heart ATP synthase was crystallized, and their structure solved to 2.8 Å resolution (Dickson *et al.*, 2006). The *b* subunit is located in the center and forms one very long α helix. This structure further supports the idea that the subunit *b* dimer does not form a rigid rod-like structure, but provides the possibility for the storage of elastic energy (Greie *et al.*, 2000). Several high-resolution structures exist for subunits of the central stalks of the ATP synthase, but details of the amino acids that form the face interactions between F_1 and F_0 are still unknown.

Structure of the ion conduction channel, F_0

No structure has so far been determined of the intact ion conduction channel F_0. The F_0 part consists of three different proteins: one subunit *a*, two subunits *b* and 10–15 subunits *c* (in chloroplasts and cyanobacteria these subunits are named IV = (*a*), I, II = (*b*, B′) and III = (*c*)). The question of why the stoichiometry of the c-ring varies among different organisms is very interesting, and provokes further questions regarding evolutionary origin of the divergence in the stoichiometry.

Subunit a

Subunit *a* contains 271 amino acids, and is the largest subunit of F_0. There is striking experimental evidence that proton translocation occurs at the interface of the oligomeric ring subunits *c* and *a* (Jiang and Fillingame, 1998; Fillingame *et al.*, 2002). The structure of subunit *a* is of outstanding importance for unraveling the molecular mechanism of ion translocation through F_0. Four, five and six transmembrane helices have been predicted in the past, but most of the experimental evidence from mutagenesis studies and cross-linking experiments, as well as chemical labeling of surface areas and lipid exposed amino acids, converge towards models in which subunit *a* consists of five transmembrane helices (Valiyaveetil and Fillingame, 1998; Wada *et al.*, 1999; Kurisu *et al.*, 2003; Vik and Ishmukhametov, 2005). Transmembrane helix IV may be in direct contact with the outer surface of the c-rotor (Fillingame *et al.*, 1998). It was proposed that helices 2–5 form a helix bundle that harbors the proton-conducting pathway through F_0. Angevine and coworkers have proposed that the aqueous access pathways at the interior of subunit *a* may be gated by a swiveling of helices in this bundle (Angevine and Fillingame, 2003). The high importance of subunit *a*, as it relates to understanding of this reaction mechanism, is contrasted by the lack of structural information on subunit *a*. So far, the only structural model of subunit *a* in the PDB data base is a theoretical model (Aksimentiev *et al.*, 2004) as no experimental structure is yet available. In 2004, a new isolation procedure was reported, leading to protein that may be suitable for structure determination by NMR (Dmitriev *et al.*, 2004a; Dmitriev *et al.*, 2004b), but no structure has yet been reported.

Subunit b

Subunit *b* is part of the stator of the ATP synthase. The enzyme from *E. coli* contains two identical subunits *b*, whereas the chloroplast and cyanobacterial enzymes contain a heterodimer of subunits I and II (*b* and B′). Each *b*-subunit has one transmembrane helix and a long extended cytoplasmic domain, which mediates the interaction between the stator parts of F_0 and the stator parts of F_1. Structural information on the transmembrane helix of subunit *b* was derived from NMR studies (Dmitriev *et al.*, 1999) and crystallization of the transmembrane helix (Del Rizzo *et al.*, 2002). Very recently, the structure of a major part of the second stalk of the ATP synthase from bovine heart was solved by X-ray crystallography (Dickson *et al.*, 2006). The structure contains the F_6 subunit, the OSPC protein (oligomycin-sensitivity conferring protein) and the amino acids 19 to 183 from subunit *b*. The structure shows that subunit *b* is located in the center of the stalk, and consists of a 160 Å-long α-helix. Dicksen *et al.* also modeled their structure into a low-resolution electron density map of the intact bovine heart enzyme, indicating a good match of the dimension of the structure with the second stalk of the ATP synthase.

Subunit c

A ring of *c*-subunits forms the rotor of the proton turbine. Each subunit *c* has a hairpin-like structure of 2 transmembrane helices, which are connected by a small hydrophilic loop. The first structural model of the monomeric subunit *c* was derived from NMR study of subunit *c* in organic solvents (Girvin *et al.*, 1998). There is a charged amino acid (Glu or Asp) present in subunit *c*, which is supposed to be directly involved in proton translocation. NMR studies in organic solvents on the *E. coli c*-ring and the *c*-ring from *Bacillus thermophilis* indicate different orientations for the outer helix (Nakano *et al.*, 2006; Fillingame *et al.*, 2000; Dmitriev *et al.*, 2002; Dmitriev and Fillingame, 2001).

The first intact ring of *c*-subunits (subunit III from the chloroplast enzyme) was isolated in 1988 and characterized by electron microscopy (Fromme *et al.*, 1987). The same isolation procedure was used later for the first AFM (atomic force microscopy) studies of the chloroplast *c*-ring (Seelert *et al.*, 2000), revealing that the ring is composed of 14 subunits III.

Further AFM and functional studies have shown that the stoichiometry of subunits in the rotor differs between different organisms (see Table 9.5). A further breakthrough was the crystallization of the F_1-*c*-ring-complex from yeast by Stock *et al.* (1999). This work represents the only structure of subunit *c* with F_1. The ring showed a stoichiometry of 10 *c*-subunits. The *c*-ring is attached to the central stalk of the ATP synthase. Due to the limitations in resolution, unfortunately, no molecular details of the interaction site between the *c*-ring and the subunits in the central stalk were discovered.

The first atomic structure of an intact *c*-ring of the F_0F_1 family of ATP synthases was solved by X-ray crystallography to 2.4-Å resolution from the Na^+ translocating ATP synthase of *Ilyobacter tartaricus* (Meier *et al.*, 2005). The structure shows that in this organism, the rotor consists of 11 subunits *c* (Figure 9.22). The Asp residues, bound to Na^+, are located in the middle of the outer (C-terminal) helix. In the same year, the structure of the rotor of the first V-type ATPase was solved (Murata *et al.*, 2005). To date no structure has been solved of any *c*-ring of the proton translocating F_0F_1 ATP synthases.

Concluding remarks and future prospects

In the last 15 years, outstanding progress on the understanding of the primary events in photosynthesis have been achieved. The determination of most of the structures of the main complexes, which catalyze light harvesting and electron transport, was a breakthrough in the understanding of the fundamental processes that provide all higher life on earth with energy. However, many burning questions on topics such as the interaction and

Table 9.5 Stoichiometry of *c*-subunits in the rotor of F_0F_1 type ATP synthases

Organism	Number of *c*-subunits	Ion transported	Method of determination	Reference
Thermophilic *Bacillus*	10	H^+	Fusion protein construction and functional assay	Mitome *et al.*, 2004
Yeast	10	H^+	X-ray structure of F_1-*c*-ring	Stock *et al.*, 1999
Escherichia coli	10	H^+	AFM	Jiang *et al.*, 2001
Ilyobacter tartaricus	11	Na^+	AFM X-ray structure	Stahlberg *et al.*, 2001, Meier *et al.*, 2005
Chlamydomonas rheinhardtii	13?	H^+	Gel electrophoresis	Meyer *et al.*, 2004
Spinach	14	H^+	AFM	Seelert *et al.*, 2003
Spirulina platensis	15	H^+	AFM	Pogoryelov *et al.*, 2005

assembly of the photosynthetic apparatus and regulation of the photosynthetic machinery still await answers. The most challenging question is still probably the mechanism of water splitting. A higher resolution structure of a non-disturbed Mn cluster must be combined with spectroscopic investigations and theoretical modeling to unravel the secret of this key process, which was invented by cyanobacteria 2.8 billion years ago and started the development of higher life on earth. By unraveling the last secrets of photosynthesis, we may form the basis for mimicking nature with artificial photosynthesis, which can provide future generations of humans with a safe and unlimited energy supply.

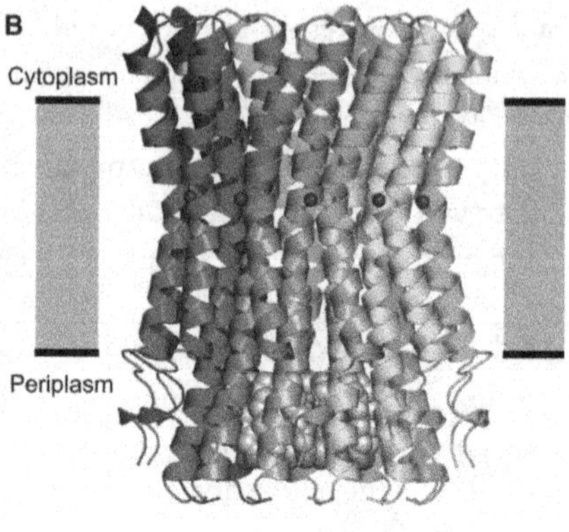

Figure 9.22 (see also Plate 9.22) Structure of the *c*11-ring of *Ilyobacter tartaricus*, PDB accession code 1YCE (Meier *et al.*, 2005).

References

Abrahams, J.P., Leslie, A.G., Lutter, R., and Walker, J.E. (1994). Structure at 2.8 Å resolution of F_1-ATPase from bovine heart mitochondria. Nature *370*, 621–628.

Adman, E.T., Siefker, L.C., and Jensen, L.H. (1976). Structure of *Peptococcus aerogenes* ferredoxin. Refinement at 2 Å resolution. J. Biol. Chem. *251*, 3801–3806.

Aksimentiev, A., Balabin, I.A., Fillingame, R.H., and Schulten, K. (2004). Insights into the molecular mechanism of rotation in the F_o sector of ATP synthase. Biophys. J. *86*, 1332–1344.

Angevine, C.M., and Fillingame, R.H. (2003). Aqueous access channels in subunit *a* of rotary ATP synthase. J. Biol. Chem. *278*, 6066–6074.

Ardelean, I., Matthijs, H.C., Havaux, M., Joset, F., and Jeanjean, R. (2002). Unexpected changes in photosystem I function in a cytochrome c_6-deficient mutant of the cyanobacterium *Synechocystis* PCC 6803. FEMS Microbiol. Lett. *213*, 113–119.

Baena-Gonzalez, E., and Aro, E.M. (2002). Biogenesis, assembly and turnover of photosystem II units. Philos. Trans. R Soc. Lond. B Biol. Sci. *357*, 1451–1459; discussion 1459–1460.

Ben-Shem A., Frolow F., and Nelson N. (2003). Crystal structure of plant photosystem I. Nature *426*: 630–635.

Bianchet, M.A., Hullihen, J., Pedersen, P.L., and Amzel, L.M. (1998). The 2.8- Å structure of rat liver F_1-ATPase: configuration of a critical intermediate in ATP synthesis/hydrolysis. Proc. Natl. Acad. Sci. USA 95, 11065–11070.

Bibby, T.S., Mary, I., Nield, J., Partensky, F., and Barber, J. (2003). Low-light-adapted *Prochlorococcus* species possess specific antennae for each photosystem. Nature *424*, 1051–1054.

Bibby, T.S., Nield, J., and Barber, J. (2001). Three-dimensional model and characterization of the iron stress-induced CP43'-photosystem I supercomplex isolated from the cyanobacterium *Synechocystis* PCC 6803. J. Biol. Chem. *276*, 43246–43252.

Biesiadka, J., Loll, B., Kern, J., Irrgang, K.D., and Zouni, A. (2004). Crystal structure of cyanobacterial photosystem II at 3.2 angstrom resolution: a closer look at the Mn-cluster. Physical Chemistry Chemical Physics 6, 4733–4736.

Blankenship, R.E. (1992). Origin and early evolution of photosynthesis. Photosynth. Res. 33, 91–111.

Blankenship, R.E., and Hartman, H. (1998). The origin and evolution of oxygenic photosynthesis. Trends Biochem. Sci. *23*, 94–97.

Boekema, E.J., Hifney, A., Yakushevska, A.E., Piotrowski, M., Keegstra, W., Berry, S., Michel, K.P., Pistorius, E.K., and Kruip, J. (2001). A giant chlorophyll-protein complex induced by iron deficiency in cyanobacteria. Nature *412*, 745–748.

Bond, C.S., Bendall, D.S., Freeman, H.C., Guss, J.M., Howe, C.J., Wagner, M.J., Wilce, M.C. (1999). The structure of plastocyanin from the cyanobacterium *Phormidium laminosum*. Acta Cryst. D 55, 414–421.

Bottcher, B., and Graber, P. (2000). The structure of the H^+-ATP synthase from chloroplasts and its subcomplexes as revealed by electron microscopy. Biochim. Biophys. Acta *1458*, 404–416.

Bovy, A., de Vrieze, G., Borrias, M., and Weisbeek, P. (1992). Transcriptional regulation of the plastocyanin and cytochrome c_{553} genes from the cyanobacterium *Anabaena* species PCC 7937. Mol. Microbiol. 6, 1507–1513.

Brejc, K., Ficner, R., Huber, R., Steinbacher, S. (1995). Isolation, crystallization, crystal structure analysis and refinement of allophycocyanin from the cyanobacterium *Spirulina platensis* at 2.3 Å resolution. J. Mol. Biol. *249*, 424–440.

Bruns, C.M., and Karplus, P.A. (1995). Refined crystal structure of spinach ferredoxin reductase at 1.7 Å resolution: oxidized, reduced and 2'-phospho-5'-AMP bound states. J. Mol. Biol. *247*, 125–145.

Burkhart, B.M., Ramakrishnan, B., Yan, H., Reedstrom, R.J., Markley, J.L., Straus, N.A., and Sundaralingam, M. (1995). Structure of the trigonal form of recombinant oxidized flavodoxin from *Anabaena*-7120 at 1.40 Angstrom resolution. Acta Crystallographica Section D-Biological Crystallography *51*, 318–330.

Chen, C., Saxena, A.K., Simcoke, W.N., Garboczi, D.N., Pedersen, P.L., and Ko, Y.H. (2006). Mitochondrial ATP synthase. Crystal structure of the catalytic F_1 unit in a vanadate-induced transition-like state and implications for mechanism. J. Biol. Chem. *281*, 13777–13783.

Chen, M., and Bibby, T.S. (2005). Photosynthetic apparatus of antenna-reaction centers supercomplexes in oxyphotobacteria: Insight through significance of Pcb/IsiA proteins. Photosynthesis Res. *86*, 165–173.

Cogdell, R.J., and Lindsay, J.G. (2000). Tansley Review No. 109—The structure of photosynthetic complexes in bacteria and plants: an illustration of the importance of protein structure to the future development of plant science. New Phytologist *145*, 167–196.

Cohen, R.O., Shen, G., Golbeck, J.H., Xu, W., Chitnis, P.R., Valieva, A.I., van der Est, A., Pushkar, Y., and Stehlik, D. (2004). Evidence for asymmetric electron transfer in cyanobacterial photosystem I: analysis of a methionine-to-leucine mutation of the ligand to the primary electron acceptor A_0. Biochemistry *43*, 4741–4754.

de la Rosa, M.A., Navarro, J.A., Diaz-Quintana, A., De la Cerda, B., Molina-Heredia, F.P., Balme, A., Murdoch Pdel, S., Diaz-Moreno, I., Duran, R.V., and Hervas, M. (2002). An evolutionary analysis of the reaction mechanisms of photosystem I reduction by cytochrome c_6 and plastocyanin. Bioelectrochemistry *55*, 41–45.

Deisenhofer, J., Epp, O., Miki, K., Huber, R., and Michel, H. (1984). X-ray structure analysis of a membrane protein complex. Electron density map at 3 Å resolution and a model of the chromophores of the photosynthetic reaction center from *Rhodopseudomonas viridis*. J. Mol. Biol. *180*, 385–398.

Deisenhofer, J., Epp, O., Sinning, I., and Michel, H. (1995). Crystallographic refinement at 2.3 Å resolution and refined model of the photosynthetic reaction center from *Rhodopseudomonas viridis*. J. Mol. Biol. *246*, 429–457.

Del Rizzo, P.A., Bi, Y., Dunn, S.D., and Shilton, B.H. (2002). The "second stalk" of Escherichia coli ATP synthase: structure of the isolated dimerization domain. Biochemistry *41*, 6875–6884.

Dickson, V.K., Silvester, J.A., Fearnley, I.M., Leslie, A.G., and Walker, J.E. (2006). On the structure of the stator of the mitochondrial ATP synthase. EMBO J. *25*, 2911–2918.

Dimroth, P., von Ballmoos, C., and Meier, T. (2006). Catalytic and mechanical cycles in F-ATP synthases. Fourth in the Cycles Review Series. EMBO Rep. *7*, 276–282.

Dmitriev, O., Jones, P.C., Jiang, W., and Fillingame, R.H. (1999). Structure of the membrane domain of subunit b of the *Escherichia coli* F_0F_1 ATP synthase. J. Biol. Chem. *274*, 15598–15604.

Dmitriev, O.Y., Abildgaard, F., Markley, J.L., and Fillingame, R.H. (2002). Structure of Ala24/Asp61 → Asp24/Asn61 substituted subunit *c* of Escherichia coli ATP synthase: implications for the mechanism of proton transport and rotary movement in the F_0 complex. Biochemistry *41*, 5537–5547.

Dmitriev, O.Y., Abildgaard, F., Markley, J.L., and Fillingame, R.H. (2004a). Backbone 1H, 15N and 13C assignments for the subunit *a* of the *E. coli* ATP synthase. J. Biomol. NMR *29*, 439–440.

Dmitriev, O.Y., Altendorf, K., and Fillingame, R.H. (2004b). Subunit *a* of the *E. coli* ATP synthase: reconstitution and high resolution NMR with protein purified in a mixed polarity solvent. FEBS Lett *556*, 35–38.

Dmitriev, O.Y., and Fillingame, R.H. (2001). Structure of Ala(20) → Pro/Pro(64) → Ala substituted subunit *c* of *Escherichia coli* ATP synthase in which the essential proline is switched between transmembrane helices. J. Biol. Chem. *276*, 27449–27454.

Durell, S.R., Labanowski, J.K., and Gross, E.L. (1990). Modeling of the electrostatic potential-field of plastocyanin. Arch. of Biochem. Biophys. *277*, 241–254.

Duysens, L.N., Kamp, B.M., and Amesz, J. (1961). Two photochemical systems in photosynthesis. Nature *190*, 510-&.

Ferreira, K.N., Iverson, T.M., Maghlaoui, K., Barber, J., and Iwata, S. (2004). Architecture of the photosynthetic oxygen-evolving center. Science *303*, 1831–1838.

Fillingame, R.H., Angevine, C.M., and Dmitriev, O.Y. (2002). Coupling proton movements to *c*-ring rotation in F_1F_0 ATP synthase: aqueous access channels and helix rotations at the *a*-*c* interface. Biochim. Biophys. Acta *1555*, 29–36.

Fillingame, R.H., Angevine, C.M., and Dmitriev, O.Y. (2003). Mechanics of coupling proton movements to *c*-ring rotation in ATP synthase. FEBS Lett. *555*, 29–34.

Fillingame, R.H., Girvin, M.E., Jiang, W., Valiyaveetil, F., and Hermolin, J. (1998). Subunit interactions coupling H^+ transport and ATP synthesis in F_1F_0 ATP synthase. Acta Physiol Scand. Suppl. *643*, 163–168.

Fillingame, R.H., Jiang, W., and Dmitriev, O.Y. (2000). The oligomeric subunit *c* rotor in the F_0 sector of ATP synthase: unresolved questions in our understanding of function. J. Bioenerg. Biomembr. *32*, 433–439.

Fish, A., Danieli, T., Ohad, I., Nechushtai, R., and Livnah, O. (2005). Structural basis for the thermostability of ferredoxin from the cyanobacterium *Mastigrocladus laminosus*. J. Mol. Biol. *350*, 599–608.

Ford, R.C., Picot, D., and Garavito, R.M. (1987). Crystallization of the photosystem-I reaction center. EMBO J. *6*, 1581–1586.

Frazao, C., Soares, C.M., Carrondo, M.A., Pohl, E., Dauter, Z., Wilson, K.S., Hervas, M., Navarro, J.A., De la Rosa, M.A., and Sheldrick, G.M. (1995). Ab initio determination of the crystal structure of cytochrome c_6 and comparison with plastocyanin. Structure 3, 1159–1169.

Fromme, P., Boekema, E.J., and Graber, P. (1987). Isolation and characterization of a supramolecular complex of subunit III of the ATP-synthase from chloroplasts. Z. Naturforsch. *42c*, 1239–1245.

Fromme, P., Jordan, P., and Krauss, N. (2001). Structure of photosystem I. Biochim. Biophys. Acta *1507*, 5–31.

Fromme, P., Kern, J., Loll, B., Biesiadka, J., Saenger, W., Witt, H.T., Krauss, N., and Zouni, A. (2002). Functional implications on the mechanism of the function of photosystem II including water oxidation based on the structure of photosystem II. Philosophical Transactions of the Royal Society of London Series B-Biological Sciences *357*, 1337–1344.

Gaballo, A., Zanotti, F., and Papa, S. (2002). Structures and interactions of proteins involved in the coupling function of the protonmotive F_0F_1-ATP synthase. Curr. Protein Pept. Sci. *3*, 451–460.

Geiss, U., Vinnemeier, J., Kunert, A., Lindner, I., Gemmer, B., Lorenz, M., Hagemann, M., and Schoor, A. (2001). Detection of the *isiA* gene across cyanobacterial strains: potential for probing iron deficiency. Appl. Environ. Microbiol. *67*, 5247–5253.

Girvin, M.E., Rastogi, V.K., Abildgaard, F., Markley, J.L., and Fillingame, R.H. (1998). Solution structure of the transmembrane H^+-transporting subunit *c* of the F_1F_0 ATP synthase. Biochemistry *37*, 8817–8824.

Glazer, A.N. (1982). Phycobilisomes—structure and dynamics. Annu. Rev. Microbiol. *36*, 173–198.

Golbeck, J.H. (2003). The binding of cofactors to photosystem I analyzed by spectroscopic and mutagenic methods. Annu. Rev. Biophys. Biomol. Structure *32*, 237–256.

Greie, J.C., Deckers-Hebestreit, G., and Altendorf, K. (2000). Subunit organization of the stator part of the F_0 complex from Escherichia coli ATP synthase. J Bioenerg. Biomembr. *32*, 357–364.

Grossman, A.R. (2003). A molecular understanding of complementary chromatic adaptation. Photosynth. Res. *76*, 207–215.

Grossman, A.R., Bhaya, D., and He, Q. (2001). Tracking the light environment by cyanobacteria and the dynamic nature of light harvesting. J. Biol. Chem. *276*, 11449–11452.

Grossman, A.R., Schaefer, M.R., Chiang, G.G., and Collier, J.L. (1993). The phycobilisome, a light-harvesting complex responsive to environmental-conditions. Microbiol. Rev. *57*, 725–749.

Groth, G., and Pohl, E. (2001). The structure of the chloroplast F_1-ATPase at 3.2 Å resolution. J. Biol. Chem. *276*, 1345–1352.

Grotjohann, I., and Fromme, P. (2004). Evolution of photosynthesis and oxygen evolution: Implications from the structural comparison of Photosystems I and II. Phys. Chem. Chem. Phys. *6*, 4743–4753.

Guergova-Kuras, M., Boudreaux, B., Joliot, A., Joliot, P., and Redding, K. (2001). Evidence for two active branches for electron transfer in photosystem I. Proc. Natl. Acad. Sci. USA *98*, 4437–4442.

Harding, M.M. (2001). Geometry of metal-ligand interactions in proteins. Acta Crystallographica Section D-Biological Crystallography *57*, 401–411.

Hart, S. (2005), PhD thesis, University of Cambridge.

Hausrath, A.C., Capaldi, R.A., and Matthews, B.W. (2001). The conformation of the epsilon- and gamma-subunits within the Escherichia coli F_1 ATPase. J. Biol. Chem. *276*, 47227–47232.

Helfrich, M., Ross, A., King, G.C., Turner, A.G., and Larkum, A.W. (1999). Identification of [8-vinyl]-protochlorophyllide *a* in phototrophic prokaryotes and algae: chemical and spectroscopic properties. Biochim. Biophys. Acta *1410*, 262–272.

Hervas, M., Navarro, J.A., Diaz, A., Bottin, H., and De la Rosa, M.A. (1995). Laser-flash kinetic analysis of the fast electron transfer from plastocyanin and cytochrome c_6 to photosystem I. Experimental evidence on the evolution of the reaction mechanism. Biochemistry *34*, 11321–11326.

Hiller, R.G., and Larkum, A.W.D. (1985). The chlorophyll-protein complexes of *Prochloron* sp. (Prochlorophyta). Biochim. Biophys. Acta *806*: 107–115.

Houmard, J., and de Marsac, N.T. (1988). Cyanobacterial genetic tools: current status. Methods Enzymol. *167*, 808–847.

Hurt, E., and Hauska, G. (1981). A cytochrome f/b6 Complex of 5 polypeptides with plastoquinol-plastocyanin-oxidoreductase activity from spinach chloroplasts. European J. Biochem. *117*, 591–599.

Ikeuchi, M., Nyhus, K.J., Inoue, Y., and Pakrasi, H.B. (1991). Identities of 4 low-molecular-mass subunits of the photosystem-I complex from *Anabaena variabilis* ATCC 29413: Evidence for the presence of the *psaI* gene-product in a cyanobacterial complex. FEBS Lett. *287*, 5–9.

Jeanjean, R., Zuther, E., Yeremenko, N., Havaux, M., Matthijs, H.C., and Hagemann, M. (2003). A photosystem 1 *psaFJ*-null mutant of the cyanobacterium *Synechocystis* PCC 6803 expresses the *isiAB* operon under iron replete conditions. FEBS Lett. *549*, 52–56.

Jiang, W., and Fillingame, R.H. (1998). Interacting helical faces of subunits *a* and *c* in the F_1F_0 ATP synthase of *Escherichia coli* defined by disulfide cross-linking. Proc. Natl. Acad. Sci. USA *95*, 6607–6612.

Jiang, W.P., Hermolin, J., and Fillingame, R.H. (2001). The preferred stoichiometry of *c* subunits in the rotary motor sector of *Escherichia coli* ATP synthase is 10. Proc. Natl. Acad. Sci. USA *98*: 4966–4971.

Jolley, C., Ben-Shem, A., Nelson, N., and Fromme, P. (2005). Structure of plant photosystem I revealed by theoretical modeling. J. Biol. Chem. *280*: 33627–33636.

Jolley, C.C., Wells, S.A., Hespenheide, B.M., Thorpe, M.F., and Fromme, P. (2006). Docking of photosystem I subunit C using a constrained geometric simulation. J. American Chem. Society *128*, 8803–8812.

Jordan, P., Fromme, P., Klukas, O., Witt, H.T., Saenger, W., and Krauß, N. (2001). Three-dimensional structure of cyanobacterial photosystem I at 2.5 Å resolution. Nature *411*, 909–917.

Kamiya, N., and Shen, J.R. (2003). Crystal structure of oxygen-evolving photosystem II from *Thermosynechococcus vulcanus* at 3.7-Å resolution. Proc. Natl. Acad. Sci. USA *100*, 98–103.

Kammel, M., Kern, J., Lubitz, W., and Bittl, R. (2003). Photosystem II single crystals studied by transient EPR: The light-induced triplet state. Biochim. Biophys. Acta *1605*, 47–54.

Karrasch, S., and Walker, J.E. (1999). Novel features in the structure of bovine ATP synthase. J. Mol. Biol. *290*, 379–384.

Keilty, A.T., Vavilin, D.V., and Vermaas, W.F.J. (2001). Functional analysis of combinatorial mutants with changes in the C-terminus of the CD loop of the D2 protein in photosystem II of *Synechocystis* sp PCC 6803. Biochemistry *40*, 4131–4139.

Kerfeld, C.A., Anwar, H.P., Interrante, R., Merchant, S., and Yeates, T.O. (1995). The structure of chloroplast cytochrome-c_6 at 1.9 Angstrom resolution—evidence for functional oligomerization. J. Mol. Biol. *250*, 627–647.

Knoetzel, H., Mant, A., Haldrup, A., Jensen, P.E., and Scheller, H.V. (2002). PST-O, a new 10-kDa subunit of eukaryotic photosystem I. FEBS Lett. *510*, 145–148.

Kohlbrenner, W.E., and Boyer, P.D. (1983). Probes of catalytic site cooperativity during catalysis by the chloroplast adenosine triphosphate and the adenosine triphosphate synthase. J. Biol. Chem. *258*, 10881–10886.

Krauss, N., Hinrichs, W., Witt, I., Fromme, P., Pritzkow, W., Dauter, Z., Betzel, C., Wilson, K.S., Witt, H.T., and Saenger, W. (1993). 3-Dimensional structure of System-I of photosynthesis at 6 Å resolution. Nature *361*, 326–331.

Krauss, N., Schubert, W.D., Klukas, O., Fromme, P., Witt, H.T., and Saenger, W. (1996). Photosystem I at 4 Å resolution represents the first structural model of a joint photosynthetic reaction center and core antenna system. Nature Structural Biol. *3*, 965–973.

Kurisu, G., Zhang, H., Smith, J.L., and Cramer, W.A. (2003). Structure of the cytochrome b_6f complex of oxygenic photosynthesis: Tuning the cavity. Science *302*, 1009–1014.

Lakshmi, K.V., Jung, Y.S., Golbeck, J.H., and Brudvig, G.W. (1999). Location of the iron–sulfur clusters F_A and F_B in photosystem I: An electron paramagnetic resonance study of spin relaxation enhancement of $P700^+$. Biochemistry *38*, 13210–13215.

Leonhardt, K.G., and Straus, N.A. (1989). Sequence of the flavodoxin gene from *Anabaena variabilis* 7120. Nucleic Acids Res. *17*, 4384–4384.

Li, Y., van der Est, A., Lucas, M.G., Ramesh, V.M., Gu, F., Petrenko, A., Lin, S., Webber, A.N., Rappaport, F., and Redding, K. (2006). Directing electron transfer within Photosystem I by breaking H-bonds in the cofactor branches. Proc. Natl. Acad. Sci. USA *103*, 2144–2149.

Loll, B., Kern, J., Saenger, W., Zouni, A., and Biesiadka, J. (2005). Towards complete cofactor arrangement in the 3.0 Å resolution structure of photosystem II. Nature *438*, 1040–1044.

Mazouni, K., Domain, F., Chauvat, F., and Cassier-Chauvat, C. (2003). Expression and regulation of the crucial plant-like ferredoxin of cyanobacteria. Mol. Microbiol. *49*, 1019–1029.

Medina, M., Lostao, A., Sancho, J., Gomez-Moreno, C., Cammack, R., Alonso, P.J., and Martinez, J.I. (1999). Electron-nuclear double resonance and hyperfine sublevel correlation spectroscopic studies of flavodoxin mutants from *Anabaena* sp. PCC 7119. Biophys. J 77, 1712–1720.

Meier, T., Polzer, P., Diederichs, K., Welte, W., and Dimroth, P. (2005). Structure of the rotor ring of F-type Na^+-ATPase from *Ilyobacter tartaricus*. Science *308*, 659–662.

Melkozernov, A.N., Bibby, T S., Lin, S., Barber, J., and Blankenship, R.E. (2003). Time-resolved absorption and emission show that the CP43′ antenna ring of iron-stressed *Synechocystis* sp. PCC6803 is efficiently coupled to the photosystem I reaction center core. Biochemistry *42*, 3893–3903.

Merchant, S., and Bogorad, L. (1986). Regulation by copper of the expression of plastocyanin and cytochrome c_{552} in *Chlamydomonas reinhardtii*. Mol. Cell Biol. 6, 462–469.

Meyer, Z., Tittingdorf, J.M., Rexroth, S., Schafer, E., Schlichting, R., Giersch, C., Dencher, N.A., and Seelert, H. (2004). The stoichiometry of the chloroplast ATP synthase oligomer III in *Chlamydomonas reinhardtii* is not affected by the metabolic state. Biochim. Biophys. Acta *1659*: 92–99.

Michel, K.P., and Pistorius, E.K. (2004). Adaptation of the photosynthetic electron transport chain in cyanobacteria to iron deficiency: The function of IdiA and IsiA. Physiol. Plant. *120*, 36–50.

Mitome, N., Suzuki, T., Hayashi, S., and Yoshida, M. (2004). Thermophilic ATP synthase has a decamer *c*-ring: indication of noninteger 10:3 H^+/ATP ratio and permissive elastic coupling. Proc. Natl. Acad. Sci. USA *101*: 12159–12164.

Morales, R., Charon, M.H., Hudry-Clergeon, G., Petillot, Y., Norager, S., Medina, M., Frey, M. (1999). Refined X-ray structures of the oxidized, at 1.3 Å, and reduced, at 1.17 Å, [2Fe-2S] ferredoxin from the cyanobacterium *Anabaena* PCC7119 show redox-linked conformational changes. Biochemistry 38, 15764–15773.

Muller, V., Lingl, A., Lewalter, K., and Fritz, M. (2005). ATP synthases with novel rotor subunits: New insights into structure, function and evolution of ATPases. J. Bioenerg. Biomembr. *37*, 455–460.

Murata, T., Yamato, I., Kakinuma, Y., Leslie, A.G., and Walker, J.E. (2005). Structure of the rotor of the V-Type Na^+-ATPase from *Enterococcus hirae*. Science *308*, 654–659.

Nakamura, Y., Kaneko, T., Sato, S., Ikeuchi, M., Katoh, H., Sasamoto, S., Watanabe, A., Iriguchi, M., Kawashima, K., Kimura, T., Kishida, Y., Kiyokawa, C., Kohara, M., Matsumoto, M., Matsuno, A., Nakazaki, N., Shimpo, S., Sugimoto, M., Takeuchi, C., Yamada, M., and Tabata, S. (2002). Complete genome structure of the thermophilic cyanobacterium *Thermosynechococcus elongatus* BP-1. DNA Res. 9, 123–130.

Nakano, T., Ikegami, T., Suzuki, T., Yoshida, M., and Akutsu, H. (2006). A new solution structure of ATP synthase subunit *c* from thermophilic *Bacillus* PS3, suggesting a local conformational change for H^+-translocation. J. Mol. Biol. *358*, 132–144.

Navarro, J.A., Hervas, M., and De la Rosa, M.A. (1997). Co-evolution of cytochrome c_6 and plastocyanin, mobile proteins transferring electrons from cytochrome b_6f to photosystem I. J. Biol. Inorg. Chem. 2, 11–22.

Noji, H., Yasuda, R., Yoshida, M., and Kinosita, K., Jr. (1997). Direct observation of the rotation of F_1-ATPase. Nature *386*, 299–302.

Partensky, F., Hess, W.R., and Vaulot, D. (1999). *Prochlorococcus*, a marine photosynthetic prokaryote of global significance. Microbiol. Mol. Biol. Rev. *63*, 106–127.

Pogoryelov, D., Yu, J., Meier, T., Vonck, J., Dimroth, P., and Muller, D.J. (2005). The c15 ring of the *Spirulina platensis* F-ATP synthase: F_1/F_0 symmetry mismatch is not obligatory. EMBO Rep. 6: 1040–1044.

Ramesh, V.M., Gibasiewicz, K., Lin, S., Bingham, S.E., and Webber, A.N. (2004). Bidirectional electron transfer in photosystem I: accumulation of A0- in A-side or B-side mutants of the axial ligand to chlorophyll A0. Biochemistry *43*, 1369–1375.

Rao, S.T., Shaffie, F., Yu, C., Satyshur, K.A., Stockman, B.J., Markley, J.L., and Sundaralingam, M. (1992). Structure of the oxidized long-chain flavodoxin from *Anabaena* 7120 at 2-Angstrom Resolution. Protein Sci. *1*, 1413–1427.

Redinbo, M.R., Cascio, D., Choukair, M.K., Rice, D., Merchant, S., and Yeates, T.O. (1993). The 1.5-Angstrom crystal-structure of plastocyanin from the green-llga *Chlamydomonas Reinhardtii*. Biochemistry 32, 10560–10567.

Rubinstein, J.L., Walker, J.E., and Henderson, R. (2003). Structure of the mitochondrial ATP synthase by electron cryomicroscopy. EMBO J. 22, 6182–6192.

Sandstrom, S., Park, Y.I., Oquist, G., and Gustafsson, P. (2001). CP43," the *isiA* gene product, functions as an excitation energy dissipator in the cyanobacterium *Synechococcus* sp. PCC 7942. Photochem. Photobiol. *74*, 431–437.

Schirmer, T., Bode, W., Huber, R., Sidler, W., and Zuber, H. (1985). X-ray crystallographic structure of the light-harvesting biliprotein C-phycocyanin from the thermophilic cyanobacterium *Mastigocladus laminosus* and its resemblance to globin structures. J. Mol. Biol. *184*, 257–277.

Schlarb, B.G., Wagner, M.J., Vijgenboom, E., Ubbink, M., Bendall, D.S., and Howe, C.J. (1999). Expression of plastocyanin and cytochrome *f* of the cyanobacterium *Phormidium laminosum* in *Escherichia coli* and *Paracoccus denitrificans* and the role of leader peptides. Gene *234*, 275–283.

Schlarb-Ridley, B.G., Bendall, D.S., and Howe, C.J. (2002). Role of electrostatics in the interaction between cytochrome *f* and plastocyanin of the cyanobacterium *Phormidium laminosum*. Biochemistry *41*, 3279–3285.

Schubert, W.D., Klukas, O., Krauss, N., Saenger, W., Fromme, P., and Witt, H.T. (1997). Photosystem I of *Synechococcus elongatus* at 4 Å resolution: Comprehensive structure analysis. J. Mol. Biol. *272*, 741–769.

Schubert, W.D., Klukas, O., Saenger, W., Witt, H.T., Fromme, P., and Krauss, N. (1998). A common ancestor for oxygenic and anoxygenic photosynthetic systems: A comparison based on the structural model of photosystem I. J. Mol. Biol. *280*, 297–314.

Seelert, H., Poetsch, A., Dencher, N.A., Engel, A., Stahlberg, H., and Muller, D.J. (2000). Structural biology. Proton-powered turbine of a plant motor. Nature *405*, 418–419.

Seelert, H., Dencher, N.A., and Muller, D.J. (2003). Fourteen protomers compose the oligomer III of the proton-rotor in spinach chloroplast ATP synthase. J. Mol. Biol. *333*: 337–344.

Sener, M.K., Park, S., Lu, D.Y., Damjanovic, A., Ritz, T., Fromme, P., and Schulten, K. (2004). Excitation migration in trimeric cyanobacterial photosystem I. J. Chem. Phys. *120*: 11183–11195.

Setif, P., Fischer, N., Lagoutte, B., Bottin, H., and Rochaix, J.D. (2002). The ferredoxin docking site of photosystem I. Biochim. Biophys. Acta-Bioenergetics *1555*, 204–209.

Shirakihara, Y., Leslie, A.G., Abrahams, J.P., Walker, J.E., Ueda, T., Sekimoto, Y., Kambara, M., Saika, K., Kagawa, Y., and Yoshida, M. (1997). The crystal structure of the nucleotide-free alpha 3 beta 3 subcomplex of F_1-ATPase from the thermophilic *Bacillus* PS3 is a symmetric trimer. Structure *5*, 825–836.

Stahlberg, H., Muller, D.J., Suda, K., Fotiadis, D., Engel, A., Meier, T., Matthey, U., and Dimroth, P. (2001). Bacterial Na^+-ATP synthase has an undecameric rotor. EMBO Rep. *2*: 229–233.

Stock, D., Leslie, A.G., and Walker, J.E. (1999). Molecular architecture of the rotary motor in ATP synthase. Science *286*, 1700–1705.

Stroebel, D., Choquet, Y., Popot, J.L., and Picot, D. (2003). An atypical haem in the cytochrome b_6f complex. Nature *426*, 413–418.

Valiyaveetil, F.I., and Fillingame, R.H. (1998). Transmembrane topography of subunit *a* in the *Escherichia coli* F_1F_0 ATP synthase. J. Biol. Chem. *273*, 16241–16247.

van Mieghem, F.J.E., Satoh, K., and Rutherford, A.W. (1991). A chlorophyll tilted 30-degrees relative to the membrane in the photosystem-II reaction center. Biochim. Biophys. Acta *1058*: 379–385.

Varley, J.P., Moehrle, J.J., Manasse, R.S., Bendall, D.S., and Howe, C.J. (1995). Characterization of plastocyanin from the cyanobacterium *Phormidium laminosum*: copper-inducible expression and SecA-dependent targeting in *Escherichia coli*. Plant Mol. Biol. *27*, 179–190.

Vik, S.B., and Ishmukhametov, R.R. (2005). Structure and function of subunit *a* of the ATP synthase of *Escherichia coli*. J. Bioenerg. Biomembr. *37*, 445–449.

Wada, T., Long, J.C., Zhang, D., and Vik, S.B. (1999). A novel labeling approach supports the five-transmembrane model of subunit *a* of the *Escherichia coli* ATP synthase. J. Biol. Chem. *274*, 17353–17357.

Walker, J.E., and Dickson, V.K. (2006). The peripheral stalk of the mitochondrial ATP synthase. Biochim. Biophys. Acta *1757*, 286–296.

Wastl, J., Purton, S., Bendall, D.S., and Howe, C.J. (2004). Two forms of cytochrome c_6 in a single eukaryote. Trends Plant Sci. *9*, 474–476.

Watenpaugh, K.D., Sieker, L.C., Jensen, L.H., Legall, J., and Dubourdieu, M. (1972). Structure of the oxidized form of a flavodoxin at 2.5-Angstrom resolution: Resolution of the phase ambiguity by anomalous scattering. Proc. Natl. Acad. Sci. USA *69*, 3185–3188.

Wilkens, S., and Capaldi, R.A. (1998). Electron microscopic evidence of two stalks linking the F_1 and F_0 parts of the *Escherichia coli* ATP synthase. Biochim. Biophys. Acta *1365*, 93–97.

Xue, Y., Okvist, M., Hansson, O., Young, S. (1998). Crystal structure of spinach plastocyanin at 1.7 Å resolution. Protein Sci. *7*, 2099–2105.

Yeremenko, N., Kouril, R., Ihalainen, J.A., D'Haene, S., van Oosterwijk, N., Andrizhiyevskaya, E.G. Keegstra, W., Dekker, H.L., Hagemann, M., Boekema, E.J., Matthijs, H.C.P., and Dekker, J.P. (2004). Supramolecular organization and dual function of the IsiA chlorophyll-binding protein in cyanobacteria. Biochemistry *43*, 10308–10313.

Yousef, N., Pistorius, E.K., and Michel, K.P. (2003). Comparative analysis of *idiA* and *isiA* transcription under iron starvation and oxidative stress in *Synechococcus elongatus* PCC 7942 wild-type and selected mutants. Arch. Microbiol. *180*, 471–483.

Zech, S.G., Kurreck, J., Eckert, H.J., Renger, G., Lubitz, W., and Bittl, R. (1997). Pulsed EPR measurement of the distance between $P680^{+\cdot,\ and}$ $Q(A)^{-\cdot}$ in photosystem II. FEBS Lett. *414*, 454–456.

Zech, S.G., Kurreck, J., Renger, G., Lubitz, W., and Bittl, R. (1999). Determination of the distance between Y(Z)ox* and Q_A-* in photosystem II by pulsed EPR spectroscopy on light-induced radical pairs. FEBS Lett. *442*, 79–82.

Zouni, A., Jordan, R., Schlodder, E., Fromme, P., and Witt, H. (2000). First photosystem II crystals capable of water oxidation. Biochim. Biophys. Acta *1457*, 103–105.

Zouni, A., Lueneberg, C., Fromme, P., Schubert, W.D., Saenger, W., and Witt, H.T. (1998). Characterization of crystals of photosystem II from the thermophilic cyanobacterium *Synechococcus elongatus*. In Photosynthesis: Mechanisms and Effects, G. Garab, ed. (Dordrecht, Kluwer), pp. 925–928.

Zouni, A., Witt, H.T., Kern, J., Fromme, P., Krauss, N., Saenger, W., and Orth, P. (2001). Crystal structure of photosystem II from *Synechococcus elongatus* at 3.8 Å resolution. Nature *409*, 739–743.

Membrane Systems in Cyanobacteria

10

Michelle Liberton and Himadri B. Pakrasi

Abstract

Cyanobacteria are photosynthetic prokaryotes with highly differentiated membrane systems. In addition to a Gram-negative-type cell envelope with plasma membrane and outer membrane separated by a periplasmic space, cyanobacteria have an internal system of thylakoid membranes where the fully functional electron transfer chains of photosynthesis and respiration reside. The presence of different membrane systems lends these cells a unique complexity among bacteria. Cyanobacteria must be able to reorganize the membranes, synthesize new membrane lipids, and properly target proteins to the correct membrane system. The outer membrane, plasma membrane, and thylakoid membranes each have specialized roles in the cyanobacterial cell. Understanding the organization, functionality, protein composition and dynamics of the membrane systems remains a great challenge in cyanobacterial cell biology.

Introduction

The oxygenic photosynthetic prokaryotes are characterized by their ability to use water as an electron donor and produce oxygen as a product of photosynthesis. The members of this group include the canonical cyanobacteria and relatives, all of which contain chlorophyll *a*; the chlorophyll *a* and *b* containing organisms, including those of the genus *Prochlorococcus*; and the chlorophyll *d* containing organisms of the genus *Acaryochloris* (Blankenship, 2002). All of these photosynthetic bacteria have been shown to form a single large phylum, which is dominated by the cyanobacteria. Cyanobacteria are a diverse group of organisms found in a wide variety of habitats, from marine, benthic, and freshwater environments to extreme conditions such as deserts, hot springs, and polar waters. Cyanobacteria can be filamentous, unicellular, or colonial, and they can vary greatly in size, from hundreds of micrometers to only a few micrometers. Concomitant with this morphological and ecological diversity is a distinct intracellular complexity. Cyanobacteria are able to perform different and often competing metabolic processes in the same cell, including protein synthesis, photosynthesis, respiration, and nitrogen fixation. To accomplish this, cyanobacteria contain numerous inclusion bodies as well as distinct membrane systems. Cyanobacteria are generally considered to have three membrane systems: outer membrane, plasma membrane, and thylakoid membranes. These membrane systems house large protein complexes, maintain proton gradients, participate in cell division, and perform many other duties vital to cellular functioning.

Protein complexes are fundamental components of cyanobacterial membranes, and indeed, can be thought to define membranes as functional units (Figure 10.1). In both the

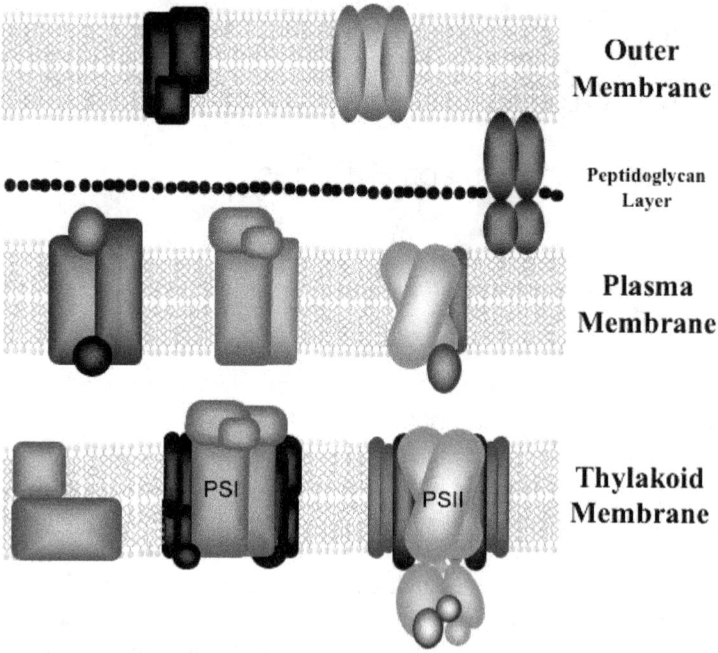

Figure 10.1 Cyanobacterial membranes contain large membrane protein complexes, including photosystem I (PSI) and photosystem II (PSII), in addition to other proteins and protein complexes of various general classes of proteins.

outer membrane and plasma membrane, the existence of porins and transporters identifies the role of these membranes in selection of materials for inclusion or exclusion from the cell, whereas thylakoid membranes are characterized by the presence of the functional photosynthetic electron transport chain and the maintenance of a lumenal proton gradient (see Chapter 9).

Recent studies using a variety of techniques have greatly elucidated the roles of cyanobacterial membranes. This chapter will focus on new findings in this area, as well as important issues that remain unresolved, and future prospects for understanding cyanobacterial membranes.

The cyanobacterial cell envelope

Overview of cell envelope components

Cyanobacteria are generally considered to have a Gram-negative cell envelope, although in many cyanobacteria there is evidence of features specific to the Gram-positive as well as the Gram-negative-type envelope, particularly in terms of lipid components, thickness of the peptidoglycan layer, and degree of crosslinking between peptidoglycan chains (Smarda *et al.*, 1979; Hoiczyk and Hansel, 2000). The cyanobacterial cell envelope consists of the outer membrane, peptidoglycan layer, and plasma or cytoplasmic membrane. A periplasmic space exists between the outer and plasma membranes and contains the peptidoglycan layer. The layers of the cyanobacterial cell envelope and organization of membrane systems and associated structures are shown in Figure 10.2.

Cyanobacterial membranes have been examined by electron microscopy for decades; thus, the organization of the membranes of the cell envelope has been well described. In the cyanobacterial cell envelope, the outer membrane is observed as a 5 nm to 10 nm membrane bilayer of medium electron density outside the peptidoglycan layer, and usually separated from the peptidoglycan layer by an electron opaque area of periplasm. Interior to the peptidoglycan layer, and separated from it by another area of periplasm, is the plasma membrane. Usually ~8 nm thick, the plasma membrane is observed as a membrane bilayer that directly encloses the cell cytoplasm. Outside the outer membrane, and likely connected to the outer membrane at variously spaced junctions, is the S-layer or surface layer.

Two recent advances have contributed greatly to our understanding of cyanobacterial membranes. These are ultra-rapid freeze-fixation followed by image processing for electron microscopy, and proteomic analysis of isolated membranes. The influence of these technical advances will be discussed in the following sections describing significant recent developments pertaining to each of the membrane systems.

S-layer

Surface layers, or S-layers, are important components of cellular ultrastructure that are not as well understood in cyanobacteria compared to other bacteria. This is likely due to the approximately twenty year delay in commencement of research in cyanobacterial S-layers

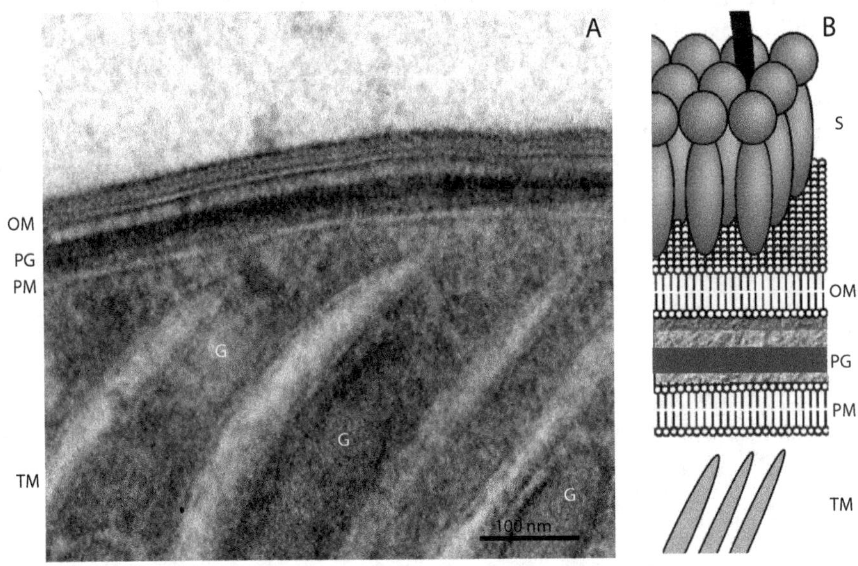

Figure 10.2 (A) Components of the cell envelope of the unicellular cyanobacterium *Cyanothece* sp. ATCC 51142. PM, plasma membrane; PG, peptidoglycan layer; OM, outer membrane. Also labeled are TM, thylakoid membrane, G, glycogen granule. B) Schematic diagram of membrane cross section showing membrane organization in a cyanobacterium. S, surface layer; OM, outer membrane; PG, peptidoglycan layer; PM, plasma membrane; TM, thylakoid membranes. Adapted from (Samuel *et al.*, 2001), with kind permission from the author.

compared to other bacteria and archaea (Smarda *et al.*, 2002). The S-layer, while not membranous in the sense of lipoprotein membranes, will be discussed here due to its importance in cell function, relevance to discussions of cyanobacterial cell envelope architecture, and close association with the outer membrane. S-layers are widely found throughout the bacteria and archaea, where they are thought to perform various functions, including but not limited to protection, cell integrity, communication with the environment, and adhesion (Engelhardt and Peters, 1998). In cyanobacteria, S-layers have been found in over 60 strains representing 16 genera, including both unicellular and filamentous families, but the occurrence of S-layers is not universal, as they are lacking in some strains of both unicellular and filamentous types (Smarda *et al.*, 2002). The S-layer is a paracrystalline array of proteinaceous material that forms a layer over the entire cell. Cyanobacteria are covered with a mucilaginous sheath on the exterior of the cell; the S-layer is between this sheath and the outer membrane. The subunits of the S-layer apparently self-assemble, forming a regular network of pores though which materials of certain sizes and charges are permitted passage (Beveridge and Graham, 1991). The S-layer is usually a monomolecular array of identical units arranged in a lattice symmetry, of which the p6 or hexagonal arrangement is the most common in cyanobacteria (Smarda *et al.*, 2002). Center to center spacing of the S-layer subunits in cyanobacteria ranges from 5.0 nm to 22.0 nm (Smarda *et al.*, 2002). While the S-layer of each cyanobacterial strain has characteristics unique to that strain, the general features of hexagonal symmetry remain the same: six protein subunits arranged around a pore (Figure 10.3).

The S-layers of cyanobacteria are amenable to study by a variety of techniques due to the regular arrangement of subunits, including transmission electron microscopy (TEM), optical diffraction, and computer reconstruction in 2D and 3D (Smarda *et al.*, 2002). Freeze-fracture preparations can show the S-layer subunits in great detail (Figure 10.4).

Recent work has described a novel function for the S-layer in swimming motility in *Synechococcus* sp. strain WH8102. The mechanisms of swimming motility in cyanobacteria are largely unknown due to the lack of visible organelles used for swimming. A mutant

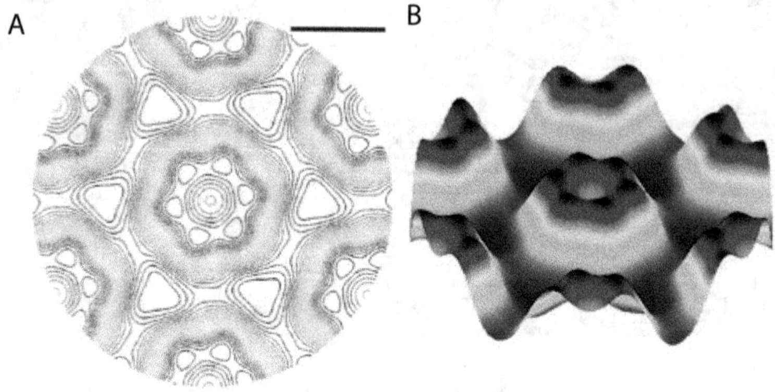

Figure 10.3 (see also Plate 10.3) Computer reconstructions of a cyanobacterial S-layer displaying a p6 symmetry. (A) Isodensity lines, and (B) space simulation. Bar = 10 nm. Adapted from (Smarda *et al.*, 2002), reprinted from Micron *33*, 257–277, copyright 2002, with permission from Elsevier.

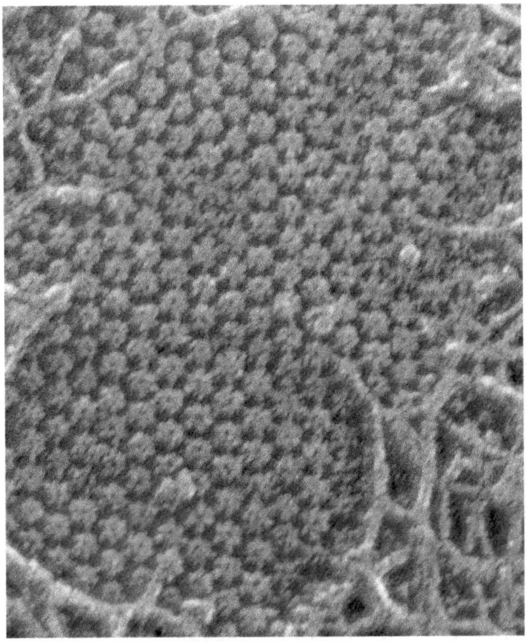

Figure 10.4 S-layer of *Synechocystis* sp. PCC 6803. Deep-etch freeze-fracture electron microscopy reveals the S-layer of *Synechocystis* to be composed of identical subunits of hexagonal symmetry. (Laboratory of John Heuser, Washington University School of Medicine)

strain in which the *swmA* gene was inactivated was known to lack swimming motility, and further analysis to understand the involvement of SwmA in swimming motility showed that this mutant displayed altered cell envelope layers (McCarren *et al.*, 2005). The cells were found by deep-etch freeze fracture as well as thin-section electron microscopy to lack the S-layer, but retain the fibrillar layer, apparently the attachment layer between the outer membrane and S-layer (McCarren *et al.*, 2005). The *swmA* mutant was still able to rotate when cells were attached to a surface, but the cells could no longer generate thrust for movement, indicating a potential involvement of the S-layer in generating propulsional force.

Notably, biosynthesis and protein composition of the S-layers of cyanobacteria remain largely unknown. Presumably, conclusions regarding the components of S-layers of other bacteria would be relevant to cyanobacteria based on the similar appearance of the S-layers from freeze-fracture studies. In particular, the SLH (surface layer homology) domain is a conserved motif of bacterial proteins, and has been proposed to act as a linker between the S-layer and the other components of the cell envelope, specifically the peptidoglycan layer (Lupas *et al.*, 1994). SLH domains have been found in the SomA and SomB outer membrane proteins of *Synechococcus* sp. PCC 6301, and in six additional *Synechocystis* sp. PCC 6803 protein sequences (Hansel *et al.*, 1998). Thus, SLH domains might have a general function in connecting and stabilizing the components of the cell envelope. Future work in this area will be needed to determine the full protein complementation of the surface layer as well as the resulting functionality.

Outer membrane

The outer membranes of cyanobacteria contain lipopolysaccharide (LPS) and other polysaccharides that provide protection for the cell, in addition to numerous proteins such as pilus proteins and porins. In general, the cyanobacterial outer membrane remains largely uncharacterized, particularly when compared to studies of the outer membranes of other Gram-negative bacteria.

The protein components of the outer membrane of *Synechocystis* 6803 have recently been explored by mass spectrometry and database identification of proteins from isolated outer membranes (Huang *et al.*, 2004). This study identified 29 different proteins, all of which contained a predicted N-terminal signal peptide required for correct targeting of outer membrane proteins. Interestingly, 26 of these proteins contained a Sec signal sequence, while the other three contained a Tat signal sequence (Huang *et al.*, 2004), suggesting that the Sec and Tat pathways both function in protein delivery to the outer membrane in cyanobacteria, and implying that the Sec pathway may be used preferentially.

A number of proteins found in this study to be localized to the *Synechocystis* outer membrane provide insight into the function of the outer membrane and its relation to the other components of the cell envelope. Toc75, the *Synechocystis* homolog to the chloroplast Toc complex protein, is found in the outer membrane, an analogous localization to the protein found in the chloroplast outer envelope. Importantly, two of the six *Synechocystis* proteins containing SLH domains were found in the outer membrane. Interestingly, the identification of some proteins in both the outer membrane proteome as well as in the plasma membrane proteome suggests that points of contact may exist between these two membrane systems (see Figure 10.1), presumably for the transfer of materials between the membranes (Huang *et al.*, 2002; Huang *et al.*, 2004). While Huang and colleagues present an important first analysis of the outer membrane proteome, undoubtedly, given the estimated size of the total *Synechocystis* proteome at nearly 3600 proteins, of which approximately 30% are proposed to be membrane proteins, many more outer membrane proteins remain to be identified in future studies. In an analysis of the *Anabaena* sp. PCC 7120 outer membrane proteome (Moslavac *et al.*, 2005), 55 proteins were identified, of which only 44 proteins (80%) have counterparts in *Synechocystis* 6803, and of these 44 proteins, only 12 were identified in studies of the *Synechocystis* membrane preparations (Fulda *et al.*, 1999; Huang *et al.*, 2002; Huang *et al.*, 2004). These data emphasize the extent to which the complete membrane proteome remains to be determined.

The susceptibility of the outer membrane to poration by EDTA has been used to shed light on the sequestration of manganese by *Synechocystis* 6803 (Keren *et al.*, 2002). Manganese is critical to Photosystem II functionality as well as a vital micronutrient and cofactor for various enzymes. When introduced into new media, *Synechocystis* cells were found to rapidly sequester large amounts of manganese from the media into a pool associated with cell envelope. This large pool is released by EDTA and by collapse of the membrane potential, indicating that it is retained by the cell in an active manner. It is likely that this pool is associated with the outer membrane, and perhaps stored in the periplasm (Keren *et al.*, 2002). The stored manganese is available for transport into the interior of the cell via unknown transporters as required by the cell, and may represent a mechanism by which cyanobacterial cells are able to sequester micronutrients to sustain cell functioning under nutrient-depleted conditions.

The location of the outer membrane on the cell exterior necessitates its function in protection of the cell, and carotenoids in the outer membrane appear to play a role in this protection. The xanthophyll glycoside myxoxanthophyll is found in the outer membrane and plasma membrane of *Synechocystis* sp. PCC 6714, and deletion of this gene in *Synechocystis* 6803 by Mohamed and coworkers resulted in cells with altered ultrastructural characteristics (Mohamed *et al.*, 2005). Strains lacking myxoxanthophyll were found to lack the S-layer, possibly due to the role of myxoxanthophyll in the outer membrane providing an anchor for the S-layer. In addition, the arrangement of thylakoid membranes is altered in the mutant, suggesting a role in thylakoid membrane stability (Mohamed *et al.*, 2005). This role in membrane stability is surprising and demonstrates the complexity of the composition and function of the cyanobacterial outer membrane.

Plasma membrane

Cyanobacterial plasma membrane characterization has also benefited from proteomics analysis. A proteomics study of isolated plasma membranes by Huang and colleagues (Huang *et al.*, 2002) identified 57 *Synechocystis* 6803 plasma membrane proteins, including 40 peripheral membrane proteins and 17 integral membrane proteins. Thirteen proteins contained a Sec signal sequence, while three proteins had a Tat pathway sequence. As in the outer membrane proteomics study, a greater number of proteins were identified that use the Sec pathway compared to the Tat pathway.

It is significant that a number of proteins from the photosynthetic and respiratory transport chains were found in the plasma membrane proteome. These include the cytochrome *c* oxidase subunit II and components of the ATP synthase, as well as proteins from photosystem II and photosystem I (Huang *et al.*, 2002), proteins that are known to function in the thylakoid membrane (discussed below). In addition, the VIPP1 protein, proposed to be involved in thylakoid biogenesis, was also found in the plasma membrane (Huang *et al.*, 2002). As in the outer membrane proteomic analysis, many more plasma membrane proteins remain to be identified. The possibility of plasma membrane proteins arranged in the membrane in functional domains is an intriguing suggestion, and analysis of membrane fractions showed that the plasma membrane has a heterogeneous protein composition in which proteins associated with specific functions were isolated together (Srivastava *et al.*, 2006).

While the thylakoid membranes are known to be the sites of active photosynthesis, an important role for the plasma membrane in photosystem biogenesis has recently been demonstrated. Initial work showed the presence of the Photosystem II proteins D1, D2 and PsbO in plasma membranes of *Synechococcus* sp. PCC 7942 (formerly known as *Anacystis nidulans* R2) purified by sucrose density centrifugation (Smith and Howe, 1993). Subsequent studies with membranes isolated by a combination of sucrose density centrifugation and aqueous two-phase partitioning resulted in highly purified isolated membranes, from which immunoblot analysis showed that the reaction core complex proteins of both Photosystem I and Photosystem II were found in the plasma membrane (Zak *et al.*, 2001). In addition, proteins necessary for complex assembly, including the D1 processing protease CtpA, were found only in the plasma membrane, while the proximal antennae proteins CP43 and CP47 were localized only to the thylakoid membranes (Zak *et al.*, 2001). These data suggested that the initial steps of photosystem assembly happened in the plasma membrane, and that the partially assembled complexes were then translocated to the thylakoid membranes by an unknown mechanism, where the assembly process was completed.

Consistent with this proposed assembly pathway, the photosystem proteins in the plasma membrane were found to be assembled into multiprotein complexes, and both Photosystem I and Photosystem II in the plasma membrane were shown to be capable of charge separation (Zak *et al.*, 2001; Keren *et al.*, 2005). Photosystem II in the plasma membrane has an assembled acceptor side, but an unassembled donor side (Keren *et al.*, 2005). However, while the complexes in the plasma membrane have been characterized and their functionality described, their method of translocation between plasma membrane and thylakoid membrane remains unknown.

In addition to the partially assembled protein complexes of the photosynthetic electron transport chain, the plasma membrane also contains other complexes, including ATP synthase, components of the respiratory chain, and various transporters, including members of the ABC group of transporters. In recent analyses, two classes of ABC transporters have emerged. X-ray crystallographic structures of ZnuA, the solute binding protein of the zinc transporter (Banerjee *et al.*, 2003), and MntC, the periplasmic binding protein of the manganese transporter (Rukhman *et al.*, 2005), both from *Synechocystis* 6803, have shown that these metal binding proteins are highly similar, although with distinct metal specificity. The recent high resolution X-ray crystallographic structure of NrtA, the solute binding lipoprotein of the NrtABCD transporter, has revealed the novel structure of this transporter and a previously unknown connection between nitrate and bicarbonate uptake in cyanobacteria (Koropatkin *et al.*, 2006a). The structure of CmpA, the solute binding component of the bicarbonate transporter (Koropatkin, *et al.*, 2006b) and NrtA are similar. How ligand binding specificity is conferred in these transporters remains to be determined.

Thylakoid membranes

Morphology and membrane arrangement

Thylakoid membranes are the most well characterized of the cyanobacterial membrane systems. All known cyanobacteria except *Gloeobacter violaceus* contain thylakoid membranes. Phylogenetic analysis has shown that *Gloeobacter* diverged within the cyanobacteria prior to the origin of plastids (Nelissen *et al.*, 1995). In *Gloeobacter*, the photosynthetic apparatus is located in the plasma membrane and the proton gradient is formed along the plasma membrane, with phycobilisomes layered along the cytoplasmic side of the membrane (Koyama *et al.*, 2006). *Gloeobacter* displays lower photosynthetic efficiency than cyanobacteria with internal thylakoid membrane systems, suggesting that a developed system of thylakoid membranes is required for high levels of photosynthesis.

Thylakoid membranes in cyanobacteria are double bilayers of approximately 15 nm to 20 nm with closed ends that allow the accumulation of a proton gradient across the membrane. The contents of the thylakoid lumen remain largely uncharacterized due to the lack of lumenal preparation techniques, although the thylakoid membranes themselves have been efficiently isolated from a number of different cyanobacterial strains. In contrast to the membranes of chloroplasts, in cyanobacteria thylakoid membranes are not arranged in grana stacks and stromal lamella, but rather in layers of approximately 3 to 8 double membrane bilayers.

Phycobilisomes are the light harvesting phycobiliprotein complexes in cyanobacteria that associate mainly with Photosystem II on the cytoplasmic side of the thylakoid membrane. The hemidiscoidal phycobilisome complexes can often be observed by electron microscopy as 50–70 nm dense structures arranged alternately between the thylakoid membranes (Mullineaux *et al.*, 1997; Anderson and Toole, 1998) (Figure 10.5). The nature of association between phycobilisomes and thylakoid membranes remains unclear. The movement of phycobilisomes and Photosystem II complexes in cyanobacterial cells has recently been explored using fluorescence techniques such as fluorescence recovery after photobleaching (FRAP). In *Dactylococcopsis salina*, photosystem II complexes were found to be immobile, while phycobilisomes diffused rapidly in the membrane (Mullineaux *et al.*, 1997).

The arrangement of thylakoid membranes in cyanobacteria has been examined by electron microscopy. It is notable that there appears to be a large variation in membrane arrangement patterns between different cyanobacteria (Figure 10.6). In *Synechococcus* 7942 and *Agmenellum quadruplicatum* PR-6 (now known as *Synechococcus* sp. PCC 7002), thylakoid membranes are arranged largely as concentric cylinders around the periphery of the cell (Figure 10.6B, C) (Nierzwicki-Bauer *et al.*, 1983; Sherman *et al.*, 1994), whereas in *Synechocystis* 6803, thylakoids are more loosely organized in loops openly arranged throughout the cell cytoplasm (Figure 10.6A) (Liberton *et al.*, 2006). *Cyanothece* sp. ATCC 51142 accumulates glycogen granules between thylakoid membranes (Figure 10.6D), an occurrence which might influence the arrangement of the radially oriented thylakoids in this organism. It has been shown that the thylakoid arrangement in *Synechococcus* 7942 coincides with the specific localization of the photosynthetic complexes, with Photosystem I found preferentially in the thylakoid membrane layers located closest to the plasma membrane, while Photosystem II proteins localized throughout the thylakoid membrane layers (Sherman *et al.*, 1994). In other cyanobacteria such as *Synechocystis* 6803, no such specific arrangement of the photosystems is observed.

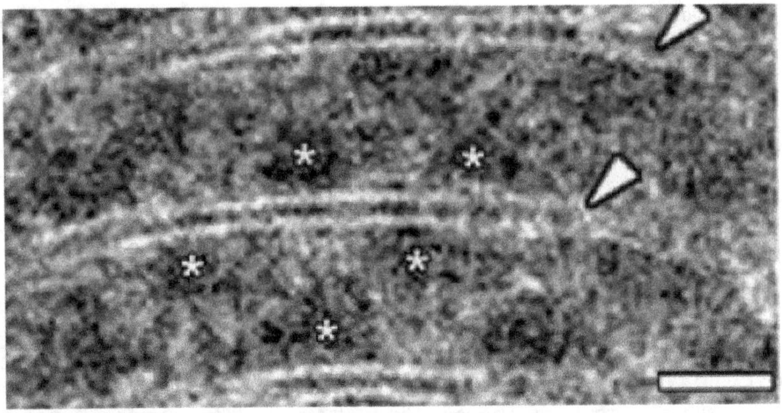

Figure 10.5 Phycobilisomes associated with thylakoid membranes of *Synechocystis* 6803. Asterisks, phycobilisomes; arrowheads, thylakoid membranes. Bar = 50 nm. From (van de Meene *et al.*, 2006), with kind permission of Springer Science and Business Media.

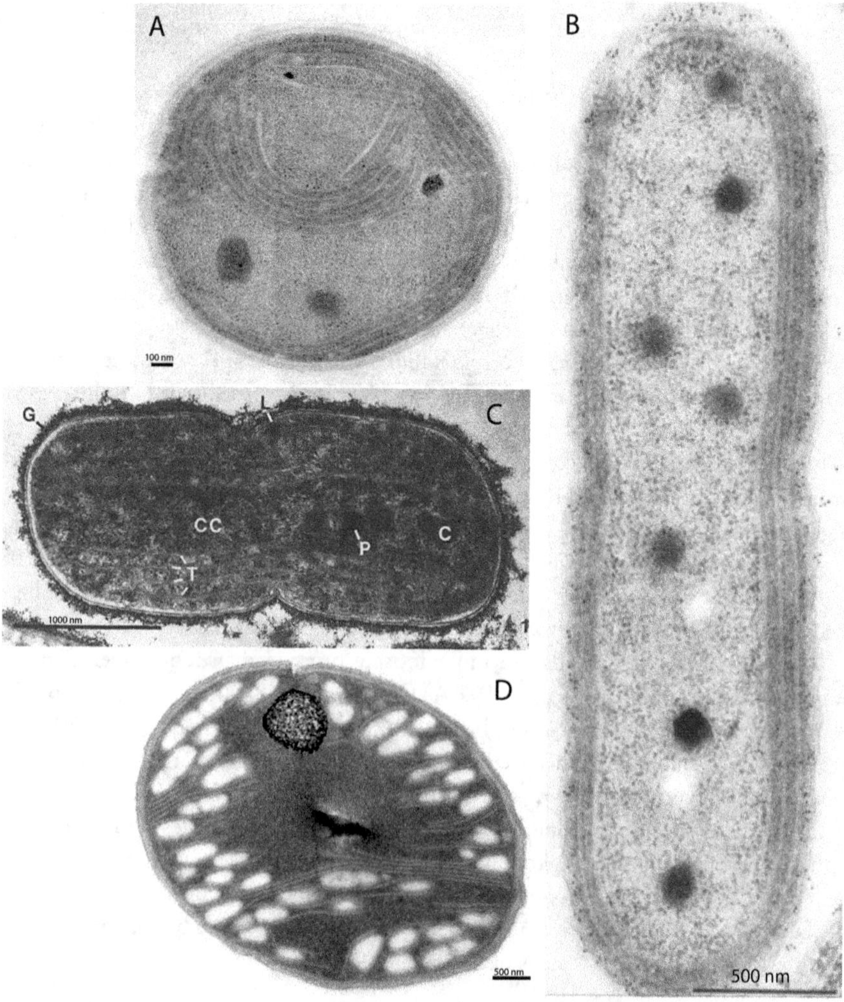

Figure 10.6 Diversity in thylakoid membrane arrangement in different cyanobacteria. (A) *Synechocystis* 6803, (B) *Synechococcus elongatus*, formerly *Synechococcus* 7942, (C) *Synechococcus* 7002, formerly *Agmenellum quadruplicatum*, and (D) *Cyanothece* 51142. (A) Adapted from (Liberton *et al.*, 2006), reproduced with permission from Springer-Verlag, (B) adapted from (Sherman *et al.*, 1994), reproduced with permission of American Society of Plant Biologists via Copyright Clearance Center, (C) adapted from (Nierzwicki-Bauer *et al.*, 1983), reproduced from The Journal of Cell Biology, 1983, 97: 713–722. Copyright 1983 The Rockefeller University Press. Labeling in panel C: G, glycocalyx; C, carboxysome; CC, central cytoplasmic region; L, lipid body; P, polyphosphate body; T, thylakoid membranes (from Nierzwicki-Bauer *et al.*, 1983).

The area of the cell cytoplasm occupied by the thylakoid membranes often forms a ribosome exclusion area, in which a distinct lack of ribosomes is noted as compared to other areas of the cell cytoplasm. The central cytoplasmic region contains various inclusion bodies, including carboxysomes, glycogen granules, polyphosphate granules, cyanophycin granules, and poly-β-hydroxybutyrate bodies (Allen, 1984).

Thylakoid membrane biogenesis

Thylakoid membrane biogenesis remains an outstanding question in cyanobacteria. Despite years of analysis and numerous studies, the mechanism by which cyanobacterial thylakoid membranes are remodeled, expanded, and reorganized remains elusive. Two basic mechanisms have been suggested for movement of lipid and protein to the thylakoid membranes, continuity with the plasma membrane and vesicle transport (Gantt, 1994; Mullineaux, 1999; Spence *et al.*, 2003; see also Chapter 11).

Importantly, the existence of a vesicle transport system in chloroplasts has been described. In isolated pea chloroplasts, vesicles are seen to accumulate after treatment with microcystin LR and after incubation at 4°C, both conditions known to interfere with vesicle fusion in eukaryotic systems (Westphal *et al.*, 2001b). While it is generally accepted that cyanobacteria are the progenitors of the chloroplasts of plants and green algae, the existence of a vesicle transport system has not been demonstrated in prokaryotes. Furthermore, the components of the chloroplast transport system are proposed to be similar to those in eukaryotes, such as dynamin (Westphal *et al.*, 2001b), which is not present in the genome of cyanobacteria such as *Synechocystis* 6803. Thus it appears that the vesicle transport system in chloroplasts is of eukaryotic origin and not derived from the prokaryotic ancestor.

The *VIPP1* gene has been proposed to be involved in thylakoid membrane formation in *Arabidopsis thaliana* (Kroll *et al.*, 2001). The T-DNA insertional mutant of this gene contained chloroplasts with severely diminished and distorted thylakoid membranes. VIPP1 is present in both the inner envelope membrane and thylakoid membranes, and is proposed to be involved in the formation of vesicles at the chloroplast inner envelope, and by extension in thylakoid membrane formation and maintenance (Kroll *et al.*, 2001). In *Synechocystis* 6803, the Vipp1 homologue is localized only to the plasma membrane, unlike in *Arabidopsis* (Westphal *et al.*, 2001a). The cyanobacterial insertional mutant did not completely segregate, but analysis of partially segregated cultures indicated an involvement of Vipp1 in thylakoid membrane biogenesis. By electron microscopy, thylakoid membranes in mutant cells were diminished in number and disorganized. Further analysis of Vipp1 from cyanobacteria and plants showed that Vipp1 forms complexes consisting solely of multiple copies of Vipp1 proteins (Aseeva *et al.*, 2004). However, the precise involvement of Vipp1 proteins in thylakoid membrane formation or maintenance, or in vesicle formation, remains unknown and controversial.

Structures called thylakoid centers have been proposed to function in thylakoid organization and biogenesis. These structures were first described in four different genera of cyanobacteria (Kunkel, 1982). Thylakoid centers were observed where thylakoid membranes converged near the plasma membrane, and thylakoid membranes associated directly with the thylakoid centers in these areas, although no plasma membrane-thylakoid membrane connectivity could be observed. In this early study, thylakoid centers were proposed to be cylindrical in nature, measuring 30 nm wide and 320 nm long, and composed of regularly organized subunits with thylakoid membranes spaced and connected along the long axis (Kunkel, 1982). Using high resolution techniques including high-pressure freezing and electron tomography, this topic was recently revisited (van de Meene *et al.*, 2006). In *Synechocystis* 6803, thylakoid membranes were also observed to converge at various sites adjacent to the plasma membrane and associate along the length of cylindrical structures identified as thylakoid centers. Thylakoid centers in *Synechocystis* were 40–50 nm in diameter and up to 1000 nm in length, larger than those observed by Kunkel (Kunkel, 1982). While clear continuity between thylakoid centers and thylakoid membranes could not be observed, thylakoid centers were reported to be contiguous with the periplasm or adjacent to the plasma membrane (van de Meene *et al.*, 2006).

While the observation of thylakoid centers as distinct structures seems clear, their function remains unknown, and their purported role in thylakoid organization, stability, or biogenesis is unsubstantiated. Another possibility is that thylakoid centers are remnants of cell division. The fate of thylakoid membranes during cell division remains unclear. Electron micrographs show the thylakoid membranes extending through the invaginating septum in *Synechocystis* 6803, and as the septum closes, the thylakoid membranes are divided into the two new daughter cells. How the membranes are apparently pinched apart and the ends closed to maintain the internal proton gradient is unknown. Nonetheless, cell division results in a change in thylakoid membrane morphology, so that the thylakoid membranes have distinct ends that appear to converge near the plasma membrane at the former division site. The thylakoid center may mark the site on the plasma membrane where division occurred.

Thylakoid membrane topology

As mentioned above, a long-standing controversy in cyanobacteria concerns the relationship between the plasma membrane and the thylakoid membranes, specifically whether physical continuity exists between these two membrane systems. It is significant that the existence of intracytoplasmic membranes in purple phototrophic bacteria is well established. Intracytoplasmic membranes can be isolated as vesicular chromatophores which contain the photochemical reaction centers and light-harvesting apparatus. In electron micrographs of purple bacteria such as *Rhodospirillum rubrum*, the continuity between the cytoplasmic membranes and intracytoplasmic membranes is apparent (Collins and Remsen, 1991). In cyanobacteria, comparable areas of obvious physical continuity between plasma membrane and thylakoid membranes are not observed in electron micrographs. In chloroplasts, connections between plasma membrane and thylakoid membranes are observed in undifferentiated proplastids but not in mature chloroplasts (Vothknecht and Westhoff, 2001).

Two recent studies have addressed the question of plasma membrane-thylakoid membrane continuity, with different conclusions. Van de Meene and coworkers, using thin section electron microscopy and electron tomography, found evidence of a connection between thylakoid membrane and plasma membrane (van de Meene *et al.*, 2006). However, such connections are not common and would not be sufficient to account for thylakoid biogenesis via transfer of lipids and proteins from the plasma membrane. Furthermore, when compared to the unequivocal evidence of plasma membrane-intracytoplasmic membrane continuity seen in purple bacteria, the proposed connection site in *Synechocystis* 6803 is markedly different in nature.

Also in *Synechocystis* 6803, a study using thin section electron microscopy, freeze-fracture, and serial sectioning reached the conclusion that areas of physical continuity between plasma membrane and thylakoid membranes could not be detected using multiple approaches. Thus, the existence of such connections is highly unlikely, and thylakoid membranes are in fact distinct intracellular compartments in cyanobacteria (Liberton *et al.*, 2006).

Protein targeting to cyanobacterial membranes

An outstanding question in cyanobacteria is how proteins are accurately targeted to the correct membrane where they function. In cyanobacteria, integral membrane proteins are likely transported by an SRP (signal recognition particle)-dependent pathway. In the case of soluble proteins, analysis of the *Synechocystis* 6803 genome shows that the genes encoding Sec and Tat proteins are found as single copies. Proteomics data indicating the involvement of the Sec pathway in outer membrane protein localization (Huang *et al.*, 2004) are important and consistent with previous studies showing that the Sec translocase component SecY is present in both plasma membrane and thylakoid membranes of *Synechococcus* 7942 (Nakai *et al.*, 1993). However, the lack of translocation mutants in cyanobacteria has impeded work in understanding the functioning and substrates of the different translocation apparatuses. Using a different approach, Spence and colleagues (Spence *et al.*, 2003) used GFP as a reporter protein linked to the Tat-specific signal peptide from the *Escherichia coli* TorA protein. They showed that the Tat-specific targeting signal resulted in specific localization of GFP to the periplasm in *Synechocystis* 6803 (Figure 10.7). This periplasmic targeting could be due to the localization of the Tat apparatus only in the plasma membrane, a possibility in view of the lack of identified Tat-targeted thylakoid lumenal proteins. But as the Tat apparatus functions in the thylakoid membranes of chloroplasts, the lack of identified thylakoid lumenal proteins in cyanobacteria may only be due to the lack of satisfactory lumenal preparations. In effect, the localization of GFP to the periplasm may reflect a default targeting pathway, in which proteins are first targeted to the periplasm and later to the thylakoid lumen if they contain a recognizable, but so far unidentified, lumenal targeting signal (Spence *et al.*, 2003).

Conclusions and future directions

A complete understanding of cyanobacterial membranes, as a step towards understanding cell physiology, remains a critical goal in cyanobacterial research. Currently, the combination of proteomics of isolated membrane fractions, advances in electron microscopy based techniques, and fluorescence-based approaches have greatly contributed to our understanding of cyanobacterial membranes. However, numerous fundamental questions remain. Foremost is the question of thylakoid membrane biogenesis. To achieve a thorough understanding of cyanobacterial membranes requires understanding how the thylakoid membranes are formed and remodeled during cell growth and division.

The small size of cyanobacteria relative to mammalian or plant cells has so far largely prohibited the use of fluorescence techniques which have proved invaluable in cell biology studies in these systems. However, the use of GFP in studies such as that of Spence and colleagues (Spence *et al.*, 2003) has demonstrated that specific fluorescence localization can be detected in cyanobacterial cells, and this work is likely to catalyze future use of these tools. In addition, improvements in optical imaging systems, such as fluorescence lifetime imaging microscopy (FLIM) are likely to benefit studies in cyanobacterial cells.

Due to the small size of these organisms, electron microscopy has long been the preferred imaging method. Recent advances in electron microscopy are increasing the

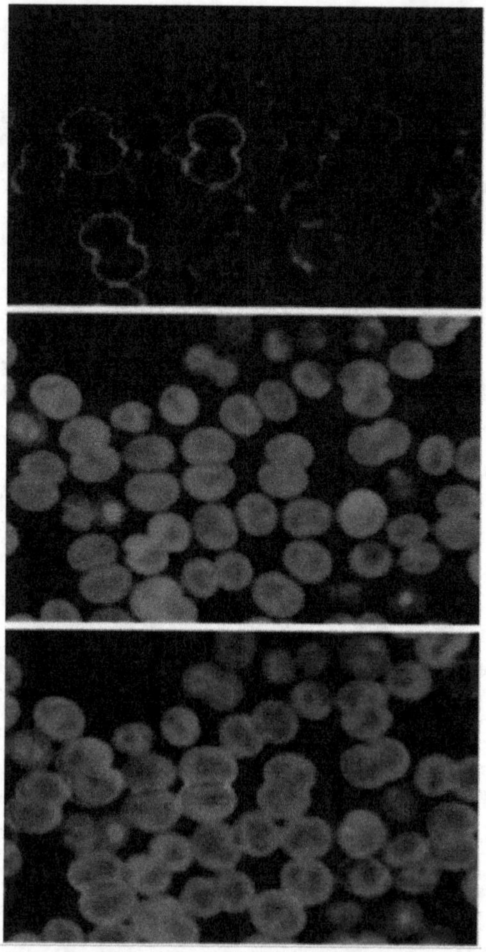

Figure 10.7 (see also Plate 10.7) The addition of a Tat-specific targeting signal to GFP results in localization of GFP only to the periplasm in *Synechocystis* 6803. Top panel, GFP; middle panel, endogenous chlorophyll; bottom panel, merged images. From (Spence *et al.*, 2003) with permission from Blackwell Publishing.

resolution to the 3–5 nm range. Most important for cyanobacterial studies is the emergence of electron tomography as a tool for high-resolution 3-D imaging of bacterial cells. Cryoelectron tomography, in which cells are preserved in a frozen-hydrated state, is expected to provide high-resolution images in 3-D, particularly once current technical challenges are overcome.

While electron microscopy will continue to be a valuable tool in studies of cyanobacterial membranes, it nonetheless presents a static view of what are undoubtedly dynamic membrane systems. Thylakoid membranes, as well as the membranes of the cell envelope, are constantly undergoing significant changes in their composition and organization during the course of cell growth and division. The greater challenge therefore remains to understand the dynamics of cyanobacterial membrane systems.

Acknowledgments

We thank members of the Pakrasi laboratory for collegial discussions and comments on the manuscript, John Heuser for his collaborative work imaging the S-layer of *Synechocystis* 6803 and R. Howard Berg for his critical reading of the manuscript. This work is part of a Membrane Biology EMSL Scientific Grand Challenge project at the W. R. Wiley Environmental Molecular Sciences Laboratory, a national scientific user facility sponsored by the U.S. Department of Energy's Office of Biological and Environmental Research (BER) program located at Pacific Northwest National Laboratory. PNNL is operated for the Department of Energy by Battelle.

References

Allen, M.M. (1984). Cyanobacterial cell inclusions. Annu. Rev. Microbiol. *38*, 1–25.

Anderson, L.K., and Toole, C.M. (1998). A model for early events in the assembly pathway of cyanobacterial phycobilisomes. Mol. Microbiol. *30*, 467–474.

Aseeva, E., Ossenbuhl, F., Eichacker, L.A., Wanner, G., Soll, J., and Vothknecht, U.C. (2004). Complex formation of Vipp1 depends on its alpha-helical PspA-like domain. J. Biol. Chem. *279*, 35535–35541.

Banerjee, S., Wei, B., Bhattacharyya-Pakrasi, M., Pakrasi, H.B., and Smith, T.J. (2003). Structural determinants of metal specificity in the zinc transport protein ZnuA from synechocystis 6803. J. Mol. Biol. *333*, 1061–1069.

Beveridge, T.J., and Graham, L.L. (1991). Surface layers of bacteria. Microbiol. Rev. *55*, 684–705.

Blankenship, R.E. (2002). Molecular Mechanisms of Photosynthesis (Blackwell Science).

Collins, M.L.P., and Remsen, C.C. (1991). The purple phototrophic bacteria, In Structure of Phototrophic Prokaryotes, J.F. Stolz, ed. (Boca Raton, FL 33431: CRC Press, Inc), pp. 49–77.

Engelhardt, H., and Peters, J. (1998). Structural research on surface layers: a focus on stability, surface layer homology domains, and surface layer-cell wall interactions. J. Struct. Biol. *124*, 276–302.

Fulda, S., Mikkat, S., Schroder, W., and Hagemann, M. (1999). Isolation of salt-induced periplasmic proteins from *Synechocystis* sp. strain PCC 6803. Arch. Microbiol. *171*, 214–217.

Gantt, E. (1994). Supramolecular Membrane Organization. In The Molecular Biology of Cyanobacteria, D.A. Bryant, ed. (Dordrecht, The Netherlands: Kluwer), pp. 119–138.

Hansel, A., Pattus, F., Jurgens, U.J., and Tadros, M.H. (1998). Cloning and characterization of the genes coding for two porins in the unicellular cyanobacterium *Synechococcus* PCC 6301. Biochim. Biophys. Acta *1399*, 31–39.

Hoiczyk, E., and Hansel, A. (2000). Cyanobacterial cell walls: news from an unusual prokaryotic envelope. J. Bacteriol. *182*, 1191–1199.

Huang, F., Hedman, E., Funk, C., Kieselbach, T., Schroder, W.P., and Norling, B. (2004). Isolation of outer membrane of *Synechocystis* sp. PCC 6803 and its proteomic characterization. Mol. Cell Proteomics 3, 586–595.

Huang, F., Parmryd, I., Nilsson, F., Persson, A.L., Pakrasi, H.B., Andersson, B., and Norling, B. (2002). Proteomics of *Synechocystis* sp. strain PCC 6803: identification of plasma membrane proteins. Mol. Cell Proteomics *1*, 956–966.

Keren, N., Kidd, M.J., Penner-Hahn, J.E., and Pakrasi, H.B. (2002). A light-dependent mechanism for massive accumulation of manganese in the photosynthetic bacterium *Synechocystis* sp. PCC 6803. Biochemistry *41*, 15085–15092.

Keren, N., Liberton, M., and Pakrasi, H.B. (2005). Photochemical competence of assembled photosystem II core complex in cyanobacterial plasma membrane. J. Biol. Chem. *280*, 6548–6553.

Koropatkin, N.M., Pakrasi, H.B., and Smith, T.J. (2006a). Atomic structure of a nitrate-binding protein crucial for photosynthetic productivity. Proc. Natl. Acad. Sci. USA *103*, 9820–9825.

Koropatkin, N.M., Koppenaal, D.W., Pakrasi, H.B., and Smith, T.J. (2006b). The structure of a cyanobacterial bicarbonate transport protein, CmpA. J. Biol. Chem., *in press*.

Koyama, K., Tsuchiya, T., Akimoto, S., Yokono, M., Miyashita, H., and Mimuro, M. (2006). New linker proteins in phycobilisomes isolated from the cyanobacterium *Gloeobacter violaceus* PCC 7421. FEBS Lett. *580*, 3457–3461.

Kroll, D., Meierhoff, K., Bechtold, N., Kinoshita, M., Westphal, S., Vothknecht, U.C., Soll, J., and Westhoff, P. (2001). VIPP1, a nuclear gene of *Arabidopsis thaliana* essential for thylakoid membrane formation. Proc. Natl. Acad. Sci. USA *98*, 4238–4242.

Kunkel, D.D. (1982). Thylakoid centers: Structures associated with the cyanobacterial photosynthetic membrane system. Arch. Microbiol. *133*, 97–99.

Liberton, M., Howard Berg, R., Heuser, J., Roth, R., and Pakrasi, H.B. (2006). Ultrastructure of the membrane systems in the unicellular cyanobacterium *Synechocystis* sp. strain PCC 6803. Protoplasma *227*, 129–138.

Lupas, A., Engelhardt, H., Peters, J., Santarius, U., Volker, S., and Baumeister, W. (1994). Domain structure of the *Acetogenium kivui* surface layer revealed by electron crystallography and sequence analysis. J. Bacteriol. *176*, 1224–1233.

McCarren, J., Heuser, J., Roth, R., Yamada, N., Martone, M., and Brahamsha, B. (2005). Inactivation of *swmA* results in the loss of an outer cell layer in a swimming *Synechococcus* strain. J. Bacteriol. *187*, 224–230.

Mohamed, H.E., van de Meene, A.M., Roberson, R.W., and Vermaas, W.F. (2005). Myxoxanthophyll is required for normal cell wall structure and thylakoid organization in the cyanobacterium *Synechocystis* sp. strain PCC 6803. J. Bacteriol. *187*, 6883–6892.

Moslavac, S., Bredemeier, R., Mirus, O., Granvogl, B., Eichacker, L.A., and Schleiff, E. (2005). Proteomic analysis of the outer membrane of *Anabaena* sp. strain PCC 7120. J Proteome Res. *4*, 1330–1338.

Mullineaux, C.W. (1999). The thylakoid membranes of cyanobacteria: structure, dynamics and function. Aust. J. Plant Physiol. *26*, 671–677.

Mullineaux, C.W., Tobin, M.J., and Jones, G.R. (1997). Mobility of photosynthetic complexes in thylakoid membranes. Nature *390*, 421–424.

Nakai, M., Sugita, D., Omata, T., and Endo, T. (1993). Sec-Y protein is localized in both the cytoplasmic and thylakoid membranes in the cyanobacterium *Synechococcus* PCC7942. Biochem. Biophys. Res. Commun. *193*, 228–234.

Nelissen, B., Van de Peer, Y., Wilmotte, A., and De Wachter, R. (1995). An early origin of plastids within the cyanobacterial divergence is suggested by evolutionary trees based on complete 16S rRNA sequences. Mol. Biol. Evol. *12*, 1166–1173.

Nierzwicki-Bauer, S.A., Balkwill, D.L., and Stevens, S.E., Jr. (1983). Three-dimensional ultrastructure of a unicellular cyanobacterium. J. Cell Biol. *97*, 713–722.

Rukhman, V., Anati, R., Melamed-Frank, M., and Adir, N. (2005). The MntC crystal structure suggests that import of Mn^{2+} in cyanobacteria is redox controlled. J. Mol. Biol. *348*, 961–969.

Samuel, A.D., Petersen, J.D., and Reese, T.S. (2001). Envelope structure of *Synechococcus* sp. WH8113, a nonflagellated swimming cyanobacterium. BMC Microbiol. *1*, 4.

Sherman, D.M., Troyan, T.A., and Sherman, L.A. (1994). Localization of membrane proteins in the cyanobacterium *Synechococcus* sp. PCC7942 (radial asymmetry in the photosynthetic complexes). Plant Physiol. *106*, 251–262.

Smarda, J., Caslavska, J., and Komarek, J. (1979). Cell wall structure of *Synechocystis aquatilis (Cyanophyceae)*. Arch. Hydrobiol. *56*, 154–165.

Smarda, J., Smajs, D., Komrska, J., and Krzyzanek, V. (2002). S-layers on cell walls of cyanobacteria. Micron *33*, 257–277.

Smith, D., and Howe, C.J. (1993). The distribution of Photosystem I and Photosystem II polypeptides between the cytoplasmic and thylakoid membranes of cyanobacteria. FEMS Microbiol. Lett. *110*, 341–348.

Spence, E., Sarcina, M., Ray, N., Moller, S.G., Mullineaux, C.W., and Robinson, C. (2003). Membrane-specific targeting of green fluorescent protein by the Tat pathway in the cyanobacterium *Synechocystis* PCC6803. Mol. Microbiol. *48*, 1481–1489.

Srivastava, R., Battchikova, N., Norling, B., and Aro, E.M. (2006). Plasma membrane of *Synechocystis* PCC 6803: a heterogeneous distribution of membrane proteins. Arch. Microbiol. *185*, 238–243.

van de Meene, A.M., Hohmann-Marriott, M.F., Vermaas, W.F., and Roberson, R.W. (2006). The three-dimensional structure of the cyanobacterium *Synechocystis* sp. PCC 6803. Arch. Microbiol. *184*, 259–270.

Vothknecht, U.C., and Westhoff, P. (2001). Biogenesis and origin of thylakoid membranes. Biochim. Biophys. Acta *1541*, 91–101.

Westphal, S., Heins, L., Soll, J., and Vothknecht, U.C. (2001a). Vipp1 deletion mutant of *Synechocystis*: a connection between bacterial phage shock and thylakoid biogenesis? Proc. Natl. Acad. Sci. USA *98*, 4243–4248.

Westphal, S., Soll, J., and Vothknecht, U.C. (2001b). A vesicle transport system inside chloroplasts. FEBS Lett. *506*, 257–261.

Zak, E., Norling, B., Maitra, R., Huang, F., Andersson, B., and Pakrasi, H.B. (2001). The initial steps of biogenesis of cyanobacterial photosystems occur in plasma membranes. Proc. Natl. Acad. Sci. USA *98*, 13443–13448.

Biogenesis and Dynamics of Thylakoid Membranes and the Photosynthetic Apparatus

11

Conrad W. Mullineaux

Abstract

Thylakoid membranes are the site of the light-reactions of photosynthesis, and they are crucial to the photosynthetic lifestyle of cyanobacteria. This chapter discusses what we know about the structure, organization and biogenesis of thylakoid membranes in cyanobacteria. There is an emphasis on the dynamics of the membrane, and the roles that protein diffusion may play in membrane biogenesis, regulation of photosynthesis and the turnover and repair of the photosynthetic apparatus. Although we have detailed knowledge of many thylakoid membrane components and some thylakoid membrane processes, much remains to be learned about the large-scale organization and biogenesis of thylakoid membranes. The chapter highlights some of the questions that are still unresolved.

Introduction to cyanobacterial thylakoid membranes

A distinctive feature of cyanobacteria is the thylakoid membrane—a complex, specialized internal membrane system that is the site of the light-reactions of photosynthesis. In cyanobacteria photosynthetic and respiratory electron transport are closely connected, and the thylakoids are also a major site of respiratory electron transport (Vermaas, 1994; Peschek *et al.*, 2004). Processes occurring in and around the thylakoid membrane are crucial to the photosynthetic lifestyle of cyanobacteria. Membranes play an essential role in photosynthetic electron transport and energy coupling. The presence of thylakoids greatly increases the membrane area, and therefore allows the cell a much higher rate of photosynthetic energy transduction. In most cyanobacteria the thylakoids account for most of the membrane area in the cell and typically account for about 90% of the lipid (Murata *et al.*, 1981). The protein and lipid composition of the thylakoid membrane is very distinct from that of the cytoplasmic membrane (Norling *et al.*, 1998). Thylakoid membranes have a much higher ratio of protein to lipid (Szalontai *et al.*, 2000). Only one known cyanobacterium, the very divergent *Gloeobacter violaceus,* lacks thylakoid membranes. In this cyanobacterium the photosynthetic complexes are located in the cytoplasmic membrane (Stanier, 1988; Nakamura *et al.*, 2003).

As well as being crucial for photosynthesis, cyanobacterial thylakoids are a beautiful and intriguing model system for biological membrane function. We know a great deal about the structure and function of the major thylakoid protein complexes (Chapter 9; see also Jordan *et al.*, 2001; Zouni *et al.*, 2001; Kurisu *et al.*, 2003). The photosynthetic complexes are naturally labeled with chromophores, making it easy to monitor their abundance, interactions and location. Many years of multidisciplinary research in photosynthesis have left a legacy of powerful, non-invasive techniques for probing the biological functions of the membrane. The Photosystem II repair cycle is a particularly beautiful experimental paradigm for biochemical repair systems, because damage to Photosystem II can be initiated rapidly and specifically with light (Barber and Andersson, 1992). Finally, the thylakoid membranes of some cyanobacteria are amenable to photobleaching measurements which provide a direct probe of membrane dynamics *in vivo* (Mullineaux and Sarcina, 2002).

This chapter summarizes what we know about the structure and function of cyanobacterial thylakoid membranes, the biogenesis of thylakoid membrane components and the dynamics of the membrane: how free are proteins and lipids to diffuse?

Architecture of cyanobacterial thylakoid membranes

The three-dimensional structure of the thylakoid membrane is complex, and varies in different cyanobacteria. However, some features are conserved in all species. Thylakoid membranes are always paired, with the membrane pair enclosing a distinct cell compartment, the thylakoid lumen (Figure 11.1) (Stanier, 1988). The separation of the lumen from the cytoplasm is important for the bioenergetics of the cell. Respiratory and photosynthetic electron transport generates a proton gradient across the thylakoid membrane, with lower pH in the lumen. Flow of protons back across the thylakoid membrane is used to power the generation of ATP by the proton-translocating ATPase. Thylakoid membrane proteins are always asymmetrically oriented. For example, the "donor" sides of Photosystem I and Photosystem II face the lumen, while the "acceptor" sides face the cytoplasm. Phycobilisomes and the F_1 part of the ATPase are found on the cytoplasmic surface of the membrane (Figure 11.1; see also Chapter 9).

Thin-section electron micrographs generally give only a crude idea of the three-dimensional organization of the thylakoid membranes. In elongated cyanobacteria (for example *Synechococcus* sp. PCC 7942) the thylakoid membrane pairs approximate to a series of concentric cylinders running the length of the cell, inside the cytoplasmic membrane (Sherman *et al.*, 1994; Mullineaux and Sarcina, 2002). This picture is likely to be

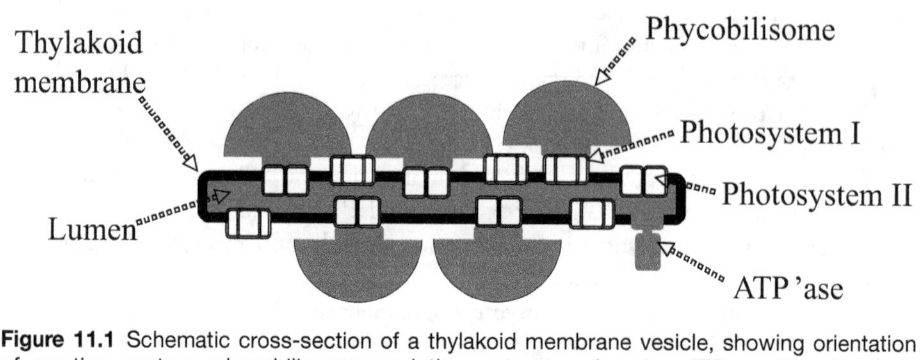

Figure 11.1 Schematic cross-section of a thylakoid membrane vesicle, showing orientation of reaction centers, phycobilisomes and the proton-translocating ATPase. The thylakoid membrane encloses an aqueous compartment, the thylakoid lumen, and separates it from the cytoplasm.

an over-simplification, however. In particular, it is clear that there must be breaks in the cylinders to allow the cytoplasm to be continuous. The nucleoid and other structures such as carboxysomes are usually found in the central cytoplasm, apparently surrounded by multiple layers of thylakoid membrane (Nierzwicki-Bauer *et al.*, 1983; Stanier, 1988). It must be essential for cell function for proteins, RNA and numerous metabolites to diffuse freely between the central cytoplasm and the cytoplasmic membrane, and this would be impossible if the thylakoid membranes were continuous cylinders. There are several possible solutions to the puzzle, which would allow the cytoplasm to be continuous whilst also keeping the lumen separate from the cytoplasm (Mullineaux, 1999). An exciting new approach to the problem uses cryo-electron tomography, in which the three-dimensional structure of the membranes in thick, frozen sections is reconstructed using information taken from transmission electron micrographs recorded at different angles. Recent studies using cryo-electron tomography in *Synechococcus* 7942 and *Microcoleus* sp. show that there are sealed "perforations" through the thylakoid membrane pairs which allow the cytoplasm to be continuous. There are also membrane connections between the different concentric membrane cylinders, possibly allowing both the thylakoid membrane and the lumen to be continuous throughout the cell (Nevo *et al.*, 2007).

The unicellular spherical cyanobacterium *Synechocystis* sp. PCC 6803 is a popular model system for studies of photosynthesis and respiration. In *Synechocystis* 6803 the thylakoids are less regularly arranged than in elongated cyanobacteria such as *Synechococcus* 7942. The membrane pairs form a series of parallel, flattened sacs radiating out into the cytoplasm from points near the cytoplasmic membrane (Figure 11.2). Recent studies using cryo-electron tomography in *Synechocystis* 6803 provides the best view yet of thylakoid structure in this cyanobacterium (van de Meene *et al.*, 2006; Liberton *et al.*, 2006). One of

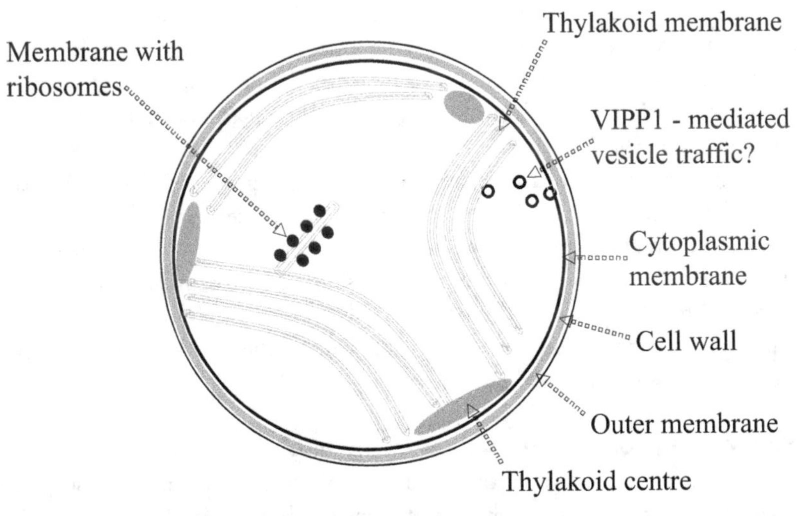

Figure 11.2 Schematic illustration of cell and thylakoid membrane organization in the cyanobacterium *Synechocystis* sp. PCC 6803. Possible sites of protein synthesis and protein entry into the thylakoid membrane and the lumen are highlighted. For further information see in particular (Kroll *et al.*, 2001; van de Meene *et al.*, 2006).

these studies highlights the role of "thylakoid centers" (van de Meene *et al.*, 2006). These are bodies whose properties are little understood, but which seem to be closely associated both with the radiating thylakoid pairs and with the cytoplasmic membrane. A key question for thylakoid membrane biogenesis and function is the nature of connections (if any) between the thylakoid membranes and the cytoplasmic membrane (see Chapter 10). The thylakoid membranes do not seem to be simple invaginations of the cytoplasmic membrane, and it now seems probable that the thylakoid membranes are completely discontinuous from the cytoplasmic membrane in *Synechocystis* 6803 (Liberton *et al.*, 2006). It is possible that there are occasional transient dynamic connections between the two membrane systems. Another possibility is that lipids and proteins may be exchanged between the cytoplasmic and thylakoid membranes by vesicle trafficking (Figure 11.2). This is suggested by the importance of Vipp1 (Vesicle Inducing Protein in Plastids 1) which in cyanobacteria is found only in the plasma membrane, but is essential for thylakoid biogenesis (Westphal *et al.*, 2001). Small membrane-bound vesicles were recently observed in the cyanobacterium *Microcoleus* sp., and these vesicles were often observed fused to the thylakoids, suggesting that they serve as mediators of transport to these membranes (Nevo *et al.*, 2007).

The best-understood photosynthetic membranes are the chromatophore membranes of purple photosynthetic bacteria, which are formed by invaginations of the plasma membrane. In these membranes the protein complexes form a dense mesh that controls the three-dimensional structure of the membrane (Bahatyrova *et al.*, 2004). In wild-type *Rhodobacter sphaeroides* the chromatophores are almost spherical, whereas in an LH2 null mutant (lacking one of the two major light-harvesting complexes) they are tubular. This can be explained by consideration of the way in which the membrane proteins pack in the wild type and in the mutant (Siebert *et al.*, 2004). In purple bacteria, the formation of chromatophores is always closely correlated with the synthesis of photosynthetic proteins. When *Rhodobacter* cells are growing under aerobic conditions they have no photosynthetic proteins and no chromatophores: the plasma membrane appears smooth, with no invaginations. Switching to anaerobic growth in the light triggers the synthesis of photosynthetic proteins and chromatophore formation. It is likely that chromatophore formation is triggered by the aggregation of photosynthetic proteins in the plasma membrane.

In cyanobacteria the situation seems to be very different. There is no close connection between membrane conformation and protein composition. Strikingly, a *Synechocystis* mutant that is almost devoid of chlorophyll-protein complexes has thylakoid membranes that appear structurally normal in thin-section electron micrographs (E. Spence, N.H. Mann, unpublished). Although the major photosynthetic proteins occupy much of the membrane area in wild-type cells (Figure 11.3), they seem to have little influence on the overall membrane configuration. The factors that control the three-dimensional structure of cyanobacterial thylakoids remain unknown.

Organization of protein complexes in thylakoid membranes

Insight into the protein content of the thylakoid and cytoplasmic membranes has come from biochemical fractionation of the membranes combined with Western blotting and proteomic studies. This approach depends on a clean, reliable separation of the thylakoid and cytoplasmic membranes. The aqueous polymer two-phase partition method (Norling *et al.*, 1998) undoubtedly gives a clean separation between the bulk membrane fractions, but a proviso is that there is always the possibility that heterogeneous areas of membrane may partition with the "wrong" fraction. The biochemical approach has shown a remarkably distinct protein content of the two membrane systems. All the mature chlorophyll-protein complexes that mediate the photosynthetic light reactions are confined to the thylakoid membranes, although some precursor complexes are found in plasma membrane fractions (Zak *et al.*, 2001).

The last few years have seen remarkable advances in knowledge of the structures of individual cyanobacterial thylakoid membrane protein complexes. However, we know much less about the way in which these complexes are organized in the intact membrane *in vivo*. The best information still comes from freeze-fracture electron micrographs (Olive *et al.*, 1997; Westermann *et al.*, 1999). These indicate a dense packing of the thylakoid membrane with protein complexes, which is consistent with the very high protein-to-lipid ratio in these membranes (Szalontai *et al.*, 2000) and with estimates of packing density based on counting the number of reaction centers per cell and estimating the thylakoid membrane area (Mullineaux, 1999) (Figure 11.3). 10–13 nm particles seen in the exoplasmic fracture face can be clearly identified as Photosystem II reaction centers. These are often seen to be arranged in rows, probably corresponding to linear arrays of Photosystem II dimers (Olive *et al.*, 1997) (Figure 11.3). It is possible, however, that the prevalence of these rows has been exaggerated by a natural tendency to publish micrographs that show recognizable

Figure 11.3 Reconstruction of a typical organization of reaction centers in cyanobacterial thylakoid membranes. Photosystem II dimers are arranged in roughly parallel rows (Olive *et al.*, 1997), with trimers of Photosystem I more randomly arranged between the rows (Westermann *et al.*, 1999). The density of complexes is estimated from measurements of the number of complexes per cell and estimates of thylakoid membrane area, assuming that the complexes are evenly dispersed throughout the thylakoid membrane system (Mullineaux, 1999).

structural features (Mustardy *et al.*, 1992). Where Photosystem II rows are present they tend to be parallel and about 50 nm apart (Olive *et al.*, 1997) (Figure 11.3). Photosystem I complexes are predominantly trimeric, and are apparently randomly distributed between the rows of Photosystem II (Westermann *et al.*, 1999) (Figure 11.3). The phycobilisome complexes are associated with the cytoplasmic surface of the membrane (Figure 11.1). They can be easily resolved in thin-section electron micrographs (Stanier, 1988), but are not seen in ordinary freeze-fracture electron micrographs. They have occasionally been resolved by freeze-etching, and one such study shows a row of phycobilisomes aligned on top of a row of Photosystem II complexes in the membrane (Giddings *et al.*, 1983). There is no high-resolution information on the localization of the many other known thylakoid membrane complexes (for example the proton-translocating ATPase, the cytochrome b_6f complex and various respiratory electron transport complexes).

In the thylakoid membranes of green plant chloroplasts, there is an obvious and striking lateral heterogeneity between the stacked, tightly appressed membranes in the grana and the unstacked stromal lamellae. The two membrane regions have quite distinct protein compositions (reviewed by Dekker and Boekema, 2005). For example Photosystem I is excluded from the grana, which contain the bulk of the Photosystem II. Cyanobacteria do not have grana and generally there is no membrane appression. A rare exception is the atypical cyanobacterium *Acaryochloris marina,* in which there is partial membrane appression, presumably accompanied by lateral heterogeneity of thylakoids in terms of composition and function (Marquardt *et al.*, 2000). In "typical" cyanobacteria there is no evidence for any dramatic, large-scale heterogeneity in the thylakoid membranes. However, a more subtle effect has been detected in one cyanobacterium. Sherman *et al.* (1994) used immuno-gold electron microscopy to show a "radial asymmetry" in the protein composition of thylakoid membranes of *Synechococcus* 7942. Photosystem II and the cytochrome b_6f complex are spread throughout the thylakoid membrane system, whereas Photosystem I and the ATP synthase are concentrated in the outermost thylakoid membrane layers adjacent to the plasma membrane. It is possible that linear photosynthetic electron transport, involving Photosystem II and Photosystem I acting in series, may occur mainly in the inner membrane cylinders. The outer membrane cylinders, where Photosystem I seems to predominate, may be specialized for alternative modes of electron transport that do not involve Photosystem II. Similar patterns were not, however, observed in the cyanobacteria *Synechocystis* 6803 and *Cyanothece*, which have a different thylakoid membrane organization (Sherman *et al.*, 1994).

It is very plausible that thylakoid membranes may play a role in localizing and organizing soluble enzymes and other cell components. Biochemical fractionation studies on the cyanobacterium *Anacystis nidulans* suggest specific association of some of the Calvin cycle enzymes with thylakoid membranes. This study suggests an almost complete thylakoid membrane-associated "photosynthetic module" capable of carrying out the "dark reactions" as well as the light reactions (Dani and Sainis, 2005)

Cryo-electron tomography promises a new approach to the problem of determining structure and possible functional heterogeneity in cyanobacterial thylakoids (van de Meene *et al.*, 2006; Liberton *et al.*, 2006). An interesting finding is that in *Synechocystis* 6803 there are ribbons of membrane in the central cytoplasm, apparently continuous with the thylakoids and densely packed with ribosomes. These could conceivably be nascent areas of thylakoid membrane, or specialized regions where thylakoid membrane protein synthesis occurs (van de Meene *et al.*, 2006).

Dynamics of thylakoid membranes

In general, biological membranes are highly dynamic systems in which lipids and many proteins are able to diffuse rather rapidly in the membrane plane. Membrane fluidity is important for many biological processes. Bacteria in general have mechanisms to maintain membrane fluidity by adjusting membrane lipid composition in response to changes in growth temperature. This phenomenon is particularly well understood in cyanobacteria (Murata, 1989), where lipid desaturase mutants have provided insights into the importance of thylakoid membrane lipid composition for processes such as the Photosystem II repair cycle. A *Synechocystis* mutant lacking lipid desaturases has a lower level of thylakoid lipid desaturation, which would be expected to lead to a higher phase transition temperature and therefore less membrane fluidity at low temperatures. This mutant is more sensitive to photoinhibition at low temperatures, suggesting that the Photosystem II repair cycle is impaired (Kanervo *et al.*, 1997). Conversely, a *Synechococcus* transformant with an additional desaturase gene can adjust the subunit composition of Photosystem II more rapidly than the wild-type (Sippola *et al.*, 1998). This transformant also shows more rapid lateral mobility of thylakoid membrane lipids (Sarcina *et al.*, 2003). These findings suggest a key role for the mobility of thylakoid membrane components in the maintenance of the photosynthetic apparatus. It is not clear, however, which membrane components need to move. Is it the reaction centers themselves, or enzymes involved in turnover and repair? It seems likely that protein diffusion is involved in many other membrane processes, including biogenesis, electron transport, and regulation.

Direct information on the mobility of thylakoid membrane components has come from Fluorescence Recovery after Photobleaching (FRAP) studies (Mullineaux *et al.*, 1997; Sarcina *et al.*, 2001). These measurements can be carried out on intact cells using a laser-scanning confocal microscope, which is used to visualize fluorescence from the photosynthetic pigments, or from Green Fluorescent Protein (GFP) tags or other fluorescent molecules. The technique is described in detail in (Mullineaux and Sarcina, 2002; Mullineaux, 2004). To observe diffusion, the laser power is increased and the highly focused confocal spot is used to bleach out fluorescence from a small region of the thylakoid membranes. The bleached area then appears as a dark spot or line on the fluorescence image. If the fluorescent pigment can diffuse the bleach spreads and fills in a characteristic way. If the cells are large enough and the membrane conformation is simple and predictable, the measurements are quantitative and diffusion coefficients can be measured. These conditions are met in some cyanobacteria, including *Synechococcus* 7942 (Sarcina *et al.*, 2001; 2003; 2006). This cyanobacterium has elongated cells in which the thylakoids are arranged approximately as concentric cylinders running the full length of the cell (see above). Quantitative FRAP measurements can be performed by bleaching a line across the centre of the cell and then following diffusion in one dimension along the long axis of the cell (Mullineaux and Sarcina, 2002; Mullineaux, 2004).

FRAP measurements have shown that Photosystem II reaction centers are normally extremely immobile—the diffusion coefficient is below 2×10^{-13} cm^2 s^{-1}, which is astonishingly slow by the normal standards of biological membrane components. However, IsiA, an integral membrane chlorophyll-protein synthesized under conditions of iron limitation, can diffuse relatively rapidly within the membrane (Sarcina and Mullineaux, 2004). This shows that not all thylakoid membrane integral protein complexes are immobile. Therefore there must be some specific factor that prevents Photosystem II from diffusing in the membrane. This factor remains unknown, but it could involve either the large domain on the lumenal side of the membrane (Zouni *et al.*, 2001), or the assembly of Photosystem II dimers into rows in the membrane (Olive *et al.*, 1997). In contrast to Photosystem II, the phycobilisomes diffuse very rapidly on the membrane surface: a phycobilisome will typically diffuse right round a cell on a timescale of a few minutes. This indicates that there is no stable connection between the phycobilisomes and the reaction centers (Mullineaux *et al.*, 1997; Sarcina *et al.*, 2001). A phycobilisome must associate only transiently with a reaction centre before detaching, diffusing on the membrane surface and re-associating with another reaction centre. The mobility of most other thylakoid membrane proteins is not yet known. However, another application of the FRAP technique has been to probe the mobility of thylakoid membrane lipids, using a fluorescent lipid analog. These studies have shown lateral lipid diffusion in the thylakoid membranes *in vivo*. As expected, lipid diffusion becomes slower when the cells are cooled below phase transition temperature, and is considerably faster in a transformant in which the membrane lipids are more desaturated (Sarcina *et al.*, 2003).

As summarized above, FRAP studies show that Photosystem II is normally immobile in the membrane, while phycobilisomes diffuse rapidly on the membrane surface (Mullineaux *et al.*, 1997; Sarcina *et al.*, 2001). This was unexpected, as it is generally assumed that light-harvesting complexes must be tightly associated with reaction centers for efficient energy transfer. A recent study has shed some light on the physiological role of phycobilisome mobility. It is essential for state transitions, a physiological adaptation mechanism that rapidly fine-tunes photosynthetic electron transport by switching the balance of light-harvesting between Photosystem I and Photosystem II (Joshua and Mullineaux, 2004). Phycobilisome mobility also plays a role in a photoprotective mechanism that quenches excess energy under iron-limited conditions (Joshua *et al.*, 2005). Thus it appears that phycobilisome mobility is necessary to confer flexibility in light-harvesting. Phycobilisomes are rather loosely coupled to reaction centers, and this allows them to rapidly redistribute in order to regulate light-harvesting. The affinity of phycobilisomes for different membrane complexes is probably controlled by covalent modification, but the biochemistry remains uncertain (van Thor *et al.*, 1998).

The physiological reasons for Photosystem II immobility are not yet certain. A plausible idea is that it is necessary to maintain some separation between Photosystem II and Photosystem I. In cyanobacterial thylakoids Photosystem II and Photosystem I are not normally widely separated as they are in green plant chloroplasts. This creates a potential problem of "spillover" of energy from Photosystem II to Photosystem I. When Photosystem II and Photosystem I are close together, excitation energy may be exchanged between their chlorophyll molecules. If this can occur, most of the energy will be trapped by Photosystem I, because it is a much faster trap for energy than Photosystem II (Trissl and Wilhelm, 1993). If this process is not controlled it will result in unbalanced photosynthetic electron flow. For efficient and safe photosynthetic electron transport it is necessary for Photosystem II and Photosystem I to turn over at similar rates. The cyanobacterial solution to the problem seems to involve oligomerization of reaction centers. Photosystem I is normally trimeric in cyanobacteria (Westermann *et al.*, 1999) in contrast to green plants (Ben-Shem *et al.*, 2003), and Photosystem II is often oligomerized into rows of dimers. One effect of oligomerization will be to minimize contact between Photosystem II and Photosystem I in a tightly packed thylakoid membrane (Mullineaux, 1999) (Figure 11.3). The immobility of Photosystem II may be a side effect of its association into rows, necessary to minimize spillover and therefore important for efficient photosynthesis.

The inability of Photosystem II to diffuse may create problems. If Photosystem II is fixed in its place in the membrane, how and where is it assembled and how does it move to its correct place in the membrane? How can photodamaged Photosystem II centers be repaired? Recent results show that the mobility of Photosystem II is under physiological control. When *Synechococcus* cells are exposed to intense red light, up to 50% of Photosystem II centers start to diffuse (Sarcina *et al.*, 2006). This is accompanied by relocation of Photosystem II to specific regions of the thylakoid membrane system. In short cells these are at the poles of the cell, whereas longer cells also have concentrations of Photosystem II at intervals down the length of the cell. There is some evidence to suggest that this phenomenon is connected with the Photosystem II repair cycle (Sarcina *et al.*, 2006). This will be discussed further below.

Biogenesis of the thylakoid membrane and the photosynthetic apparatus

We still know rather little about how and where thylakoid membranes and their protein components are synthesized. Where are lipids synthesized and thylakoid protein complexes assembled, and what regulates these processes? Is there a link between the biogenesis of the membrane and its component protein complexes? Bleaching studies on *Synechocystis* 6803 have shed some light on these questions. This cyanobacterium has the advantage that it can grow heterotrophically, and therefore can be grown (albeit very slowly) in the absence of a functioning photosynthetic apparatus. Cells can be "bleached" by prolonged growth in the absence of phosphate or nitrate (Morrison *et al.*, 2005). A similar experiment can be carried out in a less physiological, but more controlled, way using a *chlL* null mutant (Wu and Vermaas, 1995). This gene is essential for the light-independent pathway of chlorophyll biosynthesis. The null mutant is phenotypically normal when grown in the light, but when grown in the dark it gradually loses all its chlorophyll-protein complexes, as new complexes cannot be synthesized without chlorophyll biosynthesis. Bleached cells of the *chlL* null mutant retain phycobilisomes, which are assembled as normal and associated with the thylakoid membranes (Yu *et al.*, 1999). Bleaching by nitrate or phosphate starvation results in loss of phycobilisomes as well as chlorophyll complexes (Grossman *et al.*, 1993; Morrison *et al.*, 2005). However, the thylakoid membranes are retained, and there are no obvious changes in thylakoid membrane conformation. Therefore the synthesis and maintenance of the thylakoid membrane is not dependent on the presence of the photosynthetic complexes. It must not be forgotten that the thylakoid membranes are multifunctional: they are important for respiration as well as for photosynthesis (Berry *et al.*, 2002; Peschek *et al.*, 2004). Presumably the thylakoid membranes are retained in bleached cells to act as a major site for respiratory electron transport, which is essential for cell survival under these conditions. Although the details are not clear, there must, however, be physiological controls on the amount of thylakoid membrane in the cell and the membrane conformation. For example, when *Anabaena* cells differentiate to form heterocysts, they lose Photosystem II, while retaining Photosystem I and respiratory complexes. At the same time there are obvious changes in thylakoid membrane architecture (Stanier and Cohen-Bazire, 1977; Lang *et al.*, 1987).

An important recent finding is that the Vipp1 protein is essential for normal thylakoid membrane biogenesis (Westphal *et al.*, 2001). A non-segregated *Synechocystis* mutant with reduced Vipp1 content is non-photosynthetic and has no normal thylakoid membranes, with only a few dispersed membrane-like structures in the cytoplasm. Complete deletion of Vipp1 is lethal. Interestingly, the Vipp1 protein appears to be tightly associated with the cytoplasmic membrane, indicating that the cytoplasmic membrane somehow plays an essential role in thylakoid biogenesis (Westphal *et al.*, 2001). The precise mode of action of Vipp1 is unknown. However, loss of the Vipp1 ortholog in *Arabidopsis* prevents vesicle budding from the chloroplast inner envelope membrane (Kroll *et al.*, 2001). This finding strongly suggests that Vipp1 may be involved in cytoplasmic membrane vesicle formation in cyanobacteria. These vesicles could transport lipids and proteins from the cytoplasmic membrane to the thylakoids (Figure 11.2).

Where are thylakoid membrane proteins synthesized? There are three clear possibilities:

a Proteins could be synthesized at the cytoplasmic membrane and transported to the thylakoids by vesicles or possible transient connections between the thylakoids and the plasma membrane.
b Proteins could be synthesized *in situ* in the thylakoid membrane.
c Proteins could be synthesized at ribosome-rich membrane structures protruding into the central cytoplasm. These structures appear to be connected to the thylakoids (van de Meene *et al.*, 2006).

The location of ribosomes is revealed by cryo-electron tomography, and this should give a strong clue as to the location of protein synthesis sites within the cell. Ribosomes are rarely observed within or between the thylakoid membrane layers (van de Meene *et al.*, 2006), which strongly suggests that the mature thylakoids are not a major site of protein synthesis. It seems more likely that most thylakoid proteins are synthesized either at the cytoplasmic membrane (a) or at the membrane structures in the central cytoplasm (c). There is biochemical evidence for (a), since fractionation studies reveal the presence of some partially assembled Photosystem II and Photosystem I complexes in the cytoplasmic membrane fraction (Zak *et al.*, 2001). However other subunits are never detected in the cytoplasmic membrane fraction and are found only in the thylakoids (Zak *et al.*, 2001). This suggests that reaction centers are partially assembled in the cytoplasmic membrane before somehow being transferred to the thylakoid membranes, where synthesis is completed. If Photosystem II biosynthesis is not initiated *in situ* in the thylakoid membranes, how do newly assembled Photosystem II centers migrate to an appropriate location within the thylakoid membrane system? This is an interesting question, as FRAP measurements indicate that Photosystem II centers are normally completely immobile (Mullineaux *et al.*, 1997; Sarcina and Mullineaux, 2004). There are two obvious solutions to the problem:

a New reaction centers always enter newly synthesized areas of membrane. Thylakoid membrane biogenesis is concomitant with reaction centre biogenesis. In another system, the chromatophores of purple photosynthetic bacteria, this is probably the case. However, in cyanobacteria, thylakoid membrane quantity and biogenesis do not appear to be directly coupled to reaction centre quantity and biogenesis (see discussion above).
b Photosystem II reaction centers actually are mobile under some conditions. Newly synthesized reaction centers can diffuse to an appropriate location within the membrane. This view is plausible, since FRAP measurements have shown that although Photosystem II reaction centers are normally immobile, some of them do become mobile under certain conditions (Sarcina *et al.*, 2006).

A related question concerns the proteins located in the thylakoid lumen. Where are they synthesized and how are they transported to the lumen? In green plants thylakoid and lumenal proteins that are encoded by the chloroplast genome may be synthesized at thylakoid-bound ribosomes and co-translationally inserted across the membrane (Nilsson *et al.*, 1999). A similar process is plausible in cyanobacterial thylakoids, though it is not clear if there are specialized domains of the membrane where this occurs. In green plants there is another route into the lumen for nuclear-encoded polypeptides. These are synthesized in the cytoplasm and imported post-translationally across the chloroplast envelope membranes and the thylakoid membrane (Gutensohn *et al.*, 2006). There are two known systems that transport proteins across the thylakoid membrane into the lumen, the Sec and Tat pathways (reviewed by Gutensohn *et al.*, 2006). Both pathways are of bacterial origin and occur in cyanobacteria. Both pathways may therefore be involved in the translocation of proteins into the cyanobacterial thylakoid lumen, but the details remain to be determined (Spence *et al.*, 2003). Generally in bacteria the Sec and Tat pathways are involved in protein export to the periplasm (reviewed by Robinson and Bolhuis, 2001). It is not clear if the cyanobacterial Sec and Tat pathways translocate proteins to the lumen as well as to the periplasm. GFP expressed in *Synechocystis* with a Tat leader sequence is exported to the periplasm, with no sign of accumulation in the lumen (Spence *et al.*, 2003). This suggests that the Tat system in cyanobacteria is located in the cytoplasmic membrane rather than the thylakoids. However it is possible that there may be a route into the lumen via the periplasm. Proteins could be exported to the periplasm and then transferred to the lumen in the interior of vesicles (Figure 11.2) or during transient contacts between the thylakoid and cytoplasmic membranes. These possibilities remain to be tested. The accumulation of GFP in the periplasm of *Synechocystis* (Spence *et al.*, 2003) strongly suggests that there are no direct connections between the lumen and the periplasm. If there were, then GFP would be expected to diffuse into the lumen. This further suggests that there are no direct connections between the thylakoid and plasma membranes.

Turnover and repair of the photosynthetic apparatus

There are special problems connected with the repair of photosynthetic complexes. Photosystem II, in particular, is extremely susceptible to photodamage (Barber and Andersson, 1992). An active Photosystem II repair cycle is essential to sustain phototrophic growth under most conditions. Photodamage is normally concentrated at the D1 protein at the heart of Photosystem II. The repair cycle involves specific degradation of photodamaged D1 polypeptides, and insertion of newly D1 polypeptides into the Photosystem II complexes, with apparently tight coupling between degradation of the damaged subunits and translation and insertion of the new subunits (Barber and Andersson, 1992). In green plants it is well-established that Photosystem II repair takes place in the stromal lamellae. This is presumably because the grana are inaccessible to repair enzymes and ribosomes. So the first part of the repair cycle in green plants involves the migration of photodamaged Photosystem II centers from the grana to the stromal lamellae (reviewed by Aro *et al.*, 2004). The mechanisms controlling this migration are not understood. It has been suggested that protein phosphorylation is involved (Aro *et al.*, 2004), but an *Arabidopsis* mutant deficient in phosphorylation of the Photosystem II core proteins shows a normal repair cycle (Bonardi *et al.*, 2005).

Where does Photosystem II repair take place in cyanobacteria? One clue should come from the location of key repair enzymes. One such repair enzyme has been convincingly identified in cyanobacteria on the basis of mutant studies. It is a membrane-bound FtsH protease which appears to play a key role in the degradation of photodamaged D1 subunits. Most cyanobacteria contain a conserved set of four genes encoding FtsH proteases. Deletion of one of these, designated *slr0228* in *Synechocystis* 6803, leads to inactivation of the Photosystem II repair cycle (Silva *et al.*, 2003). A similar result is obtained when the nearest ortholog to *slr0228* is deleted in *Synechococcus* 7942 (N. Bouzovitis, C.W. Mullineaux, unpublished). FtsH mutants are also deficient in Photosystem II degradation during nitrate or phosphate starvation, or when chlorophyll biosynthesis is prevented in the dark (E.P. Thompson, C.W. Mullineaux, unpublished) suggesting that FtsH plays a rather general role in the turnover of the photosynthetic apparatus. FtsH (*slr0228*) has been localized in *Synechocystis* 6803 by fractionation and Western blotting, and by GFP-tagging and confocal fluorescence imaging (Komenda *et al.*, 2006). It is found in the thylakoid membranes, with no indication that it is present in the cytoplasmic membrane (Komenda *et al.*, 2006). This strongly suggests that the repair cycle also takes place in the thylakoids in *Synechocystis*. The confocal images show that concentrations of FtsH do not always coincide with concentrations of Photosystem II fluorescence, suggesting that there may be areas of the membrane specialized for Photosystem II repair (Komenda *et al.*, 2006). However, the resolution of optical images may be too low to give a clear answer on this point. FtsH has not yet been localized in *Synechococcus* 7942, but here there is some evidence for specialized repair zones within the thylakoid membranes. When cells are exposed to intense red light, a proportion of Photosystem II fluorescence becomes mobile and relocates to concentrations which are visualized as bright spots in chlorophyll fluorescence images. These are within the thylakoid membrane system. They are often located close to the poles of the cell, though in long cells they also occur at intermediate points down the length of the cell (Sarcina *et al.*, 2006). There is a correlation between the mobilization of Photosystem II and the efficiency of the repair cycle, suggesting that the relocation of

Photosystem II centers to these zones may be important for rapid Photosystem II repair (Sarcina *et al.*, 2006).

Much remains to be learned about possible functional heterogeneity in cyanobacterial thylakoids, the locations of assembly and repair processes and the roles of protein mobility. The way ahead will require a combination of biochemical and molecular biological approaches with high-resolution imaging.

Acknowledgments

Work in the author's laboratory is supported by the Biotechnology and Biological Sciences Research Council and the Wellcome Trust. I thank my research workers and collaborators.

References

Aro, E.-M., Suorsa, M., Rokka, A., Allahverdiyeva, Y., Paakkarinen, V., Saleem, A., Battchikova, N., and Rintamäki, E. (2005). Dynamics of photosystem II: a proteomic approach to thylakoid protein complexes. J. Exp. Bot. *56*, 347–356.

Bahatyrova, S., Frese, R.N., Siebert, C.A, Olsen, J.D., van der Werf, K.O., van Grondelle, R., Niederman, R.A., Bullough, P.A., Otto, C., and Hunter, C.N. (2004). The native architecture of a photosynthetic membrane. Nature 430, 1058–1062.

Barber, J., and Andersson, B. (1992). Too much of a good thing: light can be bad for photosynthesis. Trends Biochem. Sci. *17*, 61–66.

Ben-Shem, A., Frolow, F., and Nelson, N. (2003). Crystal structure of plant Photosystem I. Nature *426*, 631–638.

Berry, S., Schneider, D., Vermaas, W.F.J., and Rögner, M. (2002). Electron transport routes in whole cells of *Synechocystis* sp. strain PCC6803: the role of the cytochrome bd-type oxidase. Biochemistry *41*, 3422–3429.

Bonardi, V., Pesaresi, P., Becker, T., Schleiff, E., Wagner, R., Pfannschmidt, T., Jahns, P., and Leister, D. (2005). Photosystem II core phosphorylation and photosynthetic acclimation require two different protein kinases. Nature *437*, 1179–1182.

Dani, D.N., and Sainis, J.K. (2005). Isolation and characterization of a thylakoid membrane module showing partial light and dark reactions. Biochim. Biophys. Acta *1669*, 43–52.

Dekker, J.P., and Boekema, E. (2005). Supramolecular organization of thylakoid membrane proteins in green plants. Biochim. Biophys. Acta *1706*, 12–39.

Giddings,, T.H., Wasmann, C., and Staehelin, L.A. (1983). Structure of the thylakoids and envelope membranes of the cyanelles of *Cyanophora paradoxa*. Plant Physiol. *71*, 409–419.

Grossman, A.R., Schaefer, M.R., Chiang, G.G., and Collier, J.L. (1993). The phycobilisome: a light-harvesting complex responsive to environmental conditions. Microbiol. Rev. *57*, 725–749.

Gutensohn, M., Fan, E., Frielingsdorf, S., Hanner, P., Hou, B., Hust, B., and Klösgen, R.B. (2006). Toc, Tic, Tat *et al.*: structure and function of protein transport machineries in chloroplasts. J. Plant Physiol. *163*, 333–347.

Jordan, P., Fromme, P., Witt, H.T., Klukas, O., Saenger, W., and Krauss, N. (2001). Three-dimensional structure of cyanobacterial photosystem I at 2.5 angstrom resolution. Nature *411*, 909–917.

Joshua, S., and Mullineaux, C.W. (2004). Phycobilisome diffusion is required for light-state transitions in cyanobacteria. Plant Physiol. *135*, 2112–2119.

Joshua, S., Bailey, S., Mann, N.H., and Mullineaux, C.W. (2005). Involvement of phycobilisome diffusion in energy quenching in cyanobacteria. Plant Physiology *138*, 1577–1585.

Kanervo, E., Tasaka, Y., Murata, N., and Aro, E.-M. (1997). Membrane lipid unsaturation modulates processing of the photosystem II reaction-center protein D1 at low temperatures. Plant Physiol. *114*, 841–849.

Komenda, J., Barker, M., Kuvikova, S., De Vries, R., Mullineaux, C.W., Tichy, M., and Nixon, P.J. (2006). The FtsH protease, slr0228, is important for quality control of the thylakoid membrane of *Synechocystis* PCC6803. J. Biol. Chem. *281*, 1145–1151.

Kroll, D., Meierhoff, K., Bechtold, N., Kinoshita, M., Westphal, S., Vothknecht, U.C., Soll, J., and Westhoff, P. (2001). *VIPP1*, a nuclear gene of *Arabidopsis thaliana* essential for thylakoid membrane formation. Proc. Natl. Acad. Sci. USA *98*, 4238–4242.

Kurisu, G., Zhang, H., Smith, J.L., and Cramer, W.A. (2003). Structure of the cytochrome b_6f complex of oxygenic photosynthesis: tuning the cavity. Science *302*, 1009–1014.

Lang, N.J., Krupp, J.M., and Koller, A.L. (1987). Morphological and ultrastructural changes in vegetative cells and heterocysts of *Anabaena variabilis* grown with fructose. J. Bacteriol. *169*, 920–923.

Liberton, M., Berg, R.H., Heuser, J., Roth, R., and Pakrasi, H.B. (2006). Ultrastructure of the membrane systems in the unicellular cyanobacterium *Synechocystis* sp. strain PCC 6803. Protoplasma *227*, 129–138.

Marquardt, J., Mörschel, E., Rhiel, E., and Westermann, M. (2000). Ultrastructure of *Acaryochloris marina*, an oxyphotobacterium containing mainly chlorophyll *d*. Arch. Microbiol. *174* (2000). 181–188.

Morrison, S.S., Mullineaux, C.W., and Ashby, M.K. (2005). The influence of acetyl phosphate on DspA signalling in the cyanobacterium *Synechocystis* sp. PCC6803. BMC Microbiology 2005, 5:47.

Mullineaux, C.W., Tobin, M.J., and Jones, G.R. (1997). Mobility of photosynthetic complexes in thylakoid membranes. Nature *390*, 421–424.

Mullineaux, C.W. (1999). The thylakoid membranes of cyanobacteria: structure, dynamics and function. Aust. J. Plant Physiol. *26*, 671–677.

Mullineaux, C.W., and Sarcina, M. (2002). Probing the dynamics of photosynthetic membranes with Fluorescence Recovery after Photobleaching. Trends Plant Sci. *7*, 237–240.

Mullineaux, C.W. (2004). FRAP analysis of photosynthetic membranes. J. Exp. Bot. *55*, 1207–1211.

Murata, N., Sato, N., Omata, T., and Kuwabara, T. (1981). Separation and characterization of thylakoid and cell-envelope of the blue-green-alga (cyanobacterium) *Anacystis nidulans*. Plant Cell Physiol. *22*, 403–410.

Murata, N. (1989). Low-temperature effects on cyanobacterial membranes. J. Bioenergetics Biomembranes *21*, 61–75.

Mustardy, L., Cunningham, F.X., and Gantt, E. (1992). Photosynthetic membrane topography: quantitative *in situ* localization of photosystems I and II. Proc. Natl. Acad. Sci. USA *89*, 10021–10025.

Nakamura, Y., Kaneko, T., Sato, S., Mimuro, M., Miyashita, H., Tsuchiya, T., Sasamoto, S., Watanabe, A., Kawashima, K., Kishida, Y., Kiyokawa, C., Kohara, M., Matsumoto, M., Matsuno, A., Nakazaki, N., Shimpo, S., Takeuchi, C., Yamada, M., and Tabata, S. (2003). Complete genome structure of *Gloeobacter violaceous* PCC7421, a cyanobacterium that lacks thylakoids. DNA Res. *10*, 137–145.

Nevo, R., Charuvi, D., Shimoni, E., Schwarz, R., Kaplan, A., Ohad, I., and Reich, Z. (2007). Thylakoid membrane perforations and connectivity enable intracellular traffic in cyanobacteria. EMBO J. *26*, 1467–1473.

Nierzwicki-Bauer, S.A., Balkwill, D.L., and Stevens, S.E. (1983). Three-dimensional ultrastructure of a unicellular cyanobacterium. J. Cell Biol. *97*, 713–722.

Nilsson, R., Brunner, J. Hoffman, N.E., and van Wijk, K.J. (1999). Interactions of ribosome nascent chain complexes of the chloroplast-encoded D1 thylakoid membrane protein with cpSRP54. EMBO J. *18*, 733–742.

Norling, B., Zak, E., Andersson, B., and Pakrasi, H. (1998). 2D-isolation of pure plasma and thylakoid membranes from the cyanobacterium *Synechocystis* sp. PCC6803. FEBS Lett. *436*, 189–192.

Olive, J., Ajlani, G., Astier, C., Recouvreur, M., and Vernotte, C. (1997). Ultrastructure and light-adaptation of phycobilisome mutants of *Synechocystis* PCC6803. Biochim. Biophys. Acta *1319*, 275–282.

Peschek, G.A., Obinger, C., and Paumann, M. (2004). The respiratory chain of blue-green algase (cyanobacteria). Physiologia Plantarum *120*, 358–369.

Robinson, C., and Bolhuis, A. (2001). Protein targeting by the twin-arginine translocation pathway. Nature Rev. Mol. Cell Biol. 2, 350–355.

Sarcina, M., Tobin, M.J., and Mullineaux, C.W. (2001). Diffusion of phycobilisomes on the thylakoid membranes of the cyanobacterium *Synechococcus* 7942: effects of phycobilisome size, temperature and membrane lipid composition. J. Biol. Chem. *276*, 46830–46834.

Sarcina, M., Murata, N., Tobin, M.J., and Mullineaux, C.W. (2003). Lipid diffusion in the thylakoid membranes of the cyanobacterium *Synechococcus* sp.: effect of fatty acid desaturation. FEBS Lett. *553*, 295–298.

Sarcina, M., and Mullineaux, C.W. (2004). Mobility of the IsiA chlorophyll-binding protein in cyanobacterial thylakoid membranes. J. Biol. Chem. *279*, 36514–36518.

Sarcina, M., Bouzovitis, N., and Mullineaux, C.W. (2006). Mobilization of Photosystem II induced by intense red light in the cyanobacterium *Synechococcus* sp. PCC7942. Plant Cell *18*, 457–464.

Sherman, D.M., Troyan, T.A., and Sherman, L.A. (1994). Localization of membrane proteins in the cyanobacterium *Synechococcus* sp. PCC7942. Radial asymmetry in the photosynthetic complexes. Plant Physiol. *106*, 251–262.

Siebert, C.A., Qian, P., Fotiadis, D., Engel, A., Hunter, C.N., and Bullough, P.A. (2004). Molecular architecture of photosynthetic membranes in *Rhodobacter sphaeroides*: the role of PufX. EMBO J. *23*, 690–700.

Silva, P., Thompson, E.P., Bailey, S., Kruse, O., Mullineaux, C.W., Robinson, C., Mann, N.H., and Nixon, P.J. (2003). FtsH is involved in the early stages of repair of photosystem two in *Synechocystis* sp. PCC 6803. Plant Cell *15*, 2152–2164.

Sippola, K., Kanervo, E., Murata, N., and Aro, E.-M. (1998). A genetically engineered increase in fatty acid desaturation in *Synechococcus* sp. PCC7942 allows exchange of D1 protein forms and sustenance of photosystem II activity at low temperature. Eur. J. Biochem. *251*, 641–648.

Spence, E., Sarcina, M., Ray, N., Møller S.G., Mullineaux, C.W., and Robinson, C. (2003). Membrane-specific targeting of green fluorescent protein by the Tat pathway in the cyanobacterium *Synechocystis* PCC6803. Mol Microbiol. *48*, 1481–1489.

Stanier, R.Y., and Cohen-Bazire, G. (1977). Phototrophic prokaryotes: the cyanobacteria. Annu. Rev. Microbiol. *31*, 225–274.

Stanier, G. (1988). Fine structure of cyanobacteria. In Methods in Enzymology Vol 167, L. Packer and A.N. Glazer eds. (San Diego, USA; Academic Press), pp 157–172.

Szalontai, B., Nishiyama, Y., Gombos, Z., and Murata, N. (2000). Membrane dynamics as seen by Fourier transform infrared spectroscopy in a cyanobacterium, *Synechocystis* PCC6803. The effects of lipid unsaturation and protein-to-lipid ratio. Biochim. Biophys. Acta *1509*, 409–419.

Trissl, H.-W., and Wilhelm, C. (1993). Why do thylakoid membranes from higher plants form grana stacks? Trends Biochem. Sci. *18*, 415–419.

van de Meene, A.M.L., Hohmann-Marriott, M.F., Vermaas, W.F.J., and Roberson, R.W. (2006). The three-dimensional structure of the cyanobacterium *Synechocystis* sp. PCC6803. Arch. Microbiol. *184*, 259–270.

van Thor, J.J., Mullineaux, C.W., Matthijs, H.C.P., and Hellingwerf, K.J. (1998). Light-harvesting and state transitions in cyanobacteria. Botanica Acta *111*, 430–443.

Vermaas, W.F.J. (1994). Molecular-genetic approaches to study photosynthetic and respiratory electron transport in thylakoids from cyanobacteria. Biochim. Biophys. Acta *1187*, 181–186.

Westermann. M., Neuschaefer-Rube, O., Mörschel, E., and Wehrmeyer, W. (1999). Trimeric Photosystem I complexes exist *in vivo* in thylakoid membranes of the *Synechocystis* strain BO9201 and differ in absorption characteristics from monomeric Photosystem I complexes. J. Plant Physiol. *155*, 24–33.

Westphal, S., Heins, L., Soll, J., and Vothknecht, U.C. (2001). *Vipp1* deletion mutant of *Synechocystis*: A connection between bacterial phage shock and thylakoid biogenesis? Proc. Natl. Acad. Sci USA *98*, 4243–4248.

Wu, Q., and Vermaas, W.F.J. (1995). Light-dependent chlorophyll *a* biosynthesis upon *chl*L deletion in wild-type and photosystem 1-less strains of the cyanobacterium *Synechocystis sp.* PCC6803. Plant Mol. Biol. *29*, 933–945.

Yu, J.J., Wu, Q.Y., Mao, H.B., Zhao, N.M., and Vermaas, W.F.J. (1999). Effects of chlorophyll availability on phycobilisomes in *Synechocystis* sp PCC6803. IUBMB Life *48*, 625–630.

Zak, E., Norling, B., Maitra, R., Huang, F., Andersson, B., and Pakrasi, H.B. (2001). The initial steps of biogenesis of cyanobacterial photosystems occur in plasma membranes. Proc. Natl. Acad. Sci USA *98*, 13443–13448.

Zouni, A., Witt, H.-T., Kern, J., Fromme, P., Krauss, N., Saenger, W., and Orth, P. (2001). Crystal structure of photosystem II from *Synechococcus elongatus* at 3.8 Å resolution. Nature 409, 739–743.

Carbon Acquisition by Cyanobacteria: Mechanisms, Comparative Genomics, and Evolution

12

Aaron Kaplan, Martin Hagemann, Hermann Bauwe, Shira Kahlon, and Teruo Ogawa

Abstract

In this chapter we mainly focus on the mechanisms of inorganic carbon uptake, photorespiration, and the regulation between the metabolic fluxes involved in photoautotrophic, photomixotrophic and heterotrophic growth. We identify the genes involved, their regulation and phylogeny.

General introduction

Living in an environment where the CO_2 concentration is considerably lower than required to saturate their carboxylating enzyme, ribulose 1,5-bisphosphate carboxylase/oxygenase (Rubisco), cyanobacteria acquired the CO_2 concentrating mechanism (CCM) that enables them to accumulate CO_2 at the carboxylation site. All the cyanobacteria examined to date are able to fix CO_2 into carbohydrates. However, in addition to variance in the range of physical growth conditions, cyanobacteria also vary substantially in their ability to consume organic carbon from their surroundings. Many strains are obligate photoautotrophs where the sole carbon source is CO_2, while others are able to perform photomixotrophic or even heterotrophic growth using a wide variety of organic substances (c.f. Rippka *et al.*, 1979; Stal and Moezelaar, 1997b). Cyanobacteria constitute a unique case where the anabolic and catabolic carbohydrate metabolisms function in the same cellular compartment. In addition, the photosynthetic and respiratory electron transport pathways share components in the thylakoid membranes. Despite its importance to our understanding of cyanobacterial metabolism, little is known about the mechanisms involved in the shifts between photoautotrophic, heterotrophic and photomixotrophic modes of growth, and their regulation; between the different pathways of carbohydrate breakdown—glycolysis, fermentation, the oxidative pentose phosphate, the Krebs cycle and the photorespiratory pathways. In this chapter we shall briefly focus on recent advances in our understanding of the CCM and carbon metabolism in cyanobacteria.

The CO_2 concentrating mechanism

General description

The appearance and evolution of cyanobacteria caused a long-term reduction in atmospheric CO_2 concentration and an increase in O_2. Consequently cyanobacterial photosynthesis became CO_2-limited because of the low affinity of their Rubisco to CO_2 and the competition between CO_2 and O_2 for the cleavage of ribulose 1,5-bisphosphate (RuBP). To cope with this situation, cyanobacteria acquired the CO_2 concentrating mechanism (CCM). Thereby they can accumulate inorganic carbon (Ci) within the cells to levels 1000-fold higher (see Kaplan and Reinhold, 1999; Giordano *et al.*, 2005) than the external one and raise the concentration of CO_2 in close proximity of the carboxysomal-located Rubisco. Other enzymes of the Calvin cycle are located outside the carboxysomes. Thus, cyanobacterial photosynthesis involves a flux of RuBP into, and of 3-phosphoglycerate from, the carboxysomes. The cellular components involved in the operation of the CCM include those engaged in Ci uptake and the carboxysomes (see the schematic model in Figure 12.1), where most of the Rubisco and cytoplasmic carbonic anhydrase (CA) are confined. Two CO_2-uptake systems and three HCO_3^- transporters have been identified. CO_2 enters into the cells by diffusion and is converted to HCO_3^- by the CO_2 uptake systems that involve thylakoid membrane-located NDH-1 complexes (see below). Light energy is used to maintain the cytoplasmic concentration of the CO_2 species below chemical equilibrium by converting it to HCO_3^-. The HCO_3^- transporters are located in the cytoplasmic membrane. HCO_3^- penetrates the carboxysomes and is converted to CO_2 by catalysis of CA

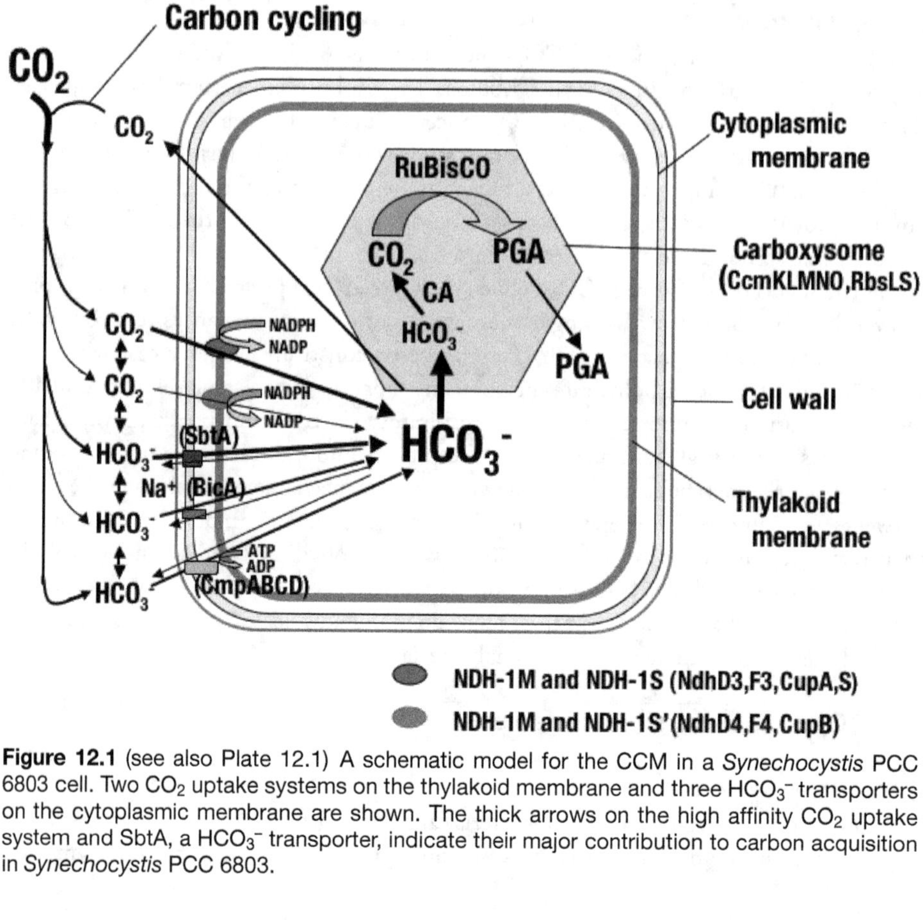

Figure 12.1 (see also Plate 12.1) A schematic model for the CCM in a *Synechocystis* PCC 6803 cell. Two CO_2 uptake systems on the thylakoid membrane and three HCO_3^- transporters on the cytoplasmic membrane are shown. The thick arrows on the high affinity CO_2 uptake system and SbtA, a HCO_3^- transporter, indicate their major contribution to carbon acquisition in *Synechocystis* PCC 6803.

in close proximity to Rubisco (Kaplan and Reinhold, 1999; So *et al.*, 2002). In addition to compensating for the relatively low affinity of Rubisco for CO_2, raising the CO_2 concentration at the carboxylation site activates the enzyme and depresses photorespiration (see section C, below). In many environmental conditions, the amount of Ci taken up by the cells exceeds that utilized in photosynthesis. The excess Ci leaks out of the cells as CO_2 and HCO_3^-, which are then taken up by the cells (Tchernov *et al.*, 1997, 2003). This cycling of Ci species and the massive transmembrane Ci fluxes involved in the operation of the CCM consume a lot of light energy, thus helping the cells to dissipate excess light energy and protect them from high light stress.

α-Cyanobacteria and β-cyanobacteria

Cyanobacteria are grouped into α-cyanobacteria and β-cyanobacteria based on the Rubisco phylogeny; α-cyanobacteria for those containing Form 1A Rubisco and β-cyanobacteria with Form 1B Rubisco. In addition to *rbcL* and *rbcS*, β-cyanobacteria possess *rbcX* that encodes a putative Rubisco chaperonin, but *rbcX* is absent in α-cyanobacteria (Badger *et al.*, 2006; Giordano *et al.*, 2005). Both α and β-cyanobacteria have carboxysomes (see below) that are protein microbodies consisting of shell proteins and soluble proteins, most of which are Rubisco (Beudeker *et al.*, 1980; McKay *et al.*, 1993). There are no homologies in many of the shell proteins between α- and β-cyanobacteria. Based on these differences, carboxysomes in β-cyanobacteria associated with Form 1B Rubisco are termed β-carboxysomes while those in α-cyanobacteria associated with Form 1A Rubisco are termed α-carboxysomes. Branching of α-cyanobacterial lineage from β-cyanobacterial might have occurred at an early stage of development by the replacement of β-carboxysomes with α-carboxysomes derived from non-photosynthetic bacteria.

Interestingly, Rubisco activase (*rca*) is present only in a few cyanobacterial species including *Anabaena* strains, *Nostoc punctiforme* strain PCC 73102 (ATCC 29133), *Trichodesmium erythraeum* strain IMS101 and *Gloeobacter violaceus* strain PCC 7421 but apparently missing in several others. Some sequence stretches within *ccmM* (see below) show homology to *rca* but the evolutionary/physiological meanings of this observation are yet to be clarified.

Carboxysomes

Role and structure

It is widely accepted that the elevated internal dissolved CO_2 concentration does not prevail throughout the cell but it is confined to the carboxysomes (Kaplan and Reinhold, 1999). The generation of CO_2 from the accumulated HCO_3^- is not catalyzed in the cytoplasm, and the Ci species do not reach equilibrium there. Diffusion of the accumulated HCO_3^- into the carboxysomes, where CO_2 formation is catalyzed at CA sites, leads to an elevated CO_2 concentration in close proximity to Rubisco. The highly organized structure of these bodies brings into close conjunction CA sites and Rubisco sites where a substantial proportion of the CO_2 is fixed. The biological advantage of the dense packing of the Rubisco molecules within carboxysomes may well lie in the substantial barrier to diffusion that this arrangement affords reducing the leak of CO_2 from the carboxysomes.

The means by which the bulk cytoplasmic Rubisco and CA are gathered and confined to the carboxysomes is poorly understood (see Kaplan and Reinhold, 1999). Application of the Hilbert differential contrast transmission electron microscopy showed that Rubisco molecules are packed within the cyanobacterial carboxysomes into paracrystalline arrays surrounded by a 5–6 nm thick monolayer shell consisting of orderly arrays of globular particles (Kaneko *et al.*, 2006). Analysis of the crystal structures of multiple carboxysomal shell proteins revealed a hexameric unit as the basic microcompartment building block that assemble to form flat facets of the polyhedral shell (Kerfeld *et al.*, 2005). Structural analysis of CsoSCA (CsoS3), the carboxysomal shell-located CA of *Halothiobacillus neapolitanus* (formerly *Thiobacillus neapolitanus*) suggested that it exists either as a homodimer or as extended filaments within the shell (Sawaya *et al.*, 2006). If, indeed, conversion of HCO_3^- to CO_2 occurs at the carboxysomal surface, it is hard to visualize how an expected massive leak of CO_2 is prevented. The fact that this CA contains only one active domain rather than the two typically found in β-CAs may indicate divergence that evolved to fulfill its function in the carboxysome (Sawaya *et al.*, 2006).

Mutants with abnormal carboxysomes

Since the first high CO_2 requiring (HCR) mutant (E1) was isolated from *Synechococcus elongatus* strain PCC 7942 (hereafter *Synechococcus* PCC 7942) in 1986 (Marcus *et al.*, 1986), many HCR mutants have been isolated or constructed from this strain and from *Synechocystis* sp. strain PCC 6803 (hereafter *Synechocystis* PCC 6803) (Ogawa *et al.*, 1987; Price *et al.*, 1989). Most of the mutants isolated from randomly mutagenized cells, including the E1 mutant, showed abnormal carboxysome structure and were unable to fix CO_2 under air levels of CO_2, although they accumulated high internal concentrations of Ci (Friedberg *et al.*, 1989; Price *et al.*, 1989). HCR-mutants defective in carboxysomal CA showed a similar phenotype as the E1 mutant but possessed normal carboxysomal structure (Fukuzawa *et al.*, 1992; Price *et al.*, 1992).

Complementation experiments of these HCR mutants enabled one to identify the genes impaired. These genes were designated *ccmK, ccmL, ccmM, ccmN* and *ccmO* that encode carboxysomal proteins (Kaplan *et al.*, 1994) and *icfA* (Fukuzawa *et al.*, 1992; Price *et al.*, 1992) that encodes carboxysomal β-CA. The gene named *ccmA* also complemented an HCR mutant defective in carboxysome (Ogawa *et al.*, 1994). This gene encodes a protein homologous to phospho-2-dehydro-3-deoxyheptonate aldolase or 3-deoxy-D-arabinoheptulosonate 7-phosphate syntase. The role of CcmA in carboxysome formation is not known and it is not clear whether the gene is directly involved in carboxysome formation.

Carboxysomal shell proteins and carbonic anhydrases

Carboxysomal shell proteins in α-cyanobacteria do not have homology with those in β-cyanobacteria, except that CsoS1 in α-carboxysomes is homologous to CcmK2 in β-carboxysomes. There are a number of homologues of *ccmK*, designated *ccmK1, ccmK2, ccmK3* and *ccmK4* in β-cyanobacteria (Figure 12.2). Many of the β-cyanobacterial strains possess all these *ccmK* genes but some of the strains contain 2 or 3 homologues. In addition to the *ccmK* genes, all the β-cyanobacterial strains possess *ccmL, ccmM, ccmN* and *ccmO* genes that constitute an operon. In *Synechococcus* PCC 7942, the *ccm* operon is located in

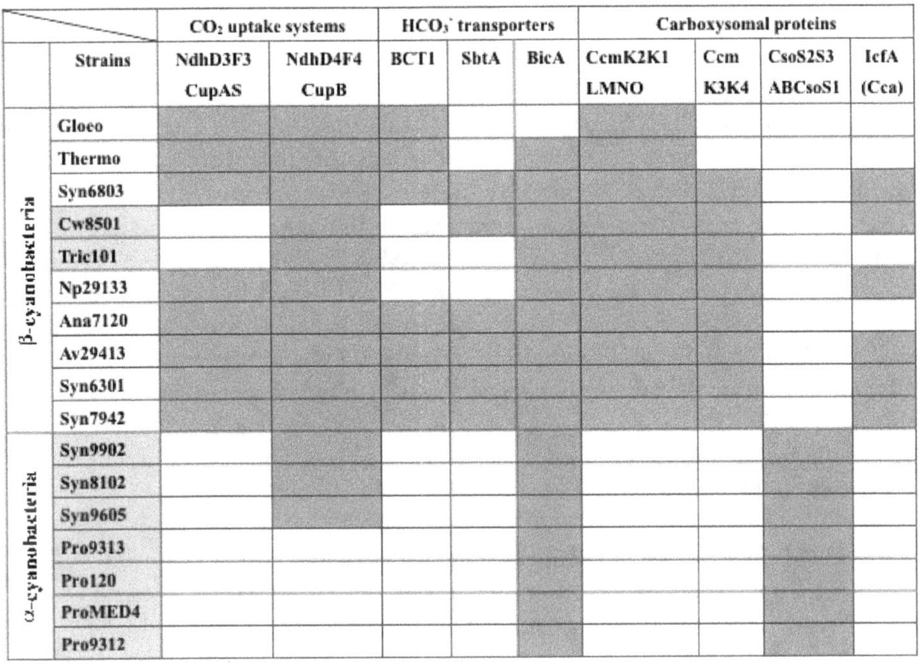

Figure 12.2 The presence (shadowed) or absence (blank) of the two CO_2-uptake systems, three HCO_3^- transporters and carboxysomal proteins in 17 cyanobacterial strains (marine strains are shadowed). Abbreviations for the strains are: Gloeo (*Gloeobacter violaceus* PCC 7421), Thermo (*Thermosynechococcus elongatus* BP-1), Syn6803 (*Synechocystis* PCC 6803), Cw8501 (*Crocosphaera watsonii* WH 8501), Tric101 (*Trichodesmium erythraeum* IMS101), Np29133 (*Nostoc punctiforme* ATCC 29133), Ana7120 (*Anabaena* sp. PCC 7120), Av29413 (*Anabaena variabilis* ATCC 29413), Syn6301 (*Synechococcus* sp. PCC 6301), Syn7942 (*Synechococcus elongatus* PCC 7942), Syn8102 (*Synechococcus* sp. WH8102), Syn9605 (*Synechococcus* sp. CC9605), Syn9902 (*Synechococcus* sp. CC9902), Pro9313 (*Prochlorococcus marinus* MIT 9313), Pro120 (*Prochlorococcus marinus* SS120), ProMED4 (*Prochlorococcus marinus* MED4), Pro9312 (*Prochlorococcus marinus* sp. MIT9312). In addition to Rubisco, carboxysomal proteins in α-cyanobacteria include CsoS1, CsoS2, CsoS3 (CsoSCA), peptide A, and peptide B; those in β-cyanobacteria are CcmK1, CcmK2, CcmK3, CcmK4, CcmL, CcmM, CcmN, CcmO and IcfA (Cca). Among them, CsoS1, CsoS2, Cca and CcmM have already been identified in carboxysome-enriched fractions.

the 5′-flanking region of the *rbcL-rbcS* operon (Kaplan *et al.*, 1994) but in other strains the *rbc* operon is located distant from the *ccm* operon. In contrast, in α-cyanobacteria, *rbcL* and *rbcS* constitute an operon with the genes encoding carboxysomal shell proteins in the order of *csoS1, rbcL, rbcS, csoS2, csoSCA,* peptide A gene, peptide B gene and *csoS1.*

Homologues of carboxysomal β-CA gene (*slr1347* in *Synechocystis* PCC 6803) are present in many but not all of the β-cyanobacterial strains, and are absent in α-cyanobacterial species. The presence of CA activity has been demonstrated in CsoSCA of *Synechococcus* sp. strain WH8102 (So *et al.*, 2004). Although CcmM in β-cyanobacteria is postulated to be a γ-class CA, experimental evidence showing CA activity in this protein is missing (Ludwig *et al.*, 2000).

CO_2-uptake systems

Mutants defective in CO_2 uptake

The presence of multiple Ci acquisition systems made it difficult to isolate mutants defective in specific systems since they complemented one another and mutation in one relevant gene did not result in a clear phenotype. However, isolation and analysis of *Synechocystis* PCC 6803 mutants, RKa and RKb, of less stringent HCR phenotype enabled us to identify the genes involved in CO_2 uptake (Ogawa, 1990, 1991a, 1991b, 1992). The *ndhB* gene which encodes one of the subunits of type-1 NAD(P)H dehydrogenase (NDH-1) complemented the RKa mutant, and *ictA* (later renamed *ndhL*, Ogawa, 1992) encoding a small protein of 80 amino acids, complemented mutant RKb. Recent proteomics analysis of NDH-1 identified the product of *ndhL* as an essential component of this enzyme (Battchikova *et al.*, 2005). These and similar studies indicated the central role of NDH-1 in CO_2 uptake in cyanobacteria and led to the isolation of several other mutants impaired in subunits of NDH-1 (see below).

Unique NDH-1 complexes involved in two CO_2 uptake systems

Cyanobacteria possess unique subunits of NDH-1 complexes that are essential to CO_2 uptake. These include the proteins encoded by *ndhD3*, *ndhD4*, *ndhF3* and *ndhF4*. Mutants constructed by inactivating *ndhD3* or *ndhF3* are unable to grow at 20 ppm CO_2 although inactivation of *ndhD4* or *ndhF4* has no effect on growth under these conditions (Ohkawa *et al.*, 1998, 2000b; Klughammer *et al.*, 1999). These results led to the findings of functionally distinct NDH-1 complexes (Ohkawa *et al.*, 2000a) and two CO_2-uptake systems (Shibata *et al.*, 2001; Maeda *et al.*, 2002). All the cyanobacterial strains listed in Figure 12.2 except the *Prochlorococcus* species possess multiple copies of *ndhD* and *ndhF* (Genome online database; http://www.genomesonline.org/). *Synechocystis* PCC 6803 possesses at least 4 *ndhD* (*D1-D4*) and 3 *ndhF* (*F1, F3* and *F4*) genes. A phylogenetic tree indicated that *ndhD* and *ndhF* are both members of a larger family and that *ndhD1/D2* and *ndhF1* show the highest similarity to *ndhD* and *ndhF*, respectively, present in a single copy in chloroplast genomes of higher plants (Shibata *et al.*, 2001). In addition to *ndhD* and *ndhF*, cyanobacteria possess unique subunits designated *cupA* (*chyY*) and *cupB* (*chyX*) that are also essential to CO_2 uptake (Shibata *et al.*, 2001; Maeda *et al.*, 2002). These genes constitute operons of *ndhF3-ndhD3-cupA* and *ndhF4-ndhD4-cupB* in many strains but not all. For example, *cupB* in *Synechocystis* 6803 is located far from the *ndhF4-ndhD4* operon. Recent proteomics analysis identified *cupS* (*sll1735*) as a subunit of NDH-1S complex that also contains NdhD3, NdhF3 and CupA (Herranen *et al.*, 2004; see 3–3). Unlike the Δ*ndhD3* mutant, a Δ*cupS* mutant was able to grow at 20 ppm CO_2 (Ohkawa *et al.*, 2000b). The *ndhD3*, *ndhF3* and *cupA* genes are expressed as an operon and *cupS* as a single gene, and the expression of these genes is induced under low CO_2. In contrast, the *ndhD4*, *ndhF4* and *cupB* genes are constitutively expressed even under high CO_2 (Ohkawa *et al.*, 1998; Shibata *et al.*, 2001; Maeda *et al.*, 2002; Wang *et al.*, 2004). Mutants constructed by inactivating *ndhD4*, *ndhF4* or *cupB* exhibited the CO_2 uptake activity of the low-CO_2-inducible system characterized by higher affinity for CO_2 and V_{max} typical of low CO_2-grown cells (Shibata *et al.*, 2001; Maeda *et al.*, 2002). Those where *ndhD3*, *ndhF3* or *cupA* were inactivated exhibited the activity of the constitutive system characterized by a relatively low affinity and low V_{max} for CO_2 uptake, regardless of the CO_2 concentrations during growth (Shibata *et al.*, 2001). When two genes were inactivated, one from each group, the mutants were unable to grow under air at pH 7.0. Thus the presence of two CO_2 uptake systems became evident; one is induced by low CO_2 and involves *ndhD3*, *ndhF3*, *cupA* and possibly *cupS*, and the other is constitutive and involves *ndhD4*, *ndhF4* and *cupB* (Shibata *et al.*, 2001, 2002). The K_m (CO_2) value of the low-CO_2-inducible system was as low as 0.8 μM and did not change, but the V_{max} of uptake increased significantly during acclimation of the cells following transfer from high to low CO_2. The above result may indicate that the rate of CO_2 to HCO_3^- conversion is dependent on the amount of the proteins involved in this reaction (Shibata *et al.*, 2001).

Inactivation of *ndhD1/D2* or *ndhF1* has no effect on CO_2 uptake but strongly inhibited the respiration and abolished the ability of cells to grow under photoheterotrophic conditions (Ohkawa *et al.*, 2000a).

Proteomics of cyanobacterial NDH-1 complexes

Proteomic analysis of cyanobacterial NDH-1 complexes has revealed the presence of three complexes: NDH-1L (large size), NDH-1M (medium size) and NDH-1S (small size) (Herranen *et al.*, 2004). NDH-1L complex contains NdhD1 and NdhF1 in addition to NdhA, NdhB, NdhC, NdhE, NdhG, NdhH, NdhI, NdhJ, NdhK, NdhL, NdhM, NdhN and NdhO (Zhang *et al.*, 2004). All these subunits, except NdhD1 and NdhF1, are present in NDH-1M. NDH-1S contains NdhD3, NdhF3, CupA and CupS and is absent in the mutant defective in *ndhD3*. High CO_2-grown cells contain predominantly NDH-1L while NDH-1M and NDH-1S are the major complexes in low CO_2-adapted cells. A complex (NDH-MS) containing both NDH-1S and NDH-1M has been isolated from a *Thermosynechococcus elongatus* strain in which the C terminus of NdhL has a 6-His tag (Zhang *et al.*, 2005). This complex is easily dissociated into NDH-1M and NDH-1S complexes (Arteni *et al.*, 2006). Inspection of NDH-1 complexes from the WT and mutants by electron microscopy showed that the complete NDH-1 complex is L-shaped, with a relatively short hydrophilic arm, and that the abundance of specific complexes was affected by the CO_2 concentration experienced during growth (Arteni *et al.*, 2006).

Analysis of various mutants of *Synechocystis* PCC 6803 for their physiological properties in relation to the presence or absence of NDH-1L, NDH-1M and NDH-1S has indicated that NDH-1L is essential to photoheterotrophic growth, while NDH-1M is involved in PSI cyclic electron transport and the NDH-1MS complex participates in the inducible CO_2-uptake system (Zhang *et al.*, 2004). It is presumed that CupA and CupB may have a CA-like activity and play a role in the conversion of CO_2 to HCO_3^- but experimental evidence is still missing. CupB has been identified in a band of about 440 kDa after native gel electrophoresis but was absent in mutant Δ*ndhD4* (M. Xu, T. Ogawa, H. Mi, unpublished). This result suggests that CupB is associated with NdhD4 and NdhF4 to form NDH-1S′, a homologue of NDH-1S, and is present as a complex of NDH-1MS′. NDH-1S′ does not contain CupS since its homologues are absent in the strains devoid of the inducible CO_2 uptake system.

The mechanism of CO_2 uptake is poorly understood but NDH-1MS and NDH-1MS′ may be involved in the conversion of CO_2 to HCO_3^- in the thylakoid membranes (Kaplan and Reinhold, 1999; Tchernov *et al.*, 2001; Ohkawa *et al.*, 2001). Clearly, the

HCO_3^- produced is released in an environment where its concentration may exceed 50 mM, i.e. against the chemical equilibrium. Such uni-directional conversion of CO_2 to HCO_3^- needs a continuous supply of OH^-, which may be produced by the reduction of $NADP^+$ to NADPH. It has been postulated that the conversion of CO_2 to HCO_3^- occurs at an "alkaline pocket" in the thylakoid membranes (Kaplan and Reinhold, 1999).

Phylogeny of the CO_2 uptake systems

The most divergent examined cyanobacterial strain, *Gloeobacter violaceus* (see Chapter 2), possesses both CO_2-uptake systems (Figure 12.2). Thus, cyanobacteria acquired the CO_2-uptake systems at an early stage of evolution before the branching of α-cyanobacterial lineage. *Prochlorococcus* strains do not possess any of the CO_2-uptake systems and marine *Synechococcus* strains possess only the constitutive CO_2 uptake. The low CO_2-inducible system might have been lost after the α-cyanobacterial lineage was branched. Then deletion of the constitutive CO_2-uptake system might have followed after the branching of the marine *Synechococcus* lineage from the *Prochlorococcus* lineage. The high-affinity CO_2-uptake system (NdhD3F3 CupAS) is absent in many marine species (shaded in left column of Figure 12.2) but not in all, since this system is present in *Synechoccoccus* sp. strain PCC 7002 (Klughammer *et al.*, 1999). It is possible that the high-affinity CO_2-uptake system has been lost from marine strains but some marine strains acquired the system by horizontal gene transfer.

Three types of HCO_3^- transporter and their phylogenies

Research on *Synechocystis* PCC 6803, *Synechococcus* PCC 7942 and *Synechococcus* PCC 7002 led to the identification of three HCO_3^- transporters. These are (1) BCT1, an ABC-type high affinity HCO_3^- transporter (Omata *et al.*, 1999); (2) SbtA, a high affinity sodium-dependent HCO_3^- transporter (Shibata *et al.*, 2002); and (3) BicA, a SulP-type low affinity HCO_3^- transporter (Price *et al.*, 2004).

BCT1, an ABC-type transporter

A 42-kD protein that is induced under low CO_2 was found in the cytoplasmic membrane of *Synechococcus* PCC 7942 (Omata and Ogawa, 1985, 1986). Cloning of the gene (designated *cmpA*) encoding this protein and sequencing of the *cmpA* region indicated that *cmpA* forms an operon with *cmpB*, *cmpC* and *cmpD* and that these genes encode subunits of an ABC-type transporter, with CmpA being a substrate-binding subunit (Omata *et al.*, 1990; Omata, 1992). The evidence that these genes encode an HCO_3^- transporter was obtained by constructing a strain of *Synechococcus* PCC 7942 in which the promoter region of the *cmp* operon was replaced with that of the *nrt* operon encoding a nitrate transporter (Omata *et al.*, 1999). When the *cmp* operon is expressed in this strain in the presence of nitrate under high CO_2 conditions, where most of other Ci-acquisition systems do not function, cells take up HCO_3^- with a K_m (HCO_3^-) value of 15 μM. The CmpA protein expressed in *E. coli* shows HCO_3^- binding capability (Maeda *et al.*, 2000). Analysis of the 3D structure of NrtA, the oxyanion-binding protein identified as the nitrate-specific receptor in *Synechocystis* PCC 6803, and identification of the nitrate-binding domain, suggested similarity to HCO_3^--binding by CmpA (Koropatkin *et al.*, 2006). These authors also proposed that transport of nitrate and bicarbonate are regulated by the cytoplasmic proteins NrtC and CmpC, so that the abundance of nitrate may affect the uptake of HCO_3^- by the CMP (BCT1) system.

A *cmpA* deletion mutant did not show an HCR phenotype because of the functioning of other Ci transporters. Contribution of this transporter to the overall Ci uptake by *Synechocystis* PCC 6803 appears to be insignificant, since deletion of *cmpA* in a Δ*ndhD3/D4* mutant has no effect on its growth under air. In contrast, a Δ*ndhD3/D4/sbtA* mutant is unable to grow under these conditions (Shibata *et al.*, 2002).

BCT1 is present in *Gloeobacter violaceus* and the *Anabena* strains, but absent in all the marine strains, listed in Figure 12.2. Thus it is likely that cyanobacteria acquired this HCO_3^- transporter at an early stage of evolution and then it was lost in most of the marine strains during their branching. It is too early to determine whether horizontal gene transfer might have been involved. It is possible that continuous presence of sodium ions outside the cells of marine organisms preclude the use of costly ABC systems for Ci uptake.

SbtA, a sodium-dependent transporter

The Δ*ndhD3/D4* mutant of *Synechocystis* PCC 6803 is unable to take up CO_2 and does not grow at pH 7.0 in air but grows at alkaline conditions. Mutants unable to grow at pH 9.0 in air were isolated after random tagging of the Δ*ndhD3/D4* mutant. These mutants had tags within *slr1512*, which was identified to encode a novel transporter involved in Na^+-dependent HCO_3^- uptake and was designated *sbtA* (sodium-bicarbonate transport) (Shibata *et al.*, 2002). Disruption of *sbtA* in the single *ndhD3* or *ndhD4* mutants, which are able to take up CO_2 either by the constitutive or by the inducible systems, respectively, does not affect their growth (Shibata *et al.*, 2001). The SbtA-dependent HCO_3^- transporter is expressed under low CO_2, has a high affinity for the substrate (K_m = 12 μM) and may function as a Na^+/HCO_3^- symporter. Its recognition may explain, at least in part, the dependence of cyanobacterial photosynthesis and growth on the present of a few millimolar sodium (Reinhold *et al.*, 1984; Miller *et al.*, 1984).

SbtA is absent in *Gleobacter violaceus, Thermosynechococcus elongatus*, α-cyanobacterial strains and several strains of β-cyanobacteria. Thus cyanobacteria might have acquired SbtA at a certain stage of evolution after the branching of α-cyanobacteria. *Prochlorococcus* strains possess distant homologues of SbtA (not listed in Figure 12.2), which have been suggested to play a role Ci acquisition (Ogawa and Kaplan, 2003). The finding of BicA (Price *et al.*, 2004), however, indicated that accumulation of Ci in these strains may be explained without assuming the distant homologue of SbtA to be a HCO_3^- transporter. Whether this homologue of SbtA functions as a HCO_3^- transporter is yet to be established.

BicA, a SulP-type low affinity transporter

This is the most recently identified and probably the last HCO_3^- transporter that can be found in cyanobacteria (Price *et al.*, 2004). Inactivation of one of the *sul* genes (*sul3/bicA*) in *Synechococcus* PCC 7002, annotated to encode a low affinity sulfate transporter, results in a lower rate of O_2 evolution at high HCO_3^- concentrations as compared with the wild type. Expression of *bicA* of *Synechococcus* PCC 7002, *Synechocystis* PCC 6803 and *Synechococcus* WH8102 in high-CO_2-grown cells of *Synechococcus* PCC 7942, using a shuttle vector, results in faster rates of O_2 evolution that are dependent on Na^+. BicA has a low affinity for its substrate with the K_m (HCO_3^-) values between 75 and 360 μM. It is induced by low CO_2 in *Synechococcus* PCC 7002 (Price *et al.*, 2004) but appears to be expressed constitutively even under high CO_2 in *Synechocystis* PCC 6803 (Wang *et al.*, 2004; H. Katoh and T. Ogawa, unpublished).

BicA is present in all the cyanobacterial strains analyzed except *Gloeobacter violaceus*. Thus it appears reasonable to assume that this HCO_3^- transporter was present before the branching of α- and β-cyanobacterial strains and was lost from *Gloeobacter violaceus*.

Energization of HCO_3^- transporters

ATP energizes BCT1 but it is not known how the two other HCO_3^- transporters are energized. Both SbtA and BicA require Na^+, which is consistent with the early findings of the requirement of Na^+ for HCO_3^- uptake by cyanobacteria (Miller *et al.*, 1984; Reinhold *et al.*, 1984; Espie *et al.*, 1994; So *et al.*, 1998) and hence for their growth, particularly at alkaline pH values. The nature of the primary pump that provides the free energy for accumulation of HCO_3^- against its electrochemical potential, via the SbtA or BicA, is not known. Although a primary sodium pump appears essential for active transport of HCO_3^- across the cytoplasmic membrane (see Kaplan and Reinhold, 1999), such a sodium pump has not yet been identified. A protein encoded by *slr1509* in *Synechocystis* PCC 6803 was once considered to be a sodium pump but it has been found to be a Na^+-dependent K^+-uptake pump (Berry *et al.*, 2003; Matsuda *et al.*, 2004).

Role of IctB

Application of an inactivation library resulted in a *Synechococcus* PCC 7942 mutant impaired in the ability to accumulate Ci within the cells (Bonfil *et al.*, 1998). The mutation responsible for this phenotype was within a gene, designated *ictB*, encoding a transmembrane protein most likely involved in transport. Despite numerous attempts in three different laboratories, it has not been possible to isolate a fully segregated mutant of *Synechocystis* PCC 6803 where the *ictB* homologue, *slr1515*, was inactivated. The role of IctB, which is highly conserved in all the cyanobacteria examined so far, is not known but expression in *Arabidopsis* and tobacco resulted in transgenic plants that show a higher photosynthetic rate under limiting CO_2 levels than the respective wild types. Growth of the transgenic plants under limiting supply of water is also enhanced (Lieman-Hurwitz *et al.*, 2003).

Acclimation to low CO_2 conditions

Many low CO_2-inducible genes have been identified by microarray analysis (Wang *et al.*, 2004) but despite intensive studies, the mechanism by which the expression of these genes is controlled by CO_2 concentrations is poorly understood. Products of the photorespiratory cycle such as phosphoglycolate, internal Ci level and CO_2 concentration as such were all implicated as possible signals (Kaplan and Reinhold, 1999; Giordano *et al.*, 2005; Woodger *et al.*, 2005). The high affinity CO_2-uptake system and two HCO_3^- transporters, SbtA and BCT1, are induced under low CO_2. The HCO_3^- transporter BicA in *Synechococcus* PCC 7002 is also induced by low CO_2 (Price *et al.*, 2004) but its homologue in *Synechocystis* PCC 6803 (Slr0834) appears to be constitutively expressed (Wang *et al.*, 2004). The LysR-type transcriptional factor, NdhR, is involved in low CO_2-induced expression of the *ndhF3/ndhD3/cupA* operon and of *sbtA*. These genes are constitutively expressed in

Δ*ndhR* mutant (Figge *et al.*, 2001; Wang *et al.*, 2004). Inactivation of *ndhR* has no effect on the expression of the *cmp* operon encoding the BCT1-type HCO_3^- transporter, which was abolished by inactivation of *cmpR* in *Synechococcus* PCC 7942. Interestingly, CmpR also regulates the expression of *psbA2* (Takahashi *et al.*, 2004). A protein kinase has recently been implicated in CO_2 sensing in guard cells of *Arabidopsis* (Hashimoto *et al.*, 2006) suggesting that a similar mechanism may be involved in cyanobacteria. The P_{II} protein may also play a role. This regulator is known as a sensor for availability of nitrogen, but its phosphorylation state changes also under different light and/or C_i conditions (Schwarz and Forchhammer, 2005; see also Chapter 13). Hypothetically, the P_{II} protein may act as a sensing molecule, which measures the ratio of inorganic nitrogen to C_i. Therefore, high C_i might mean low inorganic nitrogen and *vice versa*.

Dissipation of excess energy by carbon cycling

The uptake of CO_2 and HCO_3^- and leakage of these Ci species occur during the operation of the CCM. This carbon cycling maintains an intracellular CO_2 concentration substantially above or below that expected at chemical equilibrium with the prevailing HCO_3^- concentration in the medium (Tchernov *et al.*, 2003). It has been suggested that the cycling of Ci species may play an important role in protecting the cells from high light stress by dissipating excess light energy. Preliminary data to support this hypothesis have recently been obtained. A mutant in which all the CO_2 uptake systems and HCO_3^- transporters were inactivated was more sensitive to high light than the wild type even under high CO_2 conditions, whereas the Δ*ndhR* mutant in which genes involved in high affinity CO_2-uptake system and *sbtA* are constitutively expressed is more tolerant to high light (M. Xu, T. Ogawa, H. Mi, unpublished). Thus, in addition to CO_2 concentration, the CCM appears to have an important role in dissipating excess light energy to protect the cells from high light stress. These observations may lend support to the possibility (see Kaplan and Reinhold, 1999) that Ci transporting systems originally emerged as energy dissipation tools and were then (following the decline in atmospheric CO_2 concentration) acquired to accumulate Ci within the cells.

Function and role of photorespiration in cyanobacteria

Description of the process

In addition to the carboxylating reaction, Rubisco oxygenates RuBP to 3-phosphoglyceric acid (PGA) and 2-phosphoglycolate (PG), a compound which inhibits the Calvin cycle enzyme triosephosphate isomerase (Husic *et al.*, 1987) and is not well compatible with photosynthetic metabolism. Therefore, PG must be efficiently metabolized. In plants and other organisms that perform oxygenic photosynthesis, this occurs in the photorespiratory glycolate cycle often referred to as the C_2 cycle (Ogren, 1984; Tolbert 1997; Bauwe and Kolukisaoglu, 2003). This pathway converts PG to PGA (which is then reduced to triosephosphates in the Calvin cycle) and photorespiratory CO_2 and NH_3 are released. In addition to this primary function, the C_2 cycle interacts with other metabolic pathways, such as one-carbon metabolism, and is thought to contribute to the protection of the photosynthetic apparatus from photoinhibition (Kozaki and Takeba, 1996).

In plants, at least 10 enzymes, located in three different organelles, are involved in the C_2 cycle (Figure 12.3). Briefly, the PG is dephosphorylated by PG phosphatase (PGPase) in chloroplasts. The peroxisomal-located glycolate oxidase (GOX) catalyzes the irreversible oxidation of the glycolate formed to glyoxylate which, by the action of two aminotransferases and using glutamate and serine as amino donors, is subsequently transaminated to glycine. The combined action of two mitochondrial enzymes, the glycine decarboxylase complex (GDC) and serine hydroxymethyltransferase (SHMT), then converts two molecules of glycine to one molecule of serine, NH_3 and CO_2. Serine is exported from the mitochondria and deaminated to 3-hydroxypyruvate (HP) in a second peroxisomal aminotransferase reaction. Next, another peroxisomal enzyme, HP reductase, reduces HP to glycerate which becomes phosphorylated by the chloroplastidal glycerate kinase (GLYK) to finally yield the Calvin cycle intermediate PGA.

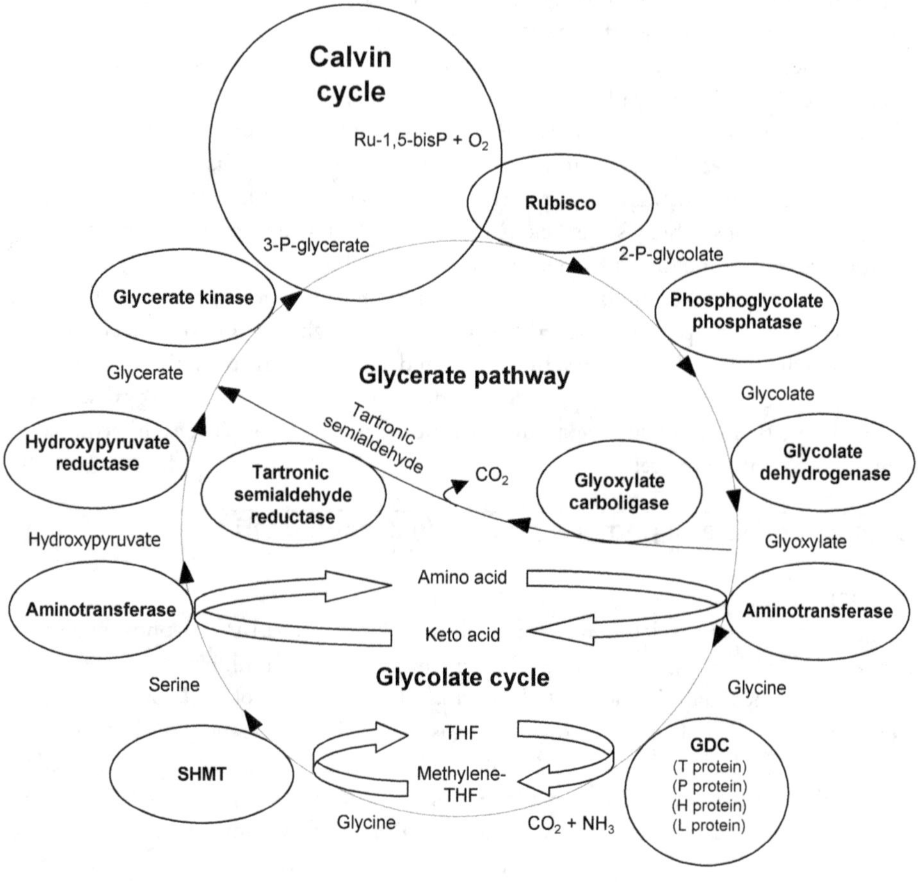

Figure 12.3 Known pathways to metabolize 2-phosphoglycolate by a plant-like glycolate cycle (outer cycle) or the glycerate pathway (inner branch). The metabolites involved, enzymatic steps and corresponding enzymes are shown. (Rubisco: ribulose-1,5-bisphosphate carboxylase/oxygenase; GDC: glycine decarboxylase complex; SHMT: serine hydroxymethyltransferase; Ru-1,5-bisP: ribulose-1,5-bisphosphate; THF: tetrahydrofolate)

It is not clear whether such a pathway exists in cyanobacteria, too. Since the discovery of the CCM (see section B, above), which raises the level of CO_2 in close proximity to Rubisco thereby inhibiting its oxygenase activity, it is generally thought that the photorespiratory C_2 cycle is negligible in cyanobacteria. However, for several reasons, including the low affinity of cyanobacterial Rubisco towards CO_2, cyanobacteria were expected to produce considerable amounts of PG. This is in agreement with the finding of photorespiratory rates comparable to those observed in green algae and C_3 plants in some cyanobacterial species (Birmingham *et al.*, 1982) and other reports suggesting the occurrence of photorespiration in cyanobacteria (Norman and Colman, 1992). Therefore, numerous attempts were made to identify a plant-like photorespiratory C_2 cycle in these organisms. Analysis of the early ^{14}C-fixation products of cyanobacteria showed clear accumulation of glycolate and other metabolites of the C_2 cycle although its proportion was rather low in comparison with Calvin cycle metabolites. Under artificial growth conditions (nearly complete O_2-atmosphere or addition of inhibitors for photorespiratory enzymes), significant accumulation of glycolate and glyoxylate was found (Norman and Colman, 1988; Renström and Bergman, 1989). Moreover, early attempts to verify the presence of a photorespiratory cycle in cyanobacteria showed the existence of corresponding enzyme reactions and metabolite fluxes (Codd and Stewart, 1973; Colman, 1989; Renström-Kellner and Bergman, 1990).

Search of cyanobacterial genomes for photorespiratory enzymes

In-silico searches of the available cyanobacterial genomes identified a number of genes which encode proteins with high similarities to enzymes involved in the higher plant C_2 cycle. Such genes were recognized in the genomes of *Synechocystis* PCC 6803, *Prochlorococcus* sp. strain SS120 and *Anabaena* PCC 7120 (Table 12.1); for some of them, the amino acid sequence similarity of the derived cyanobacterial protein to the respective plant protein is very high. For example, the phylogenetic analysis of GDC subunits P-protein and L-protein from prokaryotes and eukaryotes revealed the highest similarities between the higher plant and cyanobacterial proteins (unpublished data). This indicates that the respective plant proteins most likely originated from the genome of the cyanobacterial chloroplast ancestor (Martin *et al.*, 2002). It appears as if the genetic information provided by the cyanobacterial endosymbiont to the eukaryotic host during plant evolution not only included essential genes for oxygenic photosynthesis but also many "photorespiratory" genes.

These *in-silico* comparisons also revealed differences with respect to the nature of genes necessary for specific functions both between cyanobacteria and plants and amongst the cyanobacteria themselves. Interestingly, the enzymes necessary for the conversion of serine into PGA are not conserved between plants and cyanobacteria. This suggests that the respective plant enzymes were not derived from the cyanobacterial endosymbiont and indicates the possibility of alternative pathways for the serine-to-PGA conversion in cyanobacteria and plants. The genomes of *Prochlorococcus* SS120 and *Anabaena* PCC 7120 harbor genes for plant-type GOX and GLYK. On the other hand, glycolate dehydrogenases (GlcD) with high similarity to the corresponding *E. coli* enzyme are present in *Arabidopsis, Synechocystis* PCC 6803, and *Anabaena* PCC 7120, but not in *Prochlorococcus* SS120. *Synechocystis* PCC 6803 possesses a bacterial-type glycerate kinase instead of

Table 12.1 Selected proteins involved in 2-phosphoglycolate metabolism in *Arabidopsis thaliana* (http://www.arabidopsis.org/index.jsp) and ORFs in the cyanobacterial genomes of *Synechocystis* sp. strain PCC 6803 (6803), *Anabaena* sp. strain PCC 7120 (7120), and *Prochlorococcus* sp. strain SS120 (SS120), which show significant (e-values better than e^{-05}) similarities in amino acid sequences. Amino acid similarity searches were performed using local Blast at CyanoBase (http://www.kazusa.or.jp/cyanobase/blast.html)

Protein	ORF in *A. thaliana*	Ref.	ORF in 6803	Similarity e-value	ORF in 7120	Similarity e-value	ORF in SS120	Similarity e-value
2-phosphoglycolate phosphatase	At5g36790	(1)	none		none		none	
	At5g47760	(1)	none		none		none	
Glycolate oxidase	At3g14415	(1)	none		all0170	$1e^{-95}$	Pro1754	$3e^{-64}$
Glycolate DH	At5g06580	(2)						
subunit GlcD			sll0404	$4e^{-53}$	alr5269	$6e^{-52}$	none	
subunit GlcE			sll1189	$9e^{-29}$	all0624	$7e^{-26}$	none	
Serine-glyoxylate-aminotransferase	At2g13360	(3)	sll1559	$8e^{-59}$	alr1004	$9e^{-41}$	Pro0036	$2e^{-57}$
GDC, P-protein	At2g26080	(1)	slr0293	0.0	all4607	0.0	Pro1829	0.0
	At4g33010	(1)	slr0293	0.0	all4607	0.0	Pro1829	0.0

GDC, T-protein	At1g11860	(1)	sll0171	$2e^{-51}$	all4609	$3e^{-54}$	Pro1851	$6e^{-53}$
	At1g60990	(1)	slr0635	$3e^{-46}$	all4055	$7e^{-82}$	Pro1851	$2e^{-12}$
GDC, H-protein	At2g35120	(1)	slr0879	$2e^{-28}$	all4608	$1e^{-31}$	Pro1830	$2e^{-27}$
GDC, L-protein	At3g16950	(1)	slr1096	$1e^{-164}$	alr4745	$1e^{-174}$	Pro1372	$1e^{-151}$
Serine-hydroxymethyl-transferase, SHMT1	At4g37930	(4)	sll1931	$1e^{-93}$	alr4806	$1e^{-93}$	Pro0290	$4e^{-92}$
Glutamate:glyoxylate-aminotransferase	At1g23310	(3)	slr0036	$9e^{-14}$	alr2765	$4e^{-14}$	Pro1018	$2e^{-19}$
Hydroxypyruvate reductase	At1g68010	(1)	sll1908	$7e^{-28}$	all8087	$3e^{-28}$	Pro1436	$3e^{-29}$
			slr1556	$5e^{-26}$	alr1890	$3e^{-27}$		
			slr2123	$1e^{-23}$	alr0058	$9e^{-25}$		
Glycerate kinase	At1g80380	(5)	none		alr2873	$1e^{-22}$	Pro0730	$3e^{-13}$

Glycolate DH, glycolate dehydropgenase; GDC, glycine decarboxylate complex.

References

(1) Bauwe *et al*., unpublished.
(2) Bari *et al*., 2004.
(3) Lipman and Olsen, 2001.
(4) Voll *et al*., 2006.
(5) Boldt *et al*., 2005.

the plant-type GLYK which was found, however, in *Anabaena* PCC 7120. None of the cyanobacterial genomes encodes a plant-type PGPase, but genes similar to *cbbZ*, encoding PGPase in *Alcaligenes eutrophus* (Schäferjohann *et al.*, 1993), were identified. It needs to be mentioned that PGPase is present in many, if not all, bacteria. In contrast to plants, however, glycolate enters bacterial metabolism mostly via the so-called glycerate pathway. This pathway converts glyoxylate directly into glycerate by the cooperative action of glyoxylate carboligase and tartronic semialdehyde reductase. Interestingly, genes encoding very similar proteins are present in all the cyanobacterial genomes examined to date raising the possibility of a branched PG metabolism integrating both the C_2 cycle- and the glycerate pathway types of PG metabolism (Figure 12.3).

In summary, from the genomic point of view, all cyanobacteria including the *Prochlorococcus* strains that harbor the smallest genomes in this group are potentially able to metabolize PG. One may speculate that conservation of PG metabolism is required in all photoautotrophic cells performing oxygenic photosynthesis.

Role of photorespiration in cyanobacteria

In contrast with *Arabidopsis* or other bacteria, a functional analysis of proteins potentially involved in PG metabolism in cyanobacteria is still missing. Exceptions include SHMT (*sll1931*) and glycolate dehydrogenase (*sll0404*) from *Synechocystis* PCC 6803 where the recombinant proteins showed the respective enzymatic activities (Eisenhut *et al.*, 2006). In a complementary approach, specific mutations were introduced into the ORFs mentioned in Table 12.2 and Figure 12.3. In higher plants, mutations in photorespiratory genes are usually lethal unless the plants are grown in a CO_2-enriched atmosphere. As an example, GLYK mutants can only grow under conditions where the oxygenase reaction of Rubisco is inhibited by a high CO_2 concentration (Boldt *et al.*, 2005). Notably, *Synechocystis* PCC 6803 mutants with inactivated P-, T- or H-protein of the GDC complex were able to completely segregate and were fully viable under both high and low CO_2 conditions. These mutants accumulated glycine and lysine and showed an increased sensitivity towards external glycine (Hagemann *et al.*, 2005). The apparent tolerance of *Synechocystis* PCC 6803 to the inactivation of GDC can be interpreted in two ways. First, the efficient CCM in cyanobacteria might be able to almost completely reduce the oxygenase activity of Rubisco and hence PG synthesis. Second, there exists an additional metabolic route for PG metabolism in *Synechocystis* PCC 6803 and possibly other cyanobacteria which by-passes GDC, thus avoiding the accumulation of toxic metabolites such as glycine.

An extensive analysis of *Synechocystis* PCC 6803 mutants impaired in many genes encoding potential C_2 cycle and glycerate pathway enzyme components indeed indicated that a flexible PG metabolism exists, which is necessary for this organism even under high CO_2 conditions (Eisenhut *et al.*, 2006). However, the individual inactivation of most of these genes did not lead to a high-CO_2-requiring phenotype. When exposed to low CO_2, the GDC mutants accumulated far more glycine and lysine than the wild type and mutants with an inactivated glycerate pathway. As indicated, conversion of glycolate to glyoxylate in *Synechocystis* PCC 6803 is mediated by glycolate dehydrogenase (GlcD) rather than the plant-type GOX. The inactivation of this enzymatic step, located immediately prior to the branching of PG metabolism into a plant-like C_2 cycle and a bacterial-like glycerate pathway (see Figure 12.3), resulted in glycolate accumulation and a growth depression even

Table 12.2 Bacterial proteins involved in 2-phosphoglycolate metabolism and ORFs in the cyanobacterial genomes of *Synechocystis* sp. strain PCC 6803 (6803), *Anabaena* sp. strain PCC 7120 (7120), and *Prochlorococcus* sp. strain SS120 (SS120), which show significant (e-values better than e^{-05}) similarities in amino acid sequences. Amino acid similarity searches were done using local Blast at CyanoBase (http://www.kazusa.or.jp/cyanobase/blast.html)

Protein	Bacterium	Acc.No.	Ref.	ORF in 6803	Similarity e-value	ORF in 7120	Similarity e-value	ORF in SS120	Similarity e-value
phosphoglycolate phosphatase	*A. eutrophus*	P40852	(1)	slr0458 sll1349	$4e^{-10}$ $2e^{-05}$	all2681 alr4944	$1e^{-11}$ $9e^{-07}$	Pro0350	$2e^{-05}$
Glycolate DH	*E. coli*		(2)						
subunit GlcD		AAC76015		sll0404	$1e^{-126}$	alr5269	$1e^{-121}$	none	
subunit GlcE		AAA69145		sll1189	$8e^{-25}$	all0624	$5e^{-16}$	none	
Glycerate kinase	*E. coli*	AAB93855	(3)	slr1840	$7e^{-64}$	none		none	
Glyoxylate carboligase	*E. coli*	AAA23864	(4)	slr2088 sll1981	$4e^{-94}$ $8e^{-58}$	all4613 all3555	$2e^{-86}$ $2e^{-59}$	Pro0526	$2e^{-80}$
Tartronic semialdehyde reductase	*E. coli*	AAB93851	(3)	slr0229	$9e^{-35}$	alr3358	$2e^{-24}$	Pro1229	$1e^{-42}$

Glycolate DH, glycolate dehydropgenase; GDC, glycine decarboxylate complex.

References

(1) Schäferjohann *et al.*, 1993
(2) Pellicer *et al.*, 1996
(3) Cusa *et al.*, 1993
(4) Chang *et al.*, 1993

under high CO_2. Notably, unlike several filamentous cyanobacterial strains (Renström and Bergman, 1989), *Synechocystis* PCC 6803 does not excrete glycolate into the medium (Eisenhut *et al.*, 2006). A similar growth inhibition was found for a double mutant affected in both the C_2 cycle and the glycerate pathway cultivated at low CO_2. All these findings suggested that despite the CCM activity the oxygenase activity of Rubisco is not completely inhibited in *Synechocystis* PCC 6803 and probably also in other cyanobacteria. Consequently, PG is produced in significant amounts and metabolized via two overlapping and cooperating pathways, a plant-like C_2 cycle and a bacterial-like glycerate pathway (Eisenhut *et al.*, 2006). Similar to the photoprotective function of the CCM discussed above (B-7), in addition to its function in the avoidance of the accumulation of toxic metabolites such as PG and glycine, cyanobacterial photorespiratory activity might also contribute to the dissipation of excess light energy.

In addition to their potential role in PG metabolism, the proteins identified in the cyanobacterial genomes (Tables 12.1 and 12.2) may also be involved in other processes. Mutants in the GDC L-protein (Engels and Pistorius, 1997), in one of the genes (*slr*2088) encoding glyoxylate carboligase (Kouhen and Joset, 2002), and in the gene (*sll1931*) encoding SHMT (Hagemann *et al.*, 2005) in *Synechocystis* PCC 6803 did not fully segregate, which may indicate essential metabolic functions of the respective enzymes in this organism. The Slr2088 protein was shown to be involved in branched chain amino acid biosynthesis (Kouhen and Joset, 2002). The L-protein is most probably also interacting with the pyruvate dehydrogenase complex (Engels and Pistorius, 1997). Similar to other organisms, SHMT is likely to be essential for one-carbon metabolism. Moreover, all these enzymes could be involved in the biosynthesis of glycine and serine or other pivotal metabolic processes which may explain their occurrence even in the smallest cyanobacterial genomes.

Regulation between anabolic and catabolic carbohydrate metabolisms

Introduction

A large diversity in the ability of cyanobacteria to consume carbon from their surroundings has been reported, ranging from obligate photoautotrophs where the sole carbon source is CO_2 to photomixotrophic and even heterotrophic growth. Figure 12.4 shows a schematic view of the anabolic and catabolic processes highlighting the biochemical steps where the reductive (Calvin cycle) and the oxidative pentose phosphate cycles and glycogen formation and degradation converge. In this section, we mainly focus on the regulation between photoautotrophy, heterotrophy and photomixotrophy with emphasis on the response to glucose. We do not discuss the formation of carbohydrates via the Calvin cycle (see Tabita, 1994) and provide a very brief account on glycogen formation and breakdown, respiration and fermentation.

Glycogen

The ADP-glucose pyrophosphorylase (AGPase), highly conserved in cyanobacteria and plants, is involved in the biosynthesis of reserve carbohydrates in bacteria, green algae and plants (Ball and Morell, 2003). Ultimately, its activity results in the formation of both

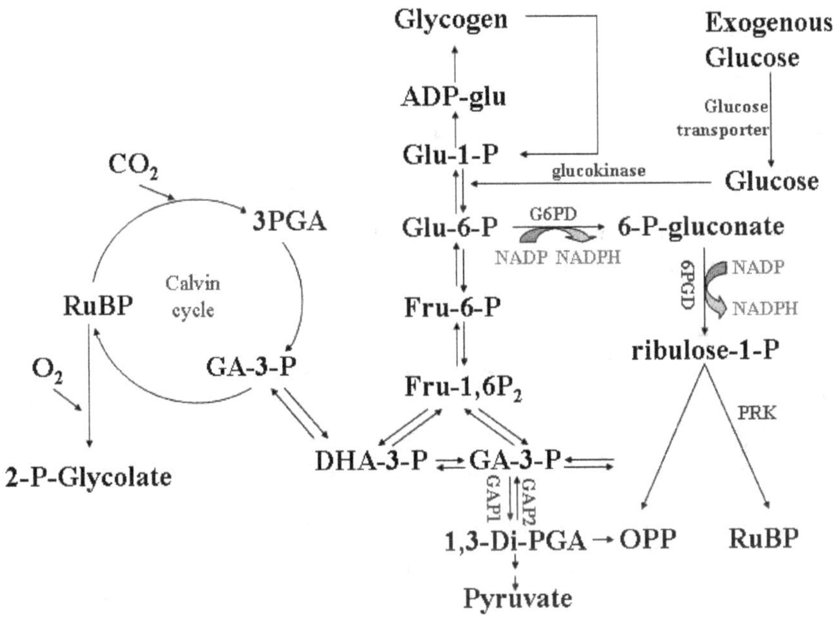

Figure 12.4 A schematic presentation of the metabolic pathways involved in formation (Calvin cycle) and breakdown (via the oxidative pentose phosphate, OPP, cycle) of carbohydrates and of glucose acquisition in cyanobacteria.

glycogen, the main reserve carbohydrate in most cyanobacteria, and glucosylglycerol, an important compatible solute in many cyanobacteria (Hagemann and Erdmann, 1994). Gearing of the rate of glycogen formation to that of CO_2 fixation in photosynthesis is largely achieved by monitoring the level of the 3-phosphogycerate (PGA) produced by the carboxylating reaction of Rubisco. The AGPase activity is strongly upregulated by a rise in the abundance of PGA, and site directed mutagenesis identified some of the amino acid residues involved in this activation (Ballicora *et al.*, 2003; Gómez-Casati *et al.*, 2003). *Synechocystis* PCC 6803 mutants where *agp*, encoding AGPase, was inactivated do not possess glycogen granules (Miao *et al.*, 2003) and produce sucrose rather than glycosylglycerol when salt-shocked. Some cyanobacteria produce other reserve carbohydrates such as semi-amylopectin type alpha-polyglucans (Nakamura *et al.*, 2005).

Respiration and fermentation

It has been reported that cyanobacteria lack 2-oxoglutarate dehydrogenase and thus do not possess a complete tricarboxylic acid cycle (Pearce *et al.*, 1969). It was therefore proposed that they rely on the oxidative pentose phosphate cycle (OPP) for carbohydrate degradation (see below) but this may not be the general case in cyanobacteria. Supply of 2-oxoglutarate to *Synechocystis* PCC 6803 resulted in a rise in the level of succinate suggesting conversion of 2-oxoglutarate to succinate (Cooley *et al.*, 2000). The flow of metabolites may operate via glutamate, γ-amino butyric acid (GABA), and succinate semialdehyde to succinate (Vermaas, personal communication). A branched Krebs cycle serving mainly to provide biochemical intermediates is not restricted to cyanobacteria; it is also found in many other bacteria where the 2-oxoglutarate dehydrogenase complex is missing. However, in those bacteria an additional route over succinic semialdehyde allows to bypass the missing step, enabling a complete metabolism of citrate (Tian *et al.*, 2005). An additional indication that cyanobacteria possess multiple means to degrade carbohydrates rests on the observation that a *Synechocystis* PCC 6803 mutant impaired in *zwf*, encoding glucose 6-phosphate dehydrogenase (G6PD), an essential step in the OPP, showed no phenotype (Scanlan *et al.*, 1995) suggesting alternative routes, the nature of which are not clear.

Glucose-6-phosphate dehydrogenase

As indicated, the OPP cycle is the main route of glucose catabolism in cyanobacteria (Hagen and Meeks, 2001; Scanlan *et al.*, 1995; Summers and Meeks, 1996; Summers *et al.*, 1995; Yang *et al.*, 2002). Therefore, activities of two key enzymes of the OPP pathway, glucose-6-phosphate dehydrogenase (G6PD) and 6-phophogluconate dehydrogenase (6PGD), encoded by *zwf* and *gnd*, respectively, is of particular importance (see Figure 12.4). The level of glucose-6-phosphate (G6P) increased following initiation of photosynthesis whereas that of 6-phosphogluconate (6PG) declined (Pelroy and Bassham, 1972), suggesting regulation of G6PD. Indeed, detailed kinetic studies showed strict regulation of G6PD activity by G6P, ATP, $NADP^+$ and NADPH and a shift in the nature of the curve relating the rate of G6PD activity to G6P concentration, from hyperbolic to sigmoidal, with a rising pH (Grossman and McGowan, 1975). Biochemical analyses indicated that G6PD was present in various aggregation states that showed different kinetic properties (Gleason, 1996; Schaeffer and Stanier, 1978; Sundaram *et al.*, 1998). It has been suggested that OpcA (OPP cycle protein A) is involved in the assembly of G6PD, its oligomerization and activation in various cyanobacteria (Min and Golden, 2000; Sundaram *et al.*, 1998). In contrast, detailed studies on G6PD from *Nostoc punctiforme* did not support involvement of OpcA in its aggregation but showed that the presence of OpcA strongly affected the activity of G6PD and its affinity towards its substrate, G6P (Hagen and Meeks, 2001). Sequence analyses indicated that both G6PD and OpcA are highly conserved but their phylogeny is not clear.

Circadian/diel regulation and SigE

By modulating gene expression in circadian/diel rhythms the cells may prepare themselves to the major change in metabolism from photosynthetic production to carbohydrate degradation following darkening. Analysis of the transcript abundance in a thermophilic *Synechococcus* cells inhabiting a microbial mat showed that *coxA* and *cydA* involved in respiration declined in the evening (Steunou *et al.*, 2006), whereas the level of transcripts of some genes potentially involved in fermentation (such as *pflB*, *pflA*, *adhE*, and *acs*) increased at the end of the day and remained high during the night (see Chapter 4). This acclimation probably reflects the specific conditions experienced in the mat where the organisms may face anaerobic conditions, due to high O_2 consumption, shortly after sunset. We draw the attention of the reader to earlier studies on fermentation (Stal and Moezelaar, 1997) and the current search for relevant genes in various cyanobacteria.

Analysis of the transcript abundance in *Synechocystis* PCC 6803 during a diel cycle (Labiosa *et al.*, 2006) showed that expression of certain genes involved in glycolysis and

OPP, such as *pgi*, *gap1*, *pykF*, *pfkA*, and a *devB* homologue peaked during the evening and early night period, while the level of transcripts involved in the CCM and carboxylation declined. Another study on the pattern of expression in *Synechocystis* PCC 6803 revealed that the abundance of transcripts originating from many of the genes encoding enzymes involved in carbohydrate degradation, such as *glgX*, *zwf*, *opcA*, *gnd*, *pfkA*, *gap1* and *tal*, peaked towards the end of the day (Kucho *et al.*, 2005). Similar pattern of circadian oscillations was observed in *Synechocystis* PCC 6803 with respect to genes encoding respiratory components. An interesting exception is those encoding the cytochrome b_6f complex. Their level does not change significantly during the day, most likely reflecting the involvement of cytochrome b_6f in both the photosynthetic and respiratory electron transport.

Interestingly, the DNA microarray analysis also indicated that the transcription of *sigE*, encoding sigma factor E, was under circadian oscillation, peaking towards the end of the day (Kucho *et al.*, 2005). A *Synechocystis* PCC 6803 mutant where *sigE* was inactivated was unable to grow under heterotrophic conditions and contained an elevated level of glycogen, implicating SigE in carbohydrate degradation (Osanai *et al.*, 2005). The transcript abundance of genes involved in glycogen breakdown, glycolysis and the OPP cycle was significantly reduced in this mutant as was the activity of rate-limiting enzymes in these pathways. A reduced level of *gap1* transcript (encoding glyceraldehyde-3-phosphate dehydrogenase, GAPDH) in a Δ*hik8* mutant of *Synechocystis* PCC 6803 lowered the ability of the cells to degrade glycogen and grow in the dark (Singh and Sherman, 2005), thus implicating this sensor kinase in the response to glucose. The level of the *hik8* transcript was not affected by inactivation of *sigE*, nor was the amount of SigE affected by the *hik8* mutation (Osanai *et al.*, 2005). Whether the Hik8 and SigE interact in modulating the expression of carbon degradation genes is yet to be elucidated.

Analysis of polypeptide patterns

Compared with photoautotrophy, heterotrophic growth of *Synechocystis* PCC 6803 led to a 4-fold rise in the rate of respiration but a 9 to 10-fold decline in PSII activity and CO_2-dependent O_2 evolution (Kurian *et al.*, 2006). Heterotrophy involved a marked decline in the abundance of both large and small subunits of Rubisco and of carboxysomal shell proteins, in agreement with the large decline in photosynthetic capability. Interestingly, the level of phosphoribulokinase, which catalyzes a committing reaction drawing carbon from the oxidative (degradation) to the reductive (Calvin) cycle did not change after transfer to heterotrophic growth. Thus, regulation of its activity between light and dark condition does not occur at the level of protein abundance.

Reduction of the level of photosynthetic proteins occurred simultaneously with a large rise in G6PD, Gap1 and also Gap2 (Kurian *et al.*, 2006). Studies by Shestakov and colleagues (Koksharova *et al.*, 1998) using *Synechocystis* PCC 6803 mutants impaired in *gap1* and *gap2* indicated that Gap2 operate in the Calvin cycle and in non-photosynthetic gluconeogenesis, whereas Gap1 is essential only for glycolytic glucose breakdown (Figure 12.4). Thus it is not clear how the *Synechocystis* PCC 6803 cells maintain the balance between autotrophy and heterotrophy if both Gap1 and Gap2 are accumulated under heterotrophy. Finally, in *Synechococcus* PCC 7942, *gap1* (but not *gap2*) is closely linked to the *glgP* gene encoding glycogen phosphorylase, which catalyzes the first step of glycogen degradation, and *gap1* and *glgP* are coexpressed (Koksharova *et al.*, 2004).

Interestingly, a marked decline in the level of Flv3 (Sll0550) was observed under heterotrophic growth conditions (Kurian *et al.*, 2006). This A-type flavoprotein is involved in the cyanobacterial Mehler reaction, i.e. the transfer of electrons from PSI, via NADPH, to O_2 forming water molecules without the release of reactive oxygen species (Helman *et al.*, 2003). The transcript abundance of *flv3* increased sharply after transfer of *Synechocystis* PCC 6803 from high to low CO_2 conditions (Wang *et al.*, 2004). Naturally, the Flv3 is not essential in the dark but the mode of regulation of this protein by light and heterotrophy is yet to be established.

The proteomic approach provided some insight into the regulation between the light and dark conditions but understanding of the regulation between the reductive and oxidative pentose phosphate cycles would have to rest on detailed analysis of the changes in the polypeptide abundances during shifts between photoautotrophic and photomixotrophic growth conditions.

Response of *Synechocystis* PCC 6803 to D-glucose

It is not clear what distinguishes obligatory photoautotrophs from those able to assimilate external organic carbon. A well known case is the glucose-sensitive (GS) and tolerant (GT) strains of *Synechocystis* PCC 6803 where a loss of glucokinase activity resulted in the inability of the GS strain to metabolize glucose (Kahlon *et al.*, 2006). Naturally, to consume extracellular organic carbon it must enter the cells. Introduction of *glcP*, encoding the glucose transporter in *Synechocystis* PCC 6803, into the chromosome of the obligatory autotroph *Synechococcus* PCC 7942 resulted in glucose sensitivity. These data suggested that lack of glucose transport could only partly explain obligatory autotrophy, at least in *Synechococcus* PCC 7942, and that glucose uptake was essential to inflict glucose sensitivity by an unknown mechanism (Zhang *et al.*, 1998). A similar conclusion was reached for the case of *Synechocystis* PCC 6803 (Kahlon *et al.*, 2006).

Exposure of *Synechocystis* PCC 6803 to glucose results in many changes at various cellular levels including carbon metabolism (Knowles and Plaxton, 2003), pigmentation (Hsiao *et al.*, 2004; Ryu *et al.*, 2004), phosphorylation patterns (Bloye *et al.*, 1992), CO_2 uptake (Kaplan and Reinhold, 1999), redox state of the plastoquinone pool and organization of the photosynthetic reaction center II (Kahlon, Ohad and Kaplan, unpublished). Nevertheless, the mechanism whereby *Synechocystis* PCC 6803 senses the presence of glucose and switches between CO_2 fixation and glucose catabolism via the reductive or oxidative pentose phosphate pathway, respectively, is poorly understood.

Despite the numerous responses of the GT strain of *Synechocystis* PCC 6803 to glucose, and with the exception of cases such as carotene biosynthesis in the dark (Ryu *et al.*, 2004), the profiles of transcript abundance and protein levels during steady state growth did not show marked changes after the addition of glucose (Kahlon *et al.*, 2006; Knowles and Plaxton, 2003; Tu *et al.*, 2004; Yang *et al.*, 2002). This may suggest that the response to glucose in *Synechocystis* PCC 6803 rests mainly on alterations at the level of post-translational modifications or modulation of enzyme activities.

Mutants of *Synechocystis* PCC 6803 impaired in their ability to grow in the presence of glucose were isolated, including mutants in *hik31*, *icfG*, *hik8* and *pmgA*. A mutant where both the chromosomal and plasmid copies of a gene encoding of a putative sensor kinase, Hik31, were inactivated, was unable to grow in the presence of glucose, particularly under high levels of CO_2 (Kahlon *et al.*, 2006). Inactivation of *glcP* in the Δ*hik31* mutant alleviated the sensitivity to glucose indicating that, like in *Synechococcus* PCC 7942 (Zhang *et al.*, 1998), entrance of glucose to the cells was required to inflict lethality. Like the GS strain, the Δ*hik31* mutant did not exhibit glucokinase activity despite the fact that transcripts originating from *glk* (*slr0329*), encoding glucokinase, were present (Kahlon *et al.*, 2006). Down-regulation of glucokinase activity is probably mediated by post-transcriptional regulation, the nature of which is not known. Inactivation of *glk* in the GT strain resulted in sensitivity to glucose indicating that the presence of glucose itself, within the cells, was lethal.

It has been reported that inactivation of *icfG*, encoding a protein-serine phosphatase in *Synechocystis* PCC 6803 (Shi *et al.*, 1999) led to inability to grow under low CO_2 in the presence of glucose (Beuf *et al.*, 1994). A considerable rise in *icfG* transcript was induced by glucose in the WT but this transcript was undetectable in the Δ*hik31* mutant, and the Δ*icfG* mutant was able to grow in the dark whereas the Δ*hik31* mutant could not. These data suggested a higher hierarchy of Hik31 in response to glucose but the interaction between Hik31 and IcfG is not clear.

A Δ*hik8* mutant could grow in the presence of glucose but not under an extended dark period, presumably due to the altered ability to express *gap1* (glyceraldehyde 3-phosphate dehydrogenase 1) and thus to utilize glycogen in the dark (Singh and Sherman, 2005). In contrast, the Δ*hik31* mutant possesses a normal level of *gap1* transcript but could not grow in the dark due to lack of glucokinase activity.

The putative antisigma factor, PmgA, is essential for the acclimation of *Synechocystis* PCC 6803 to changing light intensity (Hihara *et al.*, 1998). Growth of a Δ*pmgA* mutant is impaired when glucose is provided but the mechanism involved is not known. Interestingly, the glucose sensitivity of the Δ*pmgA* mutant could be alleviated in a suppression mutant bearing G354W mutation in *ndhF3* (Hihara, unpublished). As discussed above, the latter gene is required for the high affinity CO_2 uptake system in *Synechocystis* PCC 6803 and its entire operon is upregulated under low CO_2 conditions. This observation, and the differential effects of glucose on CO_2-dependent growth of various mutants provided strong indication to the close link between CO_2 level and response to glucose, but the nature of this interaction is not clear.

Finally, while the data summarized here support the notion that SigE, Hik31, IcfG, Hik8 and PmgA play an important role in the response of *Synechocystis* PCC 6803 to glucose their exact role and the interaction between them is not well understood. Further work on this system is expected to contribute valuable information on the basis of glucose catabolism and shifting between photoautotrophy and heterotrophy.

Acknowledgments

Research in our laboratories was supported by the DFG (Deutsche Forschungsgemeinschaft) to M.H., A.K., and H.B., The Israel Science Foundation (ISF) and the German-Israel Science Foundation (GIF) to A.K., and to T.O. by the Membrane Biology EMSL Scientific Grand Challenge project at the W. R. Wiley Environmental Molecular Sciences Laboratory, a national scientific user facility sponsored by the U.S. Department of Energy Office of Biological and Environmental Research program located at Pacific Northwest National Laboratory. Pacific Northwest National Laboratory is operated for the Department of Energy by Battelle.

References

Arabidopsis Genome Initiative (2000). Analysis of the genome sequence of the flowering plant *Arabidopsis thaliana*. Nature *408*, 796–815.

Arteni, A.A, Zhang, P., Battchikova, N., Ogawa, T., Aro, E.-M., and Boekema, E.J. (2006). Structural characterization of NDH-1 complexes of *Thermosynechococcus elongatus* by single particle electron microscopy, Biochim. Biophys. Acta *1757*, 1469–1475.

Badger, M.R., Price, G.D,. Long, B.M., and Woodger, F.J. (2006). The environmental plasticity and ecological genomics of the cyanobacterial CO_2 concentrating mechanism. J. Exp. Bot. *57*, 249–265.

Ball, S.G., and Morell, M.K. (2003). From bacterial glycogen to starch: Understanding the biogenesis of the plant starch granule. Ann. Rev. Plant. Biol. *54*, 207–233.

Ballicora, M.A., Iglesias, A.A., and Preiss, J. (2003). ADP-glucose pyrophosphorylase, a regulatory enzyme for bacterial glycogen synthesis. Microbiol. Mol. Biol. Rev. *67*, 213–225.

Bari, R., Kebeish, R., Kalamajka, R., Rademacher, T., and Peterhansel, C. (2004). A glycolate dehydrogenase in the mitochondria of *Arabidopsis thaliana*. J. Exp. Bot. *55*, 623–630.

Battchikova, N., Zhang, P., Rudd, S., Ogawa, T., and Aro, E.M. (2005). Identification of NdhL and Ssl1690 (NdhO) in NDH-1L and NDH-1M complexes of *Synechocystis* sp. PCC 6803. J. Biol. Chem. *280*, 2587–2595.

Bauwe, H., and Kolukisaoglu, U. (2003). Genetic manipulation of glycine decarboxylation. J. Exp. Bot. *54*, 1523–1535.

Berry, S., Esper, B., Karandashova, I., Teuber, M., Elanskaya, I., Rögner, M., Hagemann M., (2003). Potassium uptake in the unicellular cyanobacterium *Synechocystis* sp. strain PCC 6803 mainly depends on a Ktr-like system encoded by slr1509 (*ntpJ*). FEBS Lett. *548*, 53–58.

Beudeker, R., Cannon, G.C., Kuenen, J.G., and Shively, J.M. (1980). Relations between RuBP carboxylase, carboxysomes and CO_2 fixing capacity in the obligate chemolithotroph *Thiobacillus neapolitana* grown under different limitations in the chemostat. Arch. Microbiol. *1241*, 185–189.

Beuf, L., Bedu, S., Durand, M.C., and Joset, F. (1994). A protein involved in co-ordinated regulation of inorganic carbon and glucose metabolism in the facultative photoautrophic cyanobacterium *Synechocystis* PCC6803. Plant Mol. Biol. *25*, 855–864.

Birmingham, B.C., Coleman, J.R., and Colman, B. (1982). Measurement of photorespiration in algae. Plant Physiol. *69*, 159–262.

Bloye, S.A., Silman, N.J., Mann, N.H., and Carr, N.G. (1992). Bicarbonate concentration by *Synechocystis* PCC6803 modulation of protein phosphorylation and inorganic carbon transport by glucose. Plant Physiol. *99*, 601–606.

Boldt, R., Edner, C., Kolukisaoglu, U., Hagemann, M., Weckwerth, W., Wienkoop, S., Morgenthal, K., and Bauwe, H. (2005). D-glycerate 3-kinase, the last unknown enzyme in the photorespiratory cycle in *Arabidopsis*, belongs to a novel kinase family. Plant Cell *17*, 2413–2420.

Bonfil, D.J., Ronen-Tarazi, M., Sültemeyer, D., Lieman-Hurwitz, J., Schatz, D., and Kaplan, A. (1998). A putative HCO_3^- transporter in the cyanobacterium *Synechococcus* sp. strain PCC 7942. FEBS Lett. *430*, 236–240.

Chang, Y.Y., Wang, A.Y., and Cronan, J.E. Jr. (1993). Molecular cloning, DNA sequencing, and biochemical analyses of *Escherichia coli* glyoxylate carboligase. An enzyme of the acetohydroxy acid synthase-pyruvate oxidase family. J. Biol. Chem. *268*, 3911–3919.

Codd, G.A., and Stewart, W.D.P. (1973). Pathways of glycollate metabolism in the blue-green alga *Anabaena cylindrica*. Arch. Mikrobiol. *94*, 11–28.

Colman, B. (1989). Photosynthetic carbon assimilation and the suppression of photorespiration in the cyanobacteria. Aquatic. Bot. *34*, 211–231.

Cooley, J.W., Howitt, C.A., and Vermaas, W.F.J. (2000). Succinate:quinol oxidoreductases in the cyanobacterium *Synechocystis* sp strain PCC 6803: Presence and function in metabolism and electron transport. J. Bacteriol. *182*, 714–722.

Cusa, E., Obradors, N., Baldoma, L., Badia, J., and Aguilar, J. (1999). Genetic analysis of a chromosomal region containing genes required for assimilation of allantoin nitrogen and linked glyoxylate metabolism in *Escherichia coli*. J. Bacteriol. *181*, 7479–7484.

Douce, R., Bourguignon, J., Neuburger, M., and Rebeille, F. (2001). The glycine decarboxylase system: a fascinating complex. Trends Plant. Sci. *6*, 167–176.

Eisenhut, M., Kahlon, S., Hasse, D., Ewald, R., Lieman-Hurwitz, J., Ogawa, T., Ruth, W., Bauwe, H., Kaplan, A., and Hagemann, M. (2006). The plant-like C2 glycolate cycle and the bacterial-like glycerate pathway cooperate in phosphoglycolate metabolism in cyanobacteria. Plant. Physiol. 142, 333–342.

Engels, A., and Pistorius, E.K. (1997). Characterization of a gene encoding dihydrolipoamide dehydrogenase of the cyanobacterium *Synechocystis* sp. PCC 6803. Microbiology *143*, 3543–3553.

Espie, G.S., and Kandasamy, R.A. (1994). Monensin inhibition of Na^+-dependent HCO_3^- transport distinguishes it from Na^+-independent HCO_3^- transport and provides evidence for Na^+/HCO_3^- symport in the cyanobacterium *Synechococcus* UTEX 625. Plant Physiol. *104*, 1419–1428.

Figge, R.M., Cassier-Chauvat, C., Chauvat, F., and Cerff, R. (2001). Characterization and analysis of an NAD(P)H dehydrogenase transcriptional regulator critical for the survival of cyanobacteria facing inorganic carbon starvation and osmotic stress. Mol. Microbiol. *39*, 455–468.

Friedberg, D., Kaplan, A., Ariel, R., Kessel, M., and Seijffers, J. (1989). The 5′ flanking region of the gene encoding the large subunit of ribulose-1,5-bisphosphate carboxylase/oxygenase is crucial for growth of the cyanobacterium *Synechococcus* PCC7942 in air level of CO_2. J. Bacteriol. *171*, 6069–6076.

Fukuzawa, H., Suzuki, E., Komukai, Y., and Miyachi, S. (1992). A gene homologous to chloroplast carbonic anhydrase (*icfA*) is essential to photosynthetic carbon dioxide fixation by *Synechococcus* PCC 7942. Proc. Natl. Acad. Sci. USA *89*, 4437–4444.

Giordano, M., Beardall, J., and Raven, J.A. (2005). CO_2 concentrating mechanisms in algae: Mechanisms, environmental modulation, and evolution. Annu. Rev. Plant Biol. *56*, 99–131.

Gleason, F.K. (1996). Glucose-6-phosphate dehydrogenase from the cyanobacterium, *Anabaena* sp. PCC 7120: purification and kinetics of redox modulation. Arch. Biochem. Biophys. *334*, 277–283.

Gómez-Casati, D.F., Cortassa, S., Aon, M.A., and Iglesias, A.A. (2003). Ultrasensitive behavior in the synthesis of storage polysaccharides in cyanobacteria. Planta. *216*, 969–975.

Grossman, A., and McGowan, R.E. (1975). Regulation of glucose-6-phosphate dehydrogenase in blue-green algae. Plant Physiol. *55*, 658–662.

Hagemann, M., Vinnemeier, J., Oberpichler, I., Boldt, R., and Bauwe, H. (2005). The glycine decarboxylase complex is not essential for the cyanobacterium *Synechocystis* sp. strain PCC 6803. Plant Biol. *7*, 15–22.

Hagen, K.D., and Meeks, J.C. (2001). The unique cyanobacterial protein OpcA is an allosteric effector of glucose-6-phosphate dehydrogenase in *Nostoc punctiforme* ATCC 29133. J. Biol. Chem. *276*, 11477–11486.

Hashimoto, M., Negi, J., Young, J., Israelsson, M., Schroeder, J.I., and Iba, K. (2006). *Arabidopsis* HT1 kinase controls stomatal movements in response to CO_2. Nat. Cell Biol. *8*, 891–397.

Helman, Y., Tchernov, D., Reinhold, L., Shibata, M., Ogawa, T., Schwarz, R., Ohad, I., and Kaplan, A. (2003). Genes encoding A-type flavoproteins are essential for photoreduction of O_2 in cyanobacteria. Curr. Biol. *13*, 230–235.

Herranen, M., Battchikova, N., Zhang, P., Graf, A., Sirpio, S., Paakkarinen, V., and Aro, E.M. (2004). Towards functional proteomics of membrane protein complexes in *Synechocystis* sp. PCC 6803. Plant Physiol. *134*, 470–481.

Hihara, Y., Sonoike, K., and Ikeuchi, M. (1998). A novel gene, *pmgA*, specifically regulates photosystem stoichiometry in the cyanobacterium *Synechocystis* sp. PCC 6803 in response to high light. Plant Physiol. *117*, 1205–1216.

Hsiao, H.Y., He, Q.F., van Waasbergen, L.G., and Grossman, A.R. (2004). Control of photosynthetic and high-light-responsive genes by the histidine kinase DspA: Negative and positive regulation and interactions between signal transduction pathways. J. Bacteriol. *186*, 3882–3888.

Husic, D.W., Husic, H.D., and Tolbert, N.E. (1987). The oxidative photosynthetic carbon cycle or C2 cycle. CRC. Crit. Rev. Plant Sci. *5*, 45–100.

Igarashi, D., Miwa, T., Seki, M., Kobayashi, M., Kato, T., Tabata, S., Shinozaki, K., and Ohsumi, C. (2003). Identification of photorespiratory glutamate:glyoxylate aminotransferase (GGAT) gene in *Arabidopsis*. Plant J. *33*, 975–987.

Iwasaki, W., Miyatake, H., Ebihara, A., and Miki, K. (2004). Crystallization and preliminary X-ray crystallographic studies of the small form of glucose-inhibited division protein A from *Thermus thermophilus* HB8. Acta. Cryst. *D60*, 515–517.

Kahlon, S., Beeri, K., Ohkawa, H., Hihara, Y., Murik, O., Suzuki, I., Ogawa, T., and Kaplan, A. (2006). A putative sensor kinase, Hik31, is involved in the response of *Synechocystis* sp. strain PCC 6803 to the presence of glucose. Microbiology *152*, 647–655.

Kaneko, T., Sato, S., Kotani, H., Tanaka, A., Asamizu, E., Nakamura, Y., Miyajima, N., Hirosawa, M., Sugiura, M., Sasamoto, S., Kimura, T., Hosouchi, T., Matsuno, A., Muraki, A., Nakazaki, N., Naruo,

K., Okumura, S., Shimpo, S., Takeuchi, C., Wada, T., Watanabe, A., Yamada, M., Yasuda, M., and Tabata, S. (1996). Sequence analysis of the genome of the unicellular cyanobacterium *Synechocystis* sp. strain PCC 6803. II. Sequence determination of the entire genome and assignment of potential protein-coding regions. DNA Res. *3*, 109–136.

Kaneko, Y., Danev, R., Nagayama, K., Nakamoto, H. (2006). Intact carboxysomes in a cyanobacterial cell visualized by Hilbert differential contrast transmission electron microscopy. J. Bacteriol. *188*, 805–808.

Kaplan, A., Schwarz, R., Lieman-Hurwitz, J., Ronen-Tarazi, M., and Reinhold, L. (1994). Physiological and molecular studies on the response of cyanobacteria to changes in the ambient inorganic carbon concentration. In The Molecular Biology of Cyanobacteria, D.A. Bryant ed. (Netherlands: Kluwer, Dordrecht), pp. 469–485.

Kaplan, A., and Reinhold, L. (1999). The CO_2 concentrating mechanisms in photosynthetic microorganisms. Annu. Rev. Plant Physiol. Plant Mol. Biol. *50*, 539–570.

Kerfeld, C.A., Sawaya, M.R., Tanaka, S., Nguyen, C.V., Phillips, M., Beeby, M., and Yeates, T.O. (2005). Protein structures forming the shell of primitive bacterial organelles. Science *309*, 936–938.

Kinscherf, T.G., and Willis, D.K. (2002). Global Regulation by *gidA* in *Pseudomonas syringae*. J. Bacteriol. *184*, 2281–2286.

Klughammer, B., Sültemeyer, D., Badger, M.R., and Price, G.D. (1999). The involvement of NAD(P)H dehydrogenase subunits, NdhD3 and NdhF3, in high-affinity CO_2 uptake in *Synechococcus* sp. PCC7002 gives evidence for multiple NDH-1 complexes with specific roles in cyanobacteria. Mol. Microbiol. *32*, 1305–1315.

Knowles, V.L., and Plaxton, W.C. (2003). From genome to enzyme: Analysis of key glycolytic and oxidative pentose-phosphate pathway enzymes in the cyanobacterium *Synechocystis* sp. PCC 6803. Plant Cell Physiol. *44*, 758–763.

Koksharova, O., Schubert, M., Shestakov, S., and Cerff, R. (1998). Genetic and biochemical evidence for distinct key functions of two highly divergent GAPDH genes in catabolic and anabolic carbon flow of the cyanobacterium *Synechocystis* sp. PCC 6803. Plant Mol. Biol. *36*, 183–194.

Koksharova, O.A., Brandt, U., and Cerff, R. (2004). The *gap1* operon of the cyanobacterium *Synechococcus* PCC 7942 carries a gene encoding glycogen phosphorylase and is induced under anaerobic conditions. Microbiology. *73*, 326–329.

Koropatkin, N.M., Pakrasi, H.B., and Smith, T.J. (2006). Atomic structure of a nitrate-binding protein crucial for photosynthetic productivity. Proc. Natl. Acad. Sci. USA *103*, 9820–9825.

Kouhen, O.M., and Joset, F. (2002). Biosynthesis of the branched-chain amino acids in the cyanobacterium *Synechocystis* PCC6803: existence of compensatory pathways. Curr. Microbiol. *45*, 94–98.

Kozaki, A., and Takeba, G. (1996). Photorespiration protects C3 plants from photooxidation. Nature *384*, 557–560.

Kucho, K.-I., Okamoto, K., Tsuchiya, Y., Nomura, S., Nango, M., Kanehisa, M., and Ishiura, M. (2005). Global analysis of circadian expression in the cyanobacterium *Synechocystis* sp. strain PCC 6803. J. Bacteriol. *187*, 2190–2199.

Kurian, D., Jansen, T., and Maenpaa, P. (2006). Proteomic analysis of heterotrophy in *Synechocystis* sp. PCC 6803. Proteomics *6*, 1483–1494.

Labiosa, R.G., Arrigo, K.R., Tu, C.J., Bhaya, D., Bay, S., Grossman, A.R., and Shrager, J. (2006). Examination of diel changes in global transcript accumulation in *Synechocystis* (cyanobacteria). J. Phycol. *42*, 622–636.

Lieman-Hurwitz, J., Rachmilevitch, S., Mittler, R., Marcus, Y., and Kaplan, A. (2003). Enhanced photosynthesis and growth of transgenic plants that express *ictB*, a gene involved in HCO_3^- accumulation in cyanobacteria. Plant Biotechnol. J. *1*, 43–50.

Liepman, A.H., and Olsen, L.J. (2001). Peroxisomal alanine:glyoxylate aminotransferase (AGT1) is a photorespiratory enzyme with multiple substrates in *Arabidopsis thaliana*. Plant J. *25*, 487–498.

Ludwig, M., Sültemeyer, D., and Price, G.D. (2000). Isolation of *ccmKLMN* genes from the marine cyanobacterium, *Synechococcus* sp. PCC7002 (Cyanobacteria), and evidence that CcmM is essential for carboxysome assembly. J. Phycol. *36*, 1109–1118.

Maeda, S., Badger, M.R., and Price, G.D. (2002). Novel gene products associated with NdhD3/D4-containing NDH-1 complexes are involved in photosynthetic CO_2 hydration in the cyanobacterium, *Synechococcus* sp. PCC7942. Mol. Microbiol. *43*, 425–435.

Maeda, S., Price, G.D., Badger, M.R., Enomoto, C., and Omata, T. (2000). Bicarbonate binding activity of the CmpA protein of the cyanobacterium *Synechococcus* sp. strain PCC 7942 involved in active transport of bicarbonate. J. Biol. Chem. *275*, 20551–20555.

Marcus, Y., Schwarz, R., Friedberg, D., and Kaplan, A. (1986). High CO_2 requiring mutant of *Anacystis nidulans* R2. Plant Physiol. *82*, 610–612.

Martin, W., Rujan, T., Richly, E., Hansen, A., Cornelsen, S., Lins, T., Leister, D., Stoebe, B., Hasegawa, M., and Penny, D. (2002). Evolutionary analysis of *Arabidopsis*, cyanobacterial, and chloroplast genomes reveals plastid phylogeny and thousands of cyanobacterial genes in the nucleus. Proc. Natl. Acad. Sci. USA *99*, 2246–12251.

Matsuda, N., Kobayashi, H., Katoh, H., Ogawa, T., Futatsugi, L., Nakamura, T., Bakker, E.P., and Uozumi, N. (2004). Na^+-dependent K^+ uptake Ktr system from the cyanobacterium *Synechocystis* sp. PCC 6803 and its role in the early phases of cell adaptation to hyperosmotic shock. J. Biol. Chem. *279*, 54952–54962.

McGinn, P.J., Price, G.D., and Badger, M.R. (2004). High light enhances the expression of low-CO2-inducible transcripts involved in the CO_2-concentrating mechanism in *Synechocystis* sp. PCC 6803. Plant Cell Environ. *27*, 615–626.

Miao, X.L., Wu, Q.Y., Wu, G.F., and Zhao, N.M. (2003). Sucrose accumulation in salt-stressed cells of *agp* gene deletion-mutant in cyanobacterium *Synechocystis* sp. PCC 6803. FEMS Microbiol. Lett. *218*, 71–77.

Miller, A., Turpin, D.H., and Canvin, D.T. (1984). Na^+ requirement for growth, photosynthesis, and pH regulation in the alkalotolerant cyanobacterium *Synechococcus leopoliensis*. J. Bacteriol. *159*, 100–106.

Min, H.T., and Golden, S.S. (2000). A new circadian class 2 gene, *opcA*, whose product is important for reductant production at night in *Synechococcus elongatus* PCC 7942. J. Bacteriol. *182*, 6214–6221.

Nakamura, Y., Takahashi, J., Sakurai, A., Inaba, Y., Suzuki, E., Nihei, S., Fujiwara, S., Tsuzuki, M., Miyashita, H., Ikemoto, H., Kawachi, M., Sekiguchi, H., and Kurano, N. (2005). Some cyanobacteria synthesize semi-amylopectin type alpha-polyglucans instead of glycogen. Plant Cell Physiol. *46*, 539–545.

Norman, E.G., and Colman, B. (1988). Evidence for an incomplete glycolate pathway in cyanobacteria. J. Plant. Physiol. *132*, 766–768.

Norman, E.G., and Colman, B. (1992). Formation and metabolism of glycolate in the cyanobacterium *Coccochloris peniocystis*. Arch. Microbiol. *157*, 375–380.

Ogawa, T. (1990). Mutants of *Synechocystis* PCC 6803 defective in inorganic carbon transport. Plant Physiol. *94*, 760–765.

Ogawa, T. (1991a). A gene homologous to the subunit-2 gene of NADH dehydrogenase is essential to inorganic carbon transport of *Synechocystis* PCC6803. Proc. Natl. Acad. Sci. USA *88*, 4275–4279.

Ogawa, T. (1991b). Cloning and inactivation of a gene essential to inorganic carbon transport of *Synechocystis* PCC 6803. Plant Physiol. *96*, 280–284.

Ogawa, T. (1992). Identification and characterization of the ictA/ndhL gene product essential to inorganic carbon transport of *Synechocystis* PCC 6803. Plant Physiol. *99*, 1604–1608.

Ogawa, T., Kaneda, T., and Omata, T. (1987). A mutant of *Synechococcus* PCC7942 incapable of adapting to low CO_2 concentration. Plant Physiol. *84*, 711–715.

Ogawa, T., and Kaplan, A. (2003). Inorganic carbon acquisition systems in cyanobacteria. Photosynth. Res. *77*, 105–115.

Ogawa, T., Marco, E., and Orus, M.I. (1994). A gene (*ccmA*) required for carboxysome formation in the cyanobacterium *Synechocystis* sp. strain PCC6803. J. Bacteriol. *176*, 2374–2378.

Ogren W.L. (1984). Photorespiration: Pathways, regulation, and modification. Ann. Rev. Plant Physiol. *35*, 415–442.

Ohkawa, H., Pakrasi, H.B., and Ogawa, T. (2000a). Two types of functionally distinct NAD(P)H dehydrogenases in *Synechocystis* sp. strain PCC6803. J. Biol. Chem. *275*, 31630–31634.

Ohkawa, H., Price, G.D., Badger, M.R., and Ogawa, T. (2000b). Mutation of *ndh* genes leads to inhibition of CO_2 uptake rather than HCO_3^- uptake in *Synechocystis* sp. strain PCC 6803. J. Bacteriol. *182*, 2591–2596.

Ohkawa, H., Sonoda, M., Katoh, H., and Ogawa, T. (1998). The use of mutants in the analysis of the CCM in cyanobacteria. Can. J. Bot. *76*, 1035–1042.

Ohkawa, H., Sonoda, M., Shibata, M., and Ogawa, T. (2001). Localization of NAD(P)H dehydrogenase in the cyanobacterium *Synechocystis* sp. strain PCC 6803. J. Bacteriol. *183*, 4938–4939.

Omata, T. (1992). Characterization of the downstream region of *cmpA*: Identification of a gene cluster encoding a putative permease of the cyanobacterium *Synechococcus* PCC7942. In Research in Photosynthesis. N. Murata, ed. (Dorfrecht, Kluwer Academic Pub.), pp. 807–810.

Omata, T., Carlson, T.J., Ogawa, T., and Pierce, J. (1990). Sequencing and modification of the gene encoding the 42-kD protein in the cytoplasmic membrane of *Synechococcus* PCC 7942. Plant Physiol. *93*, 305–311.

Omata, T., and Ogawa, T. (1985). Changes in the polypeptide composition of the cytoplasmic membrane of the cyanobacterium *Anacystis nidulans* during adaptation to low CO_2 conditions. Plant Cell Physiol. *26*, 1075–1081.

Omata, T., and Ogawa, T. (1986). Biosynthesis of a 42-kD poly-peptide in the cytoplasmic membrane of the cyanobacterium *Anacystis nidulans* strain R2 during adaptation to low CO_2 concentration. Plant Physiol. *80*, 525–530.

Omata, T., Price, G.D., Badger, M.R., Okamura, M., Gohta, S., and Ogawa, T. (1999). Identification of an ATP-binding cassette transporter involved in bicarbonate uptake in the cyanobacterium *Synechococcus* sp. strain PCC 7942. Proc. Natl. Acad. Sci. USA *96*, 13571–13576.

Osanai, T., Kanesaki, Y., Nakano, T., Takahashi, H., Asayama, M., Shirai, M., Kanehisa, M., Suzuki, I., Murata, N., and Tanaka, K. (2005). Positive regulation of sugar catabolic pathways in the cyanobacterium *Synechocystis* sp. PCC 6803 by the group 2 sigma factor SigE. J. Biol. Chem. *280*, 30653–30659.

Pearce, J., Leach, C.K., and Carr, N.G. (1969). The incomplete citric acid cycle in the blue-green alga *Anabaena variabilis*. J. Gen. Microbiol. *49*, 301–313.

Pellicer, M.T., Badia, J., Aguilar, J., and Baldoma, L. (1996). Glc locus of *Escherichia coli*: characterization of genes encoding the subunits of glycolate oxidase and the Glc regulator protein. J. Bacteriol. *178*, 2051–2059.

Pelroy, R.A., and Bassham, J.A. (1972). Photosynthetic and dark carbon metabolism in unicellular blue-green algae. Arch. Mikrobiol. *86*, 303–322.

Price, G.D., and Badger, M.R. (1989). Isolation and characterization of high CO_2-requiring mutants of the cyanobacterium *Synechococcus* PCC7942: two phenotypes that accumulate carbon but are apparently unable to generate CO_2 within the carboxysome. Plant Physiol. *91*, 514–525.

Price, G.D., Coleman, J.R., and Badger, M.R. (1992). Association of carbonic anhydrase activity with carboxysomes isolated from the cyanobacterium *Synechococcus* PCC7942. Plant Physiol. *100*, 784 –793.

Price, G.D., Woodger, F.J., Badger, M.R., Howitt, S.M., and Tucker, L. (2004). Identification of a SulP-type bicarbonate transporter in marine cyanobacteria. Proc. Natl. Acad. Sci. USA *101*, 18228–18233.

Reinhold, L., Volokita, M., Zenvirth, D., and Kaplan, A. (1984). Is HCO_3^- transport in *Anabaena* a Na^+ symport? Plant Physiol. *76*, 1090–1092.

Renström, E., and Bergman, B. (1989). Glycolate metabolism in cyanobacteria. I. Glycolate excretion and phosphoglycolate phosphatase activity. Physiol. Plant. *43*, 137–143.

Renström-Kellner, E., and Bergman, B. (1990). Glycolate metabolism in cyanobacteria. IV. Uptake, growth and metabolic pathways. Physiol. Plant. *78*, 285–292.

Rippka, R., Deruelles, J., Waterbury, J.-B., Herdman, M., and Stanier, R.-Y. (1979). Genetic assignments, strain histories and properties of pure cultures of cyanobacteria. J. Gen. Microbiol. *111*, 1–61.

Romanowski, M.J., Bonanno, J.B., and Burley, S.K. (2002). Crystal structure of the *Escherichia coli* glucose-inhibited division protein B (GidB) reveals a methyltransferase fold. Proteins Structure Function and Genetics. *47*, 563–567.

Ryu, J.Y., Song, J.Y., Lee, J.M., Jeong, S.W., Chow, W.S., Choi, S.B., Pogson, B.J., and Park, Y.I. (2004). Glucose-induced expression of carotenoid biosynthesis genes in the dark is mediated by cytosolic pH in the cyanobacterium *Synechocystis* sp. PCC 6803. J. Biol. Chem. *279*, 25320–25325.

Sawaya, M.R., Cannon, G.C., Heinhorst, S., Tanaka, S., Williams, E.B., Yeates, T.O., Kerfeld, C.A. (2006). The structure of beta-carbonic anhydrase from the carboxysomal shell reveals a distinct subclass with one active site for the price of two. J. Biol. Chem. *281*, 7546–7555.

Scanlan, D.J., Sundaram, S., Newman, J., Mann, N.H., and Carr, N.G. (1995). Characterization of a *zwf* mutant of *Synechococcus* sp. strain PCC 7942. J. Bacteriol. *177*, 2550–2553.

Schaeffer, F., and Stanier, R.Y. (1978). Glucose-6-phosphate dehydrogenase of *Anabaena* sp. Arch. Microbiol. *116*, 9–19.

Schaferjohann, J., Yoo, J.G., Kusian, B., and Bowien, B. (1993). The *cbb* operons of the facultative chemoautotroph *Alcaligenes eutrophus* encode phosphoglycolate phosphatase J. Bacteriol. *175*, 7329–7340.

Schwarz, R., and Forchhammer, K., (2005). Acclimation of unicellular cyanobacteria to macronutrient deficiency: emergence of a complex network of cellular responses. Microbiology *151*, 2503–2514.

Sha, J., Kozlova, E.V., Fadl, A.A., Olano, J.P., Houston, C.W., Peterson, J.W., and Chopra, A.K. (2004). Molecular characterization of a glucose-inhibited division gene, *gidA*, that regulates cytotoxic enterotoxin of *Aeromonas hydrophila*. Infect. Immun. *72*, 1084–1095.

Shi, L., Bischoff, K.M., and Kennelly, P.J. (1999). The *icfG* gene cluster of *Synechocystis* sp. strain PCC 6803 encodes an Rsb/Spo-like protein kinase, protein phosphatase, and two phosphoproteins. J. Bacteriol. *181*, 4761–4767.

Shibata, M., Ohkawa, H., Kaneko, T., Fukuzawa, H., Tabata, S., Kaplan, A., and Ogawa, T. (2001). Distinct constitutive and low-CO_2-induced CO_2 uptake systems in cyanobacteria: genes involved and their phylogenetic relationship with homologous genes in other organisms. Proc. Natl. Acad. Sci. USA *98*, 11789–11794.

Shibata, M., Katoh, H., Sonoda, M., Ohkawa, H., Shimoyama, M., Fukuzawa, H., Kaplan, A., and Ogawa, T. (2002). Genes essential to sodium-dependent bicarbonate transport in cyanobacteria: function and phylogenetic analysis. J. Biol. Chem. *277*, 18658–18664.

Shibata, M., Ohkawa, H., Katoh, H., Shimoyama, M., and Ogawa, T. (2002). Two CO_2 uptake systems in cyanobacteria: four systems for inorganic carbon acquisition in *Synechocystis* sp. strain PCC 6803. Funct. Plant Biol. *29*, 123–129.

Singh, A.K., and Sherman, L.A. (2005). Pleiotropic effect of a histidine kinase on carbohydrate metabolism in *Synechocystis* sp. strain PCC 6803 and its requirement for heterotrophic growth. J. Bacteriol. *187*, 2368–2376.

So, A.K., Espie, G.S., Williams, E.B., Shively, J.M., Heinhorst, S., and Cannon, G.C. (2004). A novel evolutionary lineage of carbonic anhydrase (epsilon class) is a component of the carboxysome shell. J. Bacteriol. *186*, 623–630.

So, A.K., John-McKay, M., and Espie, G.S. (2002). Characterization of a mutant lacking carboxysomal carbonic anhydrase from the cyanobacterium *Synechocystis* PCC 6803. Planta. *214*, 456–467.

So, A.K., Kassam, A., and Espie, G.S. (1998). Na^+-dependent HCO_3^- transport in the cyanobacterium *Synechocystis* PCC 6803. Can. J. Bot. *67*, 1084–1091.

Stal, L.J., and Moezelaar, R. (1997). Fermentation in cyanobacteria. FEMS Microbiol. Rev. *21*, 179–211.

Steunou, A.S., Bhaya, D., Bateson, M.M., Melendrez, M.C., Ward, D.M., Brecht, E., Peters, J.W., Kuhl, M., and Grossman, A.R. (2006). In situ analysis of nitrogen fixation and metabolic switching in unicellular thermophilic cyanobacteria inhabiting hot spring microbial mats. Proc. Natl. Acad. Sci. USA *103*, 2398–2403.

Summers, M.L., and Meeks, J.C. (1996). Transcriptional regulation of *zwf*, encoding glucose-6-phosphate dehydrogenase, from the cyanobacterium *Nostoc punctiforme* strain ATCC 29133. Mol. Microbiol. *22*, 473–480.

Summers, M.L., Wallis, J.G., Campbell, E.L., and Meeks, J.C. (1995). Genetic evidence of a major role for glucose-6-phosphate dehydrogenase in nitrogen fixation and dark growth of the cyanobacterium *Nostoc* sp. strain ATCC 29133. J. Bacteriol. *177*, 6184–6194.

Sundaram, S., Karakaya, H., Scanlan, D.J., and Mann, N.H. (1998). Multiple oligomeric forms of glucose-6-phosphate dehydrogenase in cyanobacteria and the role of OpcA in the assembly process. Microbiology *144*, 1549–1556.

Tabita, F.R., (1994). The biochemistry and molecular regulation of carbon dioxide metabolism in cyanobacteria. In The Molecular Biology of Cyanobacteria, D.A. Bryant, ed. (Dordrecht, The Netherlands: Kluwer Academic Publishers), pp. 437–467.

Takahashi, Y., Yamaguchi, O., and Omata, T. (2004). Roles of CmpR, a LysR family transcriptional regulator, in acclimation of the cyanobacterium *Synechococcus* sp. strain PCC 7942 to low-CO_2 and high-light conditions. Mol. Microbiol. *52*, 837–845.

Tchernov, D., Hassidim, M., Luz, B., Sukenik, A., Reinhold, L., and Kaplan, A. (1997). Sustained net CO_2 evolution during photosynthesis by marine microorganisms. Curr. Biol. *7*, 723–728.

Tchernov, D., Helman, Y., Keren, N., Luz, B., Ohad, I., Reinhold, L., Ogawa, T., and Kaplan, A. (2001). Passive entry of CO_2 and its energy-dependent intracellular conversion to HCO_3^- in cyanobacteria are driven by a photosystem I-generated $\Delta\mu H^+$. J. Biol. Chem. *276*, 23450–23455.

Tchernov, D., Silverman, J., Luz, B., Reinhold, L., and Kaplan, A. (2003). Massive light-dependent cycling of inorganic carbon between oxygenic photosynthetic microorganisms and their surroundings. Photosynth. Res. *77*, 95–103.

Tian, J., Bryk, R., Itoh, M., Suematsu, M., and Nathan, C. (2005). Variant tricarboxylic acid cycle in *Mycobacterium tuberculosis*: Identification of alpha-ketoglutarate decarboxylase. Proc. Natl. Acad. Sci. USA *102*, 10670–10675.

Tolbert, N.E. (1997). The C-2 oxidative photosynthetic carbon cycle. Ann. Rev. Plant Physiol. Plant Mol. Biol. *48*, 1–23.

Tu, C.-J., Shrager, J., Burnap, R.L., Postier, B.L., and Grossman, A.R. (2004). Consequences of a deletion in *dspA* on transcript accumulation in *Synechocystis* sp. strain PCC6803. J. Bacteriol. *186*, 3889–3902.

Voll, L.M., Jamai, A., Renne, P., Voll, H., McClung, C.R., and Weber, A.P. (2006). The photorespiratory *Arabidopsis* shm1 mutant is deficient in SHM1. Plant Physiol. *140*, 59–66.

Volokita, M., Zenvirth, D., Kaplan, A., and Reinhold, L. (1984). Nature of the inorganic carbon species actively taken up by the cyanobacterium *Anabaena variabilis*. Plant Physiol. *76*, 599–602.

von Meyenburg, K., Jorgensen, B.B., Nielsen, J., and Hansen, F.G. (1982). Promoters of the *atp* operon coding for the membrane-bound ATP synthase of *Escherichia coli* mapped by Tn10 insertion mutations. Mol. Gen. Genet. *188*, 240–248.

Wang, H.L., Postier, B.L., and Burnap, R.L. (2004). Alterations in global patterns of gene expression in *Synechocystis* sp. PCC 6803 in response to inorganic carbon limitation and the inactivation of *ndhR*, a LysR family regulator. J. Biol. Chem. *279*, 5739–5751.

Woodger, F.J., Badger, M.R., and Price, G.D. (2005). Sensing of inorganic carbon limitation in *Synechococcus* PCC7942 is correlated with the size of the internal inorganic carbon pool and involves oxygen. Plant Physiol. *139*, 1959–1969.

Yang, C., Hua, Q., and Shimizu, K. (2002). Integration of the information from gene expression and metabolic fluxes for the analysis of the regulatory mechanisms in *Synechocystis*. Appl. Microbiol. Biotechnol. *58*, 813–822.

Zhang, C.C., Jeanjean, R., and Joset, F. (1998). Obligate phototrophy in cyanobacteria: more than a lack of sugar transport. FEMS Microbiol. Lett. *161*, 285–292.

Zhang, P., Battchikova, N., Jansen, T., Appel, J., Ogawa, T., and Aro, E.M. (2004). Expression and functional roles of the two distinct NDH-1 complexes and the carbon acquisition complex NdhD3/NdhF3/CupA/Sll1735 in *Synechocystis* sp. PCC 6803. Plant Cell *16*, 3326–3340.

Zhang, P., Battchikova, N., Paakkarinen, V., Katoh, H., Iwai, M., Ikeuchi, M., Pakrasi, H.B., Ogawa, T., and Aro, E.M. (2005). Isolation, subunit composition and interaction of the NDH-1 complexes from *Thermosynechococcus elongatus* BP-1. Biochem. J. *390*, 513–520.

Nitrogen Assimilation and C/N Balance Sensing

13

Ignacio Luque and Karl Forchhammer

Abstract

Cyanobacteria inhabit nearly all illuminated environments on Earth and play key roles in the carbon and nitrogen cycles of the biosphere. Generally, cyanobacteria are able to utilize a variety of inorganic and organic sources of combined nitrogen, like nitrate, nitrite, ammonium, urea or some amino acids. Several cyanobacterial strains are also capable of diazotrophic growth. Genome sequencing has provided a large amount of information on the genetic basis of nitrogen metabolism and its control in different cyanobacteria. Comparative genomics, together with functional studies, has led to a significant advance in this field over the past years. The first part of this review will summarize the current knowledge on the physiology, biochemistry and genetics of the utilization of inorganic and organic nitrogen sources. The second part of the review will turn towards the sophisticated mechanisms cyanobacteria have evolved to tune nitrogen metabolism to the environmental conditions. 2-Oxoglutarate has turned out to be the central signaling molecule reflecting the carbon/nitrogen balance of cyanobacteria. Central players of nitrogen control are the global transcriptional factor NtcA, which controls the expression of many genes involved in nitrogen metabolism, as well as the P_{II} signaling protein, which fine-tunes cellular activities in response to changing C/N conditions. These two proteins are sensors of the cellular 2-oxoglutarate level and have been conserved in all cyanobacteria. In contrast, the adaptation to nitrogen starvation involves heterogeneous responses in different strains.

Introduction

Cyanobacteria evolved a photoautotrophic lifestyle with minimal nutritional requirements, which enabled them to colonize almost all illuminated environments. In the course of their evolution, cyanobacteria have adapted to habitats as diverse as the deep ocean and the rocks on top of mountains, hot deserts and artic environments, they have colonized nutrient poor and eutrophic zones, aquatic and dry ecosystems. Representatives of the cyanobacteria are among the most efficient and abundant N_2-fixing organisms in the biosphere and are thus of particular importance in environments which are poor in combined nitrogen sources. Nitrogen fixation occurs in symbiotic association with plants, fungi, various eukaryotic algae as well as in free-living state. The nitrogen budget of aquatic ecosystems is largely dominated by various diazotrophic unicellular and filamentous cyanobacteria. Cyanobacteria are also able to utilize a variety of inorganic and organic combined nitrogen sources.

The capabilities for the utilization of different nitrogen sources vary among the species and reflect evolutionary adaptations to their particular habitats.

A key to their successful proliferation is the perfect adaptation of their metabolism to the physico-chemical conditions of their environments. As we will outline in the following sections, cyanobacteria have evolved complex mechanisms to tune nitrogen metabolism in response to the ambient conditions. Although the conditions to which they have to respond may be quite different, the central players of nitrogen control have been conserved in the course of evolution, in spite of diversity in peripheral reactions. An intense research in the last few decades has provided valuable information on the spectrum of genes and proteins involved in nitrogen metabolism and its control in different cyanobacteria. Sequencing of cyanobacterial genomes has allowed some global analysis on this issue. All this work has led to a rather comprehensive, thought not complete, view on how cyanobacteria sense and respond to the nitrogen regime of their environments.

Nitrogen assimilation pathways

Nitrogen in organic matter is almost always present as amino- or amido-nitrogen. The inorganic nitrogen precursor, which is assimilated to produce nitrogen-containing bioorganic compounds, is ammonium. Cyanobacteria are able to utilize various inorganic and organic nitrogen sources. They have to take up these compounds from the environment and convert them intracellularly to ammonium, which is finally assimilated. We shall first discuss the assimilation of intracellular ammonium and then turn to the utilization of various ambient nitrogen sources.

Assimilation of intracellular ammonium: the GS/GOGAT pathway

The over-all metabolic organization of ammonium assimilation

There are various biological reactions in which ammonium is converted to amino nitrogen. Enzymes catalyzing such reactions are glycine synthase as well as several amino acid dehydrogenases such as alanine dehydrogenase, aspartate dehydrogenase and glutamate dehydrogenase. The latter is supposed to play an important role in ammonium assimilation in heterotrophic bacteria under ammonium-excess conditions. Furthermore, ammonium can be used for the synthesis of the amido-nitrogen of carbamoylphosphate by carbamoylphosphate synthetase (CPS) and glutamine by glutamine synthetase (GS). However, of all these possibilities to assimilate ammonium, in cyanobacteria only glutamine synthetase is essential for ammonium assimilation, as revealed by genetic and physiological studies in various cyanobacterial strains (reviewed in Muro-Pastor *et al.*, 2005). In agreement with GS being the primary ammonium-assimilating enzyme, ^{13}N-labelling studies demonstrated that the first product of nitrogen assimilation is glutamine, followed by the appearance of glutamate (Wolk *et al.*, 1976; Meeks *et al.*, 1978).

GS incorporates ammonia into the γ-carboxyl group of glutamate in a reaction that involves initial phosphorylation of the γ-carboxyl of glutamate by ATP followed by ammonium incorporation (amidation) and release of inorganic phosphate. ATP hydrolysis drives this reaction in the direction of glutamine synthesis and allows for efficient glutamine synthesis even at low ammonium concentrations.

Ammonium assimilation via GS operates in concert with glutamate synthase, also termed glutamine-2-oxoglutarate-amido transferase (GOGAT). This enzyme uses the amide group of glutamine to reductively aminate 2-oxoglutarate (2-OG), producing two molecules of glutamate. Photosystem I-reduced ferredoxin (PSI-Fd_{red}) is generally used as reductant in this reaction (Marqués *et al.*, 1992). Overall, the GS/GOGAT cycle produces glutamate from 2-OG and ammonium, consuming one ATP and two reduction equivalents (provided by reduced ferredoxin) as shown below.

GS: Glutamate + NH_3 + ATP $\rightarrow$ Glutamine + ADP + P_i

GOGAT: Glutamine + 2-OG + 2[H] $\rightarrow$ 2 Glutamate

GS-GOGAT: 2-OG + NH_3 + ATP + 2[H] $\rightarrow$ Glutamate + ADP + P_i

An alternative route of ammonia assimilation may occur via glutamate dehydrogenase (GDH: 2-OG + NH_3 + 2 [H] $\leftrightarrows$ Glu + H_2O). Although this enzyme is present in various cyanobacteria, physiological studies could not demonstrate a role in ammonia assimilation. However, this enzyme seems to be important for survival under non-exponential growth conditions and to confer a growth advantage in the presence of ammonium (Muro-Pastor *et al.*, 2005).

Glutamate and glutamine are the major nitrogen donors for the synthesis of other nitrogen-containing molecules. GS-GOGAT activity requires a constant supply of the carbon-skeleton 2-OG, which is derived from CO_2-fixation. Cyanobacteria lack the 2-OG dehydrogenase enzyme complex, and therefore synthesis of 2-OG from isocitrate is the final step in the oxidative branch of the TCA cycle (Stanier and Cohen-Bazire, 1977). Since 2-OG is mainly consumed by GOGAT, the cellular level of this metabolite only depends on the rate of its synthesis via isocitrate dehydrogenase (IDH) and consumption by the GS-GOGAT cycle. This link makes IDH part of the nitrogen assimilatory system, which is highlighted by the fact that IDH expression is under the same mechanism of nitrogen control as GS expression (see below).

The enzymes of the GS-GOGAT pathway

With the exception of *Pseudanabaena* sp. PCC 6903, all studied cyanobacteria possess a typical bacterial GS type I (GSI) enzyme, which is encoded by the *glnA* gene (Muro-Pastor *et al.*, 2005). These enzymes are composed of 12 identical subunits of approx. 50 kDa, forming a double hexagonal ring. In addition, a few non-diazotrophic strains exhibit a second GS paralog, which is classified as GS type III (GSIII), encoded by the *glnN* gene (Reyes *et al.*, 1994). This GSIII is homologous to the GS enzymes from anaerobic bacteria of the genus *Bacteroides* and *Butyrivibrio*. GSIII is rather different from GSI since it is hexameric and composed of identical approx. 75 kDa subunits. With the exception of *Pseudanabaena* sp. PCC 6903, which contains only the GSIII, these enzymes are non-essential and only expressed under conditions of severe nitrogen starvation (see below). Regulation of *glnA* and *glnN* gene expression will be detailed in following sections. In the unicellular strain *Synechocystis* sp. PCC 6803, GS is regulated by protein-protein interaction with inhibitory peptides, IF7 and IF17 (García-Domínguez *et al.*, 1999; see also section "Other regulators related to nitrogen metabolism" below). Whether this type of GS activity regulation is more general in cyanobacteria remains to be shown.

In contrast to GOGAT from heterotrophic bacteria, but similar to the plant enzyme, the cyanobacterial GOGAT uses photosystem I-reduced ferredoxin as reductant. The so-called Fd-GOGAT (encoded by the *glsF* gene) is a large enzyme of 170 kDa, containing an iron–sulfur cluster and FMN as prosthetic groups (Muro-Pastor *et al.*, 2005). The Fd-GOGAT activity is highly influenced by the presence of the substrates 2-OG and Fd_{red}. Only in their presence, glutamine hydrolysis and amidotransfer occurs, which avoids wasteful consumption of glutamine. In addition to Fd-GOGAT, some cyanobacterial strains belonging to the genera *Synechocystis* and *Plectonema* contain a second glutamate synthase, accepting NADH as reductant. This NADH-GOGAT is a heterodimer of a large 150 kDa and a small 50 kDa subunit (encoded by *gltB* and *gltD*, respectively). The large subunit contains similar electron transfer sites as Fd-GOGAT, the small subunit contains FAD and two iron–sulfur clusters (van den Heuvel *et al.*, 2004). Under conditions of high photosynthetic activity, Fd-GOGAT seems to be the primary enzyme of the GS-GOGAT cycle (Okuhara *et al.*, 1999).

IDH catalyzes the oxidative decarboxylation of isocitrate to yield 2-OG. $NADP^+$ is the electron acceptor of cyanobacterial IDH. The enzyme has been purified and characterized from various cyanobacterial sources (reviewed in Muro-Pastor *et al.*, 2005). It is a homodimer of 55 kDa subunits, encoded by the *icd* gene. Genetic analysis indicates that *icd* is an essential gene in cyanobacteria.

Utilization of inorganic nitrogen sources

Utilization of ammonium

Ammonium is the preferred source of combined nitrogen for cyanobacteria, since it is directly used by the GS-GOGAT pathway. Ammonia diffuses through membranes, whereas the ammonium ion cannot. Since in aqueous solution, ammonia and ammonium are present in a pH-dependent equilibrium, no active transport is required when millimolar concentrations of ammonium salts are present at mildly alkaline conditions. However, in natural conditions where the environmental concentrations of ammonium are in the micromolar range or below, a high-affinity uptake system may be required. Cyanobacteria possess ammonium transporters of the widely distributed Amt family (Montesinos *et al.*, 1998; Muro-Pastor *et al.*, 2005). The crystal structure of a bacterial Amt protein has revealed a trimeric transmembrane protein, with each subunit containing at its centre a narrow hydrophobic pore through which NH_3 can diffuse (Zheng *et al.*, 2004). At the periplasmatic face of Amt, a binding site for NH_4^+ has been detected. It is proposed that NH_4^+ binds on the periplasmic face of Amt, followed by deprotonation and diffusion of NH_3 through the channel (Javelle *et al.*, 2005). Due to the availability of radiolabeled ^{14}C-methylamonium (MA) (there are no radioisotopes of nitrogen readily available), this analogue is frequently used for studies on Amt function. MA is used as a substrate by GS to form N-methylglutamine, which cannot be further metabolized and therefore accumulates inside the cells. Accurate determination of Amt activity, therefore, requires a distinction between initial uptake and subsequent metabolic conversion (Boussiba *et al.*, 1984). Although regulation of Amt activity by P_{II}-like GlnK proteins has been reported in enteric bacteria (Javelle and Merrick, 2005), activity regulation of Amt remains elusive in cyanobacteria.

Besides the acquisition of traces of ammonium from the environment which are used as N-source, an important function of the Amt transporters may be to scavenge traces of ammonium that are lost by the cells when they utilize other combined nitrogen sources and produce ammonium as an intermediate. Accordingly, the Amt proteins are highly expressed in the absence, or in the presence of very low concentrations, of external ammonium (Vázquez-Bermúdez *et al.*, 2002c).

The genes for ammonium permeases

Three genes encoding ammonium permeases, *amt1* (*sll108*), *amt2* (*sll1017*) and *amt3* (*sll0537*) have been described in *Synechocystis* sp. PCC 6803 (Table 13.1). The products of all three *amt* genes are involved in methylammonium transport although most of it is taken up through Amt1 (Montesinos *et al.*, 1998). There are also multiple *amt* genes in other cyanobacterial genera. Interestingly, in *Anabaena* sp. PCC 7120 and *Anabaena variabilis*, there are three genes encoding putative ammonium permeases arranged in tandem (Figure 13.1). Phylogenetic analyses of the cyanobacterial Amt proteins show that they cluster in 3 major groups (Figure 13.2). The most widespread is group 1 that includes Amt1 from *Synechocystis* sp. PCC 6803 and *Synechococcus elongatus* (Montesinos *et al.*, 1998; Vázquez-Bermúdez *et al.*, 2002c; Muro-Pastor *et al.*, 2005). Every cyanobacterial genome encodes at least a member of this clade except for *Nostoc punctiforme*. Group 2 includes Amt proteins from *S. elongatus* and heterocyst-forming cyanobacteria. Group 3 seems to be restricted to heterocyst-forming cyanobacteria. *Synechocystis amt2* and *amt3* and one of the *Trichodesmium erythraeum amt* (tery_Contig419_or4055) gene products cluster outside groups 1, 2 and 3. Nitrogen-responsive regulation of *amt* gene expression will be detailed below.

Utilization of nitrate and nitrite

Nitrate is the most abundant source of combined nitrogen in many environments. Accordingly, most investigated cyanobacteria have the capability to utilize this source (reviewed in Flores *et al.*, 2005). However, marine strains of the genus *Prochlorococcus* cannot utilize nitrate (Rocap *et al.*, 2003). Although this is an apparent contradiction since nitrate is the major nitrogen source in deep ocean waters where *Prochlorococcus* is abundant, the limited irradiance and the high energetic cost for its reduction to ammonium makes nitrate not a convenient source to support *Prochlorococcus* growth at depth (García-Fernández *et al.*, 2004; see Chapter 5). Generally, the utilization of nitrate requires its uptake followed by intracellular reduction by nitrate reductase (NR) and nitrite reductase (NiR) to produce ammonium, which then enters the GS-GOGAT cycle. Utilization of nitrite proceeds in a similar manner, except that no NR activity is required. The physiological electron donor for both, NR and NiR, is PSI-Fd_{red} (Flores *et al.*, 2005). Under photoautotrophic growth conditions, nitrate and nitrite reduction are coupled to photosynthetic oxygen evolution *in vivo* and thus can be considered genuine photosynthetic processes. The reduction of

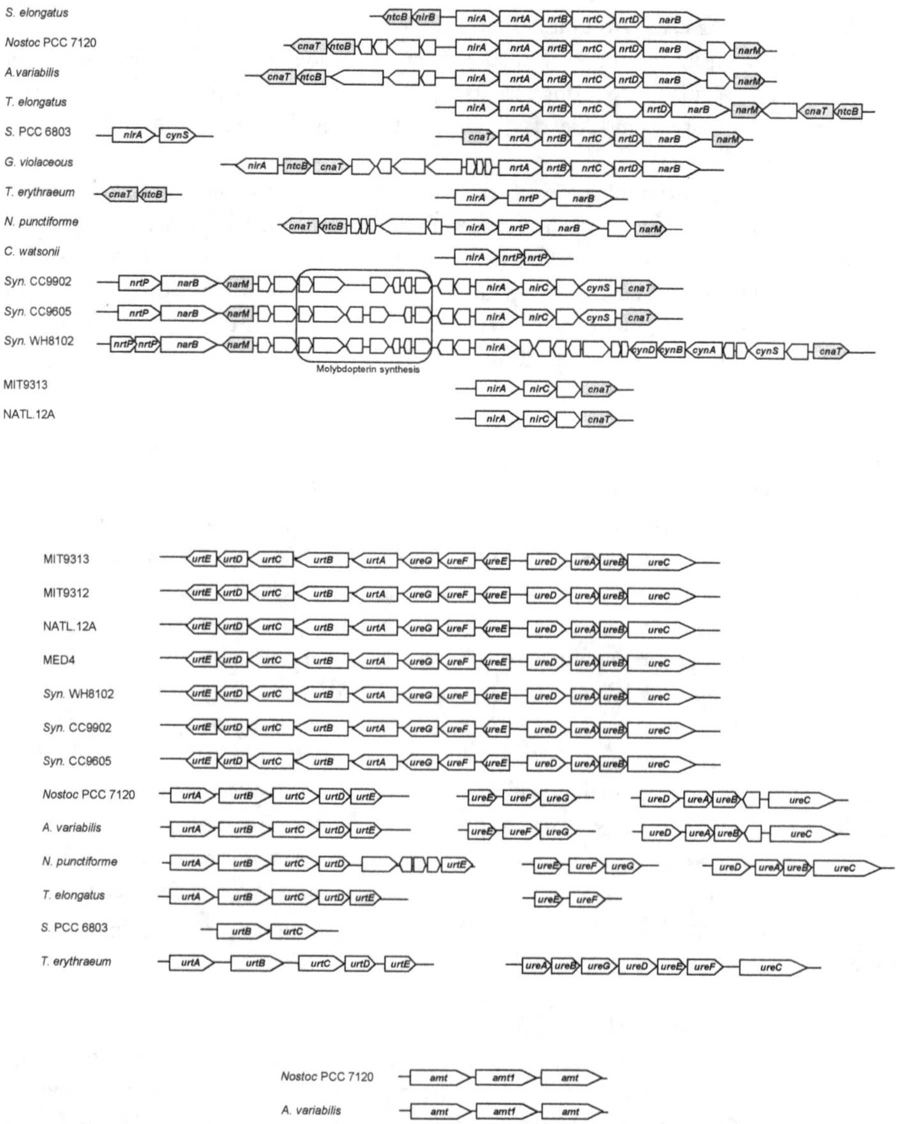

Figure 13.1 Clusters of genes involved in nitrogen assimilation in cyanobacterial genomes. Structural genes are shown as white arrows, whereas regulatory genes are shown as grey arrows. Genes are not depicted to scale. Only clustered genes are depicted.

nitrate requires two electrons, whereas six electrons are needed for nitrite reduction. Consequently, nitrate and nitrite reduction competes with CO_2-fixation for reductant under light limiting conditions. Under those conditions, the cells also have to cope with the adjustment of nitrate and nitrite reductase activities, in order to prevent wasteful accumulation of nitrite (Kloft and Forchhammer, 2005). When external ammonium becomes available, the cells immediately cease nitrate utilization, and switch to use the

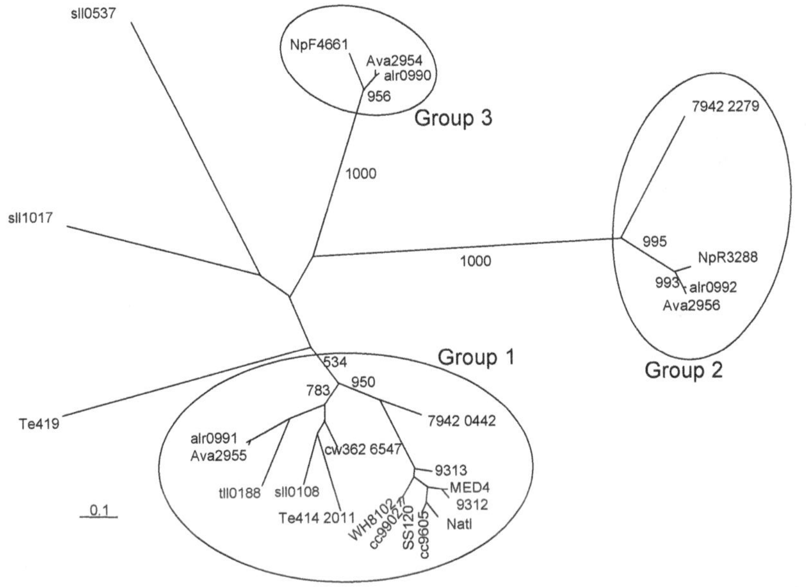

Figure 13.2 Phylogenetic analysis of Amt proteins in cyanobacteria. Sequences were aligned with the ClustalX program, followed by the selection of phylogenetically significant sequence blocks with Gblocks (Castresana, 2000). Phylogenetic trees were constructed from multiple sequence alignments by maximum likelihood using the Phyml Program (Guindon and Gascuel, 2003) with a non-parametric bootstrap analysis of 1000 replicates.

preferred nitrogen source, a process that is regulated at the level of nitrate uptake (Flores and Herrero, 1994). In addition, CO_2-fixation is required to maintain an active nitrate and nitrite uptake, a regulatory link that ensures that the product of nitrate reduction, ammonium, can be incorporated into carbon skeletons, and that accumulation of internal ammonium, which is toxic to the cells, is avoided. These regulatory processes controlling nitrate utilization are coordinated by the P_{II} signal transduction protein and will be discussed below.

Nitrate and nitrite uptake

Uptake of nitrate and nitrite operates by bispecific transporters. Two different transport systems have so far been identified in various cyanobacterial strains. Most fresh-water strains, such as *S. elongatus*, *Synechocystis* sp. PCC 6803, *Anabaena* sp. PCC 7120, *A. variabilis* or *Thermosynechococcus elongatus* and the limestone-associated *Gloeobacter violaceus* employ a binding-protein-dependent transporter (belonging to the large superfamily of ATP-binding cassette (ABC) transporters), composed of 4 different subunits, termed NrtABCD (Table 13.1; Flores *et al.*, 2005). The binding-protein-dependent transporters are characterized by high substrate affinity, optimized for uptake of substrates present in micromolar concentrations or below. Typical for bacterial ABC-transporters, the Nrt permease is composed of a periplasmic substrate binding protein, NrtA, a membrane spanning protein, NrtB, and two cytoplasmatic ATPase domains, NrtD and NrtC, coupling substrate transport to the hydrolysis of ATP. In contrast to most bacterial ATPase

Table 13.1A Some nitrogen assimilation genes in cyanobacterial genomes. An X symbol indicates the absence of the gene(s) in the corresponding genome as determined by BLAST analyses. Blank cells indicate the existence of homologues for the gene(s) that may or may not be orthologs in the corresponding genomes. Only structural genes are shown

	amt	*nrtABCD*	*nrtP*	Nitrite/formate transporter	*nirA*	*narB*	*urtABCDE*	*ureABCDEF*
Synechococcus elongatus PCC 7942	Syn_pcc79420442 (*amt1*) Syn_pcc79422279	Syn_pcc79421241 Syn_pcc79421242 Syn_pcc79421243 Syn_pcc79421244	X	X	Syn_pcc79421240	Syn_pcc79421245	X	X
Synechocystis PCC 6803	Sll0108 (*amt1*) Sll1017 (*amt2*) Sll0537 (*amt3*)	Sll1450 Sll1451 Sll1452 Sll1453	X	X	Slr0898	Sll1454	slr0447 slr1200 slr1201 sll0764 sll0374	sll1750 sll0420 slr1256 sll1639 slr1219 slr1899 sll0643
Gloeobacter violaceus	glr3061	Glr1567 Glr1568 Glr1569 Glr1570	X	X	gll1557	Glr1571	X	X
Crocosphaera watsonii	Contig362_or6547	X	cwat_Contig344_ or3684 cwat_Contig344_ or3685	X	cwat_Contig344_ or3683	cwat_Contig338_ or3495	X	X
Thermosynechococcus elongatus BP-1	tll0188	tlr1350 tlr1351 tlr1352 tlr1354	X	X	tlr1349	tlr1355	Tlr1120 Tlr1121 Tlr1122 Tlr1123 Tlr1124	tlr0005 tll0330 tlr0981 tlr0536 tll1366 tll1365 tlr0057
Nostoc PCC 7120 (*Anabaena* PCC 7120)	alr0990 alr0991 alr0992	alr0608 al0r609 alr0610 alr0611	X	X	alr0607	alr0612	All1951 All1950 All1949 All1948 All1947	alr3666 alr3667 alr3668 alr3670 alr0733 alr0734 alr0735

Anabaena variabilis	Ava2954 Ava2955 Ava2956	Ava4540 Ava4541 Ava4542 Ava4543	X	X	Ava4539	Ava4544	Ava4361 Ava4362 Ava4363 Ava4364 Ava4365	Ava3622 Ava3621 Ava3620 Ava3618 Ava4634 Ava4635 Ava4636
Nostoc punctiforme	NpR3288 NpF4661	X	NpR1527	X	NpR1528	NpR1526	NpF4615 NpF4616 NpF4617 NpF4618 NpF4623	NpF0822 NpF0823 NpF0824 NpF0825 NpR6143 NpR6142 NpR6141
Synechococcus WH8102	SYNW0253	X	SYNW2462 SYNW2463	X	SYNW2477	SYNW2464	SYNW2242 SYNW2241 SYNW2240 SYNW2239 SYNW2238	SYNW2249 SYNW2248 SYNW2247 SYNW2246 SYNW2245 SYNW2244 SYNW2243
Prochlorococcus marinus PCC 9511 (MED4)	PMM0263	X	X	X	X	X	PMM0970 PMM0971 PMM0972 PMM0973 PMM0974	PMM0963 PMM0964 PMM0965 PMM0966 PMM0967 PMM0968 PMM0969
P. marinus MIT9312	Pmt93120265	X	X	X	X	X	Pmt93120829 Pmt93120828 Pmt93120827 Pmt93120826 Pmt93120825	Pmt93120836 Pmt93120835 Pmt93120834 Pmt93120833 Pmt93120832 Pmt93120831 Pmt93120830

Table 13.1A Continued

	amt	*nrtABCD*	*nrtP*	Nitrite/formate transporter	*nirA*	*narB*	*urtABCDE*	*ureABCDEF*
P. marinus MIT9313	PMT1853	X	X	PMT2240	PMT2239	X	PMT2229 PMT2228 PMT2227 PMT2226 PMT2225	PMT2236 PMT2235 PMT2234 PMT2233 PMT2232 PMT2231 PMT2230
P. marinus NATL2A	PMN12a1629	X	X	PMN12a1299	PMN12a1298	X	PMN12a1048 PMN12a1047 PMN12a1046 PMN12a1045 PMN12a1044	PMN12a1055 PMN12a1054 PMN12a1053 PMN12a1052 PMN12a1051 PMN12a1050 PMN12a1049
P. marinus CCMP1375 (SS120)	Pro0295	X	X	X	X	X	X	X
Synechococcus sp. CC9605	Syn_cc96050247	X	Syn_cc96052642	Syn_cc96052657	Syn_cc96052656	Syn_cc96052643	Syn_cc96052620 Syn_cc96052619 Syn_cc96052618 Syn_cc96052617 Syn_cc96052616	Syn_cc96052627 Syn_cc96052626 Syn_cc96052625 Syn_cc96052624 Syn_cc96052623 Syn_cc96052622 Syn_cc96052621
Synechococcus sp. CC9902	Syn_cc99020276	X	Syn_cc99022271	Syn_cc99022285	Syn_cc99022284	Syn_cc99022272	Syn_cc99022250 Syn_cc99022249 Syn_cc99022248 Syn_cc99022247 Syn_cc99022246	Syn_cc99022257 Syn_cc99022256 Syn_cc99022255 Syn_cc99022254 Syn_cc99022253 Syn_cc99022252 Syn_cc99022251
Trichodesmium erythraeum	tery_Contig414_or2011 tery_Contig419_or4055	X	tery_Contig412_ or1340	X	tery_Contig412_ or1339	tery_Contig412_ or1338	tery_Contig404_or0294 tery_Contig404_or0295 tery_Contig404_or0296 tery_Contig404_or0297 tery_Contig404_or0298	tery_Contig413_or1693 tery_Contig413_or1694 tery_Contig413_or1695 tery_Contig413_or1696 tery_Contig413_or1697 tery_Contig413_or1698 tery_Contig413_or1699

Table 13.1B Further nitrogen assimilation genes in cyanobacterial genomes.

	cynS	*glnA*	*glnN*	*gdhA*	*glsF*	*gltB* *gltD*	*nif*	*vnf*
Synechococcus elongatus PCC 7942	Synpcc7942_2104	Syn_pcc79422156	Syn_pcc79420169	X	Syn_pcc79420890	X	X	X
Synechocystis PCC 6803	slr0899	slr1756	slr0288	slr0710	sll1499	sll1502 sll1027	X	X
Gloeobacter violaceus	X	glr1052	X	gll1133	glr1508	X	X	X
Crocosphaera watsonii		cwat_Contig361_ or5271	X	cwat_Contig359_ or4713	cwat_Contig361_ or5417	X	cwat_Contig347_or3818 cwat_Contig347_or3819 cwat_Contig347_or3820	X
Thermosynechococcus elongatus BP-1	X	tll1588	X	X	tll1368	X	X	X
Nostoc PCC 7120 (*Anabaena* PCC 7120)	all1291	alr2328	X	alr4255	alr4344	X	All1455(nifH1) All1454 All1440 Alr874(nifH2)	X
Anabaena variabilis	Ava_3015	Ava0147	X	Ava_1204 Ava_4115	Ava1294	X	Ava3916nifH1) Ava3917(nifD1) Ava3930(nifK1) Ava4247(nifH2) Ava4248(nifD2) Ava4249(nifK2) Ava4046(nifH)	Ava4054(vnfH) Ava4026(vnfDG) Ava4027(vnfK)
Nostoc punctiforme	X	NpR5387	X	X	NpR3877	X	NpR0415(nifH) NpR0414 NpR0390 NpR1035(nifH2) NpF5903(nifH3)	X
Synechococcus WH8102	SYNW2490	SYNW1073	X	X	SYNW2132	X	X	X
Prochlorococcus marinus PCC 9511 (MED4)	PMM0373	PMM0920	X	X	PMM1512	X	X	X

Table 13.1B continued

	cynS	*glnA*	*glnN*	*gdhA*	*glsF*	*gltB* *gltD*	*nif*	*vnf*
P. marinus MIT9312	X	Pmt93120880	X	X	Pmt93121604	X	X	X
P. marinus MIT9313	X	PMT0601	X	PMT0623	PMT1777	X	X	X
P. marinus NATL2A	PMN12a1390	PMN12a0141	X	X	PMN12a1078	X	X	X
P. marinus CCMP1375 (SS120)	X	Pro1038	X	X	Pro1668	X	X	X
Synechococcus sp. CC9605	Syncc9605_2659	Syn_cc96051205	X	X	Syn_cc96050326	X	X	X
Synechococcus sp. CC9902	Syncc9902_2287	Syn_cc99021265	X	X	Syn_cc99022016	X	X	X
Trichodesmium erythraeum	X	tery_Contig420_ or4519	X	tery_Contig414_ or1860	tery_Contig417_ or2955	X	Tery_Contig420_or4267 tery_Contig420_or4266 tery_Contig420_or4265	X

components of ABC transporters, NtrC bears an additional cytoplasmatic C-terminal domain with similarity to the substrate binding protein NrtA. This domain has been proposed to have a regulatory role in mediating nitrogen control over Nrt activity (Kobayashi *et al.*, 1997). The structure of the substrate binding-protein NrtA has been recently solved (Koropatkin *et al.*, 2006). NrtA is a 45 kDa lipoprotein, which is anchored to the periplasmic side of the cytoplasmic membrane (Maeda and Omata, 1997). In nitrate grown cells, it is one of the most abundant proteins in the cytoplasmic membrane. It binds nitrate and nitrite with high affinity (the K_d of the *S. elongatus* protein is about 0.3 μM) allowing the utilization of nitrate and nitrite in submicromolar concentrations.

In contrast to the situation detailed above, most marine strains, such as the marine *Synechococcus* strains, but also the filamentous strain *N. punctiforme* take up nitrate and nitrite through a permease of the major facilitator superfamily (Table 13.1; reviewed in Flores *et al.*, 2005). In general, the permease consists of one single polypeptide, termed NrtP (or NapA in some cases), a hydrophobic protein with 12 membrane spanning domains. It is not yet established whether nitrate/nitrite transport is coupled to Na^+ or H^+ symport nor has the affinity for nitrate/nitrite been reported. However, from physiological investigations, it is assumed that NrtP is a high affinity permease. In most cyanobacterial genomes, NrtP is encoded by a single gene, although in *Synechococcus* sp. WH 8102 and *Crocosphaera watsonii*, the *nrtP* ORF is split in two contiguous halves (Figure 13.1). A nitrite-specific transport activity has been described in *S. elongatus* (Maeda *et al.*, 1998), which evidences the existence of a nitrite-specific transport system, termed NIT, not characterized so far at the molecular level.

Intracellular nitrate and nitrite reduction

Reduction of nitrate to nitrite is catalyzed by an assimilatory nitrate reductase enzyme (NR). In contrast to most other bacterial assimilatory nitrate reductases, cyanobacterial NR is partially associated to the thylakoid membrane. The two electrons required for reduction of nitrate are directly delivered from PSI-Fd_{red} (Flores *et al.*, 2005). In contrast to other assimilatory nitrate reductases present in most bacteria, which are heterodimers of a ca. 45 kDa FADH-containing diaphorase subunit and a 95 kDa catalytic subunit, cyanobacterial NR is a ca. 80 kDa monomeric enzyme, containing one [4Fe-4S] cluster and a Mo-cofactor. For the biosynthesis of active NR, coordinated synthesis of the Mo-cofactor is required. Synthesis and incorporation of the Mo-cofactor in bacterial molybdoproteins requires the products of at least seven genes, homologues of which are found in the cyanobacterial genomes (Flores *et al.*, 2005).

The cyanobacterial NiR is a monomeric enzyme of 52–56 kDa subunits, containing a [4Fe-4S] cluster and siroheme as redox-active prosthetic groups. This enzyme is more closely related to nitrite reductase from eukaryotic phototrophs than to NiR from heterotrophic bacteria, which, in contrast to cyanobacterial NiR, may accept electrons from flavin and nicotinamide nucleotides (Flores *et al.*, 2005). NiR binds nitrite and releases ammonium, without the appearance of free reaction intermediates. Three successive two-electron reduction steps are involved, and from investigation of plant NiR, it seems that the reduction intermediates NO and NH_2OH are bound to siroheme (Kuznetsowa *et al.*, 2004).

The genes for nitrate utilization

In many cyanobacteria, the genes encoding nitrate uptake (*nrtABCD* or *nrtP*) are clustered in an operon with the nitrate reductase gene, *narB*, and in some cases also with the nitrite reductase gene, *nirA* (reviewed in Flores *et al.*, 2005) (Figure 13.1). As a general rule, the *narB* gene is downstream of the nitrate uptake system-encoding gene(s), *nrtABCD*, or *nrtP*, which in turn are located downstream of the *nirA* gene. This arrangement is found in *S. elongatus*, *Anabaena* sp. PCC 7120, *A. variabilis*, *T. elongatus*, *T. erythraeum* and *N. punctiforme* (Figure 13.1). In some other strains including *Synechocystis* sp. PCC 6803, *G. violaceus* and some marine *Synechococcus*, the *nrt* genes are contiguous to *narB* but not to *nirA*, whereas in *C. watsonii* it is *nirA* which is contiguous to the *nrt* genes (Figure 13.1; see also Sakamoto *et al.*, 1999; Bird and Wyman, 2003). In some unicellular marine strains, the *nirA* gene is clustered with a downstream gene (*nirC*) homologous to genes encoding formate/nitrite transporters (Figure 13.1) (see Clegg *et al.*, 2002; Rexach *et al.*, 2000). Whether the cyanobacterial *nirC* homologues encode nitrite-specific transporters remains to be demonstrated. It is remarkable that *Synechococcus* sp. CC9605 and *Synechococcus* sp. CC9902 contain an *nrtP* gene linked to *narB* and a *nirC* gene linked to *nirA*. Structural genes for nitrate and nitrite utilization are also frequently linked to regulatory genes (Figure 13.1) that will be described in section "Other regulators related to nitrogen metabolism," and with genes involved in the biosynthesis of the respective prosthetic groups.

Consistent with the inability of all *Prochlorococcus* isolates to utilize nitrate, none of the sequenced *Prochlorococcus* genomes contains a *narB* gene. *P. marinus* MED4, *P. marinus* MIT9312 and *P. marinus* CCMP1375 (SS120) also lack a *nirA* gene. The absence of nitrate/nitrite assimilation genes in *Prochlorococcus* is thought to be the result of an evolutionary process of genome compaction (García-Fernández *et al.*, 2004).

Utilization of molecular dinitrogen

Energetically, the most expensive source of nitrogen is molecular nitrogen or dinitrogen. The process of reduction of N_2 to NH_3 is termed nitrogen fixation and is catalyzed by the enzyme system known as nitrogenase. The net reaction catalyzed by nitrogenase is:

$$N_2 + 8\,[H] + 16\,ATP \rightarrow 2\,NH_3 + H_2 + 16\,ADP + 16\,P_i$$

Nitrogenase is an extremely oxygen-sensitive enzyme, composed of two components: the dinitrogenase (Mo-Fe protein) catalyzes the reduction of N_2, and the dinitrogenase reductase (Fe protein) donates electrons of very low redox potential to dinitrogenase. Dinitrogenase reductase (NifH) is a homodimer containing one central [4Fe-4S] cluster. Its function is to lower the redox potential of electrons derived from ferredoxin or flavodoxin in an ATP dependent reaction and to donate these electrons to dinitrogenase. Dinitrogenase is composed of the NifD and NifK polypeptides, arranged in an $\alpha_2\beta_2$ structure. Each α subunit contains at its active site a FeMo-cofactor, while the interface between each α–β dimer binds an [8Fe-7S] cluster (P-cluster) that is involved in the electron transfer reaction (Igarashi and Seefeldt, 2003). Under conditions of molybdate limitation, some *Anabaena* species express an alternative nitrogenase containing a vanadium-iron cofactor (Boison *et al.*, 2006; Thiel, 1993).

The ability to grow at the expense of N_2 (diazotrophic growth) is present in a subset of widely distributed cyanobacterial strains (see the distribution of *nif* genes in cyanobacterial genomes in Table 13.1). These organisms have evolved different strategies to protect nitrogenase from oxygen, including atmospheric oxygen and oxygen produced intracellularly by oxygenic photosynthesis (Bergman-Frank *et al.*, 2003; see also Chapter 15).

Utilization of organic nitrogen sources

Cyanobacteria can also utilize organic compounds as nitrogen sources, including urea and some amino acids. *Synechococcus* sp. PCC 7002 has also been reported to grow using purines as nitrogen sources (Kapp *et al.*, 1975; Neilson and Larsson, 1980). The importance of organic sources in the provision of nitrogen to natural cyanobacterial populations is difficult to estimate but may be relevant for the *Prochlorococcus* species that are unable to utilize nitrate or nitrite. This would be in line with a recent report showing higher rates of organic nitrogen uptake by *Prochlorococcus* with respect to *Synechococcus* in oligotrophic marine waters (Zubkov *et al.*, 2003).

Urea transport and metabolism

Urea is one of the most abundant forms of organic nitrogen in nature. It is excreted by many animals as a waste product and it is also formed by bacterial degradation of amino acids and nucleic acids. In the oceans urea is the dominant component of dissolved organic nitrogen compounds (Antia *et al.*, 1991). Most but not all cyanobacterial species can utilize urea as a nitrogen source. *Synechococcus* sp. WH7803, *S. elongatus, G. violaceus, C. watsonii* and *P. marinus* SS120 are among the known species that are not able to grow using urea. Genome information on the four latter strains demonstrates that they lack *ure* genes encoding urease (Table 13.1). Although yeasts and some algae use the ATP and biotin-dependent enzyme urea amidolyase (EC 3.5.1.45) to utilize urea, cyanobacteria as well as most other eubacteria rely on urease (also named urea amidohydrolase, EC 3.5.1.5). This is a nickel-dependent enzyme, which catalyzes the hydrolysis of urea rendering CO_2 and 2 molecules of ammonia (Mobley *et al.*, 1995). The cyanobacterial urease is a cytoplasmic heterotrimeric enzyme composed of two small (γ or UreA, and β or UreB) and one large (α or UreC) subunits. The native molecular mass of the *Prochlorococcus* urease has been reported to be 168 kDa, consistent with an $(\alpha\beta\gamma)_2$ structure, whereas the MW for the *Synechococcus* enzyme is 420 kDa, consistent with an $(\alpha\beta\gamma)_6$ structure (Collier *et al.*, 1999; Palinska *et al.*, 2000). In bacteria, the assembly of a catalytically active urease requires at least seven genes, three of which, *ureA, ureB* and *ureC*, encode the structural subunits, and the other four, *ureD, ureE, ureF,* and *ureG*, encode the accessory polypeptides required for the assembly of the Ni^{2+} metallocenter. The *ure* genes were originally found to be organized in two divergent clusters, one composed of *ureDABC* and the other of *ureEFG*, in *P. marinus* and *Synechococcus* sp. WH7805 (Figure 13.1; Collier *et al.*, 1999; Palinska *et al.*, 2000). These clusters, although not necessarily divergent, are also observed in a good number of cyanobacterial genomes (Figure 13.1). In *T. erythraeum*, however, all *ure* genes are arranged in tandem (Figure 13.1). Repression of urease by ammonium has been described in several cyanobacteria (Singh, 1990, 1992; Collier *et al.*, 1999). However, urease activity has been reported not to be sensitive to ammonium in some other species including *P. marinus*, *Anabaena* sp. PCC 7120 *and A. variabilis* (Ge *et al.*, 1990; Palinska *et al.*, 2000; Valladares *et al.*, 2002).

Diffusion of urea across lipid bilayers takes place readily. Thus, a passive uptake linked to consumption by cytoplasmic urease may occur for urea at concentrations in the range of the K_m observed for urease (2.5 mM for the *Anabaena* sp. PCC 7120 enzyme). However, urea is usually found in natural ecosystems at concentrations in the micromolar range (see e.g. Antia *et al.*, 1991; Cho *et al.*, 1996), making diffusion an ineffective way of uptake. An ABC-type active transport system specific for urea has recently been identified in *Synechocystis* sp. PCC 6803 and *Anabaena* sp. PCC 7120 (Valladares *et al.*, 2002). The K_s for this system (0.11 μM in *Anabaena*) would allow urea uptake at the concentrations usually found in nature. This is a concentrative permease that is supposed to accumulate urea in the cytoplasm at concentrations that can be utilized by urease. The subunits of the urea-specific transporter are encoded by 5 genes, *urtA*, *urtB*, *urtC*, *urtD* and *urtE*. In *Anabaena* sp. PCC 7120, *urtA*, *urtB* and *urtC* (and possibly *urtD* and *urtE*) are co-transcribed and subjected to metabolic repression by ammonium. Consistently with this, the *urtA* gene is expressed from an NtcA-activated canonical promoter (see section "Structure of NtcA-regulated promoters") that may direct transcription of the whole cluster (Valladares *et al.*, 2002). Clusters of genes homologous to the *Anabaena urt* genes are found in diverse cyanobacteria (Table 13.1 and Figure 13.1). Interestingly, in the genome of marine *Synechococcus* and *Prochlorococcus* species, genes putatively encoding an ABC-type permease for urea are located downstream and in the same orientation of the *ureEFG* genes of the urease accessory polypeptides (Figure 13.1).

In living organisms, cyanate (CNO^-) is produced from the decomposition of urea or carbamoyl phosphate. Cyanate is a toxic compound due to its high reactivity to nucleophilic groups in proteins. Some cyanobacteria express a cyanase (E.C.4.3.99.1), encoded by the *cynS* gene, which detoxifies cyanate generating CO_2 and ammonium. Thus cyanate has been proposed to be a potential nitrogen source. Cyanase is not ubiquitous in cyanobacteria (Table 13.1) but it is present in representatives of different phylogenetic clusters. This enzyme has been characterized in *S. elongatus* and *Synechocystis* sp. PCC 6803 (Miller and Espie, 1994; Harano *et al.*, 1997). In *S. elongatus* and *P. marinus* PCC 9511 (MED4), *cynS* is clustered with three genes *cynA*, *cynB* and *cynD* encoding a putative cyanate permease. Cyanase activity and expression of the *cyn* cluster in *S. elongatus* and *cynS* expression in *Synechocystis* sp. PCC 6803 are repressed by ammonium (Harano *et al.*, 1997).

Amino acid transport and assimilation

In contrast to the situation in other eubacteria where a large number of transport systems for individual amino acids or groups of chemically related amino acids and multiple systems with redundant functions are found, cyanobacteria seem to be equipped with a few amino acid permeases with broad specificity (see e.g. Montesinos *et al.*, 1995, 1997; Quintero *et al.*, 2001). Different cyanobacteria exhibit particular uptake capacities, both in terms of the spectrum of amino acids that each strain can transport and in the efficiency for capturing individual amino acids. For instance, *S. elongatus* only transports neutral amino acids, whereas other unicellular or filamentous cyanobacteria exhibit a much broader capacity (Montesinos *et al.*, 1997). All cyanobacteria tested seem to have at least one transport system for neutral amino acids. However, the specificity of this permease is not identical in all cyanobacteria, being restricted to the transport of neutral, non-polar amino acids in *S. elongatus*, or extended to the transport of basic and acidic amino acids in *Pseudanabaena* sp. PCC 6903 (Montesinos *et al.*, 1997).

Amino acid permeases have been characterized mainly in *Synechocystis* sp. PCC 6803 and *Anabaena* sp. PCC 7120 (Labarre *et al.*, 1987; Flores and Muro-Pastor, 1988; Flores and Muro-Pastor, 1990; Herrero and Flores, 1990; Xu and McAuley, 1990; Montesinos *et al.*, 1995, 1997; Quintero *et al.*, 2001). *Synechocystis* sp. PCC 6803 bears at least four different amino acid permeases: an ABC-type transporter for neutral amino acids, histidine and glutamine, termed Nat; another ABC-type transporter, named Bgt, for basic amino acids that also transports glutamine with low affinity; a monocomponent secondary permease for Na^+-dependent glutamate transport, named GltS; and a TRAP-like transporter, named Gtr, that also mediates Na^+-dependent glutamate transport (Labarre *et al.*, 1987; Flores and Muro-Pastor, 1990; Montesinos *et al.*, 1997; Quintero *et al.*, 2001). Since the GltS and the Gtr systems account for no more than 60% of the glutamate transport activity at least another glutamate permease must exist in *Synechocystis* sp. PCC 6803.

Anabaena sp. PCC 7120 has a minimum of five different amino acid permeases. It bears two distinct transport systems for basic amino acids, an energy-dependent, high affinity system, and a low affinity permease (Herrero and Flores, 1990; Montesinos *et al.*, 1995). Two different transporters with overlapping but not identical specificities, named N-I and N-II, mediate active transport of neutral amino acids, glutamine and histidine; and at least another permease uncharacterized so far mediates the uptake of acidic amino acids (Flores and Muro-Pastor, 1988; Montesinos *et al.*, 1995). The N-I system for neutral amino acids has been characterized at the molecular level. This system is encoded by the *natA*, *natB*, *natC*, *natD*, and *natE* genes. *natA* and *natE* encode the ATPase components, *natC* and *natD* the transmembrane proteins and *natB* the periplasmic binding protein. *natC* and *natA* form an operon whereas the other genes are scattered in the genome. The *natB* mutants exhibit reduced transport activity of neutral amino acids but increased transport of histidine and glutamine, suggesting that another periplasmic binding protein can operate with the Nat system for histidine and glutamine transport (Picossi *et al.*, 2005). As it has been observed for mutants of the neutral amino acid transport system in other cyanobacteria, mutants of the N-I system excrete hydrophobic amino acids to the extracellular medium, mainly Ala, Val, Ile, Phe and Leu. Remarkably, in *Anabaena* the *nat* mutants are also impaired in diazotrophic growth. Since the *nat* genes are only expressed in vegetative cells but not in heterocysts, it has been proposed that the N-I system could participate in the uptake of amino acids released by heterocysts, mediating thus the transfer of nitrogenous compounds from heterocysts to vegetative cells (Picossi *et al.*, 2005).

Some amino acids, including asparagine, arginine and glutamine can be utilized by cyanobacteria as nitrogen sources (Neilson and Larsson, 1980; Flores and Herrero, 1994). *Synechocystis* and *Anabaena* mutants of the basic amino acid transport system do not grow or show a reduced growth rate when supplemented with arginine as the sole nitrogen source (Flores and Muro-Pastor, 1990; Herrero and Flores, 1990) evidencing that arginine is intracellularly metabolized in these genera. *S. elongatus* can also grow at the expense of arginine as nitrogen source when supplied at millimolar concentrations. However, as a unique property to this strain, arginine is metabolized in the periplasm by an L-amino acid oxidase named L-AOX (Bockholt *et al.*, 1996). L-AOX is a flavoprotein that catalyzes the oxidative deamination of basic L-amino acids (with a preferential affinity for arginine and lysine) releasing ammonium (Engels *et al.*, 1992). Although L-AOX seems to be quite efficient in the deamination of arginine, its low affinity for its substrates (K_m for arginine 3 mM) does not support a role for this enzyme in scavenging arginine in natural environments (Flores *et al.*, 1982; Bockholt *et al.*, 1996). Intracellular catabolism of arginine in *Synechocystis* sp. PCC 6803 has been shown to involve the urea cycle and the arginase pathway rendering urea and glutamate (Quintero *et al.*, 2000). Glutamine, when utilized as a nitrogen source, seems to be mainly assimilated through GOGAT (Flores and Herrero, 1994).

Regulation of nitrogen metabolism

Nitrogen assimilation is tightly regulated in most cyanobacteria in response to the ambient nitrogen source and inorganic carbon supply (C_i: CO_2 and HCO_3^-). Ammonium, the preferred nitrogen source for cyanobacteria, exerts metabolic repression on cellular activities related to nitrogen assimilation (reviewed in Flores and Herrero, 1994). These activities, including nitrate and nitrite reductase, nitrate and nitrite transport, nitrogenase and glutamine synthetase exhibit basal levels in the presence of ammonium and high levels in its absence. This regulatory program seems to have evolved to tune the metabolic capabilities of cyanobacteria to the availability of nitrogen sources. Cells appear to sense the environmental nitrogen status and orchestrate their metabolism to assimilate first ammonium, energetically the most economical nitrogen source. The activities directed to the assimilation of alternative nitrogen sources, which are more costly to assimilate, are turned on only when ammonium has been depleted. In addition, the balance between C_i and the nitrogen source has an important impact on nitrogen regulation: limiting C_i favors the depression of nitrogen assimilatory functions. These global responses towards the nitrogen source and the C/N balance are referred to as Nitrogen control. Experiments with inhibitors of the GS/GOGAT cycle have demonstrated that ammonium incorporation into carbon skeletons is required to trigger metabolic repression. Thus, rather than the presence of extracellular ammonium, an intracellular metabolite related to the GS/GOGAT cycle seems to act as a signal for the cellular nitrogen supply (Flores and Herrero, 1994). The more recent discovery of nitrogen regulatory proteins has permitted to identify this metabolic signal as 2-OG (Forchhammer, 1999; reviewed in Forchhammer, 2004; Herrero *et al.*, 2004). Due to the lack of 2-OG dehydrogenase, the fate for 2-OG is to be combined with ammonium in the GS/GOGAT cycle. 2-OG is thus ideally suited for the signaling of the cellular C/N status: its synthesis depends on CO_2-fixation and its consumption depends on the turn-over rate of the GS-GOGAT cycle (Figure 13.3). In accord with this assumption, the cytoplasmic levels of 2-OG strictly depend on the cellular nitrogen status (Muro-Pastor *et al.*, 2001). Availability of 2-OG is necessary for the assimilation of internally generated ammonia. Since ammonia is toxic at high concentrations, its production should be avoided when the cells are depleted of 2-OG. On the other hand, an excess of 2-OG indicates that nitrogen assimilation lags behind CO_2 fixation.

Central players of nitrogen control are the transcription factor NtcA, which controls the expression of most of the genes involved in nitrogen metabolism, as well as the P_{II}

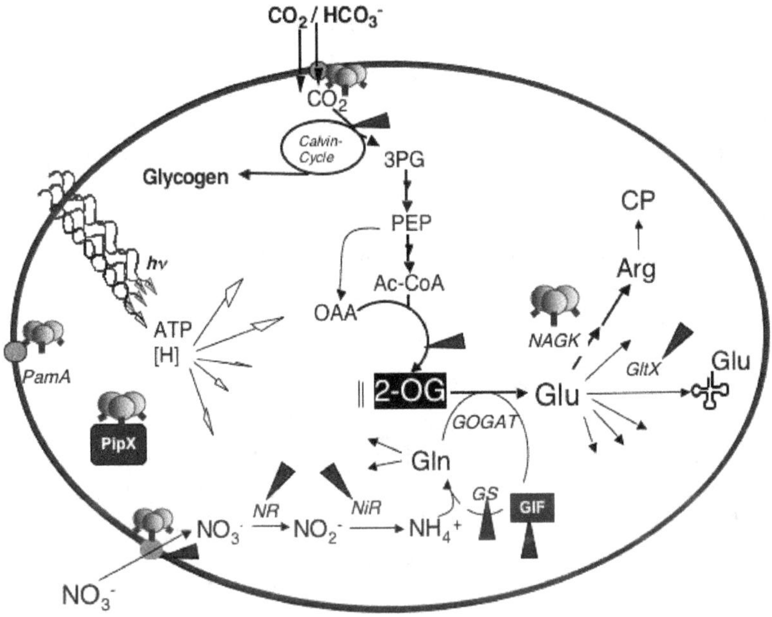

Figure 13.3 Sketch of central metabolism in cyanobacteria and its control through NtcA and P_{II}. The central position of 2-OG at the intersection between C- and N-assimilatory pathways is highlighted. Its cellular concentration is sensed by the P_{II} and NtcA regulatory proteins. P_{II} interactions, either as deduced from physiological analysis or by demonstration of protein-protein interactions, are indicated by trimeric spheres. Processes in which genes subjected to NtcA-dependent control are involved are indicated by solid arrowheads. This scheme summarizes processes in the unicellular strains *S. elongatus* and *Synechocystis* sp. PCC 6803 rather than the situation in a particular cyanobacterium, and is not representative for all cyanobacterial strains. Note that the external combined nitrogen sources are converted intracellularly to ammonium, which is then assimilated by the GS-GOGAT pathway. For details, see text.

signaling protein, which fine-tunes cellular activities in response to fluctuating C/N conditions. They will be described in some detail in the next sections.

Although nitrogen control seems to be widespread in cyanobacteria, there are some differences among strains in the functioning of the regulatory system. It is remarkable the lack of repression by ammonium for particular genes like *glnA* in *Synechococcus* sp. WH8103 (Bird and Wyman, 2003) or *amt1* in *Prochlorococcus marinus* PCC 9511 (Lindell *et al.*, 2002). These differences may reflect evolutionary adaptations of marine cyanobacteria to nutrient-poor environments that do not exhibit large fluctuations of nutrient-rich and nutrient-poor periods. Despite these differences, the P_{II} and NtcA proteins have been conserved in these oligotrophic-adapted organisms, although their precise role is not yet fully understood.

NtcA-mediated regulation

The regulatory gene *ntcA* encodes a protein of the Crp-FNR family of transcriptional regulators which includes more than 369 family members widely distributed in bacteria

(Korner *et al.*, 2003). NtcA binds to and activates the promoters of genes subjected to metabolic control by ammonium including, for instance, *glnA* encoding glutamine synthetase, the *nir* operon and its own gene (Wei *et al.*, 1993; Luque *et al.*, 1994). The *ntcA* gene is ubiquitous in cyanobacteria and exclusive of this taxonomic group (see e.g. Frías *et al.*, 1993). Mechanistically, NtcA regulates transcription acting either as an activator or as a repressor. NtcA activity has been described to respond to 2-OG, the key molecule signaling C/N status, (Vázquez-Bermúdez *et al.*, 2002b; Tanigawa *et al.*, 2002). Recent reports indicate that the NtcA regulon includes not only genes related to nitrogen metabolism but also other genes involved in a variety of functions, what broadens the spectrum of cellular processes controlled by NtcA.

The NtcA protein: structure and DNA binding

NtcA is a dimeric protein of ca. 50 kDa, composed of two identical monomers each about 220 amino acids long. NtcA from *Anabaena* sp. PCC 7120 has been crystallized and some preliminary X-ray data have been obtained (Wisén *et al.*, 2004b). Although the crystal structure of NtcA has not been solved so far, sequence conservation has allowed modeling of the *Anabaena* NtcA monomer based on the structure of Crp (Wisén *et al.*, 2004a). NtcA is composed of an N-terminal β-roll structure of about 100 amino acids, connected to the C-terminal domain by a long α-helix of ca. 30 residues that in Crp is involved in dimerization. The C-terminal domain is formed by 3 short α-helices, the two C-terminal ones forming a helix–turn–helix (HTH) motif involved in DNA binding. Although the N-terminal β-roll is characteristic of cyclic nucleotide binding proteins, it must be mentioned that only one (Arg_{82}) out of the three residues (Glu_{82}, Arg_{82}, Ser_{83}) important for cAMP binding in *E. coli* Crp is conserved in NtcA.

The sequence of the HTH motif is almost identical in the 25 NtcA proteins whose whole sequence is available (Figure 13.4). Sequence flexibility is observed at position 16 which is occupied by either Val or Ile. The high sequence similarity of the NtcA HTH motif from different cyanobacteria makes very plausible that all of them bind to identical or very similar DNA sequences. Several studies have been performed to determine the consensus DNA sequence to which NtcA binds. Early studies proposed two palindromic, partially overlapping signature sequences $GTAN_8TAC$ or $TGTN_{10}ACA$ as the consensus DNA sites for NtcA based on comparisons of natural occurring DNA targets in *S. elongatus* and *Anabaena* sp. PCC 7120, respectively (Luque *et al.*, 1994; Ramasubramanian *et al.*, 1994). Further studies were based on selection approaches of random sequence libraries (Jiang *et al.*, 2000), on the comparison of the affinity of NtcA for natural or synthetic DNA sites (Vázquez-Bermúdez *et al.*, 2002a) or on sequence comparisons of NtcA sites in cyanobacterial sequenced genomes (Su *et al.*, 2005). The most important nucleotides for NtcA binding seem to be the two external nucleotides of the GTA and TAC triplets (shown underlined), their mutation increasing largely the K_d for NtcA binding (Vázquez-Bermúdez *et al.*, 2002a). These two nucleotides are also the most conserved in the binding sites compiled from 9 cyanobacterial genomes (Su *et al.*, 2005). Nucleotides C and A are found in the second and third position downstream of the GTA triplet in the DNA sites with highest affinities for NtcA. Substitution of such nucleotides impairs NtcA binding, suggesting a role for them on NtcA binding or stabilization of the complex (Vázquez-Bermúdez *et al.*, 2002a). The presence of T or A tracks flanking the $GTAN_8TAC$ sites

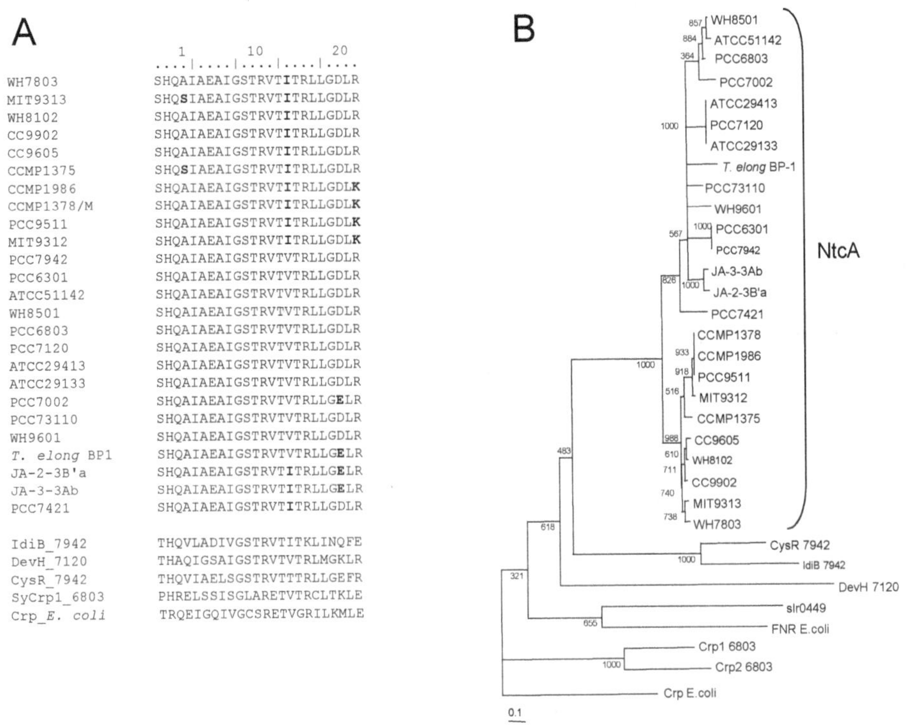

Figure 13.4 (A) Sequence alignment of the HTH motif of cyanobacterial NtcA and homologous proteins. Sequence changes between NtcA sequences are shown in bold letters. (B) Phylogenetic tree of NtcA and homologous proteins. Details for the analysis are like in Figure 13.2.

has also been shown to favor NtcA binding. T/A tracks are sequences of intrinsic DNA bending. Whether NtcA binding is favored by the sequence itself or by the secondary structure of DNA is not known. Crp bends DNA upon binding by the establishment of electrostatic interactions with phosphate groups flanking the DNA binding sites (Lawson *et al.*, 2004). Although DNA bending has not been demonstrated for NtcA, it is worth to note that some of the positively charged residues that in Crp participate in DNA bending (Lys_{48}, Lys_{172}, His_{205} and Lys_{207}) are conserved in NtcA. In the absence of any effectors, the K_d for NtcA-DNA binding has been described to vary between 27 nM for the site at the *glnA* promoter to 1.4 μM for the site at the *glnB* promoter of *S. elongatus* (Vázquez-Bermúdez *et al.*, 2002a). However, since that study was limited to a few DNA sites it is possible that the range of K_d is broader.

Some NtcA-regulated genes carry non-consensus NtcA binding sites in their promoters. Deviation from the consensus sequence may affect the sequence of the triplets. In some cases only one of the triplets is conserved and in some others both triplets deviate from the consensus, e.g. the NtcA binding site at *nifH* and the *petH* promoters in *Anabaena* sp. PCC 7120, respectively (Valladares *et al.*, 1999). Deviation of these sequences from the consensus is expected to result in a lower affinity of NtcA for these promoters (Jiang *et al.*, 2000; Vázquez-Bermúdez *et al.*, 2002a). Some sequences in which the spacer between the GTA and TAC triplets is longer or shorter than 8 nucleotides (from 7 to 9) have also been proposed to be NtcA binding sites (Jiang *et al.*, 1995, 2000; Luque *et al.*, 2001). The increase or decrease of the spacer length is supposed to result in a lower affinity for NtcA.

Modulation of NtcA function

NtcA is a transducer of nitrogen stimuli, so it must be able to directly sense the nitrogen status or it must receive this information from an element upstream in the signal transduction pathway. In several cyanobacteria *ntcA* has been shown to be an auto-regulated gene whose expression increases in the absence of ammonium as a result of activation by its own product. Thus, *ntcA* is expressed at basal levels in the presence of ammonium and at high levels in its absence (Luque *et al.*, 1994; Ramasubramanian *et al.*, 1996; Lindell *et al.*, 1998; Muro-Pastor *et al.*, 2002). In *S. elongatus* it has been reported that forcing a high cytoplasmic concentration of NtcA by genetic manipulation is not sufficient to promote activation of its target genes under non-inducing conditions. This suggests that NtcA can exist in two interconvertible states: an inactive state in the presence of ammonium that turns into an active state upon ammonium withdrawal (Luque *et al.*, 2004). As explained above, 2-OG is the molecule signaling the C/N balance of the cell. Although NtcA exhibits an intrinsic capacity for sequence-specific DNA binding *in vitro*, 2-OG at concentrations below the millimolar range has been shown to promote a five-fold increase of NtcA affinity for its DNA site in the *S. elongatus glnA* promoter (Vázquez-Bermúdez *et al.*, 2002b). Tanigawa *et al.* (2002) have also shown that, *in vitro*, 2-OG stimulates NtcA-mediated transcriptional activity in two promoters bearing the NtcA binding site centered at −40.5 (*glnA* and *ntcA* from *S. elongatus*). All these data strongly suggest that 2-OG interacts as a cofactor with NtcA enhancing its sequence-dependent affinity for DNA and its transcriptional activity. However, direct binding of 2-OG to NtcA has not been demonstrated so far, nor is known the mechanism by which 2-OG modulates NtcA activity. Consistent with the proposed role for 2-OG on the modulation of NtcA function, the nitrate reductase levels as well as the expression of two NtcA-dependent transcriptional units, the *amt1* gene and the *nir* operon, have been reported to be stimulated in a *S. elongatus* derivative strain manipulated to express an heterologous 2-OG permease (KgtP from *E. coli*) when supplied with 2-OG in the culture medium (Vázquez-Bermúdez *et al.*, 2003). As the pool of 2-OG not only depends on nitrogen availability but also on the flow of fixed carbon, availability of carbon has also an effect on the expression of genes regulated by NtcA (Vázquez-Bermúdez *et al.*, 2003).

In addition to sensing the C/N status directly from 2-OG levels, NtcA can receive physiological information from the P_{II} protein (see next section). However, activation or repression of several (but not all) NtcA-dependent genes appear to proceed normally in a P_{II}-deficient mutant of *Synechocystis* sp. PCC 6803 (Takatani and Omata, 2006).

Transcription of some NtcA-regulated genes like *glnA*, *glnB* or *ntcA* from *Synechocystis* sp. PCC 6803 is downregulated by incubation under non-phototrophic conditions (Reyes and Florencio, 1995; García-Domínguez and Florencio, 1997; Alfonso *et al.*, 2001). This effect is counteracted if the cells are provided with glucose, that *Synechocystis* can use photoheterotrophically as a carbon source (Reyes and Florencio, 1995; García-Domínguez and Florencio, 1997). Based on these data it has been proposed that NtcA activity is regulated by the photosynthetic electron transport or by redox conditions (Alfonso *et al.*, 2001). Although these studies clearly show a correlation between the redox conditions and the transcript levels of NtcA-dependent genes, those conditions are also expected to have profound effects on carbon fixation and assimilation and, in turn, on the 2-OG levels. So, the possibility remains that the effects observed are not directly induced by the redox conditions but by the changes in 2-OG levels. This is a point that deserves further investigation. Other studies have shown that DNA binding by NtcA is enhanced *in vitro* by reducing agents like DTT at concentrations from 1 to 10 mM (Jiang *et al.*, 1997; Alfonso *et al.*, 2001). Furthermore, since treatment with thiol-oxidizing or thiol-modifying reagents impair NtcA binding to DNA, it has been proposed that the oxidation state of cysteine sulphydryl radicals in NtcA would regulate DNA binding (Jiang *et al.*, 1997; Alfonso *et al.*, 2001). There is a single cysteine conserved in all NtcA sequences available located close to the N-terminus of the HTH motif, whereas a second cysteine is conserved in filamentous heterocyst-forming cyanobacteria. Modeling of the *Anabaena* sp. PCC 7120 NtcA structure does not support the formation of an intramolecular disulphide bond between both cysteines (Wisén *et al.*, 2004a). Besides, mutation of one or both cysteines have a minor effect on NtcA affinity for DNA and the resulting protein still responds to DTT treatment (Wisén *et al.*, 2004a). Further work would be required to elucidate the molecular basis for the effects of DTT on NtcA-DNA binding observed *in vitro*.

Structure of NtcA-regulated promoters

The structure of the NtcA-activated promoter was first described for some NtcA-dependent genes in *S. elongatus*. The canonical NtcA-activated promoter is composed of an NtcA-binding box centered around −40.5 with respect to the transcription start point (tsp) and a −10 Pribnow-like box with the consensus sequence TAN_3T located about 22 bp downstream (Luque *et al.*, 1994). A large and increasing number of genes in different cyanobacteria have been described to bear canonical NtcA-activated promoters (see Herrero *et al.*, 2001). Recent searches for genes bearing canonical NtcA-activated promoters with bioinformatic tools (Muro-Pastor, A.M. *et al.*, 2001; Su *et al.*, 2005; Su *et al.*, 2006) have largely increased the number of registered genes presumably activated by NtcA. Genes identified in these studies oscillate between 16 in *P. marinus* MIT9313 and 106 in *Anabaena* sp. PCC 7120, although experimental demonstration for regulation by NtcA remains to be established for most of them. The structure described above (Figure 13.5A) fits well with that of Class II Crp-activated promoters of *E. coli* (Herrero *et al.*, 2001; Lawson *et al.*, 2004). In these promoters, the activator binds around position −41.5 overlapping, and in most cases replacing, the −35 region for RNA polymerase binding. Activation is established by three contacts between the activator and the α and σ subunits of the RNA polymerase which provoke recruitment of the RNA polymerase holoenzyme (i.e. occupation of the promoter by the activator increases the affinity of the RNA polymerase for the promoter), and a post-recruitment effect (i.e. activator-RNA polymerase contacts favor the formation of the promoter open complex).

Some NtcA-activated promoters have been described to bear NtcA binding sites upstream of position −41.5 not matching the structure of Class II, but matching that of Class I regulated bacterial promoters (Figure 13.5B). Examples for such promoters can be found in recent reviews (Herrero *et al.*, 2001; Herrero *et al.*, 2004) (Figure 13.5B). In the Class I promoters,

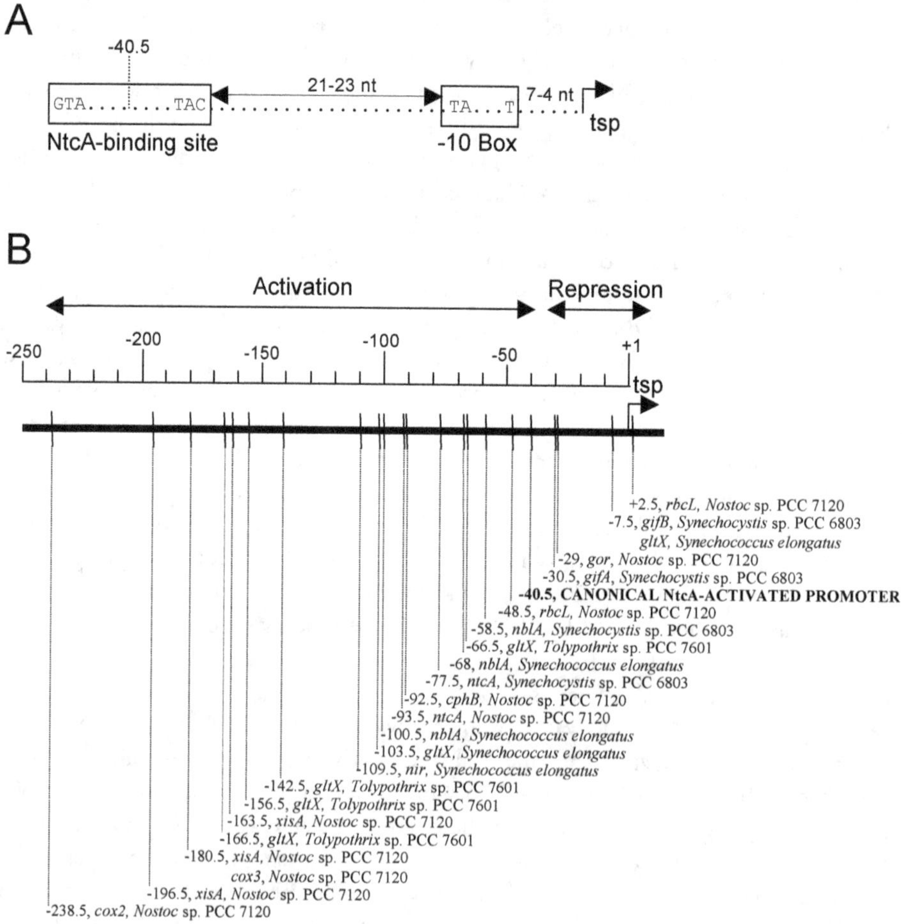

Figure 13.5 (A) Structure of the canonical NtcA-activated promoter. A vertical dotted line indicates the symmetry axis of the NtcA-binding site with respect to the tsp. A bent arrow indicates the position of the tsp. (B) Diagram of the position of NtcA binding sites in cyanobacterial promoters. Vertical lines indicate the position for the symmetry axis of each binding site. Numbers below indicate the position of the symmetry axis with respect to the tsp and are followed by the name of the corresponding gene and organism.

activation is dependent on a single contact between the activator and RNA polymerase and proceeds through a polymerase recruitment mechanism (Lawson *et al.*, 2004).

In the case of NtcA-repressed promoters, NtcA sites are centered downstream of position −40.5 (Figure 13.5B). Binding of NtcA to those sites is not productive for transcription activation. Furthermore, NtcA-binding sites in repressed promoters usually overlap with the −35 or −10 boxes, or even with the tsp, so that bound NtcA blocks the access of RNA polymerase to the promoter. Repressor NtcA bindings sites have been mapped centered at positions −30.5, −29, −7.5 or +2.5. Some genes like *gifA* and *gifB* of *Synechocystis* sp. PCC 6803 (García-Domínguez *et al.*, 2000), *gor* and *rbcL* of *Anabaena* sp. PCC 7120

(Ramasubramanian *et al.*, 1994; Jiang *et al.*, 1995), *rbcL* and *gltX* of *S. elongatus* (Luque *et al.*, 2002; Aldehni *et al.*, 2003) have been reported to be repressed by NtcA. Transcriptional repression may also occur through inhibition of promoter clearance when the repressor binds downstream of RNA polymerase (Rojo, 1999). This might be the mechanism for repression of the *hanA* gene of *Anabaena* sp. PCC 7120 in which an NtcA-binding site has been mapped centered at position +68 (Khudyakov and Wolk, 1996). The HanA protein is absent from heterocysts and its transcript has been detected in ammonium-grown cells, what is consistent with repression by NtcA (Nagaraja and Haselkorn, 1994; Khudyakov and Wolk, 1996). However, involvement of NtcA in *hanA* transcriptional repression requires further investigation. Expression of some NtcA-regulated genes may take place from multiple promoters including NtcA-controlled and NtcA-independent promoters, the latter ones allowing gene transcription under non-inducing conditions for NtcA. This is for instance the case of the *ntcA*, *glnA* and *glnB* genes in several cyanobacteria (Luque *et al.*, 1994; Reyes *et al.*, 1997; Lee *et al.*, 1999; García-Domínguez *et al.*, 2000) or *petH* from *Anabaena* sp. PCC 7120 (Valladares *et al.*, 2004).

Complex promoters

Some other NtcA-dependent promoters have a complex architecture due to the existence of binding sites for additional regulators and/or multiple sites for NtcA binding. In some promoters of the later group NtcA may act both as an activator and as a repressor. Conjoint activation of bacterial promoters by several regulators allows for synergistic activation of transcription by multiple activator-RNA polymerase contacts and for integration of multiple environmental signals (Barnard *et al.*, 2004). Promoters controlled by NtcA and other regulators like NtcB or NblR are described in section "Other regulators related to nitrogen metabolism."

Among the genes bearing a complex promoter with multiple NtcA binding sites are *xisA* and *rbcL* of *Anabaena* sp. PCC 7120 (Ramasubramanian *et al.*, 1994; Olmedo-Verd *et al.*, 2005), *gltX* of *S. elongatus* and *Tolypothrix* sp. PCC 7601 (Luque *et al.*, 2002; 2006), and *nirA* of *S. elongatus* (Luque *et al.*, 1994; Maeda *et al.*, 1998). One promoter has been described in which NtcA acts both as an activator and as a repressor. In *S. elongatus*, the *gltX* gene encodes the glutamyl-tRNA synthetase, an enzyme playing a key role in protein and pigment synthesis. Expression of *gltX* is regulated by NtcA but its regulatory pattern is unique among the NtcA-regulated genes. Remarkably, it is the only gene known to date to be activated by NtcA in the presence of ammonium. The *gltX* promoter bears three NtcA binding sites centered at –103.5, –40.5 and –7.5 with respect to the tsp. The two upstream sites are putative activator sites and show high affinity for NtcA whereas the one centered at –7.5 overlaps the –10 box and has low affinity for NtcA. A model where the occupancy of NtcA binding sites determines the transcription level of *gltX* has been proposed to explain NtcA regulation of *gltX*. In this model, low or intermediate concentrations of NtcA correlate with high expression of *gltX*, whereas a high concentration of NtcA determines repression of *gltX* (Luque *et al.*, 2002).

The NtcA regulon

The NtcA regulon includes genes involved in almost every step in nitrogen assimilation pathways (Herrero *et al.*, 2001; Flores and Herrero, 2005) (Figure 13.3). In most cyanobacteria characterized to date the genes involved in uptake of nitrogen sources including *amt1* (Montesinos *et al.*, 1998; Vázquez-Bermúdez *et al.*, 2002c), *nrtABCD* (Luque *et al.*, 1994; Frías *et al.*, 1997, 2000), *nrtP* (Sakamoto *et al.*, 1999; see also Bird and Wyman, 2003), and *urtABCDE* (Valladares *et al.*, 2002) are regulated by NtcA. All these genes are subjected to metabolic repression since they are involved in scavenging traces of ammonium or alternative nitrogen sources when ammonium is absent or scarce.

Other genes activated by NtcA are those involved in the intracellular generation of ammonium from oxidized nitrogen sources like *nirA*, *narB* (Luque *et al.*, 1994; Frías *et al.*, 1997; Sakamoto *et al.*, 1999; Frías *et al.*, 2000) and the *nifHDK* operon (Frías *et al.*, 1994; Ramasubramanian *et al.*, 1994). In some cyanobacteria, probably the *ure* operons, involved in the utilization of urea as a nitrogen source, are also under the control of NtcA (Collier *et al.*, 1999; Palinska *et al.*, 2000). Further metabolism of ammonium implies its incorporation to carbon skeletons through the GS/GOGAT cycle. Genes *glnA* and *glnN* encoding glutamine synthetase type I and type III, respectively, are activated by NtcA in several cyanobacteria (Frías *et al.*, 1994; Luque *et al.*, 1994; Wei *et al.*, 1994; Reyes *et al.*, 1997; Crespo *et al.*, 1998; Sauer *et al.*, 2000). The *icd* gene encoding isocitrate dehydrogenase which generates 2-OG to feed the GS/GOGAT cycle has been reported to be activated by NtcA in *Synechocystis* sp. PCC 6803 (Muro-Pastor *et al.*, 1996). The glutamate synthesized in the GS/GOGAT cycle is loaded to the corresponding tRNAs by the glutamyl-tRNA synthetase, encoded by *gltX*, which is under NtcA control in unicellular and filamentous cyanobacteria (Luque *et al.*, 2002, 2006). Since NtcA participates in transduction of signals of the C/N balance, it is not surprising that the *rbcL* gene encoding the large subunit of ribulose-1,5-bisphosphate carboxylase/oxygenase (Rubisco) involved in carbon fixation through the Calvin cycle is also under the control of NtcA. The *rbcL* gene has been reported to be regulated by NtcA in *S. elongatus* (Aldehni *et al.*, 2003) and *Anabaena* sp. PCC 7120 (Ramasubramanian *et al.*, 1994). Two recent publications based on global approaches have shown that many genes involved in photosynthesis also contain putative NtcA-regulated promoters suggesting a tight coordination between photosynthesis and nitrogen regulation mediated by NtcA (Su *et al.*, 2005, 2006).

Apart from its own gene, NtcA also modulates the expression of regulatory genes involved in the control of particular aspects of nitrogen metabolism, (Luque *et al.*, 1994; Ramasubramanian *et al.*, 1996; Lindell *et al.*, 1998; Lindell *et al.*, 2002). These include *glnB* (García-Domínguez and Florencio, 1997; Lee *et al.*, 1999), *ntcB* (Maeda *et al.*, 1998; Frías *et al.*, 2000; Aichi *et al.*, 2001), *gifA* and *gifB* (García-Domínguez *et al.*, 2000), and other regulatory genes involved in the control of heterocyst differentiation like *hetR* (Muro-Pastor *et al.*, 2002), *devH* (Hebbar and Curtis, 2000) or *nrrA* (*all4312*) encoding a response regulator (Ehira and Ohmori, 2006; Muro-Pastor *et al.*, 2006). Heterocyst differentiation and function is a process linked to nitrogen metabolism (see Herrero *et al.*, 2004 for a recent review; see also Chapter 14). *hetR* encodes a key regulator of heterocyst differentiation whose expression is early activated during differentiation and depends on NtcA, although no NtcA binding site has been mapped in its promoter. *ntcA* is also early upregulated during differentiation and interestingly its upregulation requires *hetR*. Thus, in *Anabaena* sp. PCC 7120, the *ntcA* and the *hetR* regulatory genes exhibit a mutual dependence for expression (Muro-Pastor *et al.*, 2002). *hetR* upregulation precedes that of *ntcA*, so that the basal levels of NtcA present at the beginning of heterocyst differentiation seem to be sufficient to activate *hetR*, whose product would then mediate *ntcA* upregulation. In diazotrophic conditions, NtcA plays a double role since it is required for activation of some genes in vegetative cells as well as for activation of heterocyst-specific genes. How heterocyst-specific genes are not activated by NtcA in vegetative cells could be explained by the requirement of heterocyst-specific regulators or the requirement of the elevated NtcA levels transiently observed during heterocyst maturation. It has been shown that some heterocyst-specific genes like *devH* or *xisA* require high levels of NtcA for activation whereas others like *coxB2* and *nifH* require additional factors (Olmedo-Verd *et al.*, 2005). Genes involved in cyanophycin synthesis and degradation are also controlled by NtcA in *Anabaena* sp. PCC 7120 (Picossi *et al.*, 2004).

NtcA also regulates other processes related to nitrogen conditions but not necessarily involved in nitrogen metabolism. For instance, cyanobacteria degrade their phycobilisomes in response to nutrient stress conditions or high light (described in section "Nitrogen specific and general stress responses"). In the particular case of nitrogen stress, NtcA participates in the activation of the *nblA* gene which plays an essential role in phycobilisome degradation (Sauer *et al.*, 1999; Luque *et al.*, 2001).

Paralogs of ntcA

Several studies have characterized regulatory proteins that are homologous but do not play the same role as NtcA, including the SYCRP1 protein from *Synechocystis* sp. PCC 6803, which is essential for pilus motility (Yoshimura *et al.*, 2002), CysR and IdiB from *S. elongatus* which are involved in the regulation of genes in response to sulfur and iron stimuli, respectively (Nicholson and Laudenbach, 1995; Michel *et al.*, 2001), and DevH from *Anabaena* sp. PCC 7120, which is required for the synthesis of the heterocyst glycolipid layer (Hebbar and Curtis, 2000; Ramirez *et al.*, 2005). A survey of the cyanobacterial genomes sequenced to date show that a cyanobacterium like *Anabaena* sp. PCC 7120 may have up to seven genes encoding proteins homologous to NtcA. These proteins also belong to the Crp/FNR family of transcriptional regulators and are predicted to bear a HTH motif in their C-terminal domain. Phylogenetic analyses (Figure 13.4B) show that proteins annotated in the cyanobacterial genomes as NtcA form a coherent cluster with two major branches, one including NtcA proteins from the *Synechococcus/Prochlorococcus* cluster of marine cyanobacteria and another including the remaining NtcA proteins. Other cyanobacterial proteins of the Crp/FNR family are out of the NtcA cluster. Deep branching for these other proteins suggests ancient gene duplication events. Sequence conservation between the HTH motifs of NtcA and other cyanobacterial proteins of the Crp/FNR family allow the prediction that these protein bind to similar DNA sequences (Figure 13.4A). Indeed, the DNA binding sequences for some of these proteins have been reported to be similar to the NtcA binding site (TGTGAN$_6$TCACA for the *Synechocystis* sp. PCC 6803 SYCRP1, and GTGTGCTGGCACAC for the *S. elongatus* IdiB). Data on NtcA mutants and mutants of other cyanobacterial proteins of the Crp/FNR family suggest that functional redundancy does not occur. Whether crosstalk between the cyanobacterial regulators of the Crp/FNR family may occur under exceptional conditions deserves further investigation.

P_{II}-mediated regulation

General properties of P_{II} signaling proteins and their occurrence in cyanobacteria
The P_{II}-like signaling proteins constitute one of the most widely distributed families of signal transduction proteins, whose members are found in bacteria, archaea and plants (for reviews, see Arcondeguy *et al.*, 2001; Forchhammer, 2004; Ninfa and Jiang, 2005). The P_{II} proteins are pivotal regulators in bacterial nitrogen assimilatory processes, where they are engaged in sensing and integrating signals of the central metabolic state of the cells. P_{II} transmits this information to P_{II}-receptors, thereby coordinating various processes of nitrogen assimilation at different levels of regulation (transcriptional, posttranslational).

The P_{II} proteins are trimers of 12–13 kDa subunits with a highly conserved three-dimensional structure, as revealed from X-ray structural analysis of various proteobacterial and non-proteobacterial P_{II} proteins, including two cyanobacterial P_{II} proteins (*Synechococcus elongatus* and *Synechocystis* sp. PCC 6803; Xu *et al.*, 2003). The three subunits form a squat barrel made up of a central core of antiparallel beta-sheets and laterally arranged α-helices and loop structures. The largest of these loops (termed T-loop) protrudes from the protein. It seems to be flexible in solution and is of particular importance for P_{II}-receptor interactions (Xu *et al.*, 2003).

There are two basic modes of signal perception by P_{II}. The first one involves the binding of the effector molecules ATP (ADP) and 2-OG to P_{II}. The adenine nucleotide binding sites of various P_{II} proteins have been resolved by crystallographic analysis (see e.g. Xu *et al.*, 2001; Benelli *et al.*, 2002; Sakai *et al.*, 2005) and are located in the clefts between the subunits of the trimers, with the ligand making contacts with residues from opposing subunits. This architecture results in three identical binding sites in the trimeric protein. The ATP binding sites of the P_{II} protein from *S. elongatus* exhibit negative cooperativity: high affinity binding to one site reduces affinity to the other sites, and this may enable sensing of the cellular energy status (Forchhammer and Hedler, 1997). In addition to binding adenine nucleotides, P_{II} proteins also bind 2-OG, with the binding of ATP and 2-OG mutually stimulating each other (Forchhammer, 2004). In the cyanobacterial and plant P_{II} proteins, which are closely related, 2-OG binding can only be detected in the presence of ATP. The three 2-OG binding sites exhibit negative cooperativity, allowing the P_{II} proteins to detect 2-OG in a wide range of concentrations (Forchhammer, 2004; Ninfa and Jiang, 2005).

A second mode of signal perception occurs by covalent modification of P_{II} proteins at the apex of the exposed T-loop. Although covalent modification is not a general property of P_{II} signaling, in those cases where it occurs, it can integrate additional signals into the P_{II} signaling process. In the case of proteobacteria, modification of P_{II} proteins occurs by reversible uridylylation of a conserved tyrosyl-residue (Tyr51). A distinct type of P_{II} modification has been demonstrated in unicellular cyanobacteria, where a P_{II} protein is phosphorylated and dephosphorylated in a seryl residue (Ser49) located in the vicinity of the conserved Tyr51, as described in more detail in section "P_{II}-regulated processes."

Effector molecule binding and covalent modification result in different signal input states. Depending on these modifications, the P_{II} binds to different P_{II} receptors and modifies the activity of these target molecules (see Figure 13.6, which refers to cyanobacterial

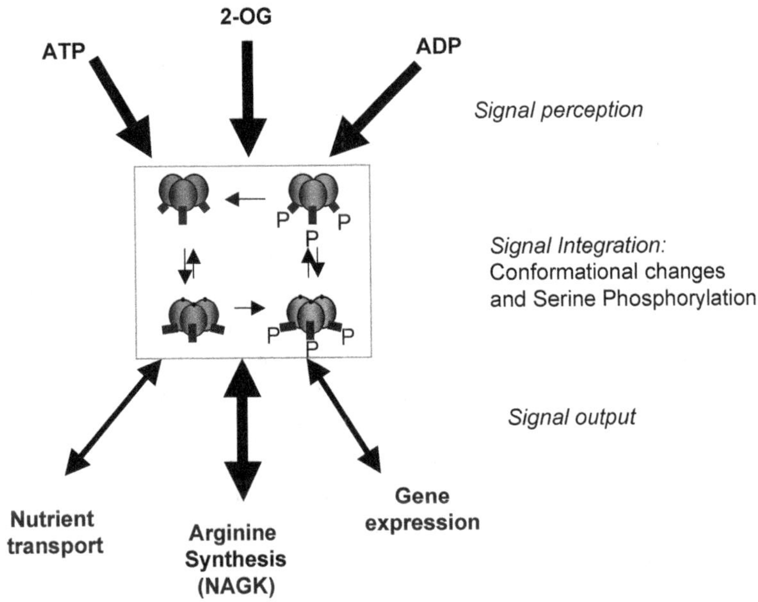

Figure 13.6 Schematic representation of the P_{II} signal integration mode operated in *S. elongatus*. Effector molecules indicating the energy (ATP/ADP) and C/N (2-OG) status of the cells are bound by P_{II} (ligands symbolized by black dots). Ligand binding may result in conformational changes of the P_{II} protein and thereby control phosphorylation/dephosphorylation of Ser49 located at the apex of a solvent exposed large loop structure (T-loop). Depending on Ser49 phosphorylation and ligand binding, P_{II} binds to various receptors, thereby controlling various cellular processes involved in nitrogen metabolism.

P_{II} targets). In most cases, these targets are key regulatory proteins, metabolic enzymes or proteins involved in nutrient transport.

The P_{II}-like signaling proteins can be classified into three subfamilies, encoded by the *glnB*, *glnK* and *nifI* genes (Arcondeguy *et al.*, 2001). All cyanobacteria analyzed so far contain homologues of the *glnB* subfamily, usually one *glnB* homologue per genome, whereas *glnK* or *nifI* homologues are absent. The occurrence of *glnB* is particularly intriguing in the species *P. marinus*, which exhibits the smallest genome of oxygenic photosynthetic organisms known to date. It lacks many genes involved in nutrient transport, intermediary metabolism, motility or acclimation processes. The conservation of a copy of *glnB* in this species indicates essential functions of P_{II} for a "minimal cyanobacterium." Genomic analysis further revealed cyanobacterial-like *glnB* genes in plants and red algae. The plant *glnB* genes are nuclear, whereas it is plastidial in the red algae *Porphyra purpurea*. The plant P_{II} proteins contain N-terminal sequences for translocation of the protein into the chloroplast (Smith *et al.*, 2003).

Comparison of cyanobacterial GlnB sequences with other GlnB and GlnK sequences identified 14 amino acids unique to the cyanobacterial sequences that constitute a cyanobacterial GlnB signature (Palinska *et al.*, 2002). Most of these amino acids, including

Ser49, are within or near the T-loop region. Interestingly, the plant GlnB homologues have conserved 5 of these 14 signature residues, including Ser49, the site of phosphorylation of *S. elongatus* GlnB.

Signal perception by P_{II} in unicellular cyanobacteria

The mechanism of P_{II} signal perception and integration has been studied in detail in the unicellular strains *S. elongatus* and *Synechocystis* sp. PCC 6803 through biochemical analyses of the protein and physiological studies of wild-type and mutant strains (reviewed in Forchhammer 2004). In these strains, P_{II} modification operates by phosphorylation of a seryl-residue (Ser49). However, it seems that phosphorylation of P_{II} proteins is absent in *Prochlorococcus* strains (Palinska *et al.*, 2002). In the presence of ammonium, P_{II} is almost completely non-phosphorylated. In nitrate grown cells, the phosphorylation state strongly depends on the growth conditions, in particular on the CO_2 supply to the cells (with low CO_2 favoring the non-modified state and *vice versa*), and to a lesser extent on the photon flux density (PPFD) (with increasing PPFD favoring P_{II} phosphorylation) (Forchhammer, 1999; Kloft and Forchhammer, 2005). Experiments with inhibitors of the GS-GOGAT cycle showed that the responses towards the C/N conditions require ammonia assimilation, and that P_{II} phosphorylation parallels the responses of nitrogen control (see above). Biochemical and physiological investigations revealed 2-OG as the main effector molecule controlling the P_{II} phosphorylation status (Forchhammer, 2004).

As detailed above, binding of 2-OG to *S. elongatus* P_{II} requires ATP, and conversely, the binding of ATP is stimulated by 2-OG. When P_{II} is fully liganded with ATP and 2-OG, each subunit is phosphoryalted at Ser49 by a poorly characterized kinase activity. As long as the effector molecules stay bound to phosphorylated P_{II} (P_{II}-P), it resists dephosphorylation. However, non-liganded (or only partially liganded) P_{II}-P is dephosphorylated by a phosphatase of the PP2C family, termed PphA, which was identified in *Synechocystis* sp. PCC 6803. Dephosphorylation of P_{II}-P by PphA is inhibited in a concentration dependent manner by 2-OG in the range between 10 and 100 μM, as long as saturating ATP concentrations are present (Ruppert *et al.*, 2002). The amount of PphA in *Synechocystis* cells depends on the growth conditions: maximum levels of PphA are observed in nitrate- or nitrite-grown cells, and the accumulation of PphA is post-transcriptionally regulated by nitrite (Kloft *et al.*, 2005). The role which ATP or ADP may play as a signal perceived by P_{II} is not entirely clear (Forchhammer *et al.*, 2004). However, a P_{II} regulated process has been discovered (the P_{II}-NAGK interaction, see below), which is sensitive to different ATP/ADP levels, pointing towards a role of the energy status, at least for some P_{II} regulated processes.

P_{II}-regulated processes

The regulatory function of P_{II} proteins is achieved by specific protein-protein interactions between P_{II} and regulatory or metabolic key enzymes, the so-called targets of P_{II} signaling. P_{II}-targets may respond to the ligand binding status of P_{II}, to the phosphorylation status or to both states (Figure 13.6). Three different cellular processes have been recognized as P_{II}-regulated in cyanobacteria: nitrate/nitrite transport, NtcA-dependent gene expression and arginine biosynthesis (Figure 13.3 and 13.6).

Nutrient uptake

Phenotypical analysis of P_{II}-deficient mutants has identified nitrate/nitrite transport to be P_{II}-regulated (Lee *et al.*, 1998; Lee *et al.*, 2000). In many cyanobacteria, nitrate transport ceases rapidly upon the addition of ammonium or removal of CO_2. In the unicellular strains *S. elongatus* and *Synechocystis* sp. PCC 6803, this regulation was shown to be P_{II} dependent. From the analysis of strains carrying different mutated versions of P_{II} mimicking non-phosphorylated or phosphorylated P_{II}, the non-modified form of P_{II} appears to be inhibitory for nitrate transport. In contrast, the forms mimicking phosphorylated P_{II} are inhibitory or non-inhibitory, depending on whether the cells are incubated in the presence or absence of ammonium, respectively (Lee *et al.*, 2000). This means that (with respect to nitrate transport) the (pseudo)phosphorylated form of P_{II} is able to switch between an inhibitory and a non-inhibitory form, likely depending on the concentration of the effector molecule 2-OG. Consistently, a P_{II}-phosphatase deficient strain (in which P_{II} remains phosphorylated upon ammonium addition) shows normal regulation of nitrate transport in response to ammonium (Kloft and Forchhammer, 2005).

Regulation of nitrate transport in *Synechocystis* sp. PCC 6803 not only prevents uneconomical nitrate utilization in the presence of ammonium, but also adjusts nitrate utilization to the availability of reductant to avoid wasteful nitrite formation under light-limiting conditions. This control depends on the phosphorylation state of P_{II} and is thus impaired in a P_{II} phosphatase-deficient strain (Kloft and Forchhammer, 2005). Apparently, nitrate uptake is under dual control by P_{II} in *Synechocystis* sp. PCC 6803: independent of P_{II}-dephosphorylation in response to ammonium and depending on P_{II} dephosphorylation in response to the availability of reductant. The molecular target(s) that mediate P_{II} control over nitrate transport have so far not been resolved and attempts to reveal binding of P_{II} to a component of the NrtABCD transport apparatus have been unsuccessful.

Arginine and cyanophycin synthesis

Yeast-two hybrid screening with P_{II} as bait identified the enzyme N-acetyl-L-glutamate-kinase (NAGK) as a potential P_{II} target (Burillo *et al.*, 2004; Heinrich *et al.*, 2004). NAGK catalyzes the first committed step in arginine biosynthesis in cyanobacteria, which is regulated by arginine feed-back inhibition at the level of enzyme activity. *Synechococcus* P_{II} interacts with NAGK and strongly activates NAGK catalytic activity (Heinrich *et al.*, 2004; Maheswaran *et al.*, 2004). Moreover, complex formation relieves arginine feedback inhibition of NAGK almost 10-fold. Only non-phosphorylated P_{II} is able to interact with NAGK. This interaction is further sensitive to P_{II} ligands: ATP in synergy with 2-OG strongly impairs complex formation, whereas these metabolites added individually have no effect. ADP is also a strong inhibitor of P_{II}-NAGK complex formation, and its effect is independent of 2-OG (Maheswaran *et al.*, 2004). Thus it seems that *in vivo* binding of P_{II} to NAGK only occurs when P_{II} is not phosphorylated and when the 2-OG and ADP levels are below a critical concentration.

The physiological significance of this regulation could be shown in *Synechocystis* sp. PCC 6803. When shifted from nitrogen-deprived to ammonium-excess conditions (conditions triggering P_{II} dephosphorylation), *Synechocystis* cells synthesize the nitrogen storage polymer cyanophycin (multi-L-arginyl-poly-L-aspartic acid) from arginine and aspartate.

Following ammonium-promoted P_{II}-dephosphorylation, P_{II} interacts with NAGK and activates its catalytic activity, resulting in a strong increase in arginine biosynthesis. The cellular arginine level increases highly, and the excess of arginine is stored as cyanophycin (Maheswaran *et al.*, 2006). The regulation of NAGK activity by P_{II} provides the mechanistic basis for the dual role of arginine as an amino acid used for protein synthesis and as a nitrogen buffer for storing excess nitrogen in form of cyanophycin: in the non-activated state, NAGK is tightly arginine-controlled, its activity being sufficient to produce arginine for the purpose of protein synthesis. In the P_{II}-activated state, excess assimilated nitrogen flows into the arginine pathway generating an arginine pool able to feed both protein synthesis and cyanophycin production.

NAGK-P_{II} interaction seems to be conserved in the eukaryotic oxygenic photosynthetic organisms (Burillo *et al.*, 2004). In *Arabidopsis thaliana* P_{II} interacts with NAGK and relieves NAGK from feedback-inhibition by arginine in a similar way as in cyanobacteria (Chen *et al.*, 2006). Furthermore, P_{II} has been shown to control arginine biosynthesis in *Arabidopsis* plastids (Ferrario-Mèry *et al.*, 2006). Since the Ser49 seems to be crucial for P_{II}-NAGK interaction, the conservation of this residue in plant P_{II} proteins could reflect conservation of this interaction, rather than conservation of P_{II} phosphorylation.

Nitrogen-controlled gene expression

Nitrogen control over gene expression is operated by NtcA, the transcription factor of Nitrogen control (see section "NtcA-mediated regulation"). P_{II} appears to have antagonistic effects on the activity of NtcA-mediated gene expression: under conditions of nitrate-dependent growth, NtcA is more active in the absence of P_{II}, such that P_{II} appears to depress NtcA (Aldehni *et al.*, 2003; Aldehni and Forchhammer, 2006). By contrast, maximum activation of NtcA under conditions of nitrogen starvation requires a functional P_{II} protein (Aldehni *et al.*, 2003; Paz-Yepes *et al.*, 2003; Aldehni and Forchhammer, 2006). A strain carrying a Ser49 to Ala mutant of P_{II} exhibits intermediate levels of NtcA-dependent gene transcripts, suggesting that P_{II} has to be phosphorylated to achieve NtcA activation (Paz-Yepes *et al.*, 2003). However, the mechanism of P_{II} control over NtcA activity remained elusive, since attempts to detect binding of P_{II} to NtcA were unsuccessful (Espinosa *et al.*, 2006a).

Recent studies aimed at the identification of potential P_{II} targets by yeast-two hybrid screening with *S. elongatus* genomic libraries revealed the protein PipX (P_{II}-interacting protein X), which is able to interact with both, P_{II} and NtcA. PipX is a small monomeric protein of 89 amino acids, conserved in all cyanobacterial genomes but lacking homologues outside the cyanobacterial phylum. PipX interaction with P_{II} or NtcA is antagonistically regulated by 2-OG: in the presence of low 2-OG, it binds preferentially to P_{II}, whereas increased 2-OG levels favor binding to NtcA (Espinosa *et al.*, 2006a). In a PipX-deficient mutant, activation of some NtcA-activated genes is severely impaired, indicating that PipX helps NtcA to activate some promoters under conditions of increased 2-OG. Although the mechanistic details of PipX function are not clear, these findings suggest that PipX is a regulatory link between P_{II} and NtcA.

Yeast two-hybrid screening with *Synechocystis* sp. PCC 6803 genomic libraries revealed P_{II} interaction with an integral membrane protein, termed PamA, exhibiting similarity to mechanosensitive channels (Osanai *et al.*, 2005a). *In vitro*, this interaction responded to 2-OG and ATP, implying that it is of physiological significance. Although the function of PamA is not known to date, a PamA deficient mutant displays impaired expression of some, but not all, NtcA-regulated genes. Furthermore, a group 2 sigma-factor (SigE) was also downregulated by the *pamA* mutation, and accordingly, SigE-dependent-transcripts (of sugar-catabolic genes such as *gap1*, *zwf*, and *gnd*) were decreased in this mutant. However, the mechanism by which PamA and its interaction with P_{II} affects gene expression is not yet clear.

Other regulators related to nitrogen metabolism

NirB and NtcB

In the *S. elongatus* genome two ORFs are located upstream from and divergently to the *nir* operon (Suzuki *et al.*, 1995). Transcription of these two genes, named *nirB* and *ntcB*, is activated by NtcA and mutants of any of them show reduced growth in medium with nitrate as a nitrogen source, evidencing their involvement in nitrate assimilation. In nitrate supplemented medium, the *nirB* mutant exhibits WT levels of the *nir* operon transcript but reduced levels of nitrite reductase activity and excretes nitrite to the medium suggesting that NirB must act post-transcriptionally to reach maximal nitrite reductase activity (Suzuki *et al.*, 1995). *nirB* is not widespread in cyanobacteria and *ntcB* is only present in those cyanobacterial genomes that contain the *nrtABCD* genes for nitrate/nitrite transport (Su *et al.*, 2005) and *N. punctiforme* (Table 13.1).

NtcB is a LysR-type transcriptional regulator that binds to a palindromic sequence, ATGCN_7GCAT, in the *nirA* promoter of *Anabaena* sp. PCC 7120 (Frías *et al.*, 2000). A sequence TGCAN_5TGCA resembling the NtcB binding sequence mapped in *Anabaena* mediates nitrate responsiveness of the *S. elongatus nirA* promoter and has been proposed to be a NtcB binding site (Maeda *et al.*, 1998). NtcA/NtcB-activated promoters are thus composed of a −10 box, an NtcA binding site centered around the −40.5 and an NtcB binding site centered around the −70 or further upstream (Frías *et al.*, 2000; Maeda *et al.*, 1998; Aichi *et al.*, 2001). This is compatible with NtcA and NtcB acting as Class II and Class I activators, respectively. Taking into account that *ntcB* is activated by NtcA, it has been proposed that NtcB acts as a booster of NtcA function upon induction. However, some differences in NtcB function have been observed in different cyanobacterial species. A recent study has highlighted the importance of NtcB in *S. elongatus*, where it has been shown to confer a selective advantage under a limiting supply of nitrate, conditions which are frequent in natural ecosystems (Aichi *et al.*, 2004).

CnaT

In *Anabaena* sp. PCC 7120 *cnaT* is located adjacent to *ntcB* but it is transcribed independently and constitutively. This gene encodes a putative glycosyl transferase that is required for transcription of the *nir* operon. The *cnaT* mutant is unable to assimilate nitrate or nitrite and, as a result, undergoes heterocyst development in nitrate-containing medium. It has been suggested that CnaT and NtcB are both elements of a single signal transduction pathway (Frías *et al.*, 2003). Interestingly, a *cnaT* mutant of *S. elongatus* PCC 7942 exhibits normal growth in nitrate-containing medium and normal levels of *nir* operon expression

(Maeda and Omata, 2004). It is therefore possible that these two homologous genes play different roles in *S. elongatus* and *Anabaena* sp. PCC 7120.

NblR

NblR is a response regulator of the PhoB/OmpR family that has been characterized in *S. elongatus* (Schwarz and Grossman, 1998). NblR is not a nitrogen-specific regulator, since it is involved in the response to general nutrient stress as well as high-light stress. The only gene known so far to be activated by NblR is *nblA* (Collier and Grossman, 1994; Schwarz and Grossman, 1998; Luque *et al.*, 2001). The NblA protein is involved in the degradation of phycobilisomes under various stress conditions (described in section "Regulation of phycobilisome degradation"). In *S. elongatus*, there is a single *nblA* gene that is transcribed from a complex promoter with at least five different tsps. Only two of these tsps, named *tspI* and *tspII*, are inducible by nutrient stress and are responsible for > 99% of total transcripts. Of these, *tspII* is strongly regulated by NtcA and NblR. The promoter corresponding to *tspII* is composed of −10 and −35 boxes for binding of the RNA polymerase and two NtcA binding sites centered at positions −68 and −100.5 (Luque *et al.*, 2001). Although NblR binding to a DNA fragment of the *nblA* promoter has been reported, the binding sequence for this factor has not been described so far. Activation of *nblA* is partially impaired in an *ntcA* mutant under conditions of nitrogen starvation, but not under sulfur stress conditions. On the other hand, a slight residual activation of *nblA* could be observed in an *nblR* mutant, suggesting that NtcA and NblR can act independently on the *nblA* promoter (Luque *et al.*, 2001). Accordingly, bleaching and cell survival is not only impaired in the *nblR* mutant, but also in an *ntcA* mutant under nitrogen starvation conditions (Sauer *et al.*, 1999; Osanai *et al.*, 2005b). The elevated *nblA* transcript levels observed in the wild type under nitrogen deprivation must result from synergistic activation by NblR and NtcA (Luque *et al.*, 2001).

Sigma factors

The cyanobacterial RNA polymerase holoenzyme is constituted by the core, a pentamer named $\alpha_2\beta\beta'\gamma$ and one of the multiple sigma factors mediating the interaction of the enzyme with specific promoter sequences in DNA. Sigma factors are classified in three distinct groups: group 1 comprises the so-called principal sigma factors that are responsible for the transcription of housekeeping genes and are essential for cell growth. Sigma factors of group 2 are similar in sequence and structure to the principal sigma factor but are not essential. Group 3 sigma factors are also non-essential and not related in sequence or structure to those of groups 1 and 2.

In the cyanobacterial genomes sequenced to date multiple genes putatively encoding group 2 and 3 sigma factors are present, some of which are differentially expressed depending on nitrogen conditions. In *Anabaena* sp. PCC 7120, the *sigB* and *sigC* genes encoding group 2 sigma factors are transiently expressed upon nitrogen starvation. However, single and double mutants of these genes are capable of heterocyst differentiation and diazotrophic growth (Brahamsha and Haselkorn, 1992). The group 2 sigma factors SigD, SigE and SigF are not induced by nitrogen deprivation, but single mutants of these factors are transiently impaired in diazotrophic growth and the *sigD sigE* double mutant is unable of diazotrophic growth and heterocyst differentiation. These data indicate that several group

2 sigma factors contribute to the expression of genes involved in heterocyst differentiation and nitrogen fixation (Khudyakov and Golden, 2001).

In *Synechococcus* sp. PCC 7002, expression of *sigB* is induced by advanced C or N starvation whereas expression of *sigC* (corresponding to *sigE* in other cyanobacteria) exhibits a transient increase (Caslake *et al.*, 1997). In *Synechocystis* sp. PCC 6803, the gene *rpoD2-V* encoding a group 2 sigma factor (RpoD2-V or SigE), bears an NtcA-dependent promoter and its expression is induced by nitrogen starvation. SigE contributes to transcription of the NtcA-dependent *glnN* gene, which is activated under conditions of severe nitrogen stress and helps to recovery after prolonged starvation. However, SigE does not contribute to transcription of some other genes of the NtcA regulon (Muro-Pastor *et al.*, 2001). In *Synechocystis*, SigE is also involved in transcription under nitrogen-replete conditions of genes involved in sugar catabolic pathways including glycolysis, oxidative pentose phosphate pathway and glycogen degradation, suggesting a possible role for SigE in coordinating carbon and nitrogen metabolisms (Osanai *et al.*, 2005a). NtcA-mediated activation seems to occur mainly with the RNA polymerase carrying SigA, the principal sigma factor. Recently, *in vitro* and *in vivo* evidences have been presented showing that other group 2 sigma factors, SigB and SigC, contribute to transcription under nitrogen starvation conditions of NtcA activated promoters including those of *amt1*, *glnA*, *glnB* and *sigE* but not *ntcA*, which seems to be transcribed mainly by the holoenzyme carrying SigA (Asayama *et al.*, 2004; Imamura *et al.*, 2006).

Given that NtcA seems to activate transcription with the RNA polymerase carrying a variety of sigma factors including the group 1 sigma factor and some group 2 ones, the emerging picture is that the use of different sigma factors to transcribe genes of the NtcA regulon serves to integrate multiple environmental signals in addition to the signal(s) transduced by NtcA. Thus, an NtcA-dependent promoter can react not only to 2-OG (C/N balance), but also to growth phase or stress signals transduced by sigma factors. Interestingly, a number of genes encoding group 1, 2 and 3 sigma factors in various cyanobacterial genomes have been shown to harbor sequences matching the canonical NtcA-activated promoter structure in their immediate upstream sequences (Su *et al.*, 2005).

IF7, IF17

In enterobacteria, glutamine synthetase activity is regulated by post-translational modification by adenylylation. Such mechanism does not exist in cyanobacteria. However, in *Synechocystis* sp. PCC 6803 glutamine synthetase activity of GSI is rapidly inactivated by ammonium by a post-translational mechanism involving protein-protein interactions (Mérida *et al.*, 1991; García-Domínguez *et al.*, 1999). Two polypeptides, termed IF7 and IF17, that are homologous to each other interact with glutamine synthetase when ammonium is added to the culture medium. GS activity decays to levels corresponding to 30% within 30 min as a result of the interaction with IF7 and IF17. The expression of *gifA* and *gifB* genes encoding IF7 and IF17, respectively, is repressed by NtcA (García-Domínguez *et al.*, 2000). Thus, when ammonium is added to the medium, IF7 and IF17 are expressed and GS is inactivated. Genes homologous of *gifA* and *gifB* have been found in some cyanobacterial genomes but seem to be absent in *Prochlorococcus*. Most cyanobacterial strains bear genes encoding IF7-like peptides (65–68 amino acids), whereas *T. elongatus* harbors two genes encoding putative IF17-like factors. Sequence comparisons have uncover a third putative factor of the IF family in *T. elongatus* and *T. erythraeum*, showing an N-terminal part similar to that of IF7 and IF17 and a divergent C-terminal part (Muro-Pastor *et al.*, 2005). *Synechocystis* IF7 has been reported to be a natively unfolded protein, extremely sensitive to proteolysis (Muro-Pastor *et al.*, 2003). Instability of IF7 (and probably of IF17) may help to prevent GS inactivation in conditions where NtcA repression is not absolute (i.e. in the presence of nitrate). Despite the presence of IF7- and IF17-like factors in several cyanobacteria, ammonium-promoted downregulation of glutamine synthetase activity has only been found in *Synechocystis* sp. PCC 6803.

Cyanobacterial adaptations to nitrogen-depletion

Nitrogen-specific and general stress responses

Depletion of combined nitrogen sources leads to expression of nitrogenase activity in diazotrophic cyanobacteria (see above), which allows them to use molecular dinitrogen as nitrogen source. However, in non-diazotrophic strains, depletion of combined-nitrogen sources imposes a severe stress and leads to the induction of general nutrient stress responses, in addition to nitrogen-specific responses. The general nutrient-stress response involves morphological and physiological adaptations that allow the cells to survive periods of starvation (reviewed in Schwarz and Forchhammer, 2005). Lack of an essential nutrient causes cessation of anabolic reactions and, as a consequence, imposes a situation of excess excitation energy. The cells respond with a progressive deactivation of photosynthetic activities and a corresponding modification of the photosynthetic apparatus, known as chlorosis, which will be described in the next section.

Gene expression responds rapidly following the onset of nitrogen starvation. Genes whose expression is specifically regulated by nitrogen depletion can be distinguished from genes whose expression is regulated in response to multiple stress and nutrient-starvation conditions. It appears that in *S. elongatus*, all the nitrogen-starvation responding genes belong to the NtcA regulon, and their expression is impaired in both NtcA- and P_{II}-deficient cells (Aldehni *et al.*, 2003). Under nitrogen-deprived conditions, the NtcA transcription factor exhibits maximum activity. It strongly activates the NtcA dependent genes, including its own gene and the P_{II}-encoding *glnB* gene. Under nitrogen-deprived conditions, maximum activation of NtcA requires P_{II} function (see above), resulting in a positive feedback loop, which finally gives raise to increased levels of activated NtcA protein. Under those conditions, promoters with low-affinity NtcA binding sites can be induced. Among the genes that are promptly upregulated by N-starvation, genes that encode proteins involved in nitrogen assimilatory functions, such as high-affinity ammonium transporters as well as glutamine synthetase, are found. Indeed, the non-diazotrophic cyanobacteria contain a GS isoenzyme belonging to the GSIII class (encoded by *glnN* gene, see above). The promoter of the *glnN* gene contains a non-canonical NtcA binding site, which becomes activated only under the conditions of severe nitrogen-starvation (Aldehni and Forchhammer, 2006).

Genes responding to multiple stress conditions encode proteins involved in oxidative stress-defense, protection of DNA or general nutrient acquisition (like the genes encoding the outer membrane proteins SomA and SomB in *S. elongatus*) (Aldehni *et al.*, 2003). Furthermore, the genes of the photosynthetic apparatus respond to nutrient limitation, especially those that encode the light harvesting pigments organized in the phycobilisome.

Regulation of phycobilisome degradation

Limitation of nutrients slows down anabolic reactions and thus, impairs the re-oxidation of PSI-Fd_{red}. In response to this, the light reactions of photosynthesis must be tuned down. Cyanobacteria adjust the photosynthetic apparatus to nutrient-limiting conditions in a process known as chlorosis or bleaching. Chlorosis is characterized by the degradation of the phycobilisomes as well as chlorophyll *a*. Limitation of different nutrients provokes chlorotic responses, which differ in the kinetics and order of pigment degradation. Furthermore, chlorosis depends on other environmental conditions such as CO_2 supply, temperature and light intensity (Collier and Grossman, 1992; Görl *et al.*, 1998; Barker-Astrom *et al.*, 2005). In addition, different cyanobacterial strains may respond quite differently to various starvation conditions, for example, *Synechocystis* sp. PCC 6803 seems to lack a rapid response to sulfur starvation (Richaud *et al.*, 2001).

Depending on the illumination and nutrient conditions, the phycobilisomes may constitute up to 50% of soluble cellular protein. Their rapid and complete degradation, which may occur within 24 hours after an onset of N-starvation, indicates the existence of a highly effective and strictly regulated degradation machinery. Studies of mutants in *S. elongatus* with a non-bleaching (Nbl) phenotype revealed several genes involved in this process. The *nblA* gene encodes a small protein of 59 amino acids (Collier and Grossman, 1994). Its expression is highly upregulated under conditions that lead to phycobiliprotein degradation, in particular, nitrogen and sulfur starvation and high light stress, It appears that an *nblA* gene is present in all organisms containing phycobilisomes (cyanobacteria and red algae), and the corresponding gene product and their physiological functions have been characterized in various cyanobacterial species (reviewed in Schwarz and Forchhammer, 2005). In heterocystous cyanobacteria, NblA is essential for the specific phycobilisome degradation that occurs during heterocyst development. An *nblA* mutant of *Anabaena* sp. PCC 7120 develops functional heterocysts that do not degrade their phycobilisomes. This indicates that phycobilisome degradation is not essential for heterocyst development or nitrogen fixation (Baier *et al.*, 2004). In the heterocystous strain *Tolypothrix* sp. PCC 7601, an NblA homologue has been detected in vegetative cells under nitrogen sufficiency, and was found to be associated with the phycobiliproteins phycocyanin and erythrophycocyanin, indicating that in this strain the NblA protein might have additional functions than those in chlorosis (Luque *et al.*, 2003).

NblA does not exhibit homology to any other protein of known function and does not display enzymatic activity *per se*. Recently, the crystal structure of NblA from *Anabaena* sp. PCC 7120 could be resolved (Bienert *et al.*, 2005). The dimeric NblA protein forms a compact four-helix bundle. According to pull-down assays and peptide library studies, NblA specifically binds to the alpha-subunits of phycocyanin and phycoerythrocyanin, which are the main constituents of the phycobilisome rods. Together, these findings imply that upon induction of the *nblA* gene, the dimeric NblA protein binds to the periphery of the phycobilisomes and thereby triggers their degradation. Artificial induction of the *nblA* gene leads to phycobiliprotein degradation in nutrient replete *S. elongatus* cells (Collier and Grossman, 1994).

A second gene whose mutation confers an Nbl phenotype is *nblB* (Dolganov and Grossman, 1999). It is widely distributed among cyanobacteria and encodes a protein with homology to the chromophore-interacting region of phycocyanin lyases, a class of enzymes that attaches a chromophore to the apo-phycobiliproteins. It is speculated that NblB is involved in the reversal of this reaction, in particular, in removing the chromophore from the pigmented phycobiliproteins.

Other *nbl* genes described so far are involved in the control of *nblA* expression. The *nblS* gene (van Waasbergen *et al.*, 2002) encodes a homologue of sensory histidine kinases, *nblR* (Schwarz and Grossman, 1998) encodes a potential response regulator of a two-component signal transduction pathway (see above) and *nblC* encodes a bacterial antisigma factor homologue (Sendersky *et al.*, 2005). The products of *nblR*, and *nblC* are required for stress-induced expression of *nblA* in *S. elongatus*. However, the interaction of these products as well as the molecular signal(s) that trigger *nblA* expression have yet to be identified. The NblS protein (whose homologue is termed DspA or Hik33 in *Synechocystis* sp. PCC 6803; see Chapter 6) appears to be involved in acclimation to a multitude of stress conditions and a PAS domain that is present in NblS suggests that light or redox signals may be involved (van Waasbergen *et al.*, 2002). Yeast-two-hybrid screening identified a protein, termed SipA, that interacts with NblS (Espinosa *et al.*, 2006b). SipA homologues are found in all genomes that contain an *nblS* homologue, and *nblS* and *sipA* seem to have co-evolved in the cyanobacterial genomes. NblS and NblR do not seem to form a two-component regulatory system. Instead they appear to belong to converging signal transduction pathways controlling *nblA* expression (Kappell *et al.*, 2006).

In *Synechocystis* sp. PCC 6803 two *nblA* genes, termed *nblA1* and *nblA2*, form an operon which is transcribed from a single promoter bearing an NtcA-binding site centered at position −58.5 (Baier *et al.*, 2001; Richaud *et al.*, 2001). This operon is not activated under conditions of sulfur starvation (Richaud *et al.*, 2001). Furthermore, no NblR homologue has been described and the role of the NblS homologue DspA in *nblA* expression is not clear yet. Phycobiliprotein degradation is unaffected in DspA deficient mutants (Morrison *et al.*, 2005).

Survival and recovery from nitrogen stress

Following the withdrawal of combined nitrogen sources, *S. elongatus* cells complete a final cell division before arresting growth. Since CO_2-fixation continues for a while after nitrogen deprivation, fixed CO_2 is stored as glycogen, and in various species poly-hydroxybutyrate (PHB) inclusions may be formed (Hai *et al.*, 2001). Prolonged starvation leads to a gradual acclimation, which has been studied in *S. elongatus* in some detail (Görl *et al.*, 1996; Sauer *et al.*, 2001). Following the initial responses as described above, including the degradation of phycobiliproteins, which was designated "phase I" of the chlorotic response, the cells start to slow down and arrest almost all metabolic processes. During this "phase II" they tune down photosynthetic activities and degrade thylakoids and chlorophyll *a* and most of the intracellular proteins. Finally the cells are almost completely de-pigmented and reside in a dormant-like state ("phase III"). As soon as a combined N-source becomes available, these cells recover and re-initiate growth. Viability in "phase III" is maintained by a low level of both PSII and PSI activity, which represents approx. 0.1% of that in vegetatively growing cells. It was proposed that the reduction equivalents generated by PSI are consumed by reduction of oxygen to water, resulting in a pseudocyclic electron flow ("water to water cycle") around PSII and PSI (Sauer *et al.*, 2001). This type of photosynthesis would generate energy required to maintain viability and allow re-initiation of growth.

Few investigations have addressed so far the issue of recovery from nitrogen depletion. When *S. elongatus* cells are in the chlorotic state, they recover within 3 days upon addition of a combined nitrogen source (Görl *et al.*, 1996). Apparently, most of the cells in the culture are able to recover. First, the photosynthetic pigments are restored, followed by the onset of growth. *Synechocystis* sp. PCC 6803 mutants of some alternative sigma factors are unable to recover from chlorosis (Huckauf *et al.*, 2000; Muro-Pastor *et al.*, 2001). Furthermore, a GSIII-deficient mutant of *S. elongatus* shows delayed recovery (Sauer *et al.*, 2000). However, the mechanistic basis of these phenotypes is not yet understood.

Future prospects on research of cyanobacterial nitrogen metabolism

An intense investigation in the last two decades has provided a large amount of information on nitrogen metabolism and nitrogen control in cyanobacteria (compare Flores *et al.*, 1983, or Guerrero and Lara, 1987, with this review). Research during the 90ies led to the discovery of the cyanobacterial nitrogen regulators NtcA and P_{II}. Further studies allowed the characterization of their mode for C/N sensing and their respective targets for regulation. The modern trends in Biology in the 21st century attempt to a global view of the physiology of organisms. The analysis of cyanobacterial genomes often combined with experimental approaches has already provided valuable information on broad aspects of nitrogen metabolism and/or regulation in recent years (see e.g. Quintero *et al.*, 2001; Rocap *et al.*, 2003; García-Fernández *et al.*, 2004; Su *et al.*, 2005, 2006). Future global experimental approaches based on transcriptomics, proteomics and metabolomics, combined with detailed molecular analyses, will undoubtedly lead to novel insights in the functioning of nitrogen metabolic pathways and regulatory mechanisms.

Acknowledgments

The work in the author's laboratories is supported by grants BMC-2003–06686 from Ministerio de Educación y Ciencia (Spain) and grants from the Deutsche Forschugnsgemeinschaft (DFG Fo195/4) and the German-Israeli-Foundation (GIF 729).

References

Aichi, M., Maeda, S., Ichikawa, K., and Omata, T. (2004). Nitrite-responsive activation of the nitrate assimilation operon in cyanobacteria plays an essential role in up-regulation of nitrate assimilation activities under nitrate-limited growth conditions. J. Bacteriol. *186*, 3224–3229.

Aichi, M., Takatani, N., and Omata, T. (2001). Role of NtcB in activation of nitrate assimilation genes in the cyanobacterium *Synechocystis* sp. strain PCC 6803. J. Bacteriol. *183*, 5840–5847.

Aldehni, M.F., and Forchhammer, K. (2006). Analysis of a non-canonical NtcA-dependent promoter in *Synechococcus elongatus* and its regulation by NtcA and P_{II}. Arch. Microbiol. *184*, 378–386.

Aldehni, M.F., Sauer, J., Spielhaupter, C. Schmid, R., and Forchhammer, K. (2003). The Signal transduction protein P_{II} is required for NtcA-regulated gene expression during nitrogen deprivation in the cyanobacterium *Synechococcus elongatus* Strain PCC 7942. J. Bacteriol. *185*, 2582–2591.

Alfonso, M., Perewoska, I., and Kirilovsky, D. (2001). Redox control of *ntcA* gene expression in *Synechocystis* sp. PCC 6803. Nitrogen availability and electron transport regulate the levels of the NtcA protein. Plant Physiol. *125*, 969–981.

Antia, N.J., Harrison, P.J., and Oliveira, L. (1991). The role of organic nitrogen in phytoplankton nutrition, cell biology and ecology. Phycologia *30*, 1–89.

Arcondeguy, T., Jack, R., and Merrick, M. (2001). P_{II} signal transduction proteins, pivotal players in microbial nitrogen control. Microbiol. Mol. Biol. Rev. *65*, 80–105.

Asayama, M., Imamura, S., Yoshihara, S., Miyazaki, A., Yoshida, N., Sazuka, T., Kaneko, T., Ohara, O., Tabata, S., Osanai, T., Tanaka, K., Takahashi, H., and Shirai, M. (2004). SigC, the group 2 sigma factor of RNA polymerase, contributes to the late-stage gene expression and nitrogen promoter recognition in the cyanobacterium *Synechocystis* sp. strain PCC 6803. Biosci. Biotechnol. Biochem. *68*, 477–487.

Baier, K., Lehmann, H., Stephan, D.P., and Lockau, W. (2004). NblA is essential for phycobilisome degradation in *Anabaena* sp. strain PCC 7120 but not for development of functional heterocysts. Microbiology *150*, 2739–2749.

Baier, K., Nicklisch, S., Grundner, C., Reinecke, J., and Lockau, W. (2001). Expression of two *nblA*-homologous genes is required for phycobilisome degradation in nitrogen-starved *Synechocystis* sp. PCC6803. FEMS Microbiol. Lett. *195*, 35–39.

Barker-Astrom, K., Schelin, J., Gustafsson, P., Clarke, A.K., and Campbell, D.A. (2005). Chlorosis during nitrogen starvation is altered by carbon dioxide and temperature status and is mediated by the ClpP1 protease in *Synechococcus elongatus*. Arch. Microbiol. *183*, 66–69.

Barnard, A., Wolfe, A., and Busby, S. (2004). Regulation at complex bacterial promoters: how bacteria use different promoter organizations to produce different regulatory outcomes. Curr. Opin. Microbiol. *7*, 102–108.

Benelli, E.M., Buck, M., Polikarpov, I., de Souza, E.M., Cruz, L.M., and Pedrosa, F.O. (2002). *Herbaspirillum seropedicae* signal transducer protein P_{II} is structurally similar to the enteric GlnK. Eur. J. Biochem. *269*, 3296–3303.

Berman-Frank, I., Lundgren, P., and Falkowski, P. (2003). Nitrogen fixation and photosynthetic oxygen evolution in cyanobacteria. Res. Microbiol. *154*, 157–164.

Bienert, R., Baier, K., Volkmer, R., Lockau, W., and Heinemann, U. (2005). Crystal structure of NblA from *Anabaena* sp. PCC 7120, a small protein playing a key role in phycobilisome degradation. J. Biol. Chem. *281*, 5216–5223.

Bird, C., and Wyman, M. (2003). Nitrate/nitrite assimilation system of the marine picoplanctonic cyanobacterium *Synechococcus* sp. strain WH 8103: Effect of nitrogen source and availability on gene expression. Appl. Environ. Microbiol. *69*, 7009–7018.

Bockholt, R., Scholten-Beck, G., and Pistorius, E.K. (1996). Construction and partial characterization of an L-amino acid oxidase-free *Synechococcus* PCC 7942 mutant and localization of the L-amino acid oxidase in the corresponding wild type. Biochim. Biophys. Acta *1307*, 111–121.

Boison G., Steingen C., Stal L.J., and Bothe H. (2006). The rice field cyanobacteria *Anabaena azotica* and *Anabaena* sp. CH1 express vanadium-dependent nitrogenase. Arch Microbiol. *186*, 367–376.

Boussiba, S., Dilling, W., and Gibson, J. (1984). Methylammonium transport in *Anacystis nidulans* R-2. J. Bacteriol. *160*, 204–210.

Brahamsha, B., and Haselkorn, R. (1992). Identification of multiple RNA polymerase sigma factor homologs in the cyanobacterium *Anabaena* sp. strain PCC 7120: cloning, expression, and inactivation of the *sigB* and *sigC* genes. J. Bacteriol. *174*, 7273–7282.

Burillo, S., Luque, I., Fuentes, I., and Contreras, A. (2004). Interaction between the nitrogen signal transduction protein PII and N-acetylglutamate kinase in organisms that perform oxygenic photosynthesis. J. Bacteriol. *186*, 3346–3354.

Caslake, L.F., Gruber, T.M., and Bryant, D.A. (1997). Expression of two alternative sigma factors of *Synechococcus* sp. strain PCC 7002 is modulated by carbon and nitrogen stress. Microbiology *143*, 3807–3818.

Castresana, J. (2000). Selection of conserved blocks from multiple alignments for their use in phylogenetic analysis. Mol. Biol. Evol. *17*, 540–552.

Chen, Y.M., Ferrar, T.S., Lohmeir-Vogel, E., Morrice, N., Mizuno, Y., Berenger, B., Ng, K.K.S., Muench, D.G., and Moorhead, G.B.G. (2006). The PII signal transduction protein of *Arabidopsis thaliana* forms an arginine-regulated complex with plastid N-Acetyl glutamate kinase. J. Biol. Chem. *281*, 5726–5733.

Cho, B.C., Park, M.G., Shim, J.H., and Azam, F. (1996). Significance of bacteria in urea dynamics in coastal surface waters. Mar. Ecol. Prog. Ser. *142*, 19–26.

Clegg, S., Yu, F., Griffiths, L., and Cole, J.A. (2002). The roles of the polytopic membrane proteins NarK, NarU and NirC in *Escherichia coli* K-12: two nitrate and three nitrite transporters. Mol. Microbiol. *44*, 143–155.

Collier, J.L., Brahamsha, B., and Palenik, B. (1999). The marine cyanobacterium *Synechococcus* sp. WH7805 requires urease (urea amidohydrolase, EC 3.5.1.5) to utilize urea as a nitrogen source: molecular-genetic and biochemical analysis of the enzyme. Microbiology *145*, 447–459.

Collier, J.L., and Grossman, A.R. (1992). Chlorosis induced by nutrient deprivation in *Synechococcus* sp. strain PCC 7942: not all bleaching is the same. J. Bacteriol. *174*, 4718–4726.

Collier, J.L., and Grossman, A.R. (1994). A small polypeptide triggers complete degradation of light-harvesting phycobiliproteins in nutrient-deprived cyanobacteria. EMBO J. *13*, 1039–1047.

Crespo, J.L., García-Domínguez, M., and Florencio, F.J. (1998). Nitrogen control of the *glnN* gene that codes for GS type III, the only glutamine synthetase in the cyanobacterium *Pseudanabaena* sp. PCC 6903. Mol. Microbiol. *30*, 1101–1112.

Dolganov, N., and Grossman, A.R. (1999). A polypeptide with similarity to phycocyanin alpha-subunit phycocyanobilin lyase involved in degradation of phycobilisomes. J. Bacteriol. *181*, 610–617.

Ehira, S., and Ohmori, M. (2006). NrrA, a nitrogen-responsive response regulator facilitates heterocyst development in the cyanobacterium *Anabaena* sp. strain PCC 7120. Mol. Microbiol. *59*, 1692–1703.

El-Shehawy, R., Lugomela, C., Ernst, A., Bergman, B. (2003). Diurnal expression of *hetR* and diazocyte development in the filamentous non-heterocystous cyanobacterium *Trichodesmium erythraeum*. Microbiology 149, 1139–1146.

Engels, D.H., Engels, A., and Pistorius, E.K. (1992). Isolation and partial characterization of an L-amino acid oxidase and of photosystem II complexes from the cyanobacterium *Synechococcus* PCC 7942. Z. Naturforsch. *47c*, 859–866.

Espinosa, J., Forchhammer, K., Burillo, S., and Contreras, A. (2006a). Interaction network in cyanobacterial nitrogen regulation: PipX, a protein that interacts in a 2-oxoglutarate dependent manner with P_{II} and NtcA. Mol. Microbiol. *61*, 457–469.

Espinosa, J., Fuentes, I., Burillo, S., Rodríguez-Mateos, F., and Contreras, A. (2006b). SipA, a novel type of protein from *Synechococcus* sp. PCC 7942, binds to the kinase domain f NblS. FEMS Microbiol. Lett. *254*, 41–47.

Ferrario-Méry, S., Besin, E., Pichon, O., Meyer, C., and Hodges M. (2006). The regulatory P_{II} protein controls arginine biosynthesis in *Arabidopsis*, FEBS Lett. *580*, 2015–2020.

Flores, E., Frías, J.E., Rubio, L.M., and Herrero, A. (2005). Photosynthetic nitrate assimilation in cyanobacteria. Photosynth. Res. *83*, 117–133.

Flores, E., Herrero, A., and Guerrero, M.G. (1982). Production of ammonium dependent on basic L-amino acids by *Anacystis nidulans*. Arch. Microbiol. *131*, 91–94.

Flores, E., and Herrero, A. (1994). Assimilatory nitrogen metabolism and its regulation. In The molecular biology of cyanobacteria. Bryant, D.A. ed. (Kluwer Academic Publishers) pp. 488–517.

Flores, E., and Muro-Pastor, M.I. (1988). Uptake of glutamine and glutamate by the dinitrogen-fixing cyanobacterium *Anabaena* sp. PCC 7120. FEMS Microbiol. Lett. 56, 127–130.

Flores, E., and Muro-Pastor, A.M. (1990). Mutational and kinetic analysis of basic amino acid transport in the cyanobacterium *Synechocystis* PCC 6803. Arch. Microbiol. *154*, 521–527.

Flores, E., and Herrero, A. (2005). Nitrogen assimilation and nitrogen control in cyanobacteria. Biochem. Soc. Trans. *33*, 164–167.

Flores, E., Ramos, J.L., Herrero, A., and Guerrero, M.G. (1983). Nitrate assimilation by cyanobacteria. In Photosynthetic Prokaryotes: Cell Differentiation and Function, G.C. Papageorgiou and L. Packer, eds. (Amsterdam, The Netherlands: Elsevier/North Holland), pp. 363–387.

Forchhammer, K. (1999). The P_{II} protein in *Synechococcus* PCC 7942 senses and signals 2-oxoglutarate under ATP-replete conditions. In The Phototrophic Prokaryotes, Peschek, G., Löffelhardt, W., and Schmetterer, G., eds. (Kluwer Academic, New York, N.Y.) pp. 549–553.

Forchhammer, K. (2004). Global carbon/nitrogen control by P_{II} signal transduction in cyanobacteria: from signals to targets. FEMS Microbiol. Rev. *28*, 319–333.

Forchhammer, K., and Hedler, A. (1997). Phosphoprotein P_{II} from cyanobacteria: Analysis of functional conservation to the P_{II} signal transduction protein from *Escherichia coli*. Eur. J. Biochem. *244*, 869–875.

Forchhammer, K., Irmler, A., Kloft, N., and Ruppert, U. (2004). P_{II} signalling in unicellular cyanobacteria: analysis of redox-signals and energy charge. Physiol. Plantarum *120*, 51–54.

Frías, J.E., Flores, E., and Herrero, A. (1994). Requirement of the regulatory protein NtcA for the expression of nitrogen assimilation and heterocyst development genes in the cyanobacterium *Anabaena* sp. PCC 7120. Mol Microbiol *14*, 823–832.

Frías, J.E., Flores, E., and Herrero, A. (1997). Nitrate assimilation gene cluster from the heterocyst-forming cyanobacterium *Anabaena* sp. strain PCC 7120. J Bacteriol *179*, 477–486.

Frías, J.E., Flores, E., and Herrero, A. (2000). Activation of the *Anabaena nir* operon promoter requires both NtcA (CAP family) and NtcB (LysR family) transcription factors. Mol Microbiol *38*, 613–625.

Frías, J.E., Herrero, A., and Flores, E. (2003). Open reading frame *all0601* from *Anabaena* sp. strain PCC 7120 represents a novel gene, *cnaT*, required for expression of the nitrate assimilation *nir* operon. J. Bacteriol. *185*, 5037–5044.

Frías, J.E., Mérida, A., Herrero, A., Martin-Nieto, J., and Flores, E. (1993). General distribution of the nitrogen control gene *ntcA* in cyanobacteria. J. Bacteriol. *175*, 5710–5713.

García-Domínguez, M., and Florencio, F.J. (1997). Nitrogen availability and electron transport control the expression of *glnB* gene (encoding P_{II} protein) in the cyanobacterium *Synechocystis* sp. PCC 6803. Plant Mol. Biol. *35*, 723–734.

García-Domínguez, M., Reyes, J.C., and Florencio, F.J. (1999). Glutamine synthetase inactivation by protein-protein interaction. Proc. Natl. Acad. Sci. USA *96*, 7161–7166.

García-Domínguez, M., Reyes, J.C., and Florencio, F.J. (2000). NtcA represses transcription of *gifA* and *gifB* genes that encode inhibitors of glutamine synthetase type I from *Synechocystis* sp. PCC 6803. Mol. Microbiol. *35*, 1192–1201.

García-Fernandez, J.M., de Marsac, N.T., and Diez, J. (2004). Streamlined regulation and gene loss as adaptive mechanisms in *Prochlorococcus* for optimized nitrogen utilization in oligotrophic environments. Microbiol. Mol. Biol. Rev. *68*, 630–638.

Ge, X., Cain, K., and Hirschberg, R. (1990). Urea metabolism and urease regulation in the cyanobacterium *Anabaena variabilis*. Can. J. Microbiol. *36*, 218–222.

Görl, M., Sauer, J., Baier, T., and Forchhammer, K. (1998). Nitrogen-starvation-induced chlorosis in *Synechococcus* PCC 7942: adaptation to long-term survival. Microbiology *144*, 2449–2458.

Guerrero, M.G., and Lara, C. (1987). Assimilation of inorganic nitrogen. In The Cyanobacteria, P. Fay and C. van Baalen, eds. (Amsterdam, The Netherlands: Elsevier Science Publishers), pp 163–186.

Guindon, S., and Gascuel, O. (2003). A simple, fast, and accurate algorithm to estimate large phylogenies by maximum likelihood. Syst. Biol. *52*, 696–704.

Hai, T., Hein, S., and Steinbuchel, A. (2001). Multiple evidence for widespread and general occurrence of type-III PHA synthases in cyanobacteria and molecular characterization of the PHA synthases from two thermophilic cyanobacteria: *Chlorogloeopsis fritschii* PCC 6912 and *Synechococcus* sp. strain MA19. Microbiology *147*, 3047–3060.

Harano, Y., Suzuki, I., Maeda, S., Kaneko, T., Tabata, S., and Omata, T. (1997). Identification and nitrogen regulation of the cyanase gene from the cyanobacteria *Synechocystis* sp. strain PCC 6803 and *Synechococcus* sp. strain PCC 7942. J. Bacteriol. *179*, 5744–5750.

Hebbar, P.B., and Curtis, S.E. (2000). Characterization of *devH*, a gene encoding a putative DNA binding protein required for heterocyst function in *Anabaena* sp. strain PCC 7120. J. Bacteriol. *182*, 3572–3581.

Heinrich, A., Maheswaran, M. Ruppert, U., and Forchhammer, K. (2004). The *Synechococcus elongatus* P_{II} signal transduction protein controls arginine synthesis by complex formation with N-acetyl-L-glutamate kinase. Mol. Microbiol. *52*, 1303–1314.

Herrero, A., and Flores, E. (1990). Transport of basic amino acids by the dinitrogen-fixing cyanobacterium *Anabaena* PCC 7120. J. Biol. Chem. *265*, 3931–3935.

Herrero, A., Muro-Pastor, A.M., and Flores, E. (2001). Nitrogen control in cyanobacteria. J. Bacteriol. *183*, 411–425.

Herrero, A., Muro-Pastor, A., Valladares, A., and Flores, E. (2004). Cellular differentiation and the NtcA transcription factor in filamentous cyanobacteria. FEMS Microbiol. Rev. *28*, 469–487.

Huckauf, J., Nomura, C., Forchhammer, K., and Hagemann, M. (2000). Stress responses of *Synechocystis* sp. strain PCC 6803 mutants impaired in genes encoding putative alternative sigma factors. Microbiology *146*, 2877–2889.

Igarashi, R.Y and Seefeldt, L.C. (2003). Nitrogen fixation: The mechanism of the Mo-dependent nitrogenase. Crit. Rev. Biochem. Mol. Biol. *38*, 351–384.

Imamura, S., Tanaka, K., Shirai, M., and Asayama, M. (2006). Growth phase-dependent activation of nitrogen-related genes by a control network of group 1 and group 2 sigma factors in a cyanobacterium. J. Biol. Chem. *281*, 2668–2675.

Javelle, A., and Merrick, M. (2005). Complex formation between AmtB and GlnK: an ancestral role in prokaryotic nitrogen control. Biochem. Soc. Trans. *33*, 170–172.

Javelle, A., Thomas, G., Marini, A.-M., Krämer, R., and Merrick, M (2005). *In vivo* functional characterisation of the *E. coli* ammonium channel AmtB: evidence for metabolic coupling of AmtB to glutamine synthetase. Biochem. J. *390*, 215–222.

Jiang, F., Hellman, U., Sroga, G.E., Bergman, B., and Mannervik, B. (1995). Cloning, sequencing, and regulation of the glutathione reductase gene from the cyanobacterium *Anabaena* PCC 7120. J. Biol. Chem. *270*, 22882–22889.

Jiang, F., Mannervik, B., and Bergman, B. (1997). Evidence for redox regulation of the transcription factor NtcA, acting both as an activator and a repressor, in the cyanobacterium *Anabaena* PCC 7120. Biochem. J. *327*, 513–517.

Jiang, F., Wisén, S., Widersten, M., Bergman, B., and Mannervik, B. (2000). Examination of the transcription factor NtcA-binding motif by *in vitro* selection of DNA sequences from a random library. J. Mol. Biol. *301*, 783–793.

Kapp, R., Stevens, S.E.J., and Fox, J.L. (1975). A survey of available nitrogen sources for growth of the blue-green alga, *Agmenellum quadruplicatum*. Arch. Microbiol. *104*, 135–138.

Kappell, A.D., Bhaya, D., and van Waasbergen, L.G. (2006). Negative control of the high light-inducible *hliA* gene and implications for the activities of the NblS sensor kinase in the cyanobacterium *Synechococcus elongatus* strain PCC 7942. Arch Microbiol. *186*, 403–413.

Khudyakov, I.Y., and Golden, J.W. (2001). Identification and inactivation of three group 2 sigma factor genes in *Anabaena* sp. strain PCC 7120. J. Bacteriol. *183*, 6667–6675.

Khudyakov, I., and Wolk, C.P. (1996). Evidence that the *hanA* gene coding for HU protein is essential for heterocyst differentiation in, and cyanophage A-4(L) sensitivity of, *Anabaena* sp. strain PCC 7120. J. Bacteriol. *178*, 3572–3577.

Kloft, N., and Forchhammer, K. (2005). Signal Transduction Protein P_{II} Phosphatase PphA Is Required for Light-Dependent Control of Nitrate Utilization in *Synechocystis* sp. strain PCC 6803. J. Bacteriol. *187*, 6683–6690.

Kloft, N., Rasch, G., and Forchhammer, K. (2005). Protein phosphatase PphA from *Synechocystis* sp. PCC 6803: the physiological framework of P_{II}-P dephosphorylation. Microbiology *151*, 1275–1283.

Kobayashi, M., Rodriguez, R., Lara, C., and Omata, T. (1997). Involvement of the C-terminal domain of an ATP-binding subunit in the regulation of the ABC-type nitrate/nitrite transporter of the cyanobacterium *Synechococcus* sp. strain PCC 7942. J. Biol. Chem. *272*, 27197–27201.

Korner, H., Sofia, H.J., and Zumft, W.G. (2003). Phylogeny of the bacterial superfamily of Crp-Fnr transcription regulators: exploiting the metabolic spectrum by controlling alternative gene programs. FEMS Microbiol. Rev. *27*, 559–592.

Koropatkin N.M., Pakrasi H.B., and Smith T.J. (2006). Atomic structure of a nitrate-binding protein crucial for photosynthetic productivity. Proc Natl Acad Sci U S A *103*, 9820–9825.

Labarre, J., Thuriaux, P., and Chauvat, F. (1987). Genetic analysis of amino acid transport in the facultatively heterotrophic cyanobacterium *Synechocystis* sp. strain 6803. J. Bacteriol. *169*, 4668–4673.

Lawson, C.L., Swigon, D., Murakami, K.S., Darst, S.A., Berman, H.M., and Ebright, R.H. (2004). Catabolite activator protein: DNA binding and transcription activation. Curr. Opin. Struct. Biol. *14*, 10–20.

Lee, H.-M., Flores, E., Forchhammer, K., Herrero, A., and Tandeau de Marsac, N. (2000). Phosphorylation of the signal transducer P_{II} protein and an additional effector are required for the P_{II}-mediated regulation of nitrate and nitrite uptake in the cyanobacterium *Synechococcus* sp. PCC 7942. Eur. J. Biochem. *267*, 591–600.

Lee, H.-M., Flores, E., Herrero, A., Houmard, J., and Tandeau de Marsac, N. (1998). A role for the signal transduction protein P_{II} in the control of nitrate/nitrite uptake in a cyanobacterium. FEBS Lett. *427*, 291–295.

Lee, H.-M., Vázquez-Bermúdez, M.F., and Tandeau de Marsac, N. (1999). The global nitrogen regulator NtcA regulates transcription of the signal transducer P_{II} (GlnB) and influences its phosphorylation level in response to nitrogen and carbon supplies in the cyanobacterium *Synechococcus* sp. strain PCC 7942. J. Bacteriol. *181*, 2697–2702.

Lindell, D., Erdner, D., Marie, D., Prasil, O., Koblizek, M., LeGall, F., Rippka, R., Partensky, F., Scanlan, D.J., and Post, A. (2002). Nitrogen stress response of *Prochlorococcus* strain PCC 9511 (oxyphotobacteria) involves contrasting regulation of *ntcA* and *amt1*. J. Phycol. *38*, 1113–1124.

Lindell, D., Padan, E., and Post, A.F. (1998). Regulation of *ntcA* expression and nitrite uptake in the marine *Synechococcus* sp. strain WH 7803. J. Bacteriol. *180*, 1878–1886.

Luque, I., Andújar, A., Jia, L., Zabulon, G., Tandeau de Marsac, N., Flores, E., and Houmard, J. (2006). Regulated expression of glutamyl-tRNA synthetase is directed by a mobile genetic element in the cyanobacterium *Tolypothrix* sp. PCC 7601. Mol. Microbiol. *60*, 1276–1288

Luque, I., Contreras, A., Zabulon, G., Herrero, A., and Houmard, J. (2002). Expression of the glutamyl-tRNA synthetase gene from the cyanobacterium *Synechococcus* sp PCC 7942 depends on nitrogen availability and the global regulator NtcA. Mol. Microbiol. *46*, 1157–1167.

Luque, I., Flores, E., and Herrero, A. (1994). Molecular mechanism for the operation of nitrogen control in cyanobacteria. EMBO J. *13*, 2862–2869.

Luque, I., Ochoa De Alda, J.A., Richaud, C., Zabulon, G., Thomas, J.C., and Houmard, J. (2003). The NblAI protein from the filamentous cyanobacterium *Tolypothrix* PCC 7601: regulation of its expression and interactions with phycobilisome components. Mol. Microbiol. *50*, 1043–1054.

Luque, I., Vázquez-Bermúdez, M F., Paz-Yepes, J., Flores, E., and Herrero, A. (2004). *In vivo* activity of the nitrogen control transcription factor NtcA is subjected to metabolic regulation in *Synechococcus* sp. strain PCC 7942. FEMS Microbiol. Lett. *236*, 47–52.

Luque, I., Zabulon, G., Contreras, A., and Houmard, J. (2001). Convergence of two global transcriptional regulators on nitrogen induction of the stress-acclimation gene *nbl*A in the cyanobacterium *Synechococcus* sp. PCC 7942. Mol. Microbiol. *41*, 937–947.

Maeda, S.-I., Okamura, M., Kobayashi, M., and Omata, T. (1998). Nitrite-specific active transport system of the cyanobacterium *Synechococcus* sp. Strain PCC 7942. J. Bacteriol. *180*, 6761–6763.

Maeda, S.I., and Omata, T. (1997). Substrate-binding lipoprotein of the cyanobacterium *Synechococcus* sp. strain PCC 7942 involved in the transport of nitrate and nitrite. J. Biol. Chem. *272*, 3036–3041.

Maeda, S., and Omata, T. (2004). A novel gene (*narM*) required for expression of nitrate reductase activity in the cyanobacterium *Synechococcus elongatus* strain PCC7942. J. Bacteriol. *186*, 2107–2114.

Maheswaran, M., Urbanke, C., and Forchhammer, K. (2004). Complex formation and catalytic activation by the P_{II} signaling protein of N-acetyl-L-glutamate kinase from *Synechococcus elongatus* strain PCC 7942. J. Biol. Chem. *279*, 55202–55210.

Maheswaran, M., Ziegler, K., Lockau, W., Hagemann, M., and Forchhammer, K. (2006). P_{II}-Regulated Arginine Synthesis Controls Accumulation of Cyanophycin in *Synechocystis* sp. strain PCC 6803. J. Bacteriol. *188*, 2730–2734.

Marqués, S., Florencio, F.J., and Candau, P. (1992). Purification and characterization of the ferredoxin-glutamate synthase from the unicellular cyanobacterium *Synechococcus* sp. PCC 6301. Eur. J. Biochem. *206*, 69–77

Meeks, J.C., Wolk, C.P., Lockau, W., Schilling, N., Shaffer, P.W., and Chien, W.S. (1978). Pathways of assimilation of $[^{13}N]N_2$ and $^{13}NH_4^+$ by cyanobacteria with and without heterocysts. J. Bacteriol. *134*, 125–130.

Mérida, A., Candau, P., Florencio, F.J. (1991). Regulation of glutamine synthetase activity in the unicellular cyanobacterium *Synechocystis* sp. strain PCC 6803 by the nitrogen source: effect of ammonium. J. Bacteriol. *173*, 4095–4100.

Michel, K.P., Pistorius, E.K., and Golden, S.S. (2001). Unusual regulatory elements for iron deficiency induction of the *idiA* gene of *Synechococcus elongatus* PCC 7942. J. Bacteriol. *183*, 5015–5024.

Miller, A.G., and Espie, G.S. (1994). Photosynthetic metabolism of cyanate by the cyanobacterium *Synechococcus* UTEX 625. Arch. Microbiol. *162*, 151–157.

Mobley, H.L., Island, M.D., and Hausinger, R.P. (1995). Molecular biology of microbial ureases. Microbiol. Rev. *59*: 451–480.

Montesinos, M.L., Herrero, A., and Flores, E. (1995). Amino acid transport systems required for diazotrophic growth in the cyanobacterium *Anabaena* sp. strain PCC 7120. J Bacteriol *177*: 3150–3157.

Montesinos, M.L., Herrero, A., and Flores, E. (1997). Amino acid transport in taxonomically diverse cyanobacteria and identification of two genes encoding elements of a neutral amino acid permease putatively involved in recapture of leaked hydrophobic amino acids. J. Bacteriol. *179*, 853–862.

Montesinos, M.L., Muro-Pastor, A.M., Herrero, A., and Flores, E. (1998). Ammonium/methylammonium permeases of a cyanobacterium. Identification and analysis of three nitrogen regulated *amt* genes in *Synechocystis* sp. PCC 6803. J. Biol. Chem. *273*, 31463–31470.

Morrison, S.S., Mullineaux, C.W, and Ashby M.K. (2005). The influence of acetyl phosphate on DspA signalling in the cyanobacterium *Synechocystis* sp. PCC6803. BCM Microbiology *5*, 47.

Muro-Pastor M.I., Barrera F.N., Reyes J.C., Florencio F.J., and Neira J.L. (2003). The inactivating factor of glutamine synthetase, IF7, is a "natively unfolded" protein. Protein Sci. *12*, 1443–1454.

Muro-Pastor, A.M., Herrero, A., Flores, E. (2001). Nitrogen-regulated group 2 sigma factor from *Synechocystis* sp. strain PCC 6803 involved in survival under nitrogen stress J. Bacteriol. *183*, 1090–1095.

Muro-Pastor, A.M., Olmedo-Verd, E., and Flores, E. (2006). All4312, an NtcA-regulated two-component response regulator in *Anabaena* sp. strain PCC 7120. FEMS Microbiol. Lett. *256*, 171–177.

Muro-Pastor, M.I., Reyes, J.C., and Florencio, F.J. (1996). The $NADP^+$-isocitrate dehydrogenase gene (*icd*) is nitrogen regulated in cyanobacteria. J. Bacteriol. *178*, 4070–4076.

Muro-Pastor, M.I., Reyes, J.C., and Florencio, F.J. (2001). Cyanobacteria perceive nitrogen status by sensing intracellular 2-OG levels. J. Biol. Chem. *276*, 38320–38328.

Muro-Pastor, M.I., Reyes, J.C., and Florencio, F.J. (2005). Ammonium assimilation in cyanobacteria. Photosynth. Res. *83*, 135–150.

Muro-Pastor, A.M., Valladares, A., Flores, E., and Herrero, A. (2002). Mutual dependence of the expression of the cell differentiation regulatory protein HetR and the global nitrogen regulator NtcA during heterocyst development. Mol. Microbiol. *44*, 1377–1385.

Nagaraja, R., and Haselkorn, R. (1994). Protein HU from the cyanobacterium *Anabaena*. Biochimie *76*, 1082–1089.

Neilson, A.H., and Larsson, T. (1980). The utilization of organic nitrogen for growth of algae: Physiologycal aspects. Physiol. Plant. *48*, 542–553.

Nicholson, M.L., and Laudenbach, D.E. (1995). Genes encoded on a cyanobacterial plasmid are transcriptionally regulated by sulfur availability and CysR. J. Bacteriol. *177*, 2143–2150.

Ninfa, A.J., and Jiang, P. (2005). P_{II} signal transduction proteins: sensors of alpha-ketoglutarate that regulate nitrogen metabolism. Current Opin. Microbiol. *8*, 168–173.

Okuhara, H., Matsumura, T., Fujita, Y., and Hase, T. (1999). Cloning and inactivation of genes encoding ferredoxin- and NADH-dependent glutamate synthases in the cyanobacterium *Plectonema boryanum*. Imbalances in nitrogen and carbon assimilations caused by deficiency of the ferredoxin-dependent enzyme. Plant. Physiol. *120*, 33–42.

Olmedo-Verd, E., Flores, E., Herrero, A., and Muro-Pastor, A.M. (2005). HetR-dependent and -independent expression of heterocyst-related genes in an *Anabaena* strain overproducing the NtcA transcription factor. J. Bacteriol. *187*, 1985–1991.

Osanai, T., Kanesaki, Y., Nakano, T., Takahashi, H., Asayama, M., Shirai, M., Kanehisa, M., Suzuki, I., Murata, N., and Tanaka, K. (2005a). Positive regulation of sugar catabolic pathways in the cyanobacterium *Synechocystis* sp. PCC 6803 by the group 2 sigma factor SigE. J. Biol. Chem. *280*, 30653–30659.

Osanai, T., Sato, S., Tabata, S., and Tanaka, K. (2005b). Identification of PamA as a P_{II}-binding membrane protein important in nitrogen-related and sugar-catabolic gene expression in *Synechocystis* sp. PCC 6803. J. Biol. Chem. *280*, 34684–34690.

Palinska, K.A., Jahns, T., Rippka, R., and Tandeau De Marsac, N. (2000). *Prochlorococcus marinus* strain PCC 9511, a picoplanktonic cyanobacterium, synthesizes the smallest urease. Microbiology *146*, 3099–3107.

Palinska, K.A., Laloui, W., Bedu, S., Loiseaux-de Goer, S., Castets, A.M., Rippka, R., and Tandeau de Marsac, N. (2002). The signal transducer P_{II} and bicarbonate aquisition in *Prochlorococcus marinus* PCC 9511, a marine cyanobacterium naturally deficient in nitrate and nitrite assimilation. Microbiology *148*, 2505–2412.

Paz-Yepes, J., Flores, E., and Herrero, A. (2003). Transcriptional effects of the signal transduction protein P_{II} (*glnB* gene product) on NtcA-dependent genes in *Synechococcus* sp. PCC 7942. FEBS Lett. *543*, 42–46.

Picossi, S., Montesinos, M.L., Pernil, R., Lichtle, C., Herrero, A., and Flores, E. (2005). ABC-type neutral amino acid permease N-I is required for optimal diazotrophic growth and is repressed in the heterocysts of *Anabaena* sp. strain PCC 7120. Mol. Microbiol. *57*, 1582–1592.

Picossi, S., Valladares, A., Flores, E., and Herrero, A. (2004). Nitrogen-regulated genes for the metabolism of cyanophycin, a bacterial nitrogen reserve polymer: expression and mutational analysis of two cyanophycin synthetase and cyanophycinase gene clusters in the heterocyst-forming cyanobacterium *Anabaena* sp. PCC 7120. J. Biol. Chem. *279*, 11582–11592.

Quintero, M.J., Montesinos, M.L., Herrero, A., and Flores, E. (2001). Identification of genes encoding amino acid permeases by inactivation of selected ORFs from the *Synechocystis* genomic sequence. Genome Res. *11*, 2034–2040.

Quintero, M.J., Muro-Pastor, A.M., Herrero, A., and Flores, E. (2000). Arginine catabolism in the cyanobacterium *Synechocystis* sp. strain PCC 6803 involves the urea cycle and arginase pathway. J. Bacteriol. *182*, 1008–1015.

Ramasubramanian, T.S., Wei, T.F., and Golden, J.W. (1994). Two *Anabaena* sp. strain PCC 7120 DNA-binding factors interact with vegetative cell- and heterocyst-specific genes. J. Bacteriol. *176*, 1214–1223.

Ramasubramanian, T.S., Wei, T.F., Oldham, A.K., and Golden, J.W. (1996). Transcription of the *Anabaena* sp. strain PCC 7120 *ntcA* gene: multiple transcripts and NtcA binding. J. Bacteriol. *178*, 922–926.

Ramirez, M.E., Hebbar, P.B., Zhou, R., Wolk, C.P., and Curtis, S.E. (2005). *Anabaena* sp. strain PCC 7120 gene *devH* is required for synthesis of the heterocyst glycolipid layer. J. Bacteriol. *187*, 2326–2331.

Reyes, J.C., and Florencio, F.J. (1994). A new type of glutamine synthetase in cyanobacteria: the protein encoded by the *glnN* gene supports nitrogen assimilation in *Synechocystis* sp. strain PCC 6803. J. Bacteriol. *176*, 1260–1267.

Reyes, J.C., and Florencio, F.J. (1995). Electron transport controls transcription of the glutamine synthetase gene (*glnA*) from the cyanobacterium *Synechocystis* sp. PCC 6803. Plant Mol. Biol. *27*, 789–799.

Reyes, J.C., Muro-Pastor, M.I., and Florencio, F.J. (1997). Transcription of glutamine synthetase genes (*glnA* and *glnN*) from the cyanobacterium *Synechocystis* sp. strain PCC 6803 is differently regulated in response to nitrogen availability. J. Bacteriol. *179*, 2678–2689.

Rexach, J., Fernández, E., and Galvan, A. (2000). The *Chlamydomonas reinhardtii Nar1* gene encodes a chloroplast membrane protein involved in nitrite transport. Plant Cell *12*, 1441–1453.

Rocap, G., Larimer, F.W., Lamerdin, J., Malfatti, S., Chain, P., Ahlgren, N.A., Arellano, A., Coleman, M., Hauser, L., Hess, W.R., Johnson, Z.I., Land, M., Lindell, D., Post, A.F., Regala, W., Shah, M., Shaw, S.L., Steglich, C., Sullivan, M.B., Ting, C.S. Tolonen, A., Webb, E.A., Zinser, E.R., and Chisholm, S.W. (2003). Genome divergence in two *Prochlorococcus* ecotypes reflects oceanic niche differentiation. Nature *424*, 1042–1047.

Richaud, C., Zabulon, G., Joder, A., and Thomas, J.C. (2001). Nitrogen or sulfur starvation differentially affects phycobilisome degradation and expression of the *nblA* gene in *Synechocystis* strain PCC 6803. J. Bacteriol. *183*, 2989–2994.

Rojo, F. (1999). Repression of transcription initiation in bacteria. J. Bacteriol. *181*, 2987–2991.

Ruppert, U., Irmler, A., Kloft, N., and Forchhammer, K. (2002). The novel protein phosphatase PphA from *Synechocystis* PCC 6803 controls dephosphorylation of the signalling protein P_{II}. Mol. Microbiol. *44*, 855–864.

Sakai, H., Wang, H., Takemoto-Hori, C., Kaminishi, T., Yamaguchi, H., Kamewari, Y., Terada, T., Kuramitsu, S., Shirouzu, M., and Yokohama, S. (2005). Crystal structures of the signal transducing protein GlnK from *Thermus thermophilus* Hb8. J. Struct. Biol. *149*, 99–110.

Sakamoto, T., Inoue-Sakamoto, K., and Bryant, D.A. (1999). A novel nitrate/nitrite permease in the marine cyanobacterium *Synechococcus* sp. strain PCC 7002. J Bacteriol. *181*, 7363–7372.

Sauer, J., Dirmeier, U., and Forchhammer, K. (2000). The *Synechococcus* strain PCC 7942 *glnN* product (glutamine synthetase III) helps recovery from prolonged nitrogen chlorosis. J. Bacteriol. *182*, 5615–5619.

Sauer, J., Görl, M., and Forchhammer, K. (1999). Nitrogen Starvation in *Synechococcus* PCC 7942: involvement of glutamine synthetase and NtcA in phycobiliprotein degradation and survival. Arch. Microbiol. *172*, 247–255.

Sauer, J., Schreiber, U., Schmid, R., Volker, U., and Forchhammer, K. (2001). Nitrogen starvation-induced chlorosis in *Synechococcus* PCC 7942. Low-level photosynthesis as a mechanism of long-term survival. Plant Physiol. *126*, 233–243.

Schwarz, R., and Forchhammer, K. (2005). Acclimation of unicellular cyanobacteria to macronutrient deficiency: emergence of a complex network of cellular responses. Microbiology *151*, 2503–2514.

Schwarz, R., and Grossman, A.R. (1998). A response regulator of cyanobacteria integrates diverse environmental signals and is critical for survival under extreme conditions. Proc. Natl. Acad. Sci. USA *95*, 11008–11013.

Sendersky E,, Lahmi R,, Shaltiel J,, Perelman A, and Schwarz R. (2005). NblC, a novel component required for pigment degradation during starvation in *Synechococcus* PCC 7942. Mol Microbiol. *58*, 659–668.

Singh, S. (1990). Regulation of urease activity in the cyanobacterium *Anabaena doliolum*. FEMS Microbiol. Lett. *67*, 79–84.

Singh, S. (1992). Regulation of urease cellular levels in the cyanobacteria *Anacystis nidulans* and *Nostoc muscorum*. Biochem. Physiol. Pflanzen *188*, 33–38.

Smith, C., Weljie, A,M,., and Moorhead, G.B.G. (2003). Molecular properties of the putative nitrogen sensor, P_{II}, from *Arabidopsis thaliana*. Plant J. *33*, 353–360.

Stanier R.Y., and Cohen-Bazire G. (1977). Phototrophic prokaryotes: the cyanobacteria. Annu. Rev. Microbiol. *31*, 225–274.

Su, Z., Mao, F., Dam, P., Wu, H., Olman, V., Paulsen, I.T., Palenik, B., and Xu, Y. (2006). Computational inference and experimental validation of the nitrogen assimilation regulatory network in cyanobacterium *Synechococcus* sp. WH 8102. Nucleic Acids Res. *34*, 1050–1065.

Su, Z., Olman, V., Mao, F., and Xu, Y. (2005). Comparative genomics analysis of NtcA regulons in cyanobacteria: regulation of nitrogen assimilation and its coupling to photosynthesis. Nucleic Acids Res. *33*, 5156–5171.

Suzuki, I., Horie, N., Sugiyama, T., and Omata, T. (1995). Identification and characterization of two nitrogen-regulated genes of the cyanobacterium *Synechococcus* sp. strain PCC7942 required for maximum efficiency of nitrogen assimilation. J. Bacteriol. *177*, 290–296.

Takatani, N., and Omata, T. (2006). Effects of P_{II} deficiency on expression of the genes involved in ammonium utilization in the cyanobacterium *Synechocystis* sp. strain PCC 6803. Plant Cell Physiol. *47*, 1182–1186

Tanigawa, R., Shirokane, M., Maeda Si, S., Omata, T., Tanaka, K., and Takahashi, H. (2002). Transcriptional activation of NtcA-dependent promoters of *Synechococcus* sp. PCC 7942 by 2-oxoglutarate *in vitro*. Proc. Natl. Acad. Sci. USA *99*, 4251–4255.

Thiel T. (1993). Characterization of genes for an alternative nitrogenase in the cyanobacterium *Anabaena variabilis*. J Bacteriol. *175*, 6276–6286.

Valladares, A., Montesinos, M.L., Herrero, A., and Flores, E. (2002). An ABC-type, high-affinity urea permease identified in cyanobacteria. Mol. Microbiol. *43*, 703–715.

Valladares, A., Muro-Pastor, A.M., Fillat, M.F., Herrero, A., and Flores, E. (1999). Constitutive and nitrogen-regulated promoters of the *petH* gene encoding ferredoxin:$NADP^+$ reductase in the heterocyst-forming cyanobacterium *Anabaena* sp. FEBS Lett. *449*, 159–164.

Valladares, A., Muro-Pastor, A.M., Herrero, A., and Flores, E. (2004). The NtcA-dependent P_1 promoter is utilized for *glnA* expression in N_2-fixing heterocysts of *Anabaena* sp. strain PCC 7120. J. Bacteriol. *186*, 7337–7343.

van den Heuvel, R.H., Curti, B., Vanoni, M.A., and Mattevi, A. (2004). Glutamate synthase: a fascinating pathway from L-glutamine to L-glutamate. Cell Mol. Life Sci. *61*, 669–681.

van Waasbergen, L.G., Dolganov, N., and Grossman, A.R. (2002). *nblS*, a gene involved in controlling photosynthesis-related gene expression during high light and nutrient stress in *Synechococcus elongatus* PCC 7942. J. Bacteriol. *184*, 2481–2490.

Vázquez-Bermúdez, M.F., Flores, E., and Herrero, A. (2002a). Analysis of binding sites for the nitrogen-control transcription factor NtcA in the promoters of *Synechococcus* nitrogen-regulated genes. Biochim. Biophys. Acta *1578*, 95–98.

Vázquez-Bermúdez, M.F., Herrero, A., and Flores, E. (2002b). 2-Oxoglutarate increases the binding affinity of the NtcA (nitrogen control) transcription factor for the *Synechococcus glnA* promoter. FEBS Lett. *512*, 71–74.

Vázquez-Bermúdez, M.F., Herrero, A., and Flores, E. (2003). Carbon supply and 2-oxoglutarate effects on expression of nitrate reductase and nitrogen-regulated genes in *Synechococcus* sp. strain PCC 7942. FEMS Microbiol. Lett. *221*, 155–159.

Vázquez-Bermúdez, M.F., Paz-Yepes, J., Herrero, A., and Flores, E. (2002c). The NtcA-activated *amt1* gene encodes a permease required for uptake of low concentrations of ammonium in the cyanobacterium *Synechococcus* sp. PCC 7942. Microbiology *148*, 861–869.

Wei, T.F., Ramasubramanian, T.S., and Golden, J.W. (1994). *Anabaena* sp. strain PCC 7120 *ntcA* gene required for growth on nitrate and heterocyst development. J. Bacteriol. *176*, 4473–4482.

Wei, T.F., Ramasubramanian, T.S., Pu, F., and Golden, J.W. (1993). *Anabaena* sp. strain PCC 7120 *bifA* gene encoding a sequence-specific DNA-binding protein cloned by *in vivo* transcriptional interference selection. J. Bacteriol. *175*, 4025–4035.

Wisén, S., Bergman, B., and Mannervik, B. (2004a). Mutagenesis of the cysteine residues in the transcription factor NtcA from *Anabaena* PCC 7120 and its effects on DNA binding *in vitro*. Biochim. Biophys. Acta *1679*, 156–163.

Wisén, S., Sjogren, T., Olin, B., and Mannervik, B. (2004b). Purification, crystallization and preliminary X-ray data of the transcription factor NtcA from the cyanobacterium *Anabaena* PCC 7120. Acta Crystallogr. D Biol. Crystallogr. *60*, 923–925.

Wolk, C.P., Thomas, J., Shaffer, P.W., Austin, S.M., and Galonsky, A. (1976). Pathway of nitrogen metabolism after fixation of ^{13}N-labeled nitrogen gas by the cyanobacterium *Anabaena cylindrica*. J. Biol. Chem. *251*, 5027–5034.

Xu, P., and McAuley, P.J. (1990). Uptake of amino acids by the cyanobacterium *Anabaena* ATCC 27893. New Phytol. *115*, 581–585.

Xu, Y., Carr, P.D., Huber, T., Vasudevan S.G., and Ollas, D.L. (2001). The structure of the P_{II}-ATP complex. Eur. J. Biochem. *268*, 2028–2037.

Xu, Y., Carr, P.D., Clancy, P., García-Domínguez, M., Forchhammer, K., Florencio, F., Tandeau de Marsac, N., Vasudevan, S., and Ollis, D. (2003). The structures of the P_{II} proteins from the cyanobacteria *Synechococcus* sp. PCC 7942 and *Synechocystis* sp. PCC 6803. Acta Cryst.D 59, 2183–2190.

Yoshimura, H., Yoshihara, S., Okamoto, S., Ikeuchi, M., and Ohmori, M. (2002). A cAMP receptor protein, SYCRP1, is responsible for the cell motility of *Synechocystis* sp. PCC 6803. Plant Cell Physiol. *43*, 460–463.

Zheng, L., Kostrewa, D., Bernèche, S., Winkler, F.K., and Li, X.-D. (2004). The mechanism of ammonia transport based on the crystal structure of AmtB of *Escherichia coli*. Proc. Natl. Acad. Sci. USA *101*, 17090–17095.

Zubkov, M.V., Fuchs, B.M., Tarran, G.A., Burkill, P.H., and Amann, R. (2003). High rate of uptake of organic nitrogen compounds by *Prochlorococcus* cyanobacteria as a key to their dominance in oligotrophic oceanic waters. Appl. Environ. Microbiol. *69*, 1299–1304.

Transcriptional and Developmental Responses by *Anabaena* to Deprivation of Fixed Nitrogen

14

Xudong Xu, Jeff Elhai, and C. Peter Wolk

Abstract

Deprivation of fixed nitrogen, to which the formation of dinitrogen-fixing cells called heterocysts is a conspicuous response, appears to initiate a switch from reliance on photosynthesis for ATP, reductant, and carbon to reliance–in developing cells–on endogenous glycogen stores, heightened utilization of the oxidative pentose phosphate pathway, and–at least in mature heterocysts–photosystem I and influx from vegetative cells. Sugars are needed to produce heterocyst envelope layers of polysaccharide and glycolipid. Decreased transcription and translation of a few highly expressed genes may make possible the increased transcription and translation of many others. Although we interpret metabolic capabilities from microarray data, we stress the hazards of doing so, and emphasize that the interpretations remain to be evaluated. Little is known of the regulation of gene expression during heterocyst differentiation. NtcA is required for the expression of *nrrA* and many other genes, NrrA is required for the full induction of *hetR*, and HetR is required for the expression of many downstream genes. Numerous regulatory-family genes are significantly induced during heterocyst differentiation and may importantly regulate the process. Some genes of unknown function increase many-fold in expression during differentiation, leading one to wonder what their roles may be.

Introduction

Anabaena sp. strain PCC 7120 (to which we will refer as *Anabaena*) is a filamentous cyanobacterium in which, under nitrogen-limiting conditions, vegetative cells differentiate into heterocysts at semi-regular intervals along its filaments. Heterocysts (Figure 14.1A) are cells that are terminally specialized for nitrogen fixation. The interior of these cells is microoxic as a result of increased respiration, inactivation of O_2-producing photosystem (PS) II, and formation of a thickened envelope outside of the cell wall. Nitrogenase, sequestered within these cells, transforms dinitrogen into ammonium at the expense of ATP and reductant—both generated by carbohydrate metabolism, a process that is supplemented, in the light, by the activity of PS I (Wolk *et al.*, 1994). Carbohydrate, probably in the form of sucrose, is synthesized in vegetative cells and moves into heterocysts. In return, nitrogen fixed in heterocysts moves into the vegetative cells, at least in part in the form of amino acids (Curatti *et al.*, 2002; Wolk *et al.*, 1994). The spatial separation of heterocysts within sequences of vegetative cells may reflect the mutual dependence of the two types of cells:

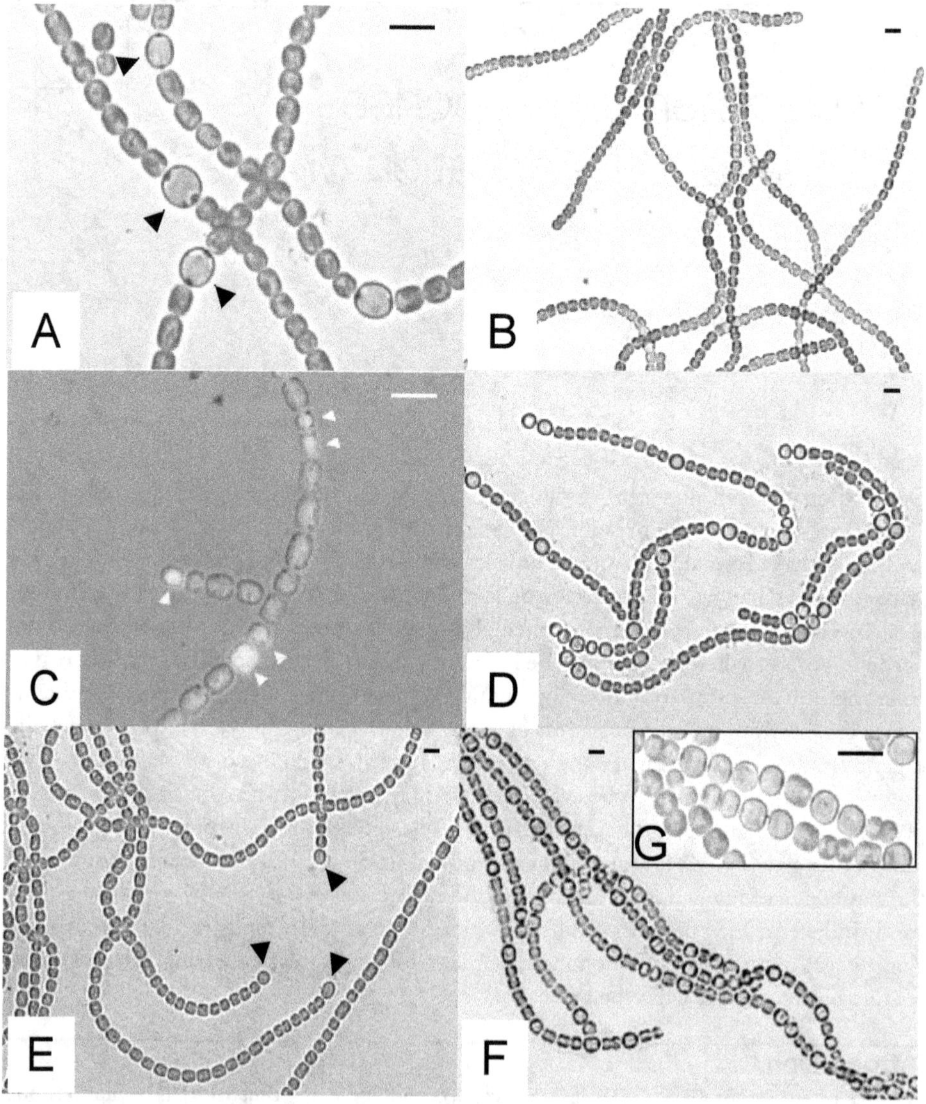

Figure 14.1 (see also Plate 14.1) Wild-type and mutant strains of *Anabaena* sp. strain PCC 7120, deprived of fixed nitrogen in medium $BG11_0$. A. The wild-type strain; arrowheads point to heterocysts. B. *hetR* mutant α41 (Ernst *et al.*, 1992), which is free of heterocysts, as described by Buikema and Haselkorn (1991). C. A *hetC* mutant, *hetC*::C.CE2, that contains a plasmid in which *gfp*, encoding green fluorescent protein, is driven by the promoter of cell-division gene *ftsZ* (Wang and Xu, 2005). *hetC* mutant cells continue to divide after initiating differentiation (arrowheads), as described by Xu and Wolk (2001). D. A fusion of the copper-regulated promoter of *petE* to *hetN* (Zhang and Xu, unpublished), showing a multiple contiguous heterocyst (Mch) phenotype in copper-free $BG11_0$, as described by Callahan and Buikema (2001). E. An insertional mutant of *patA* (Ning and Xu, unpublished), showing long filaments with terminal heterocysts (arrowheads), as described by Liang *et al.* (1992). F and G: A *patS* deletion mutant, in which the *patS*-encoding region has been replaced with the kanamycin-resistance cassette C.K2 (Xu, unpublished), showing a Mch phenotype, as described by Yoon and Golden (1998). All images are bright-field micrographs, except that in panel C, an image of GFP-based fluorescence is superimposed upon the bright-field image. Bars, 5 μm.

vegetative cells require nitrogen from heterocysts, and heterocysts require reductant from vegetative cells.

Heterocyst formation and function pose difficult questions, including the following: (1) what determines heterocyst spacing ("pattern formation"); (2) how does a heterocyst develop from a vegetative cell; (3) what are the detailed interactions between the two types of cells; and (4) what conduits and mechanisms conduct, mediate, and regulate those interactions? As discussed below, the first two of these have been yielding to investigation. In contrast, the third and fourth have proven more resistant, and invite new techniques and hypotheses (Flores *et al.*, 2006). *Anabaena* provides an excellent model for scientific study of one-dimensional pattern formation and the differentiation of prokaryotic cells.

Transcriptional changes during differentiation

Experimental technologies have recently been developed to measure the levels of a broad range of transcripts, proteins, and metabolites. A combination of these technologies with classical approaches may permit a holistic understanding of the progression of developmental processes. Analysis by microarrays, which report on transcript abundance, is the only one of these recently developed technologies that has been applied to the study of heterocyst development; we surely encourage their supplementation by so-called proteomic and metabolomic approaches. Ehira *et al.* (2003) shifted *Anabaena* from a growth medium containing nitrate and ammonium to a medium containing, at most, very small amounts of combined nitrogen, and examined changes in transcription at 1, 3, 8, and 24 hours after the shift. However, their microarrays were spotted with relatively large fragments of DNA that seldom allowed analysis at the level of individual genes. Ehira and Ohmori (2006a)[1] used microarrays consisting of gene-specific probes to examine the changes in abundance of individual transcripts at 3, 8, and 24 hours after shifting *Anabaena* from a medium containing nitrate. Under their conditions, *Anabaena* normally begins to fix dinitrogen after about 18 hours of nitrogen deprivation. Also using gene-specific arrays, Lechno-Yossef *et al.* (2006)[1] compared the effects of four mutations on transcript abundances in *Anabaena* at 14 hours after downshift from nitrate.

Ehira and Ohmori (2006a) focused their attention on the response of a mutant strain to nitrogen-deprivation, but the normal transcriptional responses of the wild-type strain were little considered. One of us (J.E.) has therefore re-analyzed their data to see what can be learned about the normal course of heterocyst differentiation in wild-type *Anabaena*. Significance, unless otherwise noted, is defined by a pairwise t-test ($p < 0.01$) that assesses the likelihood that a difference between the levels of expression in the experimental condition (n hours after nitrogen stepdown) and the control condition (continuous growth on nitrate) may have arisen by chance. Unless otherwise stated, we make note only of changes in genetic expression that appear to be at least two-fold relative to nitrogen-replete (N-replete) controls. Points were omitted from analysis if more than one experimental replicate had a signal intensity less than 1.5 times the background intensity. This procedure removed from consideration 34% to 60% of the genes examined, depending on the experimental condition. In addition, outlier measurements were discarded, according to

[1]Results deposited in the KEGG Expression Database: http://www.genome.ad.jp/dbget-bin/get_htext?Exp_DB+-n+B

the method of Grubbs (Motulsky, 2002). This procedure removed 0.5% to 1.2% of the measurements. Expression ratios after normalization (see below) are expressed as the base 2 logarithm of the ratio of the fluorescence intensity of the experimental condition and the control condition. Supplemental material that includes data for all 5336 chromosomal genes analyzed by Ehira and Ohmori (2006a) (Supplemental Table S1) and a detailed presentation of our methods of normalization and the rationale for that normalization are available at http://www.people.vcu.edu/~elhaij/microarray.

We stress that levels of transcripts do not necessarily indicate levels of proteins, and levels of protein do not necessarily correspond to protein activities. We further emphasize that interpretation of metabolic capabilities from microarray data is hazardous, and that our interpretations remain to be evaluated.

Normalization of intensities of fluorescence

The considerable variation in apparent gene expression on microarray chips must be normalized. Normalizing intensities of fluorescence by loading constant amounts of RNA per microarray cannot alleviate the considerable variation that is due in part to non-biological, chip-to-chip variation of fluorescent intensities. As in many microarray studies, that of Ehira and Ohmori (2006a) normalized signals by the assumption that total signal probed by their microarrays (confined to protein-encoding genes) was constant for every experimental condition and replicate. This assumption led to the result that during nitrogen deprivation, about 70% of the genes appeared to increase in expression (Supplementary Figure S1A). Because there is no independent experimental evidence for such an effect, we chose to normalize signals by making a different assumption, that the median expression of genes remains constant (Supplementary Figure S1B), as would be the case if the majority of genes do not change expression from one condition to another. The results presented in this chapter all rely on the latter normalization. This choice of normalization procedure, like any other, affects expression ratios and perhaps resulting conclusions. For example, both normalization by median expression and normalization by total expression would obscure a conceivable response to nitrogen deprivation: an alteration in the transcriptional apparatus to decrease transcription across the board.

Although customarily, only expression ratios are reported in spotted-array experiments, much is lost by this practice. For example, the reduced expression of a gene from a previously high level may still permit sufficient expression of a protein to allow a metabolic pathway to operate, whereas induced expression of a gene from a very low level might not. We report the expression of each gene relative to the average expression of all genes in the microarrays of Ehira and Ohmori (2006a), under N-replete conditions, set at 100 units. Coupled with the assumption that median gene expression is invariant, this measure permits inferences about the absolute level at which a gene is expressed.

Overall gene expression

Equalizing the median expression ratio (see above) leaves the total adjusted fluorescence intensity free to change from one condition to another. By this analysis, the overall abundance of mRNA decreased in all three experimental conditions, by 22%, 20%, and 6% at 3, 8, and 24 hours, respectively (Table 14.1). That decrease may be attributable, more or less, to the reduced abundance of just a few genes. The genes most highly expressed in *Ana-*

Table 14.1 Overall gene expression

	Adjusted fluorescent intensities[a]					
Category[b]	+N	−N (3 hours)	+N	−N (8 hours)	+N	−N (24 hours)
Total	537840	421388	536701	427354	537292	503183
per gene (5336)	101	79	101	80	101	94
relative to total	100%	78%	100%	80%	100%	94%
Top 1% genes	210562	104001	219648	93376	207388	146296
per gene (53)	3973	1962	4144	1762	3913	2760
relative to total	39%	25%	41%	22%	39%	29%
Phycobilisome protein genes	88894	25087	99949	17622	81993	45180
per gene (23)	3865	1091	4346	766	3565	1964
relative to total	16.5%	4.7%	18.6%	3.3%	15.2%	8.4%

[a]Fluorescent intensities were normalized to total intensity and adjusted to equalize median expression ratios as described in the text, at 0 (N-replete, denoted +N), 3, 8, or 24 hours after the removal of nitrate from the growth medium. Data are taken from Ehira and Ohmori (2006a).

[b]Intensities were summed over all 5336 genes; the 53 genes with the highest expression in the presence of nitrate; and 23 genes related to phycobilisome protein biosynthesis, 15 of which are in the top 1%. The average per gene, i.e., row 2, in the N-replete condition is not precisely 100 because outliers were removed after normalization.

baena under N-replete conditions, once subjected to 3 or 8 hours of nitrogen deprivation had, as a class, remarkably reduced levels of expression (Supplementary Figure S1C). In fact, the reduced expression of the 1% of genes that were most highly expressed under N-replete conditions was sufficient to explain the observed overall drop in mRNA abundance (Table 14.1). Genes that encode phycobiliproteins constitute the single largest subset of this group, but the decrease in expression of highly expressed genes was a general property. The trough shown in Supplementary Figure S1C for highly expressed genes is changed only minimally by removal of the phycobiliprotein genes from consideration.

Of the 5336 *Anabaena* chromosomal genes analyzed by Ehira and Ohmori (2006a), 495 increased significantly in expression at some point during the time course, 196 decreased in expression, and only 6 (4 ribosomal protein genes and two hypothetical genes) did both. A summary of the numbers of genes affected under different experimental conditions and the names of the genes significantly altered in expression may be found in Supplemental Table S2. Fifty-one percent and 71% of those genes whose transcript abundance significantly increased or decreased (respectively) at 3 hours did so at 8 hours as well, and the expression of no gene changed significantly in opposite directions. Those strong correlations suggest that the identification of genes whose expression changed significantly reflected an underlying reality. The correlation was less good between genes affected at 3 hours and 24 hours and between those affected at 8 hours and 24 hours, with similar behavior exhibited by only 2 to 7% of the genes. We conclude that the transcriptional states of *Anabaena* at 3 and 8 hours after nitrogen stepdown overlap considerably, but differ from the 24-h state when nitrogen is again being provided from the fixation of N_2.

Nitrogen metabolism

Confronted with the sudden loss of nitrate as a nitrogen source, *Anabaena* responds by drastically reducing the transcription of genes related to nitrate assimilation and by increasing the transcription of genes related to the assimilation of other nitrogen sources. (There is no conflict with the initial rapid increase of transcription of genes that are related to nitrate assimilation reported by Cai and Wolk (1997a,b) because their experimental material had been grown with ammonia.) The abundance of mRNA from the *nirA-nrtA-nrtB-nrtC-nrtD-narB* region fell at 3 hours to 2% of the N-replete value and stayed low for the duration of the time course. *cnaT*, a specific regulator of the region (Frías *et al.*, 2003), was constitutively expressed, but expression of another specific regulator, *ntcB*, increased significantly by 3 hours. Results consistent with these were reported earlier (Flores and Herrero, 1994; Frías *et al.*, 2003). A second region containing genes annotated *nrtB-nrtC-nrtA* exhibited no significant change. Expression of the genes *alr3287* and *all3035*, annotated, respectively, as (a third) *nrtB* and a nitrate-binding protein, increased significantly at 8 hours from a low level.

The expression of the *urtABCDE* region, encoding a urea transport system, increased significantly at 3 and 8 hours, although genes encoding urease did not. The apparent increase in expression, at 3 hours, of genes *alr0990* and *alr0991* that putatively encode ammonia transporters did not quite reach the level of statistical significance. The expression of both cyanophycin synthetase and cyanophycinase, encoded by *cphA1* and *cphB1*, respectively, increased significantly at 3 hours, though somewhat less than two-fold, and this increase was maintained through 8 hours. The more weakly expressed homolog, *cphA2*, decreased significantly in expression. Expression of the genes that encode the primary NH_3-assimilatory enzymes glutamine synthetase and glutamate synthase (a.k.a. GOGAT) (Flores and Herrero, 1994) changed little throughout the time course.

By 24 hours, all gene expression related to ammonium transport, urea transport, and cyanophycin metabolism was indistinguishable from the initial, N-replete values. By that time, however, the genes *nifHDK* that encode the structural proteins of nitrogenase, and are expressed only in heterocysts (Elhai and Wolk, 1990), were induced in excess of 100-fold, averaged over all cells of the filament. Other *nif* genes, *nifB-fdxN-nifS-nifU* and *nifE-nifN-nifX-nifW*, were induced less strongly but still at a high level of significance. Genes $nifV_1$ (which encodes homocitrate synthase) and *nifZ*, both of which are involved in the biosynthesis of the FeMo-co prosthetic group of dinitrogenase, were induced by 8 hours.

NtcA and GlnB are proteins involved in the regulation of nitrogen metabolism (Flores and Herrero, 2005). Muro-Pastor *et al.* (2002) reported that a clear increase in the abundance of *ntcA* transcripts was shown by a Northern blot between 3 and 6 hours of nitrogen deprivation. Similarly, the microarray data of Ehira and Ohmori (2006a) showed a statistically significant increase of *ntcA* message at 8 but not at 3 hours. *ntcA* is induced especially strongly in developing heterocysts (Olmedo-Verd *et al.*, 2006), but its expression in whole filaments at 24 hours, as judged by microarray analysis, did not differ significantly from its control level. The abundance of transcripts of *glnB*, which encodes the regulatory protein P_{II}, was not found to change significantly during heterocyst differentiation.

In the absence of an external source of nitrogen, the new protein synthesis that accompanies heterocyst differentiation (Fleming and Haselkorn, 1974) is made possible by the breakdown of endogenous proteins, largely phycobiliproteins (Wood and Haselkorn, 1979). Together with the loss of protein, the abundance of transcripts of genes that encode phycobiliproteins decreased (Table 14.1 and Figure 14.2), as described by Wealand *et al.* (1989). The chromosomal gene *nblA*, which is required for the breakdown of phycobiliproteins in *Anabaena* (Baier *et al.*, 2004), appeared to be expressed more strongly than in the N-replete control at all points in the time course, but its increase in expression first reached statistical significance at 8 hours, when it was 10-fold. In *Synechococcus* sp. strain PCC 7942, alanine dehydrogenase, encoded by *ald*, is required for the normal breakdown of phycobiliproteins in response to nitrogen deprivation (Lahmi *et al.*, 2006). In *Anabaena*,

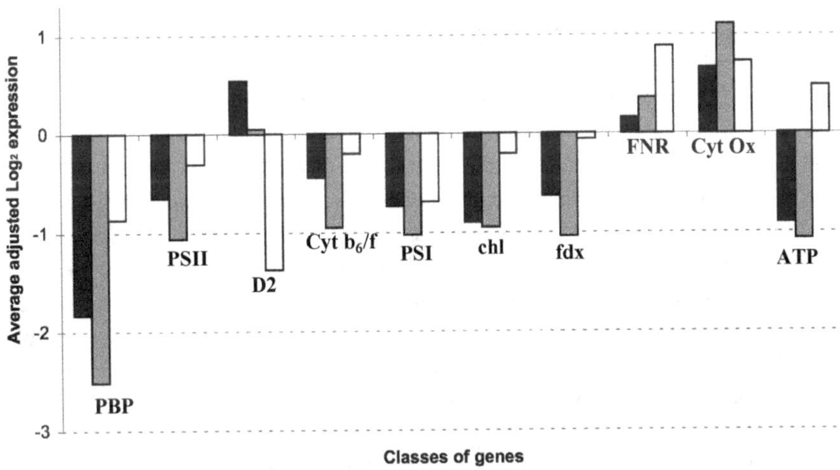

Figure 14.2 Expression of genes related to energy production. Averaged adjusted expression ratios are shown for some classes of genes related to photosynthesis, electron transport, and ATP production. The ratios are expressed as logs in base 2 after nitrogen deprivation for 3 hours (dark grey), 8 hours (light grey), and 24 hours (white). In each class of genes, only those genes with expression above background under all conditions were averaged, and (except as indicated) only one gene per gene name, the one with the greatest gene expression. The genes averaged were: (phycobilisome proteins, PBP) *apcA* (*alr0021*), *apcB*, *apcC*, *apcD*, *apcE*, *apcF*, *cpcA*, *cpcB*, *cpcD*, *cpcE*, *cpcF*, *cpcG1*, *cpcG2*, *cpcG3*, *cpcG4*, *pecA*, *pecB*, *pecC*, and *pecE*; (photosystem II, PSII) *psbB*, *psbC* (*alr4291*), *psbD*, *psbE*, *psbF*, *psbH*, *psbI*, *psbJ*, *psbL*, *psbM*, *psbN*, *psbO*, *psbP* (all3076), *psbT*, *psbU*, *psbV*, *psbW*, *psbX*, *psbY*, and *psbZ*; (photosystem II core protein, D2) *psbAI* and *psbAIV*; (cytochrome b_6/f complex, cyt b_6f) *petA*, *petB*, *petC* (*all2453*), *petD*, *petG*, and *petM*; (photosystem I, PSI) *psaA*, *psaB* (*alr5155*), *psaC*, *psaD*, *psaE*, *psaF*, *psaI*, *psaJ*, *psaK*, *psaL*, *psaM*, and *psaX*; (chlorophyll-specific biosynthesis, chl) *acsF* (*alr1358*, *alr1300*), *chlB*, *chlD*, *chlG*, *chlH* (*all4356*, *alr4773*), *chlI*, *chlL*, *chlM*, *chlN*, *por*; (ferredoxin, fdx) *petF*; (ferredoxin-NADP reductase, FNR) *petH*; (cytochrome oxidase, Cyt Ox) *coxA1*, *coxB1*, *coxC1*, *coxA2*, *coxB2*, *coxC2*, *ctaA*, *ctaB*, and *ctaC*; (ATP synthase complex, ATP) *atp1*, *atpA*, *atpB*, *atpC*, *atpD*, *atpE*, *atpF*, *atpG*, *atpH*, and *atpI*.

the expression of *ald* increased significantly at 3 and 8 hours. The role, if any, of alanine dehydrogenase in the decision of whether cells will continue to differentiate or revert to vegetative condition (see "Spatial localization of differentiation," below), remains unclear.

Carbon metabolism

The loss of phycobiliproteins and concomitant decrease in carbon fixation (Bradley and Carr, 1976) and the observations that transcripts related to PS I proteins and the synthesis of chlorophyll (antenna for PS I) were both markedly reduced early in heterocyst differentiation (Figure 14.2) raise the question of where differentiating filaments get the metabolic energy and carbon needed to make the new complement of proteins and later, the specialized polysaccharide and glycolipid layers required to complete heterocyst development. One likely source of metabolic energy is cyclic photophosphorylation by PS I, which is active in vegetative cells (Kallas, 1994) and heterocysts (Peterson and Wolk, 1978; Peterson *et al.*, 1981), and may remain active in developing cells. A source of both energy and carbon that is applicable to immature heterocysts and may be applicable earlier is an influx of carbon, probably as sucrose, from vegetative cells (Wolk, 1968; Wolk *et al.*, 1994; Curatti *et al.*, 2002). Yet another source of energy and carbon can be seen in the changes elicited by nitrogen deprivation in the pattern of transcription of genes that encode enzymes related to carbon metabolism (Figure 14.3). By 8 hours after the onset of nitrogen deprivation, the levels of expression of genes specific to the Calvin cycle (*rbcL*, *rbcS*, and *prk*) dropped to 44% of their values at 0 hours, and the expression of genes common to the Calvin cycle and gluconeogenesis also diminished by about 50%. In contrast, the expression of the four genes specific to the oxidative pentose phosphate (OPP) pathway (encoding glucose-6-phosphate dehydrogenase, 6-phosphogluconolactonase, 6-phosphogluconate dehydrogenase, and transaldolase) increased, on average, 135%. The abundance of transcripts of the two genes specific to glycogen synthesis decreased 59%, while the transcript level of the one gene specific to glycogen breakdown increased 5-fold. There is no evident pattern in the expression of genes encoding glycolytic enzymes beyond 3-phosphoglycerate.

One might consider from the foregoing that early in the process of heterocyst differentiation, to provide carbon, reductant, and ATP, *Anabaena* turns in part from a reliance on photosynthesis to reliance on glycogen-derived glucose-6-phosphate and its subsequent catabolism through the OPP pathway. However, previous experimental results do not support the idea that glycogen is broken down in response to nitrogen deprivation. Rather, measurements in three strains of *Anabaena* indicate that there is a net increase in glycogen throughout the process of differentiation (Ernst and Böger, 1985). If glycogen increases early during differentiation, as their experimental results indicate, and glycogen is simultaneously broken down, as implied by the observations summarized in Figure 14.3, either a costly futile cycle is in operation or glycogen synthesis and breakdown take place in separate cellular compartments or, more likely, in different types of cells.

Differing types of cells are well known to be present, namely, cells that express *hetR* strongly and are poised to differentiate and cells that express *hetR* weakly and will remain vegetative (Black *et al.*, 1993). We suggest that within a few hours after the onset of nitrogen deprivation, the cells in filaments of *Anabaena* have already become differentiated as to their central metabolism. Cells in which expression of *hetR* does not increase respond to nitrogen deprivation by storing, and perhaps already transferring, carbon that, in the

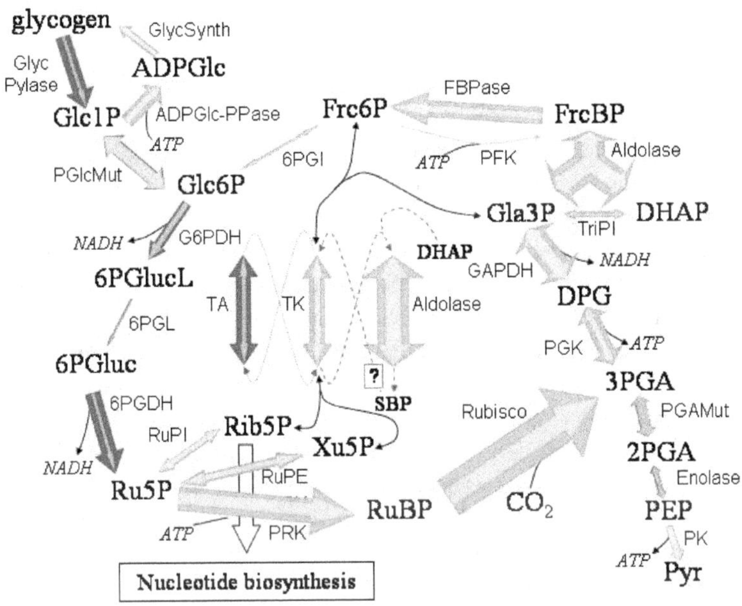

Figure 14.3 (see also Plate 14.3) Expression levels of genes related to central carbon metabolism. Expression levels at 0 vs 8 hours after nitrogen stepdown are shown for genes encoding enzymes of the Calvin cycle, the oxidative pentose phosphate (OPP) pathway, glycolysis, and glycogen metabolism. Green external arrows and orange internal arrows represent the RNA levels of genes in the presence of NO_3^-. Red external arrows and yellow-green internal arrows represent the RNA levels of genes at 8 hours after nitrogen stepdown. The widths of the two arrows for a single gene are linearly proportional to the levels of RNA under the two experimental conditions. The widths of arrows of two different genes are not linearly proportional to their RNA levels, but genes with greater levels of expression have wider arrows. A non-linear representation was necessary in order to present the full 300-fold dynamic range of RNA levels. Green dashed lines indicate the route taken by the Calvin cycle to convert Frc6P + Gla3P to Xu5P + Ery4P through TK, then (+DHAP) to SBP, through aldolase, then to Sed7P through an unknown enzyme "(?)," then (+Gla3P) to Xu5P + Rib5P through TK. Red dotted lines indicate the route taken by the OPP pathway to convert Rib5P + Xu5P to Sed7P + Gla3P through TK, then to Frc6P + Ery4P through TA, then (+ Xu5P) to Gla3P + Frc6P through TK. Compound abbreviations: ADP-Glc, ADP-glucose; Glc1P, glucose-1-phosphate; Glc6P, glucose-6-phosphate; 6PGluc, 6-phosphogluconate; Ru5P, ribulose-5-phosphate; Rib5P, ribose-5-phosphate; Xu5P, xylulose-5-phosphate; Ery4P, erythrose-4-phosphate; Sed7P, sedoheptulose-7-phosphate; SBP, sedoheptulose-1,7-bisphosphate; RuBP, ribulose-1,5-bisphosphate; Frc6P, fructose-6-phosphate; FrcBP, fructose-1,6-bisphosphate; Gla3P, glyceraldehyde-3-phosphate; DHAP, dihydroxyacetone phosphate; DPG, 1,3-diphosphoglycerate; 3PGA, 3-phosphoglycerate; 2PGA, 2-phosphoglycerate; PEP, phosphoenolpyruvate; Pyr, pyruvate. Enzyme abbreviations: (*Calvin cycle-specific*) PRK, phosphoribulose kinase; Rubisco, ribulose-bisphosphate carboxylase; (*OPP-specific*) G6PDH, glucose-6-phosphate dehydrogenase; 6PGlucL, 6-phosphogluconolactonase; 6PGDH, 6-phosphogluconate dehydrogenase; TA, transaldolase; (*glycolysis-specific*) PFK, phosphofructokinase; PGAMut, phosphoglycerate mutase; PK, pyruvate kinase; (*Calvin cycle/OPP overlap*) RuPI, ribulose phosphate isomerase; RuPE, ribulose phosphate epimerase; TK, transketolase; TriPI, triose phosphate isomerase; FBPase, fructose bisphosphatase; 6PGI, 6-phosphoglucose isomerase; (*Calvin cycle/glycolysis overlap*) PGK, phosphoglycerate kinase; GAPDH, glyceraldehyde phosphate dehydrogenase; (*glycogen metabolism*) PGlcMut, phosphoglucomutase; ADPglc-PPase, ADP-glucose-pyrophosphorylase; GlycSynth, glycogen synthase; GlycPylase, glycogen phosphorylase.

absence of a nitrogen source, cannot be used to synthesize new amino acids. Contrariwise, cells in which expression of *hetR* increases have a greater need (per cell) for sugars and ATP to allow the synthesis of heterocyst-specific proteins; to make envelope polysaccharide and later, to provide the hydrophilic head group of the envelope glycolipids; and eventually for precursors, ATP and reductant to make the acyl portion of the glycolipids (see *Formation of the heterocyst envelope*, below). They break down glycogen, using some of the sugar for synthesis and the rest as fuel.

The existence of two such cell types could also explain the difference in transcription of two genes, *alr1041* and *all4021*, that encode fructosebisphosphatase, required by both the Calvin cycle and the OPP pathway. Expression of *alr1041* decreased 3.6-fold, while *all4021*, which is part of a conserved OPP-specific operon (Hagen and Meeks, 2001; Newman *et al.*, 1995), increased in expression. The ortholog of *alr1041* in *Synechocystis* sp. strain PCC 6803 acts on sedoheptulosebisphosphate as well as fructosebisphosphate, whereas the ortholog of *all4021* acts only on fructosebisphosphate (Tamoi *et al.*, 1998). (Our usage of the word "ortholog" is based on Fitch [2000] and on considerations arising from discussion by Jensen, Gerlt, and Babbitt[2] and is given a quantitative expression in the section "Regulatory genes," below, and as a footnote to Table 14.2). If, as is not wholly clear (Crawford *et al.*, 1984), the activities of the orthologous pairs are the same, then the gene determining the two activities required for the Calvin cycle was significantly decreased by nitrogen deprivation by 3 hours, and the gene determining the activity required by the OPP pathway was significantly increased. By 24 hours after the onset of deprivation, the levels of transcripts for enzymes specific to the Calvin cycle returned to N-replete levels whereas genes related specifically to the OPP pathway remained about as highly expressed as they were at 8 hours. Part of this sustained expression may be attributable to the high level of expression of OPP genes in mature heterocysts (see *Cytoplasmic differentiation*, below).

It should be noted that exclusive focus on expression ratios, typical of microarray analyses, would confuse one's view of the metabolic response of *Anabaena* to nitrogen deprivation. For example, at 8 hours into the process, expression ratios of 7 of the 8 genes that are shared by the OPP pathway and the Calvin cycle were less than 1, ranging from 0.35 to 0.76. Unless one recognizes that the levels were initially high, one might conclude, we think erroneously, that the oxidative pentose phosphate pathway as a whole was little active.

Energy production

Anabaena invested heavily in photosynthesis: when it was growing with nitrate, more than half of the top 1% of its apparently most highly expressed genes were related to photosynthesis. Removal of fixed nitrogen from the medium, however, elicited a major change: the proteinaceous products of many of those same genes became a rich source of amino acids. The considerable decrease in RNA related to antenna phycobiliproteins has already been noted (see *Nitrogen metabolism*, above), but lesser decreases were also seen in the abundance of messages related to other components of photosynthetic electron transport and ATP production (Figure 14.2). In the cases of genes that encode proteins of PS II

[2]http://www.pubmedcentral.nih.gov/articlerender.fcgi?artid=138949

Table 14.2 Expression, during heterocyst differentiation, of previously described regulatory genes

Gene[a]	Name	Function[b]	Known Role	Family[c]	Peak[d]	Log_2 expression ratio[e] 3 h	8 h	24 h	References
all1940	*abp1*	BP; protease		L	341	0.01	0.23	–0.7	Koksharova and Wolk, 2002
all1939	*abp2*	BP, protease	Heterocyst glycolipid	L-1	69	0.26	0.00	–0.92	Koksharova and Wolk, 2002
alr3608	*abp3*	BP, partial similarity to endoglucanase	Heterocyst glycolipid	H	159	–0.25	0.14	–0.26	Koksharova and Wolk, 2002
alr4240	*abp4*	BP; transporter		G-1	44	–0.23	–0.29	–	Koksharova and Wolk, 2002
alr1010	*ccbP*	Binds Ca^{+2}	Heterocyst differentiation	H+1	53	0.02	0.12	–0.12	Zhao *et al.*, 2005
all0601	*cnaT*	Glyc transferase?	*nir* operon regulation	C-1	59	–0.11	0.17	0.29	Frías *et al.*, 2003
all0187	*conR*	TR–LytR	Heterocyst function	G+1	203	**0.8**	**1**	0.18	Fan *et al.*, 2006
alr0295	*crpA*	Binds cAMP	?	G+3	34	0.45	–0.04	–	Suzuki *et al.*, 2004
alr2325	*crpB*	Binds cAMP	?	N+2	0	–	–	–	Suzuki *et al.*, 2004
all1118	*cyaA*	Ade cyclase	?	L–2	0	–	–	–	Katayama and Ohmori, 1997
alr2266	*cyaB1*	Ade cyclase	?	F+2	0	–	–	–	Katayama and Ohmori, 1997
all1904	*cyaB2*	Ade cyclase	?	F	0	–	–	–	Katayama and Ohmori, 1997
all4963	*cyaC*	Ade cyclase	Salt response	N	60	0.36	0.3	–	Imashimizu *et al.*, 2005
all0743	*cyaD*	Ade cyclase	?	H	–	–	–	–	Katayama and Ohmori, 1997
alr3712	*devA*	ABC transporter	Heterocyst glycolipid	H+2	63	**0.98**	**4.77**	**3.05**	Cai and Wolk, 1997a; Fiedler *et al.*, 1998a
alr3952	*devH*	TR–Crp	Heterocyst glycolipid	F	964	0.1	**2**	**0.29**	Ramírez *et al.*, 2005

Table 14.2 continued

Gene[a]	Name	Function[b]	Known Role	Family[c]	Peak[d]	Log_2 expression ratio[e] 3 h	8 h	24 h	References
alr0442	*devR*$_A$	RR	Heterocyst polysaccharide	G+1	–	–	–	–	Campbell *et al.*, 1996; Zhou and Wolk, 2003
alr3858	*ftsZ*	Septum protein	Cell division	C	120	–0.11	–0.12	–0.71	Sakr *et al.*, 2006
all1691	*furA*	TR–Fur	Iron regulation	C	118	0.3	0.5	0.55	Hernández *et al.*, 2006
all2473	*furB*	TR–Fur	Modulate *furA* transcription?	C	39	–	0.1	–	López–Gomollón *et al.*, 2006
alr0957	*furC*	TR–Fur	Modulate *furA* transcription?	H+5	244	**0.8**	**0.7**	0.16	López–Gomollón *et al.*, 2006
all2319	*glnB*	Enzyme modulation	Nitrogen regulation	C	84	0	0	–0.3	Laurent *et al.*, 2004
alr0093	*hcwA*	Peptidoglycan hydrolysis	Heterocyst polysaccharide	H	94	0.42	**1.27**	0.1	Zhu *et al.*, 2001
alr1086	*henR*	RR	Heterocyst envelope	N+1	–	–	–	–	Fan *et al.*, 2006
all4496	*hepK*	HK	Heterocyst polysaccharide	G–1	387	0.2	**0.8**	**0.42**	Zhou and Wolk, 2003; Zhu *et al.*, 1998
alr0117	*hepN*	HK	Heterocyst polysaccharide	N+1	17	**1.9**	**1.4**	–	Fan *et al.*, 2006; Ning and Xu, 2004
all2760	*hepS*	STK	Heterocyst polysaccharide	N	57	–	**3.1**	–	Fan *et al.*, 2006
alr2817	*hetC*	ABC transporter	Pleiotropic; required for cessation of cell division	A	–	–	–	–	Khudyakov and Wolk., 1997; Wang and Xu, 2005; Xu and Wolk, 2001

alr3546	*hetF*	protease?	Heterocyst differentiation	F	–	–	–	–	Wong and Meeks, 2001; Aravind and Koonin, 2002
alr5358	*hetN*	ketoacyl reductase	Heterocyst differentiation	H	347	**0.8**	**1.9**	5.14	Black *et al.*, 1993; Callahan and Buikema, 2001
alr2818	*hetP*	?	Heterocyst differentiation	H	635	**0.7**	**7.1**	**2.72**	Fernández–Piñas *et al.*, 1994
alr2339	*hetR*	BP, peptidase	Heterocyst differentiation	F	462	**0.6**	**1.2**	0.09	Buikema and Haselkorn, 1991
alr2300	*hetY*	?	Heterocyst maturation	C–1	0	–	–	–	Yoon *et al.*, 2003
alr4908	*lexA*	TR–lex, peptidase	SOS function?	C–2	33	0.1	–	–0.32	Mazón *et al.*, 2004
all1964	*manR*	RR	Manganese uptake?	L	–	–	–	–	Huang and Wu, 2004
asr4517	*nblA*	Binds PBP	Phycobilisome degradation	H	82	2.7	**3.3**	–	Baier *et al.*, 2004
all4312	*nrrA*	RR	Heterocyst differentiation	L	992	**3.3**	**3.9**	**1.82**	Ehira and Ohmori, 2006a,b; Muro–Pastor *et al.*, 2006
alr4392	*ntcA*	TR–Crp	Nitrogen regulation	C	596	0.3	**2.6**	0.51	Frías *et al.*, 1993; Herrero *et al.*, 2001; Wei *et al.*, 1993
all0602	*ntcB*	TR–LysR	Nitrogen regulation	L	176	**1.41**	**1.32**	0.12	Frías *et al.*, 2000
alr3768	*orrA*	RR	Osmotic regulation	H	24	–0.94	–	–0.48	Schwartz *et al.*, 1998
all0521	*patA*	RR	Heterocyst differentiation	H	31	0.4	–0.8	–0.29	Liang *et al.*, 1992
all2512	*patB*	TR–XRE	Heterocyst differentiation	N	–	–	–	–	Jones *et al.*, 2003; Liang *et al.*, 1993
asl2301	*patS*	Binds HetR	Heterocyst differentiation	F	85	–	**3.3**	0.07	Yoon and Golden, 1998

Table 14.2 continued

Gene[a]	Name	Function[b]	Known Role	Family[c]	Peak[d]	Log_2 expression ratio[e] 3 h	8 h	24 h	References
alr2502	*pkn22*	STK	Iron regulation	A+1	67	0.4	–0.1	0.2	Xu *et al.*, 2003
alr0709	*pkn41*	STK/HK	Iron regulation	U	23	1.4	1.9	–	Cheng *et al.*, 2006
alr0710	*pkn42*	STK/HK	Iron regulation	U	–	–	–	–	Cheng *et al.*, 2006
alr4366	*pknA*	STK	Heterocyst differentiation	A+2	115	–0.2	–0.2	–0.01	Zhang, 1993
all4813	*pknC*	STK	?	H+1	38	0.7	**1.3**	–	Gonzalez *et al.*, 2001
alr4368	*pknD*	STK	Nitrogen regulation	H+1	–	–	–	–	Zhang and Libs, 1998
alr3732	*pknE*	STK	Heterocyst maturation	H	115	**1.1**	**1.6**	0.07	Zhang *et al.*, 1998
alr3731	*prpA*	STP	Heterocyst maturation	A+2	–	–	–	–	Zhang *et al.*, 1998
all1731	*prpJ*	STP	Heterocyst maturation	G+1	87	**1.2**	**1.2**	–	Jang *et al.*, 2006
alr4516	*prpS*	STP	Modification of GlnB (PII)	G+1	110	–0.1	0.1	–0.26	Laurent *et al.*, 2004
all5263	*sigA*	sigma group 1	Housekeeping	C	430	**0.1**	–0.4	–0.7	Brahamsha and Haselkorn, 1991
all1692	*sigC*	sigma group 2	?	C–3	51	**1.9**	**2.4**	–	Brahamsha and Haselkorn, 1992
alr3810	*sigD*	sigma group 2	?	C–2	194	–0.4	–0.2	–0.99	Khudyakov and Golden, 2001
alr3800	*sigE*	sigma group 2	?	G+3	256	0	0.1	0.12	Khudyakov and Golden, 2001
alr4249	*sigF*	sigma group 2	?	G+1	31	0.4	1.3	1.11	Khudyakov and Golden, 2001

[a]Genes whose function or role has been experimentally demonstrated in a strain of *Anabaena* or *Nostoc.* The role may be indirect or a demonstration that mutation of the gene leads to significant pleiotropic effects. Expression results are taken from Ehira and Ohmori (2006a)

[b]Biochemical function or structural family: BP, DNA-binding protein; HK, two-component system sensor histidine kinase; PBP, phycobiliprotein; RR, two-component system response regulator; STK, Ser/Thr protein kinase; STP, Ser/Thr protein phosphatase; TR-xxx, transcriptional regulator in the given family (xxx);

[c]Range of conservation of cyanobacterial orthologs. (An ortholog is defined here as a unique gene in the target organism that is the bidirectional [reciprocal] best Blast hit with an E value better than 10^{-6}.) The letters describe the level of conservation, derived from the phylogenetic tree of Ashby and Houmard (2006). At present, the groups are strictly nested. As more genomes are sequenced, this may no longer be the case, particularly with regard to N_2-fixing strains. Symbols: **U**: uniquely present in *Anabaena* sp. strain PCC 7120; ***A**nabaena*, the 4180 genes with common orthologs in *Anabaena* sp. strain PCC 7120 and *Anabaena variabilis* strain ATCC 29413; **H**eterocyst-forming, the 3413 genes with common orthologs in *Anabaena* spp. or *Nostoc punctiforme*; **F**ilamentous, the 2061 genes with common orthologs in a heterocyst-forming strain or in *Trichodesmium erythraeum*; **N_2**-fixing, the 1711 genes with common orthologs in a filamentous strain or in *Crocosphaera watsonii*; **G**roup, the 1461 genes with common orthologs in an N_2-fixing strain or in *Synechocystis* sp. strain PCC 6803; **L**arge, the 989 genes with common orthologs in the *Anabaena/Synechocystis* group or in *Synechococcus* sp. strain PCC 7942, *Thermosynechococcus elongatus* strain BP1, or *Gloeobacter violaceus*; **C**ore, the 744 genes with common orthologs in all available cyanobacterial genomes, including the above plus *Synechococcus* sp. strain WH8102, *Prochlorococcus marinus* strain MED4, *P. marinus* strain MIT9313, and *P. marinus* strain SS120. If the orthologs of the gene are not coextensive with the group, this is indicated by the number of orthologs by which the group is supplemented or diminished.

[d]The highest expression of the gene after removal of nitrogen during the experiment. Units are given in relation to the average expression of all genes in the presence of NO_3^-, defined as 100.

[e]The $\log_2$ ratio of expression at the indicated time compared to expression in the presence of NO_3^-. Significantly enhanced expression ($p< 0.01$) is indicated in bold.

– Gene expression not detected.

(excluding *psbA*), PS I, chlorophyll-specific biosynthesis, the cytochrome b_6/f complex (which intervenes between PS II and PS I), ATP synthase, and the most abundant ferredoxin, the patterns were similar: message abundance decreased at 3 hours, even more so at 8 hours, and returned approximately to normal levels by 24 hours.

In contrast, the abundance of transcripts of genes related to respiration, in particular those that encode cytochrome oxidase (Figure 14.2; see also *Cytoplasmic differentiation*, below), increased throughout the first 24-hour period. Ferredoxin-NADPH reductase (a.k.a. FNR, PetH, and All4121), in addition to its primary role as a recipient of electrons from PS I, may function also as an NADPH dehydrogenase (Schmetterer, 1994). Transcripts of *petH* also increased in abundance throughout that period.

Regulatory genes

It has been natural in the study of heterocyst differentiation to seek out the molecular agents that appear to direct the process. The list of known regulatory genes—taken in a broad sense—shown in Table 14.2 anticipates the discussion of our current understanding (see below) of "Mechanisms of the heterocyst developmental regulatory cascade" that underlie heterocyst differentiation. Some interesting generalizations emerge on consideration of known regulators of heterocyst differentiation. As one might expect, such regulators tend to be conserved amongst heterocyst-forming cyanobacteria. Of the 26 genes so identified in Table 14.2, 92% (all but *hetC* and *prpA*) have orthologs in *Anabaena variabilis* and *Nostoc punctiforme*, the other two heterocyst-forming strains whose genomes have been sequenced. When we write that genes in two organisms are orthologs, we mean that they are homologs with an E-value below 10^{-6} and are also bidirectional (reciprocal) best BLAST hits in those two organisms. By contrast, 43% of the total genes of *Anabaena* lack orthologs in at least one heterocyst-forming cyanobacterium. Three regulatory genes, *ntcA*, *nrrA*, and *hetR*, that act early and very pleiotropically to direct the response to nitrogen deprivation are all highly expressed: at their peaks, 6, 10, and 5 times the average level of expression of an *Anabaena* gene, respectively.

The gene *nrrA* (*all4312*) encodes a presumptive two-component response regulator that attracted the attention of two laboratories. Because there is no NtcA-binding site upstream from *hetR*, Muro-Pastor *et al.* (2006) sought a DNA-binding protein regulated by NtcA that might explain how expression of *hetR* could be regulated indirectly by NtcA. They identified a sequence 5′ from *all4312* as scoring high in a global search for NtcA binding sites upstream from conserved regulatory genes, and showed that NtcA binds upstream from *nrrA* and affects its transcription. Ehira and Ohmori (2006a) identified *all4312* as the transcriptionally most highly induced presumptive regulatory gene after 3 hours of nitrogen deprivation (Table 14.2). They determined that an *nrrA* deletion mutant of *Anabaena* exhibits normal growth with NO_3^- as a nitrogen source, but its growth on N_2 is delayed about 24 hours in the mutant relative to the wild-type strain. They then globally analyzed the relative abundances of mRNAs in strains of *Anabaena* defective in *all4312*, relative to their abundances in the wild-type strain. By 3 hours of nitrogen deprivation, 29 genes were significantly altered in expression (5 upregulated and 24 downregulated) in an *nrrA* mutant relative to the wild-type strain. Those numbers increased at 8 hours (13 upregulated and 48 downregulated) and increased further at 24 hours (177 upregulated and 71 downregulated). Northern analysis confirmed that functional *nrrA* is required for the full increase of *hetR* transcription that normally follows nitrogen stepdown (Ehira and Ohmori, 2006a), specifically at promoters affected by N-deprivation (Ehira and Ohmori, 2006b). NrrA has subsequently been shown to bind upstream from the two NtcA-regulated promoters of *hetR* and when over-expressed further increased expression of *hetR* (Ehira and Ohmori, 2006b).

These results are consistent with the idea that NtcA acts on *hetR* using NrrA as an intermediary and that the complicated phenotype of the strain lacking functional *nrrA* is a consequence of the gene's participation in *hetR* regulation. As a result, early activity of NrrA has secondary effects that can spread as heterocyst differentiation progresses. However, NrrA may directly regulate other genes besides *hetR*. For example, the normal ~3-fold increase 3 hours after nitrogen stepdown in the gene that encodes glycogen phosphorylase is blocked in the *nrrA* mutant. A failure of glycogen breakdown to provide glucose-6-phosphate (see *Carbon metabolism*, above) may itself delay heterocyst maturation. NrrA appears to be a response regulator of a two-component regulatory system. It remains to be determined which of the many histidine kinases of *Anabaena* controls its activity.

Table 14.2 may serve as a guide for those who may wish to use a list (Table 14.3) of transcriptionally activated but as yet unstudied regulatory genes to search for those that may be important in heterocyst differentiation. Table 14.2 illustrates that whereas most genes that are known to regulate the process of heterocyst differentiation were themselves transcriptionally regulated, some—such as *glnB* [see *Nitrogen metabolism*, above]—appeared not to be thusly regulated. Most of the genes were expressed sufficiently strongly to be detected, but some (*devR*, *henR*, *hetF*, *patB*, *pknD*, and *prpA*) were not. Others may not appear in Table 14.3 because, like *pknA*, they are significantly induced only transiently, at a time not sampled, or, like *patA* (Wang and Xu, 2005), specifically induced in a small subset of cells. Finally, genes may appear in a list of candidate regulatory genes only if they are recognizable as regulatory genes, and such important regulatory genes as *hetN*, *hetR*, and *patS* are not members of established classes of regulatory genes. All such genes would be difficult to identify as possibly regulatory without experimentation.

Nonetheless, potentially interesting genes are found in the list (Table 14.3) of unstudied regulators. Two putative anti-anti-sigma factors (Serrano *et al.*, 2004), encoded by *alr4329* and *alr4800*, are conserved amongst heterocyst-forming cyanobacteria, highly expressed, and significantly induced early in the course of heterocyst differentiation. Conserved genes that were significantly induced by 8 hours of nitrogen deprivation include a hybrid histidine kinase/response regulator encoded by *all3765* and several genes with lower levels of expression, including a group 3 sigma factor encoded by *alr3280*.

Unknown and hypothetical genes

When the genomic sequence of *Anabaena* was first described (Kaneko *et al.*, 2001), about 55% of the chromosomal genes of *Anabaena* were annotated "hypothetical" or "unknown." Because genes can shift between these terms depending on the stringency of the definition of similarity applied to pairs of proteins, we have chosen to pool the two groups. Given how little is known concerning the differentiation process, it appears likely that some of the genes of unknown function are highly important. Just as bioinformatics methods identified *all4312* (*nrrA*) as especially worthy of experimental consideration, perhaps such methods can bring to light additional genes that warrant special attention.

Table 14.3 Expression of unstudied regulatory genes[a]

Gene[b]	Function	Family	Peak	Log2 Expression Ratio		
				3 h	8 h	24 h
all0438	STK/WD	G	60	–	**3.83**	–
alr0548	STK	F+2	22	–0.74	**–1.07**	**–1.68**
alr0642	HK	H–1	51	–1.61	–1.77	**–1.24**
all1692	Anti-anti-sigma	N+1	54	–	**2.27**	–
alr1721	TR	H–1	37	0.98	**1.34**	0.32
alr2178	TR	U+1	39	–	**1.29**	–0.08
alr2306	RR	N–1	55	0.96	**1.29**	0.4
alr2479	TR	C–2	76	0.39	**1.05**	0.31
alr2572	HK	C	23	–	**1.67**	0.85
alr2587	TR	U+1	26	–	**2.63**	–
alr2591	TR	U	58	–	**1.66**	1.01
all2821	RR	H+3	26	**1.27**	–	–
all2874	RR	G–1	45	0.66	**1.3**	1.53
alr3280	sigma group 3	L	68	0.93	**1.12**	–
alr3646	TR	A+1	15	**1.67**	1.75	0.22
all3765	HK/RR	H	125	0.07	**2.02**	**2.09**
alr4329	anti–anti–sigma	G–1	360	**2.82**	**3.17**	1.24
all4518	STK	L	11	–	**2.12**	–1.99
alr4800	anti–anti–sigma	F	490	**2.32**	2.76	**1.18**
all4838	STK	H3	208	**–1.47**	–0.99	0.94
alr5188	RR	A+1	25	–	**1.71**	–
alr5272	HK	H	59	0.37	0.2	**–1.2**

[a]Abbreviations and conventions as in Table 14.2. Expression results are taken from Ehira and Ohmori (2006a).

[b]Genes were chosen from among those annotated as sigma factors and their modulators, transcriptional regulators (TR), two-component system sensor histidine kinases (HK) or response regulators (RR), and Ser/Thr protein kinases (STK) or phosphatases. Each gene shown has at least a two-fold, statistically significant change (bold) in RNA abundance at some point during the time course.

We used two methods to reduce the number of candidate genes to a manageable level. We considered, as a first screen, that a gene important in heterocyst differentiation would be conserved amongst the three heterocyst-forming cyanobacteria (*Anabaena*, *A. variabilis*, and *N. punctiforme*) for which genome sequences are available. Seventy percent of unknown genes significantly altered in expression during heterocyst differentiation have orthologs in all three organisms. Thus, this screen has only limited value for the current purpose. We based a second screen on our observation that many of the genes that are most highly expressed in absolute terms during heterocyst differentiation are known to be important in heterocyst differentiation (e.g. *ntcA*, *hetR*, *devH*, *nrrA*, and also *hep* genes in the cluster *alr2825*-*alr2841*). Therefore, we assembled a list of 28 functionally unidentified genes that (1) passed the first screen; (2) showed significantly increased or decreased expression at 3, 8, or 24 hours after nitrogen stepdown; and (3) exhibited at some point in the experiment a level of expression at least twice that of the average gene under N-replete conditions (a property shared with 10% of *Anabaena* genes). The resulting 43 genes (1.5% of the original list of functionally unidentified genes) are shown in Table 14.4. Six other genes are included in the list because, although not highly expressed, they have unusually high expression ratios.

Some of the genes had interesting characteristics. For example, *all4578* was the most highly expressed, presumptively protein-encoding gene over the course of heterocyst differentiation. Gene *alr0099* is in a region that has recently been experimentally linked to the control of early heterocyst differentiation (Zhang, Ning, Wolk, Fan, and Xu, unpublished; Khudyakov and Zaizev, 2006). Six genes (*alr2822* through *alr2824*, *alr2826*, *alr2829*, and *alr2833*) are in the cluster surrounding *hepA* (Ehira *et al.*, 2003) and near *hetP* and *hetC* that is the most highly induced region of the *Anabaena* chromosome during nitrogen deprivation and is further discussed, below, in the section *Formation of the heterocyst envelope*. Genes *all1424*, *asl2052*, and *alr3676*, like the six just mentioned in the region *alr2822* through *alr2833*, all showed an average increase of expression of more than 25-fold ($\log_2[25] = 4.644$) at 8 hours of nitrogen deprivation. The cluster *all2002* through *all2006* consists of four genes that lack a functional annotation and a fifth that is annotated as a cation transporter, all of which were highly expressed and significantly induced throughout the differentiation process. It will be of much interest to learn what functions all of these genes may have.

Spatial localization of differentiation

A distinctive feature of heterocyst differentiation by *Anabaena* is that normally, only well-spaced cells complete the process. 2-Oxoglutarate (2-OG, a.k.a. α-ketoglutarate), a precursor of glutamate and glutamine and typically an intermediate of the citric acid cycle, plays a critical role in determining the fate of cells in the filament deprived of nitrogen (Zhang *et al.*, 2006). This may seem surprising, but in cyanobacteria, the citric acid cycle is incomplete because of the absence of 2-oxoglutarate dehydrogenase (Kaneko *et al.*, 2001; Stanier and Cohen-Bazire, 1977). Consequently, the concentration of 2-OG can change with the level of available fixed nitrogen and therefore can represent the nitrogen status of these microorganisms (Muro-Pastor *et al.*, 2001), perhaps measuring the balance between the availabilities of nitrogen and other nutrients (see Chapter 13). In *Anabaena* expressing a 2-OG permease, addition of exogenous 2-OG promotes heterocyst differentiation (Li

Table 14.4 Expression of screened, functionally unidentified genes[a]

				Log$_2$ expression ratio		
Gene[b]	Family	+N peak[c]	–N peak[d]	3 hours	8 hours	24 hours
alr0045	C	756	764	0.07	–0.18	**–1.04**
asr0064	G+2	734	322	**–1.21**	**–1.71**	**–1.85**
alr0099	F	395	1535	**1.97**	**1.04**	**–0.66**
alr0114	G+2	339	653	–0.03	**1.09**	–0.06
asr0460	H	155	44	**–1.73**	**–3.10**	–3.05
all0565	H+2	210	425	**1.11**	0.10	–0.52
alr0600	H	223	76	–1.91	–1.81	**–1.65**
asl0845	C	217	150	**–0.54**	**–1.03**	–0.63
asr0905	H+1	34	384	**1.90**	**3.70**	**2.79**
alr1160	C–2	212	409	**0.42**	**1.02**	**0.40**
alr1356	C–1	83	204	**0.85**	**1.32**	0.24
all1424	N–1	22	487	0.61	**4.67**	**1.93**
alr1529	H	153	209	**–1.06**	**–0.65**	**0.32**
alr1713	H+1	60	694	**3.13**	**3.71**	**1.93**
asr1714	H+3	34	134	**3.16**	0.90	1.69
all1812	H	8	25	0.13	**3.05**	1.46
all1871	L	2892	9074	**1.69**	**1.75**	**–1.48**
all1872	L	450	1201	**1.23**	**1.25**	–1.28
all1873	L–1	466	955	**0.88**	**0.86**	**–1.80**
all2003	H+1	554	1637	**1.56**	**1.50**	**1.01**
all2005	H	93	771	**3.00**	**2.68**	**1.79**
all2006	H	60	386	**2.01**	**2.55**	**1.67**
all2017	H	468	1291	**1.47**	**1.30**	**0.91**
asl2052	H+1	20	1099	**5.29**	**5.85**	2.26
all2457	C	494	791	**0.76**	0.35	**–1.06**
all2705	C	1086	2501	**1.03**	**0.61**	**–1.83**
all2736	G+1	385	890	–0.03	**1.18**	–0.23
alr2790	A	21	191	0.89	**3.34**	0.48
alr2822	F+1	2	130	1.90	**5.53**	**2.74**
alr2823	F	4	272	**1.19**	**7.14**	**2.82**
alr2824	F	4	225	0.94	**6.18**	**2.58**

Table 14.4 continued

Gene[b]	Family	+N peak[c]	–N peak[d]	Log2 expression ratio 3 hours	8 hours	24 hours
alr2826	H+5	2	315	1.76	**7.68**	**3.43**
alr2829	H	4	1132	0.28	**9.25**	**10.62**
alr2833[e]	H+1	2	452	1.49	**7.41**	–
alr2887	C	370	685	**0.54**	**1.04**	–0.16
alr2888	H+3	180	551	**0.73**	**1.47**	0.12
all3173	G+1	465	1294	–0.48	0.95	**1.36**
alr3361	C–2	136	306	0.70	**1.29**	–0.22
all3520	H	128	294	0.65	**1.08**	–0.25
alr3676	N	4	58	1.51	**4.89**	1.01
alr3844	C	502	308	**–0.66**	**–1.63**	–0.84
alr4059	N–1	162	489	**1.51**	**1.52**	**–0.76**
alr4329	G–1	41	360	**2.82**	**3.17**	1.24
all4378	G+1	42	224	**1.30**	**2.40**	0.22
all4555	H+1	231	223	**–1.02**	–0.27	–0.06
all4578	H	10791	13554	**–0.65**	0.54	**–1.02**
all4962	H	40	202	**0.73**	**1.99**	0.19
all5046	H+3	178	506	–0.09	**1.48**	0.17
alr5055	G+1	129	410	**1.03**	**1.09**	0.04

[a]Abbreviations and conventions as in Table 14.2. Expression results are taken from Ehira and Ohmori (2006a)

[b]Genes shown (1) are annotated "hypothetical" or "unknown;" (2) change expression significantly at some point in the time course; (3) have orthologs in all three heterocyst-forming cyanobacteria with sequenced genomes; and either (4a) have a level of expression of at least 200 (twice the average of genes in N-replete condition) or (4b) change in expression at least 8-fold above a value that is distinguishable from background (shown in bold).

[c]Highest expression value for the three control conditions (controls for 3, 8, and 24 h)

[d]Highest expression value for the experimental condition at either 3, 8, or 24 hours.

[e]See discussion in Huang *et al*. (2005)

et al., 2003), and 2,2-difluoropentanedioic acid (DFPA), a non-metabolized, fluorinated analog of 2-OG, elicits heterocyst differentiation in the presence of ammonia (Laurent *et al.*, 2005). These findings and others described below indicate that 2-OG itself, rather than a metabolite derived from it, is the molecule that signals nitrogen deprivation and initiates heterocyst differentiation (Zhang *et al.*, 2006).

Heterocyst differentiation in response to the accumulation of 2-OG is mediated by the binding of 2-OG to NtcA (Laurent *et al.*, 2005). NtcA, a homo-dimeric transcriptional regulator, is a member of the CAP (catabolite activator protein) family (Herrero *et al.*, 2004; Wisen *et al.*, 2004) that globally controls nitrogen metabolism in cyanobacteria (Herrero *et al.*, 2004). In an *in vitro* system generated from the unicellular cyanobacterium, *Synechococcus elongatus* strain PCC 7942, 2-OG stimulates the binding activity of NtcA (Tanigawa *et al.*, 2002; Vazquez-Bermudez *et al.*, 2002), and the initiation of NtcA-dependent transcription requires the presence of 2-OG (Tanigawa *et al.*, 2002). In *Anabaena*, NtcA binds to the promoter regions of a variety of genes that are involved in nitrogen assimilation, heterocyst development, or photosynthesis (Herrero *et al.*, 2004), and its binding is stimulated by the presence of either 2-OG or DFPA (Laurent *et al.*, 2005).

In response to deprivation of fixed nitrogen, cells often begin to differentiate along the filaments of *Anabaena* in groups (clusters); one cell of each cluster eventually becomes a heterocyst, while the other cells of the cluster revert to a state of vegetative growth (Wilcox *et al.*, 1973a, b). One hypothesis for the initiation of development by grouped cells posits that competence to differentiate depends on the cell cycle, which appears often to be coordinated in neighboring cells (Meeks and Elhai, 2002). Another possibility is that neighboring cells, perhaps descended from a single cell, share the same level of sensitivity to nitrogen deprivation. The idea of clustered initiation of differentiation became more widely appreciated once a mutation in *patS* (see below) was identified that made clustering manifest.

Whereas NtcA is a master regulator of nitrogen metabolism, HetR—in coordination with NtcA—serves as a master regulator of heterocyst development (Buikema and Haselkorn, 1991; Buikema and Haselkorn, 2001; Khudyakov and Golden, 2004). Various point or insertional mutations in *hetR* abolish heterocyst development (illustrated in Figure 14.1B), whereas overexpression of *hetR* causes the formation of multiple contiguous heterocysts, the Mch phenotype (Black *et al.*, 1993; Buikema and Haselkorn, 1991; Buikema and Haselkorn, 2001; Dong *et al.*, 2000; Huang *et al.*, 2004), and leads also to heterocyst differentiation in the presence of fixed nitrogen. A different point mutation in *hetR* of *Anabaena* leads to great overproduction of heterocysts (Khudyakov and Golden, 2004). HetR, the product of *hetR*, forms homodimers that bind upstream from *hetR* and from other developmental genes (Huang *et al.*, 2004). HetR also has characteristics of a serine-type protease (Zhou *et al.*, 1998), permitting it to digest itself. The protease activity of HetR depends on calcium, whose concentration as a free ion is regulated by CcbP, a calcium-binding protein, and is crucial to heterocyst development (Zhao *et al.*, 2005; see also Torrecilla *et al.*, 2004).

Autoregulation can be a critical component of pattern-formation processes (see next paragraph). According to currently available information, the auto-regulation of *hetR* (Black *et al.*, 1993) appears to depend upon three processes: (1) the binding of a HetR dimer upstream from *hetR*, upon which transcription of *hetR* depends; (2) the mutually dependent up-regulation of *hetR* and *ntcA* (Muro-Pastor *et al.*, 2002); and (3) the calcium-dependent proteolytic activity of HetR (Zhou *et al.*, 1998), essential for heterocyst development (Dong *et al.*, 2000), which digests CcbP (Shi *et al.*, 2006), thereby promoting the increase of free calcium in developing cells (Zhao *et al.*, 2005).

A long-standing model for the formation of a pattern of spaced, differentiated cells in general (Gierer and Meinhardt, 1972; Turing, 1952) and of *Anabaena* heterocysts in particular (Fritsch, 1951; Gierer and Meinhardt, 1972; Wilcox *et al.*, 1973a; Wolk, 1967, 1979; Wolk and Quine, 1975) predicts that heterocyst development may depend on an autoregulatory activator that does not diffuse between cells and on a diffusible substance whose synthesis is dependent on the activator and whose effect is to inhibit the activator. HetR may be equated with the autoregulatory activator. Similarly, PatS, or a derivative of it, is a candidate for the diffusible inhibitor. A *patS* deletion mutant produces the Mch phenotype 24 hours after nitrogen stepdown (Figure 14.1F, G; Yoon and Golden, 1998). The gene *patS* is most strongly expressed in proheterocysts and heterocysts, and its overexpression or addition of its C-terminal pentapeptide, RGSGR (denoted PatS-5) to the medium suppresses heterocyst differentiation (Yoon and Golden, 1998). *In vitro*, binding of the HetR dimer to a fragment of the *hetR* promoter is inhibited by synthetic PatS-5 (Huang *et al.*, 2004). Because a synthetic gene encoding the pentapeptide does not restore the pattern of a Δ*patS* mutant to the wild type, Wu *et al.* (Wu *et al.*, 2004) suggested that the amino acid residues of PatS other than the C-terminal pentapeptide may be required for intercellular transport of PatS. In *Anabaena*, *patS* has two potential start codons by which it might encode a peptide of 17 or 13 amino acids (Yoon and Golden, 1998). Because its *N. punctiforme* ortholog encodes only a 13-codon protein (Meeks *et al.*, 2001), *Anabaena* PatS may also be 13 amino acids in length. The aberrant pattern of a *patS* mutant is conspicuous only early in differentiation, suggesting that *patS* may be important primarily for initial pattern formation and the resolution of clusters of differentiating cells.

Mutation of another gene, *hetN*, produces a normal pattern in nitrogen-free medium as of 24 hours and a Mch phenotype (illustrated in Figure 14.1D) only after a delay of ca. 48 hours (Black and Wolk, 1994; Callahan and Buikema, 2001). The product of this gene is predicted to be a ketoacyl reductase such as those involved in the synthesis of polyketides (Black and Wolk, 1994; Fan *et al.*, 2005), and is membrane-localized (Li *et al.*, 2002). *hetN* is positioned adjacent to a cluster of polyketide-synthase genes that are required for the synthesis of heterocyst envelope glycolipids (see "Formation of the heterocyst envelope," below). However, a *hetN* mutation does not block the synthesis of those glycolipids (Callahan and Buikema, 2001). Nonetheless, it remains possible that HetN may derivatize such a glycolipid or one of its precursors to produce an inhibitor of heterocyst formation. Only *hetN* and two other possible transcripts can, like *patS*, encode the inhibitory oligopeptide RGSGR (Li *et al.*, 2002; Wu *et al.*, 2004). Moreover, when overexpressed, HetN can block heterocyst formation, an effect possibly attributable to its possession of that peptide (Wu *et al.*, 2004). In addition, mutation of both *patS* and *hetN* leads to lethal, nearly complete differentiation of vegetative cells into heterocysts (Borthakur *et al.*, 2005). *hetN* is upregulated about 12 hours after nitrogen step-down, much later than *hetR* and *patS*. Because a *hetN* mutant shows an Mch phenotype only after about 48 hours of deprivation of nitrogen, HetN appears to be important for the maintenance of the pattern after it has been established (Callahan and Buikema, 2001). Fixed nitrogen exported from functional heterocysts may also contribute to pattern maintenance in wild type *Anabaena* (Fay *et al.*, 1968; Thiel and Pratte, 2001; Thomas *et al.*, 1977; Yoon and Golden, 2001).

The other known gene whose mutation profoundly affects the pattern of heterocysts in *Anabaena* is *patA*. This gene encodes a protein with a predicted CheY-like phosphoacceptor domain at its C-terminus and a conserved N-terminal domain, with an α-helical insertion, that may bind a ligand (Makarova *et al.*, 2006). It has no known DNA-binding domain (Liang *et al.*, 1992). A *patA* mutant of *Anabaena* produces almost exclusively terminal heterocysts (Figure 14.1E; Liang *et al.*, 1992) and so resembles members of the cyanobacterial genus, *Cylindrospermum*. According to their 16S rDNA sequences, *Anabaena* and *Cylindrospermum* are closely related (Iteman *et al.*, 2002; Wood *et al.*, 2002). Therefore, *patA* mutations may be of evolutionary as well as developmental interest. Epistasis analyses suggest that a major part of the activity of PatA involves attenuation of the inhibitory activities of PatS and HetN, perhaps by a direct effect on HetR (Orozco *et al.*, 2006).

A central question in pattern formation in *Anabaena* is how a proheterocyst can be immune to a putative diffusible inhibitor of differentiation that it produces. Two possible answers have been offered: (1) *patS* depends on *hetR* for expression, and the accumulation of a *patS*-dependent inhibitor lags behind that of HetR (Huang *et al.*, 2004); and (2) PatA reduces the inhibition of HetR by the *patS*-dependent inhibitor (Orozco *et al.*, 2006; Zhang *et al.*, 2006). Both suggested answers imply that PatS levels *in vivo* may be insufficient to inhibit differentiation. However, if so, what drives the regression of all but one in a group of proheterocysts? One of us [X.X.] has noted that the problem of "self-immunity" could be circumvented altogether by considering that PatS may elicit the synthesis of a different, diffusible, inhibitory molecule in neighboring vegetative cells. We denote such an inhibitor Iv (I for inhibitor and v for vegetative cell, its site of synthesis) and imagine that its production is proportional in quantity to the number of proheterocysts in a group. By Ih, we denote a signal substance (which could be *patS*-dependent) diffusing from proheterocysts. If Iv exists, it may both form an "inhibitory zone" (Mitchison *et al.*, 1976) in vegetative cells and diffuse within the group of developing cells, causing all but one to regress. When only one developing cell remains, the concentration of Iv might be so low that differentiation can progress (Figure 14.4).

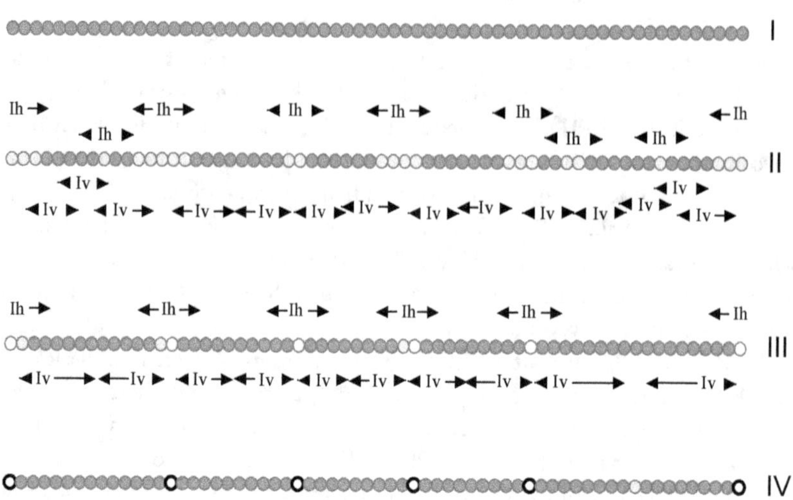

Figure 14.4 (see also Plate 14.4) A conjecture concerning heterocyst pattern formation in *Anabaena*. I to IV, different stages of pattern formation. Circles (cells) that are green, yellow, white with a thin wall, and white with a thick wall stand for vegetative cells, early developing cells, proheterocysts committed to differentiate, and mature heterocysts, respectively. Ih, heterocyst-type inhibitor (perhaps *patS*-dependent); Iv, vegetative cell-type inhibitor. Production of Iv in response to the differentiation of chosen cells may be mediated by Ih.

Although adding complexity to current models, an induced inhibitor might account also for the phenotype of a *patA* mutant, whose basis has been variously interpreted. For example, Wilcox *et al.* (1975) suggested that the *patA* phenotype might be due to the leakage of a differentiation-inhibiting molecule from the termini of the filaments, and Orozco *et al.* (2006) pointed out, as an alternative, that if (as in the induced-inhibitor model) inhibition is derived from vegetative cells, a terminal cell might receive less inhibitor than an intercalary cell and so could differentiate. Both hypotheses could account, similarly, for the observation (Fay *et al.*, 1968; Orozco *et al.*, 2006; Sutherland *et al.*, 1985; Wolk, 1967) that heterocysts produced in short filaments are most often terminal in position.

Heterocyst maturation

Cessation of cell division

Progression through the cell cycle is essential for the initiation of heterocyst differentiation (Sakr *et al.*, 2006). However, once differentiation has begun, cell division normally, and perhaps necessarily, ceases before further development takes place. FtsZ, the most critical component of bacterial cell division machinery, is a tubulin-like GTPase that polymerizes to form a ring-like structure at the division site (Errington *et al.*, 2003; Sakr *et al.*, 2006). The expression of *ftsZ* is diminished in heterocysts of wild-type *Anabaena*, compared to vegetative cells (Kuhn *et al.*, 2000; Wang and Xu, 2005), whereas *sulA*, which encodes an inhibitor of FtsZ, is upregulated (Wang and Xu, 2005). In a *hetC* mutant, the early developing cells continue to divide, resulting in smaller cells, with little autofluorescence, in which *ftsZ* is upregulated (Figure 14.1C; Wang and Xu, 2005; Xu and Wolk, 2001). HetC has been predicted to be an ATP-binding-cassette exporter of a protein or peptide (Khudyakov and Wolk, 1997). If so, knowing what it transports and how it affects the expression of *ftsZ* should help to determine how the transition to a state that lacks cell division takes place early during heterocyst differentiation.

Formation of the heterocyst envelope

The most prominent structure of a heterocyst is its envelope, which comprises a laminated layer of glycolipid packaged between an external layer of polysaccharide and the cell wall (Dunn and Wolk, 1970; Winkenbach *et al.*, 1972; Cardemil and Wolk, 1976, 1979). The glycolipids constitute the major barrier to rapid entry of oxygen into heterocysts (Haury and Wolk, 1978; Murry and Wolk, 1989; Walsby, 1985). However, ultrastructural observation of Hep^- Hgl^+ mutants has consistently shown delamination and fragmentation of the Hgl layer (Wolk, 2000; Zhou and Wolk, 2003; Fan *et al.*, 2006), leading to the interpretation that the Hep layer protects the Hgl layer and is, at least for that reason, required if nitrogen fixation is to take place under oxic conditions. The glycolipids of the cytoplasmic membranes are diacyl glycerolipids that bear one or two sugar moieties, whereas the envelope glycolipids are glycosides of polyketides (Fan *et al.*, 2005; Gambacorta *et al.*, 1995; Gambacorta *et al.*, 1996; Gambacorta *et al.*, 1998).

Most of the genes known to be required specifically for the synthesis of envelope glycolipids and polysaccharide are situated in two "expression islands" of the genome (Fan *et al.*, 2005; Huang *et al.*, 2005) that were first detected by microarray analyses (Ehira *et al.*, 2003). Genes *alr5351* through *alr5357* (*hglA* through *hglG*, but for historical reasons not in the same order) encode enzymes that may suffice for the synthesis of the glycolipid aglycones (Bauer *et al.*, 1997; Fan *et al.*, 2005). Close to those genes are gene *all5341* (denoted *hglT*), which encodes the glycosyl transferase that is responsible for glycosylation of the aglycone (Awai and Wolk, 2006) and orfs *all5345-all5347* (genes *hgdA* through *hgdC*). The phenotypes of *hgdA*, *hgdB* and *hgdC* mutants suggest that the proteins that they encode may affect the timing or positioning of glycolipid deposition (Fan *et al.*, 2005). Operon *devBCA* encodes a presumptive transporter of the glycolipids (Fiedler *et al.*, 1998a; Fiedler *et al.*, 1998b; Maldener *et al.*, 1994). Mutants affected in any gene in that operon form envelope glycolipids but do not use them to form the laminated glycolipid layer. The proteins HgdB and HgdC show similarities to DevB and DevC, respectively (Fan *et al.*, 2005). *hglK* (*all0813*), a separately located gene, also affects the extracellular deposition of glycolipids (Black *et al.*, 1995), and may possibly form a scaffold on which the glycolipids are deposited (Wolk, 2000).

At least certain (*hglE*$_A$, *hglB* [a.k.a. *hetM*]) of the genes responsible for biosynthesis of the aglycones of the heterocyst envelope glycolipids increased in transcript abundance between 8 and 24 hours of nitrogen stepdown. In contrast, genes that encode the enzymes that catalyze the biosynthesis of conventional fatty acids decreased significantly in expression by 8 hours. Despite that decrease, the average hybridization signal of the transcripts of the latter genes was more than 10-fold stronger than the hybridization signal from transcripts of the heterocyst envelope glycolipid-biosynthetic (*hgl*) genes in the hours following the onset of nitrogen deprivation, presumably reflecting, in part, the much greater number of vegetative cells than developing heterocysts.

A cluster of genes (*alr2825-alr2841*) that has been found only in heterocyst-forming cyanobacteria exhibited significantly enhanced expression at 8 and 24 hours of nitrogen deprivation (Ehira and Ohmori, 2006a). Numerous genes in the cluster are required for production of heterocyst envelope polysaccharide (Huang *et al.*, 2005). *hepA* (*alr2835*), the earliest identified gene in that cluster (Holland and Wolk, 1990; Wolk *et al.*, 1988), predicts a protein that strikingly resembles an exporter of lipid A, a constituent of lipopolysaccharide, leading to the suggestion that heterocyst envelope polysaccharide may itself be a lipopolysaccharide (Huang *et al.*, 2005). Another gene in the cluster, *hepC* (*alr2834*), resembles a UDP-galactose-lipid carrier transferase (Zhu *et al.*, 1998) that could have a role in the synthesis of a side-branch in the repeating subunit of heterocyst envelope polysaccharide (Cardemil and Wolk, 1976; Cardemil and Wolk, 1979; Cardemil and Wolk, 1981b). Many of the other genes in the cluster also predict glycosyl transferases or other enzymes that metabolize sugars or nucleotide-sugars that are detected in isolated heterocysts that synthesize envelope polysaccharide (Cardemil and Wolk, 1981a). Thus, it appears probable that these enzymes are involved directly in synthesis of heterocyst envelope polysaccharide. The abundance of transcripts from genes in this cluster was generally higher at 8 hours than at 3 or 24 hours of nitrogen deprivation, and thus decreased before the decrease in abundance of transcripts of *hglB*, *hglE*$_A$, and *hgdC*, and probably other genes, that are involved in the synthesis and deposition of heterocyst envelope glycolipids. Notably, genes *alr2819* (*hetP*) through *alr2824*, adjacent to the cluster *alr2825-alr2841*, were also highly expressed at 8 hours of nitrogen deprivation and showed expression that was enhanced, but less strongly, at 3 and 24 hours (Ehira and Ohmori, 2006a). Most of the genes *alr2819* through *alr2824* are annotated as unknown or hypothetical, but *all2821* is annotated as a two-component response regulator.

Another gene that predicts a glycosyl transferase and whose product (or that of a co-transcribed, 3′ gene) is required for synthesis of heterocyst envelope polysaccharide is *hepB* (*alr3698*), which lies distant from the region of *alr2825-alr2841*. The UV mutation that originally defined *hepB* (Wolk *et al.*, 1988) was complemented and the complementing gene sequenced (Wolk and Zarka, 1998; Wolk *et al.*, 1999). A transposon::*luxAB* fusion in the same gene was determined to be maximally luminescent at 8 hours of nitrogen deprivation (Maldener *et al.*, 2003). Accordingly, our reanalysis of the data of Ehira and Ohmori (2006a) showed a nearly 30-fold increase in the expression of *alr3698* at 8 hours. As shown by transposon mutagenesis and complementation, the gene *alr3699*, which lies immediately downstream from *hepB* and has a similar pattern of expression, also predicts a glycosyl transferase and is required for a Fox$^+$ phenotype (the ability to fix dinitrogen in the presence of oxygen) and for synthesis of heterocyst envelope polysaccharide (Wang *et al.*, 2007). Using reporter genes *luxAB* (encoding bacterial luciferase) or *gfp* (encoding green fluorescent protein), Wolk *et al.* (1993) and Wang and Xu (2005) have shown that *hepA* and *hepB* are expressed exclusively, or nearly so, in heterocysts. Therefore, *hepB* and *alr3699*, like *hepA*, may be involved directly in synthesis of heterocyst envelope polysaccharide.

An autolysin-encoding gene, *hcwA* (*alr0093*), significantly induced at 8 hours, is required for the up-regulation of *hepA* and formation of the polysaccharide layer of the heterocyst envelope (Zhu *et al.*, 2001). In addition, mutations in *rfbP* (*all4829*) and *rfbZ* (*all4830*), which affect synthesis of lipopolysaccharide in vegetative cells (Xu *et al.*, 1997), moderately affect the structure of the heterocyst envelope but result in a Fox$^-$ phenotype. A mutation in gene *pbp3* that presumptively encodes a penicillin-binding protein, with a presumptive role in synthesis of the peptidoglycan layer of cell walls, results in heterocysts that lack a polysaccharide layer but form a glycolipid layer (Leganés *et al.*, 2005). A mutation in another gene that presumptively encodes a penicillin-binding protein, *pbpB* (*pbp2*), also leads to defective heterocyst development and function, but the envelope has not been closely examined (Lázaro *et al.*, 2001). The phenotypes of these mutants suggest the likelihood that the walls of the vegetative cells that differentiate have a role(s) in the differentiation process. Therefore, Leganés *et al.* (2005) suggested that refashioning of cell wall components of the differentiating cells may be a prerequisite for the formation of the heterocyst envelope.

Of 138 genes predicted to encode glycosyl transferases according to the Kazusa website (http://www.kazusa.or.jp/cyano/Anabaena/kwd.html), many are part of large clusters of genes that encode other proteins related to sugar metabolism. Nineteen genes that putatively encode glycosyl transferases showed significantly increased expression at 8 hours of nitrogen deprivation. Three of the 19 are part of a cluster of genes, *alr3057* through *alr3074*, that contains 11 presumptive glycosyl transferase-encoding genes, most of which were significantly induced by 3 hours of nitrogen deprivation. All of those 11 genes lack orthologs in *N. punctiforme*, which forms heterocysts, and transcription of some of them was reduced in response to temperature down-shift (Ehira *et al.*, 2005), further suggesting that their role is not differentiation-specific (see also Lechno-Yossef *et al.*, 2006).

Cytoplasmic differentiation

Refractile granules of cyanophycin, a non-ribosomally synthesized co-polymer of arginine and aspartic acid, are often conspicuous in the polar pores of mature heterocysts (Lang *et al.*, 1972). Cyanophycin synthetase, which makes the polymer, is encoded by *cphA1* and *cphA2* in *Anabaena*; and cyanophycinase, which degrades it, is an exopeptidase that is encoded by *cphB1* and *cphB2* (Berg *et al.*, 2000; Picossi *et al.*, 2004; Ziegler *et al.*, 1998). *cphA1* and *cphB1* are organized in an operon and play the major role in cyanophycin metabolism. A *cphA1* mutant or a *cphA1cphA2* double mutant synthesizes no polar nodules in heterocysts, but only slightly reduces the rate of diazotrophic growth. In contrast, a *cphB1* mutant or a *cphB1cphB2* double mutant forms excessive cyanophycin granules in heterocysts and significantly reduces the rate of diazotrophic growth (Picossi *et al.*, 2004). Two interpretations of the decreased growth rate on N_2 are that (1) the large amount of cyanophycin physically restricts transport into and out of the heterocyst or (2) if cyanophycin is synthesized and degraded serving as a dynamic reservoir, as envisaged by Carr (1988), a deficiency of cyanophycinase might lead to a reduction in the rate at which fixed nitrogen reaches vegetative cells.

Heterocysts of most but not all species show much less phycobiliprotein-derived absorption and fluorescence than do vegetative cells. In some mutants, microscopically observable differentiation does not proceed beyond this phenotype (Elhai and Wolk, 1990; Xu and Wolk, 2001). The gene *nblA*, present in all phycobilisome-containing organisms, encodes a protein of approximately 7 kDa that plays a key role in phycobilisome degradation (Collier and Grossman, 1994). *Anabaena* has two *nblA* homologous genes, one on the chromosome (*nblA*) and one on the delta plasmid (*nblA*-p). Only the chromosomal *nblA* gene was upregulated upon nitrogen starvation and is required for phycobilisome degradation. A crystal structure of NblA with 1.8-Å resolution shows that NblA dimers form a four-helix bundle (Bienert *et al.*, 2006). NblA dimers bind specifically to the alpha-subunits of phycocyanin and phycoerythrocyanin, the main building blocks of the phycobilisome rod structure of (specifically) *Anabaena* (Bienert *et al.*, 2006). An *nblA* mutant, in which heterocyst biliproteins fail to bleach or to diminish in fluorescence, still fixes N_2 (Baier *et al.*, 2004).

Mature heterocysts have, as principal physiological features, nitrogenase activity and, to preserve that activity, a greatly increased rate of respiration and lack of oxygen evolution (Wolk *et al.*, 1994). The *nif* (nitrogen fixation) genes, themselves, and DNA rearrangements within *nifD*, *fdxN*, and *hupL* have been repeatedly reviewed (e.g. Golden, 1998; Haselkorn, 1992; Wolk *et al.*, 1994). The proteins that underlie the great increase in respiration are thought to be those of the oxidative pentose phosphate (OPP) pathway, whose activities are much greater in heterocysts than in vegetative cells (Winkenbach and Wolk, 1973). *all4019*, which encodes glucose 6-phosphate dehydrogenase, corresponds to the gene affected by the *N. punctiforme zwf* mutant that was described by Summers *et al.* (1995); *alr1602* presumptively encodes 6-phosphogluconolactonase, which hydrolyzes the product of G6PD producing 6-phosphogluconate, and was shown by Bauer (1994) to be required for aerobic fixation of dinitrogen; and *alr5275* encodes 6-phosphogluconate dehydrogenase. Genes *all4019* and *alr5275*, but not *alr1602*, showed significantly enhanced transcription in response to nitrogen stepdown (Ehira and Ohmori, 2006a).

Cytochrome oxidase proteins CoxB2, CoxA2 and CoxC2 (Alr2514-Alr2516) and alternative respiratory terminal oxidase (ARTO) proteins Alr2731, Alr2732 and Alr2734 (a.k.a. Cox3) are important for heterocyst respiration and aerobic fixation of dinitrogen (Jones and Haselkorn, 2002; Valladares *et al.*, 2003). Transcript levels of the *cox2* genes were not distinguishable from background at 3 hours, but were significantly increased at 8 and 24 hours of nitrogen deprivation. However, the apparent levels of expression of *coxA2* (averaged over all cells) was 4–6 fold lower than the apparent levels of the mildly induced expression of *coxA1* (*alr0951*) at both times. Transcript levels of the *cox3* genes never exceeded microarray background and the apparent increases of those of the putative ubiquinol oxidase genes *all4023* and *all4024* proved not statistically significant. In contrast, Northern blots showed, relative to nitrogen replete conditions, strong induction of *coxB3* in response to nitrogen deprivation (Valladares *et al.*, 2003). These results are not discordant: it must be understood that only 29% of the genes considered by Ehira and Ohmori had expression levels that were measurable by microarray analysis at all time points.

Mechanisms of the developmental regulatory cascade

Central regulators: NtcA and HetR

Despite identification of numerous examples of events later in differentiation that depend on particular genes, the mechanisms that underlie those dependencies remain largely unidentified. However, we will try to present what is known and some of what is unknown in the hope of stimulating work to fill in the many gaps in our knowledge.

As discussed above, 2-OG serves as a primary signal of nitrogen status in cyanobacteria and, in response to an increase in the concentration of 2-OG, the affinity of NtcA to its binding sites in DNA increases. NtcA binds upstream from *ntcA* and regulates its own transcription (Ramasubramanian *et al.*, 1996). Perhaps as a result of the 2-OG-dependent increase in that affinity, there is an autoregulatory increase in transcription of *ntcA*. That increase in transcription depends also on HetR. Conversely, two of the four transcription start sites of *hetR* are activated by an NtcA-dependent mechanism, and activation of one of those two sites depends also on HetR. The mechanisms underlying those dependencies are incompletely understood. The lack of an evident binding site for NtcA upstream from *hetR* indicates that NtcA-dependence is not direct. Rather, NrrA binds upstream from nitrogen-responsive transcription initiation sites of the *hetR* gene, and NtcA-dependence is mediated at least in part by NtcA-dependent expression of *nrrA* (Ehira and Ohmori, 2006a, b; Muro-Pastor *et al.*, 2002, 2006). The transcription start site at which an autoregulatory increase in transcription of *hetR* (Black *et al.*, 1993) depends on HetR may be responding to the HetR dimer, which binds to as yet unidentified sites in DNA. The HetR dimer also binds upstream from *patS* and *hepA* and regulates transcription of, at least, *patS* (Huang *et al.*, 2004). The role of free calcium has been discussed above, under "Spatial localization of differentiation." In *Nostoc punctiforme*, up-regulation of *hetR* depends also on HetF (Wong and Meeks, 2001), a protein that may be a protease (Aravind and Koonin, 2002).

Besides regulating its own transcription (Ramasubramanian *et al.*, 1996) and, indirectly, that of *hetR* (Muro-Pastor *et al.*, 2002), NtcA regulates the transcription of numerous genes that are involved in heterocyst differentiation, including *hetC* (Muro-Pastor *et al.*, 1999), *devBCA* (Fiedler *et al.*, 2001), *xisA*, *nifHDK*, and *rbcL* (Ramasubramanian *et al.*, 1994; Wei *et al.*, 1993). *hetC* is an example of a heterocyst-related gene that is expressed NtcA-dependently but without dependence on HetR (Muro-Pastor *et al.*, 2002). Somewhat similarly, overexpression of *ntcA* elicits expression of *devBCA* in response to withdrawal of ammonium in a *hetR* background (Olmedo-Verd *et al.*, 2005). Whether HetR can drive the expression of some genes without expression of NtcA remains unknown. Some modulators of the NtcA/HetR central regulatory machinery also play important roles in determining the fate of a cell. For example, the function of HetR is probably modulated by *patS*- and *hetN*-dependent inhibitors and PatA (see "Spatial localization of differentiation," above). Because about 70% of the cells of *patS* or *patS hetN* mutants remain vegetative as of 24 hours of nitrogen deprivation (Borthakur *et al.*, 2005), one cannot exclude the possibility that additional inhibitor(s) of differentiation may be present in *Anabaena*. The nitrogen-responsive regulator, NrrA, regulates the rate of progression of heterocyst differentiation and appears to assist in the up-regulation of *hetR* (Ehira and Ohmori, 2006a, b).

Regulation later in differentiation

One can speculate that the activation of genes such as *hetC* and *devBCA* that are upregulated at different stages of heterocyst development may require different levels of NtcA, HetR and/or cofactor(s). However, the timing of gene expression in heterocyst development may depend greatly on cascade-like regulation that may be transcriptional, such as the regulation of the gene *devH*, *alr3952* (Hebbar and Curtis, 2000), that encodes a DNA-binding protein and whose transcription was strongly upregulated at 8 hours (Ehira and Ohmori, 2006a). DevH is essential to the formation of glycolipid layer (Ramírez *et al.*, 2005). Genes *hglB*, *hglC*, *hglD*, *hglE1* (*alr5351*) and *hglE2* (*all1646*) all depend on *devH* for up-regulation (Ramírez *et al.*, 2005). However, how *devH* itself is regulated is not known. Deposition of the heterocyst envelope polysaccharide depends also on a two-component regulatory system, HepK-DevR$_A$ (Zhou and Wolk, 2003), whose activity presumably involves the sensing, communicating, and transducing of (still unknown) environmental signals.

hetC, which is directly regulated by NtcA, is required for the cessation of cell division at a very early stage of heterocyst development. Whereas up-regulation of *hepA*, *hglD*, *hglE1*, *patB*, *xisA* and *nifB* is blocked in a *hetC* mutant, the expression of *hepB*, *hepK*, *devBCA* and the *cox2* operon shows no evident dependence on *hetC* (Wang and Xu, 2005).

Diverse genes regulate the expression of *hepA* and the formation of the polysaccharide layer of the heterocyst envelope, without evidently affecting synthesis of the glycolipid layer: (1) an autolysin-encoding gene, *hcwA* (*alr0093*) (Zhu *et al.*, 2001); (2) serine-threonine kinase gene *hepS* (*all2760*) (Fan *et al.*, 2006); (3) histidine kinase/response regulator genes *hepK* (*all4496*) and *devR$_A$* (*alr0442*) whose products interact (Zhou and Wolk, 2003; Zhu *et al.*, 1998); and (4) another histidine kinase gene, *hepN* (*alr0117*) (Ning and Xu, 2004). Microarray experiments indicate that HepK and HepN regulate many of the same genes, presumably via the same response regulator, for which DevR$_A$ is a likely candidate. However, each of HepK and HepN appears also to regulate genes that the other does not (Lechno-Yossef *et al.*, 2006). Although Abp2 (All1939) and Abp3 (Alr3608) bind 5′ from *hepC* and regulate *hepA*, they show no obvious effect on the envelope polysaccharide. Surprisingly, it is the glycolipid layer that is absent from *abp2* and *abp3* mutants (Koksharova and Wolk, 2002). Response-regulator gene *henR* (*alr1086*) is required for the synthesis of both layers of the heterocyst envelope (Fan *et al.*, 2006). Because HepS regulates virtually the same set of genes as HenR, they appear to be part of the same regulatory system, even though *hepS* and *henR* mutants have differing effects on synthesis of heterocyst envelope glycolipids (Fan *et al.*, 2006; Lechno-Yossef *et al.*, 2006).

At 14 hours of nitrogen deprivation, in a *hepK* mutant relative to in the wild-type strain, the following genes that are predicted to encode the indicated kinds of regulators were up- or downregulated: *all2164*, *all4927* and *alr5241*, encoding two-component response regulators: down; *all1072* and *all2165*, encoding PATAN-domain-bearing response regulators: up and down, respectively (and *all0521* encoding PATAN-domain-bearing, response regulator *patA*, up in *hepN*, *hepS*, and *henR* mutants); *all2883*, encoding a histidine kinase: down; *hepK* itself: up; and *all0438* and *alr1869* encoding serine-threonine kinases and *asl0125*, *all1173*, and *asl2551* encoding transcriptional regulators: up (as was serine-threonine kinase *hepS* in a *hepS* mutant). These results are suggestive of transcriptional regulatory cascades, of negative autoregulation of *hepK* and *hepS*, and of inhibition of transcription of *patA* by products related to differentiation (Lechno-Yossef *et al.*, 2006).

Differentiation of the metabolism and cytoplasmic structures of the heterocyst have been less studied. Amazingly little is known of the regulation of the *nif* genes. *patB* (*all2512*), which is required for heterocyst maturation, encodes a regulator with ferredoxin and helix–turn–helix domains (Jones *et al.*, 2003; Liang *et al.*, 1993). Its product, PatB, regulates expression of *cox2* genes (Jones *et al.*, 2003). PrpA and PknE, eukaryotic-type phosphoprotein phosphatase and protein kinase proteins, are thought to ensure a correct level of phosphorylation of proteins required for maintaining the capability for nitrogen fixation (Zhang *et al.*, 1998). Although numerous studies have focused on genes that are upregulated in heterocysts, those that are downregulated in heterocysts have received much less attention. Using "segmental" microarrays, Ehira *et al.* (2003) identified DNA segments that were up- or downregulated specifically in heterocysts, and could point to strong upregulation of $hglE_A$ and down-regulation of a segment containing the *rbcLXS* (Rubisco) operon. The latter effect probably depends on the binding of NtcA to a site that overlaps the transcriptional start point of that operon (Herrero *et al.*, 2004). Now that microarrays are available for probing individual genes rather than genetic segments, microarray analyses with RNA extracted from isolated heterocysts richly deserve repetition.

Concluding words

The process of heterocyst differentiation is often described as a developmental program, a metaphor that suggests events playing out in time and space according to a set of directives. In some sense, this is an apt description. We know little of the directives themselves but can see their broad outline and the identity of many entities doing the directing. 2-OG, NtcA, and perhaps NrrA appear to be important players in the early hours following nitrogen stepdown. Later, HetR and again NtcA are important in directing events in cells that have opted for differentiation. HetR itself receives direction from PatS, HetN, and calcium. There are many other apparent directors: PatA, cell cycle, protein kinases, perhaps a large number of unstudied regulatory proteins...

...And here the metaphor of a linear set of directives breaks down. Why so many regulators? Others have proposed an alternative way of looking at the process, a chorus of signals (Zhang *et al.*, 2006), where different voices may rise up above the throng at different times. The emerging picture of heterocyst differentiation, coupled with the more mature picture of sporulation by *Bacillus subtilis* (Hilbert and Piggot, 2004), brings to mind yet another metaphor: an extended meeting, with a mixture of plenary speakers who address the body of the whole, speakers who address smaller groups, and focused conversations at the coffee table that may well include those who just spoke. There may indeed be a program that superficially directs events, but the science lurches forward not according to a fixed plan but rather in fits and starts, as a consensus arising from many quarters is formed on one matter, and then another, each allowing the next conceptual step to be taken—conversation, not direction.

The attention of those interested in heterocyst differentiation has focused primarily on the major regulators of differentiation, whereas the biochemical processes by which cells come to differ from one another has received lesser attention. Thus we have a growing (but certainly still incomplete) knowledge of the on-off switches and the signals that pass between cells to affect differentiation. However, genetic analysis rather than biochemical analysis has largely commanded our attention, in part because we have not mastered a way to get a large amount of synchronously developing cells. As a result, we still know little about the specific enzymes that build the heterocyst's special envelope. Neither do we know what regulates the expression or activity of important heterocyst-specific processes, e.g., nitrogen fixation. There are clearly many genes involved in the tasks required to build spaced heterocysts, and the global analysis of transcripts, proteins, and metabolites may help us know where to look for the genes related to end processes. Already, transcript analysis has pointed to one surprising place to look: basic metabolic processes are evidently redirected within a few hours of nitrogen deprivation, perhaps along different paths that mirror spatial differentiation. Nonetheless, even as the outline of heterocyst differentiation is coming hazily into view, one can see that there is much to be filled in.

Much to do there may be, but compare this review to those of ten or twenty years earlier and you will realize that a great deal has been learned in recent years. We are not very far from a major accomplishment: an understanding of this arguably simplest example of patterning in a multicellular organism.

Acknowledgments

We thank Shigeki Ehira and Eneas Aguirre von Wobeser for helpful discussions, Sahel Farhangi for help in microarray analysis, and Karen Bird for editorial assistance. Analysis of microarray data was done primarily in BioBIKE, a programming environment being developed with support from the National Science Foundation under grant DBI-0516378. Work in the Xu laboratory is supported by the National Natural Science Foundation of China under grants 30170009 and 30570022. Work in the Wolk laboratory is supported by the U.S. Department of Energy under grant DE-FG02–91ER20021.

References

Aravind, L., and Koonin, E.V. (2002). Classification of the caspase-hemoglobinase fold: detection of new families and implications for the origin of the eukaryotic separins. Proteins 46, 355–367

Ashby, M.K., and Houmard, J. (2006). Cyanobacterial two-component proteins: structure, diversity, distribution, and evolution. Microbiol. Mol. Biol. Rev. *70*, 472–509.

Awai, K., and Wolk, C.P. (2006). Identification of the glycosyl transferase required for synthesis of the principal glycolipid characteristic of heterocysts of *Anabaena* sp. strain PCC 7120. FEMS Microbiol. Lett. *266*, 98–102.

Baier, K., Lehmann, H., Stephan, D.P., and Lockau, W. (2004). NblA is essential for phycobilisome degradation in *Anabaena* sp. strain PCC 7120 but not for development of functional heterocysts. Microbiology *150*, 2739–2749.

Bauer, C.C. (1994). Isolation and characterization of genes involved in nitrogen fixation and heterocyst differentiation in *Anabaena* sp. strain PCC 7120. Thesis, University of Chicago, Chicago.

Bauer, C.C., Ramaswamy, K.S., Endley, S., Scappino, L.A., Golden, J.W., and Haselkorn, R. (1997). Suppression of heterocyst differentiation in *Anabaena* PCC 7120 by a cosmid carrying wild-type genes encoding enzymes for fatty acid synthesis. FEMS Microbiol. Lett. *151*, 23–30.

Bienert, R., Baier, K., Volkmer, R., Lockau, W., and Heinemann, U. (2006). Crystal structure of NblA from *Anabaena* sp. PCC 7120, a small protein playing a key role in phycobilisome degradation. J. Biol. Chem. *281*, 5216–5223.

Black, K., Buikema, W.J., and Haselkorn, R. (1995). The *hglK* gene is required for localization of heterocyst-specific glycolipids in the cyanobacterium *Anabaena* sp. strain PCC 7120. J. Bacteriol. *177*, 6440–6448.

Black, T.A., and Wolk, C.P. (1994). Analysis of a Het$^-$ mutation in *Anabaena* sp. strain PCC 7120 implicates a secondary metabolite in the regulation of heterocyst spacing. J. Bacteriol. *176*, 2282–2292.

Black, T.A., Cai, Y., and Wolk, C.P. (1993). Spatial expression and autoregulation of *hetR*, a gene involved in the control of heterocyst development in *Anabaena*. Mol. Microbiol. *9*, 77–84.

Borthakur, P.B., Orozco, C.C., Young-Robbins, S.S., Haselkorn, R., and Callahan, S.M. (2005). Inactivation of *patS* and *hetN* causes lethal levels of heterocyst differentiation in the filamentous cyanobacterium *Anabaena* sp. PCC 7120. Mol. Microbiol. *57*, 111–123.

Bradley, S., and Carr, N.G. (1976). Heterocyst and nitrogenase development in *Anabaena cylindrica*. J. Gen. Microbiol. *96*, 175–184.

Brahamsha, B., and Haselkorn, R. (1991). Isolation and characterization of the gene encoding the principal sigma factor of the vegetative cell RNA polymerase from the cyanobacterium *Anabaena* sp. strain PCC 7120. J. Bacteriol. *173*, 2442–2450.

Brahamsha, B., and Haselkorn, R. (1992). Identification of multiple RNA polymerase sigma factor homologs in the cyanobacterium *Anabaena* sp. strain PCC 7120: cloning, expression, and inactivation of the *sigB* and *sigC* genes. J. Bacteriol. *174*, 7273–7282.

Buikema, W.J., and Haselkorn, R. (1991). Characterization of a gene controlling heterocyst differentiation in the cyanobacterium *Anabaena* 7120. Genes Dev. *5*, 321–330.

Buikema, W.J., and Haselkorn, R. (2001). Expression of the *Anabaena hetR* gene from a copper-regulated promoter leads to heterocyst differentiation under repressing conditions. Proc. Natl. Acad. Sci. USA *98*, 2729–2734.

Cai, Y., and Wolk, C.P. (1997a). *Anabaena* sp. strain PCC 7120 responds to nitrogen deprivation with a cascade-like sequence of transcriptional activations. J. Bacteriol. *179*, 267–271.

Cai, Y., and Wolk, C.P. (1997b). Nitrogen deprivation of *Anabaena* sp. strain PCC 7120 elicits rapid activation of a gene cluster that is essential for uptake and utilization of nitrate. J. Bacteriol. *179*, 258–266.

Callahan, S.M., and Buikema, W.J. (2001). The role of HetN in maintenance of the heterocyst pattern in *Anabaena* sp. PCC 7120. Mol. Microbiol. *40*, 941–950.

Campbell, E.L., Hagen, K.D., Cohen, M.F., Summers, M.L., and Meeks., J.C. (1996). The *devR* gene product is characteristic of receivers of two-component regulatory systems and is essential for heterocyst development in the filamentous cyanobacterium *Nostoc* sp. strain ATCC 29133. J. Bacteriol. *178*, 2037–2043.

Cardemil, L., and Wolk, C.P. (1976). The polysaccharides from heterocyst and spore envelopes of a blue-green alga. Methylation analysis and structure of the backbones. J. Biol. Chem. *251*, 2967–2975.

Cardemil, L., and Wolk, C.P. (1979). The polysaccharides from heterocyst and spore envelopes of a blue-green alga. Structure of the basic repeating unit. J. Biol. Chem. *254*, 736–741.

Cardemil, L., and Wolk, C.P. (1981a). Isolated heterocysts of *Anabaena variabilis* synthesize envelope polysaccharide. Biochim. Biophys. Acta *674*, 265–276.

Cardemil, L., and Wolk, C.P. (1981b). Polysaccharides from the envelopes of heterocysts and spores of the blue-green algae *Anabaena variabilis* and *Cylindrospermum licheniforme*. J. Phycol. *17*, 234–240.
Carr, N.G. (1988). Nitrogen reserves and dynamic reservoirs in cyanobacteria. In Biochemistry of the Algae and Cyanobacteria (Proc. Phytochem. Soc. Eur. 1988), L.J. Rogers, and J.R. Gallon, eds. (Oxford, Clarendon Press), pp. 13–21.
Cheng, Y., Li, J.-H., Shi, L., Wang, L., Latifi, A., and Zhang, C.-C. (2006). A pair of iron-responsive genes encoding protein kinases with a Ser/Thr kinase domain and a His kinase domain are regulated by NtcA in the cyanobacterium *Anabaena* sp. strain PCC 7120. J. Bacteriol. *188*, 4822–4829.
Collier, J.L., and Grossman, A.R. (1994). A small polypeptide triggers complete degradation of light-harvesting phycobiliproteins in nutrient-deprived cyanobacteria. EMBO J. *13*, 1039–1047.
Crawford, N.A., Sutton, C.W., Yee, B.C., Johnson, T.C., Carlson, D.C., and Buchanan, B.B. (1984). Contrasting modes of photosynthetic enzyme regulation in oxygenic and anoxygenic prokaryotes. Arch. Microbiol. *139*: 124–129.
Curatti, L., Flores, E., and Salerno, G. (2002). Sucrose is involved in the diazotrophic metabolism of the heterocyst-forming cyanobacterium *Anabaena* sp. FEBS Lett. *513*, 175–178.
Dong, Y., Huang, X., Wu, X.-Y., and Zhao, J. (2000). Identification of the active site of HetR protease and its requirement for heterocyst differentiation in the cyanobacterium *Anabaena* sp. strain PCC 7120. J. Bacteriol. *182*, 1575–1579.
Dunn, J.H., and Wolk, C.P. (1970). Composition of the cellular envelopes of *Anabaena cylindrica*. J. Bacteriol. *103*, 153–158.
Ehira, S., and Ohmori, M. (2006a). NrrA, a novel nitrogen-responsive response regulator facilitates heterocyst development in the cyanobacterium *Anabaena* sp. strain PCC 7120. Mol. Microbiol. *59*, 1692–1703.
Ehira, S., and Ohmori, M. (2006b). NrrA directly regulates the expression of *hetR* during heterocyst differentiation in the cyanobacterium *Anabaena* sp. strain PCC 7120. J. Bacteriol. *188*, 8520–8525.
Ehira, S., Ohmori, M., and Sato, N. (2003). Genome-wide expression analysis of the responses to nitrogen deprivation in the heterocyst-forming cyanobacterium *Anabaena* sp. strain PCC 7120. DNA Res. *10*, 97–113.
Ehira, S., Ohmori, M., and Sato, N. (2005). Identification of low-temperature-regulated ORFs in the cyanobacterium *Anabaena* sp. strain PCC 7120: distinguishing the effects of low temperature from the effects of photosystem II excitation pressure. Plant Cell Physiol. *46*, 1237–1245.
Elhai, J., and Wolk, C.P. (1990). Developmental regulation and spatial pattern of expression of the structural genes for nitrogenase in the cyanobacterium *Anabaena*. EMBO J. *9*, 3379–3388.
Ernst, A., and Böger, P. (1985). Glycogen accumulation and the induction of nitrogenase activity in the heterocyst-forming cyanobacterium *Anabaena variabilis*. J. Gen. Microbiol. *131*, 3147–3153.
Ernst, A., Black, T., Cai, Y., Panoff, J.M., Tiwari, D.N., and Wolk, C.P. (1992). Synthesis of nitrogenase in mutants of the cyanobacterium *Anabaena* sp. strain PCC 7120 affected in heterocyst development or metabolism. J. Bacteriol. *174*, 6025–6032.
Errington, J., Daniel, R.A., and Scheffers, D.J. (2003). Cytokinesis in bacteria. Microbiol. Mol. Biol. Rev. *67*, 52–65.
Fan, Q., Huang, G., Lechno-Yossef, S., Wolk, C.P., Kaneko, T., and Tabata, S. (2005). Clustered genes required for synthesis and deposition of envelope glycolipids in *Anabaena* sp. strain PCC 7120. Mol. Microbiol. *58*, 227–243.
Fan, Q., Lechno-Yossef, S., Ehira, S., Kaneko, T., Ohmori, M., Sato, N., Tabata, S., and Wolk, C.P. (2006). Signal transduction genes required for heterocyst maturation in *Anabaena* sp. strain PCC 7120. J. Bacteriol. *188*, 6688–6693.
Fay, P., Stewart, W.D.P., Walsby, A.E., and Fogg, G.E. (1968). Is the heterocyst the site of nitrogen fixation in blue-green algae? Nature *220*, 810–812.
Fernández-Piñas, F., Leganés, F., and Wolk, C.P. (1994). A third genetic locus required for the formation of heterocysts in *Anabaena* sp. strain PCC 7120. J. Bacteriol. *176*, 5277–5283.
Fiedler, G., Arnold, M., Hannus, S., and Maldener, I. (1998a). The DevBCA exporter is essential for envelope formation in heterocysts of the cyanobacterium *Anabaena* sp. strain PCC 7120. Mol. Microbiol. *27*, 1193–1202.
Fiedler, G., Arnold, M., and Maldener, I. (1998b). Sequence and mutational analysis of the *devBCA* gene cluster encoding a putative ABC transporter in the cyanobacterium *Anabaena variabilis* ATCC 29413. Biochim. Biophys. Acta *1375*, 140–143.

Fiedler, G., Muro-Pastor, A.M., Flores, E., and Maldener, I. (2001). NtcA-dependent expression of the *devBCA* operon, encoding a heterocyst-specific ATP-binding cassette transporter in *Anabaena* spp. J. Bacteriol. *183*, 3795–3799.

Fitch, W.M. (2000). Homology a personal view on some of the problems. Trends Genet. *16*, 227–231.

Fleming, H., and Haselkorn, R. (1974). The program of protein synthesis during heterocyst differentiation in nitrogen-fixing blue-green algae. Cell *3*, 159–170.

Flores, E., and Herrero, A. (1994). Assimilatory nitrogen metabolism and its regulation. In The Molecular Biology of Cyanobacteria, D.A. Bryant, ed. (Dordrecht, The Netherlands, Kluwer Academic Publishers), pp. 487–517.

Flores, E., and Herrero, A. (2005). Nitrogen assimilation and nitrogen control in cyanobacteria. Biochem. Soc. Trans. *33*, 164–167.

Flores, E., Herrero, A., Wolk, C.P., and Maldener, I. (2006). Is the periplasm continuous in filamentous multicellular cyanobacteria? Trends Microbiol. *14*, 439–443.

Frías, J.E., Mérida, A., Herrero, A., Martín-Nieto, J., and Flores, E. (1993). General distribution of the nitrogen control gene *ntcA* in cyanobacteria. J. Bacteriol. *175*, 5710–5713.

Frías, J.E., Flores, E., and Herrero, A. (2000). Activation of the *Anabaena nir* operon promoter requires both NtcA (CAP family) and NtcB (LysR family) transcription factors. Mol. Microbiol. 38, 613–625.

Frías, J.E., Herrero, A., and Flores, E. (2003). Open reading frame *all0601* from *Anabaena* sp. strain PCC 7120 represents a novel gene, *cnaT*, required for expression of the nitrate assimilation *nir* operon. J. Bacteriol. *185*, 5037–5044.

Fritsch, F.E. (1951). The heterocyst: a botanical enigma. Proc. Linn. Soc. London *162*, 194–211.

Gambacorta, A., Soriente, A., Trincone, A., and Sodano, G. (1995). Biosynthesis of the heterocyst glycolipids in the cyanobacterium *Anabaena cylindrica*. Phytochemistry *39*, 771–774.

Gambacorta, A., Romano, I., Trincone, A., Soriente, A., Giordano, M., and Sodano, G. (1996). The heterocyst glycolipids of cyanobacteria. 5. Heterocyst glycolipids from five nitrogen-fixing cyanobacteria. Gazz. Chim. Ital. *126*, 653–656.

Gambacorta, A., Pagnotta, E., Romano, I., Sodano, G., and Trincone, A. (1998). Heterocyst glycolipids from nitrogen-fixing cyanobacteria other than Nostocaceae. Phytochemistry *48*, 801–805.

Gierer, A., and Meinhardt, H. (1972). A theory of biological pattern formation. Kybernetik *12*, 30–39.

Golden, J. (1998). Programmed DNA rearrangements in cyanobacteria. In Bacterial Genomes Physical Structure and Analysis, F.J. de Bruijn, J.R. Lupski, and G.M. Weinstock, eds. (N.Y., Chapman & Hall), pp. 162–173.

Gonzalez, L., Phalip, V., and Zhang, C.C. (2001). Characterization of PknC, a Ser/Thr kinase with broad substrate specificity from the cyanobacterium *Anabaena* sp. strain PCC 7120. Eur. J. Biochem. *268*, 1869–1875.

Hagen, K.D., and Meeks, J.C. (2001). The unique cyanobacterial protein OpcA is an allosteric effector of glucose-6-phosphate dehydrogenase in *Nostoc punctiforme* ATCC 29133. J. Biol. Chem. *276*, 11477–11486.

Haselkorn, R. (1992). Developmentally regulated gene rearrangements in prokaryotes. Annu. Rev. Genet. *26*, 113–130.

Haury, J.F., and Wolk, C.P. (1978). Classes of *Anabaena variabilis* mutants with oxygen-sensitive nitrogenase activity. J. Bacteriol. *136*, 688–692.

Hebbar, P.B., and Curtis, S.E. (2000). Characterization of *devH*, a gene encoding a putative DNA binding protein required for heterocyst function in *Anabaena* sp. strain PCC 7120. J. Bacteriol. *182*, 3572–3581.

Hernández, J.A., Muro-Pastor, A.M., Flores, E., Bes, M.T., Peleato, M.L., and Fillat, M.F. (2006). Identification of a *furA cis* antisense RNA in the cyanobacterium *Anabaena* sp. PCC 7120. J. Mol. Biol. *355*, 325–334.

Herrero, A., Muro-Pastor, A.M., and Flores, E. (2001). Nitrogen control in cyanobacteria. J. Bacteriol. *183*, 411–425.

Herrero, A., Muro-Pastor, A.M., Valladares, A., and Flores, E. (2004). Cellular differentiation and the NtcA transcription factor in filamentous cyanobacteria. FEMS Microbiol. Rev. *28*, 469–487.

Hilbert, D.W., and Piggot, P.J. (2004). Compartmentalization of gene expression during *Bacillus subtilis* spore formation. Microbiol. Mol. Biol. Rev. *68*:234–262.

Holland, D., and Wolk, C.P. (1990). Identification and characterization of *hetA*, a gene that acts early in the process of morphological differentiation of heterocysts. J. Bacteriol. *172*, 3131–3137.

Huang, W., and Wu, Q. (2004). The ManR specifically binds to the promoter of a Nramp transporter gene in *Anabaena* sp. PCC 7120: a novel regulatory DNA motif in cyanobacteria. Biochem. Biophys. Res. Commun. *317*, 578–585.

Huang, X., Dong, Y., and Zhao, J. (2004). HetR homodimer is a DNA-binding protein required for heterocyst differentiation, and the DNA-binding activity is inhibited by PatS. Proc. Natl. Acad. Sci. USA *101*, 4848–4853.

Huang, G., Fan, Q., Lechno-Yossef, S., Wojciuch, E., Wolk, C.P., Kaneko, T., and Tabata, S. (2005). Clustered genes required for the synthesis of heterocyst envelope polysaccharide in *Anabaena* sp. strain PCC 7120. J. Bacteriol. *187*, 1114–1123.

Imashimizu, M., Yoshimura, H., Katoh, H., Ehira, S., Ohmori, M. (2005). NaCl enhances cellular cAMP and upregulates genes related to heterocyst development in the cyanobacterium, *Anabaena* sp. strain PCC 7120. FEMS Microbiol. Lett *252*, 97–103.

Iteman, I., Rippka, R., Tandeau de Marsac, N., and Herdman, M. (2002). rDNA analyses of planktonic heterocystous cyanobacteria, including members of the genera *Anabaenopsis* and *Cyanospira*. Microbiology *148*, 481–496.

Jang, J., Wang, L., Arcondéguy, T., and Zhang, C.-C. (2006). PrpJ, a protein Ser/Thr and Tyr phosphatase located on the plasma membrane, is involved in heterocyst maturation in *Anabaena* PCC 7120. Paper presented at: 12th Intl. Symp. Phototrophic Prokaryotes, p. 175 (Pau, France).

Jones, K.M., and Haselkorn, R. (2002). Newly identified cytochrome *c* oxidase operon in the nitrogen-fixing cyanobacterium *Anabaena* sp. strain PCC 7120 specifically induced in heterocysts J. Bacteriol. *184*, 2941–2949.

Jones, K.M., Buikema, W.J., and Haselkorn, R. (2003). Heterocyst-specific expression of *patB*, a gene required for nitrogen fixation in *Anabaena* sp. strain PCC 7120. J. Bacteriol. *185*, 2306–2314.

Kallas, T. (1994). The cytochrome b_6f complex, In The Molecular Biology of Cyanobacteria, D.A. Bryant, ed. (Dordrecht, NL, Kluwer Academic Publishers), pp. 259–317.

Kaneko, T., Nakamura, Y., Wolk, C.P., Kuritz, T., Sasamoto, S., Watanabe, A., Iriguchi, M., Ishikawa, A., Kawashima, K., Kimura, T., Kishida, Y., Kohara, M., Matsumoto, M., Matsuno, A., Muraki, A., Nakazaki, N. Shimpo, S., Sugimoto, M., Takazawa, M. Yamada, M., Yasuda, M., and Tabata, S. (2001). Complete genomic sequence of the filamentous nitrogen-fixing cyanobacterium *Anabaena* sp. strain PCC 7120. DNA Res. *8*, 205–213.

Katayama, M., Ohmori, M. (1997). Isolation and characterization of multiple adenylate cyclase genes from the cyanobacterium *Anabaena* sp. strain PCC 7120. J. Bacteriol. *179*, 3588–3593.

Khudyakov, I.Y., and Golden, J.W. (2001). Identification and inactivation of three group 2 sigma factor genes in *Anabaena* sp. strain PCC 7120. J. Bacteriol. *183*, 6667–6675.

Khudyakov, I.Y., and Golden, J.W. (2004). Different functions of HetR, a master regulator of heterocyst differentiation in *Anabaena* sp. PCC 7120, can be separated by mutation. Proc. Natl. Acad. Sci. USA *101*, 16040–16045.

Khudyakov, I., and Wolk, C.P. (1997). *hetC*, a gene coding for a protein similar to bacterial ABC protein exporters, is involved in early regulation of heterocyst differentiation in *Anabaena* sp. strain PCC 7120. J. Bacteriol. *179*, 6971–6978.

Khudyakov, I., and Zaitzev, S. (2006). New genes involved in the regulation of heterocyst differentiation in *Anabaena* sp. PCC 7120 that act downstream of *patA*. Paper presented at: 12th Intl. Symp. Phototrophic Prokaryotes, p. 117 (Pau, France).

Koksharova, O.A., and Wolk, C.P. (2002). Novel DNA-binding proteins in the cyanobacterium *Anabaena* sp. strain PCC 7120. J. Bacteriol. *184*, 3931–3940.

Kuhn, I., Peng, L., Bedu, S., and Zhang, C.-C. (2000). Developmental regulation of the cell division protein FtsZ in *Anabaena* sp. strain PCC 7120, a cyanobacterium capable of terminal differentiation. J. Bacteriol. *182*, 4640–4643.

Lahmi, R., Sendersky, E., Perelman, A., Hagemann, M., Forchhammer, K., and Schwarz, R. (2006). Alanine dehydrogenase activity is required for adequate progression of phycobilisome degradation during nitrogen starvation in *Synechococcus elongatus* PCC 7942. J. Bacteriol. *188*, 5258–5265.

Lang, N.J., Simon, R.D., and Wolk, C.P. (1972). Correspondence of cyanophycin granules with structured granules in *Anabaena cylindrica*. Arch. Mikrobiol. *83*, 313–320.

Laurent, S., Forchhammer, K., Gonzalez, L., Heulin, T., Zhang, C.-C., and Bédu, S. (2004). Cell-type specific modification of P_{II} is involved in the regulation of nitrogen metabolism in the cyanobacterium *Anabaena* PCC 7120. FEBS Lett. *576*, 261–265.

Laurent, S., Chen, H., Bédu, S., Ziarelli, F., Peng, L., and Zhang, C.-C. (2005). Nonmetabolizable analogue of 2-oxoglutarate elicits heterocyst differentiation under repressive conditions in *Anabaena* sp. PCC 7120. Proc. Natl. Acad. Sci. USA *102*, 9907–9912.

Lázaro, S., Fernández-Piñas, F., Fernández-Valiente, E., Blanco-Rivero, A., and Leganés, F. (2001). *pbpB*, a gene coding for a putative penicillin-binding protein, is required for aerobic nitrogen fixation in the cyanobacterium *Anabaena* sp. strain PCC7120. J. Bacteriol. *183*, 628–636.

Lechno-Yossef, S., Fan, Q., Ehira, S., Sato, N., and Wolk, C.P. (2006). Mutations in four regulatory genes have interrelated effects on heterocyst maturation in *Anabaena* sp. strain PCC 7120. J. Bacteriol. *188*, 7387–7395.

Leganés, F., Blanco-Rivero, A., Fernández-Piñas, F., Redondo, M., Fernández-Valiente, E., Fan, Q., Lechno-Yossef, S., and Wolk, C.P. (2005). Wide variation in the cyanobacterial complement of presumptive penicillin-binding proteins. Arch. Microbiol. *184*, 234–248.

Li, B., Huang, X., and Zhao, J. (2002). Expression of *hetN* during heterocyst differentiation and its inhibition of *hetR* up-regulation in the cyanobacterium *Anabaena* sp. PCC 7120. FEBS Lett. *517*, 87–91.

Li, J.H., Laurent, S., Konde, V., Bedu, S., and Zhang, C.-C. (2003). An increase in the level of 2-oxoglutarate promotes heterocyst development in the cyanobacterium *Anabaena* sp. strain PCC 7120. Microbiology *149*, 3257–3263.

Liang, J., Scappino, L., and Haselkorn, R. (1992). The *patA* gene product, which contains a region similar to CheY of *Escherichia coli*, controls heterocyst pattern formation in the cyanobacterium *Anabaena* 7120. Proc. Natl. Acad. Sci. USA *89*, 5655–5659.

Liang, J., Scappino, L., and Haselkorn, R. (1993). The *patB* gene product, required for growth of the cyanobacterium *Anabaena* sp. strain PCC 7120 under nitrogen-limiting conditions, contains ferredoxin and helix–turn–helix domains. J. Bacteriol. *175*, 1697–1704.

López-Gomollón, S., Hernández, J.A., Wolk, C.P., Peleato, M.L., and Fillat, M.F. (2006). Expression of *furA* is modulated by NtcA and strongly enhanced in heterocysts of *Anabaena* sp. PCC 7120. Microbiology *153*, 42–50.

Makarova, K.S., Koonin, E.V., Haselkorn, R., and Galperin, M.Y. (2006). Cyanobacterial response regulator PatA contains a conserved N-terminal domain (PATAN) with an alpha-helical insertion. Bioinformatics *22*, 1297–1301.

Maldener, I., Fiedler, G., Ernst, A., Fernández-Piñas, F., and Wolk, C.P. (1994). Characterization of *devA*, a gene required for the maturation of proheterocysts in the cyanobacterium *Anabaena* sp. strain PCC 7120. J. Bacteriol. *176*, 7543–7549.

Maldener, I., Hannus, S., and Kammerer, M. (2003). Description of five mutants of the cyanobacterium *Anabaena* sp. strain PCC 7120 affected in heterocyst differentiation and identification of the transposon-tagged genes. FEMS Microbiol. Lett. *224*, 205–213.

Mazón, G., Lucena, J.M., Campoy, S., Fernández de Henestrosa, A.R., Candau, P., and Barbé, J. (2004). LexA-binding sequences in Gram-positive and cyanobacteria are closely related. Mol. Gen. Genomics *271*, 40–49.

Meeks, J.C., and Elhai, J. (2002). Regulation of cellular differentiation in filamentous cyanobacteria in free-living and plant-associated symbiotic growth states. Microbiol. Mol. Biol. Rev. *66*, 94–121.

Meeks, J.C., Elhai, J., Thiel, T., Potts, M., Larimer, F., Lamerdin, J., Predki, P., and Atlas, R. (2001). An overview of the genome of *Nostoc punctiforme*, a multicellular, symbiotic cyanobacterium. Photosynth. Res. *70*, 85–106.

Mitchison, G.J., Wilcox, M., and Smith, R.J. (1976). Measurement of an inhibitory zone. Science *191*, 866–868.

Motulsky, H. (2002). Detecting outliers. URL: http://www.graphpad.com/articles/outlier.htm.

Muro-Pastor, A.M., Valladares, A., Flores, E., and Herrero, A. (1999). The *hetC* gene is a direct target of the NtcA transcriptional regulator in cyanobacterial heterocyst development. J. Bacteriol. *181*, 6664–6669.

Muro-Pastor, M.I., Reyes, J.C., and Florencio, F.J. (2001). Cyanobacteria perceive nitrogen status by sensing intracellular 2-oxoglutarate levels. J. Biol. Chem. *276*, 38320–38328.

Muro-Pastor, A.M., Valladares, A., Flores, E., and Herrero, A. (2002). Mutual dependence of the expression of the cell differentiation regulatory protein HetR and the global nitrogen regulator NtcA during heterocyst development. Mol. Microbiol. *44*, 1377–1385.

Muro-Pastor, A.M., Olmedo-Verd, E., and Flores, E. (2006). All4312, an NtcA-regulated two-component response regulator in *Anabaena* sp. strain PCC 7120. FEMS Microbiol. Lett. *256*, 171–177.

Murry, M.A., and Wolk, C.P. (1989). Evidence that the barrier to the penetration of oxygen into heterocysts depends upon two layers of the cell envelope. Arch. Microbiol. *151*, 469–474.

Newman, J., Karakaya, H., Scanlan, D.J., and Mann, N.H. (1995). A comparison of gene organization in the *zwf* region of the genomes of the cyanobacteria *Synechococcus* sp. PCC 7942 and *Anabaena* sp. PCC 7120. FEMS Microbiol. Lett. *133*, 187–193.

Ning, D., and Xu, X. (2004). *alr0117*, a two-component histidine kinase gene, is involved in heterocyst development in *Anabaena* sp. PCC 7120. Microbiology *150*, 447–453.

Olmedo-Verd, E., Flores, E., Herrero, A., and Muro-Pastor, A.M. (2005). HetR-dependent and -independent expression of heterocyst-related genes in an *Anabaena* strain overproducing the NtcA transcription factor. J. Bacteriol. *187*, 1985–1991.

Olmedo-Verd, E., Muro-Pastor, A.M., Flores, E., and Herrero, A. (2006). Localized induction of the *ntcA* regulatory gene in developing heterocysts of *Anabaena* sp. strain PCC 7120. J. Bacteriol. *188*, 6694–6699.

Orozco, C.C., Risser, D.D., and Callahan, S.M. (2006). Epistasis analysis of four genes from *Anabaena* sp. strain PCC 7120 suggests a connection between PatA and PatS in heterocyst pattern formation. J. Bacteriol. *188*, 1808–1816.

Peterson, R.B., and Wolk, C.P. (1978). High recovery of nitrogenase activity and of ^{55}Fe-labeled nitrogenase in heterocysts isolated from *Anabaena variabilis*. Proc. Natl. Acad. Sci. USA *75*, 6271–6275.

Peterson, R.B., Shaw E.R. Dolan E., and Ke, B. (1981). A photochemically active heterocyst preparation from *Anabaena variabilis*. Photobiochem. Photobiophys. *2*, 79–84.

Picossi, S., Valladares, A., Flores, E., and Herrero, A. (2004). Nitrogen-regulated genes for the metabolism of cyanophycin, a bacterial nitrogen reserve polymer: expression and mutational analysis of two cyanophycin synthetase and cyanophycinase gene clusters in the heterocyst-forming cyanobacterium *Anabaena* sp. PCC 7120. J. Biol. Chem. *279*, 11582–11592.

Ramasubramanian, T.S., Wei, T.F., and Golden, J.W. (1994). Two *Anabaena* sp. strain PCC 7120 DNA-binding factors interact with vegetative cell- and heterocyst-specific genes. J. Bacteriol. *176*, 1214–1223.

Ramasubramanian, T.S., Wei, T.F., Oldham, A.K., and Golden, J.W. (1996). Transcription of the *Anabaena* sp. strain PCC 7120 *ntcA* gene: multiple transcripts and NtcA binding. J. Bacteriol. *178*, 922–926.

Ramírez, M.E., Hebbar, P.B., Zhou, R., Wolk, C.P., and Curtis, S.E. (2005). *Anabaena* sp. strain PCC 7120 gene *devH* is required for synthesis of the heterocyst glycolipid layer. J. Bacteriol. *187*, 2326–2331.

Sakr, S., Jeanjean, R., Zhang, C.-C., and Arcondeguy, T. (2006). Inhibition of cell division suppresses heterocyst development in *Anabaena* sp. strain PCC 7120. J. Bacteriol. *188*, 1396–1404.

Schmetterer, G. (1994). Cyanobacterial respiration. In The Molecular Biology of Cyanobacteria, D.A. Bryant, ed. (Dordrecht, The Netherlands, Kluwer Academic Publishers), pp. 409–435.

Schwartz, S.H., Black, T.A., Jäger, K., Panoff, J.M., and Wolk, C.P. (1998). Regulation of an osmoticum-responsive gene in *Anabaena* sp. strain PCC 7120. J. Bacteriol. *180*, 6332–6337.

Serrano, M., Neves, A., Soares, C.M., Moran, Jr, C.P., and Henriques, A.O. (2004). Role of the anti-sigma factor SpoIIAB in regulation of σ^G during *Bacillus subtilis* sporulation. J. Bacteriol. *186*, 4000–4013.

Shi, Y., Zhao, W., Zhang, W., Ye, Z., and Zhao, J. (2006). Regulation of intracellular free calcium concentration during heterocyst differentiation by HetR and NtcA in *Anabaena* sp. PCC 7120. Proc. Natl. Acad. Sci. USA *103*, 11334–11339.

Stanier, R.Y., and Cohen-Bazire, G. (1977). Phototrophic prokaryotes: the cyanobacteria. Ann. Rev. Microbiol. *31*, 225–274.

Summers, M.L., Wallis, J.G., Campbell, E.L., and Meeks, J.C. (1995). Genetic evidence of a major role for glucose-6-phosphate dehydrogenase in nitrogen fixation and dark growth of the cyanobacterium *Nostoc* sp. strain ATCC 29133. J. Bacteriol. *177*, 6184–6194.

Sutherland, J.M., Stewart, W.D.P., and Herdman, M. (1985). Akinetes of the cyanobacterium *Nostoc* PCC 7524: morphological changes during synchronous germination. Arch. Microbiol. *142*, 269–274.

Suzuki, T., Yoshimura, H., Hisabori, T., and Ohmori, M. (2004). Two cAMP receptor proteins with different biochemical properties in the filamentous cyanobacterium *Anabaena* sp. PCC 7120. FEBS Lett. *571*, 154–160.

Tamoi, M., Murakami, A., Takeda, T., Shigeoka, S. (1998). Acquisition of a new type of fructose-1,6-bisphosphatase with resistance to hydrogen peroxide in cyanobacteria: molecular characterization of the enzyme from *Synechocystis* PCC 6803. Biochim. Biophys. Acta *1383*, 232–244

Tanigawa, R., Shirokane, M., Maeda, S.-I., Omata, T., Tanaka, K., and Takahashi, H. (2002). Transcriptional activation of NtcA-dependent promoters of *Synechococcus* sp. PCC 7942 by 2-oxoglutarate *in vitro*. Proc. Natl. Acad. Sci. USA 99, 4251–4255.

Thiel, T., and Pratte, B. (2001). Effect on heterocyst differentiation of nitrogen fixation in vegetative cells of the cyanobacterium *Anabaena variabilis* ATCC 29413. J. Bacteriol. *183*, 280–286.

Thomas, J., Meeks, J.C., Wolk, C.P., Shaffer, P.W., and Austin, S.M. (1977). Formation of glutamine from [^{13}N]ammonia, [^{13}N]dinitrogen, and [^{14}C]glutamate by heterocysts isolated from *Anabaena cylindrica*. J. Bacteriol. *129*, 1545–1555.

Torrecilla, I., Leganés, F., Bonilla, I., and Fernández-Piñas, F. (2004). A calcium signal is involved in heterocyst differentiation in the cyanobacterium *Anabaena* sp. PCC 7120. Microbiology *150*, 3731–3739.

Turing, A.M. (1952). The chemical basis of morphogenesis. Phil. Trans. Roy. Soc. Lond. B *237*, 37–72.

Valladares, A., Herrero, A., Pils, D., Schmetterer, G., and Flores, E. (2003). Cytochrome *c* oxidase genes required for nitrogenase activity and diazotrophic growth in *Anabaena* sp. PCC 7120. Mol. Microbiol. *47*, 1239–1249.

Vazquez-Bermudez, M.F., Herrero, A., and Flores, E. (2002). 2-Oxoglutarate increases the binding affinity of the NtcA (nitrogen control) transcription factor for the *Synechococcus glnA* promoter. FEBS Lett. *512*, 71–74.

Walsby, A.E. (1985). The permeability of heterocysts to the gases nitrogen and oxygen. Proc. R. Soc. Lond. B *226*, 345–366.

Wang, Y., and Xu, X. (2005). Regulation by *hetC* of genes required for heterocyst differentiation and cell division in *Anabaena* sp. strain PCC 7120. J. Bacteriol. *187*, 8489–8493.

Wang, Y., Lechno-Yossef, S., Gong, Y., Fan, Q., Wolk, C.P., and Xu, X. (2007). Predicted glycosyl transferase genes located outside of the HEP island are required for formation of heterocyst envelope polysaccharide in *Anabaena* sp. strain PCC7120. J. Bacteriol. (in press).

Wealand, J.L., Myers, J.A., and Hirschberg, R. (1989). Changes in gene expression during nitrogen starvation in *Anabaena variabilis* ATCC 29413. J. Bacteriol. *171*, 1309–1313.

Wei, T.-F., Ramasubramanian, T.S., Pu, F., and Golden, J.W. (1993). *Anabaena* sp. strain PCC 7120 *bifA* gene encoding a sequence-specific DNA-binding protein cloned by *in vivo* transcriptional interference selection. J. Bacteriol. *175*, 4025–4035.

Wilcox, M., Mitchison, G.J., and Smith, R.J. (1973a). Pattern formation in the blue-green alga, *Anabaena*. I. Basic mechanisms. J. Cell Sci. *12*, 707–723.

Wilcox, M., Mitchison, G.J., and Smith, R.J. (1973b). Pattern formation in the blue-green alga, *Anabaena*. II. Controlled proheterocyst regression. J. Cell Sci. *13*, 637–649.

Wilcox, M., Mitchison, G.J., and Smith, R.J. (1975). Mutants of *Anabaena cylindrica* altered in heterocyst spacing. Arch. Microbiol. *103*, 219–223.

Winkenbach, F., and Wolk, C.P. (1973). Activities of enzymes of the oxidative and the reductive pentose phosphate pathways in heterocysts of a blue-green alga. Plant Physiol. *52*, 480–482.

Winkenbach, F., Wolk, C.P., and Jost, M. (1972). Lipids of membranes and of the cell envelope in heterocysts of a blue-green alga. Planta *107*, 69–80.

Wisen, S., Sjogren, T., Olin, B., and Mannervik, B. (2004). Purification, crystallization and preliminary X-ray data of the transcription factor NtcA from the cyanobacterium *Anabaena* PCC 7120. Acta Crystallogr. D Biol. Crystallogr. *60*, 923–925.

Wolk, C.P. (1967). Physiological basis of the pattern of vegetative growth of a blue-green alga. Proc. Natl. Acad. Sci. USA *57*, 1246–1251.

Wolk, C.P. (1968). Movement of carbon from vegetative cells to heterocysts in *Anabaena cylindrica*. J. Bacteriol. *96*, 2138–2143.

Wolk, C.P. (1979). Intercellular interactions and pattern formation in filamentous cyanobacteria. In Determinants of Spatial Organization, 37th Symp. Soc. Dev. Biol. S. Subtelny and I.R. Konigsberg, eds. (N.Y., Academic Press), pp. 247–266.

Wolk, C.P. (2000). Heterocyst formation in *Anabaena*. In Prokaryotic development, Y.V. Brun and L.J. Shimkets, eds. (Washington, D.C., ASM Press), pp. 83–104.

Wolk, C.P., and Quine, M.P. (1975). Formation of one-dimensional patterns by stochastic processes and by filamentous blue-green algae. Dev. Biol. *46*, 370–382.

Wolk, C.P., and Zarka, K. (1998). Genetic dissection of heterocyst differentiation. In Cyanobacterial Biotechnology. Proceedings of the International Symposium, Sept. 18–21, 1996, G. Subramanian, B.D. Kaushik, and G.S. Venkataraman, eds. (New Delhi, Oxford and IBH Publ. Co.), pp. 191–196.

Wolk, C.P., Cai, Y., Cardemil, L., Flores, E., Hohn, B., Murry, M., Schmetterer, G., Schrautemeier, B., and Wilson, R. (1988). Isolation and complementation of mutants of *Anabaena* sp. strain PCC 7120 unable to grow aerobically on dinitrogen. J. Bacteriol. *170*, 1239–1244.

Wolk, C.P., Elhai, J., Kuritz, T., and Holland, D. (1993). Amplified expression of a transcriptional pattern formed during development of *Anabaena*. Mol. Microbiol. *7*, 441–445.

Wolk, C.P., Ernst, A., and Elhai, J. (1994). Heterocyst metabolism and development. In The Molecular Biology of Cyanobacteria, D.A. Bryant, ed. (Dordrecht, NL, Kluwer Academic Publishers), pp. 769–823.

Wolk, C.P., Zhu, J., and Kong, R. (1999). Genetic analysis of heterocyst formation. In The Phototrophic Prokaryotes, G.A. Peschek, W. Loeffelhardt, and G. Schmetterer, eds. (New York, Kluwer Academic/Plenum Publishers), pp. 509–515.

Wong, F.C., and Meeks, J.C. (2001). The *hetF* gene product is essential to heterocyst differentiation and affects HetR function in the cyanobacterium *Nostoc punctiforme*. J. Bacteriol. *183*, 2654–2661.

Wood, N.B., and Haselkorn, R. (1979). Proteinase activity during heterocyst differentiation in nitrogen-fixing cyanobacteria. In Limited Proteolysis in Microorganisms, G.H. Cohen, and H. Holzer, eds. (Bethesda, MD, U.S. DHEW Publication No. (NIH) 79–1591), pp. 159–166.

Wood, A.M., Miller, S.R., Li, W.K.W., and Castenholz, R.W. (2002). Preliminary studies of cyanobacteria, picoplankton, and virioplankton in the Salton Sea with special attention to phylogenetic diversity among eight strains of filamentous cyanobacteria. Hydrobiol. *473*, 77–92.

Wu, X., Liu, D., Lee, M.H., and Golden, J.W. (2004). *patS* minigenes inhibit heterocyst development of *Anabaena* sp. strain PCC 7120. J. Bacteriol. *186*, 6422–6429.

Xu, X., and Wolk, C.P. (2001). Role for *hetC* in the transition to a nondividing state during heterocyst differentiation in *Anabaena* sp. J. Bacteriol. *183*, 393–396.

Xu, X., Khudyakov, I., and Wolk, C.P. (1997). Lipopolysaccharide dependence of cyanophage sensitivity and aerobic nitrogen fixation in *Anabaena* sp. strain PCC 7120. J. Bacteriol. *179*, 2884–2891.

Xu, W.-L., Jeanjean, R., Liu, Y.-D., and Zhang, C.-C. (2003). *pkn22* (*alr2502*) encoding a putative Ser/Thr kinase in the cyanobacterium *Anabaena* sp. PCC 7120 is induced by both iron starvation and oxidative stress and regulates the expression of *isiA*. FEBS Lett. *553*, 179–182.

Yoon, H.-S., and Golden, J.W. (1998). Heterocyst pattern formation controlled by a diffusible peptide. Science *282*, 935–938.

Yoon, H.S., and Golden, J.W. (2001). PatS and products of nitrogen fixation control heterocyst pattern. J. Bacteriol. *183*, 2605–2613.

Yoon, H.S., Lee, M.H., Xiong, J., Golden, J.W. (2003). *Anabaena* sp. strain PCC 7120 *hetY* gene influences heterocyst development. J. Bacteriol. *185*, 6995–7000.

Zhang, C.-C. (1993). A gene encoding a protein related to eukaryotic protein kinases from the filamentous heterocystous cyanobacterium *Anabaena* PCC 7120. Proc. Natl. Acad. Sci. USA *90*, 11840–11844.

Zhang, C.-C., and Libs, L. (1998). Cloning and characterisation of the pknD gene encoding an eukaryotic-type protein kinase in the cyanobacterium *Anabaena* sp. PCC7120. Mol. Gen. Genet. *258*, 26–33.

Zhang, C.-C., Friry, A., and Peng, L. (1998). Molecular and genetic analysis of two closely linked genes that encode, respectively, a protein phosphatase 1/2A/2B homolog and a protein kinase homolog in the cyanobacterium *Anabaena* sp. strain PCC 7120. J. Bacteriol. *180*, 2616–2622.

Zhang, C.-C., Laurent, S., Sakr, S., Peng, L., and Bédu, S. (2006). Heterocyst differentiation and pattern formation in cyanobacteria: a chorus of signals. Mol. Microbiol. *59*, 367–375.

Zhao, Y., Shi, Y., Zhao, W., Huang, X., Wang, D., Brown, N., Brand, J.J., and Zhao, J. (2005). CcbP, a calcium-binding protein from *Anabaena* sp. PCC 7120, provides evidence that calcium ions regulate heterocyst differentiation. Proc. Natl. Acad. Sci. USA *102*, 5744–5748.

Zhou, R., and Wolk, C.P. (2003). A two-component system mediates developmental regulation of biosynthesis of a heterocyst polysaccharide. J. Biol. Chem. *278*, 19939–19946.

Zhou, R., Wei, X., Jiang, N., Li, H., Dong, Y., His, K.L., and Zhao, J. (1998). Evidence that HetR protein is an unusual serine-type protease. Proc. Natl. Acad. Sci. USA *95*, 4959–4963.

Zhu, J., Kong, R., and Wolk, C.P. (1998). Regulation of *hepA* of *Anabaena* sp. strain PCC 7120 by elements 5′ from the gene and by *hepK*. J. Bacteriol. *180*, 4233–4242.

Zhu, J., Jäger, K., Black, T., Zarka, K., Koksharova, O., and Wolk, C.P. (2001). HcwA, an autolysin, is required for heterocyst maturation in *Anabaena* sp. strain PCC 7120. J. Bacteriol. *183*, 6841–6851.

Cyanobacterial Nitrogen Fixation in the Ocean: Diversity, Regulation, and Ecology

15

Lucas J. Stal and Jonathan P. Zehr

Abstract

Nitrogen is an essential and major component of biomass. While virtually all life depends on combined forms of nitrogen that are usually limited in availability, some prokaryotes, including many groups of cyanobacteria, can use the ubiquitous atmospheric dinitrogen (N_2). As photoautotrophic bacteria they can easily meet the energy demand that is required by nitrogenase, the enzyme that reduces N_2 to NH_3. However, nitrogenase is very sensitive to oxygen and the oxygenic cyanobacteria have evolved various strategies to cope with this paradox. Primary production in the ocean is generally considered to be limited by nitrogen. In recent years it has become clear that N_2-fixing cyanobacteria are important in the nitrogen budget of the surface oceans. Estimates of N_2 fixation indicate that approximately half of global N_2 fixation occurs in the sea. N_2 fixation is not distributed homogenously throughout the oceans. Pelagic diazotrophic cyanobacteria are only found in (sub)tropical oceans and are notably absent in temperate and colder seas. However, at lower salinities in estuaries and other brackish environments, N_2-fixing cyanobacteria can be abundant. N_2-fixing cyanobacteria are also abundant in benthic mats in coastal and aquatic environments all over the globe, including polar regions. This demonstrates that N_2-fixing cyanobacteria are not excluded from temperate and cold marine environments, even though they are only found in the water column of warm oceans. In this chapter we will discuss these aspects and review the existing knowledge of the diversity of N_2-fixing cyanobacteria and the factors that determine their global distribution.

Introduction

Virtually all life depends on combined forms of nitrogen as sources of nitrogen for biosynthetic purposes. Examples of nitrogen (N) sources for growth include nitrite, nitrate, ammonium, and organic nitrogen (e.g. amino acids and urea). Oxidized forms of nitrogen such as nitrite and nitrate are first reduced to ammonium before the N is assimilated and used for biosynthesis of nitrogenous compounds including proteins, nucleic acids, peptidoglycan (cell walls) and chlorophylls. Assimilatory nitrite and nitrate reduction is energetically expensive and requires a low-potential electron donor (reduced ferredoxin). These nitrogen sources are often used by algae and cyanobacteria that obtain the required energy and electrons from photosynthesis (see Chapter 13). However, combined nitrogen is often in low supply in many environments and particularly the major oceanic gyres are known to be chronically depleted in nitrogen. The most abundant form of nitrogen on Earth is gaseous molecular nitrogen, dinitrogen (N_2), the dominant constituent of the atmosphere. The triple bond between the two nitrogen atoms renders N_2 a very stable molecule that is difficult to break. Only certain prokaryotic organisms are capable of utilizing dinitrogen. These N_2-fixing (diazotrophic) microorganisms possess the enzyme complex nitrogenase that is responsible for the reduction of dinitrogen to ammonia.

The marine nitrogen cycle

Nitrogen exists in several redox states. In its most oxidized form, nitrate, the redox state of nitrogen is +5, while in its most reduced form, ammonia, it is −3. Microorganisms use this large difference in redox states for the utilization of N-compounds as energy sources or terminal electron acceptors. Mineralization of organic matter causes the liberation of ammonia (ammonification). Eventually, chemoautotrophic bacteria oxidize ammonium to nitrate. Nitrate is then denitrified to the gases N_2 and N_2O and becomes largely unavailable as a source of nitrogen for organisms. Under anaerobic conditions (e.g. in sediments, aggregates, oxygen minimum zones or other anaerobic microniches) ammonium can be converted to N_2 in a reaction involving nitrite. This pathway is known as anammox (anaerobic ammonium oxidation) (Jetten *et al.*, 1998). This process was originally hypothesized to be important in the marine environment based on the distributions of inorganic nitrogen in oxygen minimum zones in the water column (see Ward, 2003) but was first discovered in anaerobic water treatment plants. Recently, it has indeed been demonstrated in marine environments, including pelagic systems (Thamdrup and Dalsgaard, 2002; Dalsgaard *et al.*, 2003; Kuypers *et al.*, 2003).

It has been suggested that anammox is quantitatively important in comparison to denitrification. However, the relative importance of denitrification and anammox in oxygen minimum zones and sediments is still being determined (Ward, 2003). Biological N_2 fixation is the only process that ultimately compensates for the losses of nitrogen by denitrification (and anammox). It has been estimated that about 50% of global N_2 fixation occurs in the oceans (Galloway *et al.*, 2004). In addition to N_2 fixation, rivers and land run-off contribute substantially to the nitrogen budget of the oceans, particularly in coastal areas (Galloway *et al.*, 2004; Boyer *et al.*, 2006). Another important source of nitrogen is atmospheric deposition. Atmospheric deposition is in part (5% of global N_2 fixation) attributed to N_2 fixation mediated by lightning. The bulk of the atmospheric nitrogen deposition to the oceans results from anthropogenic activities (utilization of fertilizer, animal industry and internal combustion engines). Nitrogen fertilizer is produced through the Haber-Bosch process and is responsible for ~30% of global N_2 fixation (Galloway *et al.*, 2004). The chemical reduction of N_2 by the industrial Haber-Bosch method is an inefficient process that proceeds only under high pressure and high temperature.

For many years, mass calculations of the oceanic nitrogen cycle indicated that the sinks of biologically available N were greater than the sources. The estimated rates of denitrification exceeded the sum of atmospheric deposition, river discharge, land run-off and N_2 fixation. Of these sources, the estimates of the magnitude of N_2 fixation were the most uncertain. Estimates of N_2 fixation were principally derived from known distributions of *Trichodesmium*, and rate measurements from surveys at sea (Capone *et al.*, 1997). It is well-known that the oceans are seriously under-sampled and extrapolation using remote sensing detection of blooms of the marine cyanobacterium *Trichodesmium* (Subramaniam *et al.*, 2002; Coles *et al.*, 2004) holds promise for providing higher spatial coverage. Measurements are often biased, because research cruises are planned to focus on specific regions for predetermined purposes, while very little effort has been made to randomly survey other regions of the oceans, because of the cost and logistics of oceanic surveys.

Extrapolations using remote sensing to detect the size of blooms of *Trichodesmium* neglect other microorganisms involved in N_2 fixation. Limitations of remote sensing include its insensitivity for detecting non-bloom populations and the fact that it penetrates only the surface ~10 m (Hood *et al.*, 2002) of the water column while diazotrophs are known to occur and to fix N_2 up to a depth of 100 m. While blooms of *Trichodesmium* are formed when it occurs in colonies (tufts and puffs), this organism also occurs as single trichomes that are much less conspicuous but can be a significant fraction of total biomass (Orcutt *et al.*, 2001). Although it has been suggested that the formation of colonies is an adaptation necessary for *Trichodesmium* to fix N_2, experimental evidence to substantiate this is lacking. In fact, it was clear from early work on natural samples (Saino and Hattori, 1982) and cultures (Ohki and Fujita, 1988) that single trichomes of *Trichodesmium* are perfectly capable of fixing N_2, which was again confirmed by Letelier and Karl (1998). Orcutt *et al.* (2001) concluded that N_2 fixation at the Bermuda Atlantic Time-series study was quadrupled by taking into account the contribution of single trichomes. Blooms of diatom-diazotroph symbiotic associations are difficult to identify by remote sensing, although this might be possible when confirmation by groundtruth sampling is done at the same time (Carpenter *et al.*, 1999).

Biogeochemical estimates of the contribution of N_2 fixation have been derived using the N^* approach. This approach initially yielded higher estimates of putative N_2 fixation than could be explained by known rates of N_2 fixation by *Trichodesmium* (Michaels *et al.*, 1996; Gruber and Sarmiento, 1997). N^* is a parameter that gives the concentration of N in deficit or excess relative to the Redfield ratio (N:P = 16), normalized assuming that denitrification equals N_2 fixation at global $N^* = 0$. It relies on extensive databases of measured midwater inorganic nitrate and phosphate concentrations. A positive value of N^* is interpreted as indicating N_2 fixation. Using this approach putative N_2 fixation was detected in areas that were previously not known to contain diazotrophic organisms. However, a positive N^* does not necessarily mean that N_2 fixation occurred in that region, due to advection of water masses. Another problem is that Gruber and Sarmiento (1997) integrated pycnoclines over larger areas than may be reasonable (Hansell *et al.*, 2004) and might have overestimated global N_2 fixation. The N^* approach is based on several assumptions and it therefore has uncertainties. There are many uncertainties in the estimates of N_2 fixation from N^* and there is clearly need for more detailed measurements of N_2 fixation in the oceans surveying larger geographical areas. Such surveys will give us a more complete picture of the areas important for N_2 fixation and possibly identify new diazotrophic organisms.

Nitrogenase

Nitrogenase is a two-component enzyme, consisting of dinitrogenase and dinitrogenase reductase. Dinitrogenase contains the FeMo-cofactor and the P clusters metallocenters (see Chapter 13). Some organisms synthesize alternative dinitrogenases that contain vanadium (V) or Fe instead of molybdenum (Eady, 1996). Among cyanobacteria the V alternative nitrogenases are rare and the iron (Fe) only enzyme has not yet been reported. The V enzyme has only been unambiguously demonstrated in the heterocystous cyanobacterium *Anabaena variabilis* (Thiel, 1993) and recently also in *Anabaena azotica* strain FACHB-118 and *Anabaena* sp. strain CH1 (Boison *et al.*, 2006). The latter were isolated from rice fields. Organisms capable of synthesizing alternative nitrogenases all also possess the conventional Mo-containing nitrogenase. Only under Mo deficiency is the alternative nitrogenase expressed and synthesized. No marine cyanobacterium is yet known to possess vanadium nitrogenase, but γ-proteobacteria possessing V or Fe-only nitrogenases have been isolated from salt marshes (Loveless *et al.*, 1999). The ecological function of the alternative nitrogenases is obscure, particularly since it is much less efficient than the Mo enzyme. V nitrogenase evolves three times more hydrogen and has a higher energy demand (Eady, 1996), although it reduces N_2 more effectively at decreased temperature than Mo-nitrogenase.

Marine cyanobacteria

Since N_2 fixing microorganisms are scattered throughout the *Bacteria* and *Archaea*, it is difficult to detect N_2 fixing microorganisms using ribosomal RNA approaches, unless the organism of interest has been cultivated. Amplification and sequencing of *nifH* genes is a cultivation-independent approach that provides information on the phylogenetic affiliation of diazotrophs in the environment. This "functional gene" approach has been particularly useful for studying marine cyanobacteria and was initially used to amplify nitrogenase from *Trichodesmium* (Zehr and McReynolds, 1989) and to examine diazotroph diversity in marine microbial mats (Zehr *et al.*, 1995). A number of cyanobacterial *nifH* phylotypes have been reported from marine mats (Zehr *et al.*, 1995; Omoregie *et al.*, 2004), Antarctic lakes (Olsen *et al.*, 1986), estuaries (Affourtit *et al.*, 2001; Short *et al.* submitted) and the water column of tropical oceans (Zehr *et al.*, 1998, 2000, 2001; Falcon *et al.*, 2004; Mazard *et al.*, 2004; Langlois *et al.*, 2005).

Marine N_2 fixing cyanobacteria include many genera found in a variety of different habitats. Unicellular, filamentous heterocystous and filamentous non-heterocystous species are found worldwide in mats, sediments, marshes and corals (Paerl and Zehr, 2000). Fewer species are found in the plankton of estuaries and oceans (Howarth *et al.*, 1988). Filamentous heterocystous species sometimes form blooms in estuaries and rivers (Paerl, 1988). Planktonic heterocystous forms (*Nodularia, Aphanizomenon, Anabaena*) (Figure 15.1A, B) are common in only a few areas including the Baltic Sea and the Peel Harvey Estuary in Australia.

In the open ocean, the major N_2 fixing cyanobacterial genus is believed to be *Trichodesmium* (Figure 15.1C) (Capone *et al.*, 1997; LaRoche and Breitbart, 2005), although recently other cyanobacterial diazotrophs have been reported (Zehr *et al.*, 2001). Other filamentous non-heterocystous cyanobacteria are sometimes observed on *Trichodesmium* colonies (Siddiqui *et al.*, 1992). A handful of *Trichodesmium* species or strains have been described on the basis of morphology of filaments or colonies, and with genetic techniques (Janson *et al.*, 1999; Orcutt *et al.*, 2002).

A variety of eukaryotic protists, notably pelagic diatoms, are known that live in association with N_2-fixing cyanobacteria (Villareal, 1992; Foster *et al.*, 2006b). Recently, N_2 fixing symbioses between cyanobacteria and tintinnids, radiolarians and dinoflagellates

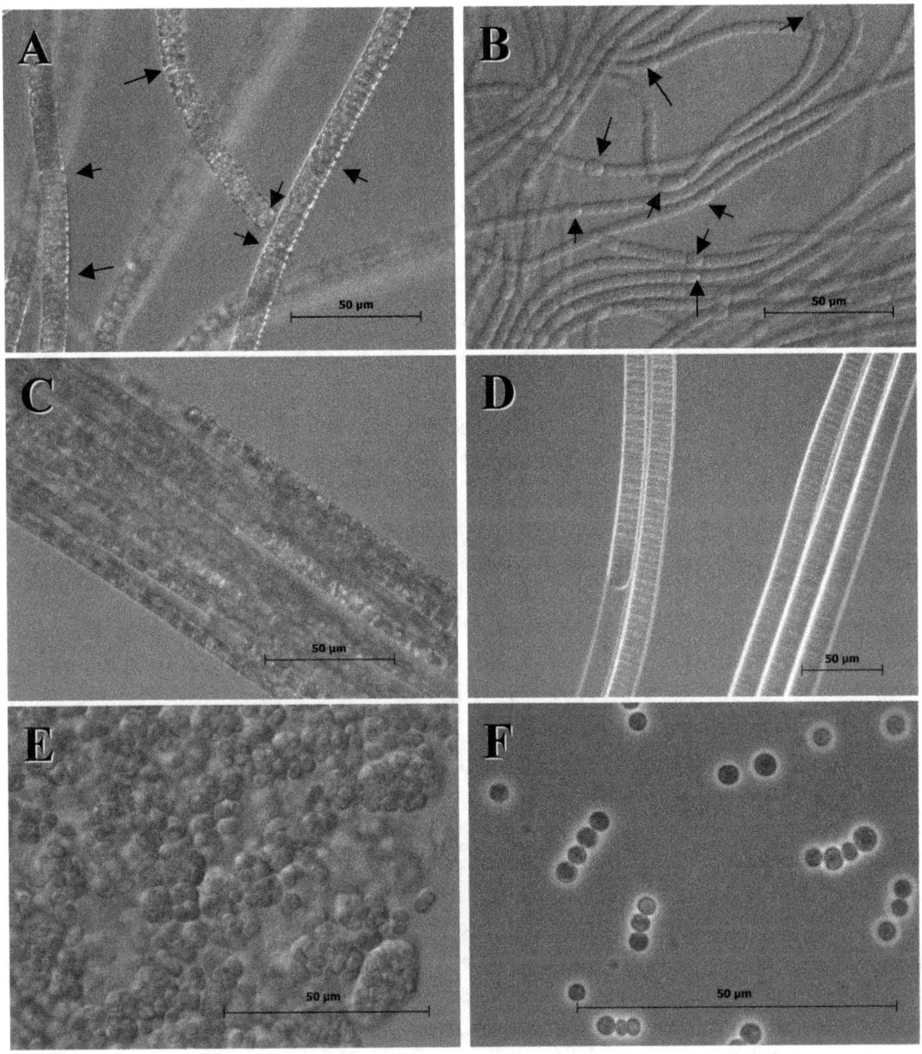

Figure 15.1 (see also Plate 15.1) Marine diazotrophic cyanobacteria. A, *Nodularia spumigena*. B, *Anabaena* sp. C, *Trichodesmium erythreum*. D, *Lyngbya aestuarii*. E, *Cyanothece* sp. F, *Crocosphaera watsonii*. Black arrows: heterocysts. Bar = 50 µm

have been described in detail (Foster *et al.*, 2006b). It appears that unicellular cyanobacteria form associations with many types of organisms. It is not yet known how many of these unicellular forms, which primarily have been observed microscopically, are diazotrophic. However, amplification of 16S rRNA genes showed that some are closely related to unicellular diazotrophic cyanobacteria (Foster *et al.*, 2006a).

Less well known are the N_2 fixing unicellular cyanobacteria that are phylogenetically related to species such as *Gloeothece* and *Cyanothece* (Figure 15.1E). Unicellular cyanobacteria can be visualized in association with a variety of eukaryotes, but have only recently been recognized in the free-living state in the pelagic waters of the open ocean using molecular approaches. Three isolates of marine unicellular cyanobacteria have been in culture for several decades: marine *Synechocystis* (now called *Crocosphaera watsonii*) (Figure 15.1F) isolated from South Atlantic waters, *Cyanothece* ATCC 51142, isolated from the Gulf of Mexico, and a *Synechococcus* (which is likely to be a strain of *Cyanothece* and not closely related to marine *Synechococcus*) of undocumented origin but believed to be isolated from a benthic environment in the Bahamas. All three of these species are in the unicellular (Group I) cyanobacteria, but are several micrometers in diameter. *Crocosphaera* is the only one thus far that has closely related species in natural assemblages in the plankton, and was discovered in the Pacific Ocean on the basis of *nifH* gene sequences. All of these unicellular cyanobacteria normally fix N_2 during the night in a light–dark cycle.

Based on *nifH* sequence analysis of samples from station ALOHA of the Hawaiian Ocean Time-series (HOT) in the Pacific Ocean, it was shown that two groups (Groups A and B) of putative unicellular diazotrophic cyanobacteria were present. *Crocosphaera watsonii* is a cultured species belonging to Group B. This group clusters with other cultivated unicellular diazotrophic cyanobacteria. Group A is known only from sequence information of *nifH* and presumed to be a cyanobacterium by the clustering of the *nifH* sequence with *Cyanothece* and others in this group. Interestingly, expression of Group A *nifH* seemed to occur exactly out of phase with the expression of Group B *nifH*. The latter was expressed mostly during the night, when the number of transcripts was 2–3 orders of magnitude higher (Church *et al.*, 2005). This pattern is in accordance with the fact that organisms belonging to or clustering close to Group B fix N_2 (mainly) during the night or in the dark phase when grown under an alternating light–dark cycle. The fact that Group A *nifH* was transcribed at a higher level during the daytime suggested that these organisms may fix N_2 during the day. Montoya *et al.* (2004) indeed measured N_2 fixation during the day in the fraction smaller than 10 μm, suggesting that this was carried out by unicellular organisms. Subsequent experiments using Pacific Ocean water showed that N_2 fixation rates were detected during the day in incubations where the only light-phase N_2 fixing cyanobacteria present were the Group A unicellular phylotypes (Zehr *et al.*, 2007). Although gene expression is not necessarily linked to enzyme activity, there are no examples yet of transcription that is out of phase with N_2 fixation activity. As yet, the Group A presumed unicellular cyanobacteria have not been cultivated.

Functional diversity of N_2-fixing cyanobacteria

Strategies avoiding O_2 inactivation of nitrogenase

Cyanobacteria are oxygenic phototrophs that usually live in aerobic environments. Roughly half of the cyanobacterial species are in principle capable of fixing N_2. Nitrogenases are extremely sensitive to O_2. In order to be able to fix N_2, cyanobacteria must avoid O_2 inactivation and therefore keep levels very low in the environment of nitrogenase. Many cyanobacteria possess the genes for nitrogenase but are unable to fix N_2 under aerobic and/or photoautotrophic conditions. These species fix N_2 only when oxygenic photosynthesis is inhibited and anaerobic conditions are provided (Rippka and Waterbury, 1977). The strategy for N_2 fixation by these organisms can be described as "avoidance" (of oxygen). For the majority of cyanobacteria that belong to this category it is unclear whether they ever thrive under conditions that allow them to fix N_2. In a few cases, cyanobacteria that occur in "sulfureta" (environments dominated by reduced sulfur; sulfide) or in marine microbial mats, may find the right conditions for diazotrophic growth (Villbrandt and Stal, 1996). Sulfureta are permanently anaerobic but in illuminated environments sulfide is known to be a potent inhibitor of oxygenic photosynthesis and sometimes acts as photosynthetic electron donor to photosystem I (Miller and Bebout, 2004). In the oceanic water column anaerobic microenvironments are sometimes present in aggregates and marine snow, but it is unknown whether these are significant sites of N_2 fixation (Paerl *et al.*, 1989a). For obvious reasons, anaerobic N_2-fixing cyanobacteria are routinely maintained in the laboratory under non-diazotrophic conditions for extended periods of time, yet these organisms do not seem to lose the capacity of N_2 fixation, indicating its remarkable conservation.

Probably the most complex adaptation that facilitates the incompatible processes of N_2 fixation and oxygenic photosynthesis is the differentiation of specialized cells, the heterocysts (sometimes also called heterocytes) (Adams and Duggan, 1999), where nitrogenase is confined in some filamentous cyanobacteria (see Chapter 14). An important property of heterocysts is their special glycolipid cell envelope that serves as a gas diffusion barrier (Walsby, 1985). It limits the diffusion of O_2 into the cell but likewise also inhibits N_2 influx. It appears that this cell envelope can be acclimated to prevailing environmental conditions, becoming more or less of a cell diffusion barrier (Staal *et al.*, 2003b).

Notwithstanding the fact that the differentiation of heterocysts seems to be the perfect adaptation for N_2-fixing cyanobacteria, several non-heterocystous species are known that fix N_2 under aerobic conditions. These include both filamentous and unicellular species (heterocystous cyanobacteria are by definition filamentous) (Bergman *et al.*, 1997).

Trichodesmium, a special case

A special case is the filamentous cyanobacterium *Trichodesmium*, which comprises a number of species, including *Katagnymene*, and often forms surface blooms in the tropical oceans. *Trichodesmium* spp. have been known for several decades as the dominant N_2 fixing species in the tropical oceans (Dugdale *et al.*, 1961; Capone *et al.*, 1997), but only recently it has been cultured in the laboratory (Ohki *et al.*, 1992; Prufert-Bebout *et al.*, 1993), allowing more in depth studies of this enigmatic organism. The genome of *Trichodesmium* sp. IMS101 has been closed and will be released by the time this chapter is published (http://www.jgi.doe.gov). Immunolocalization studies of nitrogenase in natural populations and laboratory grown cultures of *Trichodesmium* have suggested that nitrogenase is not equally distributed in the trichome or in the aggregate, but present only in a subset of cells (Lin *et al.*, 1998; Fredriksson and Bergman, 1997). These cells occur in groups of 10–15 cells and virtually all trichomes possess some cells with nitrogenase (Lin *et al.*, 1998). The cells containing the nitrogenase antigen do not appear morphologically different from the other cells, based on light microscopy or transmission electron microscopy. Similar results were reported for the benthic non-heterocystous cyanobacterium *Symploca* (Fredriksson *et al.*, 1998). It has been proposed that the nitrogenase containing cells represent physiologically differentiated cells, analogous to heterocysts and have been termed "diazocytes" (Fredriksson and Bergman, 1997). In contrast, Paerl *et al.* (1989b) found immunolabeling of *Trichodesmium* cells throughout the trichome showing that different strains, growth conditions or technical procedures may give different results. Similarly, nitrogenase antigen is present in all cells in the related benthic N_2-fixing cyanobacterium *Lyngbya aestuarii* (formerly assigned as *Oscillatoria limosa*) (Stal and Bergman, 1990) (Figure 15.1D).

Trichodesmium fixes N_2 during the daytime and it resembles therefore to some extent heterocystous cyanobacteria (Figure 15.2, compare *Trichodesmium erythreum* to *Nodularia spumigena*). It has been hypothesized that the N_2-fixing cells of *Trichodesmium* have decreased photosynthetic activity, perhaps are incapable of fixing CO_2, and may have an elevated respiratory activity (Berman-Frank *et al.*, 2001). Examination of fluorescence of photosynthetic pigments show that all cells are capable of photosynthesis (Carpenter *et al.*, 1990), although there are strings of "bright cells" that may or may not be the same cells involved in N_2 fixation (Küpper *et al.*, 2004). The evidence that they are N_2-fixing cells is only a weak (non-statistical) correlation between N_2 fixation rates and the numbers of bright cells. The bright cells have interesting fluorescence characteristics, apparently disconnecting light harvesting pigments from PSII or in state transitions over time scales of minutes. It is unclear whether this dynamic photosystem regulation would provide a mechanism for protection of nitrogenase, since nitrogenase synthesis operates on time scales of hours and destruction occurs over time scales of seconds. It should be noted that this hypothesis contradicts the diazocyte hypothesis, since the studies of localization of nitrogenase were interpreted to mean that different cells were differentiated to be N_2-fixing

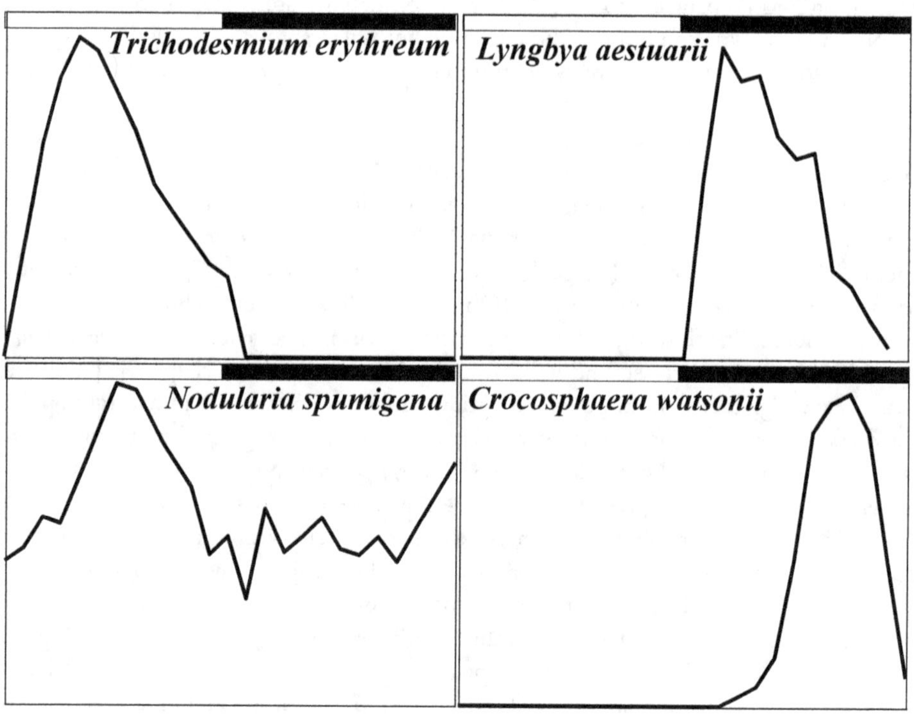

Figure 15.2 Examples of diel cycles of nitrogenase activity in cultures of marine cyanobacteria grown under alternating 12 hour light –12 hour dark regime. *Trichodesmium erythreum* (Rabouille *et al.*, 2006), *Lyngbya aestuarii* (Stal and Krumbein, 1985), *Nodularia spumigena* (Evans *et al.*, 2000) and *Crocosphaera watsonii* (Tuit *et al.*, 2004).

cells, but the traffic light hypothesis indicates that subsets of cells cycle quickly between N_2 fixation and photosynthesis. Solid proof that diazocytes are truly differentiated cells or that photosynthesis is temporally separated from nitrogen fixation is lacking.

There can be no doubt that N_2 fixation is sensitive to O_2 inactivation and it seems therefore plausible that N_2 fixation occurs in non-oxygenic photosynthetic cells or at least in cells that have a considerably lower oxygenic photosynthetic activity. However, this may be a temporary situation. One model to explain the putative spatial organization of N_2 fixation in *Trichodesmium* assumes that nitrogenase is induced when a cell reaches a state of nitrogen deficiency and returns to oxygenic photosynthesis when it becomes nitrogen replete. This strategy could be described as a combination of spatial and temporal separation of photosynthesis and N_2 fixation. It remains to be proven which model is correct.

Diel pattern of N_2 fixation

We emphasize that the resemblance of the diel pattern of N_2 fixation between heterocystous cyanobacteria and *Trichodesmium* is only superficial. The latter organism fixes N_2 exclusively during the day. At sunset nitrogenase rapidly ceases and the enzyme is hydrolyzed by proteases. At sunrise the enzyme is synthesized *de novo*. There are variations in the daily pattern of protein presence that are probably a function of growth rate and physiological status (e.g. Ohki *et al.*, 1992; Capone *et al.*, 1990). The gene expression patterns for *nif* and the photosystem reaction centers are coordinated by a circadian clock (Chen *et al.*, 1998, 1999), in patterns that are reminiscent of those seen in *Cyanothece* (Colón-López and Sherman, 1998) and *Synechococcus* (now called *Cyanothece*) sp. RF-1 (Huang and Grobbelaar, 1995). The significance of the circadian rhythm presumably would be to ensure turnover of protein prior to the N_2 fixation phase, to optimize N_2 fixation activity immediately following the light–dark or dark–light transition.

In heterocystous cyanobacteria the situation is completely different. These organisms fix N_2 during the day and night, although at a lower rate in the dark (Figure 15.2, *Nodularia spumigena*). In the dark, nitrogenase depends on aerobic respiration as a source of energy and this limits its activity (also limited by the decreased entrance of O_2 into the heterocyst). In the light, photosynthetic derived energy will increase nitrogenase activity. Under normal conditions, nitrogenase is well protected against O_2 inactivation in the heterocyst and consequently its turn-over is low. There is no evidence that nitrogenase is synthesized on a day-night cycle or that it is under any control of a circadian rhythm. In the light *de novo* synthesis of nitrogenase in the heterocyst replaces inactivated enzyme provided that O_2 and NH_4^+ are absent (Bone, 1971). It is unknown whether the actual amount of nitrogenase in heterocysts can be regulated or whether it is present at some fixed maximum level.

N_2 fixation in the dark

When grown under a light–dark regime, cultures of *Trichodesmium* exhibit nitrogenase activity exclusively during the light phase. However, cells that were taken from the light phase also exhibited activity when incubated in the dark (Staal *et al.*, 2007). This dark activity was obviously lower than in the light because of the lower ATP generation (Figure 15.3). Also, dark nitrogenase activity decreased with the length of the dark incubation in an O_2 concentration dependent manner. This was attributed to the rapid depletion of

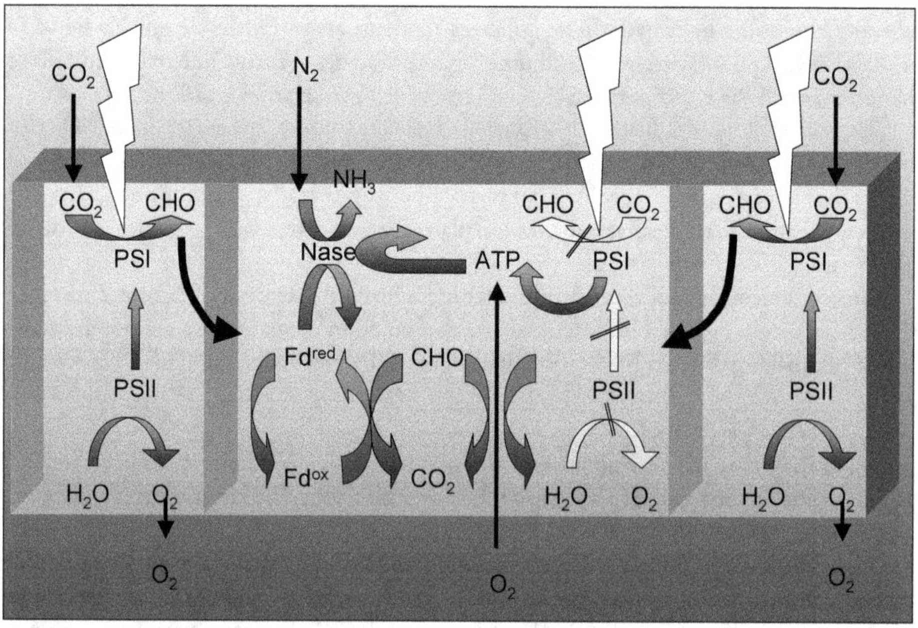

Figure 15.3 Conceptual model of a N_2-fixing cell of *Trichodesmium*. In this drawing the N_2-fixing cell is on both sides bordered by non-N_2-fixing cells. Photosystem II in the N_2-fixing cell is switched off but light is harvested and transferred to biochemical energy (ATP) through Photosystem I activity. This switch turns on respiratory electron transport. O_2 (and N_2) can freely enter the cell but is respired at the expense of stored (or imported) carbohydrate (photosynthate), also contributing to the production of ATP. Respiration competes with nitrogenase (through ferredoxin as electron carrier) for reducing equivalents. Respiration must be sufficiently fast to scavenge virtually all intracellular O_2. This depends on temperature and the flux of O_2. With increasing flux of O_2 less reducing equivalents become available and less N_2 can be fixed. When the flux of O_2 exceeds the respiratory capacity or the availability of reducing equivalents (carbohydrate), the cell cannot be maintained anoxic and nitrogenase is inactivated. This is the case in the dark when carbohydrate is not replenished. Synthesis of nitrogenase is under control of a circadian rhythm and occurs during the first half of the light period. During the second half of the light period nitrogenase activity declines due to inactivation by O_2. The cell can switch back to a normal oxygenic photosynthetic cell, depending on the nitrogen status of the cell. In that respect the N_2-fixing cell of *Trichodesmium* differs from a heterocyst, which is a permanently differentiated cell that cannot revert back to an oxygenic, CO_2-fixing cell. In addition, heterocysts have a thick glycolipid cell wall that limits the gas diffusion. Heterocystous cyanobacteria can therefore thrive in environments with higher dissolved O_2 and at lower temperatures (lower rates of respiration). In the warm tropical oceans the heterocyst cell wall would be a disadvantage: it is not necessary and it limits the flux of N_2.

carbohydrate storage. In the presence of high levels of O_2 a greater portion of carbohydrate is used for respiration. Eventually, nitrogenase activity ceases. Nitrogenase is probably inactivated when the cells are no longer capable of maintaining an anoxic interior due to the lack of carbohydrate to support respiration. Whether this mechanism is also instrumental in the regular light–dark cycle is not precisely known. Usually, graphs of nitrogenase activity during the light phase are hyperbolic, even when the light level is kept constant (Staal *et al.*, 2007). During the first half of the light period the increase of nitrogenase activity is associated with *de novo* synthesis (Zehr *et al.*, 1993; Chen *et al.*, 1998). Once present, the actual activity and inactivation of nitrogenase seems to be controlled by the availability of light and carbohydrate and by O_2 (Rabouille *et al.*, 2006; Staal *et al.*, 2007). When synthesis of nitrogenase ceases halfway through the light period, degradation or turnover of the enzyme causes a gradual loss in activity. In the dark, residual activity disappears quickly as does to some extent the nitrogenase protein, which may be degraded by proteases.

In heterocystous cyanobacteria the situation is different. These cyanobacteria seem to have a certain, fixed amount of active nitrogenase present in the heterocysts. The nitrogenase activity is a function of energy availability. There is not an indication that reducing equivalents are a limiting factor for N_2 fixation in these organisms. Hence, in the dark, lower nitrogenase activity is usually observed, which is the result of respiratory ATP formation and is dependent on the influx of O_2 into the heterocyst. In the light, ATP is generated mostly (but not exclusively) by cyclic photophosphorylation. To our knowledge there is no information to date on whether a circadian clock is involved in any of the processes related to N_2 fixation in heterocystous cyanobacteria, but the *kai* genes (see Chapter 8) are present in the genomes.

Temporal separation of N_2 fixation and photosynthesis

A number of other diazotrophic, non-heterocystous cyanobacteria have solved the problem of the incompatibility of photosynthesis and N_2 fixation by separating these processes temporally (Bergman *et al.*, 1997). While photosynthesis is confined to the light period, these organisms fix N_2 only during the dark (Figure 15.2, *Lyngbya aestuarii*, *Crocosphaera watsonii*). This is usually less efficient, because N_2 fixation completely depends on respiration for energy, using the reserve carbohydrates (glycogen) produced and stored during photosynthesis in the light period. Respiration must be sufficiently fast to ensure near anoxic conditions in the cell, which may require adaptations of the cell envelope under certain conditions, to prevent the entry of excess O_2. Typical examples of cyanobacteria that belong to this group are the filamentous *Lyngbya aestuarii*, *L. majuscula*, *Symploca* sp. and the unicellular *Cyanothece* sp., *Crocosphaera watsonii* and *Gloeothece* sp. All filamentous species listed above are marine benthic organisms. *Gloeothece* (originally assigned as *Gloeocapsa*), the first non-heterocystous diazotrophic cyanobacterium that was discovered, is a terrestrial organism, forming biofilms on carbonate cave walls. *Cyanothece* spp. (previously assigned as *Synechococcus* spp.) were reported from rice fields (Huang and Chow, 1988) and from marine pelagic and benthic environments (Mitsui *et al.*, 1987). *Cyanothece* is known from both marine and freshwater environments and *Crocosphaera watsonii* (previously called marine *Synechocystis*) is a marine pelagic species. However, the assignments in the various reports must be taken with caution. It must be emphasized that so far cyanobacteria that belong to this group are capable of diazotrophic growth in laboratory cultures under continuous light (Mitsui *et al.*, 1987; Schneegurt *et al.*, 1994) and that therefore a light–dark regime is not a prerequisite for these organisms to fix N_2. Little is known about how these organisms protect nitrogenase from O_2 inactivation when they grow in continuous light.

The strategy of temporal separation of N_2 fixation and photosynthesis in the unicellular cyanobacteria is not that simple. As mentioned above, all cyanobacteria belonging to this group are capable of diazotrophic growth under continuous light. The most obvious explanation would be that the temporal separation also functions under continuous light, i.e. when a cell reaches a state of nitrogen depletion, it will cease photosynthesis, synthesize nitrogenase and fix N_2 until replete. In a population, cells may be in different states and such cultures would appear to have simultaneous photosynthesis and N_2 fixation. Why would such cultures confine nitrogenase activity to the dark phase when grown under a light–dark regime, although it seems attractive to take advantage of the light energy? The advantage of fixing during the night of course is to avoid the inactivation by O_2 produced by the non-N_2-fixing cells. The scavenging of this O_2 might draw severely on the available reducing equivalents, decreasing the efficiency of N_2 fixation. However, a culture that has been adapted to a light–dark regime and then transferred to continuous light, may maintain its original rhythm of nitrogenase activity for some time. In this case, all cells of the culture would cease photosynthetic O_2 evolution and all would start to respire. Usually, nitrogenase activity is much higher as the result of photosystem I mediated ATP generation. The increased fixation of N_2 induced by light could lead to a nitrogen-replete situation of the cell and a stop of N_2 fixation. However, the pattern of N_2 fixation retains a daily rhythm in cultures indicating that a circadian rhythm is involved in timing N_2 fixation in the cells which are fixing N_2.

Cyanobacteria that temporally separate N_2 fixation and photosynthesis and are grown under an alternating light–dark regime may induce nitrogenase prior to the onset of the dark period, provided that this dark period is sufficiently short (Stal and Krumbein, 1985). Even if nitrogenase activity appears at the start of the dark period, transcription needs to be initiated 3 hours earlier to allow a functional enzyme to be synthesized. These cyanobacteria do not maintain active nitrogenase throughout the entire light–dark period. For every daily cycle, *de novo* synthesis seems to be necessary and the enzyme is inactivated when the cell becomes photosynthetically active and starts to evolve O_2. Nitrogenase synthesis may cease at some time and existing enzyme is turned-over more or less rapidly due to O_2 inactivation. This may be particularly rapid when the cell runs out of storage carbohydrate resulting in a lower respiratory O_2 scavenging. Some species seem to degrade nitrogenase rapidly, probably by the action of specific proteases (e.g. in *Gloeothece*), whereas in other species (e.g. *Lyngbya*) inactive nitrogenase remains for a prolonged time in the cell as determined from immunolocalization studies and Western blots (Stal and Bergman, 1990). There seems to be a fixed time required from the onset of the light period until nitrogenase activity appears (Stal, unpublished results). This is most likely the time needed to build up a stock of carbohydrate storage, which is essential for N_2 fixation (both to provide electrons for N_2 fixation as well as for respiratory O_2 scavenging and ATP generation).

Simultaneous N_2 fixation and oxygenic photosynthesis

Gloeothece sp. may represent another special case. Batch cultures of this organism exhibited the usual cycle of nitrogenase activity under an alternating light–dark regime, i.e. confined to the dark period, but when this organism was cultivated in a chemostat, nitrogenase occurred predominantly during the light (Ortega-Calvo and Stal, 1991). The calculated N_2 fixation matched the requirements for the enforced growth rate by the applied dilution rate and it was also clear that the required high nitrogenase activity could only be obtained in the light. It is unclear whether only a subset of the population fixed N_2 or that all cells fixed N_2 concomitantly with oxygenic photosynthesis. Stable continuous cultures could only be obtained with light periods of 12 hours or longer and under a 12–12 light–dark regime the maximum growth rate was low relative to cultures with longer light periods. It was anticipated that when the light period is short and/or the enforced growth rate high, the cells need a higher rate of photosynthesis in order to achieve the same 24 hour integrated growth. The resulting higher rate of O_2 evolution would then become incompatible with N_2 fixation. A satisfactory explanation of these observations is lacking but its significance is enhanced by the discovery of Group A type of oceanic putative unicellular cyanobacterial diazotrophs.

Global distribution of N_2-fixing cyanobacteria

N_2-fixing cyanobacteria exhibit a cosmopolitan distribution and can be found in almost any illuminated environment on Earth or even sometimes in dark places. This means that there are in principle few environmental conditions that would not favor the growth and activity of diazotrophic cyanobacteria. Nevertheless, N_2 fixing cyanobacteria have distinct distributions and not all types of cyanobacteria are found in all habitats. As far as we know now, diazotrophic cyanobacteria appear to have restricted distributions in the oceans. Planktonic diazotrophic cyanobacteria are only found in the warmer, (sub)tropical oceans. This is true for free-living *Trichodesmium* spp. and unicellular Groups A and B cyanobacteria but also for symbiotic cyanobacteria. Roughly, a temperature of around 20–25°C seems to be the cut-off for planktonic N_2 fixing cyanobacteria in the oceans (LaRoche and Breitbart, 2005): below this temperature diazotrophic cyanobacteria are either absent, or when present they are most certainly not fixing N_2. This happens for instance when these cyanobacteria have been transported by water currents into areas of lower surface water temperature (LaRoche and Breitbart, 2005).

In the pelagic zone of temperate and cold seas diazotrophic cyanobacteria seem to be completely absent. However, they can be found in benthic marine ecosystems, such as coastal microbial mats, all around the globe, independent of temperature (Figure 15.4B, C). This emphasizes that temperature alone is not excluding these organisms or their N_2 fixing activity. Moreover, in brackish seas such as the Baltic Sea and in estuaries, diazotrophic cyanobacteria may be an important component of the phytoplankton and even form dense blooms (Figure 15.4A). The cyanobacteria in these environments are heterocystous species, which are also common in freshwater lakes (Stal *et al.*, 2003).

Ecosystem level regulation of N_2 fixation

Factors affecting the distribution of N_2-fixing cyanobacteria

Heterocystous cyanobacteria often form blooms in lakes and sometimes in estuaries, but are not as abundant in the open ocean (Howarth *et al.*, 1988), except in symbiosis. There have been several hypotheses to describe why there are not more N_2 fixing cyanobacteria in the open ocean and estuaries and why the filamentous non-heterocystous cyanobacteria, rather than heterocystous species are dominant (Marino *et al.*, 2006; Howarth *et al.*, 1993; Howarth *et al.*, 1988). Factors that can affect the distribution and activity of N_2 fixing cyanobacteria are nutrient availability (particularly phosphorus, or the N:P ratio), trace elements (Fe), turbulence, dissolved organic matter (which can be an N source and binds trace elements), temperature, salinity and light (Paerl and Zehr, 2000).

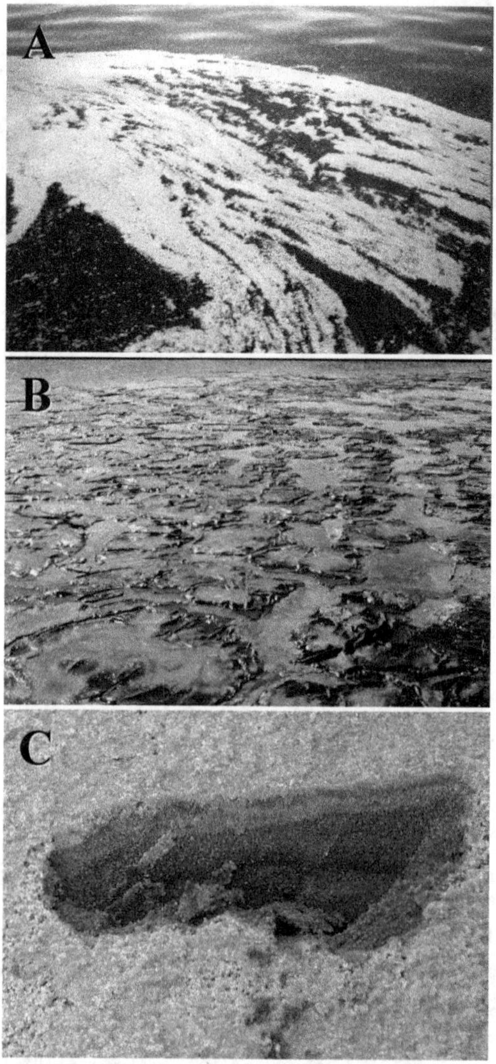

Figure 15.4 (see also Plate 15.4) A, Bloom of *Nodularia spumigena* in the Baltic Sea. B, Coastal microbial mats of *Lyngbya aestuarii* in Guerrero Negro, Baja California, Mexico. C, Cross section of a coastal microbial mat, Netherlands.

It has been shown that in heterocystous cyanobacteria from the Baltic Sea the heterocyst could actively change the relative photopigment composition over a day-night cycle, presumably in order to acclimate to spectral changes in the water column (Staal *et al.*, 2003a). It is also possible that these organisms optimize the pigmentation of the heterocyst and consequently N_2 fixation to the available part of the light spectrum, allowing them to co-exist with types that are using other parts of the spectrum as was shown for the picocyanobacteria (Stomp *et al.*, 2004).

The N_2-fixing cyanobacterial blooms in the Baltic Sea are confined to the summer, when surface water temperature is high. However, rather than temperature it is the available light in the water column that determines whether a bloom can develop or not (Stal and Walsby, 2000). Also, the development of N_2-fixing microbial mats is seasonal in most places and it seems likely that the light availability (e.g. day length and sun angle) is an important factor, although this has not been investigated in detail.

The role of temperature

The distribution of diazotrophic cyanobacteria in the marine environment requires an explanation. It is curious that free-living heterocystous cyanobacteria are absent in the pelagic of the tropical oceans, since this type of organism seems to be so excellently equipped to combine N_2 fixation with photoautotrophic growth. However, every advantage has its drawback. The glycolipid cell envelope that is formed during the differentiation of the heterocyst serves as a gas diffusion barrier. Any O_2 that diffuses into the cell needs to be scavenged by the respiratory system of the heterocyst. The efficiency of this system depends on the limitation of gas diffusion by the heterocyst cell envelope, the speed of respiration and the supply of electrons from the neighboring vegetative cells. The flux of O_2 that diffuses into the cell depends on the concentration of O_2 in the medium and on the efficiency of the heterocyst envelope. It has been long known that the heterocyst envelope adapts, for example, to the O_2 concentration in the medium, by increasing its thickness and presumably thereby its effect as a gas diffusion barrier (Murry *et al.*, 1984; Kangatharalingam *et al.*, 1990). However, another factor has been neglected. Respiration is a metabolic process, whose rate depends on temperature. Roughly, a ten degree increase or decrease in temperature will cause a doubling or halving of the respiratory rate, respectively (known as the $Q_{10} = 2$). This means that at a low temperature O_2 scavenging through respiration is relatively slow and the heterocyst envelope must be more effective as a gas diffusion barrier (i.e. presumably becomes thicker). *Vice versa,* with increasing temperature the heterocyst envelope may become thinner and ultimately no specific gas diffusion barrier would be required. Seawater is saturated at a lower concentration of O_2 than freshwater and also higher temperatures result in lower saturating concentrations of O_2 (25% less compared to freshwater). Although diffusion increases with temperature, the total effect of temperature on the diffusion influx of O_2 in the heterocyst increases hardly with a Q_{10} 1.16. Staal *et al.* (2003b) argued that in seawater in the tropics, a heterocyst envelope would not be necessary and in fact may represent a disadvantage. The differentiation of a heterocyst and the synthesis of its special envelope have cellular costs. It can be anticipated that the cost of synthesis of the heterocyst envelope is proportional to its thickness. Moreover, the heterocyst envelope not only slows down the flux of O_2 but also that of N_2. This may then limit the amount of N_2 that can be fixed. Hence, the heterocyst envelope is a trade-off between the limitation of the influx of O_2 so that the cell interior is kept anoxic and the highest possible influx of N_2 to allow maximum growth. Eventually this means that a heterocyst envelope will not be required, as is the case in *Trichodesmium* spp.

The role of respiration

Respiration has two effects on N_2 fixation. It contributes to the ATP supply and it removes electrons. It has been shown that certain levels of O_2 result in higher nitrogenase activities in heterocystous cyanobacteria, even in saturating light (Staal *et al.*, 2003c). This means that nitrogenase activity in the heterocyst is limited by energy under such conditions. However, increasing O_2 concentrations will use up an increasing portion of the reducing equivalents that are imported from the vegetative cells, increasingly limiting N_2 fixation. At the same time, respiration will provide an increasing portion of the ATP needed for the decreasing amount of N_2 fixation, until no light stimulation is observed. Above that point the heterocyst cannot be kept anoxic and nitrogenase will be quickly and irreversibly inactivated. The concentration of O_2 for optimum N_2 fixation depends on light and on the heterocyst envelope composition. At saturating light, the optimum O_2 concentration may vary from 0% in *Trichodesmium* to 10% in certain heterocystous cyanobacteria, while in the dark this is 5% to 20%, respectively. Many pelagic heterocystous cyanobacteria and *Trichodesmium* occur in colonies and it is known that the O_2 concentrations in these aggregates may vary from virtual anoxia to over-saturation (Paerl and Bebout, 1988).

Summarizing, the low saturating O_2 concentrations in warm seawater makes the possession of a heterocyst unnecessary or even a disadvantage. This explains the presence of non-heterocystous cyanobacteria and the absence of heterocystous types. It also explains why heterocystous species may be more abundant in fresh- and brackish water environments with the higher saturating concentrations of O_2. However, the absence of heterocystous cyanobacteria in the temperate and cold seas and oceans remains unexplained.

Does turbulence affect N_2 fixation?

It has been proposed that heterocystous cyanobacteria cannot cope with the turbulent conditions in estuaries and the oceans and that shearing can cause the fragmentation of the trichomes and impair the organism's metabolism (Moisander *et al.*, 2002). However, many freshwater lakes or for instance the brackish Baltic Sea also have substantial turbulence, but nevertheless extensive blooms of heterocystous cyanobacteria proliferate in these environments. These cyanobacteria often occur in aggregates that may prevent shear on single trichomes.

The effect of sulfate

The high concentration of sulfate in seawater might limit the uptake of molybdate, thereby impairing N_2 fixation in the marine environment (Cole *et al.*, 1993). Sulfate and molybdate are structural analogues, but while sulfate occurs in seawater at 28 mM, molybdate is present in the nM range. Molybdenum is a cofactor of dinitrogenase and this element occurs predominantly as molybdate in aerobic seawater. Although there is some evidence that the uptake of molybdate is impaired to some extent, it is not fully inhibited. It was suggested that, as a result of the impaired uptake of molybdate and the consequent lower growth rate, heterocystous cyanobacteria would not be able to cope with grazing and will be eliminated from the system (Marino *et al.*, 2006). However, cyanobacteria may have specific uptake systems for molybdate and sulfate (Zahalak *et al.*, 2004; Thiel *et al.*, 2002). Moreover, one could envisage that cyanobacteria may be selected that possess alternative nitrogenases that do not contain molybdenum as cofactor. Also, eukaryotic microalgae depend on nitrate and would be equally affected since nitrate reductase is also a molybdenum-containing enzyme, but no such exclusion is documented. *Trichodesmium* spp. and unicellular diazotrophic marine cyanobacteria possess the molybdenum containing nitrogenase and thrive in seawater, apparently not impaired in the uptake of this trace metal. Examination of Mo content of natural populations of *Trichodesmium* showed that they contained more than sufficient Mo (Tuit *et al.*, 2004).

Nevertheless, there is some evidence that the high sulfate concentrations in seawater may affect N_2 fixation. Experiments have shown that salinity (up to full strength seawater) alone does not impair N_2 fixation, but sulfate does, also in low-salinity medium (Stal *et al.*, 1999). In the Baltic Sea, blooms of heterocystous cyanobacteria occur in areas where the salinity is < 9. Experiments have shown that nitrogenase activity in cyanobacteria from the Baltic Sea was seriously affected by sulfate concentrations > 9 mM (Stal, unpublished results). The mechanism by which sulfate acts on N_2 fixation in heterocystous cyanobacteria is presently unresolved as is the question why the non-heterocystous cyanobacteria are unaffected. The secret may be found in the fact that heterocysts are terminally differentiated, non-growing cells, while this is not the case in non-heterocystous cyanobacteria, although diazocytes may represent a special case.

Symbiotic N_2-fixing cyanobacteria

Symbiotic heterocystous cyanobacteria living inside pelagic diatoms such as *Richelia* sp. will experience a completely different environment inside the alga. For instance, the salinity and sulfate concentrations inside the diatom may be different from the surroundings and the O_2 concentration may be higher. Moreover, the function of the diazotroph is primarily to provide nitrogen to the host and this is most conveniently done through the heterocyst, which as a non-growing cell is hardly in need of fixed nitrogen for its own metabolism. These N_2-fixing factories were designed to fix and export nitrogen and are therefore probably easy to control by the host. This might be the main reason why the majority of symbioses involving diazotrophic cyanobacteria are with heterocystous species. Whether epibiotic cyanobacteria such as the heterocystous *Calothrix* experience similar conditions as the endobiotic types is not known. Moreover, occasionally free-living heterocystous cyanobacteria have been observed (Carpenter and Janson, 2001). Whether the open ocean is their actual habitat or whether these organisms are washed into the sea from coastal environments or represent the free-living stage of otherwise symbiotic forms is not yet known.

Unicellular diazotrophic cyanobacteria

Not much is known about the ecology of the unicellular diazotrophic cyanobacteria in the tropical oceans. Their presence in the tropical oceans and their presumed absence in colder marine or freshwater systems are best explained with the same arguments as for *Trichodesium*. Putative diazotrophic unicellular cyanobacteria were found in the Arabian Sea only at temperatures above 29°C (Mazard *et al.*, 2004) but it appears that these organisms may have a greater geographic range and thrive in lower temperature higher latitude waters (Langlois *et al.*, 2005; Church *et al.* in preparation). The small size of the unicellular diazotrophs relative to *Trichodesmium* may have an impact on the affinity for nutrients, notably phosphate. A problem of the small size is that these cells are more difficult to maintain anoxic, but this may be achieved by sufficiently high rates of respiration. Experiments with laboratory cultures are needed to address these questions.

The effect of iron

Another element that may possibly be involved in ecosystem control of N_2 fixation is iron. Diazotrophic cyanobacteria have an elevated iron requirement (Kustka *et al.*, 2003). Both components of nitrogenase contain iron in metallocenters and ferredoxin is the natural electron donor, although flavodoxin may replace it when iron is short in supply. Although iron is one of the most abundant elements on Earth, in the O_2 saturated ocean water iron is quickly oxidized to ferric iron which is virtually insoluble and therefore largely unavailable. However, microorganisms have evolved various mechanisms in order to satisfy their demand for this micronutrient. It is known that iron may be bound to the extracellular sheath. Under the influence of sunlight this iron may become reduced and subsequently be taken up (Miller *et al.*, 1995). Also, a variety of chelating compounds and siderophores are produced by microorganisms, including cyanobacteria, binding iron and facilitating its uptake. The greater abundance of *Trichodesmium* in the north-western tropical Atlantic has been explained by the deposition of iron originating from dust storms of the Sahara but awaits conclusive evidence. Iron fertilization experiments have indicated that in parts of the ocean where the nutrient level is high but chlorophyll *a* content is low (HNLC areas), iron seems indeed the limiting nutrient (Boyd *et al.*, 2000).

Marine microbial mats

Contrary to the enigmatic distribution of N_2-fixing cyanobacteria in the pelagic ocean is the global distribution of benthic diazotrophic cyanobacteria. Coastal N_2-fixing microbial mats are ubiquitously found. Although the majority of such mats seem to be composed of non-heterocystous cyanobacteria, there are also ample examples of heterocystous species (Stal, 1995). Why is the global distribution of benthic species so different from that of pelagic species and what are the factors that select for heterocystous or non-heterocystous species?

Marine microbial mats are characterized by steep and fluctuating environmental conditions. For instance, in the light the microbial mat may become supersaturated with O_2 as the result of high rates of photosynthesis and a limitation of diffusion. At night, the mat turns anoxic quickly due to the high rates of respiration. Anaerobic processes then take over. Sulfate reduction is particularly important, and this activity results in enrichment of the system with sulfide. Dark anoxic conditions and/or high concentrations of the toxic sulfide may select against heterocystous cyanobacteria, although this has not been investigated extensively (Villbrandt and Stal, 1996). Alternatively, as proposed for the pelagic environment, the concentration of sulfate may pose a problem. Microbial mats are exposed to other extremes such as strong fluctuations of temperature, salinity and light but it is unknown how this affects heterocystous cyanobacteria. In some cases occasional sedimentation may cover the mat and only motile cyanobacteria may be able to reach the surface. O_2 supersaturation certainly excludes non-heterocystous cyanobacteria but may even pose a serious problem for heterocystous species. Anoxic dark conditions are likewise a problem. The only way to generate energy under such conditions is by fermentation of storage carbohydrate (Stal and Moezelaar, 1997). Cyanobacteria capable of fermentation may fix N_2 under dark anoxic conditions, notwithstanding the limited amount of energy that can be generated. Depending on the actual conditions a variety of typical daily variations of N_2 fixation may occur (Bebout *et al.*, 1987; Villbrandt *et al.*, 1990). These include a confinement of N_2 fixation to the dark period and/or with peaks at sunrise and sunset when light is available but with low O_2 saturation.

Mats containing heterocystous cyanobacteria may be found in a variety of places. It is in most cases not clear which factors allow their proliferation. In some cases freshwater input may trigger the growth of heterocystous species (Stal, unpublished). Another factor might be the lower sulfate concentration but this has not been investigated. Also mats that are less dense may allow faster diffusion of gases resulting in a lesser O_2 saturation during the light period, while such mats do not turn anoxic at night. Consequently, such mats also do not accumulate high concentrations of sulfide. Heterocystous cyanobacteria are probably more efficient diazotrophs than non-heterocystous species and when conditions allow they could out-compete the latter. However, the conditions that promote growth of heterocystous cyanobacteria in marine microbial mats are not precisely known.

Conclusions

N_2 fixation evolved early in evolution of Earth and cyanobacteria may have been diazotrophic before the diversification of heterocystous species occurred. Today, half of the global fixation of N_2 takes place in the oceans and, if not exclusively, cyanobacteria are responsible for the bulk of it. Estimations of global oceanic cyanobacterial N_2 fixation are prone with error. The oceans are seriously undersampled and remote sensing techniques detect mostly surface blooms of *Trichodesmium*. Moreover, it has also become clear that cyanobacteria other than *Trichodesmium* may contribute importantly to oceanic N_2 fixation. Higher N_2 fixation and larger geographical distributions of diazotrophs would be required to balance the anticipated larger losses of nitrogen by anaerobic ammonium oxidation in addition to denitrification but await discovery. Pelagic diazotrophic cyanobacteria are confined to the tropical oceans and this distribution seems to be dictated by temperature. However, benthic marine diazotrophic cyanobacteria can be found worldwide. This shows that temperature itself does not exclude diazotrophic cyanobacteria from the marine environment. Nitrogenase is extremely sensitive to O_2 and the cyanobacteria have diversified and evolved a variety of strategies to combine the two incompatible processes of N_2 fixation and oxygenic photosynthesis. All of this diversity is present among marine diazotrophic cyanobacteria. Although the heterocystous cyanobacteria seem to be best equipped to combine oxygenic phototrophic growth with N_2 fixation, free-living specimens are rarely found in the pelagic zones of the ocean. The possession of a heterocyst does not represent an ecological advantage in that environment. The heterocyst not only limits the influx of O_2 but also of N_2. Warm seawater contains about 25% less O_2 compared to freshwater and at the higher temperature respiration will be capable to scavenge any O_2 entering the N_2-fixing cell, making the heterocyst cell wall superfluous. The same may apply for unicellular diazotrophic cyanobacteria that have been discovered recently. These organisms may have a wider distribution but yet seem also to be absent from temperate and cold seas. Two groups of unicellular diazotrophic cyanobacteria have been identified with marked different diel patterns of *nif* gene expression. While one of these groups has cultivated representatives, the other is only known by its *nifH* sequence. Their contribution to global N_2 fixation is still uncertain but may be considerable. The specific niche of these unicellular diazotrophic cyanobacteria needs yet to be identified. Possibly, due to their small size, unicellular cyanobacteria may be better competitors for nutrients, notably phosphate, compared to *Trichodesmium*. The apparent absence of diazotrophic cyanobacteria in the plankton of temperate and cold seas is not understood. An intriguing hypothesis is that the high level of sulfate in seawater does not permit the diazotrophic growth of heterocystous cyanobacteria, which would explain their dominance in lower salinity environments such as brackish basins and estuaries. Non-heterocystous cyanobacteria are excluded from the temperate and cold marine pelagic areas due to the relatively high level of dissolved O_2 and the low respiration rates. However, they dominate benthic microbial mats. Diazotrophic cyanobacteria have evolved a range of different strategies by which they control N_2 fixation and prevent it from inactivation by O_2. These include control through circadian rhythms and delicate balances between the different metabolic processes in the cell. However, many questions about the regulation of cyanobacterial N_2 fixation and the diversity of diazotrophic organisms in the marine environment still remain open and much more to discover is ahead of us.

Acknowledgments

We thank Mrs. Ute Wollenzien for providing the micrographs in Figure 15.1. The photograph in Figure 15.4A was made by Dr. Marc Staal. This is publication nr. 3924 of NIOO-KNAW.

References

Adams, D.G., and Duggan, P.S. (1999). Heterocyst and akinete differentiation in cyanobacteria. New Phytol. *144*, 3–33.

Affourtit, J., Zehr, J.P., and Paerl, H.W. (2001). Distribution of nitrogen-fixing microorganisms along the Neuse River estuary, North Carolina. Microb. Ecol. *41*, 114–123.

Bebout, B.M., Paerl, H.W., Crocker, K.M., and Prufert, L.E. (1987). Diel interactions of oxygenic photosynthesis and N_2 fixation (acetylene reduction) in a marine microbial mat community. Appl. Environ. Microbiol. *53*, 2353–2362.

Bergman, B., Gallon, J.R., Rai, A.N., and Stal, L.J. (1997). N_2 fixation by non-heterocystous cyanobacteria. FEMS Microbiol. Rev. *19*, 139–185.

Berman-Frank, I., Lundgren, P., Chen, Y.-B., Küpper, H., Kolber, Z., Bergman, B., and Falkowski, P. (2001). Segregation of nitrogen fixation and oxygenic photosynthesis in the marine cyanobacterium *Trichodesmium*. Science *294*, 1534–1537.

Boison, G., Steingen, C., Stal, L.J., and Bothe, H. (2006). The rice field cyanobacteria *Anabaena azotica* and *Anabaena* sp. CH1 express vanadium-dependent nitrogenase. Arch. Microbiol. *186*, 367–376.

Bone, D.H. (1971). Kinetics of synthesis of nitrogenase in batch and continuous culture of *Anabaena flos-aquae*. Arch. Microbiol. *80*, 242–251.

Boyer, E.W., Howarth, R.W., Galloway, J.N., Dentener, F.J., Green, P.A., Vörösmarty, C.J. (2006). Riverine nitrogen export from the continents to the coast. Glob. Biogeochem. Cycl. *20*, GB1S91

Boyd, P.W., Watson, A.J., Law, C.S., Abraham, E.R., Trull, T., Murdoch, R., Bakker, D.C.E., Bowie, A.R., Buesseler, K.O., Chang, H., Charette, M., Croot, P., Downing, K., Frew, R., Gall, M., Hadfield, M., Hall, J., Harvey, M., Jameson, G., LaRoche, J., Liddicoat, M., Ling, R., Maldonado, M.T., McKay, R.M., Nodder, S., Pickmere, R., Rintoul, S., Safi, K., Sutton, P., Strzepek, R., Tanneberger, K., Turner, S., Waite, A., and Zeldis, J. (2000). A mesoscale phytoplankton bloom in the polar Southern Ocean stimulated by iron fertilization. Nature *407*, 695–702.

Capone, D.G., Zehr, J.P., Paerl, H.W., Bergman, B., and Carpenter, E.J. (1997). *Trichodesmium*, a globally significant marine cyanobacterium. Science *276*, 1221–1229.

Capone, D.G., O'Neil, J.M., Zehr, J., and Carpenter, E.J. (1990). Basis for diel variation in nitrogenase activity in the marine planktonic cyanobacterium *Trichodesmium thiebautii*. Appl. Environ. Microbiol. *56*, 3532–3536.

Carpenter, E.J., Chang, J., Cottrell, M., Schubauer, J., Paerl, H.W., Bebout, B.M., and Capone, D.G. (1990). Re-evaluation of nitrogenase oxygen-protective mechanisms in the planktonic marine cyanobacterium *Trichodesmium*. Mar. Ecol. Prog. Ser. *65*, 151–158.

Carpenter, E.J., and Janson, S. (2001). *Anabaena gerdii* sp. nov., a new planktonic filamentous cyanobacterium from the South Pacific Ocean and Arabian Sea. Phycologia *40*, 105–110.

Carpenter, E.J., Montoya, J.P., Burns, J., Mulholland, M.R., Subramaniam, A., and Capone, D.G. (1999). Extensive bloom of a N_2-fixing diatom/cyanobacterial association in the tropical Atlantic Ocean. Mar. Ecol. Prog. Ser. *185*, 273–283.

Chen, Y.B., Dominic, B., Mellon, M.T., and Zehr, J.P. (1998). Circadian rhythm of nitrogenase gene expression in the diazotrophic filamentous nonheterocystous cyanobacterium *Trichodesmium* sp. strain IMS 101. J. Bacteriol. *180*, 3598–3605.

Chen, Y.-B., Dominic, B., Zani, S., Mellon, M.T., and Zehr, J.P. (1999). Expression of photosynthesis genes in relation to nitrogen fixation in the diazotrophic filamentous nonheterocystous cyanobacterium *Trichodesmium* sp. IMS 101. Plant Mol. Biol. *41*, 89–104.

Church, M.J., Short, C.M., Jenkins, B.D., Karl, D.M., and Zehr, J.P. (2005). Temporal patterns of nitrogenase (*nifH*) gene expression in the oligotrophic North Pacific Ocean. Appl. Environ. Microbiol. *71*, 5362–5370.

Cole, J.J., Lane, J.M., Marino, R., and Howarth, R.W. (1993). Molybdenum assimilation by cyanobacteria and phytoplankton in freshwater and salt water. Limnol. Oceanogr. *38*, 25–35.

Coles, V.J., Wilson, C., and Hood, R.R. (2004). Remote sensing of new production fuelled by nitrogen fixation. Geophys. Res. Lett. *31*, L06301.

Colón-López, M.S., and Sherman, L.A. (1998). Transcriptional and translational regulation of photosystem I and II genes in light–dark- and continuous-light-grown cultures of the unicellular cyanobacterium *Cyanothece* sp. strain ATCC 51142. J. Bacteriol. *180*, 519–526.

Dalsgaard, T., Canfield, D.E., Petersen, J., Thamdrup, B., and Acuña-González, J. (2003). N_2 production by the anammox reaction in the anoxic water column of Golfo Dulce, Costa Rica. Nature *422*, 606–608.

Dugdale, R.C., Menzel, D.W., and Ryther, J.H. (1961). Nitrogen fixation in the Sargasso Sea. Deep Sea Res. *7*, 297–300.

Eady, R.R. (1996). Structure-function relationships of alternative nitrogenases. Chem. Rev. *96*, 3013–3030.

Foster, R.A., Collier, J.L., and Carpenter, E.J. (2006a). Reverse transcription PCR amplification of cyanobacterial symbiont 16S rRNA sequences from single non-photosynthetic eukaryotic marine planktonic host cells. J. Phycol. *42*, 243–250.

Foster, R.A., Carpenter, E.J., and Bergman, B. (2006b). Unicellular cyanobionts in open ocean dinoflagellates, radiolarians, and tintinnids: ultrastructural characterization and immuno-localization of phycoerythrin and nitrogenase. J. Phycol. *42*, 453–463.

Fredriksson, C., and Bergman, B. (1997). Ultrastructural characterization of cells specialized for nitrogen fixation in a non-heterocystous cyanobacterium, *Trichodesmium* spp. Protoplasma *197*, 76–85.

Fredriksson, C., Malin, G., Siddiqui, P.J.A., and Bergman, B. (1998). Aerobic nitrogen fixation is confined to a subset of cells in the non-heterocystous cyanobacterium *Symploca* PCC 8002. New Phytol. *140*, 531–538.

Galloway, J.N., Dentener, F.J., Capone, D.G., Boyer, E.W., Howarth, R.W., Seitzinger, S.P., Asner, G.P., Cleveland, C.C., Green, P.A., Holland, E.A., Karl, D.M., Michaels, A.F., Porter, J.H., Townsend, A.R., and Vörösmarty, C.J. (2004). Nitrogen cycles: past, present, and future. Biogeochem. *70*, 153–226

Gruber, N., and Sarmiento, J.L. (1997). Global patterns of marine nitrogen fixation and denitrification. *Glob.* Biogeochem. Cycl. *11*, 235–266.

Hansell, D.A., Bates, N.R., and Olson, D.B. (2004). Excess nitrate and nitrogen fixation in the North Atlantic Ocean. Mar. Chem. *84*, 243–265.

Hood, R.R., Subramaniam, A., May, L.R., Carpenter, E.J., and Capone, D.G. (2002). Remote estimation of nitrogen fixation by *Trichodesmium*. Deep Sea Res. *49*, 123–147.

Howarth, R.W., Marino, R., Lane, J., and Cole, J.J. (1988). Nitrogen fixation in freshwater, estuarine, and marine ecosystems. 1. Rates and importance. Limnol. Oceanogr. *33*, 669–687.

Howarth, R.W., Butler, T., Lunde, K., Swaney, D., and Chu, C.R. (1993). Turbulence and planktonic nitrogen fixation—a mesocosm experiment. Limnol. Oceanogr. *38*, 1696–1711.

Huang, T.C., and Chow, T.J. (1988). Comparative studies of some nitrogen-fixing unicellular cyanobacteria isolated from rice fields. J. Gen. Microbiol. *134*, 3089–3097.

Huang, T.C., and Grobbelaar, N. (1995). The circadian clock in the prokaryote *Synechococcus* RF-1. Microbiol. *141*, 535–540.

Janson, S., Bergman, B., Carpenter, E.J., Giovannoni, S.J., and Vergin, K. (1999). Genetic analysis of natural populations of the marine diazotrophic cyanobacterium *Trichodesmium*. FEMS Microbiol. Ecol. *30*, 57–65.

Jetten, M.S.M., Strous, M., van de Pas-Schoonen, K.T., Schalk, J., van Dongen, U.G.J.M., Van de Graaf, A.A., Logemann, S., Muyzer, G., Van Loosdrecht, M.C.M., and Kuenen, J.G. (1998). The anaerobic oxidation of ammonium. FEMS Microbiol. Rev. *22*, 421–437.

Kangatharalingam, N., Priscu, J.C., and Paerl, H.W. (1990). Heterocyst envelope thickness, heterocyst frequency and nitrogenase activity in *Anabaena flos-aquae*: influence of exogenous oxygen tension. J. Gen. Microbiol. *138*, 2673–2678.

Küpper, H., Ferimazova, N., Šetlík, I., and Berman-Frank, I. (2004). Traffic lights in *Trichodesmium*. Regulation of photosynthesis for nitrogen fixation studied by chlorophyll fluorescence kinetic microscopy. Plant Physiol. *135*, 2120–2133.

Kustka, A., Sañudo-Wilhelmy, S., Carpenter, E.J., Capone, D.G., and Raven, J.A. (2003). A revised estimate of the iron use efficiency of nitrogen fixation, with special reference to the marine cyanobacterium *Trichodesmium* spp. (Cyanophyta). J. Phycol. *39*, 12–25.

Kuypers, M.M.M., Sliekers, A.O., Lavik, G., Schmid, M., Jørgensen, B.B., Kuenen, J.G., Sinninghe Damste, J.S., Strous, M., and Jetten, M.S.M. (2003). Anaerobic ammonium oxidation by anammox bacteria in the Black Sea. Nature *422*, 608–611.

Langlois, R.J., LaRoche, J., and Raab, P.A. (2005). Diazotrophic diversity and distribution in the tropical and sub-tropical Atlantic Ocean. Appl. Environ. Microbiol. *71*, 7910–7919.

LaRoche, J., and Breitbarth, E. (2005). The importance of the diazotrophs as a source of new nitrogen in the ocean. J. Sea Res. *53*, 67–91.

Letelier R.M., and Karl, D.M. (1998). *Trichodesmium* spp. physiology and nutrient fluxes in the North Pacific subtropical gyre. Aq. Microb. Ecol. *15*, 265–276.

Lin, S., Henze, S., Lundgren, P., Bergman, B., and Carpenter, E.J. (1998). Whole-cell immunolocalization of nitrogenase in marine diazotrophic cyanobacteria, *Trichodesmium* spp. Appl. Environ. Microbiol. *64*, 3052–3058.

Loveless, T.M., Saah, J.R., and Bishop, P.E. (1999). Isolation of nitrogen-fixing bacteria containing molybdenum-independent nitrogenases from natural environments. Appl. Environ. Microbiol. *65*, 4223–4226.

Marino, R., Chan, F., Howarth, R.W., Pace, M.L., and Likens, G.E. (2006). Ecological constraints on planktonic nitrogen fixation in saline estuaries. I. Nutrient and trophic controls. Mar. Ecol. Prog. Ser. *309*, 25–39.

Mazard, S.L., Fuller, N.J., Orcutt, K.M., Bridle, O., and Scanlan,D.J. (2004). PCR analysis of the distribution of unicellular cyanobacterial diazotrophs in the Arabian Sea. Appl. Environ. Microbiol. *70*, 7355–7364.

Michaels, A.F., Olson, D., Sarmiento, J.L., Ammerman, J.W., Fanning, K., Jahnke, R., Knap, A.H., Lipschultz, F., and Prospero, J.M. (1996). Inputs, losses and transformations of nitrogen and phosphorus in the pelagic North Atlantic Ocean. Biogeochem. *35*, 181–226.

Miller, S.R., and Bebout, B.M. (2004). Variation in sulfide tolerance of photosystem II in phylogenetically diverse cyanobacteria from sulfidic habitats. Appl. Environ. Microbiol. *70*, 736–744.

Miller, W.L., King, D.W., Lin, J., and Kester, D.R. (1995). Photochemical redox cycling of iron in coastal seawater. Mar. Chem. *50*, 63–77.

Mitsui, A., Cao, S., Takahashi, A., and Arai,T. (1987). Growth synchrony and cellular parameters of the unicellular nitrogen-fixing marine cyanobacterium, *Synechococcus* sp. strain Miami B6 043511 under continuous illumination. Physiol. Plant. *69*, 1–8.

Moisander, P.H., Hench, J.L., Kononen, K., and Paerl, H.W. (2002). Small-scale shear effects on heterocystous cyanobacteria. Limnol. Oceanogr. *47*, 108–119.

Montoya, J.P., Holl, C.M., Zehr, J.P., Hansen, A., Villareal, T.A., and Capone, D.G. (2004). High rates of N_2 fixation by unicellular diazotrophs in the oligotrophic Pacific Ocean. Nature *430*, 1027–1031.

Murry, M.A., Horne, A.J., and Benemann, J.R. (1984). Physiological studies of oxygen protection mechanisms in the heterocysts of *Anabaena cylindrical*. Appl. Environ. Microbiol. *47*, 449–454.

Ohki, K., and Fujita, Y. (1988). Aerobic nitrogenase activity measured as acetylene reduction in the marine non-heterocystous cyanobacterium *Trichodesmium* spp. grown under artificial conditions. Mar. Biol. *98*, 111–114.

Ohki, K., Zehr, J.P., and Fujita, Y. (1992). Regulation of nitrogenase activity in relation to the light–dark regime in the filamentous non-heterocystous cyanobacterium *Trichodesmium* sp. NIBB 1067. J. Gen. Microbiol. *138*, 2679–2685.

Olsen, G.J., Lane, D.J. Giovannoni, S.J., and Pace, N.R. (1986). Microbial ecology and evolution: a ribosomal RNA approach. Ann. Rev. Microbiol. *40*, 337–365.

Omoregie, E.O., Crumbliss, L.L., Bebout, B.M., and Zehr, J.P. (2004). Determination of nitrogen-fixing phylotypes in *Lyngbya* sp., and *Microcoleus chthonoplastes* cyanobacterial mats from Guerrero Negro, Baja California, Mexico. Appl. Environ. Microbiol. *70*, 2119–2128.

Orcutt, K.M., Lipschultz, F., Gundersen, K., Arimoto, R., Michaels, A.F., Knap, A.H., and Gallon, J.R. (2001). A seasonal study of the significance of N_2 fixation by *Trichodesmium* spp. at the Bermuda Atlantic Time-series Study (BATS) site. Deep Sea Res. *48*, 1583–1608.

Orcutt, K.M., Rasmussen, U., Webb, E.A., Waterbury, J.B., Gundersen, K., and Bergman, B. (2002). Characterization of *Trichodesmium* spp. by genetic techniques. Appl. Environ. Microbiol. *68*, 2236–2245.

Ortega-Calvo, J.J., and Stal, L.J. (1991). Diazotrophic growth of the unicellular cyanobacterium *Gloeothece* sp. PCC 6909 in continuous culture. J. Gen. Microbiol. *137*, 1789–1797.

Paerl, H.W. (1988). Nuisance phytoplankton blooms in coastal, estuarine, and inland waters. Limnol. Oceanogr. *33*, 823–847.

Paerl, H.W., and Bebout, B.M. (1988). Direct measurements of O_2-depleted microzones in marine *Oscillatoria (Trichodesmium)*: Relation to N_2-fixing capabilities. Science *241*, 441–445.

Paerl, H.W., Bebout, B.M., and Prufert, L.E. (1989a). Bacterial associations with marine *Oscillatoria* sp. (*Trichodesmium* sp.) populations: ecophysiological implications. J. Phycol. *25*, 773–784.

Paerl, H.W., Priscu, J.C., and Brawner, D.L. (1989b). Immunochemical localization of nitrogenase in marine *Trichodesmium* aggregates: relationship to N_2 fixation potential. Appl. Environ. Microbiol. *55*, 2965–2975.

Paerl, H.W., and Zehr, J.P. (2000). Marine nitrogen fixation. In Microbial Ecology of the Oceans, D.L. Kirchman, ed. (Wiley-Liss, New York), pp. 387–426.

Prufert-Bebout, L., Paerl, H.W., and Lassen, C. (1993). Growth, nitrogen fixation, and spectral attenuation in cultivated *Trichodesmium* species. Appl. Environ. Microbiol. *59*, 1367–1375.

Rabouille, S., Staal, M., Stal, L.J., and Soetaert, K. (2006). Modeling the dynamic regulation of nitrogen fixation in the cyanobacterium *Trichodesmium* sp. Appl. Environ. Microbiol. *72*, 3217–3227.

Rippka, R., and Waterbury, J.B. (1977). The synthesis of nitrogenase by non-heterocystous cyanobacteria. FEMS Microbiol. Lett. *2*, 83–86.

Saino, T., and Hattori, A. (1982). Aerobic nitrogen fixation by the marine, non-heterocystous cyanobacterium *Trichodesmium* (*Oscillatoria*) spp.: Its protective mechanisms against oxygen. Mar. Biol. *70*, 251–254.

Schneegurt, M.A., Sherman, D.M., Nayar, S., and Sherman, L.A. (1994). Oscillating behavior of carbohydrate granule formation and dinitrogen fixation in the cyanobacterium *Cyanothece* sp. strain ATCC 51142. J. Bacteriol. *176*, 1586–1597.

Siddiqui, P.J.A., Bergman, B., and Carpenter, E.J. (1992). Filamentous cyanobacterial associates of the marine planktonic cyanobacterium *Trichodesmium*. Phycologia *31*, 326–337.

Staal, M., Stal, L.J., te Lintel Hekkert, S., and Harren, F.J.M. (2003a). Light action spectra of N_2 fixation by heterocystous cyanobacteria from the Baltic Sea. J. Phycol. 39, 668–677.

Staal, M., Meysman, F.J.R., and Stal, L.J. (2003b). Temperature excludes N_2-fixing heterocystous cyanobacteria in the tropical oceans. Nature *425*, 504–507.

Staal, M., Lintel-Hekkert, S.T., Harren, F.J.M., and Stal, L.J. (2003c). Effects of O_2 on N_2 fixation in heterocystous cyanobacteria from the Baltic Sea. Aq. Microb. Ecol. *33*, 261–270.

Staal, M., Rabouille, S., and Stal, L.J. (2007). On the role of oxygen for nitrogen fixation in the marine cyanobacterium *Trichodesmium* sp. Environmental Microbiology *9*, 727–736.

Stal, L.J. (1995). Physiological ecology of cyanobacteria in microbial mats and other communities. New Phytol. *131*, 1–32.

Stal, L.J., and Bergman, B. (1990). Immunological characterization of nitrogenase in the filamentous non-heterocystous cyanobacterium *Oscillatoria limosa*. Planta *182*, 287–291.

Stal, L.J., and Krumbein, W.E. (1985). Nitrogenase activity in the non-heterocystous cyanobacterium *Oscillatoria* sp. grown under alternating light–dark cycles. Arch. Microbiol. *143*, 67–71.

Stal, L.J., and Walsby, A.E. (2000). Photosynthesis and nitrogen fixation in a cyanobacterial bloom in the Baltic Sea. Eur. J. Phycol. *35*, 97–108.

Stal, L.J., Albertano, P., Bergman, B., von Bröckel, K., Gallon, J.R., Hayes, P.K., Sivonen, K., and Walsby, A.E. (2003). BASIC: Baltic Sea Cyanobacteria. An investigation of the structure and dynamics of water blooms of cyanobacteria in the Baltic Sea—Responses to a changing environment. Cont. Shelf Res. *23*, 1695–1714.

Stal, L.J., Staal, M., and Villbrandt, M. (1999). Nutrient control of cyanobacterial blooms in the Baltic Sea. Aq. Microb. Ecol. *18*, 165–173.

Stal, L.J., and Moezelaar, R. (1997). Fermentation in cyanobacteria. FEMS Microbiol. Rev. *21*, 179–211.

Stomp, M., Huisman, J., de Jongh, F., Veraart, A.J., Gerla, D., Rijkeboer, M., Ibelings, B.W., Wollenzien, U.I.A., and Stal, L.J. (2004). Adaptive divergence in pigment composition promotes phytoplankton biodiversity. Nature *432*, 104–107.

Subramaniam, A., Brown, C.W., Hood, R.R., Carpenter, E.J., and Capone, D.G. (2002). Detecting *Trichodesmium* blooms in SeaWiFS imagery. Deep Sea Res. *49*, 107–121.

Thamdrup, B., and Dalsgaard, T. (2002). Production of N_2 through anaerobic ammonium oxidation coupled to nitrate reduction in marine sediments. Appl. Environ. Microbiol. *68*, 1312–1318.

Thiel, T. (1993). Characterization of genes for an alternative nitrogenase in the cyanobacterium *Anabaena variabilis*. J. Bacteriol. *175*, 6276–6286.

Thiel, T., Pratte, B., and Zahalak, M. (2002). Transport of molybdate in the cyanobacterium *Anabaena variabilis* ATCC 29413. Arch. Microbiol. *179*, 50–56.

Tuit, C., Waterbury, J., and Ravizza, G. (2004). Diel variation of molybdenum and iron in marine diazotrophic cyanobacteria. Limnol. Oceanogr. *49*, 978–990.

Villareal, T.A. (1992). Marine nitrogen-fixing diatom-cyanobacteria symbioses. In Marine Pelagic Cyanobacteria: Trichodesmium and other Diazotrophs, E.J. Carpenter, D.G. Capone, and J.G. Rueter, eds. (Kluwer Academic Publishers, The Netherlands), pp. 163–175.

Villbrandt, M., Stal, L.J., and Krumbein, W.E. (1990). Interactions between nitrogen fixation and oxygenic photosynthesis in a marine cyanobacterial mat. FEMS Microbiol. Ecol. *74*, 59–72.

Villbrandt, M., and Stal, L.J. (1996). The effect of sulfide on nitrogen fixation in heterocystous and non-heterocystous cyanobacterial mat communities. Algol. Stud. *83*, 549–563.

Ward, B.B. (2003). Significance of anaerobic ammonium oxidation in the ocean. Trends Microbiol. *11*, 408–410.

Walsby, A.E. (1985). The permeability of heterocysts to the gases nitrogen and oxygen. Proc. Roy. Soc. Lond. B *226*, 345–366.

Zahalak, M., Pratte, B., Werth, K.J., and Thiel, T. (2004). Molybdate transport and its effect on nitrogen utilization in the cyanobacterium *Anabaena variabilis* ATCC 29413. Mol. Microbiol. *51*, 539–549.

Zehr, J.P., Carpenter, E.J., and Villareal, T.A. (2000). New perspectives on nitrogen-fixing microorganisms in tropical and subtropical oceans. Trends Microbiol. *8*, 68–73.

Zehr, J.P., and McReynolds, L.A. (1989). Use of degenerate oligonucleotides for amplification of the *nifH* gene from the marine cyanobacterium *Trichodesmium thiebautii*. Appl. Environ. Microbiol. 55, 2522–2526.

Zehr, J.P., Mellon, M., Braun, S., Litaker, W., Steppe, T., and Paerl, H.W. (1995). Diversity of heterotrophic nitrogen fixation genes in a marine cyanobacterial mat. Appl. Environ. Microbiol. *61*, 2527–2532.

Zehr, J.P., Mellon, M.T., and Zani, S. (1998). New nitrogen-fixing microorganisms detected in oligotrophic oceans by amplification of nitrogenase (*nifH*) genes. Appl. Environ. Microbiol. *64*, 3444–3450.

Zehr, J.P., Waterbury, J.B., Turner, P.J., Montoya, J.P., Omoregie, E., Steward, G.F., Hansen, A., and Karl, D.M. (2001). Unicellular cyanobacteria fix N_2 in the subtropical North Pacific Ocean. Nature *412*, 635–638.

Zehr, J.P., Wyman, M., Miller, V., Duguay, L., and Capone, D.G. (1993). Modification of the Fe protein of nitrogenase in natural populations of *Trichodesmium thiebautii*. Appl. Environ. Microbiol. *59*, 669–676.

Zehr, J.P., Montoya, J.P., Jenkins, B.D., Hewson, I., Mondragon, E., Short, C.M., Church, M.J., Hansen, A., and Karl, D.M. (2007). Experiments linking nitrogenase gene expression to nitrogen fixation in the North Pacific subtropical gyre. Limnol. Oceanogr. 52, 169-183

Cyanobacterial-plant Symbioses: Signaling and Development 16

Birgitta Bergman, Liang Ran, and David G. Adams

Abstract

Cyanobacteria form stable nitrogen-fixing symbioses with diverse eukaryotes. With few exceptions the cyanobacteria belong to the terrestrial and widespread genus *Nostoc*. This genus has a notable morphological plasticity which may be in part responsible for its symbiotic competence. In contrast, the symbiotic host range is wide, from mosses to angiosperms. The plant symbioses range from less intimate interactions, such as in mosses, to highly intricate symbioses, such as the intracellular symbiosis with the angiosperm *Gunnera*. In *Azolla* spp. the relationship is perpetual and maintained between generations. *Nostoc* is also one of the most developmentally advanced prokaryotes and capable of differentiating several cell types with various functions. Individual vegetative cells of *Nostoc* may differentiate into nitrogen-fixing heterocysts, and filaments may fragment into hormogonia, a motile life-stage and a prerequisite for plant infection. On internalization, the hormogonia are turned into multi-heterocystous filaments. The high frequency of heterocysts is reflected in their high nitrogen-fixing activities, and in the transfer of the fixed nitrogen to the plant. A sequence of inter-organism communication events between the partners and cellular adaptations is therefore obvious. Our current knowledge about molecular mechanisms involved in these events, and how a full integration has evolved, will be presented with emphasis on recent findings.

The cyanobacterial-plant symbioses

The formation of microbial symbiosis represents a general "macromechanism" for the origin of species, and symbioses have profoundly expanded the diversity of organisms through novel services that the microsymbionts provide their eukaryotic hosts. Cyanobacteria have played fundamental roles in the evolution of such novel life forms, being progenitors of plant chloroplasts. Also, most present day symbiotic cyanobacteria (cyanobionts) provide their plant hosts with a crucial macronutrient, i.e. combined nitrogen, via their capacity for nitrogen fixation, which has allowed niche expansion of the hosts into nitrogen-limited terrestrial environments. Cyanobacteria form symbiotic associations with a wide range of eukaryotic hosts, supplying them with fixed nitrogen or fixed carbon, or both. This chapter will concentrate on the cyanobacteria-plant symbioses, but further information on the other cyanobacterial associations is available in several reviews (Adams, 2000; Rai *et al.*, 2002; Bergman *et al.*, 2007).

Because cyanobacteria generate oxygen during photosynthesis, nitrogen-fixing forms have evolved a number of strategies for protecting the enzyme nitrogenase from oxygen inactivation (Fay, 1992; Gallon, 1992). The most complex of these is the specialized cell known as a heterocyst, which develops at regular intervals within filaments of vegetative cells and provides the anaerobic environment required for nitrogen fixation (Wolk, 1996; Adams and Duggan, 1999; Golden and Yoon, 2003; see also chapter 14). Heterocystous cyanobacteria of the families *Nostocaceae* and *Stigonemataceae* possess an additional characteristic that is crucial to the establishment of plant symbioses. The immotile parent trichomes produce specialised motile filaments known as hormogonia, which serve as the infective agents in many cyanobacterial symbioses, particularly those involving plants (Adams, 2000; Meeks *et al.*, 2002; Meeks and Elhai, 2002; Bergman *et al.*, 2007). Potential host plants improve their chances of infection by releasing chemical signals that both trigger hormogonia formation and serve as chemoattractants for hormogonia (see *What makes a successful symbiont?* and *The infection process*), and it may be for these, and additional unknown reasons, that plant cyanobionts are almost always members of the genus *Nostoc* (*Cyanobacterial diversity*).

Bryophytes (mosses, hornworts and liverworts)

The bryophytes consist of the liverworts (Hepaticae), the hornworts (Anthocerotae) and the mosses (Musci), all of which are small, non-vascular land plants, some of which form epiphytic or endophytic associations with cyanobacteria (Adams, 2002; Meeks, 2003). The moss associations are mostly epiphytic (Solheim and Zielke, 2002; Gentili *et al.*, 2005), with the exception of two species of *Sphagnum* in which the *Nostoc* spp. occupy water-filled, hyaline cells. These are thought to provide some protection from the low pH (<5) in bog habitats, which is below the usual tolerance for cyanobacterial growth (Solheim and Zielke, 2002). Similarly, the moss leaf surface is made more suitable for epiphytic cyanobacterial growth by the secretion of alkaline substances and the cation exchange capacity of the leaf (Belnap, 2001). The nitrogen-fixing epiphytic and endophytic cyanobacterial associations with mosses can provide the major input (up to 84%) of nitrogen into local ecosystems in the Arctic, the Antarctic and boreal forest regions (DeLuca *et al.*, 2002; Solheim and Zielke, 2002; Zielke *et al.*, 2002; Gentili *et al.*, 2005; Zielke *et al.*, 2005). The epiphytic cyanobacterial association with the feather moss, *Pleurozium schreberi*, may be particularly important because this moss is ubiquitous in the boreal forest regions of the world, where it can form up to 80% of the total ground cover (DeLuca *et al.*, 2002; Zackrisson *et al.*, 2004; Gentili *et al.*, 2005).

Five of the six hornwort genera form endophytic, but still extracellular, associations with cyanobacteria, but only four of over 340 genera of liverwort form associations, two of them epiphytic and two endophytic. In the field, liverworts and hornworts grow as a prostrate gametophyte thallus a few centimeters in length and symbiotic colonies can be seen as dark spots approximately 0.5 mm in diameter (Figure 16.1A). Little is known about the liverwort epiphytic associations, whereas the endophytic associations are better studied because they have proved to be amenable to laboratory experimentation. Hornworts such as *Anthoceros* and *Phaeoceros*, and liverworts such *Blasia*, can be grown in shaken liquid culture, free of their symbiotic cyanobacterial partners, and can then be re-infected with axenic cyanobacteria (Meeks, 1998; Adams, 2002; Meeks, 2003).

Figure 16.1 (see also Plate 16.1) Cyanobacterial-plant symbioses. (A) The small liverwort *Blasia* sp. colonized by cyanobacteria seen as small dark spots spread over the thallus. Inset shows a cyanobacteria-filled auricle at the surface of the thallus. (B) A branch of the small floating water-fern *Azolla* sp. The cyanobacteria are occupying cavities located in each of the leafs seen. (C) A palm-like cycad plant (*Cycas* sp.) with a prominent cone. Inset shows the blue-green cyanobacteria-filled zone located between the outer and inner cortex of a transversely sectioned coralloid root. (D) The large-leaved rhubarb-like *Gunnera manicata*. Inset shows the blue-green cyanobacterial infection sites in a cross sectioned stem.

Azolla

The free-floating aquatic fern *Azolla* is the only pteridophyte known to form symbioses with cyanobacteria. The fern is found on the surface of still or slow-moving bodies of freshwater in tropical and temperate regions (Mandal *et al.*, 1999; Adams, 2000; Rai *et al.*, 2000; Lechno-Yossef and Nierzwicki-Bauer, 2002). *Azolla* has been employed as a green manure in rice cultivation for centuries and is still in use worldwide (van Hove and Lejeune, 2002; Choudhury and Kennedy, 2004). The extensively branched main stem (rhizome) of the fern bears leaves consisting of two clearly defined lobes, approximately 1 mm in length (Figure 16.1B). The ventral lobe has little chlorophyll and functions as a float supporting the fleshy, chlorophyllous dorsal lobe above the water. The cyanobacterial symbionts are found within an extracellular cavity in the dorsal lobe (Figure 16.1B). The cyanobionts are usually referred to as *Anabaena azollae,* although their taxonomic assignment is controversial (see *Phylogeny of symbiotic cyanobacteria* – Azolla). The leaf cavities of *Azolla* also contain heterotrophic bacteria often considered a third member of the association (Lechno-Yossef and Nierzwicki-Bauer, 2002).

Gymnosperms (Cycads)

Cycads are palm-like plants, but taxonomically gymnosperms, with a trunk ranging in height from a few centimeters to 20 meters (Figure 16.1C), and a tap root from which may develop specialized apogeotropic or coralloid roots (Brenner *et al.*, 2003; Vessey *et al.*, 2005). The cycads are one of the most primitive groups of seed plants (gymnosperms) alive today, and comprise approximately 150 species found almost exclusively in the tropical and subtropical areas of the southern hemisphere, where many are endangered (Costa and Lindblad, 2002; Schneider *et al.*, 2002; Vessey *et al.*, 2005). Coralloid roots, so-called because they form dicotomously branched clusters that resemble corals, are frequently infected with heterocystous cyanobacteria (Costa and Lindblad, 2002), which can be seen as a dark blue-green band between the inner and outer cortex of the roots (Figure 16.1C; inset).

Gunnera

The only genus of flowering plant (angiosperm) known to form symbiotic associations with cyanobacteria is *Gunnera,* which consists of approximately 30–40 species, all of which can become infected. *Gunnera* is found throughout the tropical and subtropical regions of the southern hemisphere where they prefer wet areas, often at altitude (Rai *et al.*, 2000; Bergman and Osborne, 2002; Osborne and Sprent, 2002; Wanntorp *et al.*, 2002). The plants vary in size from very large rhizomatous forms with leaves two meters across, to creeping forms with leaves only 1–10 centimeters across (Figure 16.1D). The cyanobiont in *Gunnera* is found intracellularly, inside specific mucus-secreting glands on the plant stem at the base of each leaf petiole (Figure 16.1D; inset); this contrasts with the intercellular location of the cyanobionts in all cycads, bryophytes and *Azolla*. *Gunnera* has attracted particular research interest because of its status as the only flowering plant to host cyanobacterial symbioses.

Evolution of cyanobacteria and their host plants

Evolution and significance of cyanobacteria

Although cyanobacteria have existed for several billion years, our knowledge about their evolution is still in its infancy. It is however known that they have played extremely important roles, including the introduction of oxygenic photosynthesis, thereby (together with plants) increasing the level of oxygen in the atmosphere to that found today (Tomitani *et al.*, 2006). In a symbiotic context they have also made a great evolutionary impact, above all for being the prokaryotes that gave rise to the algae and land plants, via a true endosymbiotic (intracellular) event. This was one of the most successful evolutionary steps taken, considering the importance of plants in shaping the bio- and atmosphere, and a giant step in supporting the survival of animal life. This monophyletic endosymbiotic event may have occurred about two billion years ago.

Monophyly is also suggested for the cyanobacterial category comprising all plant cyanobionts, i.e. cyanobacteria capable of developing heterocysts and akinetes and belonging to Sections IV and V (Rippka *et al.*, 1979). This has been shown using 16S rDNA, *rbcL* and *hetR* analyses (Tomitani *et al.*, 2006; see also Chapters 1 and 2). Combined molecular phylogenetic and paleontological analyses suggest that cyanobacteria capable of cell differentiation diverged between 2400 and 2100 million years ago (Ma). However, the evolution of the cell cycle stage important for symbiosis, the hormogonium (see sections "What makes a successful cyanobiont?" and "The infection process"), is not known. Hormogonia development is not a general trait among Section IV and V cyanobacteria, and may also occur in Section III. Reconstitution experiments suggest that hormogonia are prerequisites for plant symbioses, but that not all hormogonia are symbiotically competent (Johansson and Bergman, 1994). Hence, the reason for the restriction of symbiotic competence to a few cyanobacterial families is not solely related to the capacity to develop hormogonia.

Evolution of cyanobacterial-plant symbioses

The age of the origin of extant cyanobacterial-plant symbioses is unknown. The fact that heterocystous cyanobacteria of Section IV arose more than 2000 Ma (Tomitani *et al.*, 2006) restricts the occurrence of any cyanobacterial-eukaryotic symbioses until after this event. The first potential land plant hosts to encounter terrestrial heterocystous cyanobacteria were probably the bryophyta which arose about 300 million years ago (Figure 16.2), while marine cyanobacterial symbioses may have predated this event. Molecular data infer that diatoms with symbiotic competence (e.g. *Hemiaulus, Rhizosolenia* and *Rophalodia*), arose about 250 million years ago, while fungal symbioses possibly arose earlier (Figure 16.2). Therefore, in view of the lack of symbiotic fossil evidence, it is not possible to establish the time of origin of any present day cyanobacterial-plant symbioses.

The only symbiotic fossils discovered so far are cyanolichen fossils from the 400 million year old Rhynie Chert in Scotland (Taylor *et al.*, 1997) and the 600 million year old Doushantou Formation in China (Taylor and Krings, 2005; Yuan *et al.*, 2005), which suggests that the symbiotic competence of some cyanobacteria existed before the evolution of land plants (Figure 16.2). Estimates suggest that lichen symbioses may have occurred not less than a hundred times (Rikkinen, 2002).

The Rhynie chert has also provided the oldest fossil evidence for a cyanobacteria-land plant endophytic association between the primitive, extinct land plant *Aglaophyton major* and an *Archaeothrix*-type filamentous cyanobacterium (Taylor and Krings, 2005). The cyanobacteria are thought to have entered the plant through stomatal pores, initially colonizing the substomatal chambers and then spreading throughout the outer cortical tissue. This method of infection has interesting similarities with that of the modern-day hornworts *Anthoceros* and *Phaeoceros* (section "Bryophytes").

Notably, all known plant hosts, according to the fossil record, are among the oldest within their respective divisions: the bryophytes, the ferns (*Azolla* spp.), the cycads and *Gunnera* spp., which may also suggest considerable antiquity of the origin of present

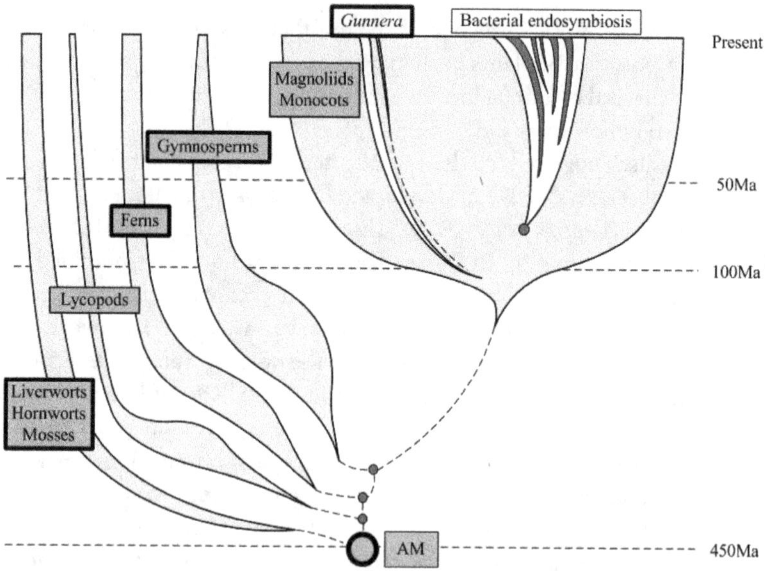

Figure 16.2 The evolution of plants with emphasis on those acting as symbiotic hosts for cyanobacteria (black-lined boxes). Cyanobacteria form symbiosis with a few representatives within each of the plant divisions boxed, ranging from mycorrhizal fungi (black ring), which appeared about 450 Ma, to bryophytes, one fern (*Azolla*), cycads (gymnosperms), and one angiosperm family, Gunneraceae, appearing about 100 Ma. Indicated in the figure are also the angiosperms acting as hosts for rhizobia. After Kistner and Parniske, 2002, with permission from Elsevier.

day cyanobacterial-plant symbioses. Assuming that most plants formed symbioses with cyanobacteria early in their evolutionary history, the date of the origin of symbioses with bryophyta and angiosperms would differ by approximately 300 million years (Figure 16.2). Are there any indications that this may have influenced the evolution of the various symbioses? Although the *Azolla* symbiosis is extracellular, a condition considered to be less advanced, it may still be the most advanced as this symbiosis is perpetually maintained, and the cyanobacterium transferred vertically. This is not the case for the likewise extracellular cyanobacterial-bryophyta symbioses which require *de novo* infection for each new generation. On the other hand, the *Gunnera* symbiosis, with the youngest host plant (Figure 16.2), is the only one in which the cyanobacterium penetrates the host cells intracellularly, which is often taken as a sign of a most advanced relationship. The evolution of plant symbioses may therefore have proceeded at different rates.

Cyanobacterial diversity

Restricted diversity

The cyanobacteria forming symbioses with plants (cyanobionts) not only have a restricted distribution within the cyanobacterial radiation, but also within the Section IV and Section V cyanobacteria (Table 16.1), the significance of which is unknown. The vast majority

Table 16.1 Partners in cyanobacterial-algal and -plant symbioses. For comparative purposes some cyanobionts of lichens are included. Genera in bold denotes the most common cyanobiont in the respective eukaryotic host

Eukaryotic host	Environment	Section	Type	Genera	N_2-fixer
Fungi/lichens[1]	Terrestrial	I/II	Unicellular	Numerous*	Some
	Terrestrial	IV	Heterocystous	***Nostoc***	+
	Terrestrial	IV	Heterocystous	*Tolypothrix*	+
	Terrestrial	IV	Heterocystous	*Dichothrix*	+
	Terrestrial	IV	Heterocystous	*Scytonema*	+
	Terrestrial	V	Heterocystous	*Stigonema*	+
	Marine	IV	Heterocystous	*Calothrix*	+
Algae[2]	Marine	IV	Heterocystous	***Richelia***	+
Bryophytes[3]	Terrestrial	IV	Heterocystous	***Nostoc***	+
				(*Calothrix*?)	
		V	Heterocystous	*Stigonema*	+
Azolla[3]	Terrestrial	IV	Heterocystous	***Nostoc*/ *Anabaena***?	+
Cycads[3]	Terrestrial	IV	Heterocystous	***Nostoc***	+
				Calothrix	+
Gunnera[3]	Terrestrial	VI	Heterocystous	***Nostoc***	+

(1) Rikkinen, 2002

**Gloeocapsa, Cyanothece, Gloeothece, Hyella, Chroococcidiopsis* etc

(2) Janson et al., 1995; 1999

(3) Rai et al., 2002

of cyanobacteria interacting with plants are affiliated to one filamentous and heterocystous genus, *Nostoc*, a genus which also comprises a range of true free-living species (Dodds *et al.*, 1995; Rai *et al.*, 2002).

The identification of cyanobionts as belonging to the filamentous and heterocystous genera had already been reported for cycads and *Gunnera* about 130 years ago (see Bergman *et al.*, 2007). A few other cyanobacterial genera have also been reported as cyanobionts, such as the likewise filamentous and heterocystous genus *Calothrix*, which enter coralloid roots of one cycad (Grobbelaar *et al.*, 1987), *Richelia*, the cyanobiont genus in marine algae (Janson *et al.*, 1999), and *Stigonema*, which associates with mosses (Gentili *et al.*, 2005). As also seen in Table 16.1, a much wider cyanobacterial diversity is found among symbioses comprising lichenized fungi in which associations with members of the genera *Nostoc*, *Calothrix* (Section IV), *Stigonema* and *Fischerella* (Section V), and with representatives of many unicellular genera (Section I and II) are well documented (Rikkinen, 2002). Marine plankton (diatoms) also harbor as cyanobionts a wider variety of both heterocystous

(Janson *et al.*, 1995; Janson *et al.*, 1999) and unicellular cyanobacteria (Foster *et al.*, 2006a) than do land plants.

Phylogeny of symbiotic cyanobacteria

Our knowledge about the taxonomy and phylogeny of cyanobacteria involved in symbioses is still in its infancy (Rasmussen and Nilsson, 2002). A more comprehensive phylogenetic analysis of symbiotic cyanobacteria, based on full-length 16S rDNA gene sequences (Svenning *et al.*, 2005), revealed well separated clades among *Nostoc* cyanobionts. For instance, *Gunnera* cyanobionts formed a separate clade, while others were inter-mixed with free-living cyanobacteria of the genus *Anabaena*. These data suggest that symbiotic competence has arisen more than once.

Liverworts/hornworts and mosses

Again, members of the genus *Nostoc* are the dominating cyanobionts in nature in liverworts and hornworts (Table 16.1; Figure 16.1A). This was demonstrated by analyzing 31 cyanobionts of the hornwort *Phaeoceros* (West and Adams, 1997) and cyanobionts of *Anthoceros* and *Blasia* (Costa *et al.*, 2001) using PCR-based molecular methods. Only one cyanobiont was found in each cavity of these bryophytes although many different strains of *Nostoc* were detected. Reconstitution experiments have however shown that other hormogonium-developing cyanobacterial genera, such as *Calothrix* (Section IV) and *Chlorogloeopsis* (Section V), may enter liverworts (West and Adams, 1997).

Azolla

The morphologial and molecular characteristics of *Azolla* cyanobionts grown under free-living conditions have always differed from those of freshly isolated cyanobiont, implying that the major symbiont(s) can not be grown in culture (Gebhardt and Nierzwicki-Bauer, 1991; Lechno-Yossef and Nierzwicki-Bauer, 2002). The cyanobiont in *Azolla* has most often been assigned to the genus *Anabaena*, although more recent molecular analyses suggest affiliation to the genus *Nostoc* (Plazinski *et al.*, 1991; Kim *et al.*, 1997; Bergman *et al.*, 2006). The latter is supported by the fact that hormogonia-like filaments enter the megasporocarps and that non-heterocystous, hormogonia-like filaments are detected in young, newly infected leaf cavities. However, Baker *et al.* (2003), using sequence comparisons of the phycocyanin intergenic spacer and a fragment of the 16S rRNA gene, concluded that the symbiont from *Azolla filiculoides* Lam. var. *rubra* was not closely related to either *Nostoc* or *Anabaena* (Table 16.1).

The cyanobacterial-*Azolla* symbiosis may be unique in that the cyanobionts are maintained through the *Azolla* generations thereby making *de novo* infection redundant (see *The infection process – invasion of the plant*). Perhaps as a consequence of this, in the seven extant *Azolla* species, one particular cyanobiont appears to be always associated with one particular *Azolla* species and co-evolution may have occurred (van Coppenolle *et al.*, 1995; Zheng *et al.*, 1999). However, further molecular refinements are needed to clarify the situation with regard to the identity of the *Azolla* cyanobionts.

Azolla symbioses are also unique in having what appears to be a third symbiotic partner, composed of numerous bacteria, which include several gram-positive and gram-negative soil bacteria, such as *Arthrobacter* spp., *Corynebacterium* sp. and *Agrobacterium* sp., but also potential nitrogen-fixers such as *Bradyrhizobium* sp. and *Rhizobium* sp. (Lechno-

Yossef and Nierzwicki-Bauer, 2002). The role of these bacteria in the *Azolla* symbiosis is not understood, although it has been proposed that they may produce the phytohormone auxin (Forni *et al.*, 1992), and that some contain nitrogenase (Lindblad *et al.*, 1991) and may therefore fix nitrogen. The intimacy of the relationship between the three partners is manifested in the fact that the bacteria are also transferred to the next generation of plants via the megasporocarps.

Cycads

Besides *Nostoc*, *Calothrix* has only once been reported as a cyanobiont in plants, namely in the roots of a cycad (*Macrozamia*; Table 16.1) (Costa and Lindblad, 2002), although these plants have a circum-global distribution in sub-tropical and tropical areas. However, a high number of *Nostoc* strains were deduced from DNA-RFLP analyses (combined with *nifH* and *glnA* probes) of cultured cyanobionts isolated from roots of five cycad species, while the cyanobionts (cultured or freshly collected) from roots of *Encephalartos altensteinii* showed identical RFLP patterns (Lindblad *et al.*, 1989). Mixed or individual strains may therefore occur in a single plant. The latter was also shown to be the case in cycads (*Zamia* and *Dioon*) cultivated in a botanical garden in Florida (Zimmerman and Rosen, 1992). Costa *et al.* (1999) later used a partial $tRNA^{Leu}$(UAA) intron as marker and reported that the cyanobionts (still within the genus *Nostoc*) of five different cycads varied, and that different roots from the same plant were colonized by the same cyanobiont. Later however, this finding was contradicted when more than one cyanobiont strain was detected in one individual coralloid root of cycads growing in China (Zheng *et al.*, 2002).

Gunnera

PCR-based DNA-fingerprinting techniques (using short tandemly repeated repetitive, STRR and long tandemly repeated repetitive, LTRR, primers) have helped to resolve affiliations of cultured cyanobiotic isolates from greenhouse grown *Gunnera* spp. and from plants growing under natural conditions (Nilsson *et al.*, 2000; Rasmussen and Svenning, 2001). Collectively, these data showed that the 23 isolates were closely related, irrespective of *Gunnera* host species, and that the same *Gunnera* species may harbor different cyanobacterial strains. A low specificity was also inferred from the analyses of 45 cultured isolates from 11 *Gunnera* species growing in geographically widely separated areas, as these showed a wide range of DNA-fingerprints, sometimes even within the same plant and within the same stem gland (the infection site). This was later confirmed using genetic methods with varying taxonomic resolution levels (RFLPs of 16S rDNA gene internal transcribed spacer, ITS, and the functional *hetR* gene), as was a lack of clustering between cyanobionts and host species (Rasmussen and Svenning, 2001). Cyanobionts collected from field grown *Gunnera tinctoria* in southern Chile and analyzed by STRR-PCR based fingerprinting, confirmed the exclusive presence of *Nostoc* strains in these *Gunnera* (Table 16.1), but pointed to a rich intra-genus diversity. In these studies, however, only one strain at a time was found to infect one specific host plant (Guevara *et al.*, 2002).

Host symbiotic structures

Unlike the bacterial symbionts in rhizobial and actinorhizal symbioses, cyanobacteria occupy existing plant structures, although they are further developed on colonization (inserts in Figure 16.1 A;C-D and Figure 16.3). All cyanobacteria are photoautotrophs, but many are also photoheterotrophs or, in some cases, chemoheterotrophs, which enables them to occupy plant structures, such as the coralloid roots of cycads, which receive little or no light.

Bryophytes

In the liverworts and hornworts the mature cyanobacterial colonies are up to 0.5 mm in diameter and occupy structures on the surface of, or within, the thallus (Figure 16.1A). In the liverwort *Blasia* the structures, known as auricles, are dome-shaped and located on the ventral surface of the thallus (insert in Figure 16.1A) (Adams, 2002). By contrast, in the hornworts *Anthoceros* and *Phaeoceros* the thallus is thicker and the cyanobacteria are found in slime cavities within the tissue, connected to the ventral surface via stomata-like pores (Meeks, 1998; Adams, 2002; Meeks, 2003). An exception to this arrangement is found in the hornwort *Leiosporoceros dussii*, in which strands of cyanobiont run parallel to the main axis of the thallus (Villarreal *et al.*, 2005; Villarreal and Renzaglia, 2006). The cyanobacteria occupy mucilage-filled "canals" formed by the separation of plant cell walls along their middle lamellae and, as these canals elongate and bifurcate, they form a integrated network throughout the plant (Villarreal and Renzaglia, 2006). This arrangement is unique to *Leiosporoceros dussii* and enables the *Nostoc* to spread through the plant without the need for multiple infections, as would be required for the isolated, globose colonies in other hornworts.

Azolla

The symbiotic cavity in the water fern *Azolla* is found in the dorsal leaf lobe (Figure 16.1B and 16.3) and is formed by an infolding of the epidermis (Lechno-Yossef and Nierzwicki-

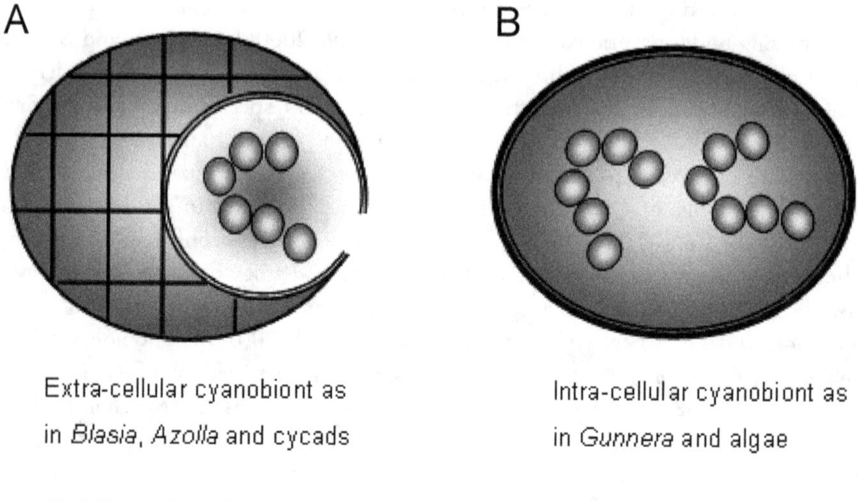

Figure 16.3 Illustration of the location of the cyanobionts in algal and plant symbioses. Cyanobacterial filaments either: (A) colonize open cavities formed in the host thallus/leaf, connected to the surrounding environment via small openings (the cyanobiont is extra-cellular); or (B) invade the host cell, although they always remain outside the plant plasmalemma (the cyanobiont is intra-cellular).

Bauer, 2002). This ellipsoidal, extracellular cavity is approximately 0.3 mm in length and is connected to the outside via a pore which narrows as the leaf matures. The pore is surrounded by teat-shaped cells that may provide a physical barrier to prevent ingress of unwanted organisms and loss of the symbionts (Veys *et al.*, 1999; Veys *et al.*, 2000, 2002). Such teat-cells are found in *Azolla* spp. from the section *Azolla*, but only under stress conditions in *Azolla* spp. from the section Rhizosperma (Veys *et al.*, 2000, 2002). Embedded in mucilage and surrounded by an envelope of plant origin, the cyanobacteria occupy only the periphery of the cavity, the centre being gaseous (Lechno-Yossef and Nierzwicki-Bauer, 2002). Within the leaf cavity are branched and simple hairs which possess transfer cell characteristics and may play a role in the exchange of nutrients between the cyanobiont and the plant (Lechno-Yossef and Nierzwicki-Bauer, 2002).

Cycads

In cycads the cyanobacterial symbionts are found in a zone between the inner and outer cortical layers of the coralloid roots (Figure 16.1C and 16.3) which are found beneath the soil surface, although they may extend above the surface in older plants (Adams, 2000; Rai *et al.*, 2000; Costa and Lindblad, 2002; Vessey *et al.*, 2005). Coralloid roots begin as precoralloids which are produced independently of the presence of cyanobacteria, although entry of the cyanobacteria irreversibly transforms them into typical coralloid roots (Ow *et al.*, 1999; Rai *et al.*, 2002).

Gunnera

In the *Gunnera-Nostoc* association the cyanobacteria enter the plant via bright red mucus secreting glands at the base of each leaf petiole (Figure 16.1D and Figure 16.4) (Rai *et al.*, 2000; Uheda and Silvester, 2001; Bergman, 2002; Bergman and Osborne, 2002). The glands consist of a central papilla, surrounded by up to nine additional papillae, which become exposed during gland development when the epidermal layer is ruptured (Figure 16.4A). Cyanobacteria enter the gland through intercellular channels between the papillae (Figure 16.4B), at least two papillae being required for successful infection (Uheda and Silvester, 2001). Glands form soon after seed germination and their development, previously thought to be constitutive, was recently shown to be induced by nitrogen starvation (Chiu *et al.*, 2005). *Gunnera* is the only plant-cyanobacterium symbiosis in which the cyanobacterium penetrates the host cell walls and becomes intracellular (Figure 16.3 and 16.4C-D). The final infected area of tissue can be from 2 mm to 2 cm in diameter depending on the size of the *Gunnera* host (Figure 16.1D, inset).

What makes a successful cyanobiont?

The sessile nature of plants may force the cyanobionts to be able to move or glide to reach target plants and the plant organs to colonize, while the more motile life style of fungi and marine plankton may allow these hosts to "capture" less motile or non-motile cyanobionts. Possibly therefore, only highly motile cyanobacteria may be capable of, or can be accepted as, plant cyanobionts. For a stable symbiosis there are additional requirements. For instance, all plant cyanobionts are nitrogen-fixers (Table 16.1) and able to support the host with its total demand of combined nitrogen. Some of the unicellular cyanobionts in lichens

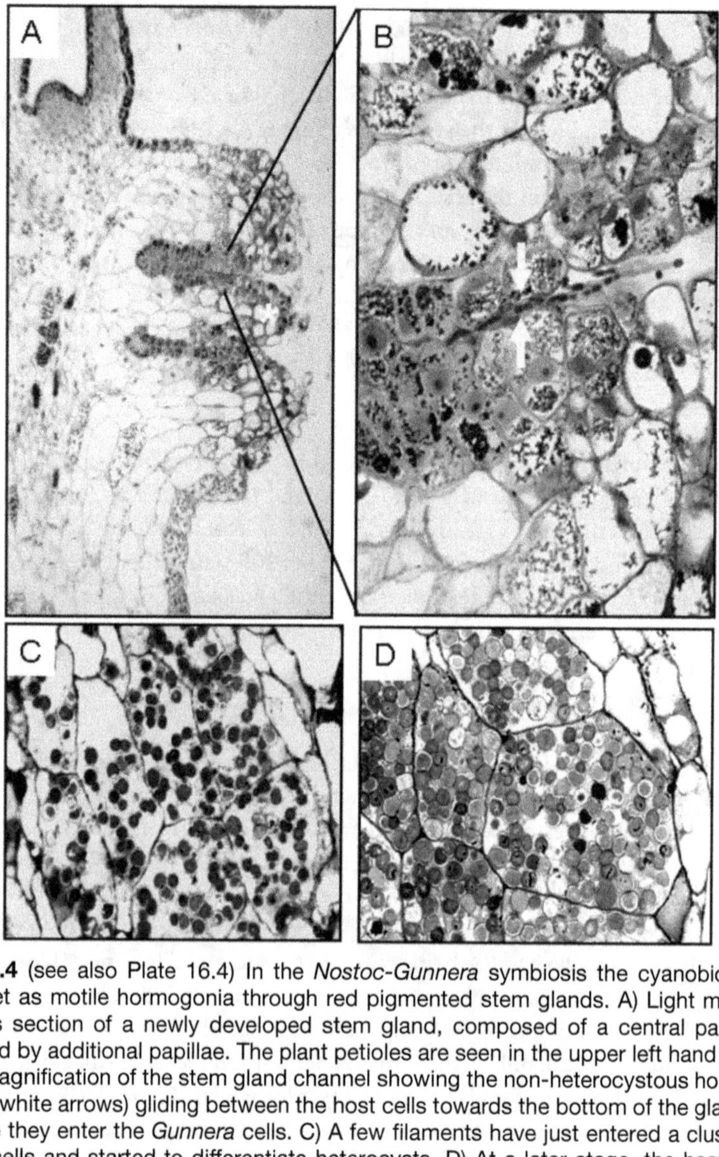

Figure 16.4 (see also Plate 16.4) In the *Nostoc-Gunnera* symbiosis the cyanobiont enters the plantlet as motile hormogonia through red pigmented stem glands. A) Light microscopy of a cross section of a newly developed stem gland, composed of a central papilla (star) surrounded by additional papillae. The plant petioles are seen in the upper left hand corner. B) A larger magnification of the stem gland channel showing the non-heterocystous hormogonial filaments (white arrows) gliding between the host cells towards the bottom of the gland (to the left) where they enter the *Gunnera* cells. C) A few filaments have just entered a cluster of the *Gunnera* cells and started to differentiate heterocysts. D) At a later stage, the host cells are filled with cyanobacteria and the heterocysts (thicker cell walls) have reached a frequency of > 50%. A) and B) from Johansson and Bergman, 1992 (Figure 16.5a and b), with permission from Springer-Verlag.

are also potential nitrogen-fixers and nitrogenase has recently been shown to be present in some unicellular cyanobionts of marine plankton (Foster *et al.*, 2006a). The pressure for the plants to gain access to a source of a macronutrient, like combined nitrogen, has probably been a strong evolutionary "driver" to enter into such symbiotic relationships (Bergman *et al.*, 1996).

Plants may in addition have evolved exclusive discrimination mechanisms to refine their selection of cyanobionts, as only certain cyanobacteria, even among closely related nitrogen-fixers (e.g. within the genus *Nostoc*), form symbioses (Table 16.1). Or conversely, only a few *Nostoc* species/strains may be able to disguise themselves enough not to be recognized as a foreign "invader." Chemotaxis may also be involved (see section "Chemotaxis") in the sense that only cyanobacteria responding positively to released plant signaling compounds may become cyanobionts (Nilsson *et al.*, 2005). As many cyanobionts are located in regions of the plant, such as the coralloid roots of cycads, that receive little or no light, they require the ability to metabolize organic carbon sources supplied by the host; indeed many cyanobionts are facultative heterotrophs. A supply of organic carbon from the host enables the cyanobiont to maintain a much higher frequency of heterocysts (which do not fix CO_2) than in the free-living state, thereby enhancing nitrogen fixation capacity (see *Colony formation and maturation*). Together, this suggests a high degree of specificity and that several perquisites are needed for a cyanobacterium to become symbiotically competent.

Hormogonia development

In the plant-cyanobacteria symbioses successful cyanobionts invariably possess the ability to develop both nitrogen-fixing heterocysts and hormogonia (Figure 16.4 and 16.5) (Bergman *et al.*, 1996; Adams, 2000; Meeks and Elhai, 2002; Meeks, 2003). However, the initial infection of the plant requires only functional hormogonia, since *Nostoc* mutants unable to form heterocysts can infect *Anthoceros punctatus*, although they are unable to form a fully functional, nitrogen-fixing symbiosis (Wong and Meeks, 2002). Hormogonia are the infective agents, possessing the necessary motility and chemotactic ability to enable what are, for most of their life cycle, immotile cyanobacterial filaments, to locate and enter the symbiotic structures of the plant. Hormogonia are formed by the heterocystous cyanobacteria by a characteristic process of reductive cell division in which all cells in a filament undergo rapid, synchronous cell division without growth, resulting in a decrease in cell size (Meeks *et al.*, 2002; Meeks and Elhai, 2002). The filament subsequently fragments at the heterocyst-vegetative cell junctions, releasing hormogonia consisting of cells much smaller than the parent trichome (Figure 16.5).

For hormogonia to successfully locate and invade the host tissue they must first attach to the plant surface and both fimbriae (pili) and extracellular polysaccharides have been suggested to have a role in this process (Adams, 2000). Pili have potential functions in gliding motility of hormogonia, in plant host recognition and in surface attachment, although none of these potential functions has been fully established. Type IV pili are required for gliding in some unicellular cyanobacteria (Bhaya, 2004), and the cell surface of hormogonia of the symbiotically competent *Nostoc punctiforme* is covered in pili (Adams and Duggan, 2005). Inactivation of the *Nostoc punctiforme* genes *pilT* and *pilD*, which encode proteins thought to be involved in pilus function, results in mutants that produce hormogonia that infect *Blasia* at a greatly reduced frequency to wild-type hormogonia, although it remains to be established if this is due to an effect on motility or some other function of the pili (Adams and Duggan, 2005). More recently, a proteomic and transcriptional analysis revealed that the expression of a homolog of *pilQ* was strongly enhanced during *Nostoc* hormogonium formation, which further supports the existence of Type IV pili in cyanobionts, and possibly their involvement in symbiotic interactions (Klint *et al.*, 2006). Moreover, upstream from *pilQ*, three adjacent ORFs are present in the *Nostoc* sp. PCC

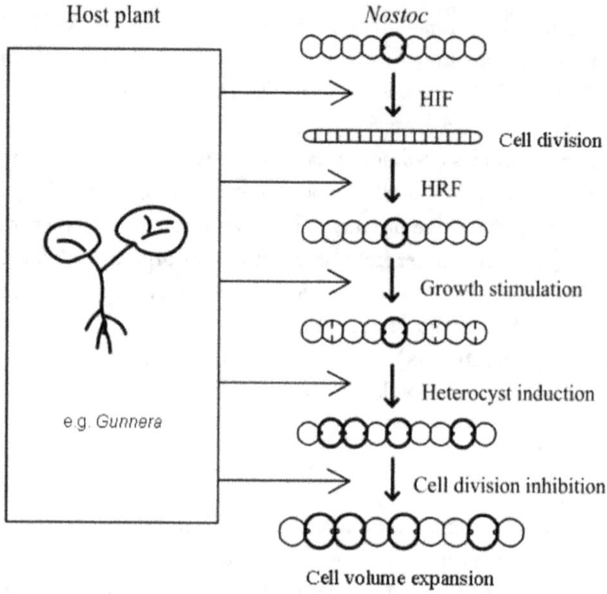

Figure 16.5 Diagrammatic representation of effects of the host plants on the phenotype of a cyanobacterium entering into a plant symbiosis. The effects illustrated apply to all plant-cyanobacteria symbioses, although the plant host given is *Gunnera*. A hormogonia-inducing factor (HIF) released by the host plant triggers a rapid cell division and hormogonia formation in a heterocystous *Nostoc* filament. The motile, small-celled hormogonia are released by fragmentation of the original filament and are attracted towards the plant by a hormogonia chemoattractant released. On entering the plant the hormogonia lose motility and, under the influence of the host, develop an elevated frequency of nitrogen-fixing heterocysts. The host also suppresses any further hormogonia formation by releasing a hormogonia repressing factor (HRF). Filament morphology is often altered in symbiosis, the cells being larger and more rounded than in the free-living state, and the rate of cell division of the cyanobiont is reduced so that it does not outgrow its host. (Drawing: J. Klint, Stockholm University, Sweden).

73102 genome, being homologous to *pilM*, *pilN*, and *pilO*, all involved in pilus formation (Meeks *et al.*, 2001). For plant infection, the hormogonia must be motile and chemotactic (section "Chemotaxis").

The infection process

Hormogonia induction

Plant hosts improve their chances of becoming infected by cyanobacteria by releasing hormogonia inducing factor (HIF) that stimulates hormogonia formation in *Nostoc* strains (Figure 16.5). The production of HIF has been demonstrated in the hornwort *Anthoceros punctatus* (Meeks, 2003), the cycad *Zamia furfuracea* (Ow *et al.*, 1999) and *Gunnera manicata* (Rasmussen *et al.*, 1994). The HIF from *Anthoceros punctatus* is a low molecular mass, heat-labile product released only when the hornwort is nitrogen starved (Cohen and Meeks, 1997; Meeks and Elhai, 2002). A mutant of the symbiotic strain *Nostoc punctiforme*, with increased sensitivity to the *Anthoceros* HIF, also showed a greater initial frequency of infection of the hornwort than the wild type (Cohen and Meeks, 1997). The HIF released by *Gunnera* is found in the mucilage secreted by the stem glands (see Figure 16.1) and, like *Anthoceros* HIF, is small and heat-labile (Bergman *et al.*, 1996). The liverwort *Blasia* also releases HIF when nitrogen starved (Babic, 1996). A mutant of the cyanobacterium *Nostoc* LBG1, which was isolated from symbiosis with the hornwort *Phaeoceros*, fails to respond to *Blasia* HIF, but still forms hormogonia on transfer from medium containing 30 mM glucose to medium lacking glucose, a procedure that induces hormogonia formation in the wild type (Babic and Adams, unpublished). This implies that there are at least two, independently controlled mechanisms for triggering hormogonia formation in this cyanobacterium.

Transcription of the *Nostoc punctiforme* gene *sigH*, which encodes an alternative sigma subunit of RNA polymerase, is induced by HIF (Campbell *et al.*, 1998; Meeks, 2003). *Nostoc* mutants with a transposon insertion in *sigH* form HIF-induced hormogonia at the same frequency as the wild type, but the hormogonia infect *A. punctatus* at a 3 to 5-fold higher frequency than the wild type, implying that the behavior of the hormogonia, as well as their frequency, is important in plant infection (Meeks and Elhai, 2002; Meeks, 2003). The gene(s) regulated by SigH are unknown. A reduced frequency of HIF-induced hormogonia formation is found in *N. punctiforme* carrying a mutation in *ntcA*, which encodes the global transcriptional regulator NtcA (Herrero *et al.*, 2004), and this results in a corresponding decrease in *Anthoceros* infection (Wong and Meeks, 2002).

Chemotaxis

Plant infection can occur when hormogonia locate and enter the plant tissue that will house the symbiotic colonies and this frequently involves movement from the light into dark areas of tissue, contrary to the natural positive phototaxis of hormogonia. For this to happen, hormogonia must be guided into the plant by chemoattraction. Indeed, the genome of the symbiotically competent cyanobacterium *Nostoc punctiforme* has 3–5 copies each of genes encoding homologues of the chemotaxis-related proteins CheA, CheB, CheW, CheD and CheR (Meeks *et al.*, 2001). When starved of combined nitrogen the liverwort *Blasia* releases extracellular signals that not only trigger hormogonia formation (Babic, 1996), but also serve as very effective chemoattractants (Knight and Adams, 1996; Watts *et al.*, 1999). The mucilage from *Gunnera* stem glands also contains a powerful chemoattractant for hormogonia from symbiotically competent *Nostoc* strains (Nilsson *et al.*, 2006). Extracts from non-host plants also contain chemoattractants; indeed an extract from the non-host plant, *Trifolium repens*, is a more effective chemoattractant than that from the natural host *Gunnera* (Nilsson *et al.*, 2006). Germinating wheat seeds have also been shown to release hormogonia chemoattractants (Knight and Adams, 1996; Watts *et al.*, 1999). The identification of these chemoattractive plant signals is now warranted.

Invasion of the plant

The infection process in liverworts and hornworts is more straightforward than that in cycads, *Gunnera* and *Azolla*, because of the ready accessibility of the symbiotic structures. Plant infection occurs when hormogonia enter *Blasia* auricles (and presumably hornwort slime cavities), after which they lose motility and differentiate heterocysts (Kimura and Nakano, 1990).

In *Azolla* the cyanobiont remains associated with the fern throughout both the asexual and sexual phases of the plant's life cycle. The inoculum for new leaf cavities resides on the apical meristem of each *Azolla* stem, in the form of a colony of heterocyst-free cyanobiont filaments, which resemble hormogonia, protected by the newly developing bi-lobed leaves (Lechno-Yossef and Nierzwicki-Bauer, 2002). In the sexual cycle the cyanobiont is packaged into the developing megasporocarps, where they differentiate into spore-like akinetes which germinate at the same time as the megasporocarp, providing the inoculum for the embryonic sporophyte.

It is not known how cyanobacteria gain entry to cycad precoralloid roots. A variety of possible entry points have been proposed, including breaks in the dermal layer, lenticels, and the papillose sheath (Costa and Lindblad, 2002; Vessey *et al.*, 2005). Prominent apical lenticels develop as a result of root invasion and once inside the root the cyanobacteria follow a continuous channel from the surface into the cortex.

Infection of *Gunnera* occurs via the channels between the papillae in the stem gland (Figure 16.1D and 16.4). In *Gunnera manicata*, removal of all, or all but one of the nine papillae, prevents infection (Uheda and Silvester, 2001). The carbohydrate-rich acidic mucus excreted by the gland gathers a population of *Nostoc*, fungi and bacteria. Under the influence of a factor in the mucilage compatible *Nostoc* strains are induced to form hormogonia which migrate down the channels between the papillae (Rai *et al.*, 2000; Bergman, 2002). In the interior of the gland the hormogonia (but not any bacteria and fungi present) penetrate the *Gunnera* cell walls and enter the cells, remaining external to the host plasmalemma (Figure 16.4). The mechanism of host cell wall penetration is not known. The hormogonia differentiate a high frequency of heterocysts and begin fixing nitrogen, while the gland channels disappear, rendering further infection impossible.

Colony formation and maturation

On entering extra- or intra-cellular symbioses with plants the phenotype of the cyanobiont is drastically changed as illustrated in Figure 16.5. Some of the changes have never been observed under any free-living conditions, suggesting that the plant environment or plant compounds may act as regulatory factors. Vital and fundamental cellular processes are affected, such as cell division, cell differentiation, carbon acquisition and nitrogen fixation and N-metabolism. In addition, the cyanobiont must be fed by the plant with all macro- and micronutrients required, except for nitrogen. This must also require cascades of regulatory molecular mechanisms to be altered and regulation must be executed at all organizational levels, from gene expression to protein synthesis and regulation of corresponding physiological activities. However, changes seen may not be executed at all organizational levels in all symbioses in the same way, and the degree of change may vary from symbiosis to symbiosis. As the ultimate outcome for all cyanobacterial-plant symbioses is always the same: high nitrogen fixation rates and the delivery of most of the nitrogen fixed by the cyanobiont to support the growth of the host plant, unifying concepts must operate.

Conditions offered *in planta*

Compared with their free-living growth state, cyanobacteria in symbiosis are subject to at least two major environmental changes: the light intensity is radically lower (in bryophytes and *Azolla*) or possibly zero (in cycads and *Gunnera*) and the oxygen tension is considerably reduced. The former requires the autotrophic cyanobacterium to adopt to a mixo- or hetero-trophic life-style, a capacity not held by all cyanobacteria. The latter is, for instance, manifested as an initially oxygen-sensitive nitrogen fixation in cyanobionts when first isolated from the host plant tissue (Rai *et al.*, 2000). The plant must also make sure that the cyanobiont is not spreading to all tissues and "out-competing" the host. Most often the cyanobiont only constitutes a small proportion of the total plant tissues and is kept strictly confined to the symbiosis-specific plant parts or tissues that mostly develop on first contact (Figure 16.1A, C, D, insets). The cyanobiont must also be fed by both Mo and Fe to sustain the nitrogen fixation process. Indeed, the cyanobionts do appear to be extremely slow growing, but are physiologically very active *in planta*. This suggests a highly efficient communication between the partners and a suite of concomittant and well-balanced physiological adaptations in both partners, to make the symbiosis function in time and space.

Changes to cyanobionts

It is now well established that the plant partners influence developmental processes and cyanobacterial functions. For instance, the development of hormogonia is radically depressed, whereas the development of heterocysts is radically stimulated (Figures 16.4 and 16.5). Additionally, cell division is first stimulated (during hormogonia development) and then slowed down *in planta* (Section "Hormogonia repression"). Photosynthesis is minimized or abolished (section "CO_2 fixation and carbohydrate transport"), while nitrogen fixation is stimulated and ammonium assimilation down-regulated (section "N_2 fixation and assimilation").

Hormogonia repression

To prevent the vegetative filaments from converting once more to the non-nitrogen-fixing, non-heterocystous hormogonial stage while inside the plant, a water-soluble hormogonium repressing factor (HRF) is administrated to the cyanobiont while in the cavities of the bryophyte *Anthoceros punctatus* (Figure 16.5) (Meeks and Elhai, 2002; Meeks, 2003). HRFs induce expression of the hormogonium repressing genes located in the *hrmUA* locus of *Nostoc punctiforme* strain ATCC 29133 (Cohen and Meeks, 1997). Regulation is achieved through the DNA binding properties of the HrmR protein, which prevent any new rounds of hormogonium development (Campbell *et al.*, 2003). The site of action is still unknown however, but HrmR binding may be regulated by a plant-derived sugar molecule. Three additional open reading frames of unknown function for symbiosis have also been discovered: *hrmI*, *hrmR* and *orfK*.

CO_2 fixation and carbohydrate transport

In two of the plant hosts, bryophytes and *Azolla*, the cyanobiont is known to receive light filtered through the chlorophyllous plant tissues surrounding the cyanobiont-filled cavities. This allows some photosynthetic activity (Adams, 2002; Lechno-Yossef and Nierzwicki-Bauer, 2002), although the vegetative cell frequency is lowered due to the high frequency of heterocysts formed. This may allow a photo-heterotrophic life-style. By contrast, light penetration into the stem glands of *Gunnera* and in the coralloid roots of cycads must be negligible (Figure 16.1C and D), and fully heterotrophic conditions prevail for the cyanobionts. An efficient transport of fixed carbon from the plant to the cyanobiont, whether extra-cellular in cavities or intra-cellular, must therefore proceed in all symbioses. The carbon compound being exported from the plant to support the cyanobionts is mostly not identified.

The primary route for fixation of carbon in cyanobacteria is, as in plants, the Calvin-cycle, with Rubisco being the major enzyme (see chapter 12). However, the Calvin-cycle may not be operative in symbiosis, and as in heterocysts, the oxidative pentose phosphate (OPP) pathway probably represents the essential catabolic route for providing reductant for nitrogen fixation and respiration (Summers *et al.*, 1995).

In spite of living in darkness in cycad coralloid roots, the cyanobionts retain a functional Rubisco and continue to fix CO_2 although at several-fold lower levels than when exposed to light (see Table 16.2, Costa and Lindblad, 2002). The fixation rates of the freshly isolated cycad cyanobionts are less than 5% of those for free-living *Nostoc* spp. when incubated under air or low O_2, the cyanobiont living heterotrophically on carbon provided by the plant.

Use of $^{14}CO_2$ has revealed that photosynthetic products synthesized in *Gunnera* are rapidly transferred to the cyanobiont as a compound of unknown composition (Söderbäck and Bergman, 1993). A distinct gradient is apparent, with actively nitrogen-fixing young glands receiving most carbohydrates. The numerous vascular bundles leading to and from the petioles to the symbiotic stem tissues are likely to function as the transfer conduits of plant-derived carbohydrates and other solutes (Bergman *et al.*, 1992). Glucose and fructose are the suggested carbohydrates transferred, as these support mixo- and heterotrophy in darkness (Wouters *et al.*, 2000; Black *et al.*, 2002). A putative cyclodextrin glycosyl transferase (CGTase), encoded by *cgt*, is present in symbiotically competent cyanobacteria and this gene is expressed in light and in darkness when the cyanobacterium is supplemented with fructose (Wouters *et al.*, 2003). CGTases belong to the α-amylase family that typically catalyses the hydrolysis of polymers of for instance α-D-glucose.

N_2 fixation and assimilation

Nitrogen fixation and assimilation are the ultimate key events and possibly the evolutionary driving force for the establishment of all cyanobacterial-plant symbioses.

The bryophyte symbiosis Nitrogen fixation is greatly enhanced in the cyanobiont entering symbioses with bryophytes, with rates being several-fold higher than for free-living cyanobacteria. The higher rates also correlate positively with the increased heterocyst frequency, which may reach 30–50% of the total cell population (Figure 16.5). The nitrogen fixed is efficiently released to the host as ammonia. As the overall glutamine synthetase (GS) levels are close to those for free-living cyanobacteria, but GS activities are reduced (see Table 16.1, Adams, 2002), some plant-dictated post-translational modifications have been proposed, although the mechanism is unknown. However, as the GS level in heterocysts, the site of primary ammonia assimilation, is specifically reduced, this may help explain the release. The ammonia released is subsequently assimilated by the plant-derived finger-like protrusions that grow into the symbiotic colony and the GS-GOGAT pathway of the host (Adams, 2002). Campbell and Meeks (1992) have shown that the oxygen tension is lower in *Anthoceros* cavities which may also permit the high nitrogen fixation activities noted.

The Azolla *symbiosis* Distinct gradients in physiological activities have been recorded for cyanobionts along *Azolla* plants (see Lechno-Yossef and Nierzwicki-Bauer, 2002). For instance, nitrogenase activity increases from the apex towards more mature leaves, as do ammonium levels, heterocyst frequencies and the percentage of vegetative cells containing cyanophycin granules, which are N-storage structures (Canini *et al.*, 1990). In addition, the cyanobiont population increases along the same axis, as does their cell size (Figure 16.5). About 50% of the nitrogen fixed is released by the cyanobiont and assimilated by the host and possibly also by the associated bacteria. It has been proposed that the primary branched hair cells stretching out into the cyanobacterial-filled cavities of *Azolla* play a central role in the assimilation of the nitrogen fixed by the cyanobionts. Indeed, immuno-labeling has shown that the GS levels are high both in the chloroplasts and the cytoplasm in the hair cells, while lower in hairs of cyanobiont-free *Azolla* plants (Uheda *et al.*, 2004).

The cycad symbiosis Lindblad and Bergman (1986) demonstrated that nitrogenase activities vary considerably from younger to older coralloid roots and that the highest activities were associated with the younger roots, or more apical parts of growing roots. Interestingly, a positive correlation between nitrogenase activity and individual heterocysts, i.e. heterocysts flanked by vegetative cells on both sides, was found, while older parts of the roots, where multiple heterocysts predominated, showed lower activity. This suggests a role for the vegetative cells in sustaining activity and that the presence of multiple heterocysts is not necessarily coupled to higher nitrogenase activity. Surprisingly, activities were considerably higher in light than in darkness in freshly isolated cyanobionts (kept at a low O_2 tension), possibly suggesting a limiting availability of energy/reductant in symbiosis (see Costa and Lindblad, 2002). As the amino acids citrulline and glutamine (in Zamiaceae) and glutamine (in Boweniaceae and Cycadaceae) are the N-solutes transported in the xylem from the *Nostoc*-colonized coralloid roots to the plant, these solutes have been inferred by Pate *et al.* (1998) as those being released from the cyanobiont, not ammonia as in other cyanobacterial-plant symbioses. This conclusion is also supported by the fact that the cyanobionts show high *in vitro* GS activity and GS protein levels similar to those in free-living cyanobacteria (Lindblad and Bergman, 1986).

The Gunnera *symbiosis* As in cycads, the nitrogenase activities are profoundly higher in cyanobionts living in *Gunnera*, at least in younger glands and at heterocyst frequencies of up to 20% (Söderbäck *et al.*, 1990). The nitrogen fixed by the cyanobiont is directly transported to the host as ammonia (Silvester *et al.*, 1996). Heterocyst differentiation is drastically stimulated (Rai *et al.*, 2000) as is also reflected in a several-fold increase in *hetR* expression (Wang *et al.*, 2004), the product of which is required for heterocyst development (Buikema and Haselkorn, 1991). Up to 70–80% heterocysts have been observed in *Nostoc* residing in older parts of the plant (Figure 16.5). Most of the fixed nitrogen is released by the cyanobiont and assimilated by the *Gunnera* plant (Silvester *et al.*, 1996) due to a reduced activity of cyanobacterial GS (Söderbäck and Bergman, 1992). As the plants are always infected in nature and as addition of combined nitrogen to cyanobacteria-infected *Gunnera tinctoria* plants does not eliminate the cyanobiont, this has been taken to suggest that the plant has a dependence on the cyanobiont and that the cyanobacterial-*Gunnera* symbiosis is evolving towards becoming an obligate symbiosis (Osborne *et al.*, 1992; Osborne and Sprent, 2002).

Changes to morphology and gene expression

Inhibition of hormogonia formation and the development of vegetative filaments with multiple heterocysts is obvious for all plant symbioses when viewed by light- and electron microscopy (Figure 16.5). Cell division is clearly restricted, but apparently not volume expansion, which leads to slow-growing inocula with larger cells (vegetative cells and heterocysts) compared with free-living *Nostoc* spp. (Meeks and Elhai, 2002). Gene expression studies of the partners in the intact symbioses are still rare, although a few recent attempts related to the cyanobiont have been mentioned above. For instance, Campbell *et al.* (1998) showed that a mutation in the alternative sigma factor *sigH* enhanced infection of the plant partner, *Anthoceros punctatus*, several-fold but led to no increase in nitrogen fixation. Three other *Nostoc punctiforme* mutants unable to differentiate heterocysts due to mutations in *ntcA*, a transcription factor regulating genes involved in N- (and C-) assimilation (Flores and Herrero, 2005), *hetR*, the master gene for heterocyst development, encoding a self-degrading serine protease (Buikema and Haselkorn, 1991), and *hetF*, possibly required for stimulating *hetR* expression (Wong and Meeks, 2001), were examined for ability to form a symbiosis with the bryophyte. Strains with mutations in *hetF* and *hetR* infected the hornwort but did not fix nitrogen, uncoupling infection from vital symbiotic performance. By contrast, the *ntcA* mutant lacked symbiotic competence and developed few hormogonia.

In addition, the expression level of genes related to N- and C-metabolism has been examined in *Nostoc*-colonized glands of the intact *Gunnera* symbioses, which represent an age gradient from newly infected to older glands (Wang *et al.*, 2004). Transcription of *ntcA* was apparent, although it was low in young, newly infected glands, but became over-expressed in cyanobionts of more mature glands. A similar expression pattern was seen for *hetR*, as was an increase in heterocyst frequency, *nifH* expression and the appearance of the NifH protein. It has also been shown that *hetR* expression is stimulated under dark heterotrophic conditions in cultured isolates (Wouters *et al.*, 2000). By contrast, the *glnB* gene and the corresponding P_{II} protein, involved in sensing the C and N status of the cell, showed a reverse pattern. Together these observations illustrate that these genes are affected in symbiosis and that the adaptations to a life in plant symbioses is executed via regulation of key-gene transcription levels. It is now of interest to understand whether these changes are driven by the conditions prevailing in the plants or by some specific plant compound(s).

Molecular mechanisms involved in the reduced cell division rate in cyanobionts are unknown, but the reduction may perhaps be merely a consequence of a plant-induced nutrient limitation, or be related to a reduced expression of the most critical cell division gene in prokaryotes, *ftsZ*. FtsZ levels decrease during the transition from vegetative cells into the terminally differentiated heterocysts (Klint *et al.*, 2007; Sakr *et al.*, 2006), the cell stage functionally dominating in symbiosis. The semi-regular spacing of single heterocysts seen in free-living cyanobacteria is also disturbed in cyanobionts, with multiple clusters of heterocysts being common. This is the phenotype of free-living cyanobacteria carrying extra copies of *hetR* (Buikema and Haselkorn, 1991) and in cyanobionts over-expressing *hetR* in *Gunnera* (Wang *et al.*, 2004). As the *hetC* gene product is involved in regulation of *ftsZ* expression (Wang and Xu, 2005) and the expression of the *patS* gene, the product of which prevents the differentiation of clustered heterocysts in free-living filaments (Golden and Yoon, 2003), the elucidation of the role of HetC in cyanobionts in plant symbioses is now of interest.

As the vegetative cells of cyanobionts have lost their function as CO_2 fixers and supplier of carbon, a role taken over by the plant, these may therefore be "allowed" to develop into heterocysts, combined nitrogen being in great demand by the plant. Together these C-replete and N-deplete conditions may induce changed C:N ratios and act as signals to initiate the over-expression of *ntcA* followed by *hetR* as noted in symbiosis (Wouters *et al.*, 2000; Wang *et al.*, 2004) and the increased heterocyst frequencies.

More recently, a proteomic approach was developed for cyanobionts to search for and identify novel proteins and metabolic pathways previously not known to be affected in cyanobacteria dwelling in plant symbiosis (Ekman *et al.*, 2006; Ran *et al.*, 2007). Initial comparative analyses of protein patterns show that the majority of the proteins synthesized under free-living conditions were expressed in the cyanobiont, indicating that most cellular processes operate in symbiosis, although a significant number of proteins were down-regulated, or were missing, in the symbiotic state, while others were more abundant or only present in symbiosis. These data are now being analyzed.

Changes to plant hosts

Compared to our knowledge about changes taking place in the cyanobionts, our knowledge about the plant hosts is rudimentary, in particular at the molecular level.

Morphological changes

Cyanobacteria potentially produce compounds involved in the initiation and establishment of symbioses. One such putative signaling compound is the recently identified phytohormone auxin, indole-3-acetic acid (Sergeeva *et al.*, 2002). Mitotic activities are induced in *Gunnera* cells at the bottom of the stem glands in the presence of compatible hormogonia (Bergman *et al.*, 1996). Other known phytohormones related to plant development may also be involved as cyanobacteria apparently are able to synthesize most (Liaimer and Bergman, 2004). Cyanobacterial phytohormones, or other compounds may also elicit the development of the finger-like hairs or protrusions extending from the cavity walls and used for exchange of metabolites in bryophytes and *Azolla* (see Rai *et al.*, 2002). Such compounds may also have a role in the development of the dichotomously dividing coralloid roots, originating from the pericycle of the main root and showing an apogeotropic growth behavior, and the full development of the cyanobacterial root zone (Figure 16.1C; Vessey *et al.*, 2005). Such findings robustly support a role for the cyanobacterium in the creation and shaping of the symbiotic plant "organs".

Changes in gene expression

So far, no host plant genes have been identified (or searched for?) as being of symbiotic significance, such as equivalents to "nodulation" genes of legumes forming symbioses with rhizobia (Udvardi *et al.*, 2005).

The symbiotic tissue/organ

After infection has been completed in bryophytes, cycads and *Gunnera*, the cyanobionts always remain distinctly enclosed and restricted to the plant structures or tissues that developed to harbor the cyanobacterium: the cavities or auricles distributed in or over the bryophyte thalli, the special cyanobacterial zone in the coralloid cycad roots, and the "stem-nodules" within the *Gunnera* stems (Figure 16.1). This is also the case in the perpetually maintained *Azolla* symbiosis, in which the cyanobacteria are restricted to the leaf cavities. The important role of these specific plant structures for the maintenance of the symbiosis is further illustrated in the development of structures that facilitate interactions, such as the exchange of nutrients, between the partners (see sections "*CO_2 fixation and carbohydrate transport*," "*N_2 fixation and assimilation*," and "*The bryophyte symbiosis*"). Additionally, in the *Gunnera* symbioses the plant plasma membrane always surrounds all invading *Nostoc* filaments, which points to a drastically enhanced plant membrane synthesis upon infection. This surface must function as the barrier between the partners and be traversed by the nutrients being exchanged (N and C), the mechanisms of which are still to be discovered.

Future directions

Further studies are needed to get a deeper insight into the evolution of cyanobacterial-plant symbioses as well as to understand ongoing life-styles of cyanobionts living in plants. Sequencing of additional cyanobacterial genomes, besides that of *Nostoc punctiforme* strain ATCC 29133, will allow genome comparisons in the search for symbiosis-specific genes or traits. A full understanding of hormogonia development and the regulation of this and related processes such as cell division and heterocyst differentiation in symbiosis is required, as is identification of mechanisms involved in the enhanced heterocyst differentiation. A better understanding of signaling and of communication events between the cyanobacterium and plants on first contact is required to better understand mechanisms involved in processes from recognition to adaptations in the cyanobiont and in the plant. Another critical issue is to identify mechanisms related to the release of the fixed nitrogen, *the* central event in all cyanobacterial-plant symbioses.

Acknowledgments

Financial support from the Swedish Research Council (to B. B.) is acknowledged.

References

Adams, D.G. (2000). Symbiotic interactions. In Ecology of Cyanobacteria: Their Diversity in Time and Space, B. Whitton and M. Potts, eds. (Dordrecht: Kluwer Academic Publishers), pp. 523–561.

Adams, D.G. (2002). Cyanobacteria in symbiosis with hornworts and liverworts. In Cyanobacteria in Symbiosis, A.N. Rai, B. Bergman, and U. Rasmussen, eds. (Dordrecht: Kluwer Academic Publishers), pp. 117–135.

Adams, D.G., and Duggan, P.S. (1999). Heterocyst and akinete differentiation in cyanobacteria. New Phytol. *144*, 3–33.

Adams, D.G., and Duggan, P.S. (2005). Symbiosis between cyanobacteria and the liverwort *Blasia* as a model system. In Abstracts of the 1st International Conference on Plant-Microbe Interactions: Endophytes and Biocontrol Agents (Finland), pp. 32.

Babic, S. (1996). Hormogonia formation and the establishment of symbiotic associations between cyanobacteria and the bryophytes *Blasia* and *Phaeoceros* (PhD thesis, University of Leeds, England).

Baker, J.A., Entsch, B., and McKay, D.B. (2003). The cyanobiont in an *Azolla* fern is neither *Anabaena* nor *Nostoc*. FEMS Microbiol. Lett. *229*, 43–47.

Belnap, J. (2001). Factors influencing nitrogen fixation and nitrogen release in biological soil crusts. In Biological Soil Crusts: Structure, Function, and Management, J. Belnap and O.L. Lange, eds. (New York: Springer-Verlag), pp. 241–261.

Bergman, B. (2002). The *Nostoc-Gunnera* symbiosis. In Cyanobacteria in Symbiosis, A.N. Rai, B. Bergman, and U. Rasmussen, eds. (Dordrecht: Kluwer Academic Publishers), pp. 207–232.

Bergman, B., and Osborne, B. (2002). The *Gunnera-Nostoc* symbiosis. In Commentaries on cyanobacterial symbioses, B. Osborne, ed. (Dublin: Royal Irish Academy), pp. 35–39.

Bergman, B., Johansson, C., and Söderbäck, E. (1992). The *Nostoc-Gunnera* symbiosis. New Phytol. *122*, 379–400.

Bergman, B., Matveyev, A., and Rasmussen, U. (1996). Chemical signalling in cyanobacterial-plant symbioses. Trends Plant Sci. *1*, 191–197.

Bergman, B., Rasmussen, U., and Rai, A.N. (2007). Cyanobacterial Associations. In Associative and Endophytic Nitrogen-fixing Bacteria and Cyanobacterial Associations. C. Elmerich, and W.E. Newton, eds. (Dordrecht: Kluwer Academic Publishers). Vol. 5, pp. 257–301.

Bhaya, D. (2004). Light matters: phototaxis and signal transduction in unicellular cyanobacteria. Mol. Microbiol. *53*, 745–754.

Black, K.G., Parsons, R., and Osborne, B.A. (2002). Uptake and metabolism of glucose in the *Nostoc-Gunnera* symbiosis. New Phytol. *153*, 297–305.

Brenner, E.D., Stevenson, D.W., and Twigg, R.W. (2003). Cycads: evolutionary innovations and the role of plant-derived neurotoxins. Trends Plant Sci. *8*, 446–452.

Buikema, W.J., and Haselkorn, R. (1991). Characterization of a gene controlling heterocyst differentiation in the cyanobacterium *Anabaena* PCC 7120. Genes Dev. *5*, 321–330.

Campbell, E.L., and Meeks, J.C. (1992). Evidence for plant-mediated regulation of nitrogenase expression in the *Anthoceros-Nostoc* symbiotic association. J. Gen. Microbiol. *138*, 473–480.

Campbell, E.L., Brahamsha, B., and Meeks, J.C. (1998). Mutation of an alternative sigma factor in the cyanobacterium *Nostoc punctiforme* results in increased infection of its symbiotic plant partner, *Anthoceros punctatus*. J. Bacteriol. *180*, 4938–4941.

Campbell, E.L., Wong, F.C.Y., and Meeks, J.C. (2003). DNA binding properties of the HrmR protein of *Nostoc punctiforme* responsible for transcriptional regulation of genes involved in the differentiation of hormogonia. Mol. Microbiol. *47*, 573–582.

Canini, A., Caiola, M.G., and Mascini, M. (1990). Ammonium content, nitrogenase activity and heterocyst frequency within the leaf cavities of *Azolla-filiculoides* Lam. FEMS Microbiol. Lett. *71*, 205–210.

Chiu, W.L., Peters, G.A., Levieille, G., Still, P.C., Cousins, S., Osborne, B., and Elhai, J. (2005). Nitrogen deprivation stimulates symbiotic gland development in *Gunnera manicata*. Plant Physiol. *139*, 224–230.

Choudhury, A.T.M.A., and Kennedy, I.R. (2004). Prospects and potentials for systems of biological nitrogen fixation in sustainable rice production. Biology and Fertility of Soils *39*, 219–227.

Cohen, M.F., and Meeks, J.C. (1997). A hormogonium regulating locus, *hrmUA*, of the cyanobacterium *Nostoc punctiforme* strain ATCC 29133 and its response to an extract of a symbiotic plant partner *Anthoceros punctatus*. Mol. Plant. Microbe Interact. *10*, 280–289.

van Coppenolle, B., McCouch, S.R., Watanabe, I., Huang, N., and Van Hove, C. (1995). Genetic diversity and phylogeny analysis of *Anabaena-Azollae* based on RFLPs detected in *Azolla-Anabaena azollae* DNA complexes using nif gene probes. Theor. Appl.Gen. *91*, 589–597.

Costa, J.L., and Lindblad, P. (2002). Cyanobacteria in symbiosis with cycads. In Cyanobacteria in Symbiosis, A.N. Rai, B. Bergman, and U. Rasmussen, eds. (Dordrecht: Kluwer Academic Publishers), pp. 195–206.

Costa, J. L., Paulsrud, P., and Lindblad, P. (1999). Cyanobiont diversity within coralloid roots of selected cycad species. FEMS Microbiol. Ecol. *28*, 85–91.

Costa, J.L., Paulsrud, P., Rikkinen, J., and Lindblad, P. (2001). Genetic diversity of *Nostoc* symbionts endophytically associated with two bryophyte species. Appl. Environ. Microbiol. *67*, 4393–4396.

DeLuca, T.H., Zackrisson, O., Nilsson, M.C., and Sellstedt, A. (2002). Quantifying nitrogen-fixation in feather moss carpets of boreal forests. Nature *419*, 917–920.

Dodds, W.K., Gudder, D.A., and Mollenhauer, D. (1995). The ecology of *Nostoc*. J. Phycol. *31*, 2–18.

Ekman, M., Tollbäck, P., Klint, J., and Bergman, B. (2006). Protein expression profiles in an endosymbiotic cyanobacterium revealed by a proteomic approach. Mol. Plant Microb. Interact. *19*, 1251-1261.

Fay, P. (1992). Oxygen relations of nitrogen fixation in cyanobacteria. Microbiol. Rev. *56*, 340–373.

Flores, E., and Herrero, A. (2005). Nitrogen assimilation and nitrogen control in cyanobacteria. Biochem. Soc. Trans. *33*, 164–167.

Forni, C., Riov, J., Caiola, M.G., and Telor, E. (1992). Indole-3-Acetic-Acid (IAA) production by *Arthrobacter* species isolated from *Azolla*. J. Gen. Microbiol. *138*, 377–381.

Foster, R.A., Collier, J.L., and Carpenter, E.J. (2006a). Reverse transcription PCR amplification of cyanobacterial symbiont 16S rRNA sequences from single non-photosynthetic eukaryotic marine planktonic host cells. J. Phycol. *42*, 243–250.

Foster, R.A., Carpenter, E.J., and Bergman, B. (2006b). Unicellular cyanobionts in open ocean dinoflagellates, radiolarians, and tintinnids: Ultrastructural characterization and immuno-localization of phycoerythrin and nitrogenase. J. Phycol. *42*, 453–463.

Gallon, J.R. (1992). Reconciling the incompatible – N_2 fixation and O_2. New Phytol. 122, 571–609.

Gebhardt, J.S., and Nierzwicki-Bauer, S.A. (1991). Identification of a common cyanobacterial symbiont associated with *Azolla* spp. through molecular and morphological characterization of free-living and symbiotic cyanobacteria. Appl. Environ. Microbiol. *57*, 2141–2146.

Gentili, F., Nilsson, M.C., Zackrisson, O., DeLuca, T.H., and Sellstedt, A. (2005). Physiological and molecular diversity of feather moss associative N_2-fixing cyanobacteria. J. Exp. Bot. *56*, 3121–3127.

Golden, J.W., and Yoon, H.S. (2003). Heterocyst development in *Anabaena*. Curr. Opin. Microbiol. *6*, 557–563.

Grobbelaar, N., Scott, W.E., Hattingh, W., and Marshall, J. (1987). The identification of the coralloid root endophytes of the Southern African cycads and the ability of the isolates to fix dinitrogen. South Afr. J. B. *53*, 111–118.

Guevara, R., Armesto, J.J., and Caru, M. (2002). Genetic diversity of *Nostoc* microsymbionts from *Gunnera tinctoria* revealed by PCR-STRR fingerprinting. Microb. Ecol. *44*, 127–136.

Herrero, A., Muro-Pastor, A.M., Valladares, A., and Flores, E. (2004). Cellular differentiation and the NtcA transcription factor in filamentous cyanobacteria. FEMS Microbiol. Rev. *28*, 469–487.

van Hove, C., and Lejeune, A. (2002). Applied aspects of *Azolla-Anabaena* symbiosis. In Cyanobacteria in Symbiosis, A.N. Rai, B. Bergman, and U. Rasmussen, eds. (Dordrecht: Kluwer Academic Publishers), pp. 179–193.

Janson, S., Rai, A.N., and Bergman, B. (1995). Intracellular Cyanobiont *Richelia intracellularis* ultrastructure and immuno-localization of phycoerythrin, nitrogenase, Rubisco and glutamine synthetase. Mar. Biol. *124*, 1–8.

Janson, S., Wouters, J., Bergman, B., and Carpenter, E.J. (1999). Host specificity in the *Richelia*-diatom symbiosis revealed by *hetR* gene sequence analysis. Environ. Microbiol. *1*, 431–438.

Johansson, C., and Bergman, B. (1994). Reconstitution of the symbiosis of *Gunnera manicata* Linden-cyanobacterial specificity. New Phytol. *126*, 643–652.

Kim, J.J.H., Krawczyk, K., Lorentz, W.P., and Zimmerman, W.J. (1997). Fingerprinting cyanobionts and hosts of the *Azolla* symbiosis by DNA amplification. World J. Microbiol. & Biotechnol. *13*, 97–101.

Kimura, J., and Nakano, T. (1990). Reconstitution of a *Blasia-Nostoc* symbiotic association under axenic conditions. Nova Hedwigia *50*, 191–200.

Kistner, C., and Parniske, M. (2002). Evolution of signal transduction in intracellular symbiosis. Trends in Plant Science *7*, 511–518.

Klint, J., Rasmussen, U., and Bergman, B. (2007). FtsZ may have dual roles in the filamentous cyanobacterium *Nostoc/Anabaena* sp. strain PCC 7120. J. Plant Physiol. *164*, 11–18.

Klint, J., Ran, L., Rasmussen, U., and Bergman, B. (2006). Identification of developmentally regulated proteins in cyanobacterial hormogonia using a proteomic approach. Symbiosis *41*, 87–95.

Knight, C.D., and Adams, D.G. (1996). A method for studying chemotaxis in nitrogen fixing cyanobacterium-plant symbioses. Physiol. Mol. Plant Pathol. *49*, 73–77.

Lechno-Yossef, S., and Nierzwicki-Bauer, S.A. (2002). *Azolla-Anabaena* symbiosis. In Cyanobacteria in Symbiosis, A.N. Rai, B. Bergman, and U. Rasmussen, eds. (Dordrecht: Kluwer Academic Publishers), pp. 153–178.

Liaimer, A., and Bergman, B. (2004). Phytohormones in cyanobacteria: occurrence and perspectives. In Biology of Plant-Microbe Interactions; International Society for Molecular Plant-Microbe Interactions, I. Tikhonovich, B. Lugtenberg, and N. Provorov, eds (St Paul Minnesota USA), pp. 394–397.

Lindblad, P., and Bergman, B. (1986). Glutamine synthetase: activity and localization in cyanobacteria of the cycads *Cycas revoluta* and *Zamia skinneri*. Planta *169*, 1–7.

Lindblad, P., Bergman, B., and Nierzwicki-Bauer, S.A. (1991). Immunocytochemical localization of nitrogenase in bacteria symbiotically associated with *Azolla* spp. Appl. Environ. Microbiol. *57*, 3637–3640.

Lindblad, P., Haselkorn, R., Bergman, B., and Nierzwicki-Bauer, S.A. (1989). Comparison of DNA restriction fragment length polymorphisms of *Nostoc* strains in and from cycads. Arch. Microbiol. *152*, 20–24.

Mandal, B., Vlek, P.L.G., and Mandal, L.N. (1999). Beneficial effects of blue-green algae and *Azolla*, excluding supplying nitrogen, on wetland rice fields: a review. Biol. Fertil. Soils *28*, 329–342.

Meeks, J.C. (1998). Symbiosis between nitrogen-fixing cyanobacteria and plants: The establishment of symbiosis causes dramatic morphological and physiological changes in the cyanobacterium. Bioscie. *48*, 266–276.

Meeks, J.C. (2003). Symbiotic interactions between *Nostoc punctiforme*, a multicellular cyanobacterium, and the hornwort *Anthoceros punctatus*. Symbiosis *35*, 55–71.

Meeks, J.C., and Elhai, J. (2002). Regulation of cellular differentiation in filamentous cyanobacteria in free-living and plant-associated symbiotic growth states. Microbiol. Mol. Biol. Rev. *66*, 94–121.

Meeks, J.C., Campbell, E.L., Summers, M.L., and Wong, F.C. (2002). Cellular differentiation in the cyanobacterium *Nostoc punctiforme*. Arch. Microbiol. *178*, 395–403.

Meeks, J.C., Elhai, J., Thiel, T., Potts, M., Larimer, F., Lamerdin, J., Predki, P., and Atlas, R. (2001). An overview of the genome of *Nostoc punctiforme*, a multicellular, symbiotic cyanobacterium. Photosyn. Res. *70*, 85–106.

Nilsson, M., Bergman, B., and Rasmussen, U. (2000). Cyanobacterial diversity in geographically related and distant host plants of the genus *Gunnera*. Arch. Microbiol. *173*, 97–102.

Nilsson, M., Rasmussen, U., and Bergman, B. (2005). Competition among symbiotic cyanobacterial *Nostoc* strains forming artificial associations with rice (*Oryza sativa*). FEMS Microbiol. Lett. *245*, 139–144.

Nilsson, M., Rasmussen, U., and Bergman, B. (2006). Cyanobacterial chemotaxis to extracts of host and nonhost plants. FEMS Microbiol. Ecol. 55, 382–390.

Osborne, B.A., and Sprent, J.I. (2002). *Nostoc-Gunnera* symbiosis: ecology. In Cyanobacteria in Symbiosis, A.N. Rai, B. Bergman, and U. Rasmussen, edss (Dordrecht: Kluwer Academic Publishers), pp. 233–251.

Osborne, B.A., Cullen, A., Jones, P.W., and Campbell, G.J. (1992). Use of nitrogen by the *Nostoc-Gunnera tinctoria* (Molina) mirbel symbiosis. New Phytol. *120*, 481–487.

Ow, M.C., Gantar, M., and Elhai, J. (1999). Reconstitution of a cycad-cyanobacterial association. Symbiosis *27*, 125–134.

Pate, J.S., Lindblad, P., and Atkins, C.A. (1988). Pathways of assimilation and transfer of the fixed nitrogen in coralloid roots of cycad-*Nostoc* symbioses. Planta *176*, 461–471.

Plazinski, J., Croft, L., Taylor, R., Zheng, Q., Rolfe, B.G., and Gunning, B.E.S. (1991). Indigenous plasmids in *Anabaena-Azollae*: their taxonomic distribution and existence of regions of homology with symbiotic genes of *Rhizobium*. Can. J. Microbiol. *37*, 171–181.

Rai, A.N., Söderbäck, E., and Bergman, B. (2000). Cyanobacterium-plant symbiosis. New Phytol. *116*, 449–481.

Rai, A.N., Bergman, B., and Rasmussen, U. (2002). Cyanobacteria in Symbiosis. (Dordrecht: Kluwer Academic Publishers).

Ran, L., Huang, F., Ekman, M., Klint, J., and Bergman, B. (2007). Proteomic analyses of the photoauto- and diazotrophically grown cyanobacterium *Nostoc* sp. PCC 73102. Microbiology 153, 608–618.

Rasmussen, U., and Svenning, M.M. (2001). Characterization by genotypic methods of symbiotic *Nostoc* strains isolated from five species of *Gunnera*. Arch. Microbiol. *176*, 204–210.

Rasmussen, U., and Nilsson, M. (2002). Cyanobacterial diversity and specificity in plant symbioses. In Cyanobacteria in Symbiosis, A.N. Rai, B. Bergman, and U. Rasmussen, eds. (Dordrecht: Kluwer Academic Publishers), pp. 329–347.

Rasmussen, U., Johansson, C., and Bergman, B. (1994). Early communication in the *Gunnera-Nostoc* symbiosis: plant-induced cell-differentiation and protein-synthesis in the Cyanobacterium. Mol. Plant. Microbe Interact. 7, 696–702.

Rikkinen, J. (2002). Cyanolichens: an evolutionary overview. In Cyanobacteria in Symbiosis, A.N. Rai, B. Bergman, and U. Rasmussen, eds. (Dordrecht: Kluwer Academic Publishers), pp. 31–72.

Rippka, R., Deruells, J., Waterbury, J.B., Herdman, M., and Stanier, R. (1979). Generic assignment, strain histories and properties of pure cultures of cyanobacteria. J. Gen. Microbiol. *111*, 1–61.

Sakr, S., Jeanjean, R., Zhang, C.C., and Arcondeguy, T. (2006). Inhibition of cell division suppresses heterocyst development in *Anabaena* sp, strain PCC 7120. J. Bacteriol. *188*, 1396–1404.
Schneider, D., Wink, M., Sporer, F., and Lounibos, P. (2002). Cycads: their evolution, toxins, herbivores and insect pollinators. Naturwissenschaften *89*, 281–294.
Sergeeva, E., Liaimer, A., and Bergman, B. (2002). Evidence for production of the phytohormone indole-3-acetic acid by cyanobacteria. Planta *215*, 229–238.
Silvester, W.B., Parsons, R., and Watt, P.W. (1996). Direct measurement of release and assimilation of ammonia in the *Gunnera-Nostoc* symbiosis. New Phytologist *132*, 617–625.
Solheim, B., and Zielke, M. (2002). Associations between cyanobacteria and mosses. In Cyanobacteria in Symbiosis, A.N. Rai, B. Bergman, and U. Rasmussen, eds. (Dordrecht: Kluwer Academic Publisher), pp. *137*–152.
Summers, M.L., Wallis, J.G., Campbell, E.L., and Meeks, J.C. (1995). Genetic-evidence of a major role for glucose-6-phosphate dehydrogenase in nitrogen-fixation and dark growth of the cyanobacterium *Nostoc* sp. strain ATCC 29133. J. Bacteriol. *177*, 6184–6194.
Svenning, M.M., Eriksson, T., and Rasmussen, U. (2005). Phylogeny of symbiotic cyanobacteria within the genus *Nostoc* based on 16S rDNA sequence analyses. Arch. Microbiol. *183*, 19–26.
Söderbäck, E., and Bergman, B. (1992). The *Nostoc-Gunnera magellanica* symbiosis: phycobiliproteins, carboxysomes and Rubisco in the cyanobiont. Physiol. Plant. 84, 425–432.
Söderbäck, E., and Bergman, B. (1993). The *Nostoc-Gunnera* symbiosis: carbon fixation and translocation. Physiol. Plant. *89*, 125–132.
Söderbäck, E., Lindblad, P., and Bergman, B. (1990). Developmental patterns related to nitrogen-fixation in the *Nostoc-Gunnera magellanica* Lam symbiosis. Planta *182*, 355–362.
Taylor, T.N., and Krings, M. (2005). Fossil microorganisms and land plants: Associations and interactions. Symbiosis *40*, 119–135.
Taylor, T.N., Hass, H., and Kerp, H. (1997). A cyanolichen from the lower devonian rhynie chert. Am. J. Bot. *84*, 992–1004.
Tomitani, A., Knoll, A.H., Cavanaugh, C.M., and Ohno, T. (2006). The evolutionary diversification of cyanobacteria: molecular-phylogenetic and paleontological perspectives. Proc. Natl. Acad. Sci. USA *103*, 5442–5447.
Udvardi, M.K., Tabata, S., Parniske, M., and Stougaard, J. (2005). *Lotus japonicus*: legume research in the fast lane. Trends Plant Sci. *10*, 222–228.
Uheda, E., and Silvester, W.B. (2001). The role of papillae during the infection process in the *Gunnera-Nostoc* symbiosis. Plant Cell Physiol. *42*, 780–783.
Uheda, E., Maejima, K., and Shiomi, N. (2004). Localization of glutamine synthetase isoforms in hair cells of *Azolla* leaves. Plant Cell Physiol. *45*, 1087–1092.
Vessey, J.K., Pawlowski, K., and Bergman, B. (2005). Root-based N_2-fixing symbioses: legumes, actinorhizal plants, *Parasponia* sp. and cycads. Plant Soil *274*, 51–78.
Veys, P., Lejeune, A., and Van Hove, C. (2000). The pore of the leaf cavity of *Azolla*: interspecific morphological differences and continuity between the cavity envelopes. Symbiosis *29*, 33–47.
Veys, P., Lejeune, A., and Van Hove, C. (2002). The pore of the leaf cavity of *Azolla* species: teat cell differentiation and cell wall projections. Protoplasma *219*, 31–42.
Veys, P., Waterkeyn, L., Lejeune, A., and Van Hove, C. (1999). The pore of the leaf cavity of *Azolla*: morphology, cytochemistry and possible functions. Symbiosis *27*, 33–57.
Villarreal, J.C., and Renzaglia, K.S. (2006). Structure and development of *Nostoc* strands in *Leiosporoceros dussii* (Anthocerotophyta): a novel symbiosis in land plants. Am. J. Bot. *93*, 693–705.
Villarreal, J.C., Duff, R.J., and Renzaglia, K.S. (2005). Anatomical and ultrastructural innovations in *Leiosporoceros dussii* (Steph.) Hässel. In Abstract of XVII International Botanical Congress (Vienna, Austria).
Wang, C.M., Ekman, M., and Bergman, B. (2004). Expression of cyanobacterial genes involved in heterocyst differentiation and dinitrogen fixation along a plant symbiosis development profile. Mol. Plant. Microbe Interact. *17*, 436–443.
Wang, Y., and Xu, X.D. (2005). Regulation by *het*C of genes required for heterocyst differentiation and cell division in *Anabaena* sp strain PCC 7120. J. Bacteriol. *187*, 8489–8493.
Wanntorp, L., Wanntorp, H.E., and Kallersjo, M. (2002). Phylogenetic relationships of *Gunnera* based on nuclear ribosomal DNA ITS region, *rbcL* and rps16 intron sequences. Syst. Bot. *27*, 512–521.

Watts, S., Knight, C.D., and Adams, D.G. (1999). Characterization of plant exudates inducing chemotaxis in nitrogen-fixing cyanobacteria. In The Phototrophic Prokaryotes, G.A. Peschek, W. Löffelhardt, and G. Schmetterer, eds. (New York: Kluwer Academic/Plenum Publishers), pp. 679–684.

West, N.J., and Adams, D.G. (1997). Phenotypic and genotypic comparison of symbiotic and free-living cyanobacteria from a single field site. Appl. Environ. Microbiol. *63*, 4479–4484.

Wolk, C.P. (1996). Heterocyst formation. Annu. Rev. Genet. *30*, 59–78.

Wong, F.C.Y., and Meeks, J.C. (2001). The *hetF* gene product is essential to heterocyst differentiation and affects HetR function in the cyanobacterium *Nostoc punctiforme*. J. Bacteriol. *183*, 2654–2661.

Wong, F.C.Y., and Meeks, J.C. (2002). Establishment of a functional symbiosis between the cyanobacterium *Nostoc punctiforme* and the bryophyte *Anthoceros punctatus* requires genes involved in nitrogen control and initiation of heterocyst differentiation. Microbiol. *148*, 315–323.

Wouters, J., Janson, S., and Bergman, B. (2000). The effect of exogenous carbohydrates on nitrogen fixation and *hetR* expression in *Nostoc* PCC 9229 forming symbiosis with *Gunnera*. Symbiosis *28*, 63–76.

Wouters, J., Bergman, B., and Janson, S. (2003). Cloning and expression of a putative cyclodextrin glucosyltransferase from the symbiotically competent cyanobacterium *Nostoc* sp. PCC 9229. FEMS Microbiol. Lett. *219*, 181–185.

Yuan, X.L., Xiao, S.H., and Taylor, T.N. (2005). Lichen-like symbiosis 600 million years ago. Science *308*, 1017–1020.

Zackrisson, O., DeLuca, T.H., Nilsson, M.C., Sellstedt, A., and Berglund, L.M. (2004). Nitrogen fixation increases with successional age in boreal forests. Ecology *85*, 3327–3334.

Zheng, W.W., Nilsson, M., Bergman, B., and Rasmussen, U. (1999). Genetic diversity and classification of cyanobacteria in different *Azolla* species by the use of PCR fingerprinting. Theor. Appl. Gen. *99*, 1187–1193.

Zheng, W.W., Song, T.Y., Bao, X.D., Bergman, B., and Rasmussen, U. (2002). High cyanobacterial diversity in coralloid roots of cycads revealed by PCR fingerprinting. FEMS Microbiol. Ecol. *40*, 215–222.

Zielke, M., Solheim, B., Spjelkavik, S., and Olsen, R.A. (2005). Nitrogen fixation in the high arctic: Role of vegetation and environmental conditions. Arctic Antarctic and Alpine Res. *37*, 372–378.

Zielke, M., Ekker, A.S., Olsen, R.A., Spjelkavik, S., and Solheim, B. (2002). The influence of abiotic factors on biological nitrogen fixation in different types of vegetation in the High Arctic, Svalbard. Arctic Antarctic and Alpine Res. *34*, 293–299.

Zimmerman, W.J., and Rosen, B.H. (1992). Cyanobiont diversity within and among cycads of one field site. Can. J. Microbiol. *38*, 1324–1328.

Index

Gene index

Species index

A

C

D

F

G

H

K

L

M

N

O

P

R

S

Subject index

Q

R

S

T

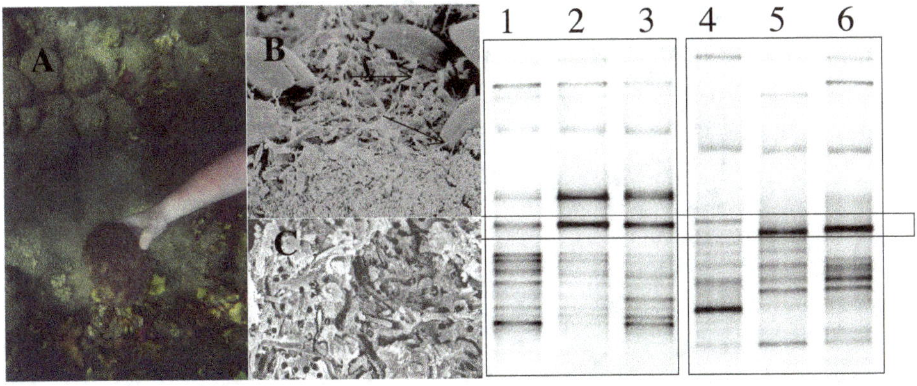

Plate 4.1 (see also Figure 4.1) Fingerprinting of cyanobacterial communities. DGGE fingerprints specific for oxygenic phototrophs done on calcified communities of cyanobacteria found as rolling oncolites (stromatolites) on the bottom of Rio Mezquites, Cuatro Ciénegas, Mexico. Each lane in the gel corresponds to one oncolite, such as that being picked up from the bottom of the river in A. Each band in this community "bar-code" represents one allele of the 16S rRNA found in the complex mixture that has been PCR-amplified from DNA extracts. The bands are separated on the gel on the basis of sequence differences. Strong bands correspond to abundant alleles. The fingerprints allow the immediate realization that oncolites collected at the same site (lanes 1, 2, 3), just meters apart, possess a common pattern of community structure. But shifts in the presence and intensity of particular alleles are also conspicuous as oncolites kilometers downstream are analyzed (lanes 4, 5 and 6). A microscopic investigation of these communities will reveal the presence of surface diatoms (B) and two major morphotypes of cyanobacteria encrusted in the carbonate matrix: one *Calothrix*-like heterocystous filament (B), only the terminal hairs are visible here (arrows) sticking out of the surface, and a thin filamentous, non-heterocystous *Phormidium/Leptolyngbya* (C). Excision, re-amplification and sequencing of the DGGE bands identify those in the upper half of the gel as diatom plastids, the intense band marked by the rectangle as a heterocystous cyanobacterium, and most of the rest as diverse sequences matching best a variety of thin-filamentous cyanobacteria. Data and photographs are from Garcia-Pichel *et al.* (2004) and Elser *et al.* (2005); the DGGE fingerprint is from Wade, B., and Garcia-Pichel, F., unpublished.

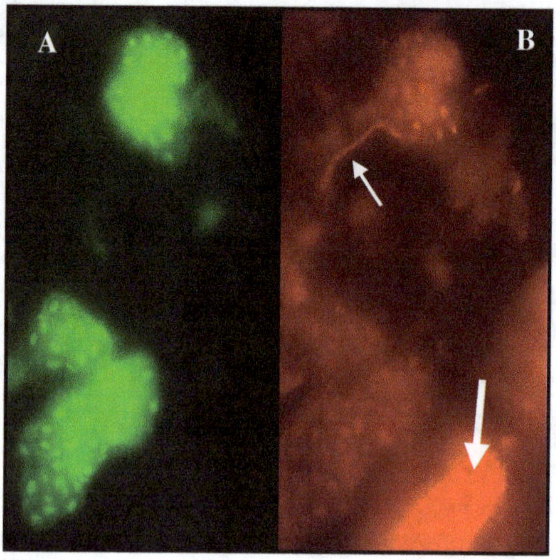

Plate 4.2 (see also Figure 4.2) FISH analysis of a cryptic picobenthic cyanobacterium from microbial mats. A: Fluorescence microscopy of a tiramide-enhanced FISH preparation. B: same frame under autofluorescence by photosynthetic pigments. Note the low levels of fluorescence compared to that of *Microcoleus chthonoplastes* filaments (thick arrow), and the negative hybridization of the probe to small and large filamentous cyanobacteria (thin arrow). Sequences for this previously unidentified cyanobacterium were retrieved from DGGE community fingerprints similar to those in Plate 4.1, and used to design specific oligonucleotide probes. After testing the specificity of the oligonucleotide probes as Southern probes on the DGGE gels, a group of very small unicellular cyanobacteria with low autofluorescence per cell, with the morphology of "*Aphanothece*" could be easily identified as the "missing" cyanobacterium. This picobenthic cyanobacterium has so far eluded cultivation. Data modified from Abed and colleagues (Abed *et al.*, 2002c)

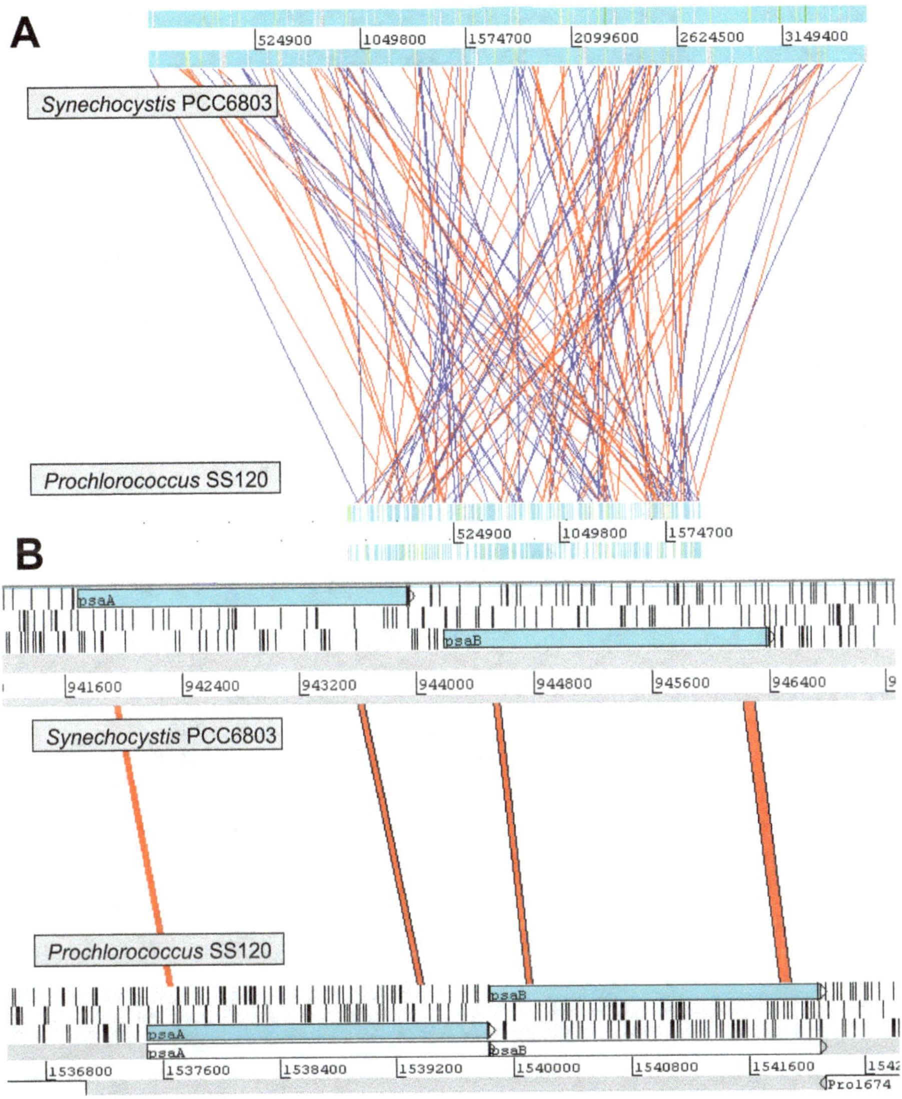

Plate 5.1 (see also Figure 5.1) A. Alignment of the *Synechocystis* sp. PCC 6803 genomic sequence (top) and *Prochlorococcus marinus* SS120 sequence (bottom). The lines connect sequences with a BLASTN E value < 10^{-2}, however, except very few stretches in ultra-conserved genes and some tRNA genes, there is no sequence similarity at the DNA level. There is clearly no synteny in the organization of the two genomes. **B.** Detail from A, photosynthesis genes such as *psaA-psaB* contain short stretches of conserved nucleotide sequence similarity.

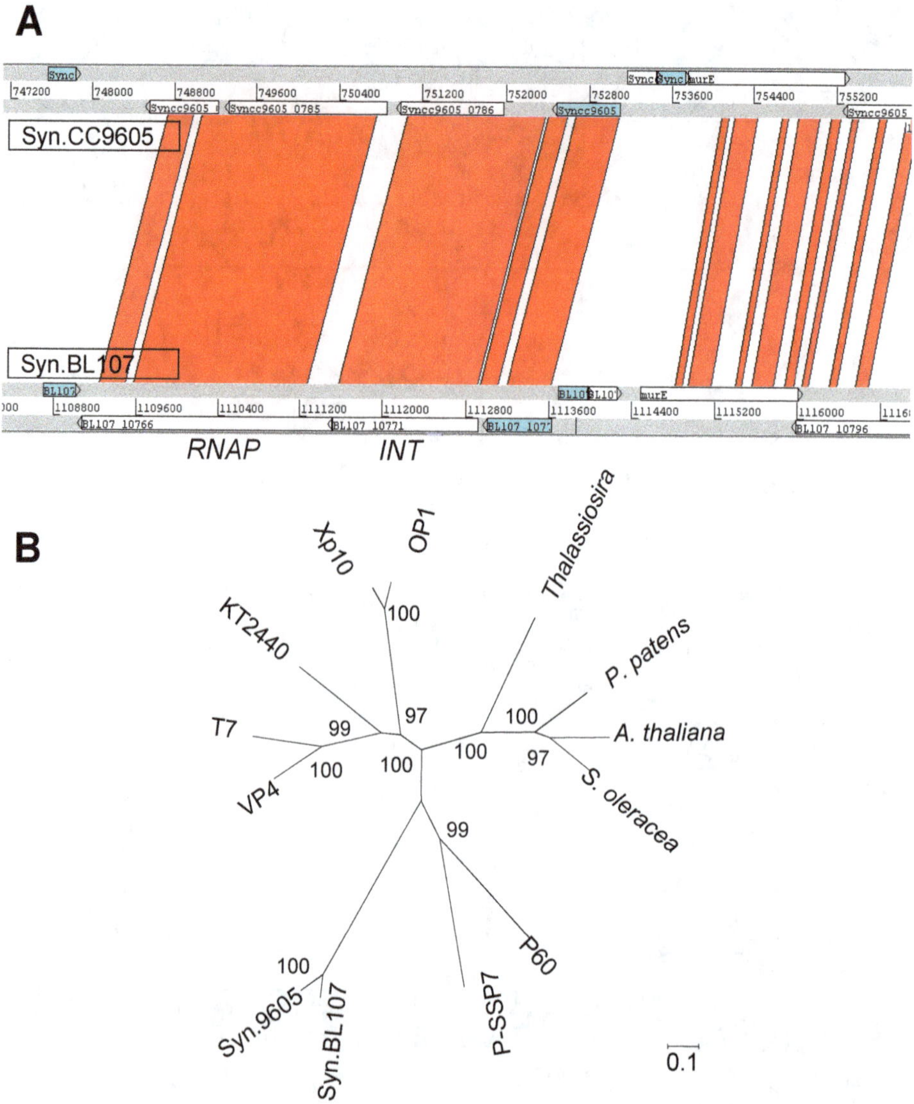

Plate 5.3 (see also Figure 5.3) A. Alignment of part of the *Synechococcus* CC9605 genomic sequence (top) and *Synechococcus* BL107 (bottom) containing genes for a T7-phage-like RNA polymerase (RNAP) and an integrase (INT). The lines connect DNA sequences with a BLASTN E value < 10^{-2}. The split of the RNAP gene in two parts in the CC9605 genome is visible, however, the DNA sequence similarity is similar for both parts with the respective gene segment in BL107. **B.** Phylogenetic analysis of RNAP sequences from cyanobacteria *Synechococcus* CC9605 (Syn.9605) and BL107 (Syn.BL107) and podoviral phages P60 and P-SSP7, non-cyanobacterial phages VP4 (Vibriophage VP4, AAY46276.1), T7 (coliphage, gi|3891671), KT2440 (*Pseudomonas putida* bacteriophage KT2440, NP_744415.1), XP10 (*Xanthomonas oryzae* bacteriophage Xp10, NP_858979.1), OP10 (*Xanthomonas oryzae* phage OP1, YP_453591.1) and eukaryotic RNAPs of the diatom *Thalassiosira pseudonana*, the moss *Physcomitrella patens* (*P. patens*) and the two higher plants *Arabidopsis thaliana* (*A. thaliana*) and *Spinacia oleracea* (*S. oleracea*).

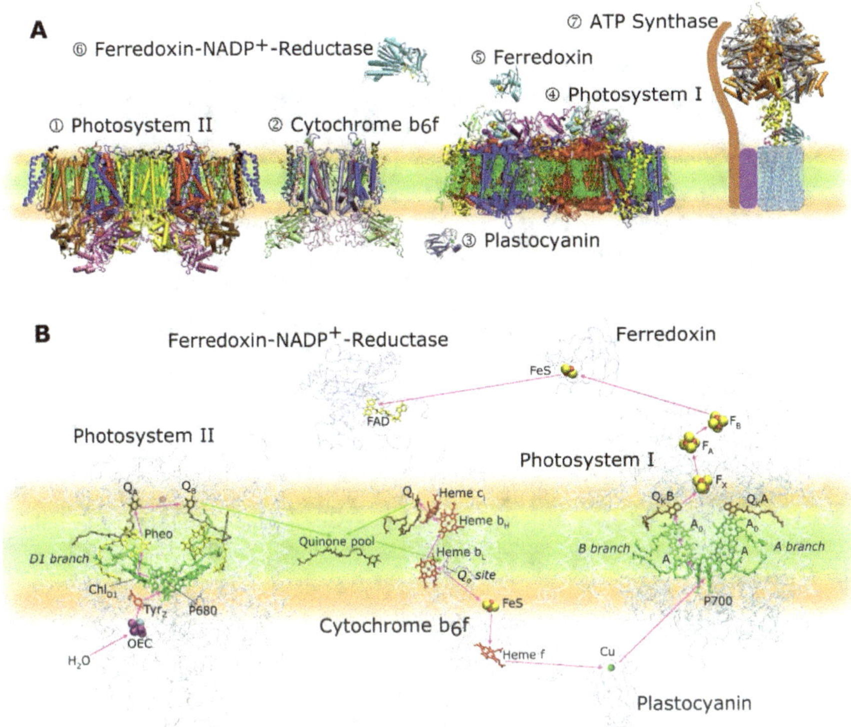

Plate 9.1 (see also Figure 9.1) Structures of the protein complexes involved in the primary reactions of oxygenic photosynthesis, with corresponding Protein Data Bank (PDB) accession codes. A, Structure of the complexes. B, Cofactors of the electron transport chain. Left to right: photosystem II (Ferreira *et al.,* 2004, 1S5L), the cytochrome b_6f complex (Kurisu *et al.,* 2003 1VF5, Stroebel *et al.,* 2003, 1Q90), photosystem I (Fromme *et al.,* 2001; Jordan *et al.,* 2001 1JB0) and the ATP synthase (Abrahams *et al.,* 1994, 1BMF; Stock *et al.,* 1999, 1QO1) as well as the soluble electron carriers plastocyanine, (Redinbo *et al.,* 1993 2PLT), cytochrome c_6 (Kerfeld *et al.,* 1995, 1CYI; Frazao *et al.,* 1995, 1CTJ), ferredoxin (Adman *et al.,* 1976, no PDB code; Morales *et al.,* 1999, 1CZP), flavodoxin (Watenpaugh *et al.,* 1972, 1FX1) and the ferredoxin: NADP$^+$ oxidoreductase (Bruns and Karplus, 1995, 1FNB).

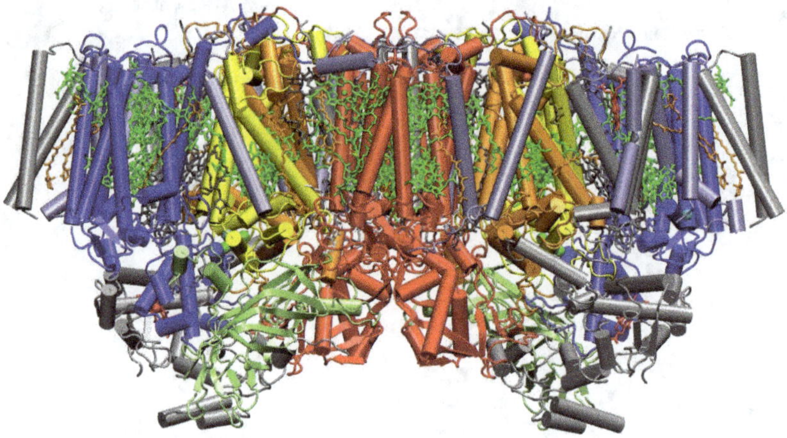

Plate 9.5 (see also Figure 9.5) Structure of photosystem II, picture prepared from PDB file 2AXT (Loll *et al.,* 2005). Yellow—D1; orange—D2; red—CP47; blue—CP43; green—PsbO; all other subunits shown in grey. Cofactors: green—chlorophyll; orange—carotenoids; grey—lipids and detergents.

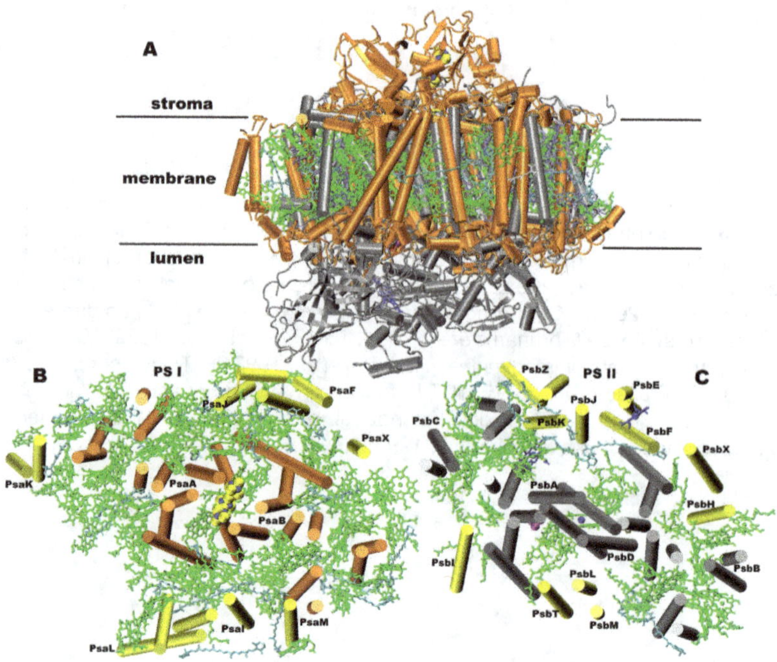

Plate 9.6 (see also Figure 9.6) Comparison of the structure of photosystems I and II. The pictures are derived from the PDB files for PSI, 1JB0 (Jordan *et al.,* 2001) and PSII, 1S6L (Ferreira *et al.,* 2004). A, view parallel to the membrane plane. PSI proteins are shown in gold, PSII in silver, chlorophylls are depicted in green and carotenoids in blue. B, view of PSI from the stromal side onto the membrane plane, the subunits are labeled according to their genes. C, view of PSII from the stromal side of the membrane, the subunits are labeled according to their genes. Note that assignment of small subunits of PSII is still under discussion (see Loll *et al.,* 2005). For clarity only transmembrane helices are shown in B and C.

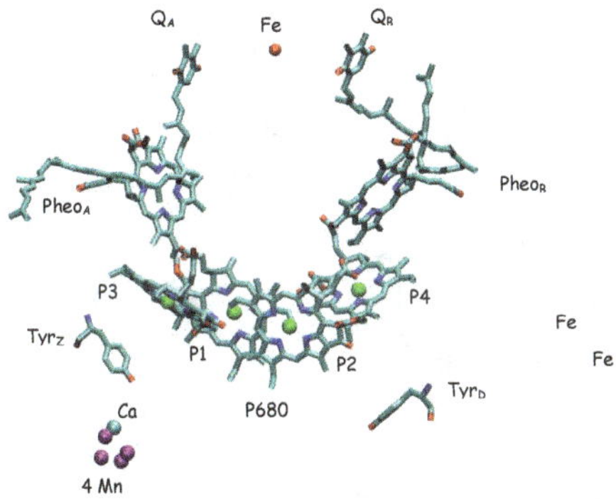

Plate 9.7 (see also Figure 9.7) Electron transfer chain of photosystem II as seen at 3.57 Å resolution, from the PDB file 1S6L (Ferreira *et al.*, 2004).

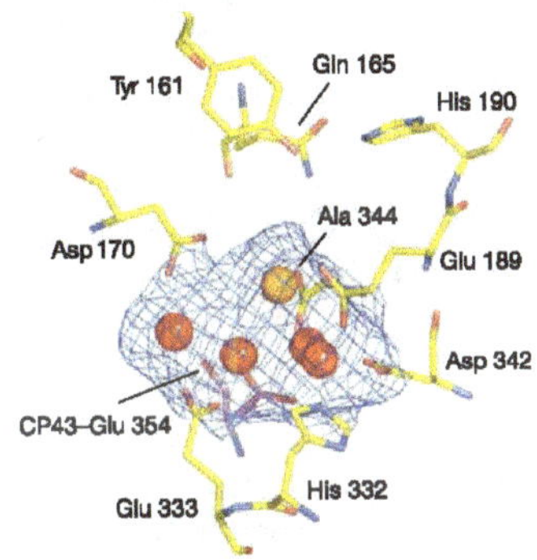

Plate 9.8 (see also Figure 9.8) Electron density of the Mn cluster at 3 Å resolution, from PDB file 2AXT (Loll *et al.*, 2005).

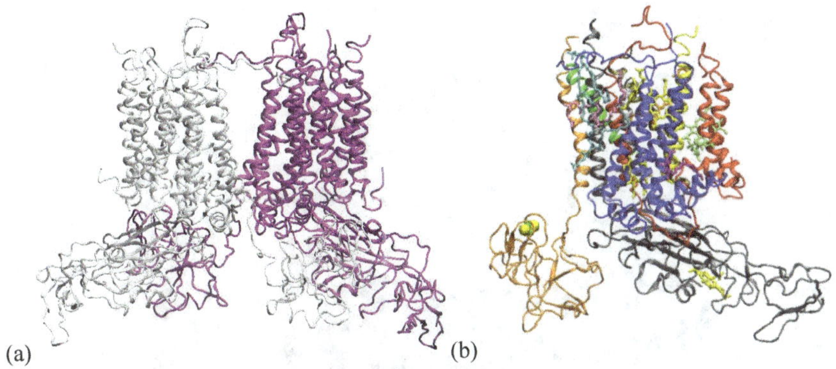

Plate 9.9 (see also Figure 9.9) (a) The dimeric nature of the cytochrome b_6f complex. Note that the Rieske protein is actually located on the opposite side of the axis of symmetry from the rest of its partners. This is a case of "domain swapping," and is thought to contribute to stability of the dimer. (b) The multisubunit nature of a cytochrome b_6f monomer, along with the cofactors belonging to each subunit. The monomer is rotated slightly with relation to its purple analogue in picture (a), to better show the complex assembly of subunits and cofactors. Subunits, coded by color—Blue: cytochome b_6. Red: subunit IV. Gray: cytochrome *f*. Orange: Rieske iron–sulfur protein. Yellow: PetL. Tan: PetM. Silver: PetG. Green: PetN. Cofactors, coded by color—Chl*a*: green. Carotenoid: orange. 2Fe-2S cluster: yellow and green spheres. Lipids: cyan. Hemes: yellow. The structure given here is drawn from the RCSB/PDB coordinate file 1VF5, for the cytochrome b_6f from *Mastigocladus laminosus* (Kurisu *et al.*, 2003).

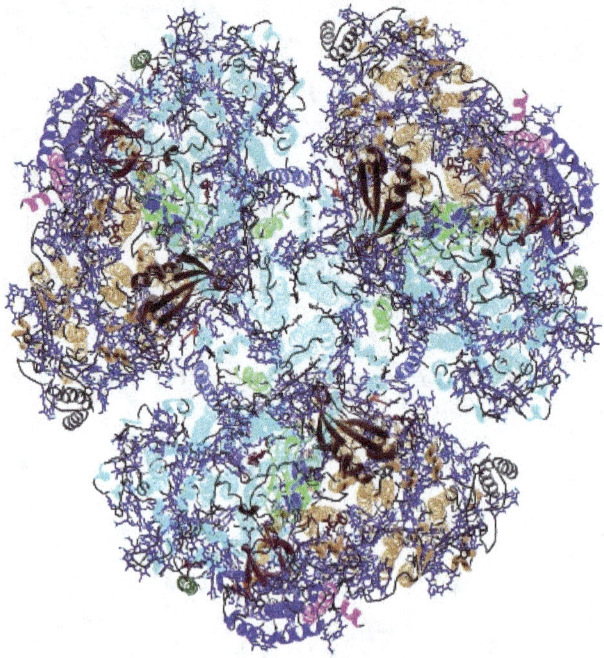

Plate 9.13 (see also Figure 9.13) PSI trimer from the cyanobacterium *Thermosynechococcus elongatus*, seen from the stromal side onto the membrane plane, PDB 1JB0 (Jordan *et al.*, 2001).

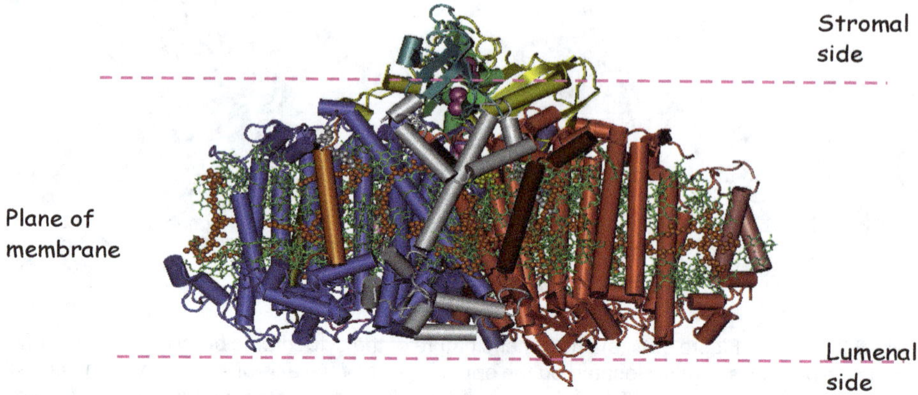

Plate 9.14 (see also Figure 9.14) A model of photosystem I from *Thermosynechococcus elongatus*, drawn from the coordinates of the structure deposited with the Brookhaven Protein Data Bank under accession code 1JB0 (Jordan *et al.*, 2001). Ribbons color-coded by subunits. PsaA—red, PsaB—blue, PsaC—green, PsaD—yellow, PsaE—cyan, PsaF—gray, PsaI—mauve, PsaJ—ochre, PsaK—pink, PsaL—periwinkle, PsaM—tan, PsaX—orange. Co-factors: chlorophyll—green, carotenoids—orange bead chains, phylloquinones—yellow bead chains, lipids—gray bead chains, iron–sulfur clusters—large magenta spheres.

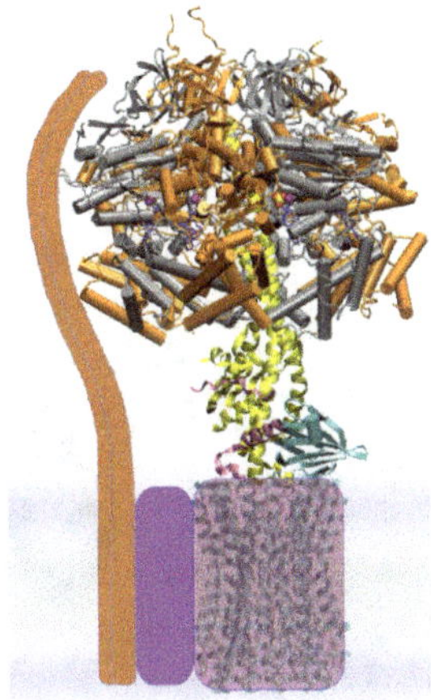

Plate 9.20 A schematic structural diagram of ATP synthase. Structure for F1 was modified after 2F43 (Chen *et al.,* 2006). *c*-ring represents the structure of the *c*11 ring from *Ilyobacter tartaricus* taken from Meier *et al*. (2005); PDB accession code 1YCE.

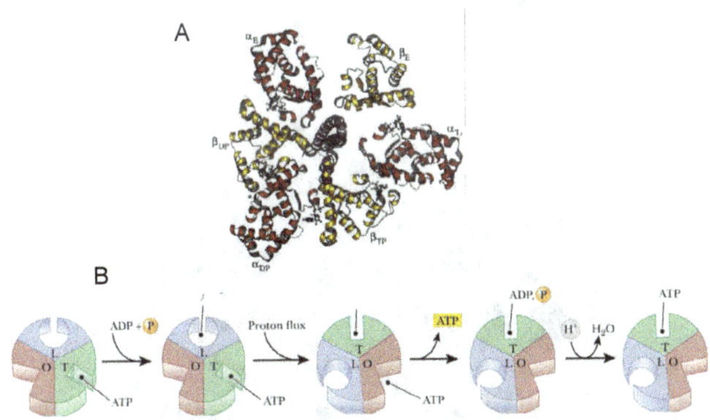

Plate 9.21 (see also Figure 9.21) A, structure of the F_1 part of the ATP synthase from beef heart mitochondria as determined by Abrahams *et al*. in 1994 (Figure from Abrahams *et al.,* 1994). B, binding change mechanism as proposed by Kohlbrenner and Boyer (1983).

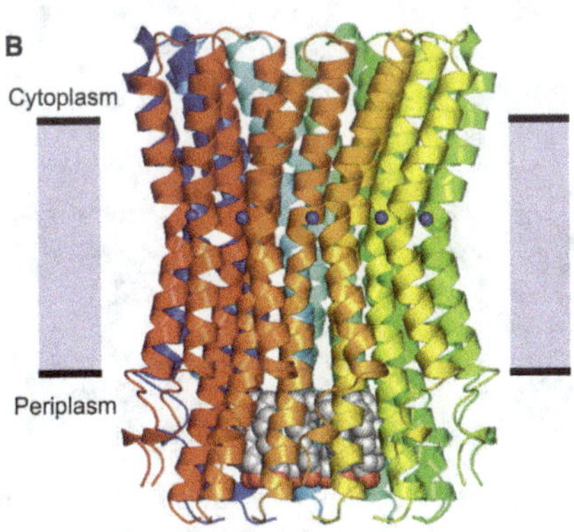

Plate 9.22 (see also Figure 9.22) Structure of the c11-ring of *Ilyobacter tartaricus*, PDB accession code 1YCE (Meier *et al.*, 2005).

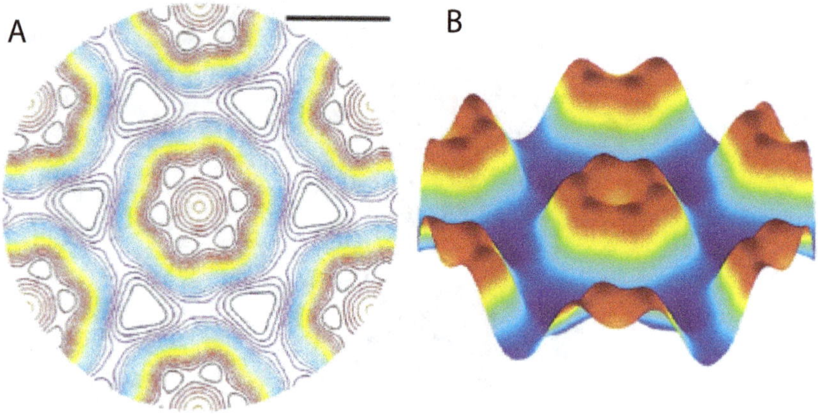

Plate 10.3 (see also Figure 10.3) Computer reconstructions of a cyanobacterial S-layer displaying a p6 symmetry. (A) Isodensity lines, and (B) space simulation. Bar = 10 nm. Adapted from (Smarda *et al.*, 2002), reprinted from Micron *33*, 257–277, copyright 2002, with permission from Elsevier.

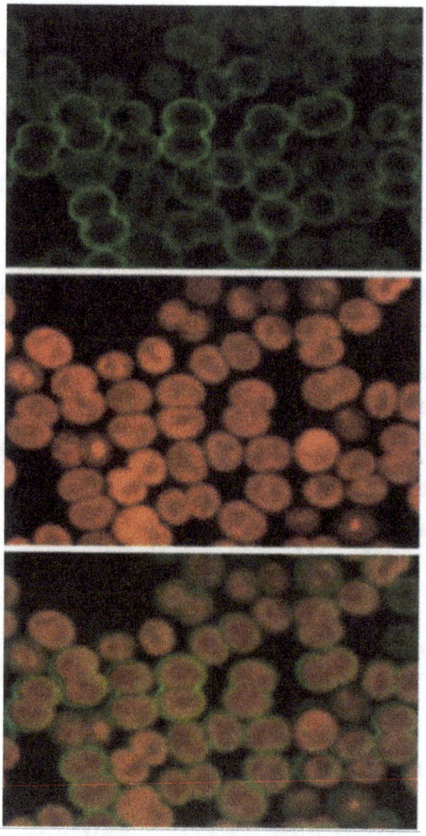

Plate 10.7 (see also Figure 10.7) The addition of a Tat-specific targeting signal to GFP results in localization of GFP only to the periplasm in *Synechocystis* 6803. Top panel, GFP; middle panel, endogenous chlorophyll; bottom panel, merged images. From (Spence *et al.*, 2003) with permission from Blackwell Publishing.

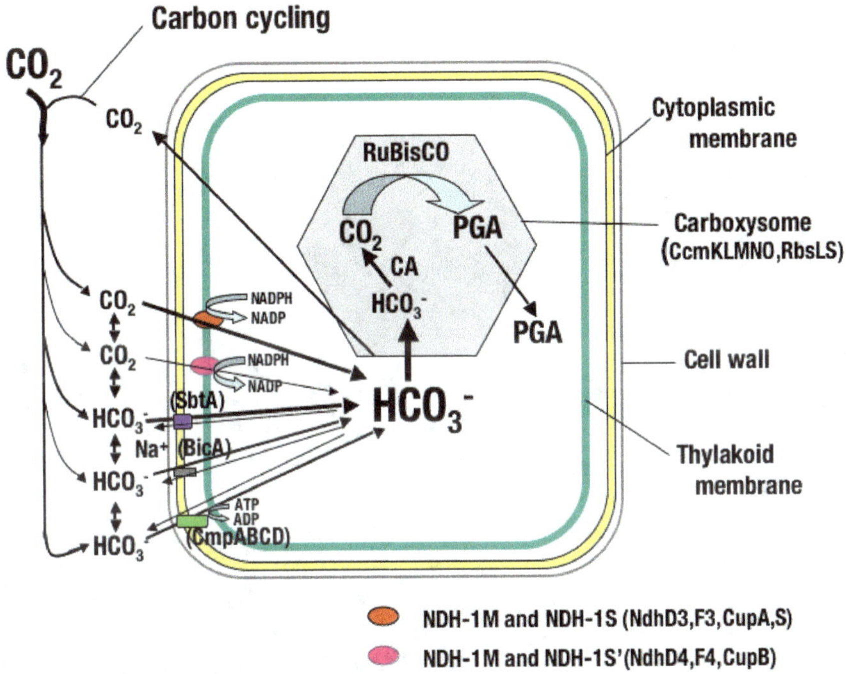

Plate 12.1 (see also Figure 12.1) A schematic model for the CCM in a *Synechocystis* PCC 6803 cell. Two CO_2 uptake systems on the thylakoid membrane and three HCO_3^- transporters on the cytoplasmic membrane are shown. The thick arrows on the high affinity CO_2 uptake system and SbtA, a HCO_3^- transporter, indicate their major contribution to carbon acquisition in *Synechocystis* PCC 6803.

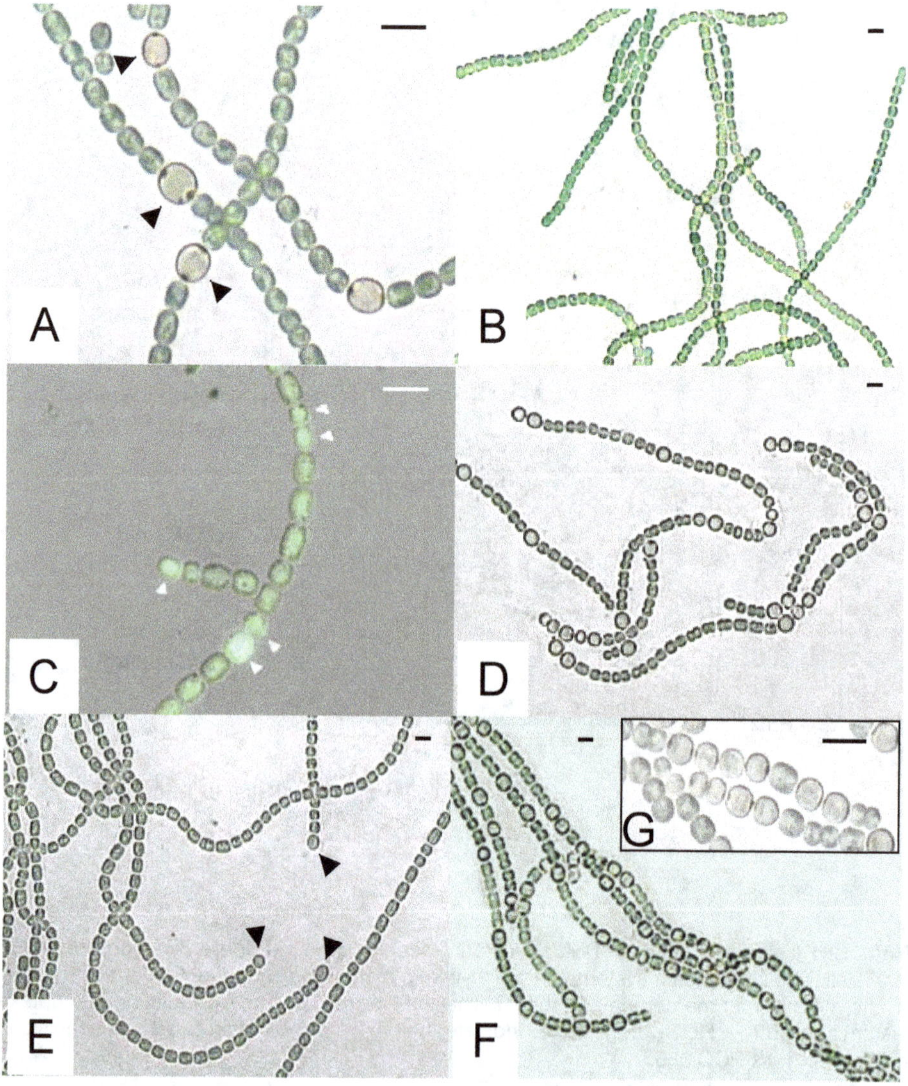

Plate 14.1 (see also Figure 14.1) Wild-type and mutant strains of *Anabaena* sp. strain PCC 7120, deprived of fixed nitrogen in medium BG11$_0$. A. The wild-type strain; arrowheads point to heterocysts. B. *hetR* mutant α41 (Ernst *et al.*, 1992), which is free of heterocysts, as described by Buikema and Haselkorn (1991). C. A *hetC* mutant, *hetC*::C.CE2, that contains a plasmid in which *gfp*, encoding green fluorescent protein, is driven by the promoter of cell-division gene *ftsZ* (Wang and Xu, 2005). *hetC* mutant cells continue to divide after initiating differentiation (arrowheads), as described by Xu and Wolk (2001). D. A fusion of the copper-regulated promoter of *petE* to *hetN* (Zhang and Xu, unpublished), showing a multiple contiguous heterocyst (Mch) phenotype in copper-free BG11$_0$, as described by Callahan and Buikema (2001). E. An insertional mutant of *patA* (Ning and Xu, unpublished), showing long filaments with terminal heterocysts (arrowheads), as described by Liang *et al.* (1992). F and G: A *patS* deletion mutant, in which the *patS*-encoding region has been replaced with the kanamycin-resistance cassette C.K2 (Xu, unpublished), showing a Mch phenotype, as described by Yoon and Golden (1998). All images are bright-field micrographs, except that in panel C, an image of GFP-based fluorescence is superimposed upon the bright-field image. Bars, 5 μm.

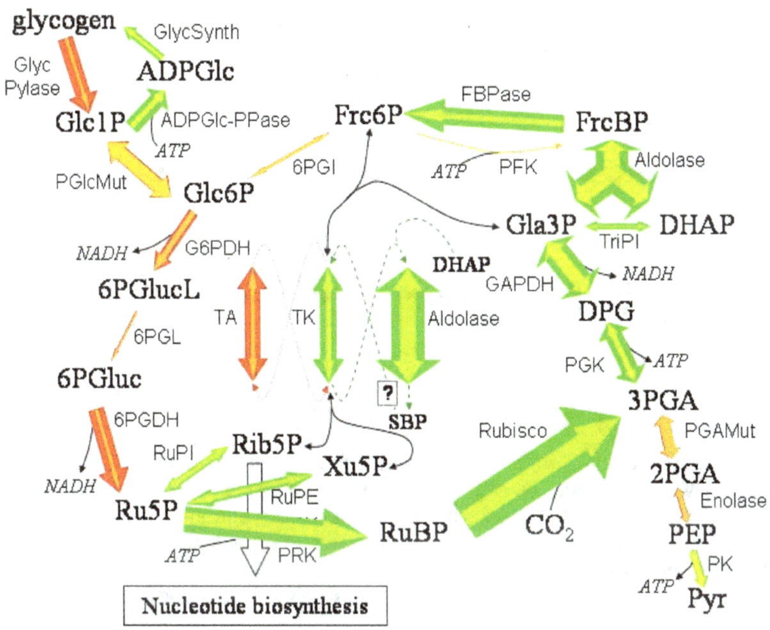

Plate 14.3 (see also Figure 14.3) Expression levels of genes related to central carbon metabolism. Expression levels at 0 vs 8 hours after nitrogen stepdown are shown for genes encoding enzymes of the Calvin cycle, the oxidative pentose phosphate (OPP) pathway, glycolysis, and glycogen metabolism. Green external arrows and orange internal arrows represent the RNA levels of genes in the presence of NO_3^-. Red external arrows and yellow-green internal arrows represent the RNA levels of genes at 8 hours after nitrogen stepdown. The widths of the two arrows for a single gene are linearly proportional to the levels of RNA under the two experimental conditions. The widths of arrows of two different genes are not linearly proportional to their RNA levels, but genes with greater levels of expression have wider arrows. A non-linear representation was necessary in order to present the full 300-fold dynamic range of RNA levels. Green dashed lines indicate the route taken by the Calvin cycle to convert Frc6P + Gla3P to Xu5P + Ery4P through TK, then (+DHAP) to SBP, through aldolase, then to Sed7P through an unknown enzyme "(?)," then (+Gla3P) to Xu5P + Rib5P through TK. Red dotted lines indicate the route taken by the OPP pathway to convert Rib5P + Xu5P to Sed7P + Gla3P through TK, then to Frc6P + Ery4P through TA, then (+ Xu5P) to Gla3P + Frc6P through TK. Compound abbreviations: ADP-Glc, ADP-glucose; Glc1P, glucose-1-phosphate; Glc6P, glucose-6-phosphate; 6PGluc, 6-phosphogluconate; Ru5P, ribulose-5-phosphate; Rib5P, ribose-5-phosphate; Xu5P, xylulose-5-phosphate; Ery4P, erythrose-4-phosphate; Sed7P, sedoheptulose-7-phosphate; SBP, sedoheptulose-1,7-bisphosphate; RuBP, ribulose-1,5-bisphosphate; Frc6P, fructose-6-phosphate; FrcBP, fructose-1,6-bisphosphate; Gla3P, glyceraldehyde-3-phosphate; DHAP, dihydroxyacetone phosphate; DPG, 1,3-diphosphoglycerate; 3PGA, 3-phosphoglycerate; 2PGA, 2-phosphoglycerate; PEP, phosphoenolpyruvate; Pyr, pyruvate. Enzyme abbreviations: (*Calvin cycle-specific*) PRK, phosphoribulose kinase; Rubisco, ribulose-bisphosphate carboxylase; (*OPP-specific*) G6PDH, glucose-6-phosphate dehydrogenase; 6PGlucL, 6-phosphogluconolactonase; 6PGDH, 6-phosphogluconate dehydrogenase; TA, transaldolase; (*glycolysis-specific*) PFK, phosphofructokinase; PGAMut, phosphoglycerate mutase; PK, pyruvate kinase; (*Calvin cycle/OPP overlap*) RuPI, ribulose phosphate isomerase; RuPE, ribulose phosphate epimerase; TK, transketolase; TriPI, triose phosphate isomerase; FBPase, fructose bisphosphatase; 6PGI, 6-phosphoglucose isomerase; (*Calvin cycle/glycolysis overlap*) PGK, phosphoglycerate kinase; GAPDH, glyceraldehyde phosphate dehydrogenase; (*glycogen metabolism*) PGlcMut, phosphoglucomutase; ADPglc-PPase, ADP-glucose-pyrophosphorylase; GlycSynth, glycogen synthase; GlycPylase, glycogen phosphorylase.

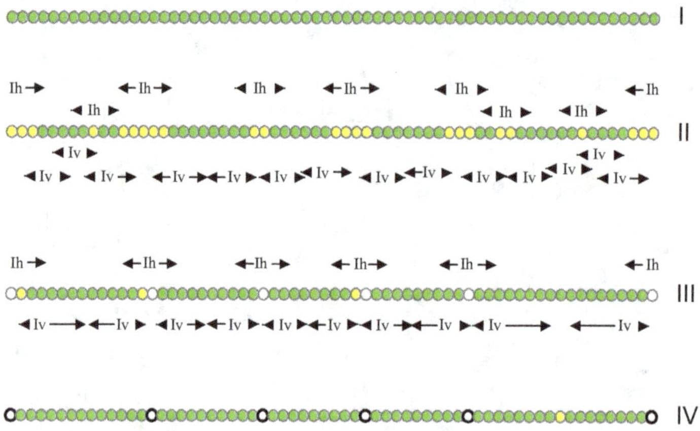

Plate 14.4 (see also Figure 14.4) A conjecture concerning heterocyst pattern formation in *Anabaena*. I to IV, different stages of pattern formation. Circles (cells) that are green, yellow, white with a thin wall, and white with a thick wall stand for vegetative cells, early developing cells, proheterocysts committed to differentiate, and mature heterocysts, respectively. Ih, heterocyst-type inhibitor (perhaps *patS*-dependent); Iv, vegetative cell-type inhibitor. Production of Iv in response to the differentiation of chosen cells may be mediated by Ih.

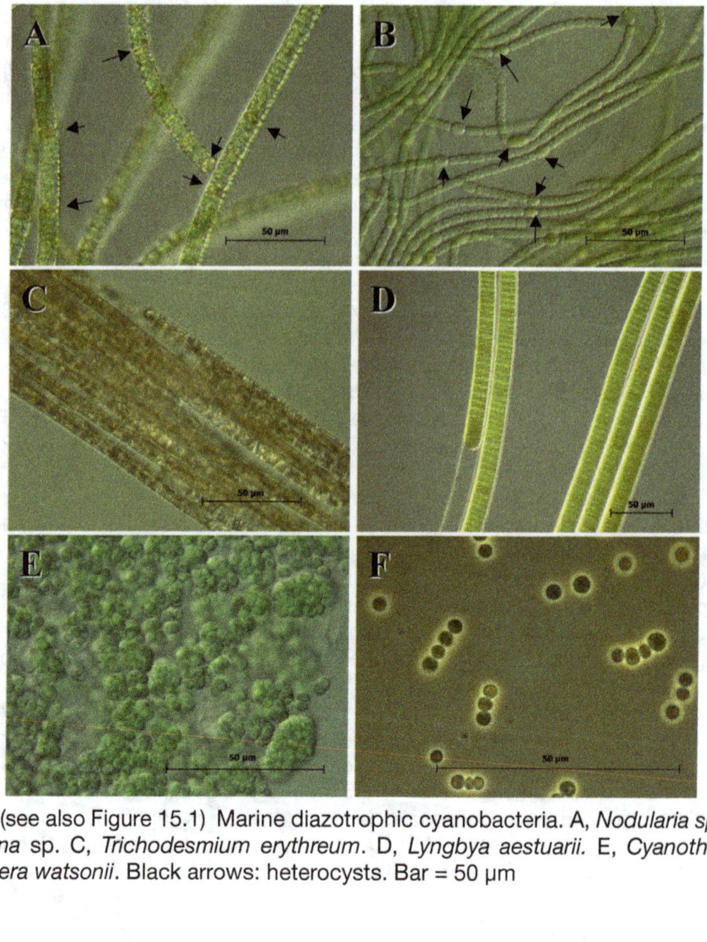

Plate 15.1 (see also Figure 15.1) Marine diazotrophic cyanobacteria. A, *Nodularia spumigena*. B, *Anabaena* sp. C, *Trichodesmium erythreum*. D, *Lyngbya aestuarii.* E, *Cyanothece* sp. F, *Crocosphaera watsonii*. Black arrows: heterocysts. Bar = 50 μm

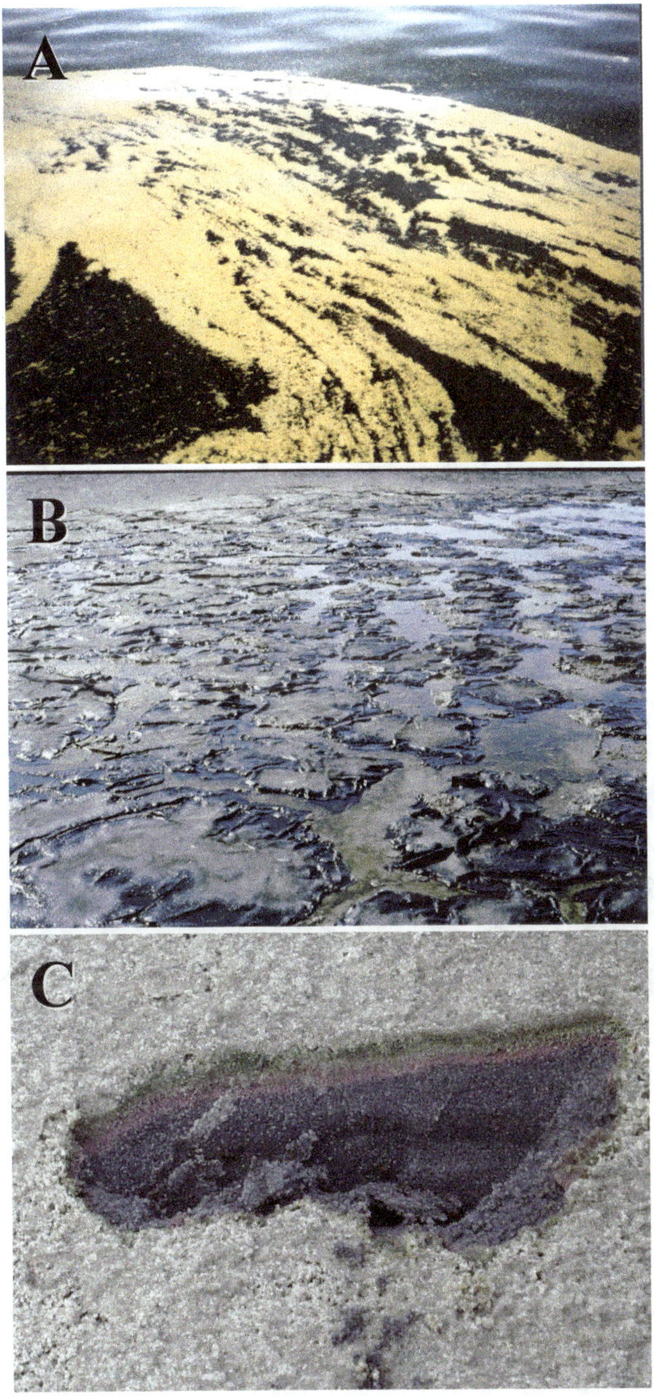

Plate 15.4 (see also Figure 15.4) A, Bloom of *Nodularia spumigena* in the Baltic Sea. B, Coastal microbial mats of *Lyngbya aestuarii* in Guerrero Negro, Baja California, Mexico. C, Cross section of a coastal microbial mat, Netherlands.

Plate 16.1 (see also Figure 16.1) Cyanobacterial-plant symbioses. (A) The small liverwort *Blasia* sp. colonized by cyanobacteria seen as small dark spots spread over the thallus. Inset shows a cyanobacteria-filled auricle at the surface of the thallus. (B) A branch of the small floating water-fern *Azolla* sp. The cyanobacteria are occupying cavities located in each of the leafs seen. (C) A palm-like cycad plant (*Cycas* sp.) with a prominent cone. Inset shows the blue-green cyanobacteria-filled zone located between the outer and inner cortex of a transversely sectioned coralloid root. (D) The large-leaved rhubarb-like *Gunnera manicata*. Inset shows the blue-green cyanobacterial infection sites in a cross sectioned stem.

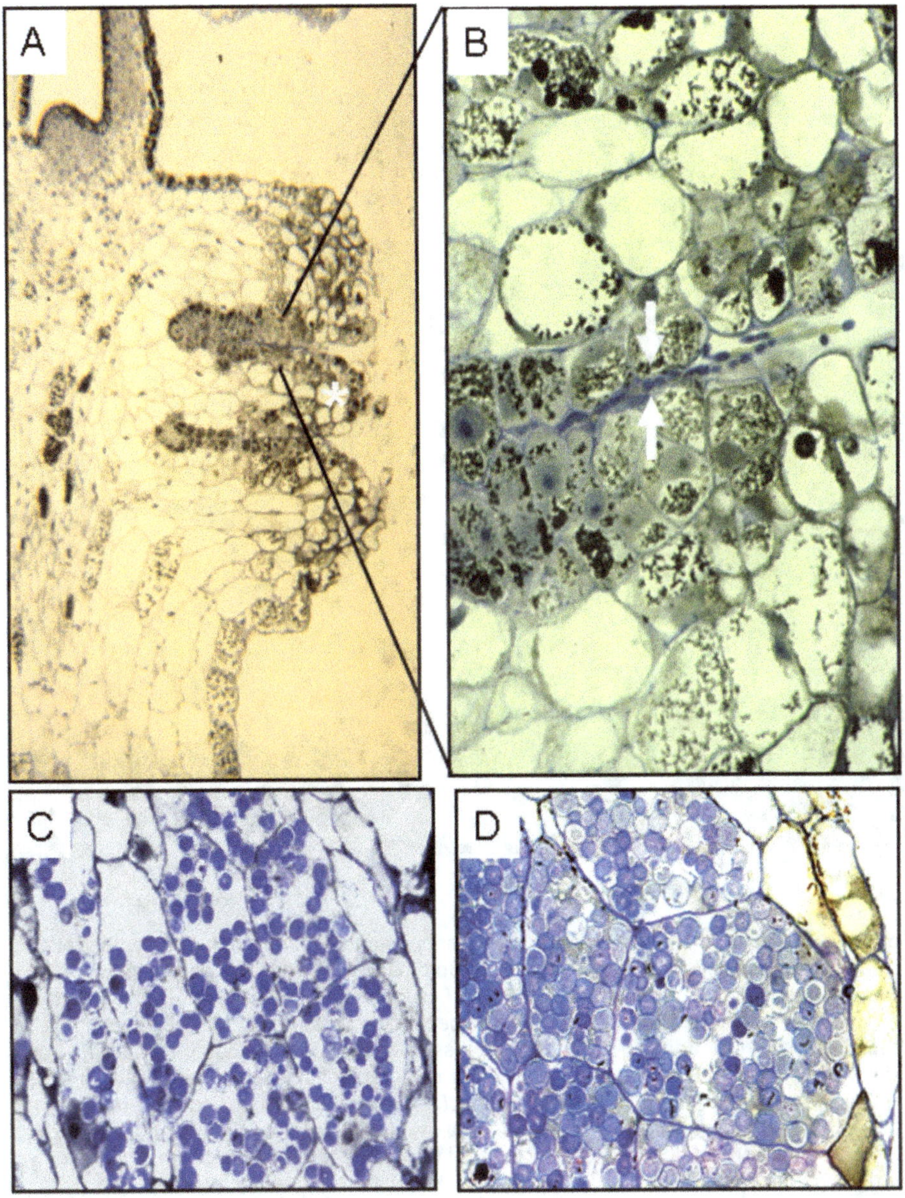

Plate 16.4 (see also Figure 16.4) In the *Nostoc-Gunnera* symbiosis the cyanobiont enters the plantlet as motile hormogonia through red pigmented stem glands. A) Light microscopy of a cross section of a newly developed stem gland, composed of a central papilla (star) surrounded by additional papillae. The plant petioles are seen in the upper left hand corner. B) A larger magnification of the stem gland channel showing the non-heterocystous hormogonial filaments (white arrows) gliding between the host cells towards the bottom of the gland (to the left) where they enter the *Gunnera* cells. C) A few filaments have just entered a cluster of the *Gunnera* cells and started to differentiate heterocysts. D) At a later stage, the host cells are filled with cyanobacteria and the heterocysts (thicker cell walls) have reached a frequency of > 50%. A) and B) from Johansson and Bergman, 1992 (Figure 16.5a and b), with permission from Springer-Verlag.

Cover Background: a surface bloom of the cyanobacterium *Nodularia* in the southern Baltic Sea in 2004 (courtesy of S. Jansson and B. Bergman). Left, top: electron micrograph of a transversal section of *Synechococcus* sp. PCC 7942 (courtesy of R. Nevo, Z. Reich, I. Ohad and A. Kaplan). Left, middle: light micrograph of *Synechococcus* sp. PCC 9321 (courtesy of A. M. Muro-Pastor). Left, bottom: Mesoproterozoic fossil akinete from the 1500 Ma Bil'yakh Group, Siberia (courtesy of A. H. Knoll). Right, top: light micrograph of a natural isolate of a heterocyst-forming cyanobacterium from Mkindo, Tanzania (courtesy of J. E. Frías and A. Herrero). Right, middle: *Trichodesmium thiebautii* colony (courtesy of B. Bergman). Right, bottom: electron micrograph of a longitudinal section of a filament of *Anabaena* sp. PCC 7120 with a heterocyst (far right) and vegetative cells (courtesy of I. Maldener). Images are not to scale.

www.ingramcontent.com/pod-product-compliance
Lightning Source LLC
LaVergne TN
LVHW011550200926
848484LV00001B/300